ENTOMOPATHOGENIC NEMATODES: SYSTEMATICS, PHYLOGENY AND BACTERIAL SYMBIONTS

ENTOMOPATHOGENIC NEMATODES: SYSTEMATICS, PHYLOGENY AND BACTERIAL SYMBIONTS

Khuong B. Nguyen and David J. Hunt (Editors)

David J. Hunt and Roland N. Perry (Series Editors)

NEMATOLOGY MONOGRAPHS AND PERSPECTIVES

VOLUME 5

BRILL
LEIDEN-BOSTON
2007

This book is printed on acid-free paper.

Library of Congress Cataloging-in-Publication Data

The Library of Congress Cataloging-in-Publication Data is available from the Publisher.

Cover illustration: *Heterorhabditis* (infective juvenile anterior region), *Steinernema* (female head) and part of a phylogenetic tree

ISBN-10: 90 04 15293 8
ISBN-10: 978 90 04 15293 9

Contents

Foreword

The number of species of entomopathogenic nematodes in the Families Steinernematidae and Heterorhabditidae has expanded almost exponentially in the last few years. Identifying a species belonging to either the genus *Steinernema* or *Heterorhabditis* has become much more complex, requiring morphometrics, life cycles, DNA analyses, and possibly even mating studies.

Thus, there is a great need for this monograph to bring together, under one cover, up-to-date information on this group of nematodes.

Drs K.B. Nguyen and D.J. Hunt, as editors and contributing authors of this monograph, have assembled a team of distinguished scientists who have done an outstanding job of collecting and synthesising the information on these nematodes. While the emphasis is necessarily on taxonomy and morphology, included is a very helpful treatise on the phylogeny of the group and the use of these nematodes as biological control organisms, and a very significant treatise on the bacterial associates of both the Steinernematidae and the Heterorhabditidae. Not only are the editors competent in the taxonomy and morphology of all nematodes, but they are also world authorities on the morphology and taxonomy of entomopathogenic nematodes.

This monograph is a must for all scientists who work with these nematodes, for students, especially graduates, who wish to become knowledgeable about the group, and for all who want to use entomopathogenic nematodes as biological control organisms.

Grover C. Smart Jr
Professor Emeritus
Entomology & Nematology Department,
University of Florida, Gainesville, Florida, USA

Acknowledgements

Khuong Nguyen would like to express his gratitude to Dr G.C. Smart Jr, Professor Emeritus of the University of Florida, for agreeing to write the Foreword for the book, and to Mrs P. Howell for help in reading and giving comments for several descriptions of the nematode species. He also wishes to thank contributors who made this book possible and the many journal editors and publishers who allowed us to use their illustrations and photographs in this book.

David Hunt would like to express particular thanks to Professor Gian Battista Vai, University of Bologna, Italy, for providing a copy of his book on Aldrovandi and for scanning an image of the great man; Professor Stefano Maini, University of Bologna, Italy, for his help with obtaining information on Aldrovandi; Dr Marion Seier, CABI Europe-UK, for her help in translating the Goeze description of emerging mermithids; and to the late Professor R.T. Leiper FRS for assiduously collecting so many pertinent antiquarian publications.

List of Contributors

Byron J. ADAMS, Microbiology & Molecular Biology Department, and Evolutionary Ecology Laboratory, Brigham Young University, Provo, UT 84602-5253, USA.

Adler R. DILLMAN, Microbiology & Molecular Biology Department, and Evolutionary Ecology Laboratory, Brigham Young University, Provo, UT 84602-5253, USA.

David J. HUNT, CABI Europe-UK, Bakeham Lane, Egham, Surrey TW20 9TY, UK.

Heather S. KOPPENHÖFER, Department of Entomology, Rutgers University, New Brunswick, NJ 08901-8524, USA.

Zdeněk MRÁČEK, Laboratory of Insect Pathology, Institute of Entomology, Czech Academy of Sciences, Branišovská 31, 370 05 České Budějovice, Czech Republic.

Khuong B. NGUYEN, Entomology and Nematology Department, Institute of Food and Agricultural Sciences, University of Florida, Gainesville, FL 32611-0620, USA.

Scott M. PEAT, Microbiology & Molecular Biology Department, and Evolutionary Ecology Laboratory, Brigham Young University, Provo, UT 84602-5253, USA.

Nematology Monographs & Perspectives, 2007, Vol. 5, 1-26

Chapter 1

Introduction

David J. HUNT

CABI Europe-UK, Bakeham Lane, Egham, Surrey TW20 9TY, UK

Early beginnings: of dragons and serpents

In his intriguingly titled treatise *The natural history of serpents and dragons*, Ulisse Aldrovandi (1522-1605) recorded and named a small dragon that had been discovered along a farm road near Bologna in northern Italy (Aldrovandi, 1640). The unfortunate dragon, which was found on 13 May 1572 by Baptista of Camaldulus, a herdsman, hissed like a snake and bravely held its ground, but was quickly felled by a blow from Baptista's staff. The body of the dragon was obtained by Aldrovandi for his museum and was described as having a long neck, fat body, only two legs and a long tail. Figure 1 shows an engraving of Aldrovandi's *Draco bipes apteros*.

Ulisse Aldrovandi (also known as Ulysses Aldrovandus) had been born into a well-connected family (Pope Gregory XIII was his cousin) in Bologna, Italy, in the autumn of 1522 and was so named after the eponymous hero in Homer's *The Iliad* and *The Odyssey*. Somewhat predictably, perhaps, his brother Achille was named after the other of Homer's two heroic figures. It is said that his far-fetched tales, of which those concerning the cock and the bull were particular favourites, gave rise to the phrase 'cock and bull story'. Many of his over 400 lavishly illustrated manuscripts, including the one dealing with dragons and serpents, were published posthumously in Bologna.

Whilst Aldrovandi (Fig. 2) was a remarkable and celebrated scientist of his time, he was clearly fascinated by the bizarre and unusual. To our eyes, the more uncritical aspects of his industry could be interpreted as a little naive, although this must be balanced against his prodigious and wide-ranging achievements in so many fields. In addition to this predilection for dragons and other mythical beings, Aldrovandi was an avid collector of specimens for scientific research, his 'curiosity

Fig. 1. *Engraving of a small dragon described by Ulisse Aldrovandi (1643). Adapted from* Mythical monsters *by Charles Gould (1884).*

cabinet', which embraced over 18 000 curios, being famed throughout Europe. His multi-folio work on natural history, his contributions to geology and the establishment of a botanic garden rank high amongst the extraordinary achievements of this wide-ranging intellectual.

Perhaps not surprisingly for a man endowed with such an insatiable, multi-facetted curiosity, even the history of nematodes associated with insects can be traced back to him when, nearly 400 years ago, he recorded the presence of long, thread-like worms on the bodies of dead grasshoppers (Aldrovandi, 1623). The worms recorded by Aldrovandi were, from their description, almost certainly what we now refer to as *Mermis nigrescens* Dujardin, 1842, a common mermithid parasite of orthopterans.

After Aldrovandi, the next record of an insect-associated nematode was by Lister (1672) who described similar, again presumably mermithid, worms from a plant in his garden. Lister compared his observations to those of Aldrovandi.

René Anthoine Ferchault de Réaumur (1683-1757) was another scientist of manifold and wide ranging talents that included metallurgy, slate-working, porcelain manufacture, egg incubation and preservation, temperature measurement (he invented the Réaumur thermometer, a device with an 80 degree range between the freezing and boiling point of water), invertebrate locomotion, insect behaviour, and parthenogenesis in aphids. Réaumur also noted that polydactyly, a rare genetic condi-

Fig. 2. *Ulisse Aldrovandi in his 78th year. Frontispiece from* Memorie della vita di Ulisse Aldrovandi, medico e filosofo Bolognese, *Museo Geologico Giovanni Capellini, by courtesy of Prof. Gian Battista Vai, University of Bologna and Director of the Museo Geologico Giovanni Capellini.*

tion, could be passed on to the descendants of an individual so afflicted. In addition to innumerable other achievements, he published a six volume treatise, *Mémoires pour servir à l'histoire des insectes*, between 1734 and 1742. As befitting a man endowed with such a diverse intellectual compass, Réaumur also made observations pertaining to nematodes and, in 1742, described an unusual worm from bees (de Réaumur, 1742). Nearly a century later, this species of nematode, a common parasite of bumble bees, was described as *Sphaerularia bombi* Dufour, 1837.

Later in the same decade, Gould (1747) graphically described the emergence of worms, probably mermithids, from ants. William Gould, a man of the cloth from Exeter College, Oxford, and often referred to as the father of British myrmecology, is most famous for his book *An account of English ants* in which he refuted, not without some criticism, the Old Testament assertion that ants stored grain:

> "*Go to the ant, thou sluggard; consider her ways and be wise; which having no chief, overseer, or ruler, provideth her bread in the summer, and gathereth her food in the harvest.*" *Proverbs.*

As a taxonomist, Gould clearly had a strong pragmatic bent, recognising the following species of pismire: 'hill ants', 'jet ants', 'red ants', 'common yellow ants' and 'small black ants'.

Moving on several more decades into the 18th century, Johann August Ephraim Goeze (1731-1793) published an account of what could only have been ovipositing female *M. nigrescens* crawling over plants and box hedges in a garden. The worms appeared in their hundreds in June 1781 after a heavy thunderstorm, a phenomenon characteristic of this parasite, the mature females of which emerge from the damp soil in order to lay eggs on the wet foliage of plants from where they are subsequently ingested by phytophagous insects (Goeze, 1782). The title page of this important early work (which mostly deals with cestodes, ascarids and acanthocephalans) is shown in Figure 3. Goeze (1782) described these worms as a species of *Gordius*, although from his account (Fig. 4) it is clear that he was well aware that they differed from 'normal' gordians, a group of animals with which he was very familiar.

At the beginning of the 19th century, Carolo Asmundo Rudolphi (1771-1832) published two major catalogues, *Entozoorum sive vermium intestinalium historia naturalis* (Rudolphi, 1809) and *Entozoorum synopsis cui accedunt mantissa duplex et indices lociepletissimi* (Rudolphi, 1819). In his lists the insect-parasitic nematodes were attributed to *As-*

Versuch
einer
Naturgeschichte
der
Eingeweidewürmer thierischer Körper

von

Johann August Ephraim Goeze,
Pastore an der Kirche S. Blasii zu Quedlinburg; Ehrenmitgliede der Gesellschaft Naturforschender Freunde zu Berlin; der Herzoglich deutschen Gesellschaft zu Helmstedt, der Naturforschenden Gesellschaft zu Halle, der Königlich Schwedischen Gesellschaft der Wissenschaften zu Stockholm, und der physiographischen Gesellschaft zu Lund in Schonen.

Mit 44 Kupfertafeln.

Auf Kosten des Verfassers: bey Ihm selbst, und in der Buchhandlung der Gelehrten zu Leipzig zu haben.

Blankenburg,
gedruckt bey Philipp Adam Pape, Fürstl. privilegirtem Buchdrucker 1782.

Fig. 3. *Title page from volume 2 of* Entozoorum sive vermium intestinalium historia naturalis *by Goeze (1782).*

caris and *Filaria*, genera with a decidedly wider diagnosis then than now (Figs 5, 6). At this time the distinction between filariids, gordiids and mermithids had yet to be clearly defined and it was not until the monograph of Schneider (1866) that mermithids and gordiids, for example, were clearly separated.

124 Zweeter Abschnitt. Beschreibung und Abbildung

Indessen zeigt uns doch die Erfahrung Beyspiele von Intestinalgordien bey verschiedenen Arten von Thieren, die wir wegen ihrer Oekonomie nicht wohl zu den Accessoriis zählen können, ob sie gleich im ganzem Bau ihres Körpers viel ähnliches mit den Wassergordien haben.

Die beste Beschreibung vom *Gordio* hat Pallas *) gegeben:

> *Corpus* filiforme æquale lævissimum. *Os* dilatatum labio subrotundo concavo. *Cauda uncinulo terminata.*

Diesen leztern Karakter vermisse ich bey den Intestinalgordien gänzlich.

Ich habe Wassergordien genug gesehen, und beobachtet; auch die rare dunkelbraune cauda bifida **). Man darf nur im Frühjahr unter Mittage, die stehenden Regenpfützen, wenn die Sonne darauf scheint, betrachten; so siehet man sie wimmeln. Auch unter der Lemna hab' ich sie häufig gefunden. So gar im Magen der Frösche, die sie zu ihrer Nahrung verschluckt hatten, und wo schon der Anfang der Verdauung an ihnen zu sehen war. Noch kürzlich im Junius 1781, nach starken Gewitterregen, lagen sie in den Gärten, auf den frisch gegrabenen Beeten, und erst gelegtem niedrigem Buchsbaum, wie zarte Fäden, bey hunderten herum. Allein dies alles sind mir noch keine eigentliche Intestinalgordien. Diese leztere sterben gar zu bald in kaltem Wasser, da ich die erstern wohl acht Tage, und länger, im Wasser erhalten habe.

Wie ich die Gordien, die in vielen Insektengeschlechtern, inwendig in ihren Feuchtigkeiten leben, betrachten soll, weiß ich nicht eigentlich zu sagen. Ueberhaupt ist mir dies Geschlecht, der Gordius *intestinalis* unter allen das dunkelste.

Fig. 4. Entozoorum sive vermium intestinalium historia naturalis, *volume 2, by Goeze (1782). Description of mermithid nematodes emerging from the soil after a thunderstorm in June 1781.*

By the time that Karl Moriz Diesing (1800-1867), an Austrian helminthologist, published his catalogue of helminths, the number of insect-associated nematodes had increased dramatically, 174 entries being listed for such nematodes in volume 2 of his *Systema helminthum* (Diesing, 1851). The title page of Diesing's 1851 work is shown in Figure 7.

The list of insect-associated nematodes was further expanded by the German Otto Friedrich Bernhard von Linstow (1842-1916) in his *Compendium der helminthologie* (von Linstow, 1878, 1889). Figure 8,

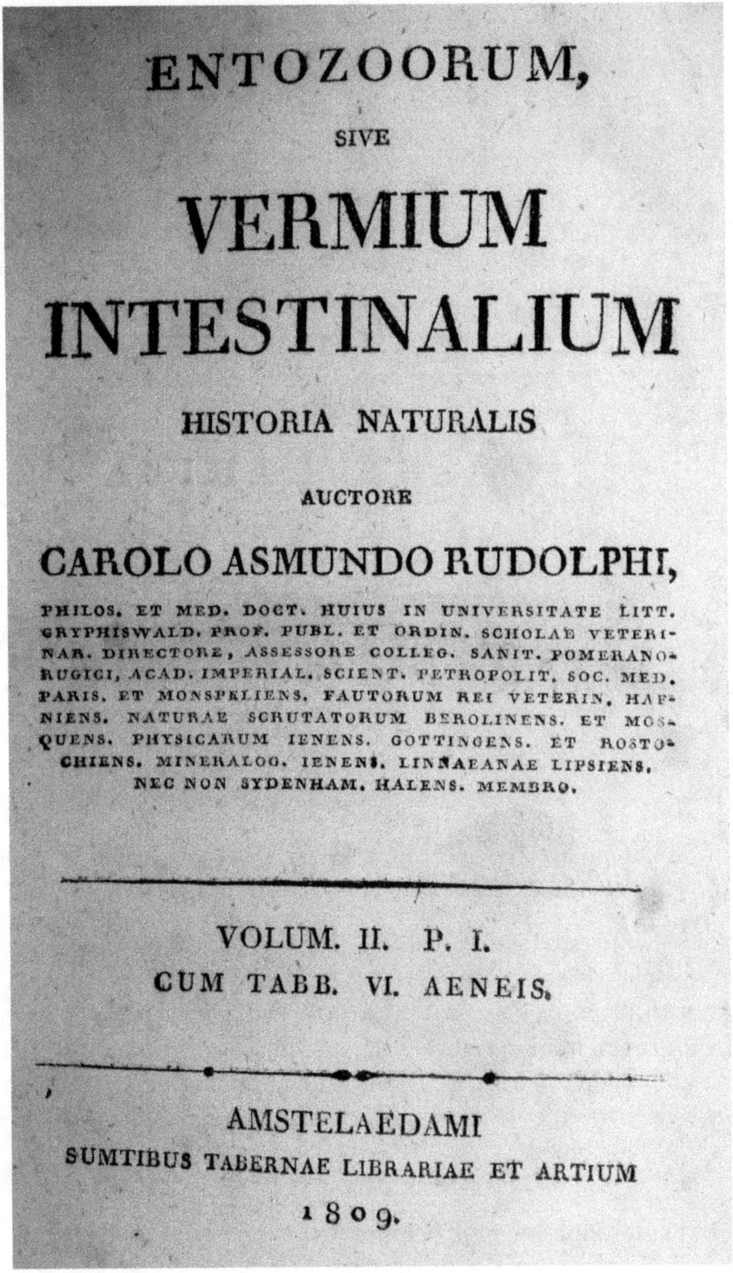

ENTOZOORUM,

SIVE

VERMIUM

INTESTINALIUM

HISTORIA NATURALIS

AUCTORE

CAROLO ASMUNDO RUDOLPHI,

PHILOS. ET MED. DOCT. HUIUS IN UNIVERSITATE LITT. GRYPHISWALD. PROF. PUBL. ET ORDIN. SCHOLAE VETERINAR. DIRECTORE, ASSESSORE COLLEG. SANIT. POMERANO-RUGICI, ACAD. IMPERIAL. SCIENT. PETROPOLIT. SOC. MED. PARIS. ET MONSPELIENS. FAUTORUM REI VETERIN. HAFNIENS. NATURAE SCRUTATORUM BEROLINENS. ET MOSQUENS. PHYSICARUM IENENS. GOTTINGENS. ET ROSTOCHIENS. MINERALOG. IENENS. LINNAEANAE LIPSIENS. NEC NON SYDENHAM. HALENS. MEMBRO.

VOLUM. II. P. I.

CUM TABB. VI. AENEIS.

AMSTELAEDAMI

SUMTIBUS TABERNAE LIBRARIAE ET ARTIUM

1809.

Fig. 5. *Title page of volume 2 of* Entozoorum sive vermium intestinalium historia naturalis *by Rudolphi (1809).*

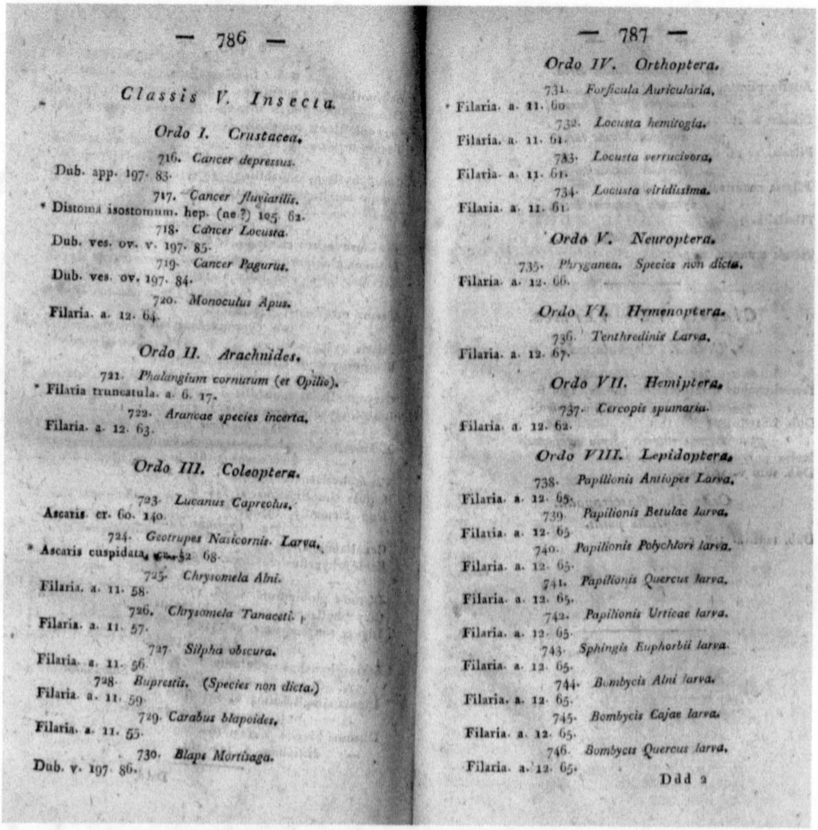

— 786 —

Classis V. Insecta.

Ordo I. Crustacea.

716. *Cancer depressus.*
Dub. app. 197. 83.

717. *Cancer fluviatilis.*
* Distoma isostomum. hep. (ne ?) 105. 62.

718. *Cancer Locusta.*
Dub. ves. ov. v. 197. 85.

719. *Cancer Pagurus.*
Dub. ves. ov. 197. 84.

720. *Monoculus Apus.*
Filaria. a. 12. 64.

Ordo II. Arachnides.

721. *Phalangium cornutum (et Opilio).*
* Filaria truncatula. a. 6. 17.

722. *Araneae species incerta.*
Filaria. a. 12. 63.

Ordo III. Coleoptera.

723. *Lucanus Capreolus.*
Ascaris. cr. 60. 140.

724. *Geotrupes Nasicornis. Larva.*
* Ascaris cuspidata, [illegible] 52. 68.

725. *Chrysomela Alni.*
Filaria. a. 11. 58.

726. *Chrysomela Tanaceti.*
Filaria. a. 11. 57.

727. *Silpha obscura.*
Filaria. a. 11. 56.

728. *Buprestis. (Species non dicta.)*
Filaria. a. 11. 59.

729. *Carabus blapoides.*
Filaria. a. 11. 55.

730. *Blaps Mortisaga.*
Dub. v. 197. 86.

— 787 —

Ordo IV. Orthoptera.

731. *Forficula Auricularia.*
* Filaria. a. 11. 60

732. *Locusta hemitogia.*
Filaria. a. 11. 61.

733. *Locusta verrucivora.*
Filaria. a. 11. 61.

734. *Locusta viridissima.*
Filaria. a. 11. 61.

Ordo V. Neuroptera.

735. *Phryganea. Species non dicta.*
Filaria. a. 12. 66.

Ordo VI. Hymenoptera.

736. *Tenthredinis Larva.*
Filaria. a. 12. 67.

Ordo VII. Hemiptera.

737. *Cercopis spumaria.*
Filaria. a. 12. 62.

Ordo VIII. Lepidoptera.

738. *Papilionis Antiopes Larva.*
Filaria. a. 12. 65.

739. *Papilionis Betulae larva.*
Filaria. a. 12. 65

740. *Papilionis Polychlori larva.*
Filaria. a. 12. 65.

741. *Papilionis Quercus larva.*
Filaria. a. 12. 65.

742. *Papilionis Urticae larva.*
Filaria. a. 12. 65.

743. *Sphingis Euphorbii larva.*
Filaria. a. 12. 65.

744. *Bombycis Alni larva.*
Filaria. a. 12. 65.

745. *Bombycis Cajae larva.*
Filaria. a. 12. 65.

746. *Bombycis Quercus larva.*
Filaria. a. 12. 65.

Ddd 2

Fig. 6. Entozoorum synopsis cui accedunt mantissa duplex et indices lociepletissimi *(Rudolphi, 1819). Part of the list of nematodes from insects.*

from his 1878 work, depicts the first page of his extensive catalogue of insect-associated nematodes.

Although many of the early records of insect-associated nematodes dealt with larger, more spectacular worms such as mermithids, it is clear that by the second half of the 19th century smaller parasites, such as oxyurids from the gut of various arthropods (see Fig. 8), were also being discovered and named.

The first entomopathogenic nematodes

Most of the early records of insect-parasitic nematodes referred to rather large and impressive animals measured in centimetres rather than

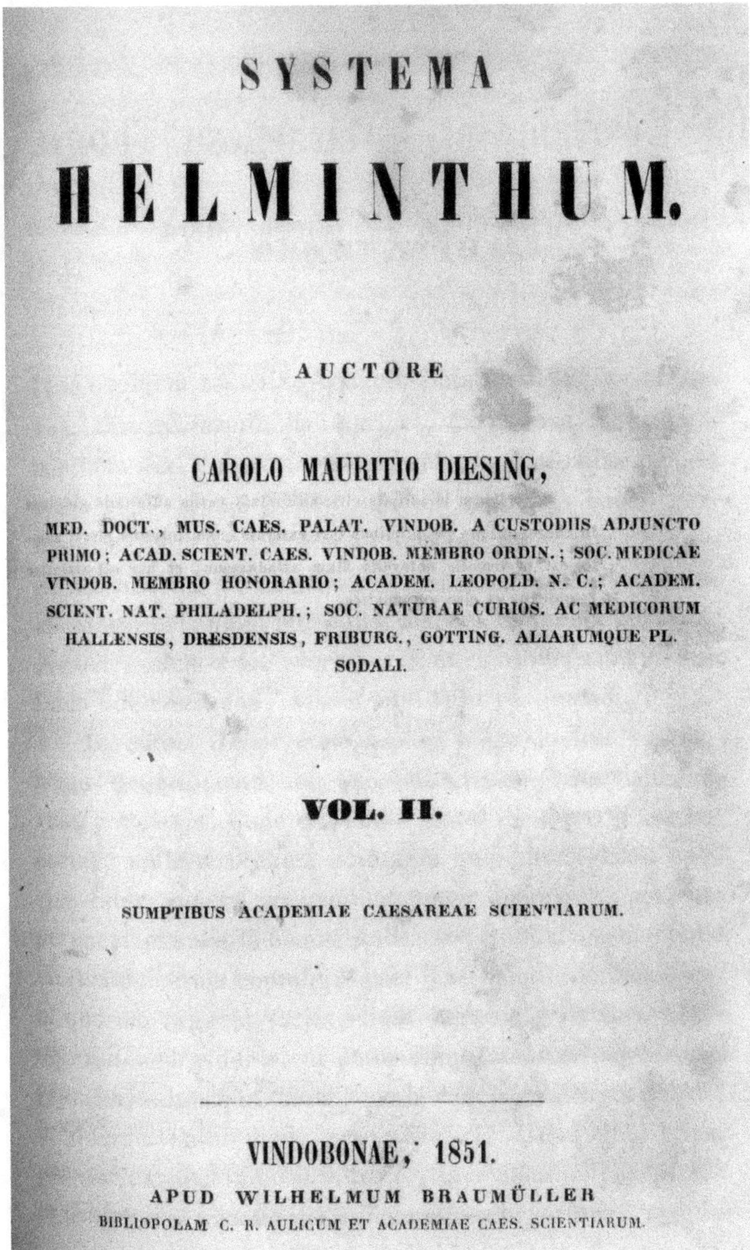

SYSTEMA

HELMINTHUM.

AUCTORE

CAROLO MAURITIO DIESING,

MED. DOCT., MUS. CAES. PALAT. VINDOB. A CUSTODIIS ADJUNCTO PRIMO; ACAD. SCIENT. CAES. VINDOB. MEMBRO ORDIN.; SOC. MEDICAE VINDOB. MEMBRO HONORARIO; ACADEM. LEOPOLD. N. C.; ACADEM. SCIENT. NAT. PHILADELPH.; SOC. NATURAE CURIOS. AC MEDICORUM HALLENSIS, DRESDENSIS, FRIBURG., GOTTING. ALIARUMQUE PL. SODALI.

VOL. II.

SUMPTIBUS ACADEMIAE CAESAREAE SCIENTIARUM.

VINDOBONAE, 1851.

APUD WILHELMUM BRAUMÜLLER

BIBLIOPOLAM C. R. AULICUM ET ACADEMIAE CAES. SCIENTIARUM.

Fig. 7. *Title page from volume 2 of the* Systema helminthum *by Diesing (1851).*

B. Evertebrata — Wirbellose Thiere.

I. Arthropoda — Gliederthiere.

a. Insecta — Insekten.

α. Orthoptera – Gradflügler.

1555. Termes lucifugus Rossi.

Isacis migrans Lespés, Ann. d. sc. natur. 4 sér. V p. 335—336, tab. VIII fig. 10—15. Abdom.

***1556. Periplaneta orientalis L. Schabe.**

Oxyuris appendiculata Leidy (= Anguillula appendiculata Leidy = Oxyuris Blattae orientalis Bütschli). Diesing, Revis. d. Nemat. p. 630—631; Bütschli, Z. f. w. Zool. XXI p. 252—293, tab. XXII fig. 20—23, 28—29; v. Linstow, Troschel's Archiv 1877, I pag. 7—9, tab. I fig. 12. Intest. crass. et ten.

Oxyuris Diesingi Hammerschmidt (= Anguillula mucrura Dies.). Diesing, Revis. d. Nemat. p. 630; Bütschli, Z. f. w. Zool. XXI p. 252—293, tab. XXI fig. 1—16, tab. XXII fig. 17—19, 24—27. Intest. crass. et ten.

Gordius Blattae orientalis v. Siebold, Stettin. entomol. Zeit. 1842 pag. 153. Abdom.

Mermis Blattae orientalis Diesing. Deslongchamps, Encyclop. méthod. II pag. 396.

1557. Blatta aegyptiaca Fabr.

Oxyuris Diesingi Hammerschmidt (v. 1556).

Oxyuris Blattae Graeffe, Beob. über Radiat. u. Würmer Nizza's pag. 51, tab. X fig. 13. Intest.

19*

Fig. 8. Compendium der helminthologie *(von Linstow, 1878). First page of the list of nematodes from insects.*

Fig. 9. *Some major figures in the history of entomopathogenic nematology. They are from right to left: Gotthold Steiner, Lauro Travassos and Ivan Filipjev (top row); Prosper Bovien, Rudolph Glaser and George Poinar Jr (bottom row).*

millimetres and, perhaps not surprisingly, it was not until the 20th century that the nematodes we now refer to as steinernematids were discovered when Steiner (1923) described *Aplectana kraussei* Steiner, 1923, a species that had actually been collected by Dr Krausse (hence the patronym) in 1917.

Gotthold Steiner (1886-1961) had been born in Switzerland but emigrated to the USA where he worked with Nathan Augustus Cobb, the father of American nematology. Steiner (Fig. 9), an outstanding nematologist, was undeniably one of the foremost scientists working with these nematodes and, in addition to proposing the genus *Neoaplectana* Steiner, 1929, also described *Neoaplectana glaseri* Steiner, 1929, the first such species to be employed in biological control of white grubs. The species was named after Rudolph Glaser, a major innovator and contributor in the development of biological control with these nematodes.

Although Steiner had originally placed his species in the genus *Aplectana* Railliet & Henry, 1916, this was a palpable misjudgement. Travassos (Fig. 9) quickly realised this and erected the new genus *Steineria* Travassos, 1927, a patronym in honour of Steiner, with *kraussei* as type species (Travassos, 1927a). Because of homonymy, Travassos (1927b) quickly renamed this genus as *Steinernema* Travassos, 1927. Lauro Pereira Travassos (1890-1970), who was born in Angra dos Rios, Rio de Janeiro, Brazil, was primarily a helminthologist, although he also published extensively on the taxonomy and systematics of nematodes parasitic in insects and other arthropods, such as diplopods. Having received his doctorate in 1913, Lauro spent most of his working life at the renowned Instituto Oswaldo Cruz in Brazil. His prolific output included over 430 publications, although strangely only two are related to entomopathogenic nematodes – his main contribution being the erection of the genus *Steinernema* referred to above, although he also proposed *Neoaplectana menozzii* Travassos, 1929, a taxon that is currently regarded as a *species inquirenda*.

Two other early pioneers in the field must be mentioned, namely Filipjev and Bovien. Ivan Nikolaevich Filipjev (1889-1940), perhaps better known as a systematist, was responsible for erecting the family Steinernematidae and also described *Neoaplectana feltiae* Filipjev, 1934 (now known as *Steinernema feltiae*), one of the commonest steinernematids (Fig. 9).

Prosper Louis Bovien, who was born in 1894 in Copenhagen, Denmark, was primarily a zoologist (Fig. 9) and is notable for his fastidious descriptions of two species of *Neoaplectana*; *N. affinis* Bovien, 1937 and *N. bibionis* Bovien, 1937. Bovien worked overseas at the Java Sugar Experiment Station from 1920-1924 before returning to Denmark and being appointed Chief Zoologist at the Phytopathological Institute. In 1936 he was also appointed as Lecturer in Zoology at the Royal Agricultural College, Copenhagen. Bovien's delightfully written doctoral thesis, entitled *Some types of association between nematodes and insects*, runs to some 114 pages (Bovien, 1937). It was in this work that he proposed *N. bibionis* and *N. affinis* and discussed in some depth the known members of the genus. Bovien was also the first to link these nematodes with the bacterium that we now know as *Xenorhabdus*.

For many decades the only genus of entomopathogenic nematode described was *Steinernema* (= *Neoaplectana*). It was not until 1976 that Poinar described *Heterorhabditis* Poinar, 1976 and a further two decades were to pass before *Neosteinernema* Nguyen & Smart, 1994 was proposed (Nguyen & Smart, 1994).

The discovery of *Steinernema carpocapsae*

In 1955, J. Weiser described a nematode obtained from larvae of the codling moth, *Carpocapsa pomonella* (Weiser, 1955). This species, which he called *Neoaplectana carpocapsae* Weiser, 1955, was appropriately named after the genus of its host and was destined to become the most heavily researched member of the genus. Coincidentally, in the same year that Weiser described *N. carpocapsae*, S.R. Dutky and W.S. Hough, working in the USA, independently found the same nematode in the same host species (Dutky & Hough, 1955). The isolate they found was named DD-136 and for a long time was the commonest isolate to be used for biological control purposes.

The re-establishment of *Steinernema kraussei*

Although *Steinernema kraussei* was the first entomopathogenic nematode to be discovered and formally named, its taxonomic status remained obscure for many decades, the later named *Neoaplectana* holding sway. The eventual recognition that *Neoaplectana* was in fact nothing more than a junior synonym of *Steinernema*, together with the establishment of *S. kraussei* as type species, were major advances that led to much-needed taxonomic stability within the group. Most of the credit for this must be given to Mráček (Mráček, 1977, 1992; Mráček & Weiser, 1979; Mráček *et al.*, 1981, 1991) for his long fight to re-establish the type species, and to Wouts *et al.* (1982) for producing what turned out to be the definitive taxonomic paper on the situation (see Chapter 2 for more detail on this aspect).

A rose by any other name?

The earliest records of insect-associated nematodes referred to them as filariids or gordiids and it was not until the 19th century that mermithids were distinguished from the true gordians. A number of different terms have been coined to refer to nematodes associated with and/or parasitic on insects. Popular for some time was the term *entomophilic*, literally meaning 'insect-loving'. This was deemed inappropriate for those nematodes that parasitised and often killed their hosts, and Poinar (1975), in an important monograph on nematodes associated with insects, used the term *entomogenous* in preference to entomophilic, presumably in an attempt to encompass the parasitic forms. 'Entomogenous', a term much used in mycology, incidentally, means 'growing on insects'. The term *entomoparasitic* was subsequently used to describe the parasitic forms of nematode and holds sway at present, whilst steinernematids and heterorhabditids, nematodes that actually kill their host by means of a specific symbiotic bacterium they deliver to the insect haemocoel, are referred to as *entomopathogenic* because the host is actually killed by a disease, rather than directly by the nematode. The term *entomopathogenic nematode*, often abbreviated to EPN, has now become universal when referring to such organisms.

Biogeography and distribution

Initial data seemed to suggest that *Steinernema* (= *Neoaplectana*) was more temperate in distribution whilst *Heterorhabditis* favoured tropical regions (Griffin *et al.*, 1991; Hominick *et al.*, 1995). This supposition was partly due to sampling bias, *Heterorhabditis*, in fact, often preferring sandy soils close to the littoral zone whilst *Steinernema* is now known to be highly speciose and widely distributed in tropical soils. The association of *Heterorhabditis* spp. with the littoral zone has been variously explained: Akhurst and Bedding (1986) suggested introduction with ship ballast, Poinar (1993) speculated that it was a reflection of their marine origin (simultaneously hypothesising – correctly as it turned out – that steinernematids and heterorhabditids were, despite their similarities, only distantly related), whilst Hominick *et al.* (1996) thought their presence was influenced by large populations of insects in marine detritus. Hominick (2002) provided a comprehensive account of the biogeography of entomopathogenic nematodes, the lists of species being arranged by nematode name and by country.

Although heterorhabditids are often associated with tropical or temperate coastal regions (*e.g.*, Griffin *et al.*, 1991; Hara *et al.*, 1991; Poinar, 1993; Amarasinghe *et al.*, 1994; Hominick *et al.*, 1995), few ecological data are available for their natural environments. Constant *et al.* (1998) found that *H. indica*, the commonest tropical species of the genus, was located in coastal areas of Guadeloupe whilst Mauléon *et al.* (2006) reported on the spatial and temporal distribution of *H. indica* in Guadeloupe. The latter authors found that *H. indica* was located on beaches and limestone cliffs where the soils were well drained, calcium-rich and neutral to slightly alkaline in pH. In such environments, the potential hosts, in addition to insects, were often small crustaceans such as sand hoppers. Mauléon *et al.* (2006) came to the conclusion that the main factor influencing distribution, at least of *H. indica*, was soil type and that the species persisted for up to 10 years (*i.e.*, the period of the study) in small patches under the canopy of trees.

Establishment of protocols for species descriptions

Descriptions of new species of EPN were often done by individuals who were not professional taxonomists and who rarely followed a consistent descriptive narrative. With the increasing number of nominal species, the lack of an overall standard was hindering research and species identification, and so minimum standards for species description were agreed by the participants in an international workshop. These protocols were subsequently published by Hominick *et al.* (1997). It was recommended that, in addition to certain morphological and morphometric data, cross-breeding tests should be performed between the putative new species and those closest to it, and that molecular characterisation, initially in the guise of RFLP profiles of ITS regions, should be provided in order to overcome the problems inherent in segregating species in genera displaying highly conserved morphology. Although the molecular methodologies have moved on since Hominick *et al.* (1997), the emphasis now being on sequencing of various genes (*e.g.*, 18S rDNA and D2D3 expansion region) rather than RFLP profiles, much of the protocol remains relevant, particularly at a time when the rise in species descriptions is almost exponential.

Morphological and morphometric characterisation

Classical identification/descriptive strategies have been employed since the first species was described in 1923. As the number of species increased it became apparent that such methodologies had difficulty in discriminating between such morphologically conserved and morphometrically variable parasites. Attempts to overcome such difficulties resulted in the creation of a number of ratios additional to the usual de Manian ones. These usually incorporated various proportions involving tail length, excretory pore and nerve ring position, and pharynx length in the infective juveniles (where morphometric variability was less because it is a non-feeding stage), whereas in males, measurements of 'hard' structures, such as the spicule and gubernaculum, were combined with other data in various ways. Many of these ratios are too variable to be of unambiguous utility, although they are *de rigueur* in formal descriptions, perhaps more from desperation than for illumination.

Some workers focused on the characters of the infective juvenile (IJ), particularly such features as the lateral field (Mráček & Bednarek, 1991), tail shape and length of the terminal clear or hyaline area, head contour and presence or absence of 'cephalic horns', and form of the bacterial pouch. Amongst these, the characters of the lateral field have proven to be most useful, the number of lines or incisures (not to be confused with the number of ridges, which is always one fewer) and their pattern becoming an essential requisite, although often requiring SEM studies to elucidate. Nguyen and Smart (1995) proposed that the status of the lateral field be assessed at various points along the length of the IJ in order to express the variation in lateral line number towards either extremity. A small group of six species bear cephalic horn-like structures, although these may be readily overlooked – they were not mentioned in the original descriptions of either *S. abbasi* or *S. riobrave*, for example (see Nguyen & Adams, 2003).

Useful characters of the adult stages focus mainly on the male. They include the detailed form of the spicules and gubernaculum as revealed by SEM, Nguyen being a leading exponent of this approach (see Nguyen & Smart, 1990, 1997), presence or absence of a caudal mucron, disposition of the copulatory papillae (not always recorded with the accuracy that they deserve), spermatozoon morphology (see Spiridonov *et al.*, 1999), and presence or absence of small cuticular projections, the epiptygmata, guarding the entrance to the female vagina. At the present time morphological characteristics can be used to recognised several species of *Steinernema* but for species of *Heterorhabditis* this approach is more difficult, molecular characterisation being imperative.

Illustrated keys to steinernematids and heterorhabditids have been published by Nguyen and Smart (1996) and Adams and Nguyen (2002), more up-to-date versions being included in Chapter 3 of this book.

Molecular methodologies and phylogeny

SPECIES IDENTIFICATION

The molecular approach to identification and phylogeny of steinernematids and heterorhabditids is dealt with in detail in Chapter 6. It is the development of these readily available methodologies that has largely facilitated the burgeoning of new species descriptions – although even now there are those who swim against the tide and publish ill-founded names reliant only on morphology and morphometrics. The discriminatory power of the molecular toolbox should not be underestimated and has become the method of choice for diagnosing these animals in an objective, rapid and repeatable manner.

The first practical diagnostic technique for identifying species of steinernematids and heterorhabditids was the development of RFLP profiling of ITS regions (see Reid & Hominick, 1992, 1993; Joyce *et al.*, 1994a, b; Reid *et al.*, 1997). This technique proved to be both cheap and reliable; a single infective juvenile, even when fixed in alcohol, yielding sufficient DNA for unequivocal species diagnosis. A database of RFLP profiles was soon established and made clear that there were many, hitherto unsuspected, new species awaiting description. RAPD profiles have been used by a number of workers, including Liu and Berry (1995, 1996).

RFLP profiling is now a little outmoded: current methodologies involving nucleotide sequence data of, for example, the ITS and D2D3 regions of rDNA are usually used for species diagnoses whilst mtDNA 16S-COII, 12S, *cox* 1 genes of the mitochondrial genome (Adams, 1998; Szalanski *et al.*, 2000; Nadler *et al.*, 2006) and 18S rDNA sequences (Blaxter *et al.*, 1998) are usually employed in assessing phylogenetic relationships. Fortunately, it is now rare (not to mention unwise) for a new species of steinernematid or heterorhabditid to be proposed without molecular data as this is a crucial contributor to the diagnostic character suite. Indeed, any new species proposed without such data is inviting immediate relegation to the status of *species inquirenda*.

With sequences of such a high proportion of the valid species now available, a robust phylogeny for *Steinernema* and *Heterorhabditis* has become increasingly apparent (Adams *et al.*, 1998; Nguyen *et al.*, 2001; Stock *et al.*, 2001; Spiridonov *et al.*, 2004; Stock & Reid, 2004; Nadler *et al.*, 2006).

PHYLOGENY OF THE STEINERNEMATIDAE AND HETERORHABDITIDAE

Initially both steinernematids and heterorhabditids were regarded as closely related genera within the Rhabditida. This mainly stemmed from the fact that both genera shared many morphological features (but crucially not the form of the male tail region) and had very similar life cycles involving a highly evolved, symbiotic relationship with specific, yet different, genera of bacteria.

Spiridonov and Belostotskaya (1983) first suggested that *Steinernema* (or *Neoaplectana* as it was then known) had affinities with alloionematids and strongyloidids. Some 10 years later, Poinar (1993), in a detailed morphological comparison, also emphasised clear differences between the two genera, concluding that the two families were polyphyletic in origin, that similarities were attributable to convergence, and that heterorhabditids were probably marine in origin.

The view of Spiridonov and Belostotskaya (1983) was recently supported by analyses of 18S rDNA sequences by Blaxter *et al.* (1998), *Steinernema* being nested within the strongyloidid clade, a position it retains in subsequent, higher resolution, tree topologies (De Ley & Blaxter, 2000, 2004).

Vermiculture and application technology

Rudolph William Glaser (1888-1941; Fig. 9) and Norman Stoll (1892-1976), although unaware of the presence of the bacterial symbiont, pioneered the artificial culture of entomopathogenic nematodes and their application for biological control. Much of their work involved the control of white grubs (the larvae of the scarabaeid beetle, *Popillia japonica*) in turf.

Bedding, working in Australia, pioneered the use of sponge as a storage matrix for EPN infective juveniles (Bedding, 1981, 1984).

Sponge formulations have been, and still are, widely used, although they are not appropriate to large scale commercial applications.

Modern mass production techniques involve the use of large capacity stainless steel biofermenters, the resulting infective juveniles being formulated in special gels to enhance shelf life of the product, whilst the polyether-polyurethane sponge-based formulations are utilised by smaller scale cottage industries catering for niche markets such as gardens. Excellent reviews of production, formulation and application methodologies for EPN have been published by Friedman (1990), Gaugler and Han (2002) and Grewal (2002).

Major modern monographs

Although George Poinar (Fig. 9) is without doubt one of the foremost names in EPN research – indeed his contributions to all aspects of insect-associated nematodes are on a colossal scale – he is perhaps more widely famous for his work on the extraction of DNA from insects fossilised in amber.

His first major monograph, *Entomogenous nematodes* (Poinar, 1975) produced an excellent overview of all insect-associated nematodes, including steinernematids, and he later published a more specialised volume, *Nematodes for biological control of insects*, which covered the topic of nematodes as biocontrol agents (Poinar, 1979). In his earlier volume Poinar (1975) compiled a comprehensive and extremely useful list of nematode/insect associations supported by a full bibliography – a major contribution to the study of insect nematodes and still a valuable resource today.

As to other monographs dealing with entomopathogenic nematodes, particular mention must be made of *Entomopathogenic nematodes in biological control* (Gaugler & Kaya, 1990) and *Entomopathogenic nematology* (Gaugler, 2002). A recent volume, *Nematodes as biological control agents* (Grewal *et al.*, 2005) has a broader canvas, yet still provides much useful data on entomopathogenic nematodes.

Perspective

The study of entomopathogenic nematodes, which can be traced back to the discovery of *S. kraussei* in 1917 (although not described until

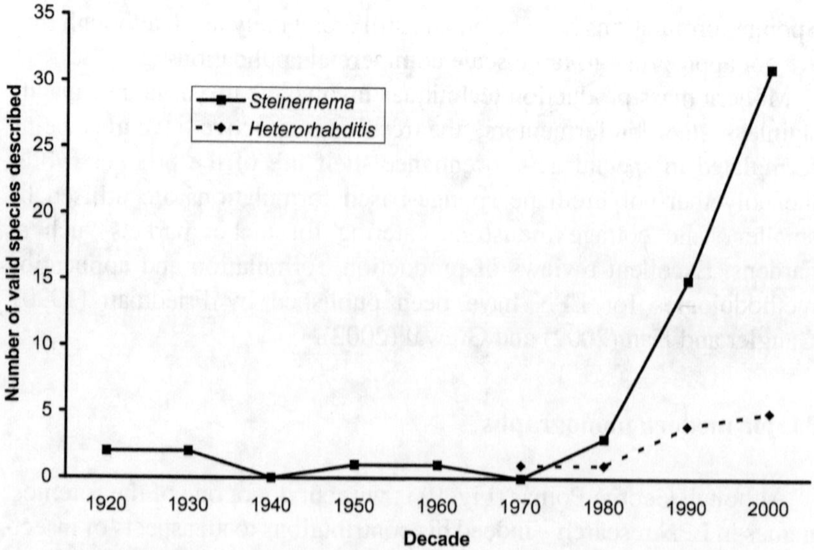

Fig. 10. *Number of valid species of* Steinernema *and* Heterorhabditis *described in each decade since 1920.*

1923), is now approaching its centenary. From a slow beginning, the pace of discovery has increased dramatically, progress being driven by the perceived advantages of biological control or management of insect pests and facilitated by the development of robust molecular techniques for species recognition.

The advent of molecular methodologies introduced a powerful discriminatory tool that has facilitated a rapid increase in the proposal of new species, particularly of *Steinernema*. Indeed, no fewer than 46 of the 55 valid species of *Steinernema* have been described since the beginning of the 1990s (Fig. 10). *Steinernema* has been shown to be far more speciose than previously suspected, an inference that would not have been possible to sustain in such morphologically-conserved organisms by relying on classical taxonomy alone. Not surprisingly, molecular methodologies have also posed some intriguing challenges to species concept in the two families (after all, what exactly *is* a species? There are almost as many opinions as there are taxonomists!), but doubtless as the science matures a consensus will emerge on which sequence(s) to use for which purpose and the relevance of small differences in base pair number to taxonomic units.

The genus *Heterorhabditis* seems to be far less speciose than *Steinernema*. The reasons for this have not been fully elucidated but probably relate to the fact that the first generation in the former is hermaphroditic, whereas in *Steinernema*, with the notable exception of one species (Stock *et al.*, 2002), it is amphimictic, thereby increasing the latter's potential for genetic diversity. Nevertheless, of the 11 valid species, nine have been described since 1990 (Fig. 10).

References

ADAMS, B.J. (1998). Species concepts and the evolutionary paradigm in modern nematology. *Journal of Nematology* 30, 1-21.

ADAMS, B.J. & NGUYEN, K.B. (2002). Taxonomy and systematics. In: Gaugler, R. (Ed.). *Entomopathogenic nematology*. Wallingford, UK, CABI Publishing, pp. 1-33.

ADAMS, B.J., BURNELL, A.M. & POWERS, T.O. (1998). A phylogenetic analysis of *Heterorhabditis* (Nemata: Rhabditidae) based on internal transcribed spacer 1 DNA sequence data. *Journal of Nematology* 30, 22-39.

AKHURST, R.J. & BEDDING, R.A. (1986). Natural occurrence of insect pathogenic nematodes (Steinernematidae and Heterorhabditidae) in soil of Australia. *Journal of the Australian Entomological Society* 25, 241-243.

ALDROVANDI, U. (1623). *De animalibus insectis libri septum*. Frankfurt, p. 271.

ALDROVANDI, U. (1640). *Serpentum et draconum historiae.* Bologna, Italy, Mark Antony Bernia, p. 402.

AMARASINGHE, L.D., HOMINICK, W.M., BRISCOE, B.R. & REID, A.P. (1994). Occurrence and distribution of entomopathogenic nematodes in Sri Lanka. *Journal of Helminthology* 68, 277-286.

BEDDING, R.A. (1981). Low cost *in vitro* mass production of *Neoaplectana* and *Heterorhabditis* species (Nematoda) for field control of insect pests. *Nematologica* 27, 109-114.

BEDDING, R.A. (1984). Large scale production, storage and transport of the insect parasitic nematodes *Neoaplectana* spp. and *Heterorhabditis* spp. *Annals of Applied Biology* 104, 117-120.

BLAXTER, M.L., DE LEY, P., GAREY, J.R., LIU, L.X., SCHELDMAN, P., VIERSTRAETE, A., VANFLETEREN, J.R., MACKEY, L.Y., DORRIS, M., FRISSE, L.M., VIDA, J.T. & THOMAS, W.K. (1998). A molecular evolutionary framework for the phylum Nematoda. *Nature* 392, 71-75.

BOVIEN, P. (1937). Some types of association between nematodes and insects. *Videnskabelige Meddelelser fra Dansk Naturhistorisk Forening, København* 101, 1-114.

CONSTANT, P., MARCHAY, L., FISCHER-LE SAUX, M., BRIAND-PANOMA, S. & MAULÉON, H. (1998). Natural occurrence of entomopathogenic nematodes (Heterorhabditidae and Steinernematidae) in Guadeloupe islands. *Fundamental and Applied Nematology* 21, 667-672.

DE LEY, P. & BLAXTER, M.L. (2002). Systematic position and phylogeny. In: Lee, D.L. (Ed.). *The biology of nematodes*. London, UK, Taylor & Francis, pp. 1-30.

DE LEY, P. & BLAXTER, M.L. (2004). A new system for Nematoda: combining morphological characters with molecular trees, and translating clades into ranks and taxa. In: Cook, R. & Hunt, D.J. (Eds). *Proceedings of the fourth international congress of nematology, Tenerife, Spain, 8-13 June 2002, Nematology monographs and perspectives, volume 2*. Leiden, The Netherlands, Brill, pp. 633-653.

DIESING, K.M. (1851). *Systema helminthum, volume 2*. Vienna, Academiae Caesareae Scientarum, 588 pp.

DUTKY, S.R. & HOUGH, W.S. (1955). Note on a parasitic nematode from codling moth larvae, *Carpocapsa pomonella* (Lepidoptera, Olethreutidae). *Proceedings of the Entomological Society of Washington* 57, 244.

FILIPJEV, I.N. (1934). Miscellanea Nematologica 1. Eine neue Art der Gattung *Neoaplectana* Steiner nebst Bermerkungen über die systematische Stellung der letzteren. *Parazitologichesky* 4, 229-240.

FRIEDMAN, M.J. (1990). Commercial production and development. In: Gaugler, R. & Kaya, H.K. (Eds). *Entomopathogenic nematodes in biological control*. Boca Raton, FL, USA, CRC Press, pp. 153-172.

GAUGLER, R. (ED.) (2002). *Entomopathogenic nematology*. Wallingford, UK, CABI Publishing, 388 pp.

GAUGLER, R. & HAN, R. (2002). Production technology. In: Gaugler, R. (Ed.). *Entomopathogenic nematology*. Wallingford, UK, CABI Publishing, pp. 289-310.

GAUGLER, R. & KAYA, H.K. (EDS) (1990). *Entomopathogenic nematodes in biological control*. Boca Raton, FL, USA, CRC Press, 365 pp.

GOEZE, J.A.E. (1782). *Entozoorum sive vermium intestinalium historia naturalis, volume 2*. Germany, Blankenburg, 471 pp. + 44 plates.

GOULD, W. (1747). *An account of English ants*. London, A. Millar, 109 pp.

GREWAL, P.S. (2002). Formulation and application technology. In: Gaugler, R. (Ed.). *Entomopathogenic nematology*. Wallingford, UK, CABI Publishing, pp. 265-287.

GREWAL, P.S., EHLERS, R.-U. & SHAPIRO-ILAN, D.I. (2005). *Nematodes as biological control agents*. Wallingford, UK, CABI Publishing, 505 pp.

GRIFFIN, C.T., MOORE, J.F. & DOWNES, M.J. (1991). Occurrence of insect-parasitic nematodes (Steinernematidae, Heterorhabditidae) in the Republic of Ireland. *Nematologica* 37, 92-100.

HARA, A.H., GAUGLER, R., KAYA, H.K. & LEBECK, L.M. (1991). Natural populations of entomopathogenic nematodes (Rhabditida: Heterorhabditidae, Steinernematidae) from the Hawaiian Islands. *Environmental Entomology* 20, 211-216.

HOMINICK, W.M. (2002). Biogeography. In: Gaugler, R. (Ed.). *Entomopathogenic nematodes*. Wallingford, UK, CABI Publishing, pp. 115-143.

HOMINICK, W.M., REID, A.P. & BRISCOE, B.R. (1995). Prevalence and habitat specificity of steinernematid and heterorhabditid nematodes isolated during soil surveys of the UK and The Netherlands. *Journal of Helminthology* 69, 27-32.

HOMINICK, W.M., REID, A.P., BOHAN, D.A. & BRISCOE, B.R. (1996). Entomopathogenic nematodes: biodiversity, geographical distribution and the convention on biological diversity. *Biocontrol Science and Technology* 6, 317-331.

HOMINICK, W.M., BRISCOE, B., DEL PINO, F.G., HENG, J., HUNT, D.J., KOZODOY, E., MRÁČEK, Z., NGUYEN, K.B., REID, A.P., SPIRIDONOV, S., STOCK, P., STURHAN, D., WATURU, C. & YOSHIDA, M. (1997). Biosystematics of entomopathogenic nematodes: current status, protocols and definitions. *Journal of Helminthology* 71, 271-298.

JOYCE, S.A., REID, A.P., DRIVER, F. & CURRAN, J. (1994a). Application of polymerase chain reaction (PCR) methods to the identification of entomopathogenic nematodes. In: Burnell, A.M., Ehlers, R.-U. & Masson, J.P. (Eds). *COST 812 Biotechnology: Genetics of entomopathogenic nematode-bacterium complexes. Proceedings of symposium & workshop, St Patrick's College, Maynooth, Co, Kildare, Ireland*. Luxembourg, DGXII, European Commission, pp. 178-187.

JOYCE, S.A., BURNELL, A.M. & POWERS, T.O. (1994b). Characterisation of *Heterorhabditis* isolates by PCR amplification of segments of mtDNA and rDNA genes. *Journal of Nematology* 26, 260-270.

VON LINSTOW, O.F.B. (1878). *Compendium der helminthologie*. Hanover, Han'sche Buchhandlung, 382 pp.

VON LINSTOW, O.F.B. (1889). *Compendium der helminthologie*. Hanover, Nachtrag, 151 pp.

LISTER, M. (1672). An extract of a letter written from York, April 12, 1672, concerning animated horse hairs, rectifying a vulgar error. *Philosophical Transactions of the Royal Society of London* 83, 4064-4065.

LIU, J. & BERRY, R.E. (1995). Determination of PCR conditions of RAPD analysis in entomopathogenic nematodes (Rhabditida: Heterorhabditidae and Steinernematidae). *Journal of Invertebrate Pathology* 65, 79-81.

LIU, J. & BERRY, R.E. (1996). Phylogenetic analysis of the genus *Steinernema* by morphological characters and randomly amplified polymorphic DNA fragments. *Fundamental and Applied Nematology* 19, 463-469.

MAULÉON, H., DENON, D. & PANOMA, S. (2006). Spatial and temporal distribution of *Heterorhabditis indica* in their natural habitats of Guadeloupe. *Nematology* 8, 603-617.

MRÁČEK, Z. (1977). *Steinernema kraussei*, a parasite of the body cavity of the sawfly, *Cephaleia abietis* in Czechoslovakia. *Journal of Invertebrate Pathology* 30, 87-94.

MRÁČEK, Z. (1992). *Steinernema kraussei* (Steiner, 1923) (Nematoda: Rhabditida: Steinernematidae): redescription of its topotype from Westphalia. *Folia Parasitologica* 41, 59-64.

MRÁČEK, Z. & WEISER, J. (1979). The head papillae of Steinernematidae. *Journal of Invertebrate Pathology* 34, 310-311.

MRÁČEK, Z., WEISER, J. & GERDIN, S. (1981). Head and cuticular structures of some species in the family Steinernematidae (Nematoda). *Nematologica* 27, 443-448.

MRÁČEK, Z., WEISER, J., BURES, M. & KAHOUNOVA, L. (1991). *Steinernema kraussei* (Steiner, 1923) (Nematoda: Rhabditida): Rediscovery of its type locality in Germany. *Folia Parasitologica* 39, 181-182.

NADLER, S.A., BOLOTIN, E. & STOCK, S.P. (2006). Phylogenetic relationships of *Steinernema* Travassos, 1927 (Nematoda: Cephalobina: Steinernematidae) based on nuclear, mitochondrial and morphological data. *Systematic Parasitology* 63, 161-181.

NGUYEN, K.B. & ADAMS, B.J. (2003). SEM and systematic studies on *Steinernema abbasi* Elawad *et al.*, 1997 and *S. riobrave* Cabanillas *et al.*, 1994. *Zootaxa* 179, 1-10.

NGUYEN, K.B. & SMART JR, G.C. (1990). *Steinernema scapterisci* n. sp. (Rhabditida: Steinernematidae). *Journal of Nematology* 22, 187-199.

NGUYEN, K.B. & SMART JR, G.C. (1994). *Neosteinernema longicurvicauda* n. gen., n. sp. (Rhabditia: Steinernematidae), a parasite of the termite *Reticulitermes flavipes* (Koller). *Journal of Nematology* 26, 162-174.

NGUYEN, K.B. & SMART JR, G.C. (1995). Scanning electron microscope studies of *Steinernema glaseri* (Nematoda: Steinernematidae). *Nematologica* 41, 183-190.

NGUYEN, K.B. & SMART JR, G.C. (1996). Identification of entomopathogenic nematodes in the Steinernematidae and Heterorhabditidae (Nemata: Rhabditida). *Journal of Nematology* 28, 286-300.

NGUYEN, K.B. & SMART JR, G.C. (1997). Scanning electron microscopic studies of spicules and gubernacula for *Steinernema* spp. (Nemata: Steinernematidae). *Nematologica* 43, 465-480.

NGUYEN, K.B., MARUNIAK, J. & ADAMS, B.J. (2001). Diagnostic and phylogenetic utility of the rDNA internal transcribed spacer sequences of *Steinernema*. *Journal of Nematology* 33, 73-82.

POINAR JR, G.O. (1975). *Entomogenous nematodes*. Leiden, The Netherlands, E.J. Brill, 317 pp.

POINAR JR, G.O. (1976). Description and biology of a new insect parasitic rhabditoid, *Heterorhabditis bacteriophora* n. gen. n. sp. (Rhabditida; Heterorhabditidae n. fam.). *Nematologica* 21, 463-470.

POINAR JR, G.O. (1979). *Nematodes for biological control of insects*. Boca Raton, FL, USA, CRC Press, 277 pp.

POINAR JR, G.O. (1993). Origins and phylogenetic relationships of the entomophilic rhabditids, *Heterorhabditis* and *Steinernema*. *Fundamental and Applied Nematology* 16, 332-338.

DE RÉAUMUR, R.A.F. (1742). *Mémoires pour servir à l'histoire des insectes. Tome sixième. Suite de l'histoire des mouches à quatre aîles, avec un supplément à celle des mouches à deux aîles*. Paris, France, 608 pp. + 48 plates.

REID, A.P. & HOMINICK, W.H. (1992). Restriction fragment length polymorphisms within the ribosomal repeat unit of British entomopathogenic nematodes (Rhabditida: Steinernematidae). *Parasitology* 105, 317-323.

REID, A.P. & HOMINICK, W.H. (1993). Cloning of the rDNA repeat unit from a British entomopathogenic nematode (Steinernematidae) and its potential for species identification. *Parasitology* 107, 529-536.

REID, A.P., HOMINICK, W.H. & BRISCOE, B.R. (1997). Molecular taxonomy and phylogeny of entomopathogenic nematode species (Rhabditida: Steinernematidae) by RFLP analysis of the ITS region of rDNA repeat unit. *Systematic Parasitology* 37, 187-193.

RUDOLPHI, C.A. (1809). *Entozoorum sive vermium intestinalium historia naturalis, volume 2*. Amsterdam, Taberni Librariae et Artium, 386 pp. + 6 plates.

RUDOLPHI, C.A. (1819). *Entozoorum synopsis cui accedunt mantissa duplex et indices lociepletissimi*. Berlin, Augusti Rücker, 811 pp. + 3 plates.

SCHNEIDER, A.F. (1866). *Monographie der Nematoden*. Berlin, 357 pp. + 28 plates.

SPIRIDONOV, S.E. & BELOSTOTSKAYA, F.N. (1983). [New data about morphology and biology of the nematodes of the genus *Neoaplectana* and their position in the system of rhabditids.] *Parasitologyia* 17, 119-125.

SPIRIDONOV, S.E., HOMINICK, W.M. & BRISCOE, B.R. (1999). Morphology of amoeboid cells in the uterus of *Steinernema* species (Rhabditida: Steinernematidae). *Russian Journal of Nematology* 7, 39-42.

SPIRIDONOV, S.E., REID, A.P., PODRUCKA, K., SUBBOTIN, S.A. & MOENS, M. (2004). Phylogenetic relationships within the genus *Steinernema* (Nematoda: Rhabditida) as inferred from analyses of sequences of the ITS1-5.8S-ITS2 region of rDNA and morphological features. *Nematology* 6, 547-566.

STEINER, G. (1923). *Aplectana kraussei* n. sp., eine in der Blattwespe *Lyda* sp. parasitierende Nematodenform, nebst Bemerkungen über das Seitenorgan der parasitischen Nematoden. *Zentralblatt für Bakteriologie, Parasitenkunde, Infektions krankheiten und Hygiene, Abteilung II* 59, 14-18.

STOCK, S.P. & REID, A.P. (2004). Biosystematics of entomopathogenic nematodes (Steinernematidae, Heterorhabditidae): current status and future directions. In: Cook, R. & Hunt, D.J. (Eds). *Proceedings of the fourth international congress of nematology, Tenerife, Spain, 8-13 June 2002, Nematology monographs and perspectives, volume 2.* Leiden, The Netherlands, Brill, pp. 435-446.

STOCK, S.P., CAMPBELL, J.F. & NADLER, S.A. (2001). Phylogeny of *Steinernema* Travassos, 1927 (Cephalobina: Steinernematidae) inferred from ribosomal DNA sequences and morphological characters. *Journal of Parasitology* 87, 877-889.

STOCK, S.P., GRIFFIN, C.T. & BURNELL, A.M. (2002). Morphological characterisation of three isolates of *Heterorhabditis* Poinar, 1976 from the 'Irish group' (Nematoda: Rhabditida: Heterorhabditidae) and additional evidence supporting their recognition as a distinct species, *H. downesi* n. sp. *Systematic Parasitology* 51, 95-106.

SZALANSKI, A.L., TAYLOR, D.B. & MULLIN, P.G. (2000). Assessing nuclear and mitochondrial DNA sequence variation within *Steinernema* (Rhabditida: Steinernematidae). *Journal of Nematology* 32, 229-233.

TRAVASSOS, L. (1927a). Sobre o genero *Oxysomatium. Boletim Biologico, Sao Paulo* 5, 20-21.

TRAVASSOS, L. (1927b). Uma nova *Capillaria* parasita de peixes be agua doce: *Capillaria sentinosa* n. sp. *Boletim Biologico, Sao Paulo* 10, 215-217.

WEISER, J. (1955). *Neoaplectana carpocapsae* n. sp. (Anguillulata, Steinernematinae), novy cizopasník housenek obaleče jablečného, *Carpocapsa pomonella* L. *Věstník Československé Zoologické Společynosti* 19, 44-52.

WOUTS, W.M., MRÁČEK, Z., GERDIN, S. & BEDDING, R.A. (1982). *Neoaplectana* Steiner, 1929 a junior synonym of *Steinernema* Travassos, 1927 (Nematoda: Rhabditida). *Systematic Parasitology* 4, 147-154.

Nematology Monographs & Perspectives, 2007, Vol. 5, 27-57

Chapter 2

Overview of taxonomy and systematics

David J. HUNT

CABI Europe-UK, Bakeham Lane, Egham, Surrey TW20 9TY, UK

The family Steinernematidae

HISTORY

The first species of Steinernematidae to be described was *Aplectana kraussei* Steiner, 1923. This was collected by Dr Krausse in Germany in 1917 and came from *Lyda potrophica* Hartig (= *Cephaleia abietis* L.), a hymenopterous sawfly (Krausse, 1917). The nematode was subsequently described by Steiner (1923). Steiner tentatively placed his species in the oxyurid genus *Aplectana* Railliet & Henry, 1916, a group of nematodes otherwise known from amphibians, and perhaps a rather unlikely repository for an insect nematode. It did not remain long in that taxon as a few years later, Travassos (1927a), working in Brazil, transferred the species to his new genus *Steineria* Travassos, 1927, nominating the former *A. kraussei* as the type species. Travassos named *Steineria* in honour of Steiner, but unfortunately overlooked the fact that this name was already occupied by *Steineria* Micoletzky, 1922 and was therefore an objective junior homonym. According to the rules of zoological nomenclature, a new name was therefore required. On realising his error, Travassos (1927b) immediately proposed *Steinernema* Travassos, 1927 as the replacement epithet, thereby neatly solving the problem yet at the same time maintaining the patronymic attribution to Steiner. For many years, *Steinernema kraussei*, the type species of the genus, remained little known, the only record being the original description from 1923. Skrjabin *et al.* (1951) even transferred the species to the genus *Oxysomatium*, a move which was ignored by other authors and was subsequently repudiated by Veremchuk (1964).

The original description of *S. kraussei* was based on five males and 18 females. The morphometrics were given as the usual Cobb formula although Steiner also included certain of the de Man indices, namely α, β and γ (nowadays universally referred to as a, b and c). The detailed description was illustrated by a rather good drawing of the mature female (Fig. 11) and a series of drawings depicting the anterior end, tail region and male copulatory organs. An *en face* view depicted the lateral sensory organs (the amphids) plus four submedian papillae, two of which were subdorsal and the other two subventral in location. No labial papillae were illustrated or mentioned in the description.

Steiner (1929) erected the genus *Neoaplectana* Steiner, 1929 and differentiated it from *Steinernema* on the basis that: *i*) the former had two circles of six cephalic papillae whereas *Steinernema* was described as having but a single ring of four; and *ii*) that *Neoaplectana* had more male copulatory papillae in a different arrangement. In fact, Steiner misinterpreted the sense organs; *Neoaplectana* actually has an outer circle of four cephalic papillae plus two large amphidial apertures and an inner circle of six labial papillae. The supposed difference between the two genera in the number of papillae on the cephalic extremity was not corrected for over 50 years and resulted in confusion and uncertainty about the status of the genera.

The type species of *Neoaplectana* was *N. glaseri* Steiner, 1929. This species was discovered in New Jersey, USA, where it attacked white grubs, the larvae of the scarabaeid beetle, *Popillia japonica*. *Neoaplectana glaseri* (now known, of course, as *Steinernema glaseri*) was the first nematode to be mass-produced and used for biological control of an insect pest under field conditions.

Neoaplectana menozzii Travassos, 1932, the next species to be described (Travassos, 1932), was found inside *Conorrhynchus* (= *Cleonus*) *mendicus* Gyll., a curculionid beetle, in Italy by C. Menozzi. This species was described from preserved material only, is not well known and has not been re-identified since the original discovery. It is possible that *N. menozzii* is conspecific with *Neoaplectana feltiae* Filipjev, 1934, a species originally discovered in 1916. Other suggestions include the possibility that *menozzii* is conspecific with either *affine* or *carpocapsae*. All things considered, it is probably best to regard the poorly known *menozzii* as a *species inquirenda*.

The description of *N. feltiae* by Filipjev (1934) was based on a nematode first isolated in Russia in 1926. *Neoaplectana feltiae* came from the cutworm, *Feltia segetum* (Schiff.) Herbst (Insecta: Lepidoptera), the specific epithet being derived from the genus of

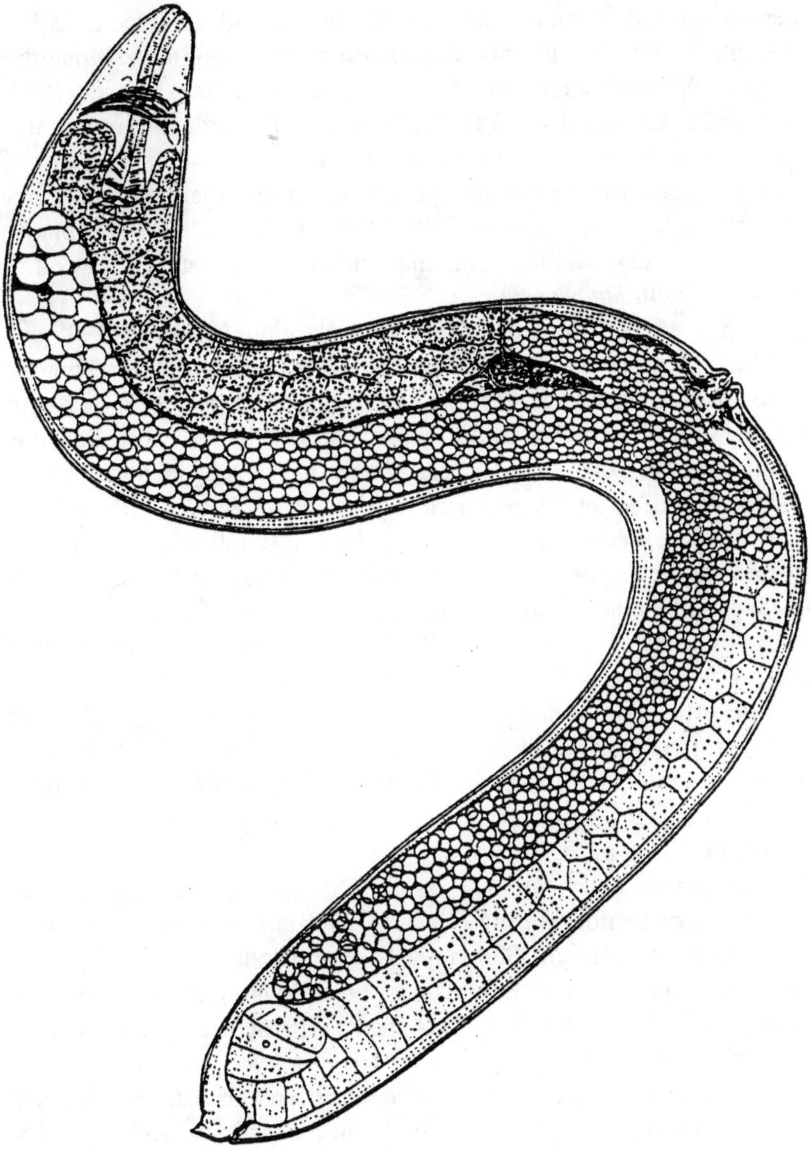

Fig. 11. Aplectana kraussei. *After original illustration by Steiner (1923).*

its insect host. In the same paper, Filipjev proposed the subfamily Steinernematinae Filipjev, 1934 (actually he called it Steinerneminae, but this name is improperly formed and must be corrected), the type genus being *Steinernema* Travassos, 1927. The subfamily was later recognised at family rank by Chitwood and Chitwood (1937), although the family authority should still be cited as Steinernematidae Filipjev, 1934 (rather than the oft-seen Steinernematidae (Filipjev, 1934) Chitwood & Chitwood, 1937) as the Principle of Coordination states that ranks created at one level in the family group (*i.e.*, subfamily, family and superfamily) are simultaneously created at the other ranks and carry the same author and date (see ICZN, 1999). Filipjev (1934) also placed *Neoaplectana* in the Steinernematinae and indicated that it could well be congeneric with *Steinernema*.

The next species to be described were *N. bibionis* Bovien, 1937 and *N. affinis* Bovien, 1937, both recorded from bibionid flies (Diptera: Bibionidae) from several localities in Denmark (Bovien, 1937). Prosper Bovien, in a delightfully crafted work that is still well worth reading, commented that one of his species may have been identical to the inadequately described *N. menozzii*. The quality of Bovien's illustrations comfortably exceed that of many modern descriptions; his care and attention to the disposition of the genital papillae, for example, serves as an example that later authors would have done well to follow (Fig. 12). Of the two species described by Bovien, *N. bibionis* was synonymised under *S. feltiae* whilst Poinar (1988) redescribed *N. affinis*.

Little progress was made in the next 15 years or so until Sobolev (1953) proposed the family Neoaplectanidae Sobolev, 1953. This family, of course, subsequently became a junior synonym of Steinernematidae when *Neoaplectana*, its type genus, was eventually recognised as being a junior synonym of *Steinernema*.

Neoaplectana carpocapsae Weiser, 1955, perhaps the most famous of all steinernematids, was described by Weiser (1955) from the larvae of the codling moth in Germany. The nematode was more or less simultaneously discovered by Dutky and Hough (1955), also from codling moth, but in the USA. This species (particularly the Dutky and Hough isolate which was known as DD-136) was widely used as a potential biocontrol agent and a number of different isolates were named and maintained under laboratory conditions. Weiser and Köhler (1955) also described *N. janickii* Weiser & Köhler, 1955, although this species is now regarded as a *species inquirenda*.

In the 1950s, 1960s and 1970s, a number of species of *Neoaplectana* were described from the then USSR (Kirjanova & Putschkova, 1955; Kakuliya & Veremchuk, 1965; Veremchuk, 1969; Veremchuk & Litvinchuk, 1971; Rubtsov & Polevik, 1979) and others were named, but not

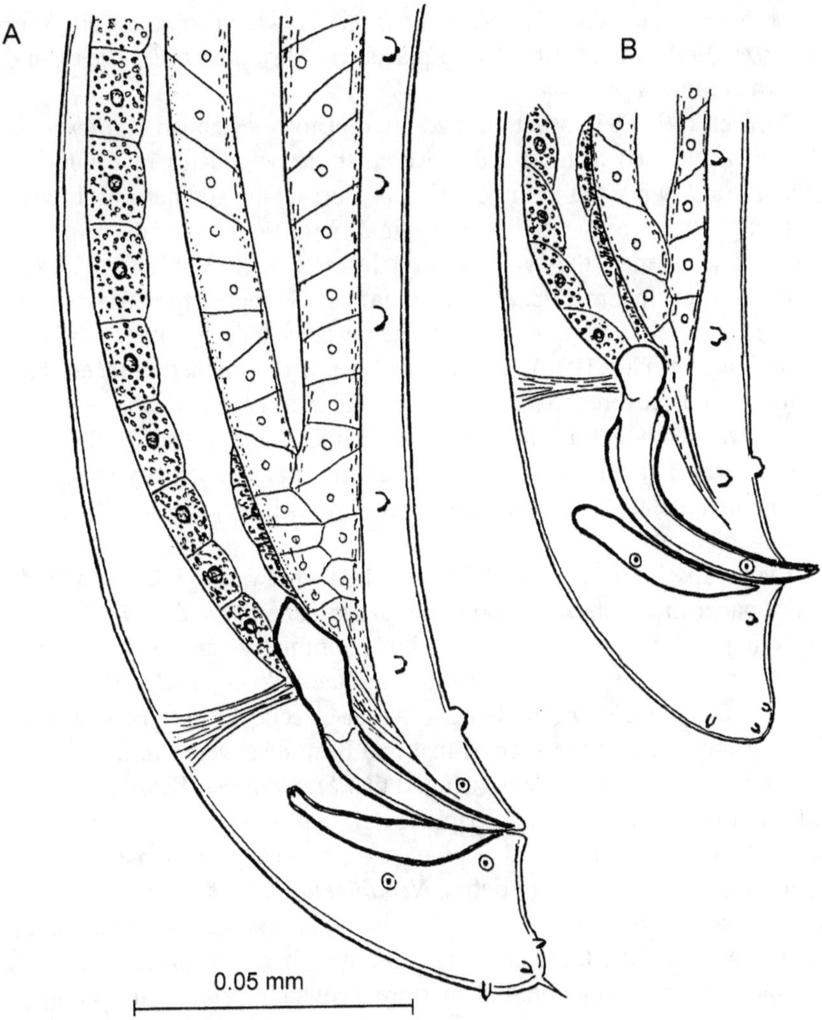

Fig. 12. Neoaplectana bibionis *(A) and* N. affinis *(B). Male tail, lateral view. After original illustration by Bovien (1937).*

formally described (Veremchuk, 1966, 1970). As these latter names have no formal description they are regarded as *nomina nuda*, or naked names, and have no official standing. Many of these 'species' would appear to be synonymous with either *S. feltiae* or *S. carpocapsae*, although it is impossible to be sure and the names are best left in peace. Veremchuk (1964) regarded Steinernematidae as invalid and replaced it with the later Neoaplectanidae. *Steinernema kraussei* was also removed from *Oxysomatium*, where it had been placed by Skrjabin *et al.* (1951), and returned to *Neoaplectana*.

Mráček (1977), in a determined attempt to re-establish *Steinernema kraussei*, claimed to have rediscovered the type species in *Cephaleia abietis*, a sawfly found in coniferous forests in southern Bohemia. Although the type host of the original *S. kraussei* was the same as Mráček's material, the type locality differed, with Mráček's material coming from Bohemia (Czechoslovakia) and Steiner's from Westphalia, Germany. His claim was met with some resistance by other workers, including Poinar (1986), as it was disputed whether Mráček had actually rediscovered *S. kraussei* or had merely found a species of *Neoaplectana* infecting the same host. Mráček and Weiser (1979) and Mráček *et al.* (1981) continued to study the cephalic sensory organs of Steinernematidae, the latter paper using the SEM to elucidate papillal arrangement.

Eventually Wouts *et al.* (1982) published what was to become a landmark paper in which they proposed that *Steinernema* and *Neoaplectana* be synonymised on the basis that both nominal genera had the same arrangement of cephalic and labial papillae. Wouts *et al.* studied the population of *Steinernema kraussei* from Czechoslovakia plus surviving type material of the species that had been discovered in the USDA collection. They clearly demonstrated that *Steinernema kraussei* had, in addition to the four cephalic papillae originally mentioned by Steiner, an inner circle of six labial papillae, *i.e.*, precisely the same arrangement that had been used to define *Neoaplectana*. This discovery meant that *Neoaplectana*, which was largely based on the presence of an outer ring of six (actually there are only four, the other two corresponding to the rather prominent amphidial apertures) cephalic papillae and an inner ring of six labial papillae, *vs* only a single ring of four cephalic papillae in *Steinernema*, became a junior synonym of *Steinernema*, the older and therefore senior genus.

Although this move gained an almost immediate general acceptance, it was not universal: Poinar, a determined champion of *Neoaplectana*, in particular, refused to accept that the true *S. kraussei* had been rediscovered in Czechoslovakia and opined that this material did not correspond to *S. kraussei* and was, in fact, a *Neoaplectana* – hence the undeniable SEM demonstration of the occurrence of four cephalic and six labial papillae, the diagnostic character of the latter genus. Poinar

(1986) seemed to overlook or dismiss the fact that the type specimen of *S. kraussei* also showed the presence of labial papillae (the material was poorly preserved, thereby making morphological observations difficult and any conclusions contentious). This dispute rumbled on for the next decade and a half with both generic combinations circulating in the literature to the confusion of one and all. Eventually, Poinar (1990), without further explanation, renounced *Neoaplectana* in favour of *Steinernema*, although he suggested that *S. kraussei* was not validly designated as type species of the genus and proposed that *S. glaseri*, the second species to be described, be used in its stead. Thankfully, this proposal was not followed, and the taxonomy of the genus finally settled down according to the proposal by Wouts *et al.* (1982).

Subsequently, Mráček *et al.* (1991) successfully re-identified the type locality of *S. kraussei* near Neuenherrsee, Westphalia, Germany, whilst Mráček (1992) provided a full description of topotype material of the type species, thus stabilising the systematics of the group beyond all reasonable contention. There now only remained to be resolved some minor skirmishing over the appropriate endings of various specific epithets (the International Commission on Zoological Nomenclature (1999) states that the gender of the species name must match that of the genus name, *Neoaplectana* being feminine and *Steinernema* neuter in gender).

The proposal to adjust the endings of the species names to conform to the Rules was made by Hominick *et al.* (1997) and reiterated by Stock *et al.* (1997). Minor tussles over the decision to change the ending of certain names were soon resolved, the situation being summarised, apparently to everyone's satisfaction, by Adams and Nguyen (2002).

Species descriptions in the genus *Steinernema* have risen at an impressive rate in the last decade, partly as a result of increased interest in the group as potential biocontrol agents, particularly in previously unresearched geographic areas such as Asia and Africa, but also because of the utility of molecular techniques in reliably characterising species in a group that is increasingly difficult to access using classical methodologies. Molecular characterisation, at first with RFLP profiles of the ITS region but later with DNA sequences and D2D3 expansion region, have provided a ready tool to probe beyond the conserved morphology of this fascinating group of entomopathogenic nematodes, and have revealed a hitherto unsuspected complexity in their taxonomy – 55 species are now accepted as valid, many of which could not have been sustained by reliance on traditional descriptions. Of the valid species, 46 have been proposed since 1990 and of these 31 since 2000. Compare this with the fact that only two currently valid species were described in the 1920s, two more in the 1930s, one in the 1950s, one in the 1960s and four in the 1980s – progress indeed!

TAXONOMY AND SYSTEMATICS

Family Steinernematidae Filipjev, 1934
= Neoaplectanidae Sobolev, 1953

DIAGNOSIS

Panagrolaimomorpha: Strongyloidoidea. Parasites of insects and other arthropods. Infective juveniles non-feeding, carrying symbiotic entomopathogenic bacteria. First and second generation adults amphimictic. Oviparous to ovoviviparous.

Adults

Medium to large sized, fairly plump nematodes, Cephalic region with six labial and four cephalic papillae. Lateral fields indistinct or absent. Cheilorhabdions forming sclerotised ring, posterior stoma funnel-shaped. Pharynx rhabditoid, comprising cylindroid procorpus with slight metacorpal swelling, short isthmus and rounded basal bulb with small valve plates. Nerve ring located in isthmus region, excretory pore distinct, usually located anterior to nerve ring. Male gonad monorchic, reflexed. Spicules paired, robust, capitulum distinct. Gubernaculum large, bifurcate in ventral view. Bursa absent. Copulatory papillae typically comprising single midventral, precloacal papilla plus 10-14 paired papillae. Tail short, terminus rounded. Female vulva in form of median transverse slit. Female genital system amphididelphic, reflexed. Tail either short, conoid (*Steinernema*), or considerably longer than anal body diam. (*Neosteinernema*).

Infective juvenile

Third-stage infective juvenile slender, vermiform. Cuticle finely annulated. Mouth and anus closed, stoma collapsed, pharynx non-functional, reduced; symbiotic bacteria usually visible in intestinal pouch just posterior to basal bulb. Lateral field prominent with up to nine incisures. Excretory pore located anterior to nerve ring. Tail medium, conoid, tapering to pointed terminus (*Steinernema*), or longer, tapering to a filiform terminus (*Neosteinernema*).

TYPE GENUS

Steinernema Travassos, 1927
= *Steineria* Travassos, 1927 *nec Steineria* Micoletzky, 1922
= *Neoaplectana* Steiner, 1929
= *Patanodontus* de Villalobos & Camino, 1997

OTHER GENUS

Neosteinernema Nguyen & Smart, 1994

Genus *Steinernema* Travassos, 1927

= *Steineria* Travassos, 1927 *nec Steineria* Micoletzky, 1922
= *Neoaplectana* Steiner, 1929
= *Patanodontus* de Villalobos & Camino, 1997

DIAGNOSIS (MODIFIED AFTER ADAMS & NGUYEN, 2002)

Male

Medium sized, fairly plump nematodes, heat relaxed habitus usually J- or C-shaped. Cephalic region with six labial and four cephalic papillae. Perioral disc usually present. Pharynx comprising cylindroid procorpus with slight metacorpal swelling, short isthmus and rounded basal bulb with small valve plates. Nerve ring located in isthmus region, excretory pore distinct, usually located anterior to nerve ring. Testis monorchic, reflexed. Spicules paired, robust, often yellowish to brownish in colour, capitulum distinct. Gubernaculum large, bifurcate in ventral view. Bursa absent. Copulatory papillae typically comprising single midventral, precloacal papilla plus 10-14 (usually 10-11) paired papillae. Precloacal subventral series numbering 7-10 pairs. Postcloacal papillae may be subventral, lateral or subdorsal in position. Tail short, terminus rounded, mucron present or absent.

Female

Large, rather plump nematodes. Heat-relaxed habitus often C-shaped or spiralled. Cephalic region with six partially or completely fused lips, each bearing a papilla; four cephalic papillae, and two small amphidial apertures. Cuticle appearing smooth or annulated, lateral fields absent. Cheilorhabdions sclerotised, forming a ring, posterior stoma funnel-shaped. Pharynx similar to male. Vulva in form of median transverse slit; epiptygmata present or absent. Genital system amphididelphic, reflexed; oviparous or ovoviviparous. Tail short, conoid, *ca* one or less anal body diam. long in first generation but longer in second generation. Phasmids may be inconspicuous.

Infective juvenile

Third-stage infective juvenile slender, vermiform, non-feeding. Heat relaxed habitus usually straight or slightly ventrally arcuate. Cuticle finely annulated, retained sheath from second-stage moult may be present. Mouth and anus closed, stoma collapsed, pharynx non-functional, reduced, basal bulb usually directed dorsad; symbiotic bacteria usually visible in intestinal pouch just posterior to basal bulb. Lateral field well-developed with up to nine incisures. Excretory pore prominent, located anterior to nerve ring. Tail medium conoid, tapering to pointed terminus. Phasmids located about midpoint of tail.

TYPE SPECIES

Steinernema kraussei (Steiner, 1923) Travassos, 1927
= *Aplectana kraussei* Steiner, 1923
= *Steineria kraussei* (Steiner, 1923) Travassos, 1927
= *Oxysomatium kraussei* (Steiner, 1923) Skrjabin, Shikhobalova & Mozgovoi, 1951

OTHER SPECIES

S. abbasi Elawad, Ahmad & Reid, 1997
= *S. thermophilum* Sudershan & Singh, 2000 n. syn.
S. aciari Qiu, Yan, Zhou, Nguyen & Pang, 2005
S. affine (Bovien, 1937) Wouts, Mráček, Gerdin & Bedding, 1982
= *Neoaplectana affinis* Bovien, 1937
S. akhursti Qiu, Hu, Zhou, Mei, Nguyen & Pang, 2005
S. anatoliense Hazir, Stock & Keskin, 2003
S. apuliae Triggiani, Mráček & Reid, 2004
S. arenarium (Artyukhovsky, 1967) Wouts, Mráček, Gerdin & Bedding, 1982
= *Neoaplectana arenaria* Artyukhovsky, 1967

= *Neoaplectana anomali* Kozodoi, 1984
= *Steinernema anomalae* (Kozodoi, 1984) Curran, 1989
S. ashiuense Phan, Takemoto & Futai, 2006
S. asiaticum Anis, Shahina, Reid & Rowe, 2002
S. backanense Phan, Spiridonov, Subbotin & Moens, 2006
S. beddingi Qiu, Hu, Zhou, Pang & Nguyen, 2005
S. bicornutum Tallosi, Peters & Ehlers, 1995
S. carpocapsae (Weiser, 1955) Wouts, Mráček, Gerdin & Bedding, 1982
= *Neoaplectana carpocapsae* Weiser, 1955
= *Neoaplectana feltiae sensu* Stanuszek, 1974a, *nec* Filipjev, 1934
= *Neoaplectana feltiae pieridarum* Stanuszek, 1974b
= *Neoaplectana pieridarum* Stanuszek, 1974b
= *Steinernema feltiae pieridarum* (Stanuszek, 1974b) Wouts, Mráček, Gerdin & Bedding, 1982
= *Steinernema pieridarum* (Stanuszek, 1974b) Wouts, Mráček, Gerdin & Bedding, 1982
= *Neoaplectana carpocapsae pieridarum* Stanuszek, 1974b
= *Neoaplectana dutkyi* Turco, Thames & Hopkins, 1971
= *Steinernema dutkyi* (Turco, Thames & Hopkins, 1971) Wouts, Mráček, Gerdin & Bedding, 1982
S. caudatum Xu, Wang & Li, 1991
S. ceratophorum Jian, Reid & Hunt, 1997
S. cubanum Mráček, Hernandez & Boemare, 1994
S. cumgarense Phan, Spiridonov, Subbotin & Moens, 2006
S. diaprepesi Nguyen & Duncan, 2002
S. eapokense Phan, Spiridonov, Subbotin & Moens, 2006
S. feltiae (Filipjev, 1934) Wouts, Mráček, Gerdin & Bedding, 1982
= *Neoaplectana feltiae* Filipjev, 1934
= *Neoaplectana bibionis* Bovien, 1937
= *Steinernema bibionis* (Bovien, 1937) Wouts, Mráček, Gerdin & Bedding, 1982
= *Neoaplectana leucaniae* Hoy, 1954
= *Steinernema leucaniae* (Hoy, 1954) Wouts, Mráček, Gerdin & Bedding, 1982
S. glaseri (Steiner, 1929) Wouts, Mráček, Gerdin & Bedding, 1982
= *Neoaplectana glaseri* Steiner, 1929
S. guangdongense Qiu, Fang, Zhou, Pang & Nguyen, 2004
S. hebeiense Chen, Li, Yan, Spiridonov & Moens, 2006
S. hermaphroditum Stock, Griffin & Chaerani, 2004

S. intermedium (Poinar, 1986) Mamiya, 1988
= *Neoaplectana intermedia* Poinar, 1986
S. jollieti Spiridonov, Krasomil-Osterfeld & Moens, 2004
S. karii Waturu, Hunt & Reid, 1997
S. khoisanae Nguyen, Malan & Gozel, 2006
S. kushidai Mamiya, 1988
S. leizhouense Nguyen, Qiu, Zhou & Pang, 2006
S. litorale Yoshida, 2005
S. loci Phan, Nguyen & Moens, 2001
S. longicaudum Shen & Wang, 1992
= *S. serratum* Liu, 1992 [= *nomen nudum*]
S. monticolum Stock, Choo & Kaya, 1997
S. neocurtillae Nguyen & Smart, 1992
S. oregonense Liu & Berry, 1996
S. pakistanense Shahina, Anis, Reid, Rowe & Maqbool, 2001
S. puertoricense Román & Figueroa, 1994
S. rarum (de Doucet, 1986) Mamiya, 1988
= *Neoaplectana rara* de Doucet, 1986
S. riobrave Cabanillas, Poinar & Raulston, 1994
S. ritteri de Doucet & Doucet, 1990
S. robustispiculum Phan, Subbotin, Waeyenberge & Moens, 2005
S. sangi Phan, Nguyen & Moens, 2001
S. sasonense Phan, Spiridonov, Subbotin & Moens, 2006
S. scapterisci Nguyen & Smart, 1990
S. scarabaei Stock & Koppenhöfer, 2003
S. siamkayai Stock, Somsook & Reid, 1998
S. sichuanense Mráček, Nguyen, Tailliez, Boemare & Chen, 2006
S. silvaticum Sturhan, Spiridonov & Mráček, 2005
S. tami Luc, Nguyen, Reid & Spiridonov, 2000
S. thanhi Phan, Nguyen & Moens, 2001
S. websteri Cutler & Stock, 2003
S. weiseri Mráček, Sturhan & Reid, 2003
S. yirgalemense Nguyen, Tesfamariam, Gozel, Gaugler & Adams, 2004

SPECIES INQUIRENDAE

Steinernema acaudatum (de Villalobos & Camino, 1997) Sudhaus & Fürst von Lieven, 2003
= *Patanodontus acaudatum* de Villalobos & Camino, 1997

Steinernema agriotos (Veremchuk, 1969) Wouts, Mráček, Gerdin & Bedding, 1982
= *Neoaplectana agriotos* Veremchuk, 1969
Steinernema belorussica (Veremchuk, 1969) Wouts, Mráček, Gerdin & Bedding, 1982
= *Neoaplectana belorussica* Veremchuk, 1969
Steinernema bothynoderi (Kirjanova & Putschkova, 1955) Wouts, Mráček, Gerdin & Bedding, 1982
= *Neoaplectana bothynoderi* Kirjanova & Putschkova, 1955
Steinernema georgica (Kakuliya & Veremchuk, 1965) Wouts, Mráček, Gerdin & Bedding, 1982
= *Neoaplectana georgica* Kakuliya & Veremchuk, 1965
Steinernema janickii (Weiser & Köhler, 1955) Wouts, Mráček, Gerdin & Bedding, 1982
= *Neoaplectana janickii* Weiser & Köhler, 1955
Steinernema kirjanovae (Veremchuk, 1969) Wouts, Mráček, Gerdin & Bedding, 1982
= *Neoaplectana kirjanovae* Veremchuk, 1969
Steinernema masoodi Ali, Shaheen, Pervez & Hussain, 2005
Steinernema melolonthae (Weiser, 1958) Wouts, Mráček, Gerdin & Bedding, 1982
= *Neoaplectana melolonthae* Weiser, 1958
Steinernema menozzii (Travassos, 1932) Wouts, Mráček, Gerdin & Bedding, 1982
= *Neoaplectana menozzii* Travassos, 1932
Steinernema octodontum (Camino & de Villalobos, 1996) Sudhaus & Fürst von Lieven, 2003
= *Diplogaster octodontus* Camino & de Villalobos, 1996
Steinernema seemae Ali, Shaheen, Pervez & Hussain, 2005
Steinernema semiothisae (Veremchuk & Litvinchuk, 1971) Wouts, Mráček, Gerdin & Bedding, 1982
= *Neoaplectana semiothisae* Veremchuk & Litvinchuk, 1971
Steinernema solanense Walia & Bajaj, 2005
Steinernema tabanivorum (Rubstov & Polevik, 1979) Wouts, Mráček, Gerdin & Bedding, 1982
= *Neoaplectana tabanivora* Rubstov & Polevik, 1979

Nomina nuda

Neoaplectana brevilarvalis pieridarum Sandner & Stanuszek, 1972

Neoaplectana chresima Steiner *in* Glaser, McCoy & Girth, 1942
Neoaplectana dutkii Welch, 1963
= *N. dutkyi* Turco, Thames & Hopkins, 1971 [= a junior synonym of *S. carpocapsae*]
Neoaplectana elateridicola Veremchuk, 1970
Neoaplectana titovi Veremchuk, 1966
Steinernema serratum Liu, 1992

NOTES

1) Poinar (1990) has been regarded as being the combining authority for a number of *Steinernema* species formally placed in *Neoaplectana* (*S. affine*, *S. anomalae*, *S. carpocapsae*, *S. intermedium* and *S. rarum*), but this is not correct. When Wouts *et al.* (1982) formally proposed *Neoaplectana* as a junior synonym of *Steinernema*, it is clear that they regarded all nominal species formerly in *Neoaplectana*, whether valid or not, as belonging to *Steinernema*. Although they did not formally propose the combination '*Steinernema carpocapsae*', for example, the action nevertheless occurred, notwithstanding the possibility that the authors themselves did not realise that such was the case. This applies to all the species placed in *Neoaplectana* (but not previously combined under *Steinernema*) prior to the paper by Wouts *et al.* (1982). Combinations made subsequent to that date take the combining authority as normal. A similar situation prevailed with combinations under the plant-parasitic genus *Globodera* and discussion with The Secretariat of the International Commission on Zoological Nomenclature (see Loof & Bakker, 1992) confirmed this point.

2) Many of the species listed as *species inquirenda* have been proposed as synonyms of several other species, notably the widespread and common *S. carpocapsae* and *S. feltiae*. Although this is a convenient way of disposing of the names and of tidying up the nomenclature, the fact remains that these species are too poorly characterised to allow such conclusions, particularly now that many more species have been and are being proposed and the importance of molecular data is being realised. Accordingly, it is preferable to follow a more objective course whilst realising and accepting that the ultimate fate of these binomina may never be unambiguously resolved.

Of the *species inquirendae* or *nomina nuda* listed above, Mráček (pers. comm., 2006) indicates from his research that: *i*) *S. agriotos*, *S.*

belorussica, *S. elateridicola* and *S. semiothisae* are likely to be junior synonyms of *S. carpocapsae*; *ii*) *S. kirjanovae* is probably a junior synonym of *S. kraussei*; and *iii*) *S. bothynoderi* and *S. tabanivora* are probably junior synonyms of *S. feltiae*.

Two recently proposed species, *S. masoodi* and *S. seemae* (Ali *et al.*, 2005), are compromised by exceptionally poor and inaccurate descriptions and also lack molecular data.

3) The species *menozzii* may be a senior synonym of *affine* (see Wouts, 1980), *carpocapsae* (Wouts, unpublished letter) or *feltiae*.

4) The endings of some of the specific epithets combined under *Steinernema* have been corrected to reflect either the fact that *-nema* is neuter in gender, or that they were originally wrongly formed by their author(s).

5) In 1963 Welch, in an unpublished thesis, used the binomen *Neoaplectana dutkii* for the nematode then known as DD-136 and subsequently shown to be *N. carpocapsae*. The name was also employed by Jackson (1965), although it has no validity under the Code. Turco *et al.* (1971), cited the species as '*N. dutkyi* Jackson, 1965' in a key – despite the fact that Jackson had not validly characterised the species. In differentiating the species in a key, however, Turco *et al.* (1971) can be regarded as validly establishing the binomen and hence the correct citation is *N. dutkyi* Turco, Thames & Hopkins, 1971.

6) According to Sudhaus and Fürst von Lieven (2003), two species described from Argentina as diplogastrids actually belong to *Steinernema*, namely *Diplogaster octodontus* Camino & de Villalobos, 1996 and *Patanodontus acaudatum* de Villalobos & Camino, 1997. Apparently, both species have all the features of *Steinernema* and were therefore combined under that genus by Sudhaus and Fürst von Lieven (2003). These authors also synonymised under *Steinernema* the genus *Patanodontus* de Villalobos & Camino, 1997, which was proposed with *P. acaudatum* as type species (de Villalobos & Camino, 1997).

7) *Steinernema caudatum* and *S. ritteri* lack molecular data, but are nevertheless regarded as valid species herein.

8) *Steinernema thermophilum* is regarded as a junior synonym of *S. abbasi* herein because the sequences of the ITS region of the two nematodes are identical (Nguyen, pers. comm., 2006). Interestingly, an extensive and detailed study (Tailliez *et al.*, 2006) on the bacterial

associates of *Steinernema* spp. found that both *S. abbasi* and *S. thermophilum* shared the same symbiont, namely *Xenorhabdus indica* Somvanshi, Lang, Ganguly, Swiderski, Saxena & Stackebrandt, 2006, a species that had been described from material identified as *S. thermophilum*.

Genus *Neosteinernema* Nguyen & Smart, 1994

DIAGNOSIS (AFTER ADAMS & NGUYEN, 2002)

Male

Similar to *Steinernema* in general respects. Smaller than female. Spicule foot-shaped with dorsal hump or swelling, gubernaculum almost as long as spicule. Genital papillae arranged as single midventral papilla located just anterior to cloacal aperture and 13-14 pairs of which eight pairs pre-cloacal. Tail conoid with digitate terminus, phasmids conspicuous.

Female

Similar to *Steinernema*. Head with four cephalic and six labial papillae. Ovoviviparous. Tail longer than anal body diam. ($c' = 1.1$-1.68), phasmids prominent, located on protuberance in posterior half of tail.

Infective juvenile

Similar in general respects to *Steinernema*. Head slightly swollen, tail long ($c = ca\ 5.5$) elongate to filiform, as long as pharynx, usually curved near extremity, phasmid conspicuous.

TYPE AND ONLY SPECIES

N. longicurvicauda Nguyen & Smart, 1994

Family Heterorhabditidae

HISTORY

The first species of heterorhabditid to be described may have been *Rhabditis hambletoni* Pereira, 1937 (Pereira, 1937), although this is currently regarded as a *species inquirenda*. The first definitive description, however, came relatively late in the history of entomopathogenic nematology when Poinar (1976) proposed *Heterorhabditis bacteriophora* Poinar, 1976. The description of this species occurred in the same year as the proposal of the new genus and new species *Chromonema heliothidis* Khan, Brooks & Hirschmann, 1976 by Khan *et al.* (1976). The latter binomen was apparently based on the same original isolate as Poinar's species but because Poinar published first, *Heterorhabditis bacteriophora* has priority and the latter becomes a junior synonym.

The family Heterorhabditidae Poinar, 1976 was proposed by Poinar (1976) at the same time as he described *Heterorhabditis*, the type genus. Although the family was initially regarded as belonging to the Rhabditida (together with the biologically similar Steinernematidae), recent molecular evidence has supported the contention of Poinar (1993) that the two families have evolved separately. Similarities in life cycle, and to some extent morphology, having therefore evolved in parallel. Current molecular evidence places the Heterorhabditidae with the strongylids (Blaxter *et al.*, 1998; De Ley & Blaxter, 2002, 2004).

The number of nominal *Heterorhabditis* species (currently fewer than 15) is rather small when compared to the 50 or so valid species of *Steinernema*. This may be a feature of under-reporting, although with the widespread search for new entomopathogenic nematodes that has taken place in the last decade or so, that explanation no longer seems viable. More likely it is due to the different life cycle, the first generation being hermaphroditic whereas in *Steinernema* it is amphimictic, thereby opening up the opportunity for greater genetic diversity in the ensuing progeny. As with *Steinernema*, the number of new species proposed has increased in recent years, molecular methodologies once again playing a substantial role in diagnostics.

TAXONOMY AND SYSTEMATICS

Family Heterorhabditidae Poinar, 1976

DIAGNOSIS

Rhabditomorpha: Strongyloidea. Obligate parasites of insects and other arthropods. Infective juveniles non-feeding, carrying symbiotic entomopathogenic bacteria. First generation adults hermaphroditic, second generation adults amphimictic. Oviparous to ovoviviparous.

Adults

Cephalic region bearing six, separate, conoid lips, each with terminal papilla. Amphidial apertures small. Stoma wide, shallow, cheilsotom forming a ring, other parts fused. Pharyngeal collar present. Pharynx short, with cylindroid procorpus, no metacorpal swelling, isthmus weak, basal bulb spheroid, with small valve plates. Nerve ring located in isthmus region, excretory pore usually opening posterior to basal bulb. Vulva median, transverse. Hermaphrodite with two opposed, reflexed, ovotestes; amphimictic female with amphididelphic, reflexed, ovaries. Tail short conoid. Male spicules paired, separate, slender, more or less straight with small, offset, capitulum. Gubernaculum linear in lateral view, plate-like in ventral view. Bursa present. Usually nine pairs of genital papillae or rays, sometimes fewer. Tail medium conoid, tapering to pointed terminus.

Infective juvenile

Third-stage infective juvenile usually enclosed in longitudinally ridged sheath of second-stage. Sheath tessellate anteriorly. Infective juvenile vermiform, non-feeding, bearing apical dorsal tooth on cephalic region. Lateral field with two ridges although central incisure not easily visible with LM so that area between bands appears smooth. Mouth and anus closed, stoma and pharynx reduced. Excretory pore located posterior to nerve ring. Symbiotic bacteria retained in intestine. Tail medium conoid, tapering to pointed terminus.

TYPE AND ONLY GENUS

Heterorhabditis Poinar, 1976
= *Chromonema* Khan, Brooks & Hirschmann, 1976

Genus *Heterorhabditis* Poinar, 1976

= *Chromonema* Khan, Brooks & Hirschmann, 1976

DIAGNOSIS (MODIFIED AFTER ADAMS & NGUYEN, 2002)

Hermaphroditic female

Large, plump, nematodes. Head truncate to slightly rounded, lips six, well-developed, separate, each with terminal papilla. One or two small raised structures may be present at base of each lip. Amphidial aperture small. Stoma wide, shallow, cheilorhabdions forming a cuticularised refractive ring, other stoma parts fused to form collapsed posterior portion. Posterior part of stoma surrounded by pharyngeal collar. Pharynx lacking metacorpus, isthmus slender, basal bulb swollen with reduced valve plates. Nerve ring located at middle of isthmus, excretory pore usually located posterior to pharynx. Vulva median, slit-like, surrounded by elliptical rings. Ovotestes two, opposed, reflexed. Oviparous at first, ovoviviparous later. Tail conoid, longer than anal body diam., postanal swelling usually present.

Amphimictic female

Similar to, but usually smaller than, hermaphrodite. Labial papillae prominent. Reproductive tract amphididelphic reflexed. Vulva not functional for egg laying (juveniles hatch in body) but functional for mating.

Male

Genital tract monorchic, reflexed. Spicules paired, separate, slightly curved ventrally. Spicule capitulum short, offset from lamina by constriction. Gubernaculum usually *ca* half of spicule length. Bursa well-developed, peloderan, usually nine pairs of ray-like genital papillae (occasionally fewer) present.

Infective juvenile

Third-stage infective juvenile usually enclosed in sheath formed from cuticle of second stage. Sheath with longitudinal ridges and tessellated anteriorly. Cuticle of infective juvenile striated, lateral field with two ridges, although central incisure not easily visible with LM so that area between bands appears smooth. Head with prominent dorsal tooth. Mouth and anus closed, non-functional. Stoma collapsed with parallel walls. Pharynx and intestine reduced. Excretory pore located posterior to nerve ring. Symbiotic bacteria retained within intestine. Tail conoid, tapering to pointed terminus.

TYPE SPECIES

Heterorhabditis bacteriophora Poinar, 1976
= *Chromonema heliothidis* Khan, Brooks & Hirschmann, 1976
= *Heterorhabditis heliothidis* (Khan, Brooks & Hirschmann, 1976) Poinar, Thomas & Hess, 1977
= *H. argentinensis* Stock, 1993

OTHER SPECIES

H. amazonensis Andaló, Nguyen & Moino, 2007
H. baujardi Phan, Subbotin, Nguyen & Moens, 2003
H. downesi Stock, Griffin & Burnell, 2002
H. floridensis Nguyen, Gozel, Koppenhöfer & Adams, 2006
H. indica Poinar, Karunakar & David, 1992
= *H. hawaiiensis* Gardner, Stock & Kaya, 1994
H. marelatus Liu & Berry, 1996
= *H. hepialius* Stock, Strong & Gardner, 1996
H. megidis Poinar, Jackson & Klein, 1987
H. mexicana Nguyen, Shapiro-Ilan, Stuart, McCoy, James & Adams, 2004
H. taysearae Shamseldean, Abou El-Sooud, Abd-Elgawad & Saleh, 1996
H. zealandica Poinar, 1990
= *H. heliothidis apud* Wouts, 1979

SPECIES INQUIRENDAE

H. brevicaudis Liu, 1994
H. egyptii Abd-Elgawaad & Ameen, 2005
H. hambletoni (Pereira, 1937) Poinar, 1976
= *Rhabditis hambletoni* Pereira, 1937
H. hoptha (Turco, 1970) Poinar, 1979
= *Neoaplectana hoptha* Turco, 1970
H. poinari Kakuliya & Mikaia, 1997

NOMEN NUDUM

Heterorhabditis downesi Hass, Downes & Griffin, 2001 *nec Heterorhabditis downesi* Stock, Griffin & Burnell, 2002

NOTES

Hass *et al.* (2001) published the combination *Heterorhabditis downesi* before the species was formally described by Stock *et al.* (2002), the former paper being published on 27 December 2001 and the latter in February 2002. Hass *et al.* (2001) did not provide sufficient information to diagnose the species and hence their regrettable use of the name prior to its official publication must be regarded as a *nomen nudum*. In view of this situation, the subsequent proposal of the binomen *Heterorhabditis downesi* by Stock *et al.* (2002) must be regarded as valid.

References

ABD-ELGAWAAD, M.M.M. & AMEEN, H.H. (2005). *Heterorhabditis egyptii* n. sp. (Rhabditida: Heterorhabditidae) from Egypt. *Egyptian Journal of Agricultural Research-National Research Center* 2, 855-870.

ADAMS, B.J. & NGUYEN, K.B. (2002). Taxonomy and systematics. In: Gaugler, R. (Ed.). *Entomopathogenic nematology*. Wallingford, UK, CABI Publishing, pp. 1-33.

ALI, S.S., SHAHEEN, A., PERVEZ, R. & HUSSAIN, M.A. (2005). *Steinernema masoodi* sp. n. and *S. seemae* sp. n. (Nematoda: Rhabditidae [*sic*]: Steinernematidae) from India. *International Journal of Nematology* 15, 89-99.

ANDALÓ, V., NGUYEN, K.B. & MOINO, A. (2007). *Heterorhabditis amazonensis* n. sp. (Rhabditida: Heterorhabditidae) from Amazonas, Brazil. *Nematology* 8 (2006), 853-867.

ANIS, M., SHAHINA, F., REID, A.P. & ROWE, J. (2002). *Steinernema asiaticum* sp. n. (Rhabditida: Steinernematidae) from Pakistan. *International Journal of Nematology* 12, 220-231.

ARTYUKHOVSKY, A.K. (1967). [*Neoaplectana arenaria* nov. sp. (Steinernematidae, Nematoda) inducing nematode disease in chafers of the Voronezh region.] *Trudy Voroneszhskogo Gosudarstvennogo Zapovednika* 15, 94-100.

BLAXTER, M.L., DE LEY, P., GAREY, J.R., LIU, L.X., SCHELDMAN, P., VIERSTRAETE, A., VANFLETEREN, J.R., MACKEY, L.Y., DORRIS, M., FRISSE, L.M., VIDA, J.T. & THOMAS, W.K. (1998). A molecular evolutionary framework for the phylum Nematoda. *Nature* 392, 71-75.

BOVIEN, P. (1937). Some types of association between nematodes and insects. *Videnskabelige Meddelelser fra Dansk Naturhistorisk Forening, København* 101, 1-114.

CABANILLAS, H.E., POINAR JR, G.O. & RAULSTON, J.R. (1994). *Steinernema riobravis* sp. nov. (Rhabditida: Steinernematidae) from Texas. *Fundamental and Applied Nematology* 17, 123-131.

CAMINO, N.B. & DE VILLALOBOS, L.C. (1996). *Diplogaster octodontus* n. sp. (Rhabditida, Diplogasteridae) parasito de ninfas de *Gryllodes laplatae* (Orthoptera, Gryllidae) en la Argentina. *Iheringia Serie Zoologia* 89, 183-186.

CHEN, S., LI, X., YAN, A., SPIRIDONOV, S.E. & MOENS, M. (2006). A new entomopathogenic nematode, *Steinernema hebeiense* sp. n. (Rhabditida: Steinernematidae) from North China. *Nematology* 8, 563-574.

CHITWOOD, B.G. & CHITWOOD, M.B. (1937). *An introduction to nematology. Section 1, Anatomy*. Baltimore, MD, USA, Monumental Printing Co., 213 pp.

CURRAN, J. (1989). Chromosome numbers of *Steinernema* and *Heterorhabditis* species. *Revue de Nématologie* 12, 145-148.

CUTLER, G.C. & STOCK, S.P. (2003). *Steinernema websteri* sp. n. (Rhabditida: Steinernematidae), a new entomopathogenic nematode from China. *Nematologia Mediterranea* 31, 215-224.

DE DOUCET, M.M.A. (1986). A new species of *Neoaplectana* Steiner, 1929 (Nematoda: Steinernematidae) from Cordoba, Argentina. *Revue de Nématologie* 9, 317-323.

DE DOUCET, M.M.A. & DOUCET, M.E. (1990). *Steinernema ritteri* n. sp. (Nematoda: Steinernematidae) with a key to the species of the genus. *Nematologica* 36, 257-265.

DE LEY, P. & BLAXTER, M.L. (2002). Systematic position and phylogeny. In: Lee, D.L. (Ed.). *The biology of nematodes*. London, UK, Taylor & Francis, pp. 1-30.

DE LEY, P. & BLAXTER, M.L. (2004). A new system for Nematoda: combining morphological characters with molecular trees, and translating clades into ranks and taxa. In: Cook, R. & Hunt, D.J. (Eds). *Proceedings of the fourth international congress of nematology, Tenerife, Spain, 8-13 June 2002, Nematology monographs and perspectives, volume 2*. Leiden, The Netherlands, Brill, pp. 633-653.

DUTKY, S.R. & HOUGH, W.S. (1955). Note on a parasitic nematode from codling moth larvae, *Carpocapsa pomonella* (Lepidoptera, Olethreutidae). *Proceedings of the Entomological Society of Washington* 57, 244.

ELAWAD, S., AHMED, W. & REID, A.P. (1997). *Steinernema abbasi* sp. n. (Nematoda: Steinernematidae) from the Sultanate of Oman. *Fundamental and Applied Nematology* 20, 435-442.

FILIPJEV, I.N. (1934). Miscellanea Nematologica 1. Eine neue Art der Gattung *Neoaplectana* Steiner nebst Bermerkungen über die systematische Stellung der letzteren. *Parazitologichesky* 4, 229-240.

GARDNER, S.L., STOCK, S.P. & KAYA, H.K. (1994). A new species of *Heterorhabditis* from the Hawaiian islands. *Journal of Parasitology* 80, 100-106.

GLASER, R.W., MCCOY, E.E. & GIRTH, H.B. (1942). The biology and culture of *Neoaplectana chresima*, a new nematode parasitic in insects. *Journal of Parasitology* 28, 122-136.

HASS, B., DOWNES, M.J. & GRIFFIN, C.T. (2001). Correlation between survival in water and persistence of infectivity in soil of *Heterorhabditis* spp. isolates. *Nematology* 3, 573-579.

HAZIR, S., STOCK, S.P. & KESKIN, N. (2003). A new entomopathogenic nematode, *Steinernema anatoliense* n. sp. (Rhabditida: Steinernematidae), from Turkey. *Systematic Parasitology* 55, 211-220.

HOMINICK, W.M., BRISCOE, B., DEL PINO, F.G., HENG, J., HUNT, D.J., KOZODOY, E., MRÁČEK, Z., NGUYEN, K.B., REID, A.P., SPIRIDONOV, S., STOCK, P., STURHAN, D., WATURU, C. & YOSHIDA, M. (1997). Biosystematics of entomopathogenic nematodes: current status, protocols and definitions. *Journal of Helminthology* 71, 271-298.

HOY, J.M. (1954). The biology and host range of *Neoaplectana leucaniae*, a new species of insect-parasitic nematode. *Parasitology* 44, 392-399.

INTERNATIONAL COMMISSION ON ZOOLOGICAL NOMENCLATURE (1999). *International Code of Zoological Nomenclature*, 4th edition. London, UK, International Trust for Zoological Nomenclature, 306 pp.

JACKSON, G.J. (1965). Differentiation of three species of *Neoaplectana* (Nematoda: Rhabditida), grown axenically. *Parasitology* 55, 571-578.

JIAN, H., REID, A.P. & HUNT, D.J. (1997). *Steinernema ceratophorum* n. sp. (Nematoda: Steinernematidae), a new entomopathogenic nematode from north-east China. *Systematic Parasitology* 37, 115-125.

KAKULIYA, G. & MIKAIA, N. (1997). [New species of the nematode *Heterorhabditis poinari* sp. nov. (Rhabditida; Heterorhabditidae).] *Soobshcheniya Akademii Nauk Gruzii SSR* 155, 457-459.

KAKULIYA, G.A. & VEREMCHUK, G.V. (1965). [*Neoaplectana georgica* n. sp. (Nematoda: Steinernematidae) in *Amphimallon solstitialis*.] *Soobshcheniya Akademii Nauk Gruzii SSR* 40, 713-718.

KHAN, A., BROOKS, W.M. & HIRSCHMANN, H. (1976). *Chromonema heliothidis* n. gen., n. sp. (Steinernematidae, Nematoda), a parasite of *Heliothis zeae* (Noctuidae, Lepidoptera), and other insects. *Journal of Nematology* 8, 159-168.

KIRJANOVA, E.S. & PUTSCHKOVA, L.V. (1955). [A new parasite of the beet weevil – *Neoaplectana bothynoderi* Kirjanova et Putschkova n. sp. (Nematoda).] *Trudi Zoologicheskogo Instituta Akademiya Nauk SSR* 18, 53-62.

KOZODOI, E.M. (1984). [A new entomophagous nematode *Neoaplectana anomali* sp. n. (Rhabditida: Steinernematidae) and its biology.] *Zoologichesky Zhurnal* 63, 1605-1909.

KRAUSSE, A. (1917). Forstentomologische Exkursionen ins Eggegebirge zum Studium der Massenvermehrung der *Cephaleia abietis* L. *Archiv für Naturgeschichte Berlin* A6, 46-49.

LIU, J. (1992). Taxonomic study of the genus: *Steinernema* Travassos and *Heterorhabditis* Poinar. In: *Proceedings of the XIX International Congress of Entomology, 28 June-4 July 1992, Beijing, China*, p. 318.

LIU, J. (1994). [A new species of the genus *Heterorhabditis* from China (Rhabditida: Heterorhabditidae).] *Acta Zootaxonomica Sinica* 19, 268-272.

LIU, J. & BERRY, R.E. (1996a). *Heterorhabditis marelatus* n. sp. (Rhabditida: Heterorhabditidae) from Oregon. *Journal of Invertebrate Pathology* 67, 48-54.

LIU, J. & BERRY, R.E. (1996b). *Steinernema oregonensis* n. sp. (Rhabditida: Steinernematidae) from Oregon, U.S.A. *Fundamental and Applied Nematology* 19, 375-380.

LOOF, P.A.A. & BAKKER, J. (1992). Authorities of specific names in, and transfers to, the nominal genus *Globodera* Skarbilovich, 1959. *Nematologica* 38, 385-386.

LUC, P.V., NGUYEN, K.B., REID, A.P. & SPIRIDONOV, S.E. (2000). *Steinernema tami* sp. n. (Rhabditida: Steinernematidae) from Cat Tien Forest, Vietnam. *Russian Journal of Nematology* 8, 33-43.

MAMIYA, Y. (1988). *Steinernema kushidai* n. sp. (Nematoda: Steinernematidae) associated with scarabaeid beetle larvae from Shizuoka, Japan. *Applied Entomology and Zoology* 23, 313-320.

MICOLETZKY, H. (1922). Die freilebenden Erd-Nematoden mit besonderer Berücksichtigung der Steiermark und der Bukowina, zuglich mit einer Revision sämtlicher nicht mariner, freilebender Nematoden in Form von Genus – Beschreibungen und Bestimmungsschlüsseln. *Archiv für Naturgeschichte Berlin* 87, 1-320 (p. 119).

MRÁČEK, Z. (1977). *Steinernema kraussei*, a parasite of the body cavity of the sawfly, *Cephaleia abietis* in Czechoslovakia. *Journal of Invertebrate Pathology* 30, 87-94.

MRÁČEK, Z. (1992). *Steinernema kraussei* (Steiner, 1923) (Nematoda: Rhabditida: Steinernematidae): redescription of its topotype from Westphalia. *Folia Parasitologica* 41, 59-64.

MRÁČEK, Z. & WEISER, J. (1979). The head papillae of Steinernematidae. *Journal of Invertebrate Pathology* 34, 310-311.

MRÁČEK, Z., WEISER, J. & GERDIN, S. (1981). Head and cuticular structures of some species in the family Steinernematidae (Nematoda). *Nematologica* 27, 443-448.

MRÁČEK, Z., WEISER, J., BURES, M. & KAHOUNOVA, L. (1991). *Steinernema kraussei* (Steiner, 1923) (Nematoda: Rhabditida): Rediscovery of its type locality in Germany. *Folia Parasitologica* 39, 181-182.

MRÁČEK, Z., HERNANDEZ, E.A. & BOEMARE, N.E. (1994). *Steinernema cubana* sp. n. (Nematoda: Rhabditida: Steinernematidae) and the preliminary characterisation of its associated bacterium. *Journal of Invertebrate Pathology* 64, 123-129.

MRÁČEK, Z., STURHAN, D. & REID, A. (2003). *Steinernema weiseri* n. sp. (Rhabditida, Steinernematidae), a new entomopathogenic nematode from Europe. *Systematic Parasitology* 56, 37-47.

MRÁČEK, Z., NGUYEN, K.B., TAILLIEZ, P., BOEMARE, N. & SHULONG-CHEN (2006). *Steinernema sichuanense* n. sp. (Rhabditida, Steinernematidae) a new species of entomopathogenic nematode from the province of Sichuan, east Tibetan Mts, China. *Journal of Invertebrate Pathology* 93, 157-169.

NGUYEN, K.B. & DUNCAN, L.W. (2002). *Steinernema diaprepesi* n. sp. (Rhabditida: Steinernematidae), a parasite of the citrus root weevil *Diaprepes abbreviatus* (L.) (Coleoptera: Curculionidae). *Journal of Nematology* 34, 159-170.

NGUYEN, K.B. & SMART JR, G.C. (1990). *Steinernema scapterisci* n. sp. (Rhabditida: Steinernematidae). *Journal of Nematology* 22, 187-199.

NGUYEN, K.B. & SMART JR, G.C. (1992). *Steinernema neocurtillis* n. sp. (Rhabditida: Steinernematidae) and a key to species of the genus *Steinernema*. *Journal of Nematology* 24, 463-477.

NGUYEN, K.B. & SMART JR, G.C. (1994). *Neosteinernema longicurvicauda* n. gen., n. sp. (Rhabditia: Steinernematidae), a parasite of the termite *Reticulitermes flavipes* (Koller). *Journal of Nematology* 26, 162-174.

NGUYEN, K.B., SHAPIRO-ILAN, D.I., STUART, R.J., MCCOY, C.W., JAMES, R.R. & ADAMS, B.J. (2004a). *Heterorhabditis mexicana* n. sp. (Rhabditida: Heterorhabditidae) from Tamaulipas, Mexico, and morphological studies of the bursa of *Heterorhabditis* spp. *Nematology* 6, 231-244.

NGUYEN, K.B., TESFAMARIAM, M., GOZEL, U., GAUGLER, R. & ADAMS, B.J. (2004b). *Steinernema yirgalemense* n. sp. (Rhabditida: Steinernematidae) from Ethiopia. *Nematology* 6, 839-856.

NGUYEN, K.B., GOZEL, U., KOPPENHÖFER, H.S. & ADAMS, B.J. (2006a). *Heterorhabditis floridensis* n. sp. (Rhabditida: Heterorhabditidae) from Florida. *Zootaxa* 1177, 1-19.

NGUYEN, K.B., MALAN, A.P. & GOZEL, U. (2006b). *Steinernema khoisanae* n. sp. (Rhabditida: Steinernematidae), a new entomopathogenic nematode from South Africa. *Nematology* 8, 157-175.

NGUYEN, K.B., QIU, L., ZHOU, Y. & PANG, Y. (2006c). *Steinernema leizhouense* sp. n. (Nematoda: Steinernematidae), a new entomopathogenic nematode from southern China. *Russian Journal of Nematology* 14, 101-118.

PEREIRA, C. (1937). *Rhabditis hambletoni* n. sp., nema apparentemente semiparasito da 'bróca do algodoeiro' (*Gasterocercodes brasiliensis*). *Archivo do Instituto Biologico* 8, 215-230.

PHAN, K.L., NGUYEN, N.C. & MOENS, M. (2001a). *Steinernema loci* sp. n. and *Steinernema thanhi* n. sp. (Rhabditida: Steinernematidae) from Vietnam. *Nematology* 3, 503-514.

PHAN, K.L., NGUYEN, N.C. & MOENS, M. (2001b). *Steinernema sangi* sp. n. (Rhabditida: Steinernematidae) from Vietnam. *Russian Journal of Nematology* 9, 1-7.

PHAN, K.L., SUBBOTIN, S.A., NGUYEN, N.C. & MOENS, M. (2003). *Heterorhabditis baujardi* sp. n. (Rhabditida: Heterorhabditidae) from Vietnam and morphometric data for *H. indica* populations. *Nematology* 5, 367-382.

PHAN, K.L., SUBBOTIN, S.A., WAEYENBERGE, L. & MOENS, M. (2005). A new entomopathogenic nematode, *Steinernema robustispiculum* n. sp. (Rhabditida: Steinernematidae), from Chumomray National Park in Vietnam. *Systematic Parasitology* 60, 23-32.

PHAN, K.L., SPIRIDONOV, S.E., SUBBOTIN, S.A. & MOENS, M. (2006a). Four new species of *Steinernema* Travassos, 1928 with short infective juveniles from Vietnam. *Russian Journal of Nematology* 14, 11-29.

PHAN, K.L., TAKEMOTO, S. & FUTAI, K. (2006b). *Steinernema ashiuense* sp. n. (Nematoda: Steinernematidae), a new entomopathogenic nematode from Japan. *Nematology* 8, 681-690.

POINAR JR, G.O. (1976). Description and biology of a new insect parasitic rhabditoid, *Heterorhabditis bacteriophora* n. gen. n. sp. (Rhabditida; Heterorhabditidae n. fam.). *Nematologica* 21, 463-470.

POINAR JR, G.O. (1986). *Neoaplectana intermedia* n. sp. (Steinernematidae: Nematoda) from South Carolina. *Revue de Nématologie* 8 (1985), 321-327.

POINAR JR, G.O. (1988). Redescription of *Neoaplectana affinis* Bovien (Rhabditida: Steinernematidae). *Revue de Nématologie* 11, 143-147.

POINAR JR, G.O. (1989). *Nematodes for biological control of insects*. Boca Raton, FL, USA, CRC Press, 277 pp.

POINAR JR, G.O. (1990). Taxonomy and biology of Steinernematidae and Heterorhabditidae. In: Gaugler, R. & Kaya, H.K. (Eds). *Entomopathogenic nematodes in biological control*. Boca Raton, FL, USA, CRC Press, pp. 23-61.

POINAR JR, G.O. (1993). Origins and phylogenetic relationships of the entomophilic rhabditids, *Heterorhabditis* and *Steinernema*. *Fundamental and Applied Nematology* 16, 333-338.

POINAR JR, G.O., HANSEN, E.L., YARWOOD, E.A. & WEISER, J. (1972). Clarification of the status of the DD-136 strain of *Neoaplectana carpocapsae* Weiser. *Nematologica* 18, 288-290.

POINAR JR, G.O., THOMAS, G.M. & HESS, R. (1977). Characteristics of the specific bacterium associated with *Heterorhabditis bacteriophora* (Heterorhabditidae: Rhabditida). *Nematologica* 23, 97.

POINAR JR, G.O., JACKSON, T. & KLEIN, M. (1987). *Heterorhabditis megidis* sp. n. (Heterorhabditidae: Rhabditida), parasitic in the Japanese beetle, *Popillia japonica* (Scarabaeidae: Coleoptera), in Ohio. *Proceedings of the Helminthological Society of Washington* 54, 53-59.

POINAR JR, G.O., KARUNAKAR, G.K. & DAVID, H. (1992). *Heterorhabditis indicus* n. sp. (Rhabditida, Nematoda) from India: separation of *Heterorhabditis* spp. by infective juveniles. *Fundamental and Applied Nematology* 15, 467-472.

QIU, L., FANG, Y., ZHOU, Y., PANG, Y. & NGUYEN, K.B. (2004). *Steinernema guangdongense* sp. n. (Nematoda: Steinernematidae), a new entomopathogenic nematode from southern China with a note on *S. serratum* (*nomen nudum*). *Zootaxa* 704, 1-20.

QIU, L., HU, X., ZHOU, Y., MEI, S., NGUYEN, K.B. & PANG, Y. (2005a). *Steinernema akhursti* n. sp. (Nematoda: Steinernematidae) from Yunnan, China. *Journal of Invertebrate Pathology* 90, 151-160.

QIU, L., HU, X., ZHOU, Y., PANG, Y. & NGUYEN, K.B. (2005b). *Steinernema beddingi* n. sp. (Nematoda: Steinernematidae), a new entomopathogenic nematode from Yunnan, China. *Nematology* 7, 737-749.

QIU, L., YAN, X., ZHOU, Y., NGUYEN, K.B. & PANG, Y. (2005c). *Steinernema aciari* sp. n. (Nematoda: Steinernematidae), a new entomopathogenic nematode from Guangdong, China. *Journal of Invertebrate Pathology* 88, 58-69.

ROMÁN, J. & FIGUEROA, W. (1994). *Steinernema puertoricensis* n. sp. (Rhabditida: Steinernematidae), a new entomopathogenic nematode from Puerto Rico. *Journal of Agriculture of the University of Puerto Rico* 78, 167-175.

RUBSTOV, I.A. & POLEVIK, N.L. (1979). [First occurrence of *Neoaplectana tabanivora* (Nematoda: Steinernematidae) parasitic in gad flies.] *Vestnik Zoologii* 3, 16-19.

SANDER, H. & STANUSZEK, S. (1972). Differentiation of nematode populations of the genus *Neoaplectana* Steiner, 1929 (Nematoda; Rhabditoidea; Steinernematidae) into generations of 'giant' and small forms. In: *11th International Symposium of Nematology, European Society of Nematologists, Reading, UK, 3-8 September, 1972*, p. 60. [Abstr.]

SHAHINA, F., ANIS, M., REID, A.P., ROWE, J. & MAQBOOL, M.A. (2001). *Steinernema pakistanense* sp. n. (Rhabditida: Steinernematidae) from Pakistan. *International Journal of Nematology* 11, 124-133.

SHAMSELDEAN, M.M., ABOU EL-SOOUD, A.B., ABD-ELGAWAD, M.M. & SALEH, M.M. (1996). Identification of a new heterorhabditid species from Egypt, *Heterorhabditis taysearae* n. sp. (Rhabditida: Heterorhabditidae). *Egyptian Journal of Biological Pest Control* 6, 129-138.

SHEN, C.P. & WANG, G.H. (1992). [Description of an entomopathogenic nematode, *Steinernema longicaudum* sp. nov. and its application.] In: *Proceedings of the XIX International Congress of Entomology, Beijing, China*, pp. 220-231.

SKRJABIN, K.I., SHIKHOBALOVA, N.P. & MOZOGOVOI, A.A. (1951). [*Descriptive catalogue of parasitic nematodes, volume 2. Oxyurata and Ascaridata.*] Moscow, 631 pp.

SOBOLEV, A.A. (1953). [To define more accurately the information on Rhabditata (superfamilies Rhabditoidea and Aphelenchoidea), parasitic in insects.] In: [*Papers on helminthology presented to Academician K.I. Skryabin on his 75th birthday.*] Moscow, USSR, Izdatelstvo Akademii Nauk SSSR, pp. 676-684.

SPIRIDONOV, S.E., KRASOMIL-OSTERFELD, K. & MOENS, M. (2004). *Steinernema jollieti* sp. n. (Rhabditida: Steinernematidae), a new entomopathogenic nematode from the American Midwest. *Russian Journal of Nematology* 12, 85-95.

STANUSZEK, S. (1974a). [*Neoaplectana feltiae* complex (Nematoda: Rhabditoidea, Steinernematidae) its taxonomic position within the genus *Neoaplectana* and intraspecific structure.] *Zeszyty Problemowe Postepow Nauk Rolniczych* 154, 331-360.

STANUSZEK, S. (1974b). [*Neoaplectana feltiae pieridarum*, n. ecotype (Nematoda: Rhabditoidea, Steinernematidae) – a parasite of *Pieris brassicae* L. and *Mamestra brassicae* L. in Poland, morphology and biology.] *Zeszyty Problemowe Postepow Nauk Rolniczych* 154, 361-393.

STEINER, G. (1923). *Aplectana kraussei* n. sp., eine in der Blattwespe *Lyda* sp. parasitierende Nematodenform, nebst Bemerkungen über das Seitenorgan der parasitischen Nematoden. *Zentralblatt für Bakteriologie, Parasitenkunde, Infektions krankheiten und Hygiene, Abteilung II* 59, 14-18.

STEINER, G. (1929). *Neoaplectana glaseri* n. g., n. sp. (Oxyuridae), a new nemic parasite of the Japanese beetle (*Popillia japonica* Newm.). *Journal of the Washington Academy of Science* 19, 436-440.

STOCK, S.P. (1993). A new species of the genus *Heterorhabditis* Poinar, 1976 (Nematoda: Heterorhabditidae) parasitizing *Graphognathus* sp. larvae (Coleoptera: Curculionidae) from Argentina. *Research and Reviews in Parasitology* 53, 103-107.

STOCK, S.P. & KOPPENHÖFER, A.M. (2003). *Steinernema scarabaei* n. sp. (Rhabditida: Steinernematidae), a natural pathogen of scarab beetle larvae (Coleoptera: Scarabaeidae) from New Jersey, USA. *Nematology* 5, 191-204.

STOCK, S.P., STRONG, D. & GARDNER, S.L. (1996). Identification of *Heterorhabditis* (Nematoda: Heterorhabditidae) from California with a new species isolated from larvae of the ghost moth *Hepialus californicus* (Lepidoptera: Hepialidae) from the Bodega Bay Natural Reserve. *Fundamental and Applied Nematology* 19, 585-592.

STOCK, S.P., CHOO, H.Y. & KAYA, H.K. (1997). An entomopathogenic nematode, *Steinernema monticolum* sp. n. (Rhabditida: Steinernematidae) from Korea with a key to other species. *Nematologica* 43, 15-29.

STOCK, S.P., SOMSOOK, V. & REID, A.P. (1998). *Steinernema siamkayai* n. sp. (Rhabditida: Steinernematidae), an entomopathogenic nematode from Thailand. *Systematic Parasitology* 41, 105-113.

STOCK, S.P., GRIFFIN, C.T. & BURNELL, A.M. (2002). Morphological characterisation of three isolates of *Heterorhabditis* Poinar, 1976 from the 'Irish group' (Nematoda: Rhabditida: Heterorhabditidae) and additional evidence supporting their recognition as a distinct species, *H. downesi* n. sp. *Systematic Parasitology* 51, 95-106.

STOCK, S.P., GRIFFIN, C.T. & CHAERANI, R. (2004). Morphological and molecular characterisation of *Steinernema hermaphroditum* n. sp. (Nematoda: Steinernematidae), an entomopathogenic nematode from Indonesia, and its phylogenetic relationships with other members of the genus. *Nematology* 6, 401-412.

STURHAN, D., SPIRIDONOV, S.E. & MRÁČEK, Z. (2005). *Steinernema silvaticum* sp. n. (Rhabditida: Steinernematidae), a new entomopathogenic nematode from Europe. *Nematology* 7, 227-241.

SUDERSHAN, G. & SINGH, L.K. (2000). *Steinernema thermophilum* sp. n. (Rhabditida: Steinernematidae) from India. *International Journal of Nematology* 10, 183-191.

SUDHAUS, W. & FÜRST VON LIEVEN, A. (2003). A phylogenetic classification and catalogue of the Diplogastridae (Secernentea, Nematoda). *Journal of Nematode Morphology and Systematics* 6, 43-90.

TAILLIEZ, P., PAGÈS, S., GINIBRE, N. & BOEMARE, N. (2006). New insight into diversity in the genus *Xenorhabdus*, including the description of ten novel species. *International Journal of Systematic and Evolutionary Microbiology* 56, 2805-2818.

TALLOSI, B., PETERS, A. & EHLERS, R.-U. (1995). *Steinernema bicornutum* sp. n. (Rhabditida: Nematoda) from Vojvodina, Yugoslavia. *Russian Journal of Nematology* 3, 71-80.

TRAVASSOS, L. (1927a). Sobre o genero *Oxysomatium*. *Boletim Biologico, Sao Paulo* 5, 20-21.

TRAVASSOS, L. (1927b). Uma nova *Capillaria* parasita de peixes be agua doce: *Capillaria sentinosa* n. sp. *Boletim Biologico, Sao Paulo* 10, 215-217.

TRAVASSOS, L. (1932). Uma nova especie do genero *Neoaplectana* Steiner, 1929 (Nematoda). *Boletim Biologico, Sao Paulo* 19, 150-154.

TRIGGIANI, O., MRÁČEK, Z. & REID, A. (2004). *Steinernema apuliae* sp. n. (Rhabditida: Steinernematidae): a new entomopathogenic nematode from southern Italy. *Zootaxa* 460, 1-12.

TURCO, C.P. (1970). *Neoaplectana hoptha* sp. n. (Neoaplectanidae: Nematoda), a parasite of the Japanese beetle, *Popillia japonica* Newm. *Proceedings of the Helminthological Society of Washington* 37, 119-121.

TURCO, C.P., THAMES, W.H. & HOPKINS, S.H. (1971). On the taxonomic status and comparative morphology of species of the genus *Neoaplectana* Steiner (Neoaplectanidae: Nematoda). *Proceedings of the Helminthological Society of Washington* 38, 68-79.

VEREMCHUK, G.V. (1964). [On the systematic position of Steinernematidae Chitwood & Chitwood, 1937 (Nematoda: Rhabditida).] *Materialy Nauchnoi Konferentsii Vsesoiuznogo Obshchestva Gel'mintologov* 1, 55-58. [Abstr.]

VEREMCHUK, G.V. (1966). [Entomopathogenic nematodes of the genus *Neoaplectana*, family Steinernematidae Chitwood & Chitwood, 1937.] In: *Tez. Doklady IX Mezdunarodnogo Kongressa po Mikrobiologii [Abstr. IX International Congress on Microbiology], Moscow*, pp. 326-327.

VEREMCHUK, G.V. (1969). [New species of *Neoaplectana* (Rhabditida: Steinernematidae), nematodes pathogenic to Elateridae.] *Materialy Nauchnoi Konferentsii Vsesoiuznogo Obshchestva Gel'mintologov* 1, 44-50.

VEREMCHUK, G.V. (1970). [A new species of *Neoaplectana* (Rhabditida: Steinernematidae) pathogenic to insects.] *Parazitologiya* 3, 249-252.

VEREMCHUK, G.V. & LITVINCHUK, L.N. (1971). [Nematodes of the genus *Neoaplectana* (Rhabditida: Steinernematidae) parasites of the larch looper moth, *Semiothisa pumila* Kusn. (Lepidoptera; Geometridae).] *Novye i maloizvestnye vidy fauny Sibiri* 4, 92-96.

DE VILLALOBOS, C. & CAMINO, N.B. (1997). *Patanodontus acaudatum* gén. n. et sp. n. (Nematoda-Diplogasteridae) un parasito de larves de *Diloboderus abderus* Sturm. (Coleoptera-Scarabaeidae) en Argentina. *Estudo de Biologia (Curitiba)* 42, 33-39.

WALIA, K.K. & BAJAJ, H.K. (2005). *Steinernema solanense* – a new entomopathogenic nematode species from Himachal Pradesh, India. In: *Sustainable plant protection strategies: health and environmental concerns. National symposium held at BSKKV, Dapoli, Maharashtra, India, 15-17 October, 2005*, p. 108.

WATURU, C.N., HUNT, D.J. & REID, A.P. (1997). *Steinernema karii* sp. n. (Nematoda: Steinernematidae), a new entomopathogenic nematode from Kenya. *International Journal of Nematology* 7, 68-75.

WEISER, J. (1955). *Neoaplectana carpocapsae* n. sp. (Anguillulata, Steinernematinae), novy cizopasník housenek obaleče jablečného, *Carpocapsa pomonella* L. *Věstník Československé Zoologické Společynosti* 19, 44-52.

WEISER, J. (1958). Ein neuer Nematode als Parasit der Engerlinge des Maikäfers, *Melolontha melolontha* in der Tschechoslowakei. *Transactions of the 1st International Conference of Insect Pathology and Biological Control, Prague, Czech Republic*, pp. 331-336.

WEISER, J. & KOHLER, W. (1955). *Neoaplectana janickii* n. sp. novy cizopasnik larvy ploskohrbetky *Acantholyda nemoralis* Thoms. v Polsku. *Roczniki Nauk Lesnych* 11, 93-110.

WELCH, H.E. (1963). Unpublished thesis.

WOUTS, W.M. (1979). The biology and life cycle of a New Zealand population of *Heterorhabditis heliothidis* (Heterorhabditidae). *Nematologica* 25, 191-202.

WOUTS, W.M. (1980). Biology, life cycle and redescription of *Neoaplectana bibionis* Bovien, 1937 (Nematoda: Steinernematidae). *Journal of Nematology* 12, 62-72.

WOUTS, W.M., MRÁČEK, Z., GERDIN, S. & BEDDING, R.A. (1982). *Neoaplectana* Steiner, 1929 a junior synonym of *Steinernema* Travassos, 1927 (Nematoda: Rhabditida). *Systematic Parasitology* 4, 147-154.

XU, Z., WANG, G. & LI, X. (1991). [A new species of the genus *Steinernema* (Rhabditida: Steinernematidae).] *Zoological Research* 12, 17-20.

YOSHIDA, M. (2005). *Steinernema litorale* n. sp. (Rhabditida: Steinernematidae) a new entomopathogenic nematode from Japan. *Nematology* 6 (2004), 819-838.

Nematology Monographs & Perspectives, 2007, Vol. 5, 59-119

Chapter 3

Methodology, morphology and identification

Khuong B. NGUYEN

Entomology and Nematology Department, Institute of Food and Agricultural Sciences, University of Florida, Gainesville, FL 32611-0620, USA

Methodology

NEMATODE PREPARATION FOR LIGHT MICROSCOPY

Nematode rearing and collection

For taxonomic studies, ten *Galleria mellonella* larvae are exposed, in a Petri dish (100 × 15 mm) lined with two moistened filter papers, to 2000 infective juveniles (IJ) of a *Steinernema* species or 200-300 IJ of a *Heterorhabditis* species. First and second generation adult nematodes are obtained by dissecting infected insects 2-4 days and 5-7 days, respectively, after the host has died. Dissection should be done in Ringer's solution or 1% saline (NaCl) solution so as to avoid excessive osmotic stress, particularly to the adult stages. Third-stage infective juveniles are obtained when they emerge from the cadavers after 7-10 days.

The recommended infection densities for *Steinernema* and *Heterorhabditis* are based on personal experience of the effect of initial inoculation density on morphometrics of the harvested stages (Nguyen, unpubl.). High inoculation densities can result in effects on nematode morphometrics that are significant enough to impact upon accurate species diagnosis.

Observations

All observations and measurements should be performed within a week after collection. For light microscope observation, at least 20 males and females and infective juveniles are examined live. Additional specimens of different stages may be killed in hot water (60°C) and fixed in TAF (Courtney *et al.*, 1955) or in lactophenol (Franklin & Goodey, 1949). These nematodes can be used further when more observation is needed to confirm the morphology or variation of some structures. Nematodes preserved in different fixatives will assist observation of certain characters. Nematodes fixed in TAF are processed to glycerin by the Seinhorst method (Seinhorst, 1959). Type specimens should always be processed to, and mounted in, glycerin. Cover glass supports are mandatory so as to avoid squashing and flattening the nematodes.

For a microscope with differential interference contrast optics, nematodes processed to glycerin result in the best observation and photographs. Structures to be observed are mentioned in the Morphology section (see p. 68).

Measurements

Either fixed or living nematodes may be measured, although it is essential to state which. A device such as a drawing tube is useful for this step. Experience has shown that it is better to work with fixed adults and living infective juveniles. Living infective juveniles can be mounted in tap water on a glass slide. Cover glass supports must be used to avoid flattening specimens. The slide is sealed with paraffin wax, nail polish or other sealants. The nematodes become immobile after about 20-30 min. If a nematode was actually collected from a host insect (*i.e.*, not by baiting with *Galleria*), measurements of the nematode reared from the natural host should be collected in order to allow comparison with measurements of the same nematode reared in *Galleria*.

NEMATODE PREPARATION FOR SCANNING ELECTRON MICROSCOPY

Working with fixed nematodes

Adults and IJ fixed by any method can be used for SEM. Transfer nematodes to a working chamber (Fig. 13) in a glass container with 2 ml of 0.1 M sodium cacodylate buffer, or 0.2 M sodium phosphate buffer. Nematodes must be washed five times with this buffer by adding and removing the solution from the glass container. Nematodes should be kept in each change of buffer solution for 10 min. They are post-fixed with 2% osmium tetroxide solution for 12 h at 25°C, washed in five changes of 0.1 M sodium cacodylate buffer for 10 min each change.

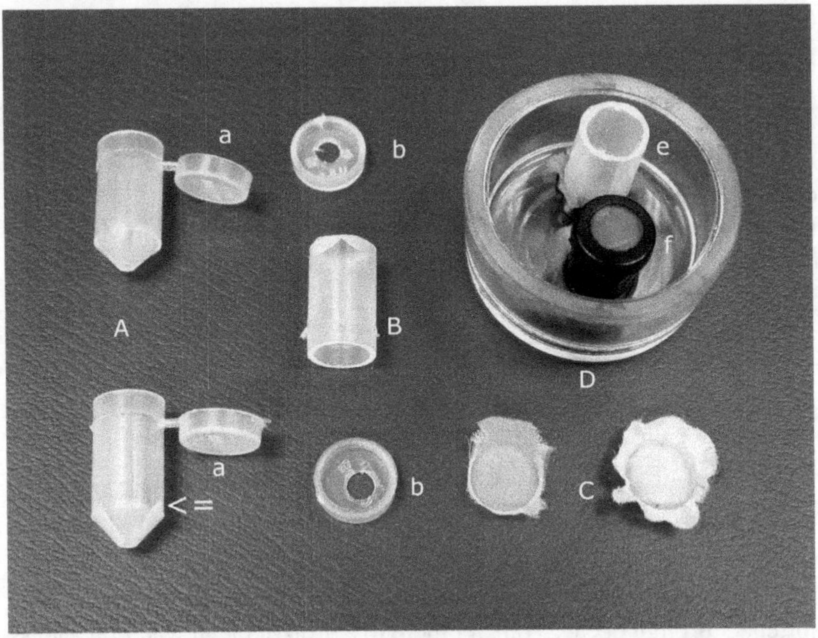

Fig. 13. *Photograph showing construction of working chamber. A: Beem capsules (cat. no. 130, Ted Pella, Inc., Redding, CA, USA). Caps (a) are removed from the capsules and a hole made to form cap (b). B: The capsule without a cap is cut at the arrow to form the cylinder B; C, D: A piece of nylon screen (C) is placed over one end of the cylinder and a cap is snapped into place to form the cylinder (e, in a dish in D) that will be used as the working chamber. The chamber containing processed specimens (f, in D) with the second cap on is now ready for critical-point drying. Modified after Eisenback (1985).*

Specimens are then dehydrated in a graded ethanol series (10-100%), with 15 min for each change/step, passed through three changes of 100% ethanol, critical point dried with liquid CO_2, mounted on SEM stubs, and coated with gold.

Caution! Sodium cacodylate, glutaraldehyde and osmium tetroxide are toxic, and should be handled with appropriate regard to recommended safety precautions. Always use gloves and carry out the process in a fume hood.

Working with live nematodes

Fix females, males and infective juveniles by the following method: first transfer the clean nematodes into a vial containing 1 ml of

0.1 M sodium cacodylate buffer (8°C, pH 7.2), then add 0.3 ml 6% glutaraldehyde in 0.1 M sodium cacodylate buffer (8°C). After 1 and 2 h, add an additional 0.3 and 0.4 ml, respectively, of the latter solution to give a final volume of 2.0 ml. All steps are done under refrigeration at 8°C. After 24 h, wash the nematodes in five changes of 0.1 M sodium cacodylate buffer for 10 min each change. Post-fix in 2% osmium tetroxide for 12 h at 25°C, wash in five changes of 0.1 M sodium cacodylate buffer for 10 min each change. Dehydrate the nematodes in a graded ethanol series (10-100%) at 25°C, then transfer them to a mixture of 50% amyl acetate and 50% ethyl alcohol for 10 min, and finally to 100% amyl acetate. Critical point dry with liquid CO_2, mount on SEM stubs, and coat with gold (see Nguyen & Smart, 1993).

EXAMINING THE SPICULES AND GUBERNACULA

This method is used to prepare spicules and gubernacula for SEM. Heat kill first generation males in a Petri dish containing water and store at room temperature. After 4-5 days, the bodies of the specimens become softened through decay. Transfer them to a drop of clean water on a glass slide. Open the posterior end of each male with the help of two small needles, dissect out the spicule and gubernaculum, and wash them free of debris by gently moving them about in water. Then place them on a SEM stub close to a short piece of hair acting as a marker (Nguyen & Smart, 1990, 1994).

A modification of this method (Nguyen & Smart, 1994) consists of dissecting the spicules and gubernacula from fixed males in lactophenol, glycerin or 40% lactic acid. For this purpose, the posterior part of a male is cut and transferred to a drop of lactic acid. After 1 h, two small needles are used to dissect the spicule and gubernaculum which are transferred to a cover glass (12 mm diam.) sitting on a glass slide. The spicules and/or gubernacula can be arranged near a marker, such as a short piece of hair, which has previously been glued near the centre of the cover glass. The cover glass is then transferred to a SEM stub previously sprayed with metallic enamel. After 1 h, the spicules and gubernacula are coated with gold and are then ready to be examined.

CROSS HYBRIDISATION STUDIES

Reproductive compatibility of different species can be conducted using the method reported by Nguyen and Duncan (2002) using

G. mellonella haemolymph. A drop of haemolymph of *G. mellonella* is placed in a sterile Petri dish (35 × 10 mm) and a single IJ of nematode species A and one IJ of nematode species B are added. Control crosses are done between IJ of the same species. All treatments are replicated 30 times. All Petri dishes are kept in closed plastic bags containing 2-3 layers of moistened facial tissue to prevent the haemolymph from drying. After 3-5 days, a single male and single female will develop in a proportion of the dishes. Observe them regularly to see if they mate and produce offspring.

EXAMINING THE BURSA OF *HETERORHABDITIS*

Transfer living males into a small drop of lactophenol (containing 0.002% acid fuchsin) on a hot plate at 65-70°C. After 30 min, transfer one of the males to a drop of lactophenol on a glass slide. Sever the anterior three-quarters of the body and remove. Place a cover glass on top of the drop of lactophenol containing the posterior part of the nematode (the drop must be big enough so that the cover glass can be easily moved). Move the cover glass slowly by means of a needle in order to rotate the nematode posterior region so as to present a ventral view. At least ten such bursas should be observed and the disposition of the copulatory papillae or rays recorded (Nguyen *et al.*, 2004).

MOLECULAR TECHNIQUES

DNA can be extracted successfully either from live nematodes or from those preserved in 95% alcohol.

Extraction of DNA

For live nematodes, DNA is extracted from a single female (or juvenile) using the modified method reported by Hominick *et al.* (1997). The nematode is crushed in 20 μl (10 μl for juvenile) of lysis buffer (50 mM KCl, 10 mM Tris pH 8.3, 2.5 mM $MgCl_2$, 0.45% NP40, 0.45% Tween 20 and 60 μg ml^{-1} proteinase K) on a sterilised hanging-drop glass slide and transferred to a sterilised 0.5 ml microcentrifuge tube on ice. The tube is frozen at −80°C (or −20°C if a deep freezer is not available) for 10 min or more if needed, incubated at 65°C for 60-90 min, followed by a 95°C incubation for 8 min to completely lyse the cells, digest the proteins and inactivate proteinase K, respectively. Subsequently, the tube is cooled on ice, and centrifuged at 11 600 *g* for 2 min. The supernatant containing the DNA is collected and kept at −20°C for future use (or at −80°C if longer storage is needed).

For nematodes from alcohol, transfer specimens to a Petri dish containing distilled water and keep them in water for 5-12 h. Other steps are similar to the method for live nematodes except the crushed nematodes should be incubated at 65°C for 5-12 h.

DNA can also be extracted from a single female using a DNeasy tissue kit (Qiagen Inc., Valencia, CA, USA). Methods of extraction are in the instruction book.

Internal transcribed spacer (ITS) regions

1. PCR amplification

The ITS region of the ribosomal DNA is amplified by the polymerase chain reaction (PCR) in a 25 μl reaction. All tubes are set up on ice and the following added to each tube: 2.5 μl of 10× PCR buffer, 1 μl of dNTP mixture (10 mM each), 1 μl of 10 pM forward primer, 1 μl of 10 pM reverse primer, 1.5 μl $MgCl_2$ (50 mM), 0.25 μl of Taq DNA polymerase (5 U μl^{-1}), 15 μl of distilled water, and 2.5 μl of DNA. The primers used in this process were reported by Vrain *et al.* (1992): 18S: 5′-TTGATTACGTCCCTGCCCTTT-3′ (forward) and 26S: 5′-TTTCACTCGCCGTTACTAAGG-3′ (reverse) corresponding to nucleotide position 2503-2523 and 3774-3794, respectively, of the sequence of the rDNA tandem unit from *Caenorhabditis elegans* (GenBank: accession no. X03680). If these two primers do not work for a nematode, the following pair of primers should be used: TW81: 5′-GTTTCCGTAGGTGAACCTGC-3′ (forward) and AB28: 5′-ATATG CTTAAGTTCAGCGGGT-3′ (reverse) (Hominick *et al.*, 1997). All PCR reactions are run in a Thermocycler with the cycling profile suggested by Nguyen *et al.* (2004): 1 cycle of 94°C for 7 min followed by 35 cycles of 94°C for 60 s, 50°C for 60 s, 72°C for 60 s. The last step is 72°C for 10 min. *Note*: For *Heterorhabditis* spp. the annealing temperature should be 45°C.

To determine if the PCR process is successful, 5 μl of PCR product is run on a 1% agarose gel in 1× TAE buffer or 0.5 TBE buffer for 30-60 min. The presence of PCR product can be visualised by ethidium bromide staining (Maniatis *et al.*, 1989).

At this point a worker can use the RFLP (restriction fragment length polymorphism) method for species differentiation (Hominick *et al.*,

1997). I prefer the sequencing and phylogeny methods, so in the following I will mention these two methods.

2. Sequencing

PCR products are purified with a QIAquick PCR purification kit (Qiagen Inc., Santa Clarita, CA, USA). Purified DNA is sequenced in both directions. The primers used for sequencing are the 18S and 26S, or TW81 and AB28 described earlier and two internal primers. The two internal primers are used to obtain a complete sequence of both strands of the PCR product. These primers are variable depending on nematode species. In general, the following two pairs are the best for *Steinernema* species: KN58 = 5′-GTATGTTTGGTTGAAGGTC-3′ (forward) and KNRV = 5′-CACGCTCATACAACTGCTC-3′ (reverse), and S58F = 5′-CGTTATTTGGCGTGAATTGCAG-3′ and S58R = 5′-CTCAATAATGATCCTTCCGCAGG-3′ (Nguyen *et al.*, 2004).

For *Steinernema* species with paired horn-like structures on the labial region, the following pair of internal primers should be used: BIC58 = 5′-GATCACTTGGTTCGTAGATCG-3′ (forward) and BICRV = 5′-GC TCAATAATGATCCTTCCGC-3′ (Nguyen, Tesfamariam, Gozel, Gaugler & Adams, 2004).

The purified PCR product should be sent to a DNA core for sequencing.

For *Heterorhabditis* the following internal primers should be used: H58P: 5′-ACGAATTGCAGACGCTTAG-3′ (forward), and H58R: 5′-GTGCGTTCAAAACTTCACC-3′ (reverse) (Nguyen *et al.*, 2004).

Important notes

To save cost and obtain a good result, sequences with external primers 18S and 26S should be done first. Next, get a consensus sequence from those two sequences, then use BLAST comparison from the GenBank http://www.ncbi.nlm.nih.gov/ to see if the studied nematode is a new species or not. If the nematode appears to be a new species (the percentage of similarity is 95% or less for *Steinernema* or 97% or less for *Heterorhabditis*), then internal primers can be used to obtain a complete sequence. In order to choose the best internal primers for sequencing, the workers should use the FIND device in Sequencher (software) to see if the above mentioned internal primers are present in the consensus sequence. When a pair of primers is found, use them for sequencing.

Sequencher^{TM} (Gene Code Corporation, Ann Arbor, MI, USA) can be used for sequence editing and verifying base-calls. At least two sequencing reactions should be performed per strand.

3. Multiple alignments

Sequences of studied species are aligned using the default parameters of Clustal X (Thompson *et al.*, 1997), then optimised manually in MacClade 4.0 (Maddison & Maddison, 2002).

4. Phylogenetic analysis

Sequences of the internal transcribed spacer regions (ITS) and the partial 28S ribosomal DNA (D2/D3 regions) of *Steinernema* species have been used by different authors (Nguyen *et al.*, 2001; Stock *et al.*, 2001; Nguyen & Duncan, 2002; Nguyen & Adams, 2003; Spiridonov *et al.*, 2004; Nadler *et al.*, 2006) in taxonomic and phylogenetic studies. These two regions are mentioned here.

Molecular phylogenetic relationships were obtained by maximum parsimony using PAUP 4.0b8 (Swofford, 2002). For the phylogenetic analysis using the ITS regions, *S. intermedium* (accession no. AF122016) or *Caenorhabditis elegans* (X03680) is used as the outgroup taxon for resolving relationships among the rest of the *Steinernema* species (Nguyen *et al.*, 2001). For D2/D3 regions, *Panagrellus redivivus* (AF331910) is used as the outgroup taxon (Stock *et al.*, 2001). Models of sequence evolution were evaluated using ModelTest 3.7 (Posada & Crandall, 1998). The ModelTest will choose a model that will be used to correct for genetic distances in the construction of the trees, and to search for the maximum likelihood solution (heuristic search, starting tree gained stepwise, stepwise addition sequence as-is, and TBR branch swapping; these options can be changed by researchers).

Branch support was estimated by bootstrap analysis (1000 replicates for maximum parsimony and 100 replicates for maximum likelihood).

For further interpretation of the data from the above methods, consult Nguyen *et al.* (2001, 2004, 2006), Stock *et al.* (2001), Nguyen and Duncan (2002), Nguyen and Adams (2003), Spiridonov *et al.* (2004) and Nadler *et al.* (2006).

D2D3 regions

All processes are similar to those for ITS regions with the following exceptions: the primers used for PCR process are: the primers D2F, 5′-CCTTAGTAACGGCGAGTGAAA-3′ (forward) and 536, 5′-CAGCT

ATCCTGAGGAAAC-3′ (reverse) are used as external primers to amplify and sequence the D2D3 regions; primers 502, 5′-CAAGTACCGTG AGGGAAAGTTGC-3′ (forward), and 503, 5′-CCTTGGTCCGTGTTT CAAGACG-3′ (reverse) are used as internal primers for additional sequencing. The primers 502, 503 and 536 were reported by Stock *et al.* (2001). Primer D2F was reported by Nguyen *et al.* (2006).

Fixative preparation and fixing process

- Using TAF fixative (Courtney *et al.*, 1955)

  Preparation

Formalin	7 ml
Triethanolamine	2 ml
Water	91 ml

Transfer nematodes to a container. Gently kill them with heat at 50-60°C. Add fixative to the container. Keep nematodes in fixative for 24 h or more before processing further.

- Using lactophenol (Franklin & Goodey, 1949)

  Preparation

Lactic acid	100 ml
Phenol	100 ml
Glycerin	200 ml
Water	100 ml

Acid fuchsin or methyl blue can be added in water before mixing the ingredients.

Caution! Phenol is toxic, avoid contact with skin.

Adults dissected from cadavers are transferred immediately to hot lactophenol in a small dish on a hot plate at 60-70°C. For health and safety reasons, this process should be done in a fume cupboard. After 20 min, remove the dish from the hot plate and keep at room temperature. The specimens are now ready for observation or measurements.

Processing specimens to glycerin

- Modified Seinhorst (1959) technique

  Preparation

  Solution 1: 20 parts of ethanol 95% + 1 part of glycerin + 79 parts of water

  Solution 2: 95 parts of ethanol 95% + 5 parts of glycerin

Step 1 – Place fixed nematodes in a 10 ml glass or plastic dish; add 7 ml of solution 1.

Step 2 – Place this dish into a larger closed container (for slow evaporation) filled with 95% ethanol up to 1/10 of its volume, then place it in an oven at 35-40°C for at least 12 h.

Step 3 – Remove dish from oven; almost fill the dish containing specimens with solution 2. Place dish into the closed container and return to the oven at 40°C. After 3-5 h, the nematodes are ready for mounting.

If many specimens are collapsed, repeat step 3.

- Slow evaporation method

A simpler method favoured by some workers is the slow-evaporation technique whereby fixed nematodes are placed into a 5% solution of glycerin in an excavated glass block (=embryo dish). The dish is partially covered so as to control the rate of evaporation and hence the rate at which the nematodes are processed to pure glycerin. For delicate nematodes, this process should take at least a week. The nematodes can then be mounted on permanent slides.

Identification

Morphology of the Family Steinernematidae Filipjev, 1934

Morphological characters are important in the taxonomy of *Steinernema* as in other nematodes and should be observed and described carefully. Individual taxonomists describe a nematode differently, some structures being reported by one author, but not by another. Workers need to choose which protocol to follow.

The following is an example of a species description and is illustrated by photographs of the important structures. All the characters mentioned should be reported in the description of a new species.

Male

Male morphological characters are the most important in the description of a new species of Steinernematidae. The range of characters utilised may occasionally differ but they are mostly as follows: *Body* always curved posteriorly, sometimes C-shaped, sometimes J-shaped. Cuticle smooth under a light microscope but with striation under SEM.

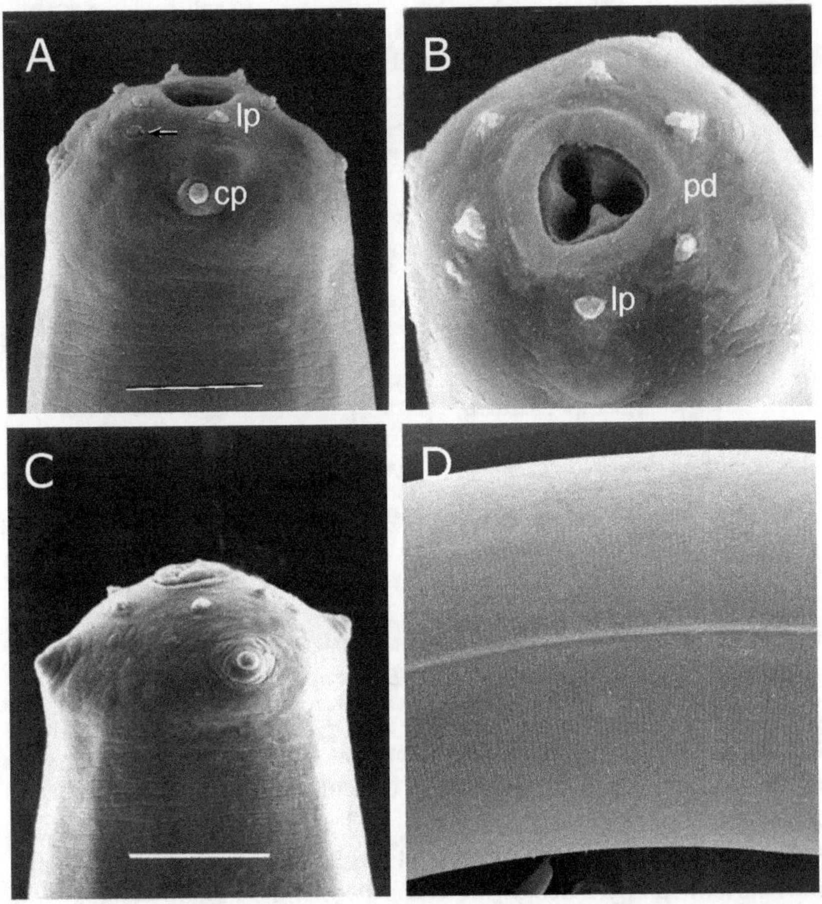

Fig. 14. Steinernema *male. A: Anterior region showing six labial papillae (lp), three of four cephalic papillae (cp) and an amphidial aperture (arrow); B: Face view showing stoma opening, six labial papillae (lp) and a perioral disc (pd); C: Anterior region, somewhat swollen, with three of four cephalic papillae. The cephalic papillae are larger than labial the papillae and the perioral disc is prominent. D: Body showing lateral field with one ridge. (Scale bars: A, B: A = 8.8 μm; B = 6 μm. C, D: C = 10 μm; D = 20 μm.)*

Lateral field absent or with one ridge (Fig. 14D). *Cephalic extremity* swollen or truncate (Fig. 14A, C) with four cephalic papillae and six labial papillae. Cephalic papillae are usually larger than labial papillae. Amphidial apertures pore- or slit-like, but may be absent or inconspicuous. Cephalic extremity with or without perioral disc. *Stoma* deep, shal-

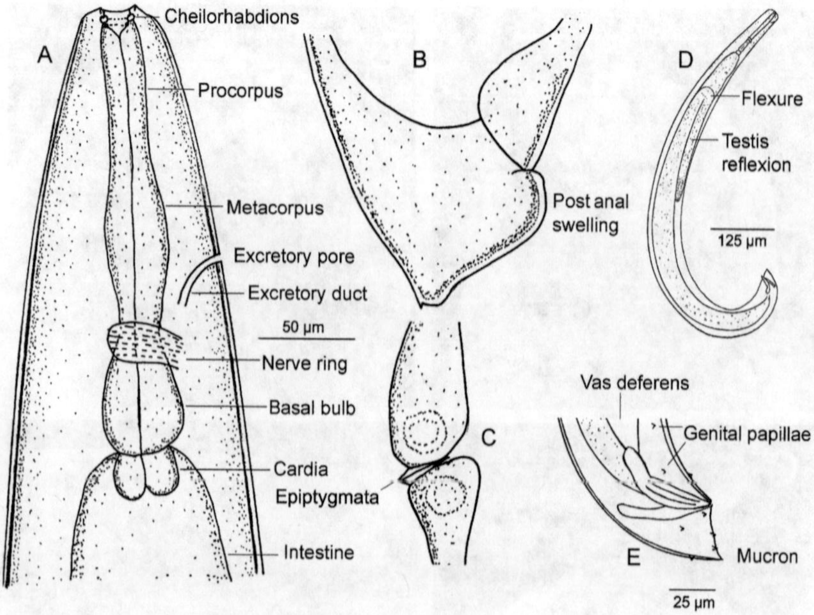

Fig. 15. *Diagrammatic morphology of* Steinernema. *A: Pharyngeal region; B: Female tail region; C: Vulval region; D: Entire male; E: Male tail region.*

low or reduced, cheilorhabdions sometimes prominent, posterior part of stoma funnel-shaped, moderately cuticularised or not cuticularised. *Pharynx* either with cylindrical corpus and swollen basal bulb, or mostly with cylindrical procorpus, somewhat swollen metacorpus, narrow isthmus surrounded by a nerve ring and a swollen basal bulb containing a valve (Fig. 15). *Nerve ring* usually surrounding isthmus or anterior part of basal bulb, or just anterior to basal bulb. *Excretory pore* position variable, either anterior or posterior to nerve ring, or close to end of pharynx; excretory duct well (or weakly) cuticularised; excretory gland with prominent nucleus (or excretory gland not observed). *Cardia* prominent or inconspicuous. *Genital system* monorchic, reflexed. Testis reflexed, comprising germinal zone, growth zone and *vas deferens*. Testis may be reduced or collapsed, the percentage of nematodes with such an atypical testis should be reported. Distance from base of pharynx to anterior end of testis may be constant or variable. In *posterior region*, the number of genital papillae is important, for example, in Figure 16, the male on the left has ten pairs of genital papillae anterior to the cloacal aperture and the tail on the right has six pairs (the normal number is six or

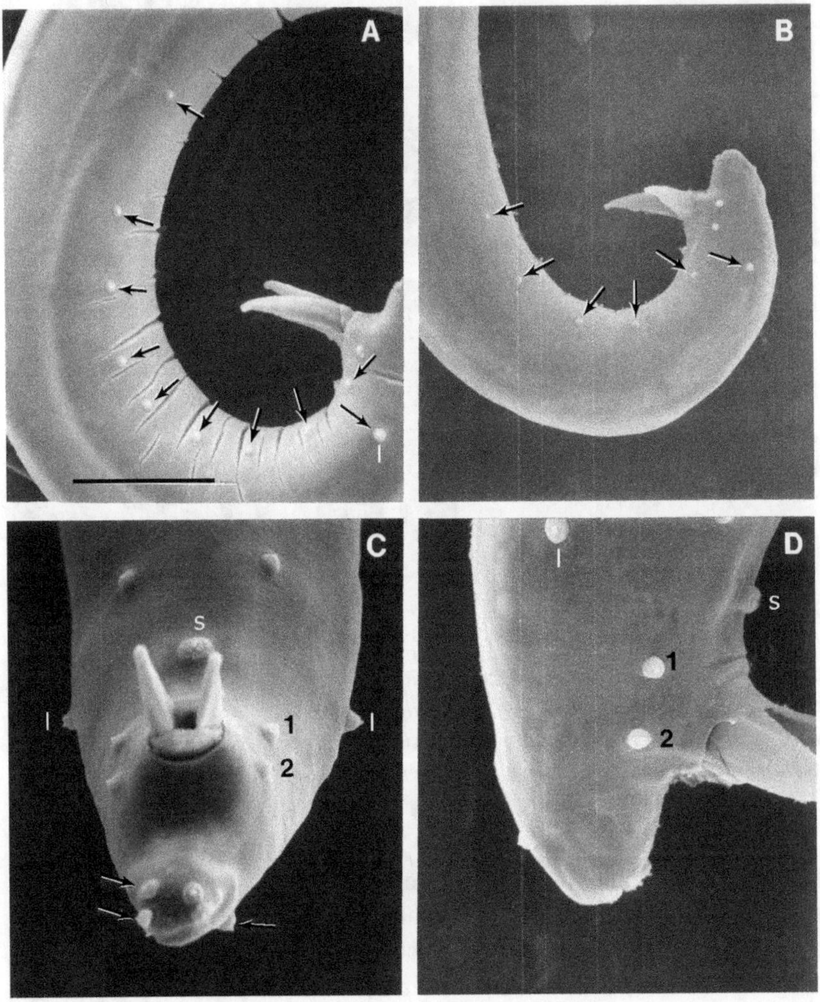

Fig. 16. *Posterior region of* Steinernema *male. A: Posterior region showing ten pairs of genital papillae anterior to cloaca; B: Posterior region showing six pairs of genital papillae anterior to cloaca; C: Ventral view of posterior region showing single papilla (s), lateral pair (l), adanal pairs (1.2), postcloacal subterminal pairs (left arrows) and postcloacal subdorsal papilla (right arrow); D: Lateral view of posterior region showing single papilla, one lateral and two adanal papillae, one subdorsal and two subterminal papillae. (Scale bar: A = 50 μm; B = 60 μm; C, D = 20 μm.)*

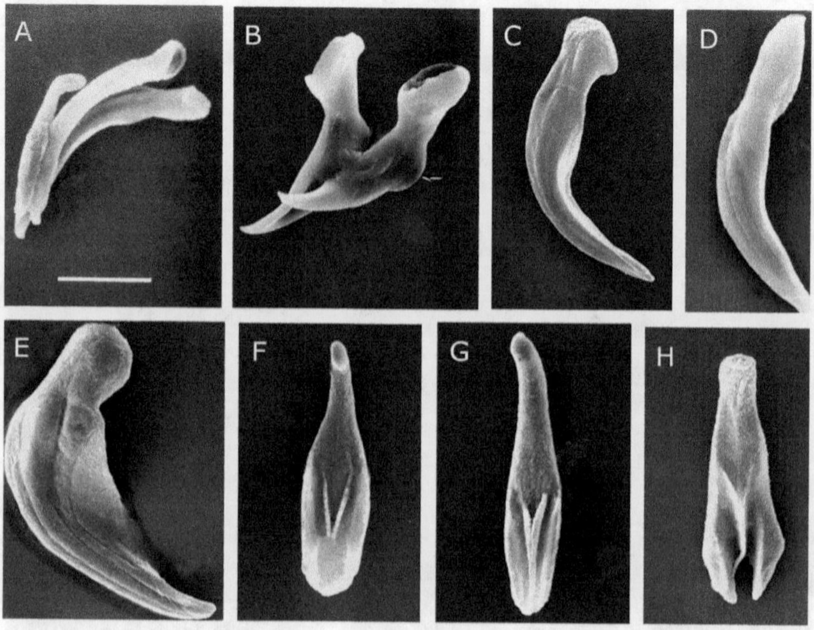

Fig. 17. Steinernema *spicules and gubernacula. A: Spicules and gubernaculum as occurring* in situ*; B: Foot-shaped spicules of* Neosteinernema longicurvicauda*; C: Spicule with head wider than long; D: Spicule with head longer than wide; E: Large spicule with round anterior end, prominent rostrum and wide velum; F: Gubernaculum with V-shaped cuneus; G: Gubernaculum with Y-shaped cuneus; H: Gubernaculum with arrow-shaped cuneus. (Scale bar: A-E = 25 μm; F-H = 18 μm.)*

seven). Additionally, there is a single precloacal midventral papilla, two pairs adcloacal (Fig. 16C, D), one pair lateral, two pairs subterminal (left arrows), and one pair subdorsal (right arrow). *The spicule* is large, foot-shaped (Fig. 17B) in *Neosteinernema.* Spicule head is either longer than wide, as long as wide, or wider than long, the ratio head length/head width must be reported (Fig. 17C-E); shaft present, either long or short, or absent; blade almost straight, moderately curved or well curved; rostrum present or absent (Fig. 17C, E); velum may be prominent, thin, or absent; posterior end may or may not reach tail tip. In SEM photograph, spicule blade has three lobes, dorsal, lateral and ventral. Terminology of spicule and gubernaculum is presented in Figure 18. *Gubernaculum*: boat-shaped in lateral view, anterior end curved, pointed or enlarged anteriorly or rounded. In ventral view, tapering anteriorly abruptly or grad-

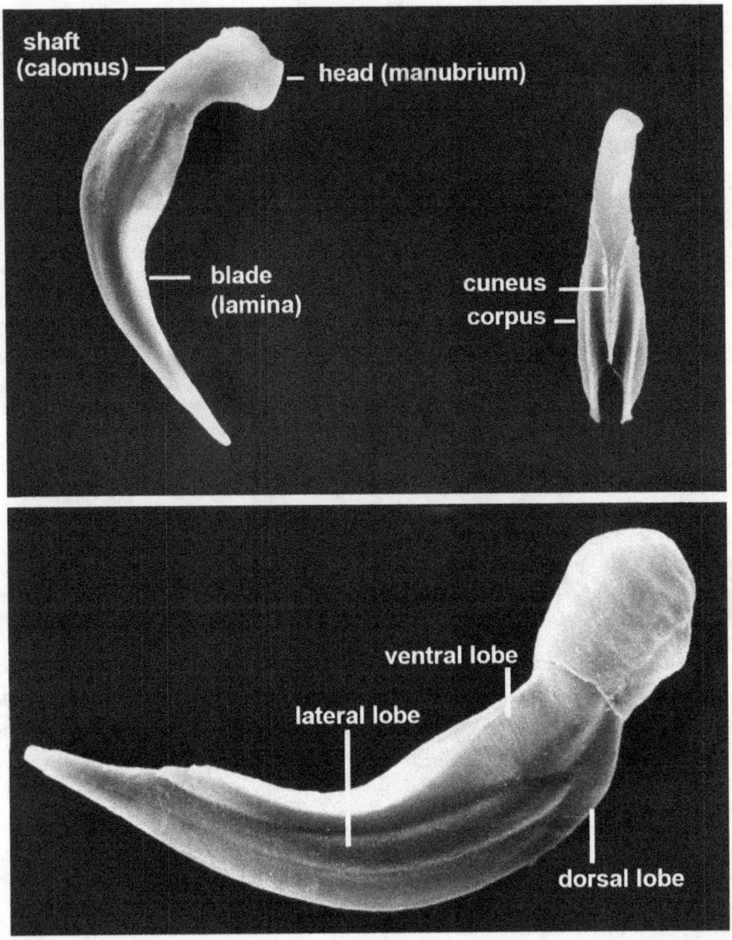

Fig. 18. *SEM of spicules and gubernaculum showing terminology used in taxonomy.*

ually (Fig. 17F-H); cuneus present or absent, when present it is either V-shaped, Y-shaped, needle-shaped or arrowhead-shaped. Gubernaculum enlarging posteriorly, abruptly or gradually to form corpus wings. Corpus is either open or closed posteriorly (Fig. 17F-H). *Tail*: conoid, without bursa, with or without phasmid. *Tail tip*: either with or without mucron or digitate.

The second-generation male should be described and differences in morphology and morphometrics between it and the first generation recorded.

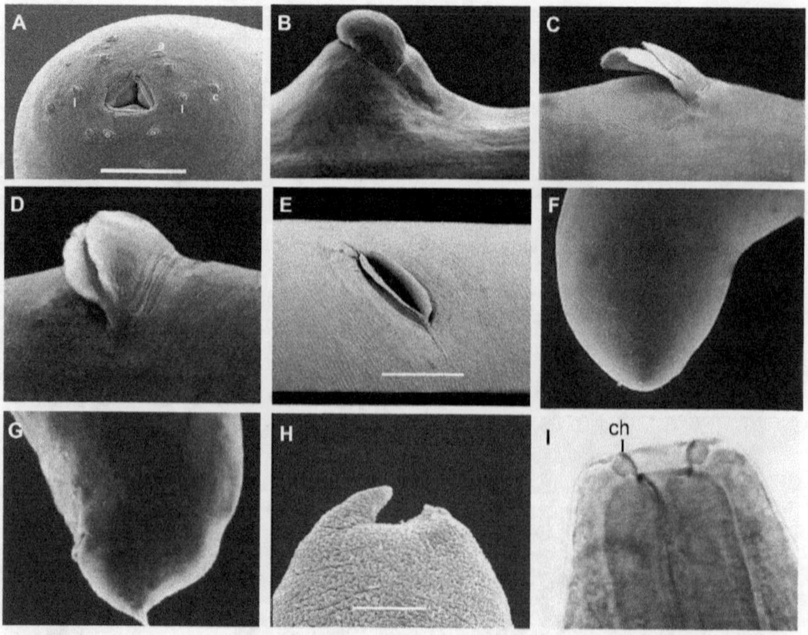

Fig. 19. Steinernema *female. A: Anterior region showing subtriangular stoma, six labial papillae (l), four cephalic papillae (c) and two amphidial apertures (a); B-E: Variation in shape of epiptygmata; F: Tail tip rounded; G: Tail tip with mucron; H: tail tip with more than one projection; I: Stoma showing prominent cheilorhabdions (ch). (Scale bars: A = 10 μm; B = 30 μm; C-E = 15 μm; F, G = 38 μm; H = 3 μm; I = 25 μm. Use scale bar in A for figures without a scale.)*

Female

Description is similar to that of male. There are some female characters that are important in taxonomy of Steinernematidae. Anterior end, as presented in Figure 19A, is almost similar for all species. The appearance, presence or absence of epiptygmata (Fig. 19B-E), and the size of cheilorhabdions in lateral view (Fig. 19I) are important. Tail may be shorter or longer than anal body diam. Shape of tail tip should be reported, *e.g.*, round with mucron or papilla-like structures, or bluntly pointed (Fig. 19F-H). These features can sometimes be used as diagnostic characters (Fig. 19B-H). Lateral field is mostly absent but sometimes present with one ridge as in male (Fig. 14D).

The second-generation female should also be described and the differences between the two generations recorded.

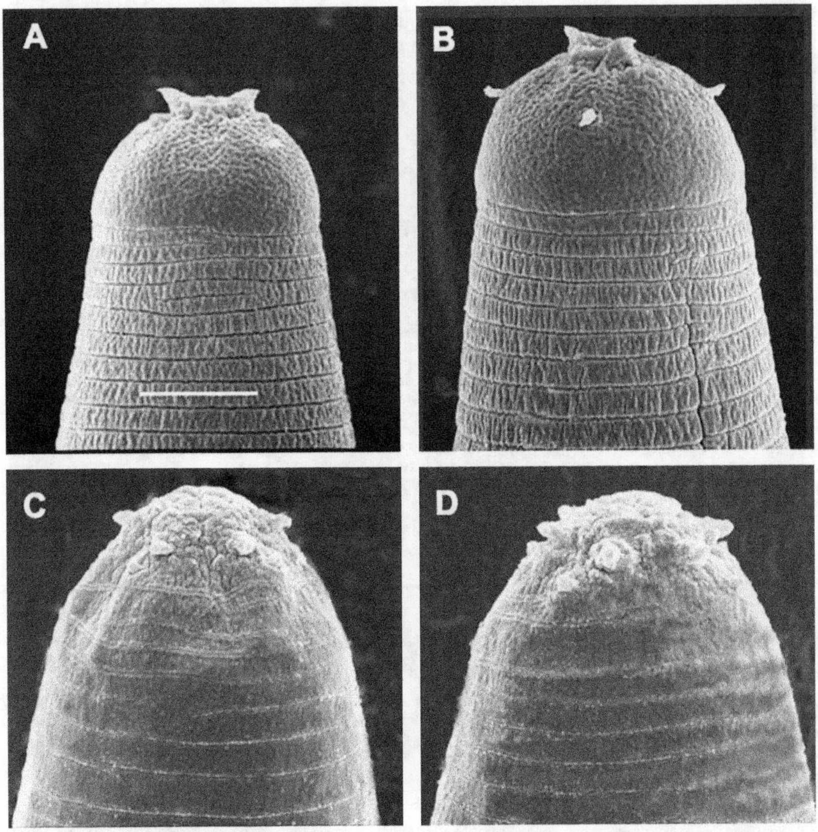

Fig. 20. Steinernema *showing exsheathed infective juveniles with horn-like structures on the labial region (A, B), and ensheathed infective juveniles showing elevated labial region with six cephalic papillae, but without horn-like structures (C, D). (Scale bar = 4.3 μm.)*

Infective juvenile

Body of heat-relaxed specimens slender, almost straight or slightly curved. Cuticle striated. Cephalic region truncate, continuous with body or offset. Stoma closed. Head smooth or annulated with four cephalic papillae and six labial papillae (in ensheathed IJ, Fig. 20C, D) or with four cephalic papillae and no labial papillae (in exsheathed IJ, Fig. 21A). Horn-like structures (Fig. 20) have been found on the cephalic region of some species. Sometimes these structures cannot be seen on ensheathed infective juvenile (Fig. 20C, D) and therefore exsheathed infective juvenile should be observed (newly harvested and 1-month-

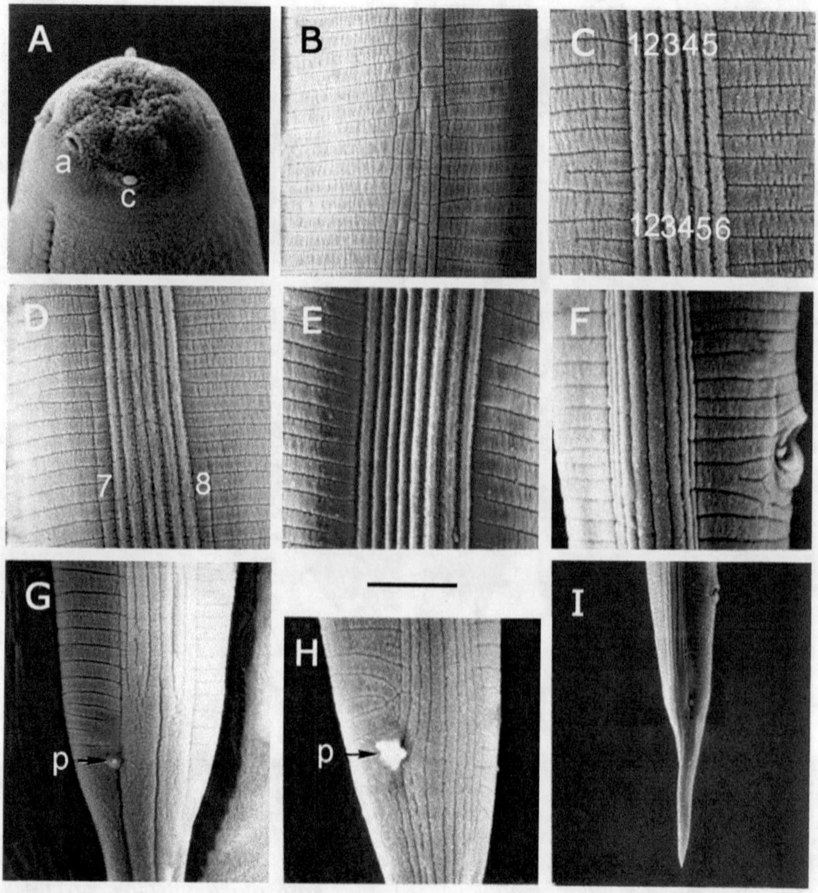

Fig. 21. Steinernema *infective juvenile showing lateral field pattern with arrangement of lateral ridges (2, 5, 6, 8, 2), and shape and position of phasmid (p). A: Anterior region showing four cephalic papillae (c), one of two amphidial apertures (a) and beginning of lateral field; B: Anterior portion of lateral field at point where two ridges become five; C: Lateral field showing central ridge splitting into two to give a total of six ridges; D: Lateral field at level where central two ridges have not separated fully, and two additional bands are formed exteriorly; E: Lateral field with maximum number of eight ridges; F: Lateral field at anal level showing enlargement of four central ridges; G: Lateral field at phasmid level showing eight ridges becoming two large ridges. The phasmid (p), with exudates in central pore, is located ventral to lateral field; H: Lateral field at phasmid level showing less frequent condition where eight ridges are not reduced to two, phasmid (p) is covered with exudates; I: Tail region showing lateral field, anus and phasmid near mid-tail.*

old nematodes should be observed to see more detail of morphological structures). Amphidial apertures prominent on SEM photographs of several species (Fig. 21A) but may be obscure on others. Lateral field pattern (Fig. 21): the arrangement of ridges in lateral field from head to tail is important, the formula for the pattern in Figure 21 is 2, 5, 6, 8, 2 ridges. If an electron microscope is available, this pattern should be described. A light microscope with DIC optics can be used to observe the lateral field pattern. Characteristics of the pharynx is usual in descriptions: pharynx with thin corpus, with almost uniform diameter, or with a slightly swollen metacorpus; nerve ring surrounding isthmus; basal bulb usually elongate with a valve. Position of excretory pore is an important character, at mid-pharynx, in anterior third or near basal bulb. Cardia small, prominent or inconspicuous. Bacterial chamber prominent or obscure, give the size (length × diam.) if possible. Intestine lumen narrow, rectum and anus distinct. Genital primordium prominent or obscure. Hemizonid (anterior to excretory pore) or hemizonion (posterior to excretory pore) should be observed and reported. Phasmid present or absent; when present, it is small or large and located either just ventral to lateral field or interrupting the ventral-most lateral ridge (Fig. 21G, H) and usually near mid-tail. Hyaline region comprises either more or less than 50% of tail length. Tail with dorsal or ventral depression, tail end with or without internal needle-like structure.

MORPHOLOGY OF THE FAMILY HETERORHABDITIDAE POINAR, 1976

The description of *Heterorhabditis* species is similar to *Steinernema* with the following exceptions:

Male

Bursa usually with nine papillae (Fig. 22) but some variation is normal. The distribution of bursal ribs is as follow: pair 1 well anterior, extending to, or beyond the bursa rim; pairs 2 and 3 in a group, immediately anterior to the cloacal aperture. Pairs 4, 5, 6 form a group just posterior to cloaca, pair 4 usually curved laterally. The arrangement of pairs 1-6 is not variable in *Heterorhabditis*. Pairs 7, 8

Fig. 21. *(Continued.) (Scale bar: A, F, H = 6 μm; B, D, E, G = 7.5 μm; C = 5 μm; I = 25 μm.) After Nguyen and Smart (1995), courtesy of* Nematologica.

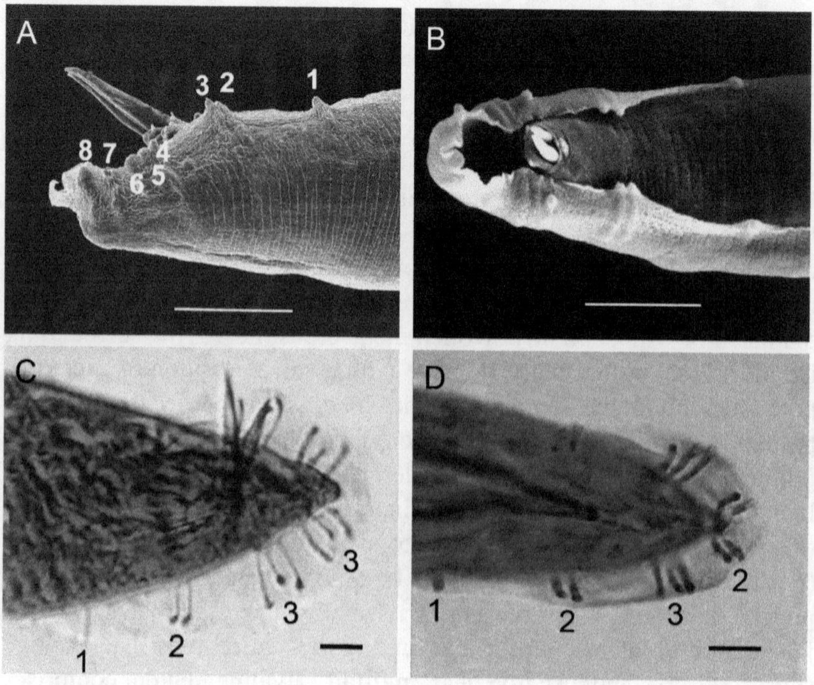

Fig. 22. Heterorhabditis *bursa. A, B: SEM micrographs of posterior regions showing bursas with ribs; C, D: Light microscope pictures showing nine and eight pairs of bursal ribs with terminal group with three pairs of ribs (C) and two pairs of ribs (D). (Scale bars: A, B = 15 μm; C, D = 20 μm.)*

and 9, forming the terminal group, are the most variable and can be used for species differentiation. Several species, such as *Heterorhabditis megidis*, *H. zealandica*, *H. marelatus*, *H. bacteriophora*, have three pairs of ribs in the terminal group; *H. mexicana*, *H. floridensis* have either one pair, two pairs, or three ribs on one side, two ribs on the other side; for *H. indica*, the arrangement of the bursa ribs is variable and the occurrence of three pairs in the terminal group is very rare (Fig. 23). At least ten bursas should be observed and the variation, if any, should be reported, *e.g.*, nine bursas with three pairs of rib; one with two pairs in the terminal group. Spicules (Fig. 24E) and ventral view of gubernaculum (Fig. 24F) can be used as diagnostic characters.

Female

Head region with six anteriorly-directed papillae (Fig. 24A); almost similar for all species except that labial papillae are curved outward

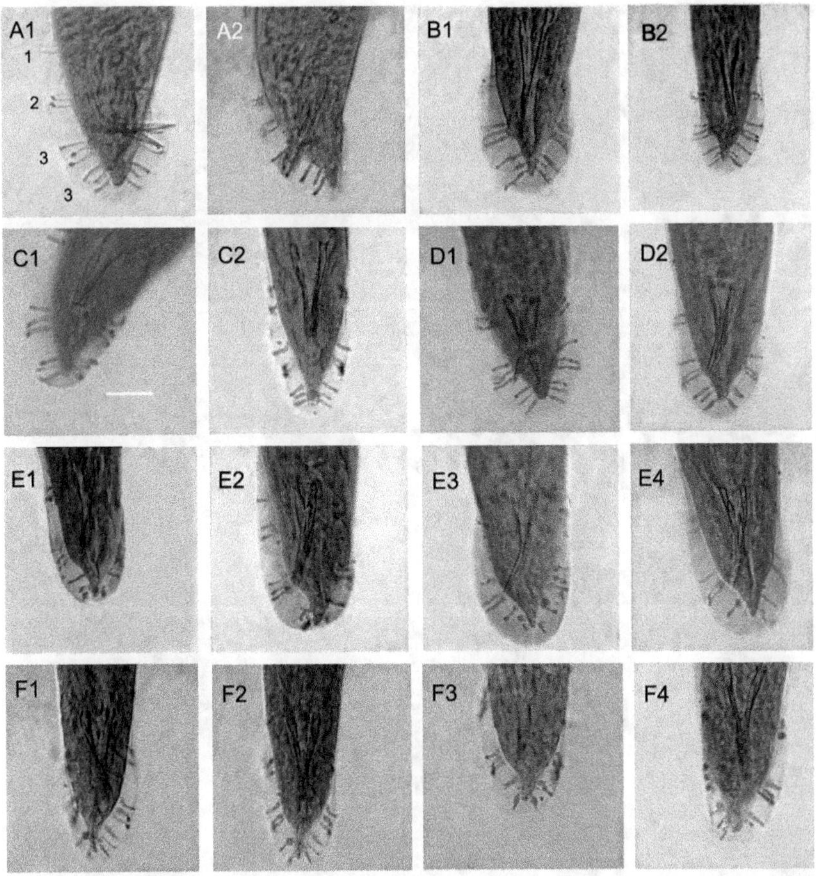

Fig. 23. *Bursa of* Heterorhabditis *spp., ventral view, showing bursal papillae. A1, A2:* H. megidis, *ventral and lateral views showing four groups of bursal papillae (1, 2, 3, 3) and three papillae in terminal group; B1, B2:* H. zealandica *strain Florida and NZH3, respectively, ventral views showing three papillae in terminal group with central papillae much shorter than others; C1, C2:* H. marelatus, *ventral views with little variation; D1, D2:* H. bacteriophora *strain GA2 and HP88, respectively, with three pairs of papillae in terminal group; E, F:* H. indica *(E1-E4 = strain India, F1-F4 = strain Mexico) with number of papillae in terminal group of bursal papillae variable. (Scale bar = 20 μm.)*

in *Heterorhabditis indica* (see Fig. 225). Amphidial apertures present, but cephalic papillae not observed with SEM. The form of the vulva and surrounding pattern (see Fig. 222) appear to be good diagnostic characters.

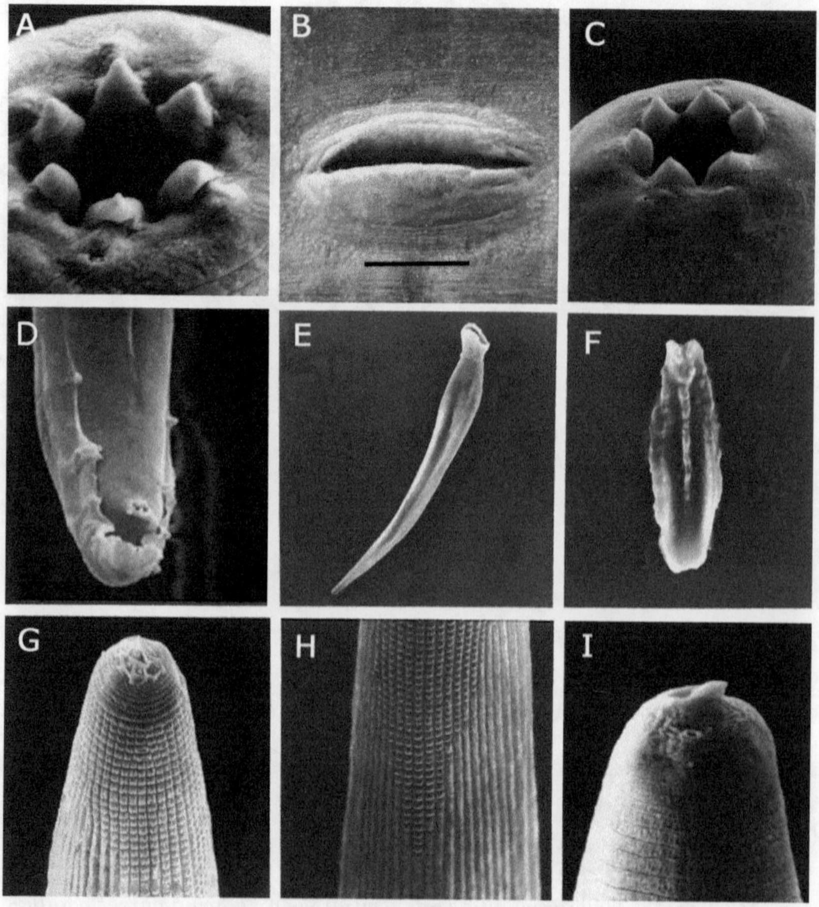

Fig. 24. *SEM micrographs of* Heterorhabditis. *A: Hermaphroditic female face view showing anteriorly directed lips, labial papillae and amphidial apertures; B: Vulva; C: Face of amphimictic female; D: Bursa, ventral view showing bursa ribs; E: Spicule, lateral view; F: Gubernaculum, ventral view; G, H: Anterior region of third-stage infective juvenile ensheathed in cuticle of second-stage juvenile. Note tessellate pattern and longitudinal ridges; I: Anterior region of infective juvenile showing dorsal tooth and prominent amphidial aperture. (Scale bar: A = 5 μm; B, F, G = 8.6 μm; C, D = 6 μm; E = 15 μm; H = 10 μm; I = 3.8 μm.)*

Infective juvenile

Morphological characters of infective juveniles can be seen in Figure 24G-I. At present, no morphological characters can be used for species diagnosis, except for the presence of a membranous ring, a diagnostic character, around the mouth of *H. megidis* (see Fig. 230F). The phasmid of *Heterorhabditis* is located just dorsal to the lateral field.

Important note: Morphological and morphometrical characteristics are insufficient to describe or identify a new species of *Heterorhabditis*. Molecular characterisation is always necessary to confirm species identity.

MEASURING AND MORPHOMETRICS

All measurement parameters for *Steinernema* species are listed in Table 1 and those for *Heterorhabditis* are listed in Table 2.

Measurement of spicule and gubernaculum length should be performed as illustrated in Figure 25. The spicule should be measured both along the longer curved median line (or arc) and along the shorter, straight line (or chord) from tip to tip. These values will differ, depending on the curvature of the spicule, and it is important to differentiate the measurements so that accurate comparisons can be made with published data.

In the case of the infective juvenile, measurements should be taken during the first week after the IJ start to emerge. The tail length of both ensheathed and exsheathed juveniles of *Steinernema* and *Heterorhabditis* should also be reported. The c value, L/tail, must be calculated using total body length divided by tail length with sheath.

IDENTIFICATION KEYS

At the time of writing, 55 valid species of *Steinernema* and 11 species of *Heterorhabditis* have been described. As this number increases the identification of individual species becomes progressively more difficult. Identification of species based on only one stage of the life cycle may not be accurate because there are few differentiating morphological characteristics between species and morphometrical ranges of species may overlap. Characteristics of males, females and infective juveniles must be used for accurate identification. Diagnostic characteristics of various stages of the different species of entomopathogenic nematodes are presented in this chapter so as to assist in identification. Morphological and morphometrical characteristics can be used to identify species of *Steinernema*, but they are less reliable for species differentiation in

Table 1. *Morphometric characters recommended (√) for describing* Steinernema *species. Measurements should be in the form: mean ± SD (range).*

Character	First generation		Second generation		Infective juvenile
	Male	Female	Male	Female	
n	20	20	20	20	20
L	√	√	√	√	√
a					√
b					√
c					√
c′					√
V		√		√	
Body diam.	√	√	√	√	√
EP	√	√	√	√	√
NR	√	√	√	√	√
ES	√	√	√	√	√
Testis reflexion	√		√		
Tail length (T)	√	√	√	√	√
Anal body diam. (ABD)	√	√	√	√	√
Spicule length (SL)	√		√		
Spicule width	√		√		
Gubernaculum length (GL)	√		√		
Gubernaculum width	√		√		
D% = EP/ES × 100	√	√	√	√	√
E% = EP/T × 100					√
SW% = SL/ABD × 100	√		√		
GS% = GL/SL × 100	√		√		
H% = Hyaline/T × 100					√

EP = distance from anterior end to excretory pore; NR = distance from anterior end to nerve ring; ES = distance from anterior end to end of pharynx; H = hyaline portion length.

Heterorhabditis. Molecular characterisation is necessary for final confirmation of identity.

Identification of *Steinernema* species

The diagnostic characters for the identification of species of *Steinernema* are presented in the polytomous key (Table 3). The nematode

Table 2. *Morphometric characters recommended (√) for describing* Heterorhabditis *species. Measurements should be in the form: mean ± SD (range).*

Character	Male	Hermaphrodite	Female	Infective juvenile
n	20	20	20	20
L	√	√	√	√
a				√
b				√
c (L/T)				√
V		√	√	
Body diam.	√	√	√	√
EP	√	√	√	√
NR	√	√	√	√
ES	√	√	√	√
Testis reflexion	√			
Tail length with sheath (T)				√
Tail length without sheath	√	√	√	√
Anal body diam. (ABD)	√	√	√	√
Spicule length (SL)	√			
Gubernaculum length (GL)	√			
D% = EP/ES × 100	√	√	√	√
E% = EP/T × 100				√
SW% = SL/ABD × 100	√			
GS% = GL/SL × 100	√			

EP = distance from anterior end to excretory pore; NR = distance from anterior end to nerve ring; ES = distance from anterior end to end of pharynx.

morphometrics are arranged in the descending order of body length of infective juveniles and important characters are in bold-face.

Ratios and abbreviations used in the keys

L = body length, MBD = maximum body diam., EP = distance from anterior end to excretory pore, NR = distance from anterior end to nerve ring, ES = pharynx length, T = tail length, a = L/MBD, b = L/ES, c = L/T, D% = EP/ES × 100, E% = EP/T × 100, ABD = anal body diam., SL = spicule length, GL = gubernaculum length, IJ = infective juvenile; SW% = spicule length/ABD ×100;

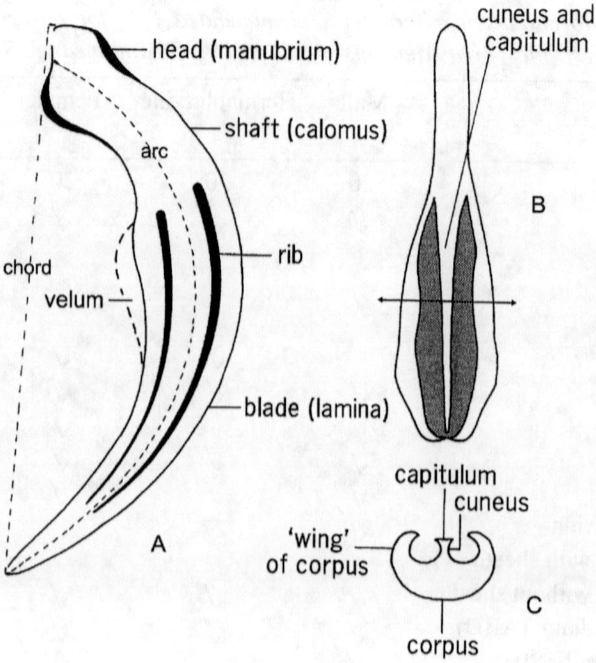

Fig. 25. *A: Spicule anatomy and measuring spicule length. Both chord and arc measurements, as indicated by broken lines, should be reported. B: Gubernaculum, ventral view; C: Gubernaculum in transverse section. Modified after Nguyen and Smart (1993).*

GS% = gubernaculum/spicule ×100; MUC = mucron, A = absent, P = present; na = not available.

How to use the table for identification

To identify a nematode, the following steps should be followed:

- The nematode should be reared in *Galleria mellonella*.
- Make measurements of ten IJ and ten males of the target nematode.
- Match the IJ body length of the nematode with nematodes that have a similar body length in the key.
- Compare the collected data with the measurements in **bold-face** letters and **specific characters** of selected nematodes.
- Decide which nematodes in the key are most similar to that being identified.
- Compare with the original descriptions to confirm the identity.

Table 3. *Tabular key for the identification of valid* Steinernema *species (arranged in descending order of IJ body length). All measurements are in μm. Abbreviations as given on pages 83 and 84.*

Species	Morphometric characters (mean, range)										
	L	MBD	EP	NR	ES	T	a	b	c	D%	E%
S. cubanum											
Infective juvenile											
	1283	37	**106**	116	148	**67**	35	8.6	19.2	70	**160**
	1149-1508	33-46	101-114	106-130	135-159	61-77	na	na	na	na	na

Male **SL = 58** (50-67); GL = 39 (37-42); MBD = 97 (77-117); **D% = 70**; SW% = 141; GS% = 70; **MUC = A**.

Spicule and gubernaculum shapes as in the micrographs:

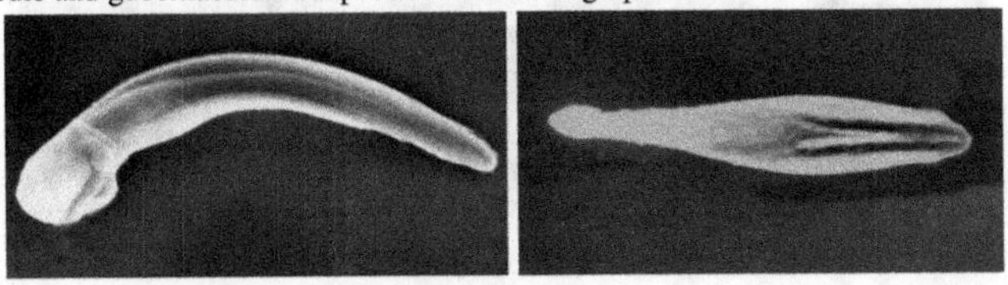

Specific character: Spicule without notch at tip.

Species	L	MBD	EP	NR	ES	T	a	b	c	D%	E%
S. puertoricense											
Infective juvenile											
	1171	51	**95**	117	143	**94**	23	8.2	12.4	66	**101**
	1057-1238	47-54	90-102	111-121	138-147	88-107	20-24	7.4-8.6	11.6-13.6	62-74	88-108

Male **SL = 78** (71-88); GL = 40 (36-45); MBD = 101 (67-148); **D% = 77**; SW% = 152; GS% = 50; **MUC = A**.

Spicule and gubernaculum shapes as in the micrographs:

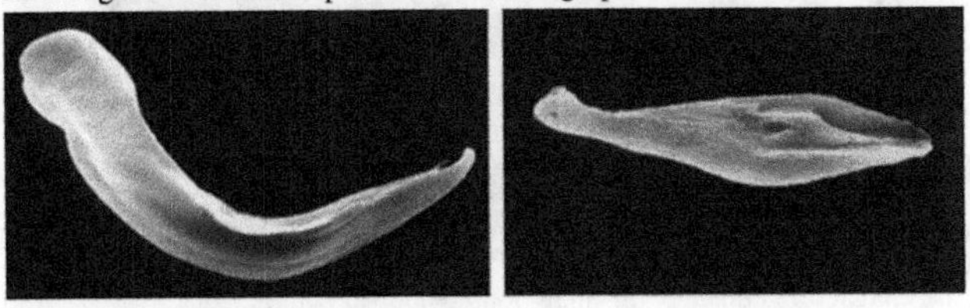

Specific character: Female with epitygmata.

Table 3. *(Continued).*

Species	Morphometric characters (mean, range)										
	L	MBD	EP	NR	ES	T	a	b	c	D%	E%
S. glaseri											
Infective juvenile											
	1130	43	**102**	120	162	**78**	29	7.3	14.7	65	**131**
	864-1448	31-50	87-110	112-126	158-168	62-87	26-35	6.3-7.8	13.6-15.7	58-71	122-138

Male **SL = 77** (64-90); GL = 55 (44-59); MBD = 72 (54-92); **D% = 70** (60-78); **SW% = 210** (164-243); GS% = 70 (60-85); **MUC = A.**

Spicule and gubernaculum shapes as in the micrographs:

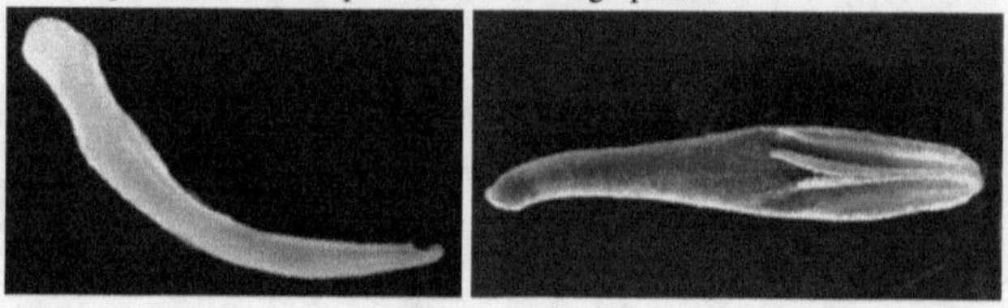

Specific characters: Spicule tip with a notch; Lateral field formula: 2, 5, 6, 8, 2.

Species	L	MBD	EP	NR	ES	T	a	b	c	D%	E%
S. aciari											
Infective juvenile											
	1113	47	**95**	114	146	**78**	24	7.6	14.4	65	**123**
	975-1250	42-53	87-100	106-125	135-157	68-88	28-33	7.0-8.3	14-16	60-70	113-134

Male **SL = 86** (75-95); GL = 56 (48-65); MBD = 104 (88-125); **D% = 76** (69-88); **SW% = 204** (180-240); GS% = 65 (57-77); **MUC = A.**

Spicule and gubernaculum shapes as in the micrographs:

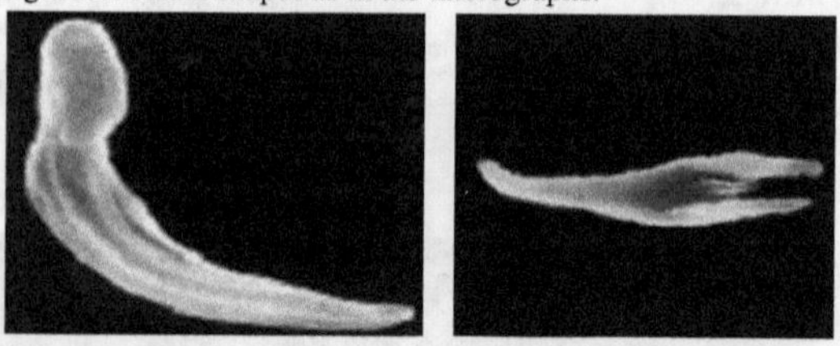

Specific character: Female with small double flapped epitygmata.

Table 3. *(Continued).*

Species	Morphometric characters (mean, range)										
	L	MBD	EP	NR	ES	T	a	b	c	D%	E%
S. caudatum											
Infective juvenile											
	1106	36	**82**	109	156	**88**	31	7.1	12.6	52	**94**
	933-1269	34-41	76-89	104-126	149-175	80-100	na	na	na	na	87-100

Male **SL = 75**; GL = 52; MBD = 91; **D% = 71**; SW% = 222; GS% = 70; **MUC = A**.

Spicule and gubernaculum shapes as in the illustration (redrawn from the original description):

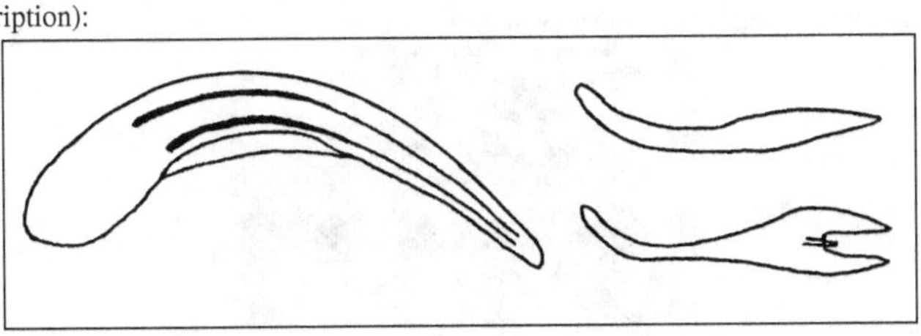

Specific characters: EP and T short; Spicule tip not hook-like in shape.

Species	L	MBD	EP	NR	ES	T	a	b	c	D%	E%
S. longicaudum											
Infective juvenile											
	1063	40	**81**	107	145	**95**	27	7.3	11.2	56	**85**
	(No ranges available)										

Male **SL = 77**; GL = 48; MBD = 155; **D% = 62**; SW% = 160; GS% = 62; **MUC = A**.

Spicule and gubernaculum shapes as in the micrographs:

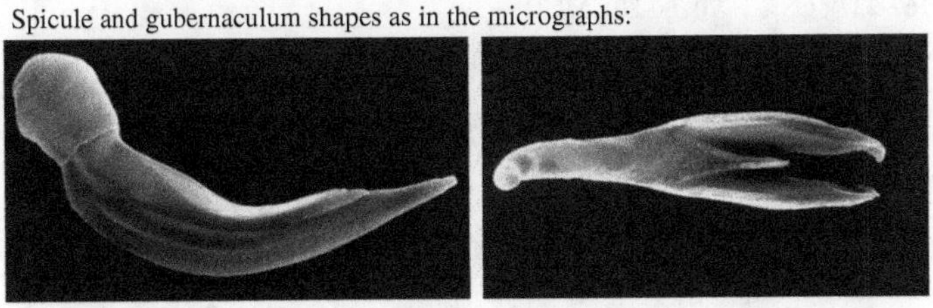

Specific characters: IJ tail long; Spicule reduced in width suddenly to form a tip with flattened terminus.

Table 3. *(Continued).*

Species	L	MBD	EP	NR	ES	T	a	b	c	D%	E%
S. apuliae											
Infective juvenile											
	1064	37	**95**	111	145	**71**	29	7.3	15.0	**66**	**130**
	945-1212	34-41	86-102	103-124	137-159	63-80	26-31	6.8-7.8	14-17	63-70	120-150

Male **SL** = **72** (64-80); GL = 50 (46-54); MBD = 104 (86-124); **D%** = **81** (75-92); **SW%** = **160** (150-180); GS% = 70 (70-80); **MUC** = **A.**

Spicule and gubernaculum shapes as in the micrograph:

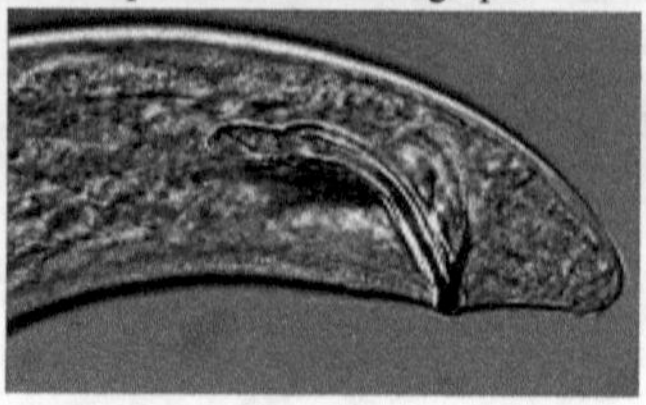

Specific characters: H% = 42 (38-47); V > 50 (50-58).

Species	L	MBD	EP	NR	ES	T	a	b	c	D%	E%
S. khoisanae											
Infective juvenile											
	1062	31	**95**	109	140	**85**	**35**	7.6	**12.5**	**68**	**112**
	994-1159	27-34	87-101	101-118	130-155	76-97	32-37	6.6-8.4	10.6-13.8	60-73	95-128

Males **SP** = **85** (70-88); GL = 56 (50-63); MBD = 108 (90-132); **D%** = **85** (71-99); **SW%** = **203** (167-227); GS% = 70 (60-80); **MUC** = **A.**

Spicule and gubernaculum as in the micrographs:

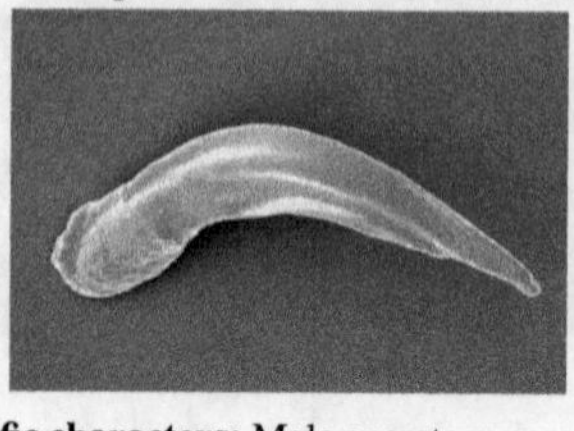

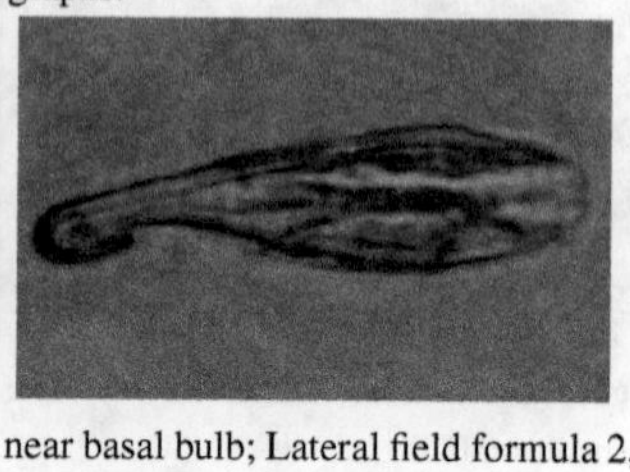

Specific characters: Male excretory pore near basal bulb; Lateral field formula 2, 7, 8, 6, 4, 2.

Table 3. *(Continued).*

Species	Morphometric characters (mean, range)										
	L	MBD	EP	NR	ES	T	a	b	c	D%	E%
S. guangdongense											
Infective juvenile											
	1055	42	**80**	102	134	**91**	25	7.9	11.6	**59**	**88**
	987-1145	30-48	71-85	88-111	123-144	82-103	22-35	7.3-8.5	10.2-12.9	54-65	74-100

Male **SL** = **86** (80-94); **GL** = **64** (47-73); **D%** = **70** (67-78); **SW%** = **175** (152-216); GS% = 75 (59-82); **MUC** = **A**.

Spicule and gubernaculum shapes as in the micrographs:

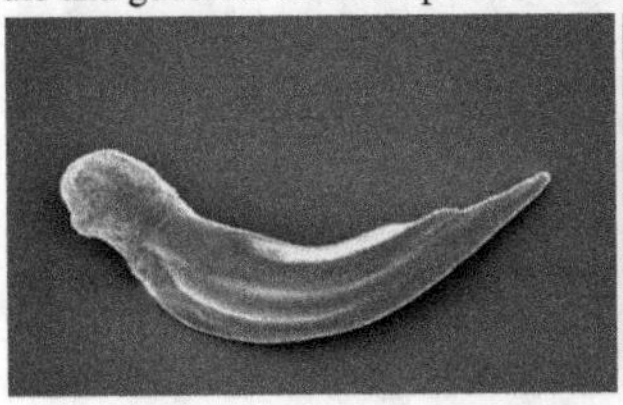
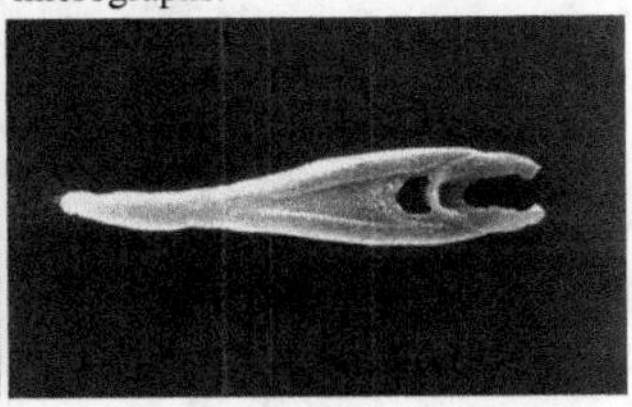

Specific characters: Closer to *S. longicaudum* but spicule longer; Lateral field formula: 2, 7, 8, 7, 4, 2.

Species	L	MBD	EP	NR	ES	T	a	b	c	D%	E%
S. arenarium											
Infective juvenile											
	1034	46	**83**	109	138	**75**	26	7.6	13.8	55	**119**
	724-1408	28-77	76-86	100-120	123-160	64-84	17-34	5.9-10.8	9.4-16.9	52-59	106-130

Male **SL** = **84** (81-91); GL = 55 (49-60); MBD = 188 (184-219); **D%** = **93** (88-102); SW% = 210; GS% = 65 (60-66); **MUC** = **A**.

Spicule and gubernaculum shapes as in the micrographs:

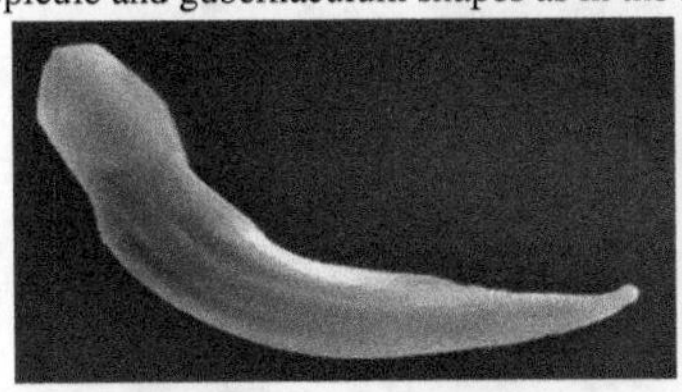
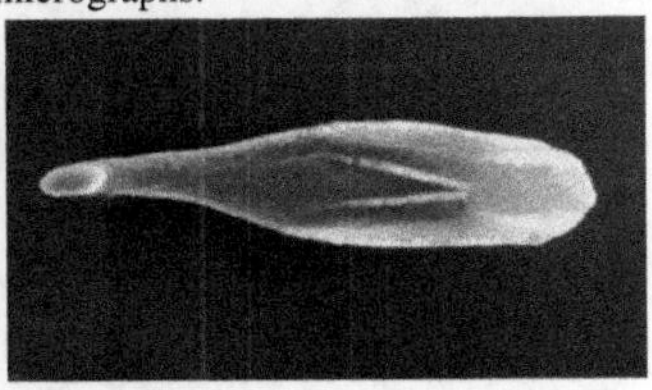

Specific characters: Spicule tip ball-shaped; Cuneus V-shaped.

Table 3. *(Continued).*

Species	L	MBD	EP	NR	ES	T	a	b	c	D%	E%
S. diaprepesi											
Infective juvenile											
	1002	34	**74**	102	138	**83**	30	7.3	12.1	**54**	**90**
	880-1133	30-42	66-83	74-109	111-152	65-91	23-35	6.5-8.3	10.4-13.2	30-70	78-114

Male **SL = 79** (71-90); GL = 54 (45-61); **D% = 80** (68-86); **SW% = 180** (150-200); GS% = 69 (59-79); **MUC = A.**

Spicule and gubernaculum shapes as in the micrographs:

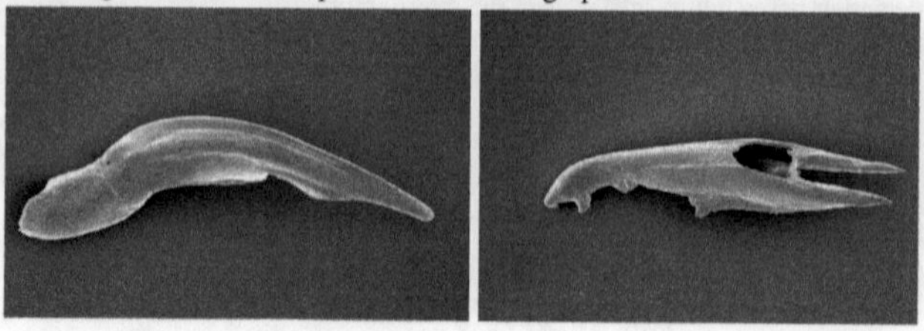

Specific characters: Epitygmata present; Female tail terminus usually with five papilla-like structures; Lateral field formula 2, 6,7, 8, 4, 2.

Species	L	MBD	EP	NR	ES	T	a	b	c	D%	E%
S. loci											
Infective juvenile											
	986	37	**80**	102	141	**75**	27	7.0	13.0	**57**	**107**
	896-1072	30-45	71-86	89-111	126-155	66-83	24-30	6-8	11-15	52-63	94-120

Male **SL = 71** (60-80); GL = 46 (40-52); **D% = 73** (61-80); SW% = 190 (170-210); GS% = 70 (50-80); **MUC = A.**

Spicule, gubernaculum and lateral field as in the micrographs:

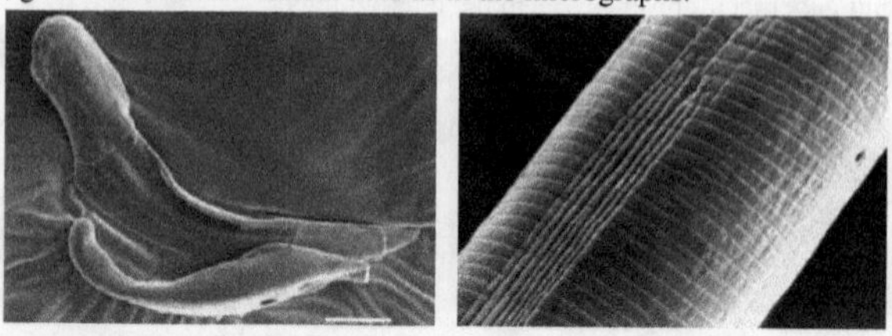

Specific characters: Lateral field with 2 ridges anteriorly then changing to 6, 7.

Table 3. *(Continued).*

Species	Morphometric characters (mean, range)										
	L	MBD	EP	NR	ES	T	a	b	c	D%	E%
S. leizhouense											
Infective juvenile											
	983	38	**83**	101	133	**66**	28	7.5	**13.9**	**59**	**120**
	900-1050	35-43	80-88	95-108	125-143	58-75	23-38	6.8-8.9	11.0-16.8	54-63	105-140

Male **SL = 82** (75-93); GL = 55 (50-60); MBD = 104 (75-143); **D% = 75** (67-83); **SW% = 213** (177-251); GS% = 67 (60-74); **MUC = A**.

Spicule and gubernaculum shapes as in the micrographs:

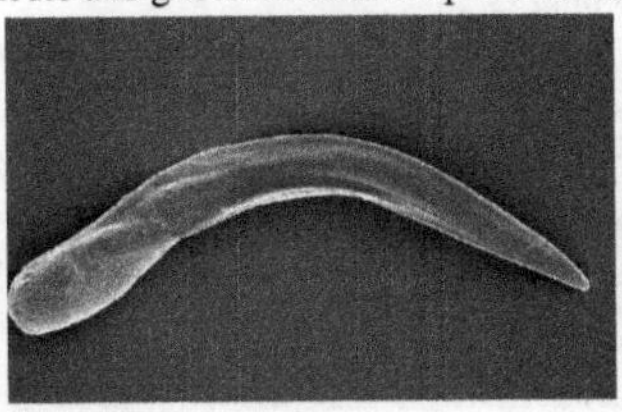

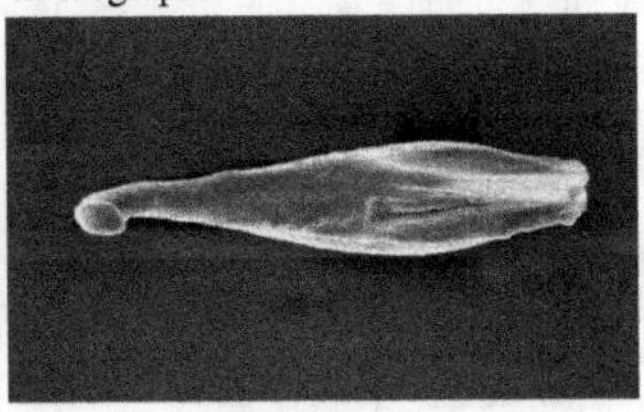

Specific characters: Hyaline portion of IJ 38%; Lateral field formula 2, 6, 8, 4, 2.

Species	L	MBD	EP	NR	ES	T	a	b	c	D%	E%
S. oregonense											
Infective juvenile											
	980	34	**66**	na	132	**70**	30	7.6	14.0	50	**100**
	820-1110	28-38	60-72	na	116-148	64-78	24-37	6.0-8.0	12.0-16.0	40-60	90-110

Male **SL = 71** (65-73); GL = 56 (52-59); MBD = 138 (105-161); **D% = 73** (64-75); SW% = 151; GS% = 79; **MUC = A**.

Spicule and gubernaculum shapes as in the micrographs:

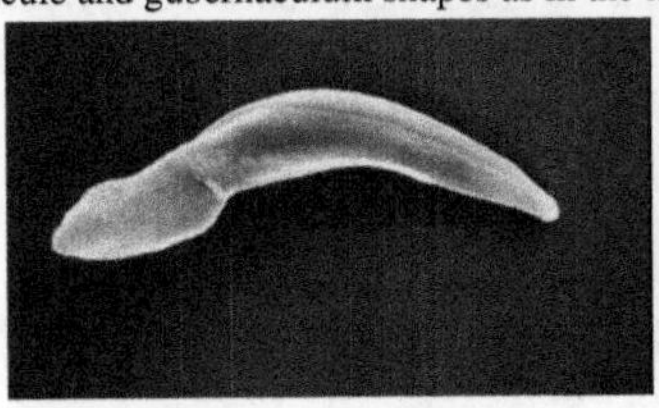

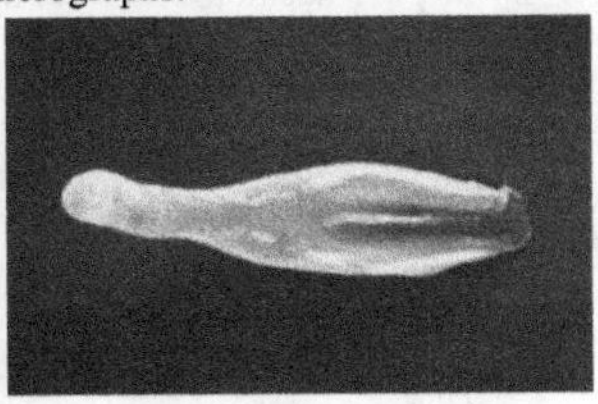

Specific character: This nematode is very similar to *Steinernema feltiae* except male without a mucron and IJ longer.

Table 3. *(Continued).*

Species	L	MBD	EP	NR	ES	T	a	b	c	D%	E%
	Morphometric characters (mean, range)										
S. kraussei											
Infective juvenile											
	951	33	**63**	105	134	**79**	29	7.1	12.1	47	**80**
	797-1102	30-36	50-66	99-111	119-145	63-86	na	na	na	na	na

Male: **SL = 55** (52-57); GL = 33 (23-38); MBD = 128 (110-144); **D% = 53**; SW% = 110; GS% = 71; **MUC = P**.

Spicule and gubernaculum shapes as in the micrographs:

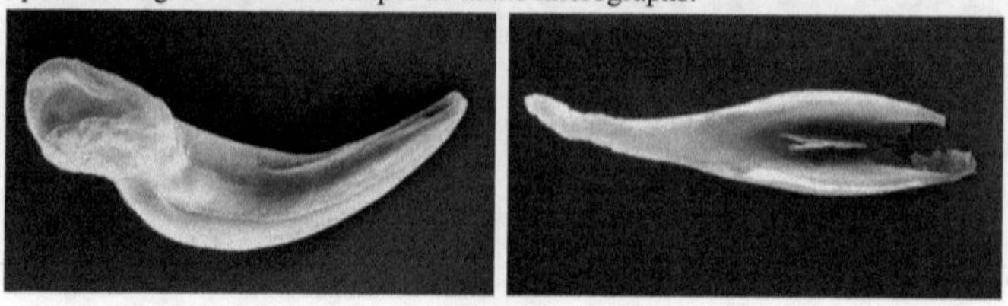

Specific characters: Spicule short; D% low. Develops at low temperature.

Species	L	MBD	EP	NR	ES	T	a	b	c	D%	E%
S. karii											
Infective juvenile											
	932	33	**74**	105	134	**74**	29	6.8	12.6	57	**96**
	876-982	31-35	68-80	97-112	122-147	64-80	na	na	na	na	na

Male: **SL = 83** (73-91); GL = 57 (42-64); MBD = 136 (107-166); **D% = 66** (57-78); **MUC = A**.

Spicule and gubernaculum shapes as in the drawing (redrawn after Waturu *et al.*, 1997):

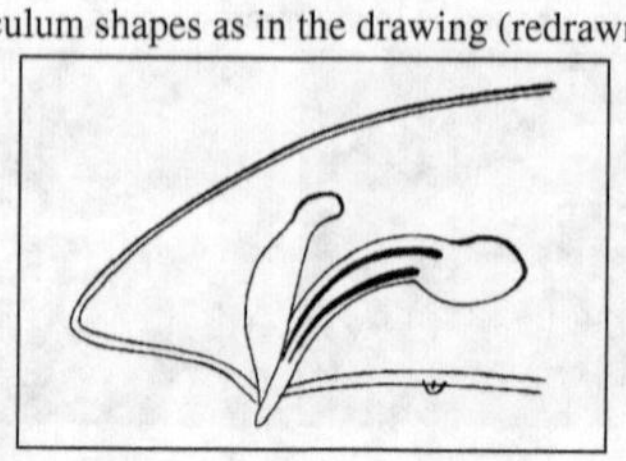

Specific character: Spicule long.

Table 3. *(Continued).*

Species	L	MBD	EP	NR	ES	T	a	b	c	D%	E%
	Morphometric characters (mean, range)										
S. hermaphroditum											
Infective juvenile											
	928	–	**65**	102	–	**77**	29	6.2	10.6	**50**	**85**
	700-110		63-68	95-113		65-83	23-35	5.6-7.1	9.3-11.7	47-55	76-100

Male **SL** = **68** (65-70); **GL** = **48** (47-50); **D%** = **48** (44-50); SW% = 150 (140-160); GS% = 71 (70-75); **MUC** = **A**.

Spicule and gubernaculum shapes as in the micrographs:

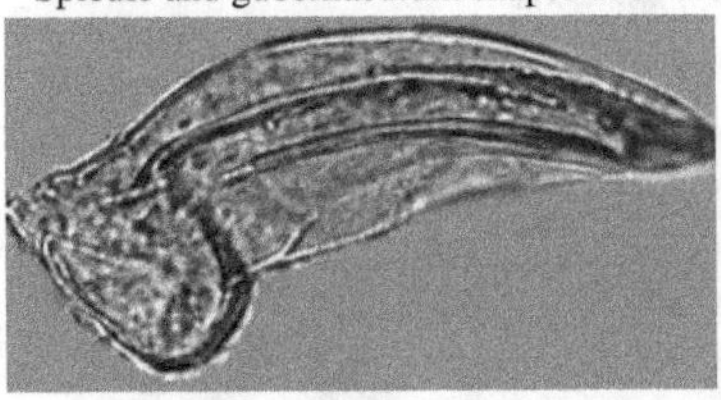

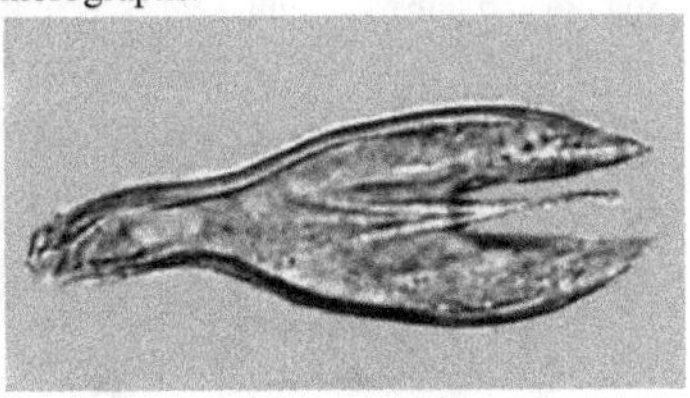

Specific character: Hermaphrodites present.

Species	L	MBD	EP	NR	ES	T	a	b	c	D%	E%
S. scarabaei											
Infective juvenile											
	918	31	**77**	120	156	**76**	31	6.6	12.0	**60**	**100**
	890-959	25-37	72-82	116-127	145-171	71-80	26-36	5.4-7.6	39034.0	50-75	90-110

Male **SL** = **75** (67-83); **GL** = **44** (36-50); MBD = 96 (55-120); **D%** = **66** (53-77); **SW%** = **170** (150-200); GS% = 60 (50-65); **MUC** = **P**.

Spicule and gubernaculum as in the micrographs:

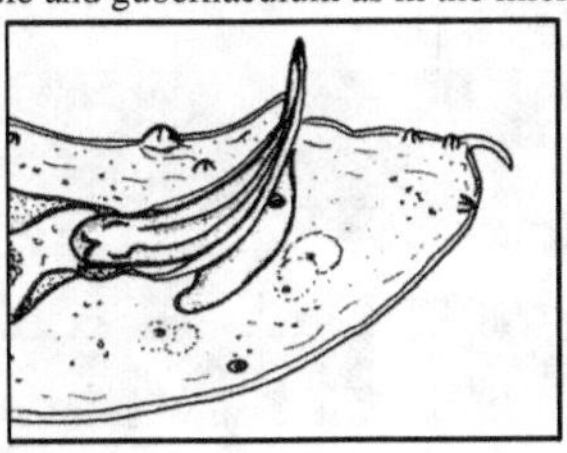

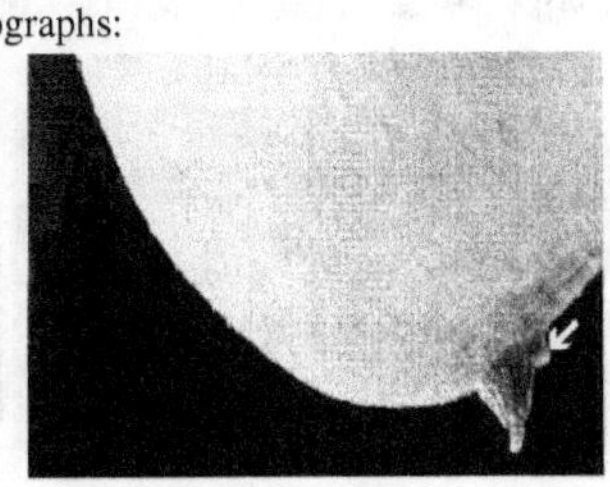

Specific character: First generation female tail with bifurcate mucron.

Table 3. *(Continued).*

Species	Morphometric characters (mean, range)										
	L	MBD	EP	NR	ES	T	a	b	c	D%	E%
S. litorale											
Infective juvenile	**909**	31	**61**	96	125	**83**	30	**7.3**	11.0	**49**	**73**
	834-988	28-33	54-69	89-104	114-133	72-91	27-31	6.7-7.9	9.7-11.9	44-56	68-84

Male **SL = 75** (67-89); GL = 53 (44-64); MBD = 96 (82-111); **D% = 65** (55-72); **SW% = 174** (154-200); GS% = 71 (62-81); **MUC = P**.

Spicule and gubernaculum shapes as in the micrographs:

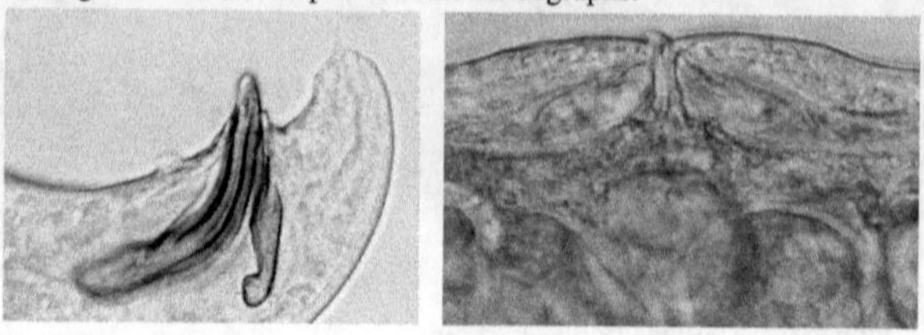

Specific characters: Spicule head twice as long as wide; Vulva with epitygmata.

Species	L	MBD	EP	NR	ES	T	a	b	c	D%	E%
S. neocurtillae											
Infective juvenile	**885**	34	**18**	107	144	**80**	26	6.1	11.2	**12**	**23**
	741-988	28-42	14-22	100-119	130-159	64-97	22-29	5.4-6.7	9.1-14.2	10.0-15.0	18-30

Male **SL = 58** (52-64); GL = 52 (44-59); MBD = 111 (77-144); **D% = 19** (13-26); SW% = 143 (118-164); **GS% = 89** (82-93); **MUC = P**.

Spicule and gubernaculum shapes as in the micrographs:

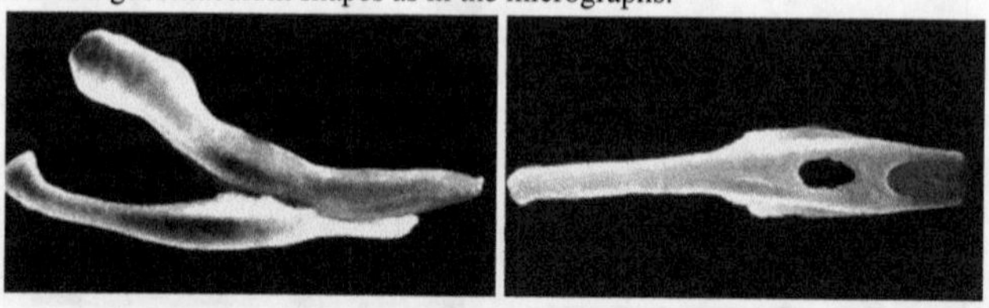

Specific characters: EP extremely short; GS very high; Cannot be produced in *Galleria*.

Table 3. *(Continued).*

Species	L	MBD	EP	NR	ES	T	a	b	c	D%	E%
S. silvaticum											
Infective juvenile											
	860	30	**62**	96	121	**75**	29	7.1	11.4	**50**	**83**
	670-975	26-35	51-73	75-109	100-141	63-86	23-33	6.3-7.7	9.9-13.1	46-56	–

Male **SL** = **51** (42-64); GL = 37 (30-43); MBD = 65 (52-78); D% = 60 (45-63); SW% = 155; GS% = 73; **MUC** = **P**.

Spicule and gubernaculum as in the micrograph:

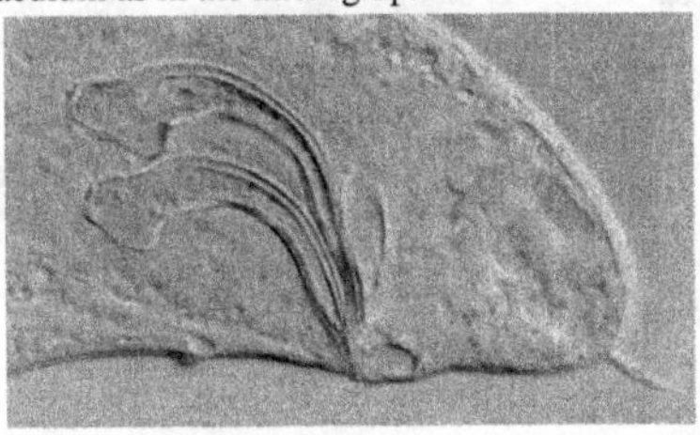

Specific characters: Male tail with prominent mucron; Spicule head wide.

Species	L	MBD	EP	NR	ES	T	a	b	c	D%	E%
S. thanhi											
Infective juvenile											
	851	31	**75**	96	129	**63**	27	7.0	**14.0**	**58**	**119**
	720-960	27-39	68-84	83-112	115-146	52-72	23-33	6.0-8.0	12.0-16.0	52-67	101-138

Male **SL** = **72** (67-78); GL = 49 (40-56); MBD = 97 (76-124); **D%** = **73** (64-82); **SW%** = **180** (150-210); GS% = 70 (50-80); **MUC** = **A**.

Spicule and lateral field as in the micrographs:

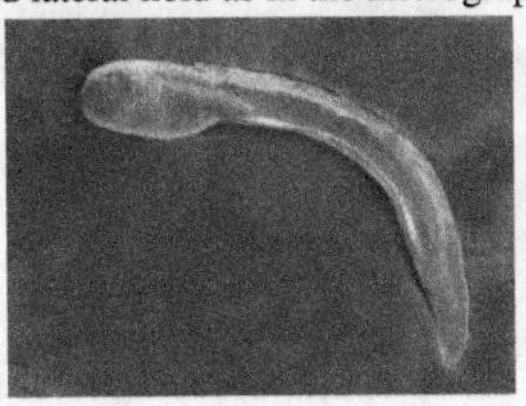

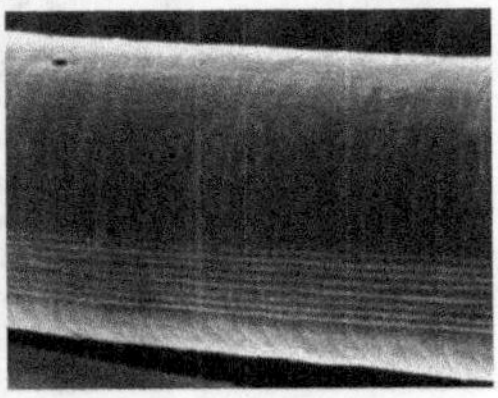

Specific character: The presence of 13 pairs of genital papillae on first generation male.

Table 3. *(Continued).*

Species	L	MBD	EP	NR	ES	T	a	b	c	D%	E%
	Morphometric characters (mean, range)										
S. feltiae											
Infective juvenile	**849**	26	**62**	99	136	**81**	31	6.0	10.4	45	**78**
	736-950	22-29	53-67	88-112	115-150	70-92	29-33	5.3-6.4	9.2-12.6	42-51	69-86

Male **SL** = **70** (65-77); GL = 41 (34-47); **MBD** = 75 (60-90); **D%** = **60**; SW% = 113 (99-130); GS% = 59 (52-61); **MUC** = **P**.

Spicule and gubernaculum shapes as in the micrographs:

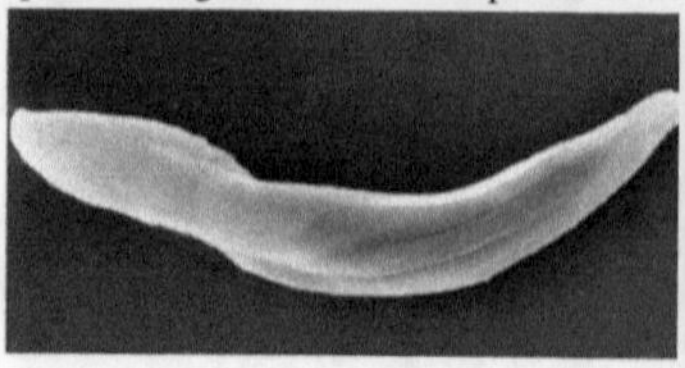
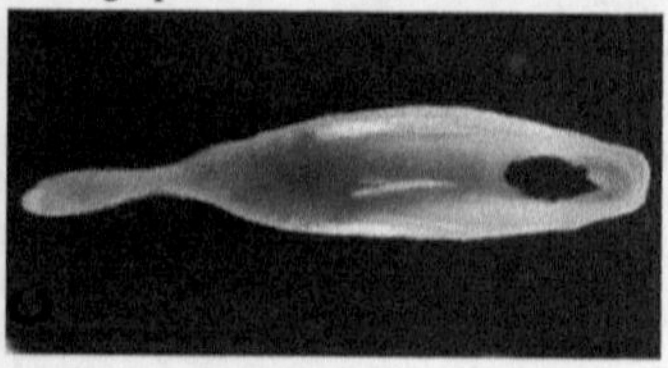

Specific characters: Spicule head very long; Male with long mucron; Lateral field formula 2, 7, 8, 7, 4, 2.

Species	L	MBD	EP	NR	ES	T	a	b	c	D%	E%
S. akhursti											
Infective juvenile	**812**	**33**	**59**	90	119	**73**	**24**	6.8	11.0	47	**77**
	770-835	33-35	55-60	83-95	115-123	68-75	23-26	6.6-7.2	10.0-12.0	45-50	73-86

Male **SL** = **90** (85-100); GL = 64 (58-68); MBD = 131 (115-150); D% = 56 (52-61); **SW%** = **180** (140-200); GS% = 71 (65-77); **MUC** = **P**.

Spicule and gubernaculum shapes as in the micrographs:

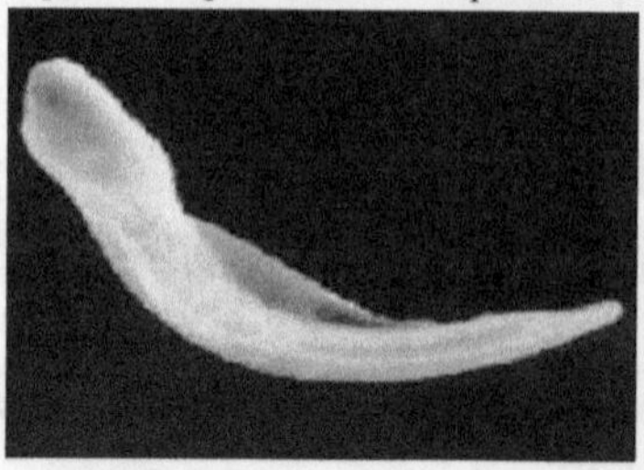
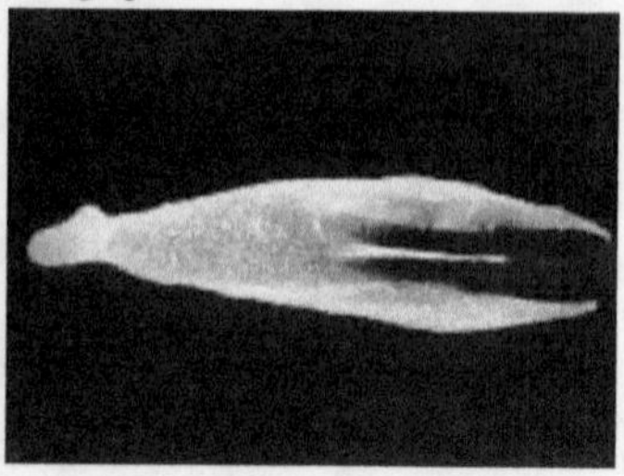

Specific characters: Lateral field formula of IJ: 2, 6, 2; Male with the longest spicule in *S. feltiae* group.

Table 3. *(Continued).*

Species	L	MBD	EP	NR	ES	T	a	b	c	D%	E%
	Morphometric characters (mean, range)										
S. bicornutum											
Infective juvenile											
	769	29	**61**	92	124	**72**	27	6.2	10.7	50	**84**
	648-873	25-33	53-65	88-100	113-135	63-78	23-29	5.6-6.9	9.7-12.0	40-60	80-100

Male **SL = 65** (53-70); GL = 48 (38-50); MBD = 109 (80-127); **D% = 52** (50-60); **SW% = 222** (218-226); GS% = 72; **MUC = A.**

Spicule and gubernaculum as in the micrographs:

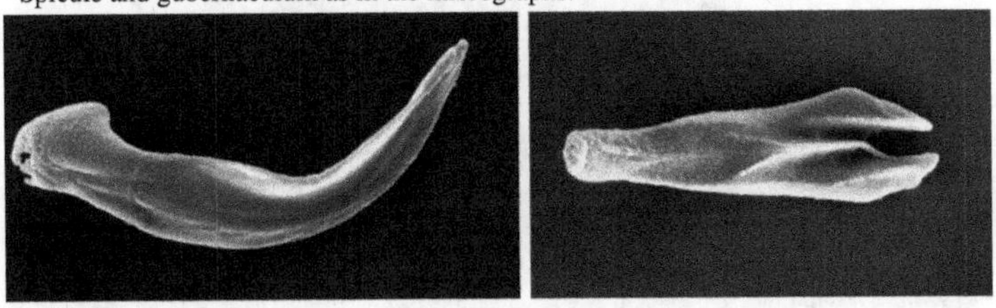

Specific characters: 2 horn-like structures on lip region of IJ; Spicule head wider than long; Cuneus arrow-head shaped.

Species	L	MBD	EP	NR	ES	T	a	b	c	D%	E%
S. ashiuense											
Infective juvenile											
	768	30	**55**	86	119	**71**	25	6.0	11.0	**46**	**78**
	720-800	28-33	51-59	77-91	113-128	66-76	24-27	6.0-7.0	10.0-12.0	43-50	70-85

Male **SL = 59** (50-65); GL = 37 (25-43); MBD = 106 (80-125); **D% = 50** (44-56); **SW% = 149** (128-167); GS% = 63 (43-73); **MUC = P.**

Spicule and gubernaculum as in the micrographs:

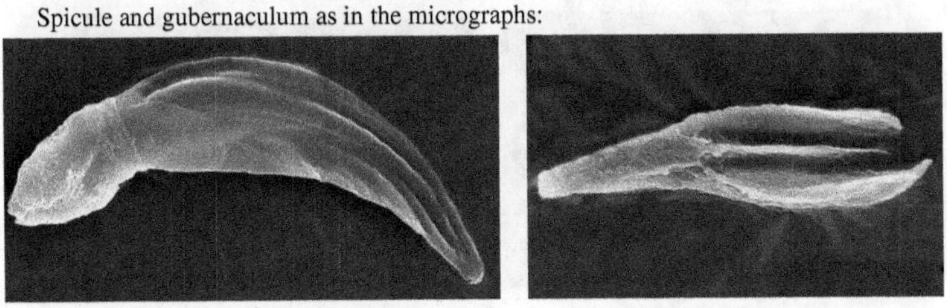

Specific characters: Lateral field with 5 ridges; Hyaline portion = 1/3 of tail.

Table 3. *(Continued).*

Species	L	MBD	EP	NR	ES	T	a	b	c	D%	E%
	Morphometric characters (mean, range)										
S. sangi											
Infective juvenile	**753**	35	**51**	91	127	**81**	**22**	5.9	9.3	**40**	**62**
	704-784	30-40	46-54	78-97	120-138	76-89	19-25	5.6-6.3	8.7-10.2	36-44	56-70

Male **SL = 63** (58-80); GL = 40 (34-46); MBD = 159 (120-225); D% = 49 (42-63); **SW% = 150** (120-160); GS% = 60 (50-70); **MUC = P**.

Spicule shape as in the micrograph:

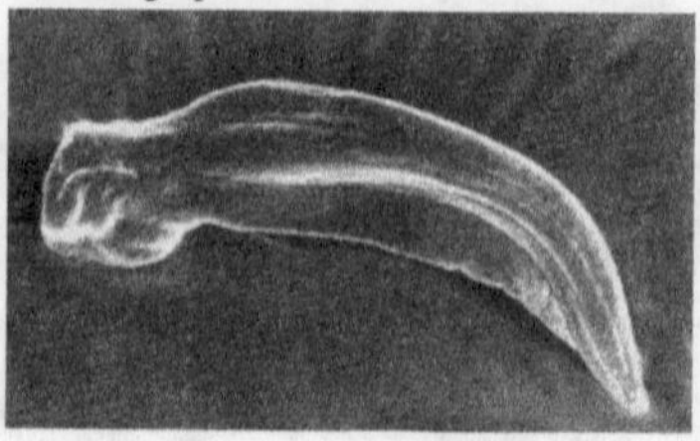

Specific characters: This species is closer to *S. kraussei* but IJ body is shorter and SW% is larger.

Species	L	MBD	EP	NR	ES	T	a	b	c	D%	E%
S. beddingi											
Infective juvenile	**743**	–	**70**	96	125	**77**	**26.7**	5.8	10.1	53	91
	700-790		64-75	80-113	113-130	72-83	24-30	5.0-6.8	9.6-10.8	44-64	80-108

Male **SL = 67** (63-73); **GL = 40** (35-45); **D% = 59** (53-62); SW% = 122 (110-131); GS% = 60 (56-66); **MUC = A**.

Spicules and gubernaculum shapes as in the micrographs:

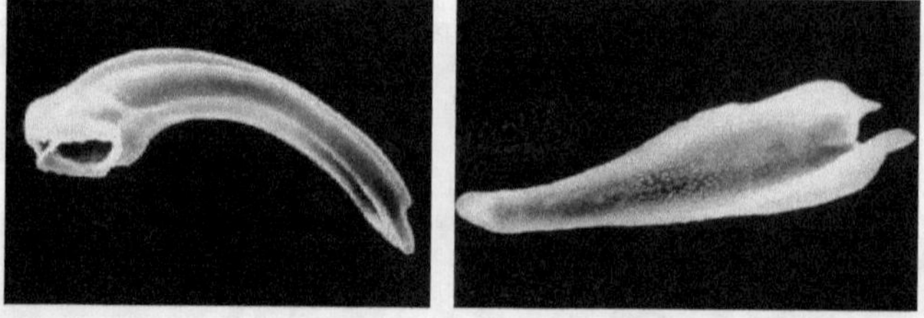

Specific characters: Spicule shape is unique; Gubernaculum without cuneus; Lateral field of IJ has 4 ridges.

Table 3. *(Continued).*

Species	Morphometric characters (mean, range)										
	L	MBD	EP	NR	ES	T	a	b	c	D%	E%
S. weiseri											
Infective juvenile											
	740	25	**57**	84	113	**60**	**29**	6.6	12.0	51	**95**
	586-828	24-29	43-65	72-92	95-119	49-68	25-33	5.7-7.2	10.0-14.0	44-55	–

Male **SL = 68** (62-72); GL = 53 (46-57); MBD = 112 (84-138); **D% = 49** (39-60); **SW% = 180** (150-240); GS% = 80 (70-85); **MUC = A.**

Spicule and gubernaculum of the second generation as in the drawing:

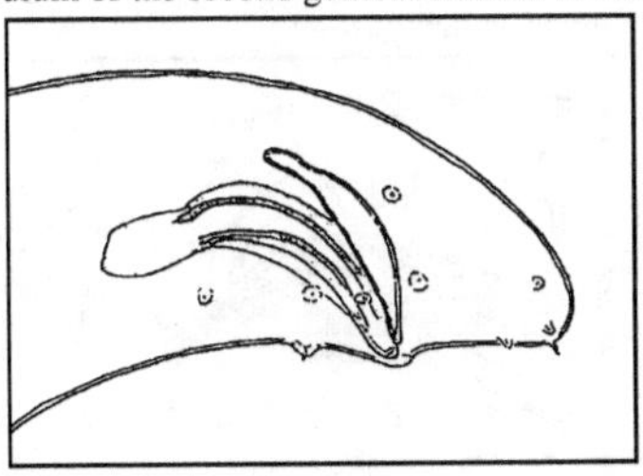

Specific character: IJ has very short hyaline portion, 1/3 of tail length.

Species	L	MBD	EP	NR	ES	T	a	b	c	D%	E%
S. robustispiculum											
Infective juvenile											
	712	28	56	84	120	**75**	25	6.0	10.0	46	**75**
	642-778	26-35	50-68	80-100	115-152	68-92	18-29	3.6-6.3	38879.0	43-59	67-87

Male **SL = 58** (51-65); GL = 41 (36-44); MBD = 127 (105-150); **D% = 56** (50-63); SW% = 129 (111-150); GS% = 70 (64-79); **MUC = P.**

Spicule large (redrawn after Phan *et al.*, 2005):

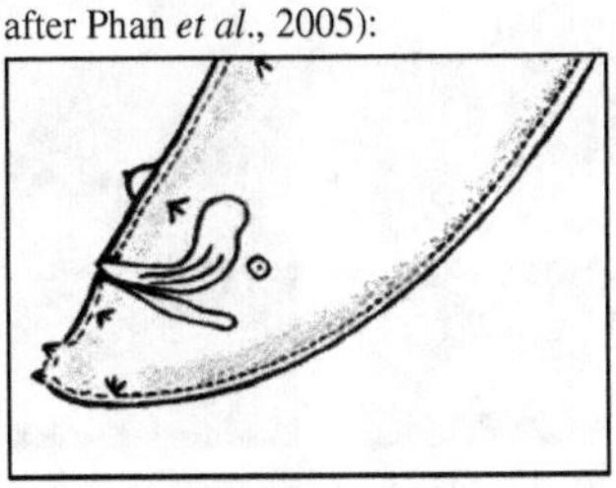

Specific character: Spicule length/spicule width = 4.5 (3.8-5.6).

Table 3. *(Continued).*

Species	Morphometric characters (mean, range)										
	L	MBD	EP	NR	ES	T	a	b	c	D%	E%
S. jollieti											
Infective juvenile											
	711	23	**60**	–	123	**68**	31	5.7	11.0	**48**	88
	625-820	20-28	53-65		115-135	60-73	25-34	4.9-6.4	9.0-12.0	46-50	–

Male **SL** = **64** (55-70); GL = 54 (45-60); MBD = 115 (98-135); **D%** = **64** (53-83); SW% = 145; GS% = 84; **MUC** = **A**.

Spicule and gubernaculum as in the drawing:

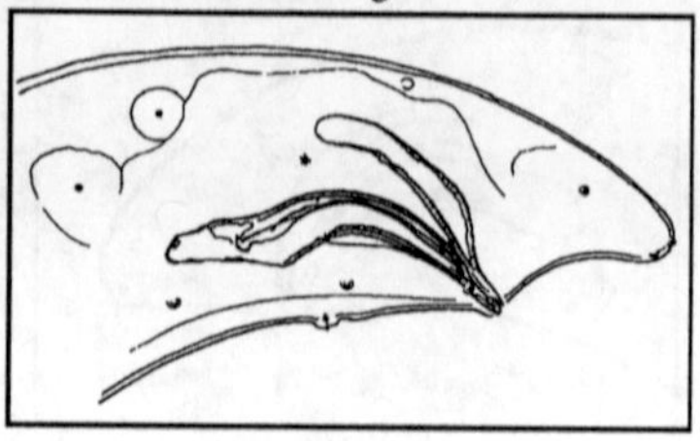

Specific character: Lateral field with 6 ridges.

Species	L	MBD	EP	NR	ES	T	a	b	c	D%	E%
S. sichuanense											
Infective juvenile											
	710	28	**64**	99	131	**72**	25	5.4	10.0	**49**	**89**
	606-787	24-32	57-68	91-108	121-142	64-76	23-28	4.9-6.0	8.9-11.0	46-53	–

Male **SL** = **88** (83-101); GL = 47 (40-51); MBD = 122 (91-141); D% = 51 (45-56); SW% = 160 (140-180); GS% = 60; **MUC** = **A**.

Spicule and gubernaculum as in the micrographs:

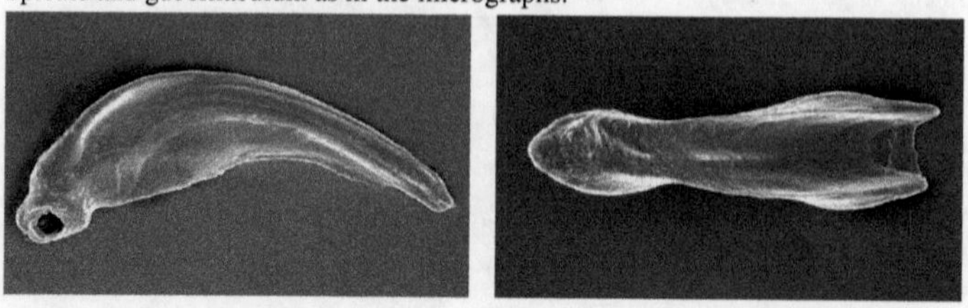

Specific characters: Lateral field with six ridges; Lateral field formula 2, 5, 6, 4, 2.

Table 3. *(Continued).*

Species	Morphometric characters (mean, range)										
	L	MBD	EP	NR	ES	T	a	b	c	D%	E%
S. monticolum											
Infective juvenile											
	706	37	**58**	88	124	**77**	19	5.7	9.3	47	**76**
	612-821	32-46	54-62	81-93	120-131	71-95	14-22	5.0-6.4	7.6-11.1	44-50	63-86

Male **SL = 70** (61-80); GL = 45 (35-54); MBD = 160 (117-206); **D% = 55** (49-61); SW% = 140 (120-150); GS% = 60 (50-70); **MUC = P**.

Spicule and gubernaculum shapes as in the micrographs:

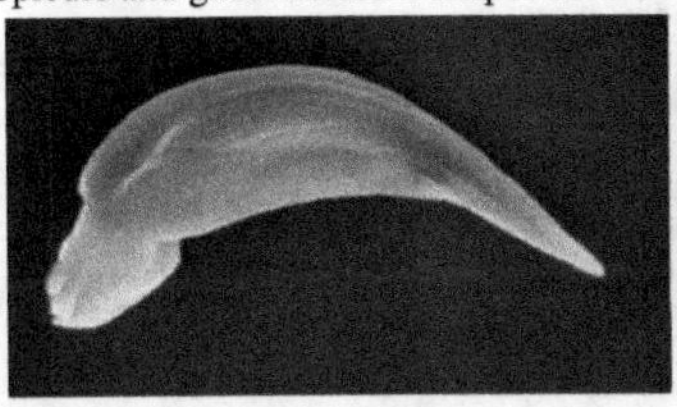
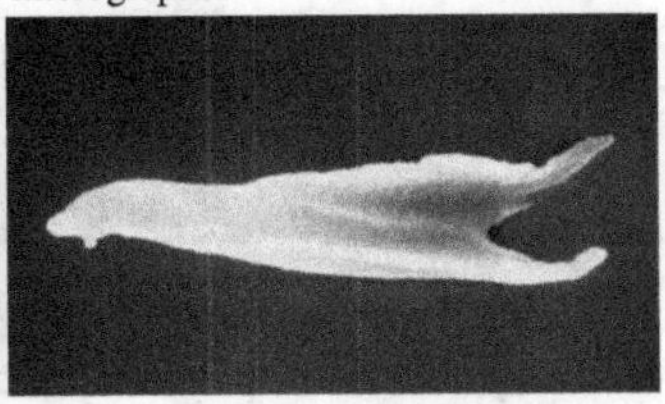

Specific characters: IJ tail long; E% low.

Species	L	MBD	EP	NR	ES	T	a	b	c	D%	E%
S. ceratophorum											
Infective juvenile											
	706	27	**55**	92	123	**66**	26	na	10.6	45	**84**
	591-800	23-34	47-70	79-103	108-144	56-74	24-28	na	8.8-12.9	40-56	74-96

Male **SL = 71** (54-90); GL = 40 (25-45); MBD = 146 (104-185); **D% = 51** (33-65); **SW% = 140** (100-200); GS% = 60 (40-80); **MUC = A**.

Spicule and gubernaculum shapes as in the micrographs:

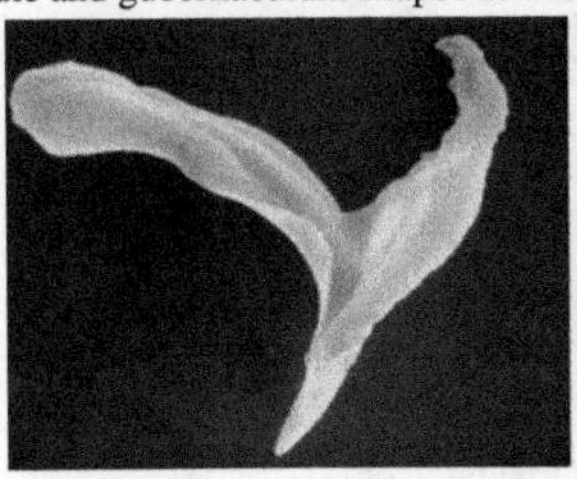
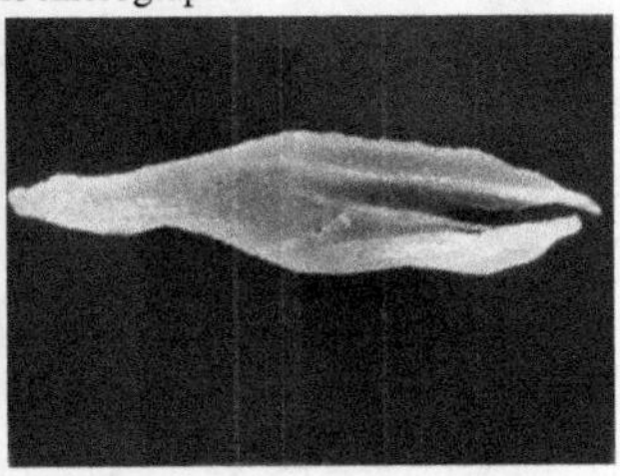

Specific character: 2 horn-like structures on lip region of IJ.

Table 3. *(Continued).*

Species	Morphometric characters (mean, range)										
	L	MBD	EP	NR	ES	T	a	b	c	D%	E%
S. affine											
Infective juvenile											
	693	30	**62**	95	126	**66**	23	5.5	10.5	49	**94**
	608-880	28-34	51-69	88-104	115-134	64-74	21-28	5.1-6.0	9.5-11.5	43-53	74-108
Male	**SL** = **70** (67-86); GL = 46 (37-56); MBD = 118 (95-164); **D%** = **61**; SW% = 117; GS% = 66; **MUC** = **P**. Spicule and gubernaculum shapes as in the micrographs:										

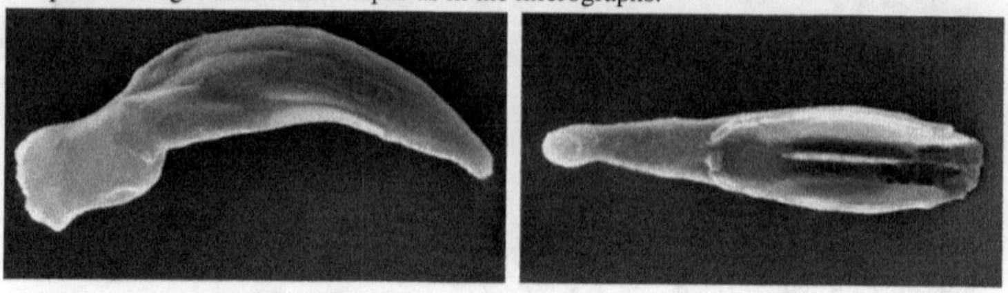

Specific character: Spine-like structure in IJ tail tip.

Species	L	MBD	EP	NR	ES	T	a	b	c	D%	E%
S. pakistanense											
Infective juvenile											
	683	27	**54**	80	113	58	24	6.0	11.0	**47**	**91**
	649-716	24-29	49-58	76-83	108-122	53-62	21-27	5.0-6.0	10.0-12.0	42-53	87-102
Male	**SL** = **68** (62-73); GL = 41 (36-45); MBD = 102 (80-128); D% = 60 (50-60); SW% = 180 (100-220); GS% = 60 (50-60); MUC = A. Spicule and gubernaculum shapes as in the drawings:										

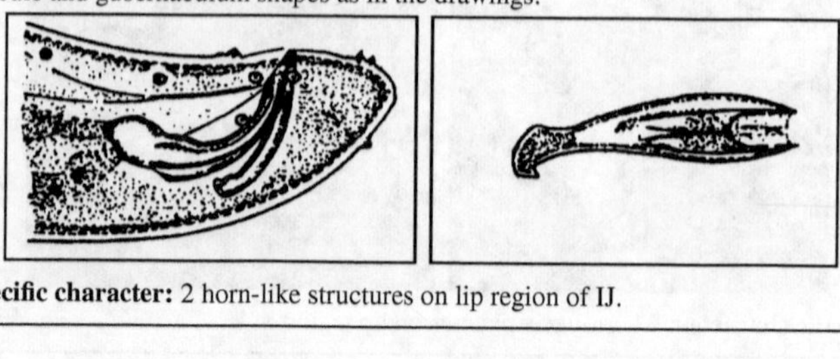

Specific character: 2 horn-like structures on lip region of IJ.

Table 3. *(Continued).*

Species	L	MBD	EP	NR	ES	T	a	b	c	D%	E%
S. intermedium											
Infective juvenile											
	671	29	**65**	93	123	**66**	23	5.3	10.0	51	**96**
	608-800	25-32	59-69	85-99	110-133	53-74	20-26	5.0-6.0	9.3-10.8	48-58	89-108

Male **SL = 91** (84-100); GL = 64 (56-75); MBD = 168 (113-207); **D% = 67** (58-76); SW% = 124 (103-139); GS% = 69 (62-77); MUC = A.

Spicule and gubernaculum shapes as the micrographs:

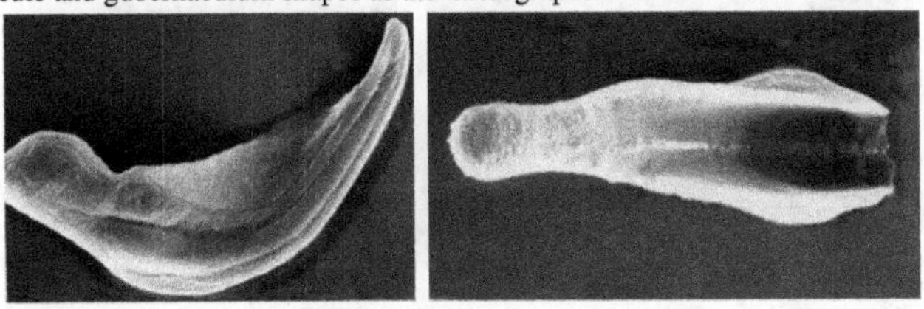

Specific character: Spicule is the largest in *Steinernema*.

Species	L	MBD	EP	NR	ES	T	a	b	c	D%	E%
S. hebeiense											
Infective juvenile											
	658	26	**48**	78	107	**66**	26	6.2	10.0	**45**	**72**
	610-710	23-28	43-51	73-83	100-111	63-71	24-28	5.7-6.7	9.4-11.0	40-50	65-80

Male **SL = 57** (51-63); GL = 46 (38-50); MBD = 86 (74-98); **D% = 51** (48-59); SW% = 140 (120-170); GS% = 80 (60-90); **MUC = A**.

Spicule and gubernaculum as in the micrographs:

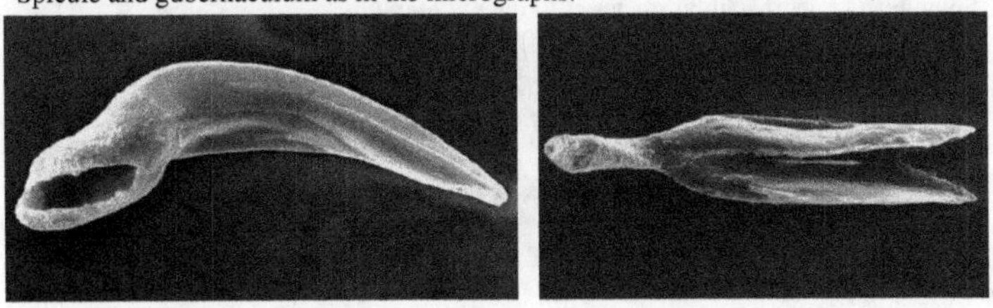

Specific character: Gubernaculum *ca* 80% of spicule.

Table 3. *(Continued).*

Species	L	MBD	EP	NR	ES	T	a	b	c	D%	E%
	Morphometric characters (mean, range)										
S. yirgalemense											
Infective juvenile											
	635	29	**51**	88	121	**62**	21	5.2	10.3	**42**	83
	548-693	24-33	45-59	82-93	115-128	57-67	20-25	4.8-5.9	9.2-11.2	38-48	67-98

Male **SL** = **64** (51-77); GL = 48 (42-54); MBD = 112 (97-138); D% = 58 (50-66); **SW%** = **171** (121-213); GS% = 74 (65-85); **MUC** = **A**.

Spicule and gubernaculum shapes as in the micrographs:

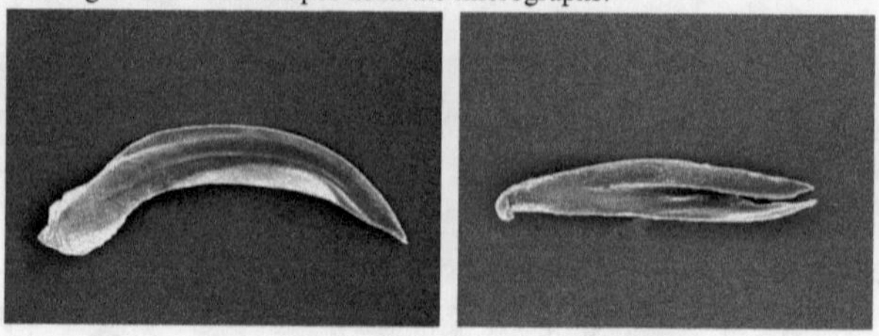

Specific characters: 2 horn-like structure present on IJ head; The presence of 12 pairs of genital papillae in male.

Species	L	MBD	EP	NR	ES	T	a	b	c	D%	E%
S. riobrave											
Infective juvenile											
	622	28	**56**	87	114	**54**	23	5.4	11.6	49	**105**
	561-701	26-30	51-64	84-89	109-116	46-59	20-24	4.9-6.0	10.1-12.4	45-55	93-111

Male **SL** = **67** (62.5-75.0); GL = 51 (47.5-56.2); MBD = 133 (116-159); **D%** = **71** (60-80); SW% = 114; GS% = 76; **MUC** = **A**.

Spicule and gubernaculum shapes as in the micrographs:

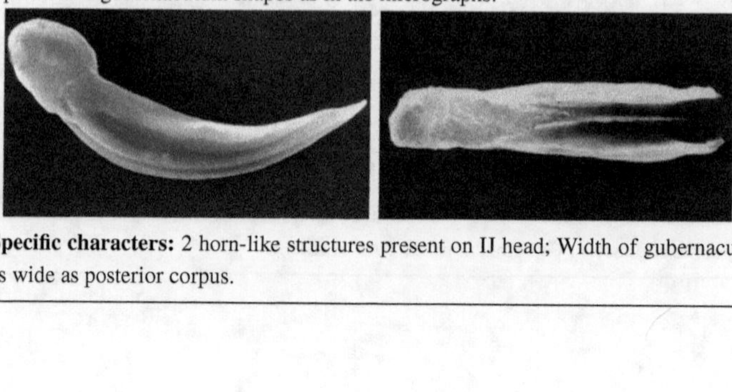

Specific characters: 2 horn-like structures present on IJ head; Width of gubernaculum neck as wide as posterior corpus.

Table 3. *(Continued).*

Species	L	MBD	EP	NR	ES	T	a	b	c	D%	E%
	Morphometric characters (mean, range)										
S. kushidai											
Infective juvenile	**589**	26	**46**	76	111	**50**	22.5	5.3	11.7	41	**92**
	524-662	22-31	42-50	70-84	106-120	44-59	19-25	4.9-5.9	9.9-12.9	38-44	84-95

Male **SL** = **63**; GL = 44; MBD = 97; **D%** = **51**; SW% = 150; GS% = 70; **MUC** = **A**.

Spicule and gubernaculum shapes as in the micrographs:

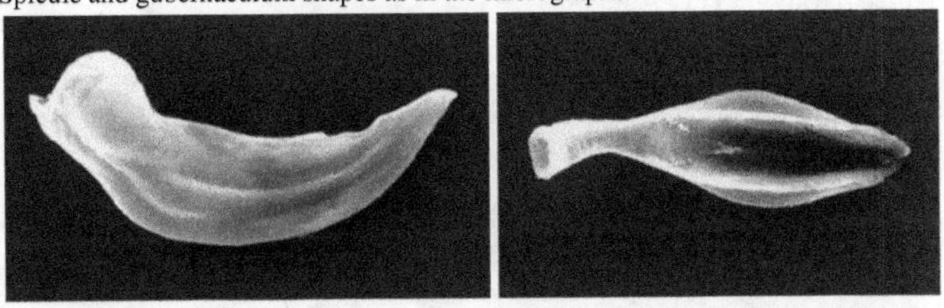

Specific characters: SL/spicule width about 3.9; Cannot reproduce in *Galleria*.

Species	L	MBD	EP	NR	ES	T	a	b	c	D%	E%
S. scapterisci											
Infective juvenile	**572**	24	**39**	97	127	**54**	24	4.5	10.7	31	**73**
	517-609	18-30	36-48	83-106	113-134	48-60	20-31	4.0-4.6	9.2-11.7	27-40	60-80

Male **SL** = 83 (72-92); GL = 65 (59-75); MBD = 156 (97-213); **D%** = **38** (32-44); **SW%** = **252** (204-280); GS% = 78; **MUC** = **P**.

Spicule and gubernaculum shapes as in the micrographs:

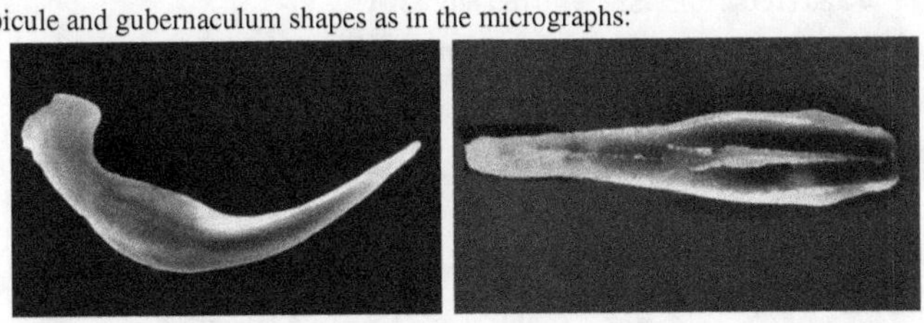

Specific characters: Female: Epitygmata very large; Excretory duct with ellipse-shaped structure; male with long spicule.

Note: Develops poorly in *Galleria*.

Table 3. *(Continued).*

Species	Morphometric characters (mean, range)										
	L	MBD	EP	NR	ES	T	a	b	c	D%	E%
S. carpocapsae											
Infective juvenile	**558**	25	**38**	85	120	**53**	21	4.4	10.0	26	**60**
	438-650	20-30	30-60	76-99	103-190	46-61	19-24	4.0-4.8	9.1-11.2	23-28	54-66

Male **SL = 66** (58-77); GL = 47 (39-55); MBD = 101 (77-130); **D% = 41** (27-55); **SW% = 172** (140-200); GS% = 71 (59-88); **MUC = P**.

Spicule and gubernaculum shapes as in the micrographs:

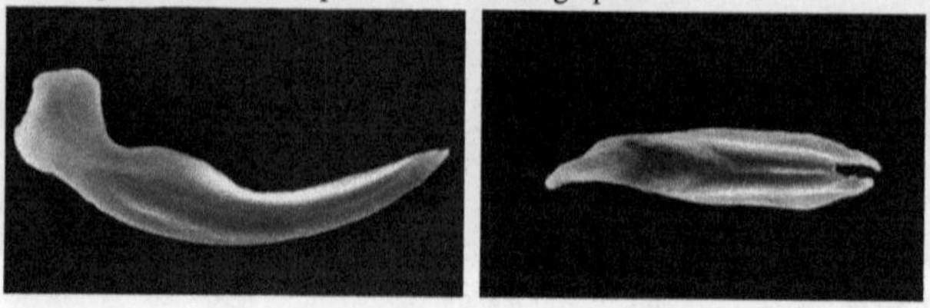

Specific characters: Spicule head wider than long; Gubernaculum with short anterior part.

Species	L	MBD	EP	NR	ES	T	a	b	c	D%	E%
S. websteri											
Infective juvenile	**584**	21	**36**	88	115	**47**	28	5.1	12.6	31	**77**
	553-631	17-25	29-40	83-95	107-122	37-56	24-35	4.8-5.6	11.0-15.5	24-34	62-102

Male **SL = 68** (64-72); GL = 49 (42-56); MBD = 147 (119-175); **D% = 40** (30-50); **SW% = 180** (160-210); GS% = 70 (60-80); **MUC = P**.

Spicule and gubernaculum shapes as in the micrographs:

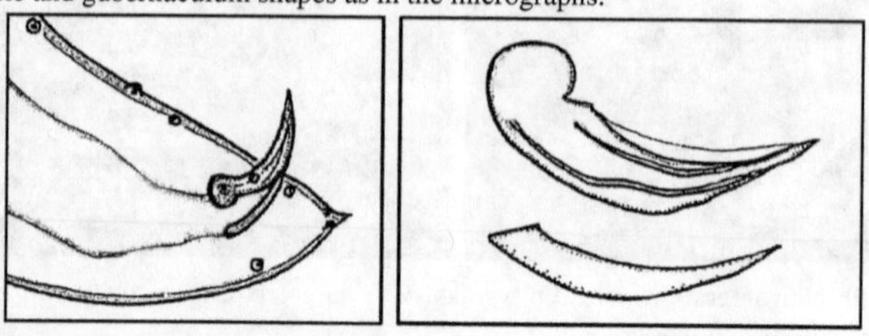

Specific character: Spicule shape is unique.

Table 3. *(Continued).*

Species	Morphometric characters (mean, range)										
	L	MBD	EP	NR	ES	T	a	b	c	D%	E%
S. anatoliense											
Infective juvenile	**545**	24.5	37	76	107	**52**	22	5.1	10.5	35	**72**
	507-580	21-28	36-39	71-82	97-124	46-58	19-27	4.6-5.7	9.4-11.8	32-39	64-82

Male **SL = 74** (68-84); GL = 47 (42-59); MBD = 109 (87-150); D% = 49 (47-55); SW% = 175 (160-190); GS% = 55 (40-65); **MUC = P.**

Spicule and gubernaculum shapes as in the micrograph:

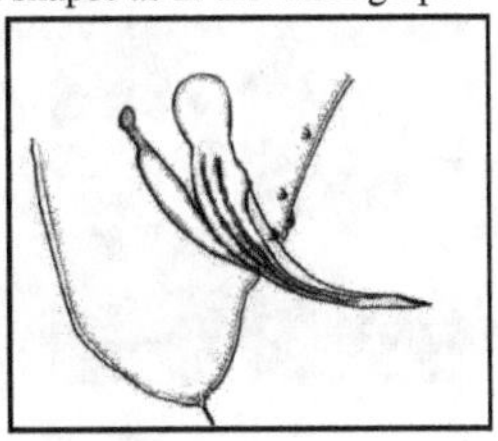

Specific characters: Spicule longer than that of *S. carpocapsae*; Spicule head elongate.

Species	L	MBD	EP	NR	ES	T	a	b	c	D%	E%
S. abbasi											
Infective juvenile	**541**	29	**48**	68	89	**56**	18	6.0	9.8	53	**86**
	496-579	27-30	46-51	64-72	85-92	52-61	17-20	5.5-6.6	8.1-10.8	51-58	79-94

Male **SL = 65** (57-74); GL = 45 (33-50); MBD = 87 (82-98); **D% = 60** (51-68); SW% = 156 (107-187); GS% = 70 (58-85); **MUC = A.**

Spicule and gubernaculum shapes as in the micrographs (redrawn from Elawad *et al.*, 1997):

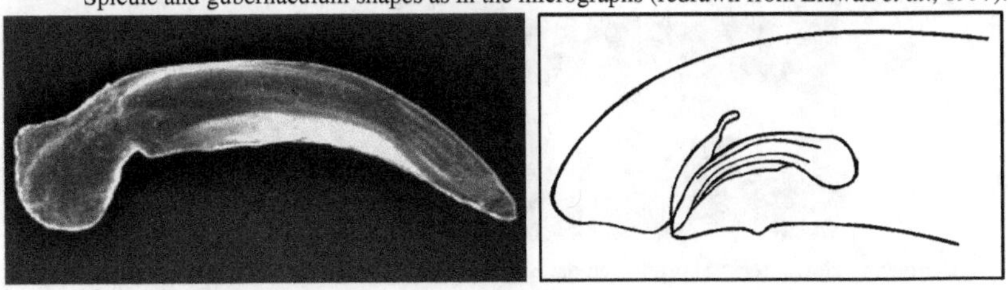

Specific characters: 2 horn-like structures present on IJ head; Mucron in male absent.

Table 3. *(Continued).*

Species	L	MBD	EP	NR	ES	T	a	b	c	D%	E%
S. tami											
Infective juvenile											
	530	23	**36**	na	117	**50**	23	5.0	10.0	31	**73**
	400-600	19-29	34-41	na	110-123	42-57	19-28	3.7-5.1	9.0-11.0	28-34	67-86

Male **SL = 77** (71-84); GL = 48 (38-55); MBD = 129 (98-161); **D% = 44** (30-60); **SW% = 200** (140-300); GS = 160; **MUC = P.**

Spicule and gubernaculum shapes as in the micrographs:

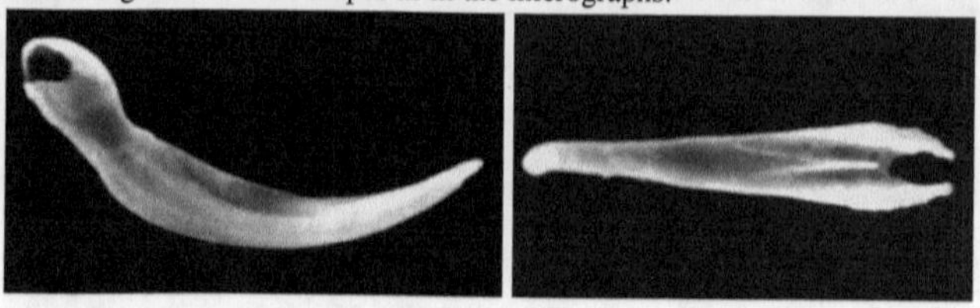

Specific characters: Short nematode with long spicule; Shaft prominent; SW greater than that of other short nematodes.

Species	L	MBD	EP	NR	ES	T	a	b	c	D%	E%
S. rarum											
Infective juvenile											
	511	23	**38**	70	102	**51**	23	4.7	9.8	35	**72**
	443-573	18-26	32-40	60-88	89-120	44-56	20-26	4.1-5.6	8.7-11.0	30-39	63-80

Male **SL = 47** (42-52); GL = 34 (23-38); MBD = 123 (100-142); **D% = 50** (44-51); SW% = 94 (91-105); GS% = 71 (55-73); **MUC = P.**

Spicule and gubernaculum shapes as in the micrographs:

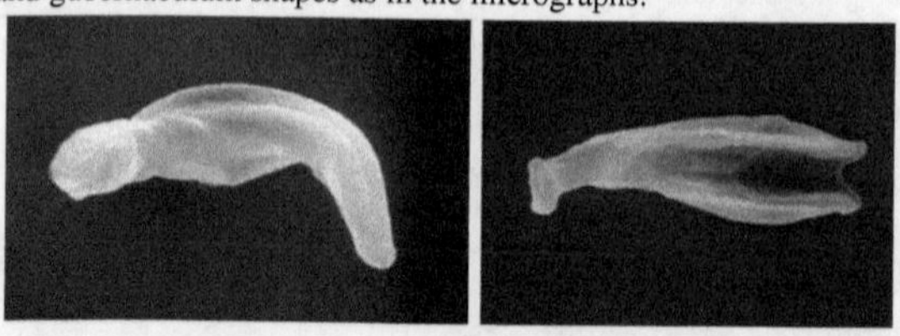

Specific characters: Short nematode; Spicule short, strongly curved posteriorly; Male has 2 openings posterior to edge of cloaca.

Table 3. *(Continued).*

Species	Morphometric characters (mean, range)										
	L	MBD	EP	NR	ES	T	a	b	c	D%	E%
S. ritteri											
Infective juvenile											
	510	22	**43**	73	92	**49**	24	5.5	10.6	46	**88**
	470-590	19-24	40-46	68-85	85-95	44-54	19-31	4.9-6.3	9.2-13.1	44-50	79-97

Male **SL** = **69** (58-75); GL = 44 (33-50); MBD = 130 (110-176); **D%** = **47** (44-50); SW% = 156 (144-157); GS% = 64 (57-67); **MUC** = **A**.

Spicule and gubernaculum as in the drawing (redrawn from Doucet & Doucet, 1990):

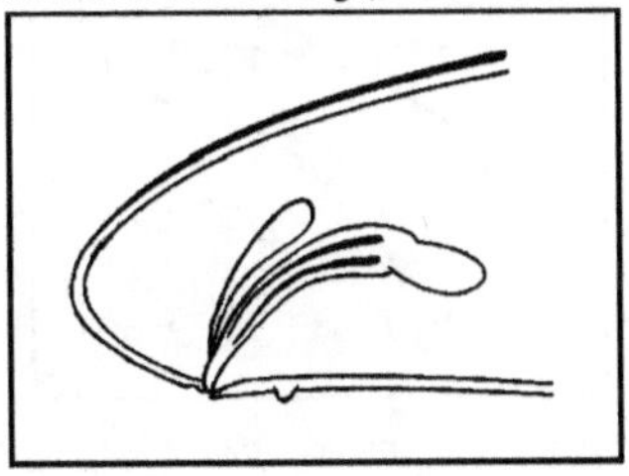

Specific characters: Short nematode; Mucron absent in first generation male.

Species	L	MBD	EP	NR	ES	T	a	b	c	D%	E%
S. sasonense											
Infective juvenile											
	486	22	**32**	69	100	**47**	22	4.8	10.0	**32**	**69**
	456-516	19-23	30-34	65-72	96-104	42-52	21-24	4.6-5.1	9.4-11.0	30-34	61-76

Male **SL** = **71** (66-78); GL = 53 (50-56); MBD = 113 (105-120); D% = 40 (37-45); **SW%** = **250** (219-306); GS = 75 (69-82); **MUC** = **P**.

Spicule and gubernaculum as in the drawing (after Phan *et al.*, 2006):

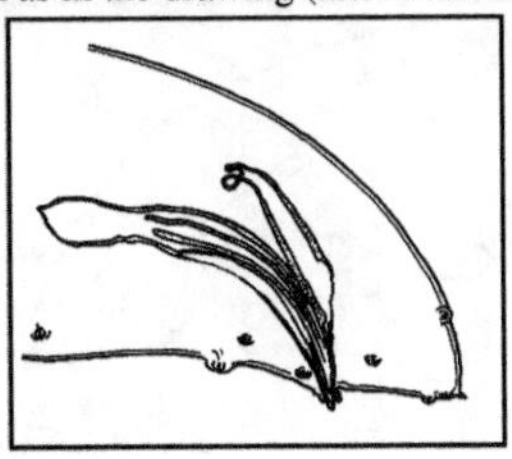

Specific character: Very short nematode.

Table 3. *(Continued).*

Species	L	MBD	EP	NR	ES	T	a	b	c	D%	E%
	Morphometric characters (mean, range)										
S. siamkayai											
Infective juvenile											
	446	21	**35**	72	95	**36**	21	4.7	11.3	**37**	**96**
	398-495	18-24	29-38	68-80	80-107	31-41	19-23	4.0-6.1	10.3-14.8	31-43	95-112

Male **SL = 77.5** (75-80); GL = 54 (47-65); MBD = 140 (107-159); D% = 42 (35-49); SW% = 170 (140-220); GS% = 70 (60-80); **MUC = P.**

Spicule and gubernaculum as in drawing (redrawn from Stock *et al.*, 1999):

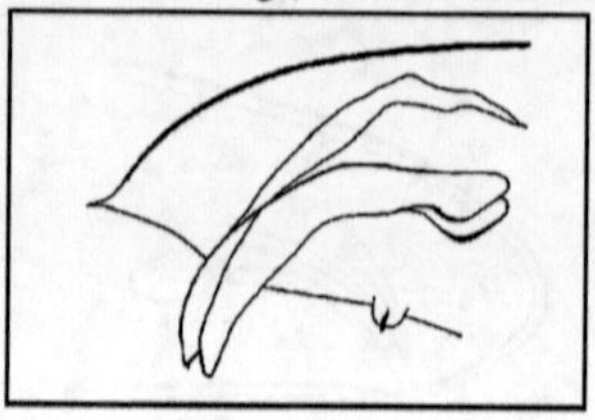

Specific characters: Very short nematode; Spicule long for a short nematode; Epitygmata present in females.

Species	L	MBD	EP	NR	ES	T	a	b	c	D%	E%
S. backanense											
Infective juvenile											
	438	20	31	62	91	**42**	22	4.8	10.0	34	**73**
	391-473	18-22	28-33	58-65	87-97	39-46	20-27	4.5-5.2	9.8-11.0	30-37	64-83

Male **SL = 58** (53-62); **GL = 43** (36-48); MBD = 101 (75-135); D% = 41 (30-48); **SW% = 219** (184-256); GS% = 74 (66-79); **MUC = P.**

Spicule and gubernaculum as in the drawing (after Phan *et al.*, 2006):

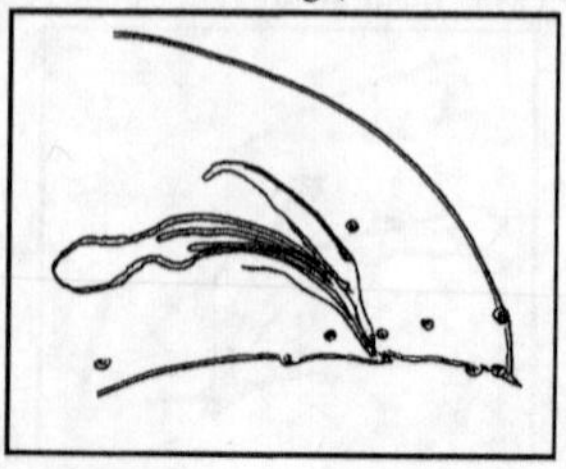

Specific character: Short nematode.

Table 3. *(Continued).*

Species	Morphometric characters (mean, range)										
	L	MBD	EP	NR	ES	T	a	b	c	D%	E%
S. asiaticum											
Infective juvenile											
	425	23	**32**	56	96	**55**	**32**	5.2	16.1?	**32**	**78**
	360-450	20-25	28-34	49-61	82-104	52-57	25-42	4.7-6.1	13.9-20.0	30-36	60-90

Male **SL** = **68** (61-74); GL = 53 (46-62); MBD = 104 (86-130); SW% = 200; GS% = 76.

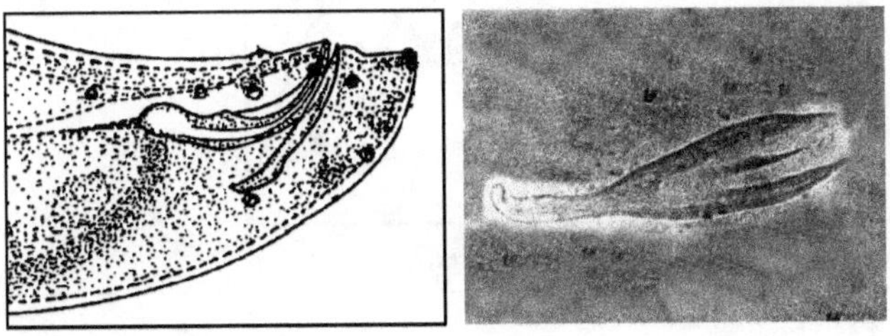

Specific character: This is a very small nematode.

Species	L	MBD	EP	NR	ES	T	a	b	c	D%	E%
S. eapokense											
Infective juvenile											
	402	19	29	58	84	**39**	21	4.8	10.2	34	73
	370-434	18-21	26-31	54-62	79-88	35-45	18-23	4.3-5.3	9.6-11.4	32-37	66-80

Male **SL** = **69** (63-74); GL = 50 (47-56); MBD = 87 (75-90); D% = 43 (37-48); **SW%** = **246** (221-282); GS% = 73 (69-80); **MUC** = **P**.

Spicule and gubernaculum as in the drawing (after Phan *et al.*, 2006):

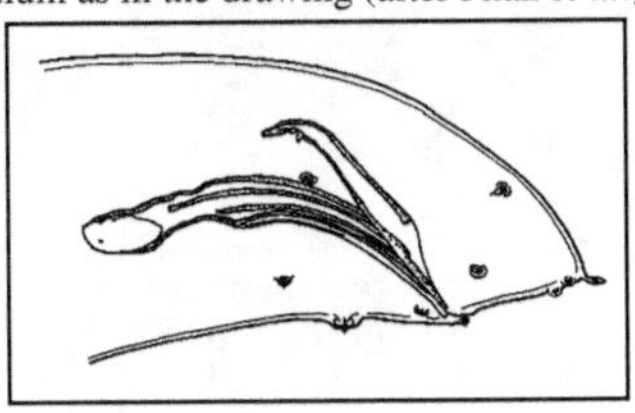

Specific character: IJ body extremely short.

Table 3. *(Continued).*

Species	Morphometric characters (mean, range)										
	L	MBD	EP	NR	ES	T	a	b	c	D%	E%
S. cumgarense											
Infective juvenile											
	402	22	**35**	58	86	36	18	4.7	11.0	**45**	**99**
	384-432	21-25	33-38	54-60	80-91	34-38	17-19	4.4-4.9	10.5-12.1	35-52	87-103
Male	**SL = 76** (68-81); GL = 58 (48-66); MBD = 96 (90-120); D% = 45 (35-52); **SW% = 277** (250-313); GS% = 76 (71-83); **MUC = P**. Spicule and gubernaculum as in the drawing (after Phan *et al.*, 2006): 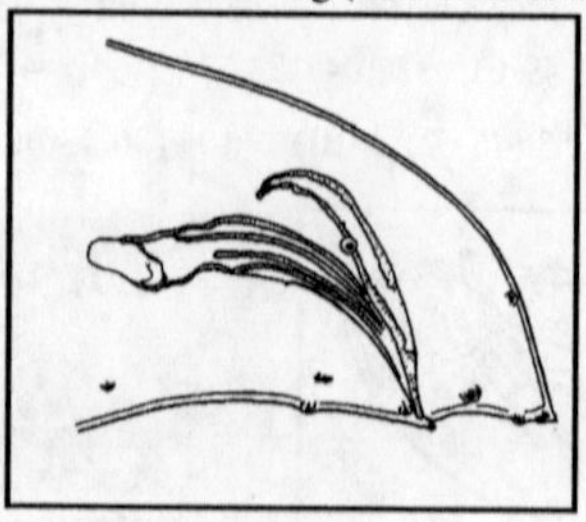 **Specific character:** IJ body extremely short.										

Identification of *Heterorhabditis* species

The diagnostic characters for the identification of species of *Heterorhabditis* are presented in the following polytomous key (Table 4).

Ratios and abbreviations used in the key are similar to those of *Steinernema* species with the addition of TR = testis reflexion.

Table 4. *Tabular key for the identification of valid* Heterorhabditis *species (arranged in descending order of IJ body length). All measurements are in μm. Abbreviations as defined on pages 83 and 84.*

Species	Morphometric characters (mean, range)										
	L	MBD	EP	NR	ES	T	a	b	c	D%	E%
H. megidis											
Infective juvenile											
	768	29	**131**	109	155	**119**	26	5.0	**6.5**	85	**110**
	736-800	27-32	123-142	104-115	147-160	112-128	23-28	4.6-5.9	6.1-6.9	81-91	103-120

Male **SL** = **49** (46-54); GL = 21 (17-24); MBD = 47 (44-50); **D%** = **122**; **TR** = **128**; **SW%** = **135** (130-140); GS% = 43.

Spicule, gubernaculum and bursa as in micrographs:

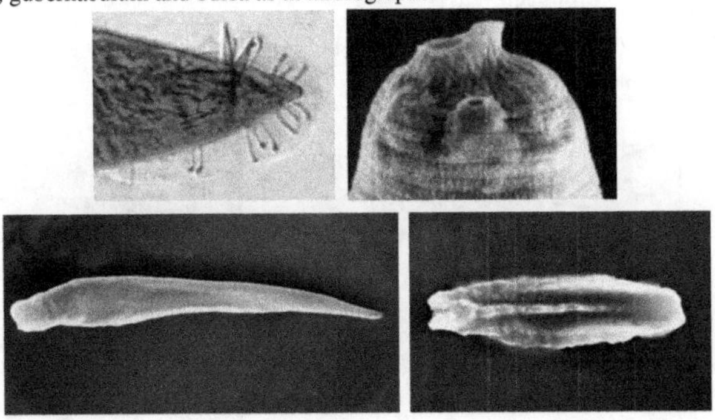

Specific character: Long IJ with membranous ring around oral aperture.

Species	L	MBD	EP	NR	ES	T	a	b	c	D%	E%
H. zealandica											
Infective juvenile											
	685	27	**112**	100	140	**102**	25	4.9	**6.6**	80	**108**
	570-740	22-30	94-123	90-107	135-147	87-119	24-26	4.2-5.0	6.2-6.7	70-84	103-109

Male **SL** = **51** (48-55); GL = 22 (19-25); MBD = 41 (36-45); **D%** = **118**; **TR** = **132** (88-173); SW% = 181 (160-209); GS% = 43.

Vulva pattern, spicule and bursa as in the micrographs:

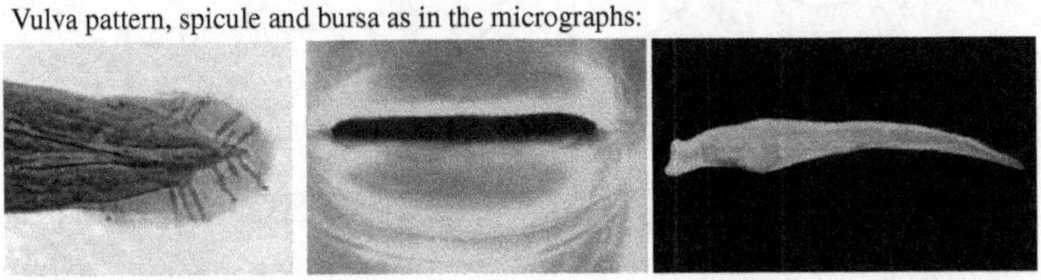

Specific characters: Spicule with prominent rostrum; Infected *Galleria* larvae usually turn green.

Table 4. *(Continued).*

Species	L	MBD	EP	NR	ES	T	a	b	c	D%	E%
H. marelatus											
Infective juvenile											
	654	28	**102**	99	133	**107**	24	4.9	**6.1**	77	**96**
	588-700	24-32	81-113	83-113	121-139	99-117	21-29	4.7-5.4	5.5-6.6	60-86	89-110

Male **SL** = **45** (42-50); GL = 19 (18-22); MBD = 51 (48-56); D% = 113; **TR** = **91** (67-136); SW% = 175 (173-178); GS% = 41.

Syn. *H. hepialius*

Bursa, spicule as in the micrographs:

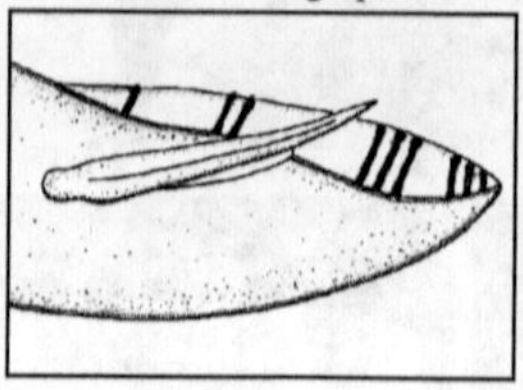

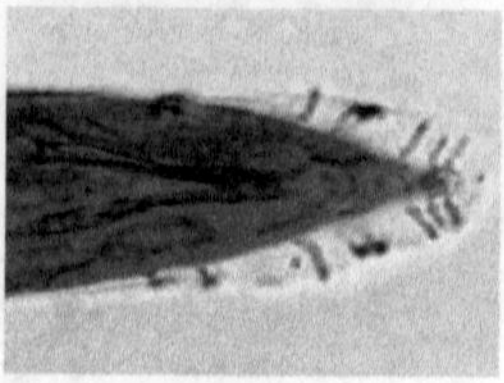

Specific character: With 4 pairs of ribs in bursal terminal group.

Species	L	MBD	EP	NR	ES	T	a	b	c	D%	E%
H. downesi											
Infective juvenile											
	637	18	**115**	101	135	**68**	35	4.7	9.5	**85**	**170**
	588-692	15-22	96-128	96-105	126-141	62-74	29-42	4.4-5.3	8.5-10.5	76-98	160-180

Male **SL** = **43** (41-47); GL = 18 (17-20); MBD = 36 (33-40); D% = 88; TR = 150 (108-181); SW% = 180 (170-220); GS% = 43 (36-47).

Hermaphrodite and male tail as in the micrographs:

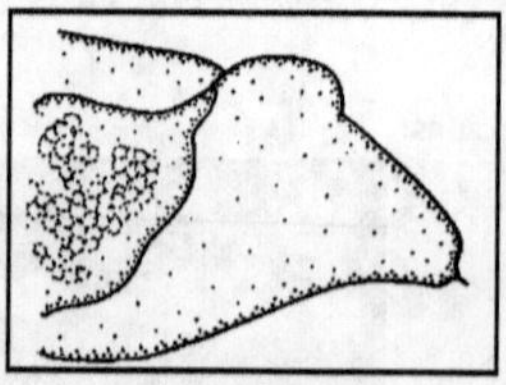

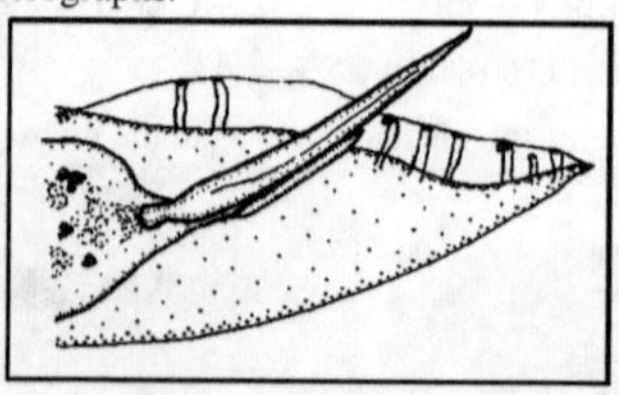

Table 4. *(Continued).*

Species	L	MBD	EP	NR	ES	T	a	b	c	D%	E%
H. amazonensis											
Infective juvenile											
	589	23	**107**	85	121	**107**	26	4.9	5.5	**88**	**100**
	567-612	20-24	89-115	76-93	107-132	98-115	24-29	4.4-5.5	5.1-6.1	83-92	89-109

Male **SL** = **41** (35-45); GL = 21 (19-23); MBD = 41 (36-43); **D%** = **103** (95-109); **TR** = **59** (42-85); **SW%** = **152** (120-187); GS% = 51 (44-56).

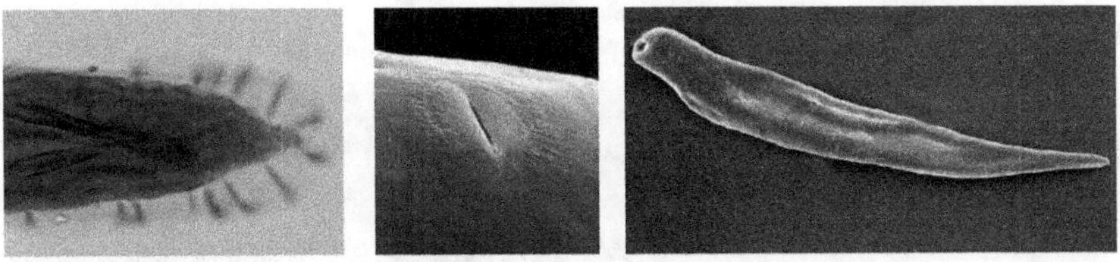

Specific characters: Bursa with 2 pairs of ribs or less in terminal group; Vulval pattern smooth, elliptical.

Species	L	MBD	EP	NR	ES	T	a	b	c	D%	E%
H. bacteriophora											
Infective juvenile											
	588	23	**103**	85	125	98	25	4.5	**6.2**	84	**112**
	512-671	18-31	87-110	72-93	100-139	83-112	17-30	4.0-5.1	5.5-7.0	76-92	103-130

Male **SL** = **40** (36-44); GL = 20 (18-25); MBD = 43 (38-46); **D%** = **117**; **TR** = **76** (61-89); **SW%** = **128** (120-134); GS% = 50.

Syn. *H. argentinensis*

Bursa, vulva pattern and spicule shape as in the micrographs:

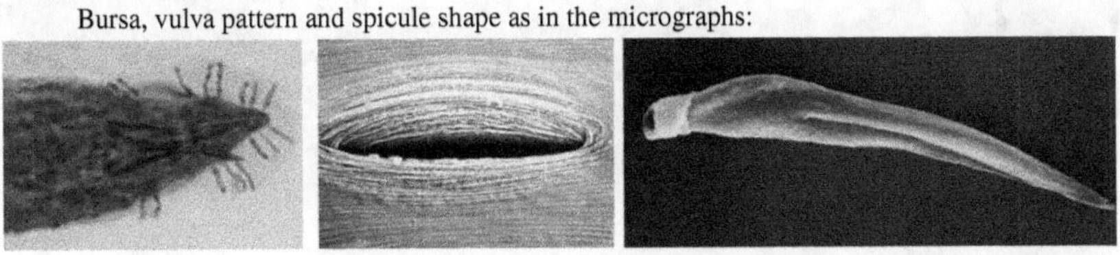

Table 4. *(Continued).*

Species	Morphometric characters (mean, range)										
	L	MBD	EP	NR	ES	T	a	b	c	D%	E%
H. mexicana											
Infective juvenile											
	578	23	**102**	81	122	**99**	25	4.6	5.9	**81**	104
	530-620	20-24	83-109	74-88	104-142	91-106	24-28	4.2-5.1	5.5-6.3	72-86	87-111

Male **SL** = **41** (30-47); GL = 23 (18-32); MBD = 42 (38-47); D% = 129 (114-149); TR = 96 (65-130); SW% = 167 (130-196); GS% = 56 (43-70).

Bursa, vulva pattern and spicule as in the micrographs:

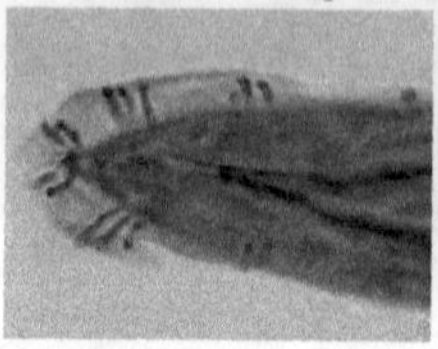

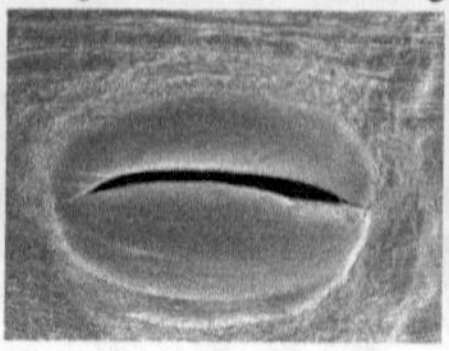

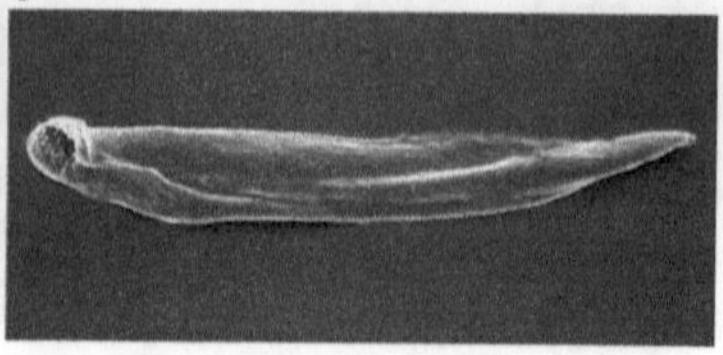

Species	L	MBD	EP	NR	ES	T	a	b	c	D%	E%
H. floridensis											
Infective juvenile											
	562	21	**109**	86	**135**	**103**	28	4.3	5.6	**81**	**105**
	554-609	19-23	101-122	68-107	123-142	91-113	25-32	3.9-4.9	5.3-6.6	71-90	95-134

Male **SL** = **42** (36-46); **GL** = **23** (17-30); MBD = 48 (43-50); **D%** = **112** (105-119); TR = 93 (78-116); SW% = 157 (133-209); GS = 54 (47-65).

Bursa and vulva pattern as in the micrographs:

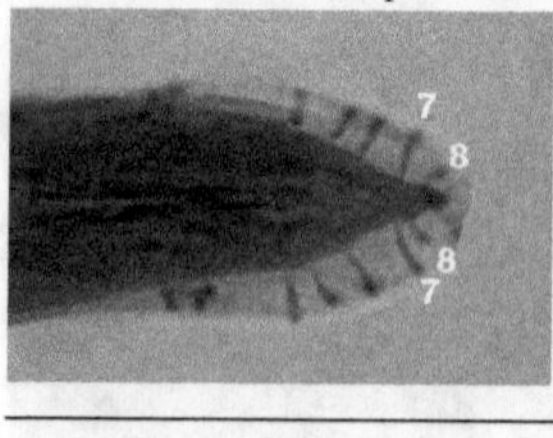

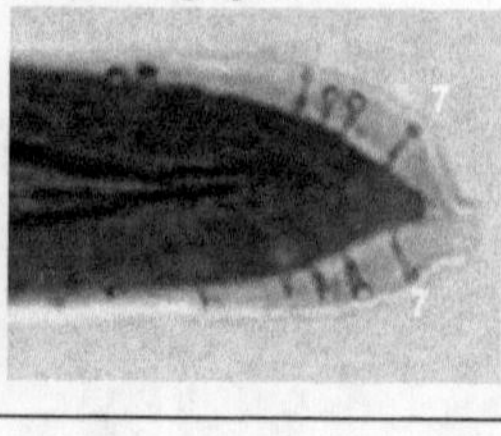

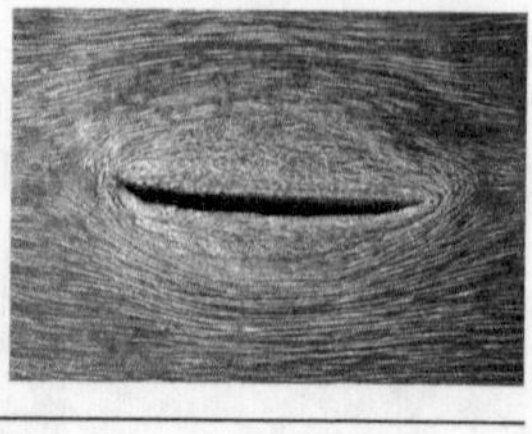

Table 4. *(Continued).*

Species	L	MBD	EP	NR	ES	T	a	b	c	D%	E%
	Morphometric characters (mean, range)										
H. baujardi											
Infective juvenile	**551**	20	**97**	81	115	**90**	28	4.8	6.0	**84**	**108**
	497-595	18-22	91-103	75-86	107-120	83-97	26-30	4.5-5.1	6.0-6.7	78-88	98-114

Male **SL** = **40** (33-45); GL = 20 (18-22); MBD = 49 (43-53); **D%** = **79**; TR = 91 (63-106); SW% = 182 (138-208); GS% = 50 (44-61).

Bursa, spicule and gubernaculum as in the micrographs:

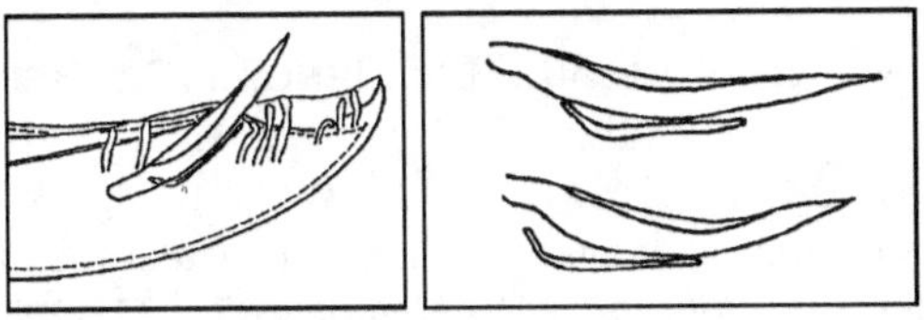

Species	L	MBD	EP	NR	ES	T	a	b	c	D%	E%
H. indica											
Infective juvenile	**528**	20	**98**	82	117	**101**	26	4.5	**5.3**	84	**94**
	479-573	19-22	88-107	72-85	109-123	93-109	25-27	4.3-4.8	4.5-5.6	79-90	83-103

Male **SL** = **43** (35-48); GL = 21 (18-23); MBD = 42 (35-46); **D%** = **122**; **TR** = **106** (78-132); SW% = 190 (180-200); GS% = 49.

Syn. *H. hawaiiensis*

Vulva, spicule and gubernaculum shape as in the micrographs:

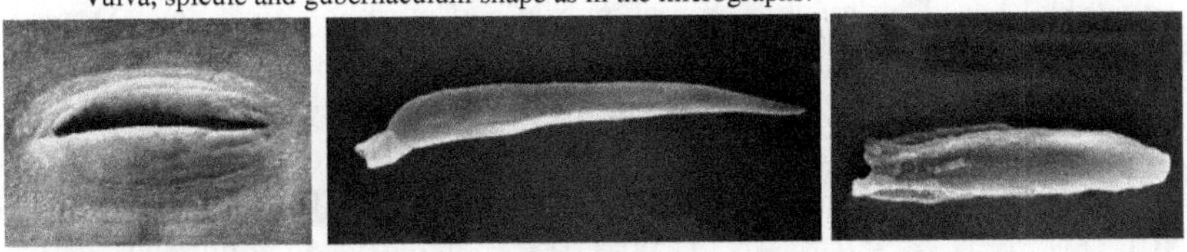

Species	L	MBD	EP	NR	ES	T	a	b	c	D%	E%
H. taysearae											
Infective juvenile	**418**	20	**90**	-	110	**55**	21	3.8	7.7	**82**	**180**
	332-499	17-23	74-113		96-130	44-70	18-27	2.4-4.2	6.5-8.7	71-96	110-230

Male **SL** = **39** (30-42); GL = 18 (14-21); MBD = 44 (38-48); TR = 122 (100-146); SW% = 127; GS% = 46.

Note: The tail looks short. The authors may have measured tail length without the sheath, making E% very high.

Specific character: IJ body is very short.

References

COURTNEY, W.D., POLLEY, D. & MILLER, V.I. (1955). TAF, an improved fixative in nematode technique. *Plant Disease Reporter* 39, 570-571.

EISENBACK, J.D. (1985). Techniques for preparing nematodes for scanning electron microscopy. In: Barker, K.R., Carter, C.C. & Sasser, J.N. (Eds). *An advanced treatise on* Meloidogyne, *volume II: Methodology*. A cooperative publication of the Department of Plant Pathology and the United State Agency for International Development, pp. 79-105.

FRANKLIN, M. & GOODEY, J.B. (1949). A cotton blue lactophenol technique for mounting plant-parasitic nematodes. *Journal of Helminthology* 23, 175-178.

HOMINICK, W.M., BRISCOE, B., DEL PINO, F.G., HENG, J., HUNT, D.J., KOZODOY, E., MRÁČEK, Z., NGUYEN, K.B., REID, A.P., SPIRIDONOV, S., STOCK, P., STURHAN, D., WATURU, C. & YOSHIDA, M. (1997). Biosystematics of entomopathogenic nematodes: current status, protocols and definitions. *Journal of Helminthology* 71, 271-298.

MADDISON, W.P. & MADDISON, D.R. (2002). *MacClade version 4.0*. Sunderland, MA, USA, Sinauer Associates.

MANIATIS, T., FRITCH, E.F. & SAMBROOK, J. (1989). *Molecular cloning: a laboratory manual*, 2nd edition. Cold Spring Harbor, NY, USA, Cold Spring Harbor Laboratory Press, 1659 pp.

NADLER, S.A., BOLOTIN, E. & STOCK, S.P. (2006). Phylogenetic relationships of *Steinernema* Travassos, 1927 (Nematoda: Cephalobina: Steinernematidae) based on nuclear, mitochondrial and morphological data. *Systematic Parasitology* 63, 161-181.

NGUYEN, K.B. & ADAMS, B.J. (2003). SEM and systematic studies of *Steinernema abbasi* Elawad *et al.*, 1997 and *S. riobrave* Cabanillas *et al.*, 1994 (Rhabditida: Steinernematidae). *Zootaxa* 179, 1-10.

NGUYEN, K.B. & DUNCAN, L.W. (2002). *Steinernema diaprepesi* n. sp. (Rhabditida: Steinernematidae), a parasite of the citrus weevil *Diaprepes abbreviatus* (L.) (Coleoptera: Curculionidae). *Journal of Nematology* 34, 159-170.

NGUYEN, K.B. & SMART JR, G.C. (1990). *Steinernema scapterisci* n. sp. (Rhabditida: Steinernematidae). *Journal of Nematology* 22, 187-199.

NGUYEN, K.B. & SMART JR, G.C. (1993). Scanning electron microscope studies of *Steinernema anomali* Kozodoi, 1984. *Journal of Nematology* 25, 486-492.

NGUYEN, K.B. & SMART JR, G.C. (1994). *Neosteinernema longicurvicauda* n. gen. n. sp. (Rhabditida: Steinernematidae), a parasite of the termite *Reticulitermes flavipes* (Koller). *Journal of Nematology* 26, 162-174.

NGUYEN, K.B., MARUNIAK, J. & ADAMS, B.J. (2001). The diagnostic and phylogenetic utility of the rDNA internal transcribed spacer sequences of *Steinernema*. *Journal of Nematology* 33, 73-82.

NGUYEN, K.B., SHAPIRO-ILAN, D.I., STUART, R.J., MCCOY, C.W., JAMES, R.R. & ADAMS, B.J. (2004). *Heterorhabditis mexicana* n. sp. (Rhabditida: Heterorhabditidae) from Tamaulipas, Mexico, and morphological studies of the bursa of *Heterorhabditis* spp. *Nematology* 6, 231-244.

NGUYEN, K.B., TESFAMARIAM, M., GOZEL, U., GAUGLER, R. & ADAMS, B.J. (2004). *Steinernema yirgalemense* n. sp. (Rhabditida: Steinernematidae) from Ethiopia. *Nematology* 6, 839-856.

NGUYEN, K.B., MALAN, A.P. & GOZEL, U. (2006). *Steinernema khoisanae* n. sp. (Rhabditida: Steinernematidae), a new entomopathogenic nematode from South Africa. *Nematology* 8, 157-175.

POSADA, D. & CRANDALL, K.A. (1998). Model test: testing the model of DNA substitution. *Bioinformatics* 14, 817-818.

SEINHORST, J.W. (1959). A rapid method for the transfer of nematodes from fixative to anhydrous glycerin. *Nematologica* 4, 67-69.

SPIRIDONOV, S.E., REID, A.P., PRODRUCKA, K., SUBBOTIN, S.A. & MOENS, M. (2004). Phylogenetic relationship within the genus *Steinernema* (Nematoda: Rhabditida) as inferred from analyses of the ITS1-5.8S-ITS2 region of rDNA and morphological features. *Nematology* 6, 547-566.

STOCK, S.P., CAMPBELL, J.F. & NADLER, S.A. (2001). Phylogeny of *Steinernema* Travassos, 1927 (Cephalobina: Steinernematidae) inferred from ribosomal DNA sequences and morphological characters. *Journal of Parasitology* 87, 877-889.

SWOFFORD, D.L. (2002). *PAUP* Phylogenetic Analysis Using Parsimony (*and Other Methods)*. Sunderland, MA, USA, Sinauer Associates.

THOMPSON, J.D., GIBSON, T.J., PLEWNIAK, F., JEANMOUGH, F. & HIGGINS, D.G. (1997). The CLUSTAL_X windows interface: flexible strategies for multiple sequence alignment aided by quality analysis tools. *Nucleic Acid Research* 25, 4876-4882.

VRAIN, T.C., WAKARCHUK, D.A., LEVESQUE, A.C. & HAMILTON, R.I. (1992). Intraspecific rDNA restriction fragment length polymorphisms in the *Xiphinema americanum* group. *Fundamental and Applied Nematology* 15, 563-574.

Nematology Monographs & Perspectives, 2007, Vol. 5, 121-609

Chapter 4

Steinernematidae: species descriptions

Khuong B. NGUYEN [1], David J. HUNT [2] and Zdeněk MRÁČEK [3]

[1] *Entomology and Nematology Department, Institute of Food and Agricultural Sciences, University of Florida, Gainesville, FL 32611-0620, USA*

[2] *CABI Europe-UK, Bakeham Lane, Egham, Surrey TW20 9TY, UK*

[3] *Laboratory of Insect Pathology, Institute of Entomology, Czech Academy of Sciences, Branišovská 31, 370 05 České Budějovice, Czech Republic*

Descriptions and illustrations of all species regarded as valid herein are presented in alphabetical order under their respective family and genus. Morphological and morphometric data are largely based on the original description (and hence differ to some extent depending on the detail originally provided) but have been expanded, corrected and/or updated where necessary. Illustrations have been taken from the original description wherever possible, although in some cases, particularly for the older species, re-descriptions have had to be employed. In some cases, illustrations have had to be re-drawn. The author responsible for producing each of the species data sheets presented below is indicated in the footnotes. Other data provided include the etymology of the specific epithet, type locality, type host (when known), deposition of types, an up-to-date geographical distribution for each species, accession codes for DNA sequences, a brief account of the bionomics, *etc.*, of each species plus relevant major references. Descriptions of *species inquirendae* (see Chapter 2 for a list) are not provided.

In the genus *Steinernema*, the following four 'species groups', which are based on body length of the infective juvenile (IJ), may be cited in the relationships section of each species description in order to indicate assemblages of similar species: *glaseri*-group (IJ L $\geqslant$ 1000 μm); *feltiae*-group (IJ L = 700-1000 μm); *intermedium*-group (IJ L = 600-700 μm); *carpocapsae*-group: (IJ L $\leqslant$ 600 μm). In addition to these, there is also the *bicornutum*-group, a small morphology-based cluster of species, the

IJ bearing two distinctive horn-like structures on the cephalic region. It should be noted that the above groups are not necessarily congruent with groups based on molecular characters.

The genus *Steinernema* Travassos, 1927

Steinernema abbasi[*,**] Elawad, Ahmad & Reid, 1997
= *S. thermophilum* Sudershan & Singh, 2000, n. syn.
(Figs 26-29)

MEASUREMENTS

See Table 5.

DESCRIPTION

Male (first generation)

Body shape and anterior region similar to males of other species of *Steinernema*. Lateral field inconspicuous. Labial region slightly swollen, bearing six labial and four cephalic papillae. Amphidial apertures small. Stoma shallow. Pharynx with cylindrical procorpus, slightly swollen metacorpus, isthmus present, nerve ring around isthmus, basal bulb distinct. Pharyngo-intestinal valve present. Excretory pore always located anterior to nerve ring and near end of metacorpus. Distance from anterior end to excretory pore always more than body diam. at excretory pore. Genital system monorchic, reflexed. About 60% of males with normal testis and 40% with reduced or collapsed testis; distance from base of pharynx to anterior end of testis always greater than distance from anterior end of nematode to base of pharynx. Spicules paired, golden dark yellow in colour. Spicule head (manubrium) 12-15 μm long, wider than long, *ca* 20% of spicule length, ventrally directed; shaft (calomus) almost absent; blade (lamina) thick, gradually tapering, *ca* three or four times longer than head; blade terminus pointed; velum present. Each spicule with two internal ribs; shape of spicule, width of head, and degree of curvature variable, spicule tip conoid, bluntly pointed. Gubernaculum length *ca* 70% of spicule, boat-shaped, ventrally curved, slightly swollen in middle and gradually narrowing distally. Bursa absent. Complement of 23 genital papillae present, with single, large, midventral precloacal papilla *ca* 22 μm anterior to cloacal aperture; six pairs of subventral precloacal papillae; a pair of sublateral papillae located almost at level of cloacal aperture and four pairs of

---

[*] Named after M.S. *Abbas*, one of the individuals who isolated the nematode in Oman.

[**] Description after Elawad *et al.* (1997). Revised by Khuong B. Nguyen.

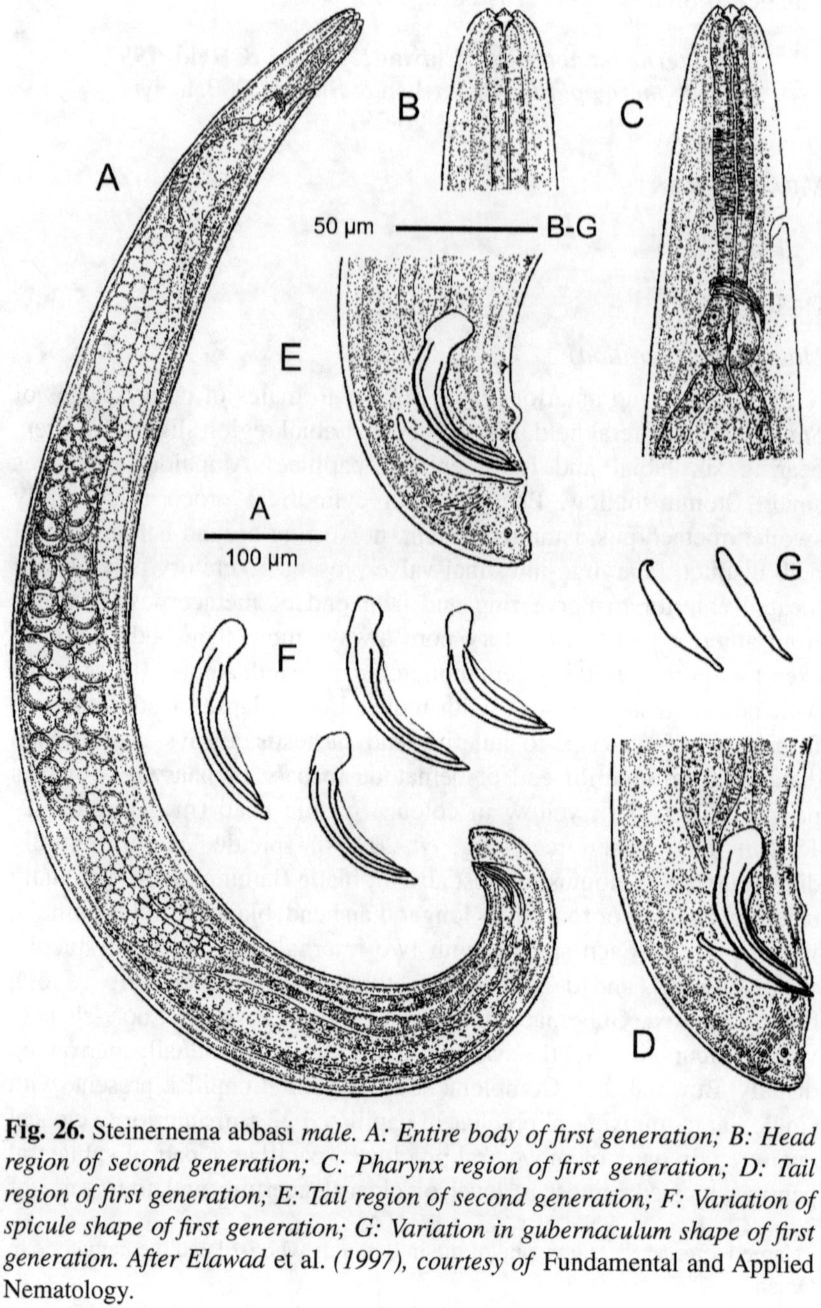

Fig. 26. Steinernema abbasi *male. A: Entire body of first generation; B: Head region of second generation; C: Pharynx region of first generation; D: Tail region of first generation; E: Tail region of second generation; F: Variation of spicule shape of first generation; G: Variation in gubernaculum shape of first generation. After Elawad* et al. *(1997), courtesy of* Fundamental and Applied Nematology.

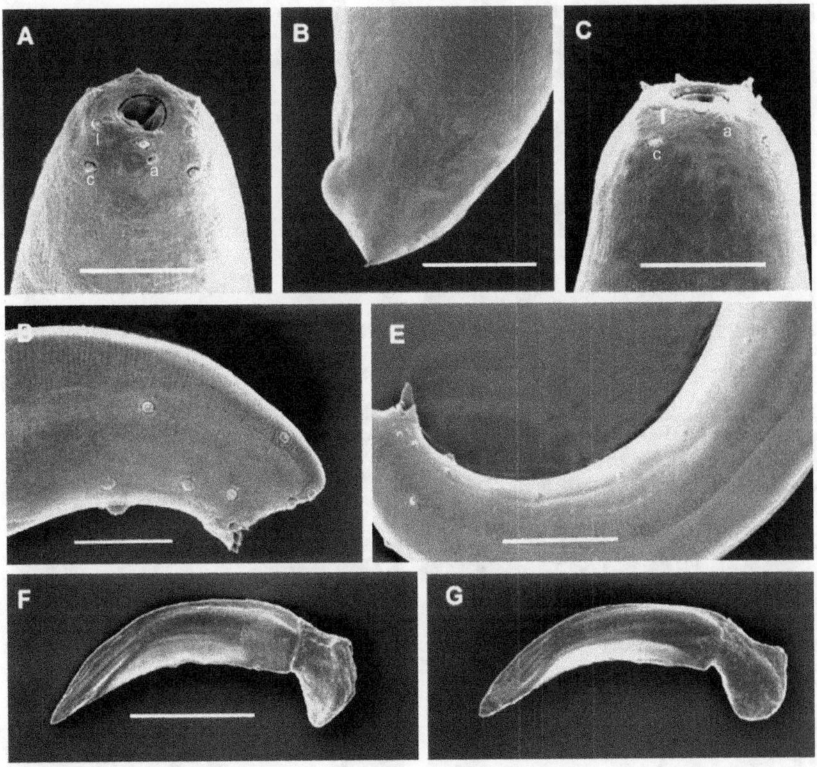

Fig. 27. Steinernema abbasi. *SEM of female and male. A: Anterior region of female showing mouth, labial papillae (l), amphidial aperture (a) and cephalic papillae (c); B: Posterior region of female showing postanal swelling; C: Anterior region of male showing mouth, labial papillae (l), amphidial aperture (a), and cephalic papillae (c); D: Posterior region of male showing genital papillae; E: Posterior region of male showing preanal papillae; F, G: Variation in shape of spicules. (Scale bars: A = 10 μm; B = 30 μm; C = 10 μm; D = 20 μm; E = 37.5 μm; F, G = 23 μm.) After Nguyen and Adams (2003), courtesy of* Zootaxa.

caudal papillae. Tail short, conoid, *ca* 60% of anal body diam. long with rounded terminus; mucron absent.

Male (second generation)

Similar to first generation except smaller and thinner, with collapsed testis, spicules and gubernaculum slightly shorter and thinner. Shape of spicule and gubernaculum not different from those of first generation male, but slightly variable between individuals of same generation.

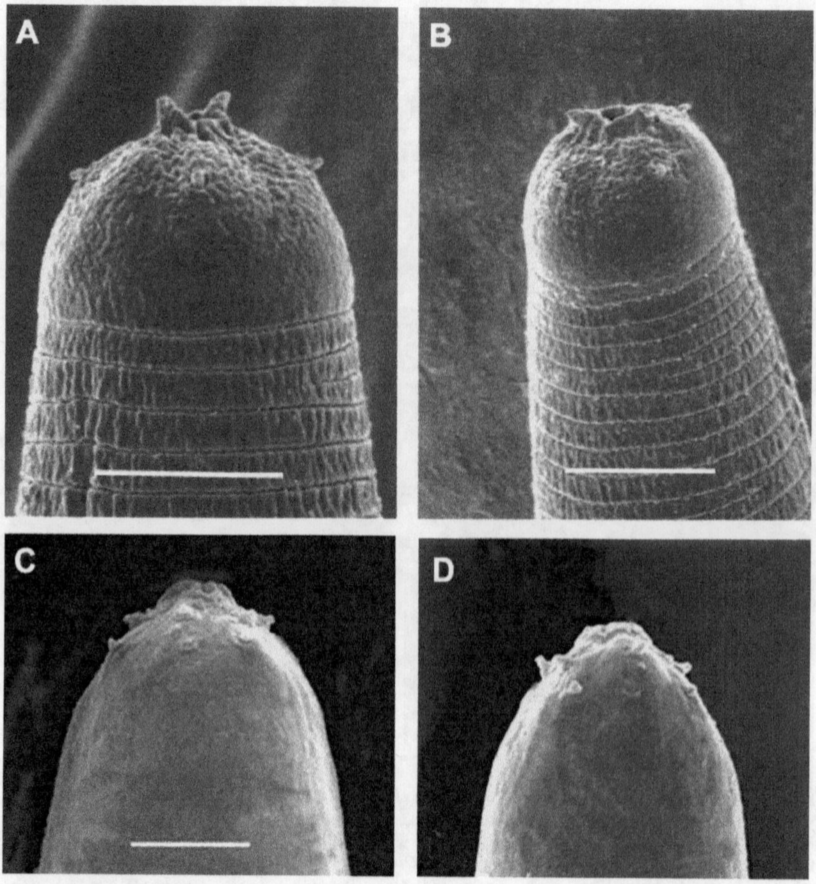

Fig. 28. Steinernema abbasi. *SEM of anterior region of infective juvenile (IJ). A, B: Labial region of exsheathed IJ showing two horn-like structures in labial region; C, D: Labial region of ensheathed infective juvenile showing labial region lacking horn-like structures. (Scale bars: A = 4.3 μm; B = 5.0 μm; C, D = 3.0 μm.) After Nguyen and Adams (2003), courtesy of* Zootaxa.

Female (first generation)

Body shape and morphology structures similar to other species of *Steinernema.* Lateral fields and phasmids inconspicuous. Labial region rounded, continuous with body; SEM face view with six labial and four cephalic papillae; amphidial apertures small, pore-like. Stoma shallow, triangular at base; oral aperture circular. Excretory pore located at level of metacorpus. Distance of excretory pore from anterior end always

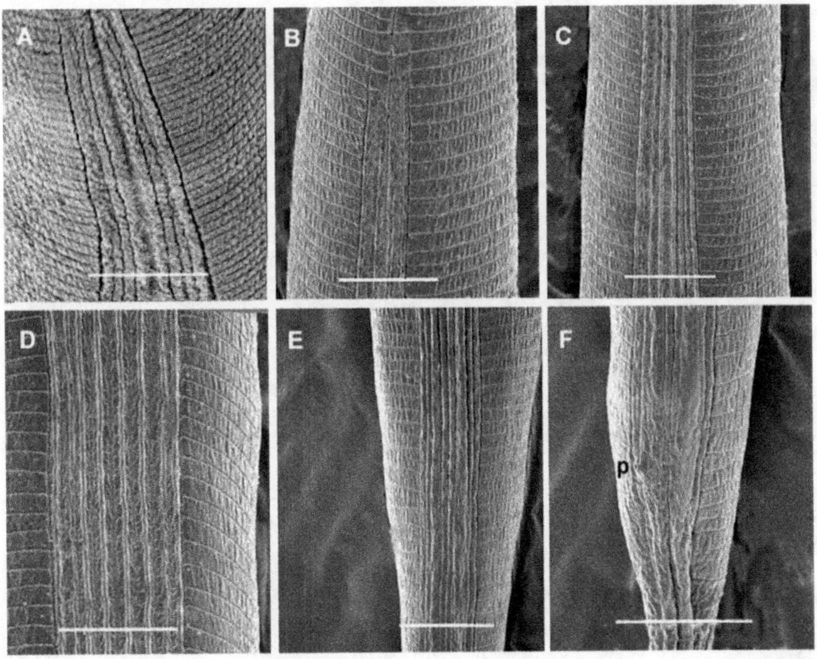

Fig. 29. Steinernema abbasi. *SEM of lateral field of infective juveniles. A: Anterior region of infective juvenile showing beginning of lateral field; B: Anterior region of infective juvenile showing change in lateral field from two to six, then to eight, ridges; C: Lateral field with eight ridges although second ridge from left is faint; D: Lateral field with eight ridges; E: Lateral field with eight ridges then seven; F: Posterior part of lateral field showing phasmid (p) and change of ridge number from six to two. (Scale bars: A = 7.5 μm; B = 5 μm; C = 7.5 μm; D = 6.6 μm; E, F = 6.7 μm.) After Nguyen and Adams (2003), courtesy of* Zootaxa.

more than body diam. at excretory pore. Pharynx as in male. Genital system amphidelphic, reflexed, often containing many eggs. Vulva a transverse slit; epiptygmata present. Vagina sclerotised, 16 μm deep and *ca* 9% of corresponding body diam. long. Tail short, conoid, with pointed tip, *ca* 65% of anal body diam. long; ventral postanal swelling always present.

Female (first generation, giant form)

Body of giant female in form of spiral on heat-relaxation, *ca* three times longer than normal female but morphologically similar.

Table 5. *Morphometrics of type population of* Steinernema abbasi. *Measurements are in μm and in the form: mean ± SD (range).*

Character	First generation			Second generation		Infective juvenile
	Holotype (male)	Male	Female	Male	Female	
n	–	15	10	15	10	15
L	1142	1252 ± 189 (999-1534)	3510 ± 638 (2453-4477)	861 ± 121 (606-1035)	2609 ± 622 (1897-3917)	541 ± 24 (496-579)
a	–	–	–	–	–	18 ± 0.9 (17-20)
b	–	–	–	–	–	6 ± 0.3 (5.5-6.6)
c	–	–	–	–	–	9.8 ± 0.8 (8.1-10.8)
V	–	–	55 ± 2.4 (52-60)	–	55 ± 2.2 (50-77)	–
Body diam.	80	87 ± 6.7 (82-98)	159 ± 12 (143-181)	70 ± 4.6 (64-80)	130 ± 5 (123-148)	29 ± 1 (27-30)
EP	68	80 ± 7.8 (68-89)	71 ± 11 (58-91)	66 ± 5 (62-79)	66 ± 4 (61-73)	48 ± 1.5 (46-51)
NR	107	103 ± 6.5 (99-123)	125 ± 5.8 (120-137)	99 ± 4 (93-106)	114 ± 2.5 (110-118)	68 ± 2.4 (64-72)
ES	134	133 ± 6 (121-144)	165 ± 7.5 (155-176)	121 ± 5 (112-130)	146 ± 6.7 (136-157)	89 ± 1.8 (85-92)
Testis reflexion	259	274 ± 33 (234-319)	–	241 ± 20	–	–
Tail length (T)	24	26 ± 3 (20-31)	37 ± 3 (31-40)	21 ± 1.7 (17-24)	36 ± 2.5 (32-39)	56 ± 3.2 (52-61)
Anal body diam. (ABD)	39	43 ± 5 (37-55)	62 ± 6.8 (51-70)	39 ± 1.6 (36-42)	50 ± 2 (47-54)	–
Spicule length (SL)	69	65 ± 5.7 (57-74)	–	61 ± 4.8 (51-69)	–	–
Spicule width	13	12 ± 1.30 (10-14)	–	10 ± 1.6 (8-13)	–	–
Gubernaculum length (GL)	50	45 ± 4.3 (35-50)	–	43 ± 3.3 (35-48)	–	–
Gubernaculum width	8	7 ± 0.1 (6-8.5)	–	7 ± 0.7 (6-8)	–	–
D% = EP/ES × 100	51	60 ± 5 (51-68)	–	56 ± 7 (50-70)	–	53 ± 2 (51-58)
E% = EP/T × 100	–	–	–	–	–	86 ± 5 (79-94)
SW% = SL/ABD × 100	177	156 ± 22 (107-187)	–	159 ± 15 (128-188)	–	–
GS% = GL/SL × 100	72	70 ± 7 (58-85)	–	70 ± 7 (58-81)	–	–

EP = distance from anterior end to excretory pore; NR = distance from anterior end to nerve ring; ES = distance from anterior end to end of pharynx.

Female (second generation)

Similar to first generation females but smaller and tail sharply pointed. Tail always with postanal ventral swelling.

Infective juvenile

Body thin, elongate. Labial region continuous with body. Exsheathed infective juvenile with two horn-like structures on cephalic extremity and four cephalic papillae. Lateral field pattern beginning anteriorly with one line usually at third or fourth annule. At annules 14 or 15 two other lines appearing to form two ridges. Near excretory pore, lateral field changing gradually from two to six ridges. A short distance posteriorly, two external ridges dividing into two making a total of eight ridges (*i.e.*, nine incisures, although only eight incisures reported in original description), the maximum number in the lateral field. Number of ridges in lateral field may not always be very clear. Short distance anterior to anus, number of ridges reducing to seven. From that point, two submarginal ridges becoming larger then, just anterior to phasmid, number of ridges becoming six. Posterior to phasmid, lateral field changing into two large ridges. In general, formula of lateral field pattern is 2, 6, 8, 7, 6, 2. Excretory pore always weak, situated near posterior end of metacorpus. Distance from anterior end to excretory pore always more than body diam. at same level. Pharynx with cylindrical procorpus and slightly swollen metacorpus. Nerve ring located just anterior to basal bulb. Tail elongate, annulated, gradually tapering, dorsally curved at tip with slight ventral depression.

TYPE HOST AND LOCALITY

Type host unknown but likely to be a bollworm. The type locality was a sandy soil in alfalfa fields in the Sultanate of Oman (latitude 16N, longitude 54E).

TYPE MATERIAL

Holotype male, two paratype males, two paratype females and two paratype juveniles deposited at the CABI Europe-UK Nematode Collection, Egham, Surrey, UK.

Note

In the original description of *S. abbasi* the authors overlooked the presence of the two horn-like structures on the cephalic region. Nguyen and Adams (2003) demonstrated that these structures were actually present and also described in detail the lateral field pattern of the infective juvenile.

Relationships

Steinernema abbasi is characterised by morphological structures of the male and infective juvenile. For the male, shape and size of spicule (65 μm), and ratio SW% (156) are diagnostic characters. For the infective juvenile, the presence of two horn-like structures on the cephalic region (Fig. 28), short body length (541 μm) and lateral field pattern (2, 6, 8, 7, 6, 2) are typical for this species.

Steinernema abbasi can be separated from all other species (except for members of the *bicornutum*-group) by the presence of two horn-like structures on the cephalic region of the infective juvenile. The male of *S. abbasi* is similar to that of *S. riobrave* because of the golden dark yellow coloured spicules and the absence of a terminal mucron, but has a shorter body length. Spicule and gubernaculum length of *S. abbasi* (65 and 45 μm, respectively) are comparable to those of *S. carpocapsae* (66 and 47 μm, respectively), *S. kushidai* (63 and 44 μm, respectively), and *S. riobrave* (67 and 51 μm, respectively) but shorter than those of *S. scapterisci* (83 and 65 μm, respectively). The first generation female of *S. abbasi* is similar to that of *S. riobrave* but the tail of the second generation is wider with a rounded, wedge-shaped, projection on the tip *vs* the sharp, V-shaped, tail in *S. riobrave*.

Molecular characterisation

This species was differentiated by comparing its RFLP pattern with those of 33 other steinernematid species/isolates (Elawad *et al.*, 1997). The sequence of the ITS regions is characterised by its sequence length, 988 bp, with ITS1 = 268 bp, ITS2 = 314 bp and its composition, A = 0.25000, C = 0.16599, G = 0.22975 and T = 0.35425 (accession no. AY248749), and the length of D2D3 regions, 870 bp and its composition, A = 0.25287, C = 0.16437, G = 0.30230, T = 0.28046 (accession no. AF331890).

Additional DNA information was provided by Nguyen and Adams (2003).

BIONOMICS

The life cycle of *S. abbasi* is comparable to other species of *Steinernema*. The infective juvenile enters the haemocoel of insects to deliver the associated bacteria and at least two generations are completed before infective juveniles emerge from the cadaver. In *Galleria mellonella*, adults develop in 48 h at 25°C and in 36 h at 30°C. *Steinernema abbasi* appears to be more active at higher temperatures than *S. riobrave* as it produces more infective juveniles in *G. mellonella* and *Spodoptera littoralis* larvae at 35°C. The LD50 for *S. abbasi* is slightly superior to that of *S. riobrave* against *G. mellonella* but the temperature profile (thermal niche breadth for establishment) for *S. abbasi* is similar to that of *S. riobrave* (Elawad *et al.*, 1996). *Steinernema abbasi* and *S. riobrave* are clearly nematodes of semi-arid and subtropical regions.

Steinernema riobrave was recovered from pre-pupae and pupae of *Helicoverpa zea* and *Spodoptera frugiperda* in Texas, USA (Raulston *et al.*, 1992). *Steinernema abbasi* is associated with bollworms but it has not been isolated from a known host; however, it reproduces well in *S. littoralis* in the laboratory. It is likely that *S. abbasi* could be developed to control pre-pupae and pupae of bollworms in the Middle East or elsewhere where bollworms are a major problem, particularly under irrigation (Elawad *et al.*, 1997).

DISTRIBUTION

Steinernema abbasi is known from the Sultanate of Oman (Elawad *et al.*, 1997). The species has also been found in India and Egypt (Nguyen, pers. comm., 2006).

References

ELAWAD, S., AHMAD, W. & REID, A.P. (1997). *Steinernema abbasi* sp. n. (Nematoda: Steinernematidae) from the Sultanate of Oman. *Fundamental and Applied Nematology* 20, 433-442.

ELAWAD, S.A., ABBAS, M.S. & HAGUE, N.G.M. (1996). The establishment, reproduction and pathogenicity of a new species of *Steinernema* from the

Sultanate of Oman in *Galleria mellonella. Afro-Asian Journal of Nematology* 6, 40-45.

NGUYEN, K.B. & ADAMS, B.J. (2003). SEM and systematic studies of *Steinernema abbasi* Elawad *et al.*, 1997 and *S. riobrave* Cabanillas *et al.*, 1994 (Rhabditida: Steinernematidae). *Zootaxa* 179, 1-10.

RAULSTON, J.R., PAIR, S.D., LOERA, J. & CABANILLAS, H.I. (1992). Prepupal and pupal parasitism of *Helicoverpa zea* and *Spodoptera frugiperda* (Lepidoptera: Noctuidae) by *Steinernema* sp. in corn field in the Lower Rio Grande Valley. *Journal of Economic Entomology* 85, 1666-1670.

SUDERSHAN, G. & SINGH, L.K. (2000). *Steinernema thermophilum* sp. n. (Rhabditida: Steinernematidae) from India. *International Journal of Nematology* 10, 183-191.

Steinernema aciari[*,**] Qiu, Yan, Zhou, Nguyen & Pang, 2004
(Figs 30-32)

MEASUREMENTS

See Table 6.

DESCRIPTION

Male (first generation)

Body C-shaped, curved posteriorly when heat-relaxed. Cuticle smooth. Cephalic extremity truncate and slightly swollen anteriorly. Pharynx similar to that of female. Excretory pore opening located slightly anterior to nerve ring. Genital system monorchic, reflexed. Spicules paired, strongly curved. Ratio between spicule length (measured along arc) and maximum spicule width = *ca* 5.2. Spicule head longer than wide with length/width ratio of 1.6-2 : 1. Blade long and thick. Lateral and dorsal lobes distinct, beginning at head and extending to, and close to, tip, respectively. Ventral lobe inconspicuous, ending well anterior to tip. Distal tip of spicule blunt, bearing a ventral aperture making it appear hook- or notch-like. Velum thin, sometimes not observed in mature adults but distinct in young males. Gubernaculum boat-shaped in lateral view. Slimmer and tapering anteriorly in ventral view. Cuneus Y-shaped, short, ending well anterior to tip of corpus that is well separated posteriorly. Spicules and gubernacula of young males slightly different from those of mature males. Genital papillae comprising 11 pairs and a single precloacal papilla distributed in similar pattern to that of most other *Steinernema* species. Tail conoid, mucron absent.

Male (second generation)

Similar to first generation male in general morphology, except that excretory pore located more anterior than in first generation, and most measurements are smaller (Table 6). Tail mucron not observed.

Female (first generation)

Body C-shaped when heat-relaxed. Cuticle smooth or with faint annules. Cephalic extremity rounded and slightly truncate anteriorly. Six

[*] Epithet formed from the initial letters of the Australian Center for International Agricultural Research (*ACIAR*).

[**] After Qiu *et al.* (2004). Revised by Khuong B. Nguyen.

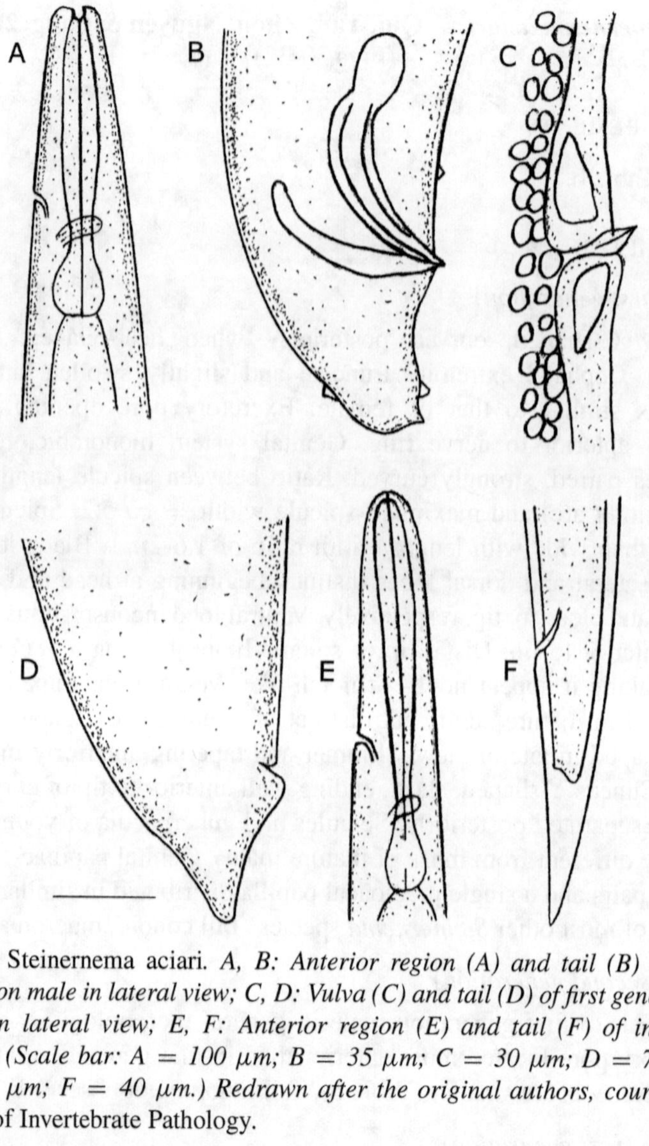

Fig. 30. Steinernema aciari. *A, B: Anterior region (A) and tail (B) of first generation male in lateral view; C, D: Vulva (C) and tail (D) of first generation female in lateral view; E, F: Anterior region (E) and tail (F) of infective juvenile. (Scale bar: A = 100 μm; B = 35 μm; C = 30 μm; D = 75 μm; E = 50 μm; F = 40 μm.) Redrawn after the original authors, courtesy of* Journal of Invertebrate Pathology.

fused lips, each with a papilla. Four cephalic papillae occurring further posteriorly on head. Amphidial apertures two, distinct, located posterior to lateral labial papillae. Stoma shallow, subtriangular anteriorly and triradiate internally. Pharyngeal tissue extending almost to oral aperture. Cheilorhabdions and prorhabdions small but distinct. Pharynx muscu-

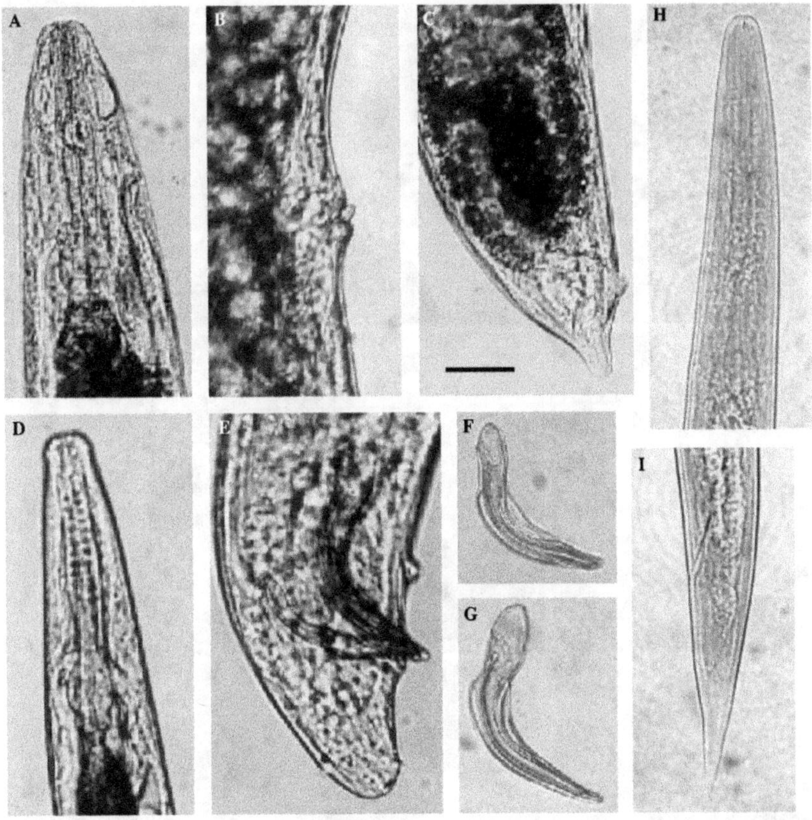

Fig. 31. Steinernema aciari. *Light micrographs. A, B, C: Anterior region (A), vulva (B) and tail (C) of first generation female in lateral view; D, E: Anterior region (D) and tail (E) of first generation male in lateral view; F-I: Spicules (F, G), anterior region (H) and tail (I) of infective juvenile. (Scale bar: A = 70 μm; B = 60 μm; C = 58 μm; D = 47 μm; E = 20 μm; F = 23 μm; G = 25 μm; H = 28 μm; I = 26 μm.) After the original authors, courtesy of* Journal of Invertebrate Pathology.

lar, procorpus, metacorpus and isthmus indistinct, basal bulb slightly enlarged. Nerve ring usually surrounding anterior portion of basal bulb. Excretory pore opening anterior to nerve ring. Lateral fields and phasmids not observed. Genital system didelphic, reflexed. Vulva a transverse slit, slightly protruding from body surface and with conspicuous double flapped epiptygmata. Vagina muscular. Tail conoid, with a long, mamillate terminus and distinct postanal swelling.

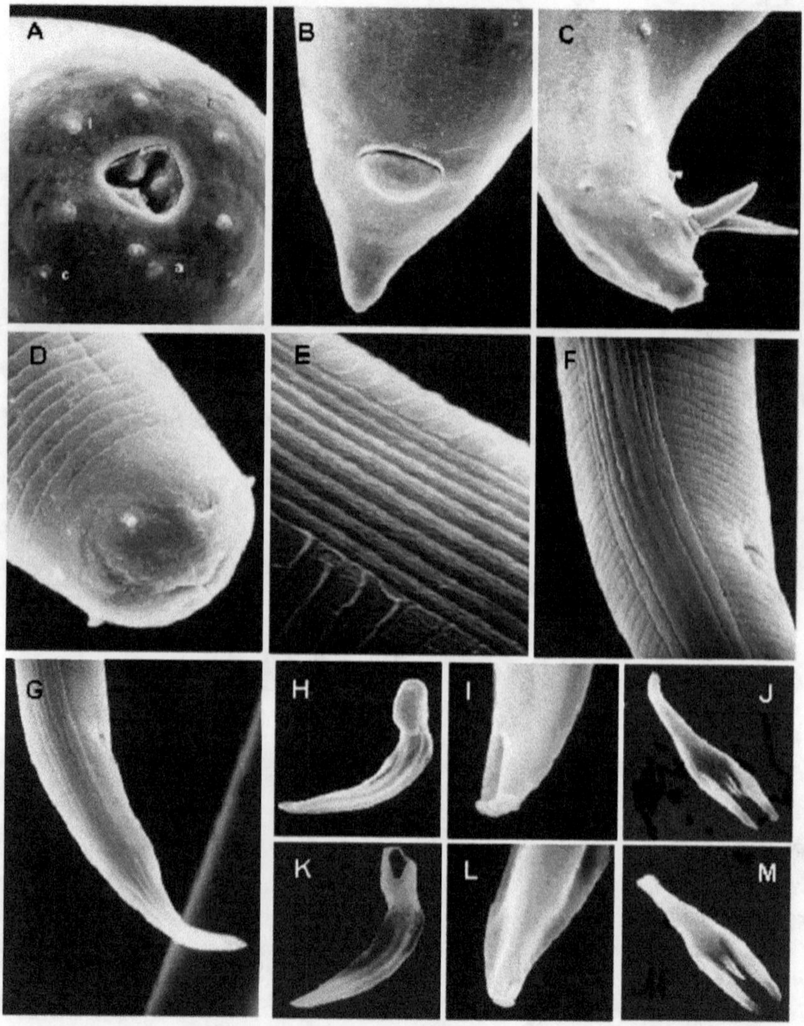

Fig. 32. Steinernema aciari. *SEM micrographs. A: Face view of first generation female showing six labial papillae, four cephalic papillae and two amphidial apertures; B: Ventral view of tail of first generation female showing postanal swelling; C: Lateral view of tail of first generation male; D: Anterior portion of infective juvenile (IJ); E, F, G: Middle and anal portion and tail of IJ showing lateral field pattern; H, I, J: Spicule, spicule tip and gubernaculum of mature male; K, L, M: Spicule, spicule tip and gubernaculum of young male. (Scale bars: A = 10 μm; B, C = 33 μm; D = 5.1 μm; E = 8.2 μm; F = 13 μm; G = 22 μm; H = 35 μm; I = 6 μm; J = 27 μm; K = 28 μm; L = 22 μm; M = 23 μm.) After the original authors, courtesy of* Journal of Invertebrate Pathology.

Table 6. *Morphometrics of* Steinernema aciari. *Measurements are in μm and in the form: mean ± SD (range).*

Character	First generation			Second generation		Infective juvenile
	Holotype (male)	Male	Female	Male	Female	
n	–	20	20	20	20	20
L	150	1597 ± 93 (1400-1750)	7233 ± 1013 (5500-8850)	1133 ± 98 (1000-1300)	2022 ± 247 (1675-2425)	1113 ± 68 (975-1250)
a	–	–	–	–	–	24 ± 2.1 (20-27)
b	–	–	–	–	–	7.6 ± 0.4 (7.0-8.3)
c	–	–	–	–	–	14.4 ± 0.5 (14-16)
c′	–	–	–	–	–	2.5 ± 0.2 (2.3-2.7)
V	–	–	52 ± 5.1 (47-57)	–	53 ± 5.8 (48-60)	–
Body diam.	95	104 ± 7.7 (88-125)	231 ± 23 (188-262)	62 ± 2.4 (58-65)	102 ± 4.2 (98-110)	47 ± 2.5 (42-53)
EP	127	135 ± 10 (115-155)	179 ± 11 (160-195)	113 ± 12 (93-135)	124 ± 7.9 (113-135)	95 ± 3.7 (87-100)
NR	132	144 ± 13 (118-165)	195 ± 18 (163-241)	145 ± 12 (113-165)	164 ± 8.3 (150-175)	114 ± 5.3 (106-125)
ES	168	178 ± 16 (140-200)	262 ± 24 (220-307)	177 ± 14 (148-205)	219 ± 6.7 (205-225)	146 ± 5 (135-157)
Tail length (T)	37	40 ± 4.4 (31-47)	75 ± 12 (50-92)	39 ± 4 (33-48)	74 ± 4.6 (70-85)	78 ± 5.2 (68-88)
Anal body diam. (ABD)	35	42 ± 4.5 (33-50)	83 ± 69 (64-91)	40 ± 2.7 (35-43)	42 ± 3.2 (38-48)	30 ± 1.3 (28-33)
Spicule length (SL)	85	86 ± 6.3 (75-95)	–	79 ± 4.3 (73-85)	–	–
Gubernaculum length (GL)	57	56 ± 5.2 (48-65)	–	46 ± 3.1 (42-51)	–	–
D% = EP/ES × 100	76	76 ± 5 (69-88)	–	65 ± 4 (61-72)	–	65 ± 3 (60-70)
E% = EP/T × 100	–	–	–	–	–	123 ± 7 (113-134)
SW% = SL/ABD × 100	243	204 ± 15 (180-240)	–	197 ± 21 (184-247)	–	–
GS% = GL/SL × 100	67	65 ± 6 (57-77)	–	59 ± 8 (53-64)	–	–
H% = Hyaline/T × 100	–	–	–	–	–	44 ± 2.7 (41-49)

EP = distance from anterior end to excretory pore; NR = distance from anterior end to nerve ring; ES = distance from anterior end to end of pharynx.
After the original authors.

Female (second generation)

Similar to first generation but much smaller. Vulva more protuberant than in first generation. Postanal swelling absent. Tail conical, pointed, longer than anal body diam., no mamillate terminus or mucron.

Infective juvenile

Body slender, enclosed in sheath formed from moulted cuticle of previous stage. Oral aperture and anus closed. Cephalic region smooth, rounded and slightly truncate; labial papillae not seen. Amphidial apertures slit-shaped but inconspicuous. Four prominent cephalic papillae present. Pharynx long and narrow; isthmus distinct, surrounded by nerve ring. Cuticle with prominent transverse striations. Lateral field beginning anteriorly with single line at fourth annule and increasing to nine lines slightly posterior to base of pharynx, making a total of eight, nearly identical, ridges. Nine-line pattern extending posteriorly to close to anus where number of lines reduces to five, then three, before disappearing. Tail attenuate and tapering evenly. Hyaline region comprising 40-50% of tail length. Phasmid inconspicuous.

TYPE HOST AND LOCALITY

The type host of this nematode in nature is unknown as it was recovered from soil using *Galleria* larvae as bait (Bedding & Akhurst, 1975). The soil sample was collected from hillside grassland, *ca* 500 m from the sea at Haimen town (N23.13, E116.38), Shantou district, Guangdong province, People's Republic of China.

TYPE MATERIAL

Holotype (first generation male), and allotype (first generation female) (ZSUNC 0001), ten paratype first generation males and females in five slides (ZSUNC 0002~0006, one male and one female per slide), *ca* 50 infective juveniles (ZSUNC 0007~0010) and other population slides deposited in the State Key Laboratory for Biocontrol, College of Life Sciences, Zhongshan University, Guangzhou 510275, People's Republic of China. Living infective juveniles preserved in liquid nitrogen in the Nematode Collection of SKLB, Zhongshan University, People's Republic of China. Some paratype specimens deposited in the University of California-Davis Nematode Collection, Davis, CA, USA.

RELATIONSHIPS

Morphologically it is characterised by a combination of features of the various developmental stages: IJ with a body length of 1113 ± 68 μm and a tail 78 ± 5.2 μm long; first generation female with a conoid tail with a long mamillate terminus and a conspicuous postanal swelling; first generation male with a unique combination of spicule and gubernaculum shape (spicule with hook-like tip, gubernaculum with a short cuneus and well separated corpus), and tail lacking a mucron.

Both molecular (either 28S rDNA or ITS rDNA) and morphological characters (IJ body length = 1113 μm) show that *S. aciari* belongs to the *glaseri*-group. Morphologically it is most similar to *S. puertoricense*, *S. glaseri* and the recently described *S. apuliae*, but several characters of various stages differentiate them. Compared to *S. puertoricense*, the tail of *S. aciari* IJ is shorter (78 ± 5.2 *vs* 94 ± 6 μm), c and D% are higher (14.4 and 123 *vs* 12.5 and 101, respectively), the excretory pore position of the first generation male of *S. aciari* is more anterior to the nerve ring, cuneus short (does not reach the end of the corpus) and the corpus is clearly separated posteriorly (Fig. 32J, M), whilst that of *S. puertoricense* is long (reaches the end of corpus) and the corpus is not separated posteriorly (Adams & Nguyen, 2002). Compared to *S. glaseri*, the excretory pore position of *S. aciari* IJ is anterior to the nerve ring while that of *S. glaseri* is at about the same level. These two nematodes can also be separated by the morphology of the tail of the first generation females and by the spicule and gubernaculum of the first generation males. Compared to *S. apuliae*, *S. aciari* IJ has a greater body diam. and anal body diam. (47±2.5 and 30±1.3 μm *vs* 37±21.9 and 23±1.4 μm, respectively), and smaller a and c′ (24 ± 2.1 and 2.5 ± 0.2 *vs* 29 ± 1.4 and 3.0±0.2, respectively). They can also be separated by the spicule of the first generation male, that of *S. aciari* being longer and with a greater SW value (86 ± 6.3 μm and 2.04 ± 0.15 *vs* 72 ± 5 μm and 1.6 ± 0.03, respectively) and with a distinct hook-like structure on the tip which is not present in *S. apuliae* (Fig. 32I, L).

Steinernema aciari can be distinguished from the remainder of the species in the *glaseri*-group by morphometrics of the IJ, as well as by the spicule shape of the first generation male and the tail morphology of the first generation female.

MOLECULAR CHARACTERISATION

Steinernema aciari is characterised by the length of ITS regions, 998bp (ITS1 = 283 bp, ITS2 = 308), and its composition (A = 0.24048, C = 0.20541, G = 0.25351 and T = 0.30060) (accession no. AY787660) and those of partial 28S rDNA, 463 bp (A = 0.24622, C = 0.18790, G = 0.30885, T = 0.25701) (accession no. AY787661).

DISTRIBUTION

The nematode has only been recorded from the Shantou district on the eastern coast of Guangdong province, People's Republic of China (Qui *et al.*, 2004).

References

ADAMS, B.J. & NGUYEN, K.B. (2002). Taxonomy and systematics. In: Gaugler, R. (Ed.). *Entomopathogenic nematology*. Wallingford, UK, CABI Publishing, pp. 1-33.

BEDDING, R.A. & AKHURST, R.J. (1975). A simple technique for the detection of insect parasitic nematode in soil. *Nematologica* 21, 109-110.

QIU, L., YAN, X., ZHOU, Y., NGUYEN, K.B. & PANG, Y. (2004). *Steinernema aciari* sp. n. (Nematoda: Steinernematidae), a new entomopathogenic nematode from Guangdong, China. *Journal of Invertebrate Pathology* 88, 58-69.

Steinernema affine[*,**] (Bovien, 1937) Wouts, Mráček, Gerdin & Bedding, 1982

= *Neoaplectana affinis* Bovien, 1937
(Figs 33-35)

MEASUREMENTS

See Table 7.

DESCRIPTION

Male (first generation)

Cuticle smooth, head rounded, not offset from rest of body, six lips united at base but distinct at tips. Each lip bearing a single labial papilla forming an inner circle of six. Outer circle of four cephalic papillae located at base of dorsolateral and ventrolateral papillae. Paired amphidial apertures located at base of lateral papillae. Lateral fields and phasmids inconspicuous. Stoma partially collapsed, posterior portion surrounded by pharyngeal tissue. Cheilorhabdions represented by thick ring of sclerotised material lining fused area of lips. Posteriorly is a second sclerotised ring, probably representing prorhabdions. Mesorhabdions thin walled, partially collapsed. Metarhabdions vestigial, represented by a small swollen area. Pharynx muscular with cylindrical procorpus followed by a slightly swollen, non-valvate, metacorpus. Isthmus joining nearly spherical basal bulb containing a reduced valve with refractive ridges. Genital system monorchic, reflexed. Seminal vesicle with secondary spermatocytes dividing into spermatids and spermatids shedding their residual bodies. Spicules variable in shape, tending to be strongly curved and not coloured, tips narrowed but blunt. Capitulum (manubrium) broad, rounded, generally with small arch (rostrum) at its base, or at top of calomus. Velum extending obliquely from rostrum almost to tip of lamina. Width and curvature variable. Gubernaculum forked with proximal portions curved upwards and overlapping the spicule laminae. Distal portion varying from simple to hook-shaped. Tip of tail bearing minute mucron, usually protruding only as

[*] From the Latin *affinis* = related to or neighbouring. This species was collected in the same host, *Bibio* sp., as *Steinernema feltiae* (= *Neoaplectana bibionis*) and is closely related to it as stated by Bovien (1937).

[**] Description after Poinar (1988). Revised by Zdeněk Mráček.

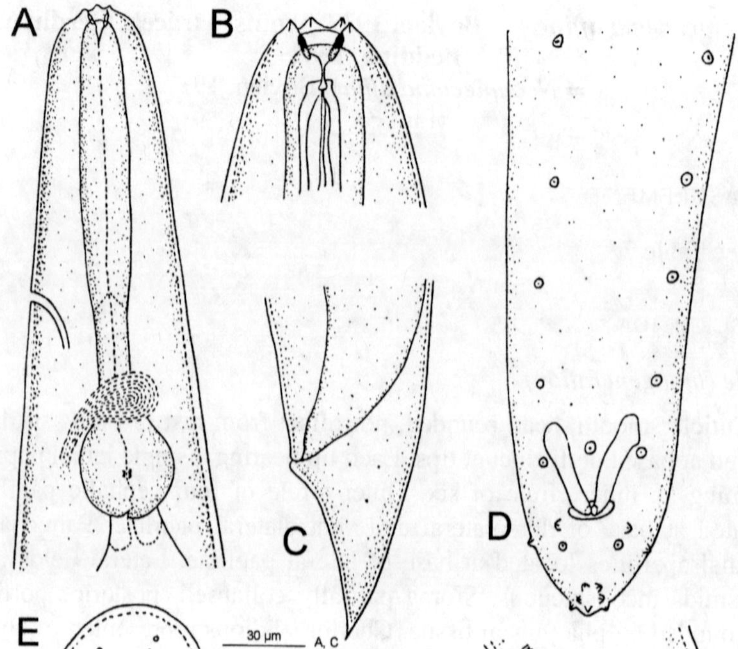

Fig. 33. Steinernema affine. *A, D, G, H: First generation male. A: Lateral view of anterior; D: Ventral view of tail portion; G: Lateral view of spicules and gubernaculum; H: Lateral view of male tail. C, E, F: Females. C: Lateral view of second generation tail; E:* En face *view of first generation female; F: Lateral view of first generation tail. After Poinar (1988), courtesy of* Revue de Nématologie.

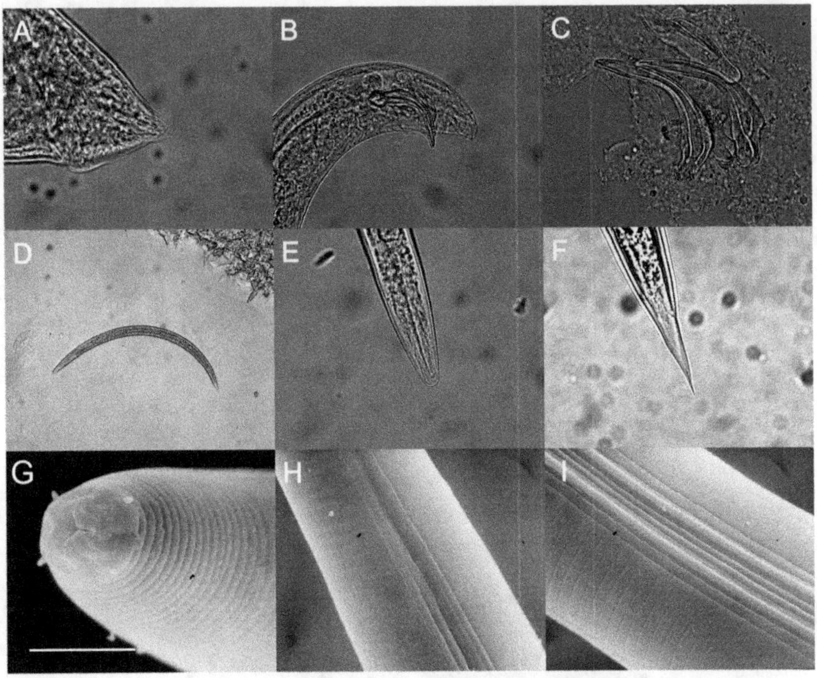

Fig. 34. Steinernema affine. *A: First generation female, tail with papilla-like structures. B, C: First generation male. B: Male tail; C: Spicules. D-I: Infective juvenile. D: Entire body; E: Head; F: Tail with characteristic refractile spine; G: Head with cephalic papillae; H: Lateral field in neck portion; I: Lateral field in mid-body, six ridges. (Scale bar: A = 70 μm; B = 30 μm; C, E, F = 50 μm; D = 350 μm; G-I = 15 μm.) After Z. Mráček.*

far as surrounding genital papillae. Tail mucron may be confused with genital papillae but latter are normally paired. Rarely, an extra-genital papilla occurring on one side and sometimes two midventral papillae occur in series. Basic complement of 23 genital papillae consisting of 11 pairs and single precloacal midventral papilla. Three pairs clustered at tail tip surrounding mucron, three pairs in region posterior to gubernaculum. Row of five subventral pairs located anterior to cloacal aperture.

Male (second generation)

Similar to first generation, but smaller size, mucron absent. Ratios E% and SW% much higher than in first generation.

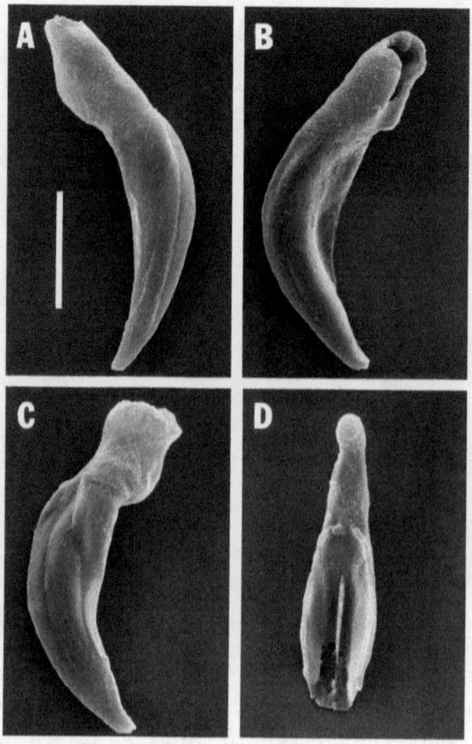

Fig. 35. Steinernema affine. *A-D: SEM of spicules and gubernaculum. A-C: Spicules; D: Gubernaculum. (Scale bar = 20 μm.) After K.B. Nguyen.*

Table 7. *Morphometrics of* Steinernema affine. *Measurements are in μm and in the form: mean (range).*

Character	First generation		Second generation		Infective juvenile
	Male*	Female*	Male*	Female*	
n	10	10	10	10	20** and 15*
L	1800 (1400-2100)	3100 (2066-3600)	1300 (1200-1500)	1700 (1600-1800)	712 ± 9.1** (626-788)
a	15	19	19	18	23* (21-28)
b	12	15	8.4	10	5.5* (5.1-6.0)
c	35	58	33	31	10.5* (9.5-12)
c′	0.9	0.7	0.9	1.4	3.6 (3.3-4.2)

Table 7. *(Continued).*

Character	First generation		Second generation		Infective juvenile
	Male*	Female*	Male*	Female*	
V	–	55 (50-60)	–	57 (54-61)	–
Body diam.	118 (95-164)	161 (139-208)	69 (63-76)	96 (88-107)	30* (28-34)
EP	94 (82-114)	102 (82-133)	93 (86-104)	103 (93-107)	62 (51-69)
NR	105 (89-117)	131 (114-152)	119 (107-128)	111 (96-136)	95* (88-104)
ES	153 (136-174)	203 (178-228)	155 (146-168)	170 (158-179)	126* (115-134)
Testis reflexion	715 (485-926)	–	220 (189-378)	–	–
Tail length (T)	51 (45-56)	53 (47-63)	40 (30-42)	55 (51-64)	64 ± 0.9** (52-68)
Hyaline (H)	–	–	–	–	25 ± 0.6** (19-30)
Anal body diam. (ABD)	60 (53-72)	73 (54-95)	44 (40-50)	39 (34-42)	18 ± 0.2** (16-19)
Spicule length (SL)	70 (67-86)	–	69 (62-72)	–	–
Spicule width	14 (11-19)	–	10 (8.0-11)	–	–
Gubernaculum length (GL)	46 (37-56)	–	36 (30-42)	–	–
Gubernaculum width	9.6 (8.0-10)	–	7.0 (4.8-8.0)	–	–
$D\% = EP/ES \times 100$	61	50	60	60	49* (43-53)
$E\% = EP/T \times 100$	180	192	233	187	94* (74-108)
$SW\% = SL/ABD \times 100$	117	–	157	–	–
$GS\% = GL/SL \times 100$	66	–	52	–	–
$H\% = H/T \times 100$	–	–	–	–	40 ± 0.6** (36-45)

EP = distance from anterior end to excretory pore; NR = distance from anterior end to nerve ring; ES = distance from anterior end to end of pharynx.

* Data and recalculated according to Poinar (1988).

** According to Půža (2001).

Female (first generation)

Cephalic extremity and cuticle and cuticular structures similar to male. Nerve ring located more towards the centre of isthmus, whereas in male it is usually positioned on anterior portion of basal bulb. Excretory pore more anteriorly located in female than in male. Genital system amphidelphic with opposed reflexed ovaries. Vulval aperture located posterior to mid-body, lips protruding by 6-25 μm. Terminal knob located at base of tail, 6-19 μm long. Postanal swelling reduced; if present, only slightly developed. Terminal projections or papilla-like mucrons on tail tip present.

Female (second generation)

Vulval aperture located posterior to mid-body, lips protruding by 0-5 μm. Tail straight, conoid, often ending in fine mucron. Postanal swelling absent or slightly developed.

Infective juvenile

Often still enclosed in second-stage cuticle. Cephalic extremity finely rounded, offset from rest of body. Oral aperture and anus closed, intestine collapsed. Bacterial cells contained in modified vesicle in anterior portion of intestinal lumen. Lateral field bearing six ridges (*i.e.*, seven lines), comprising less distinct marginal and central pair, and prominent submarginal pair. Excretory pore located anterior to nerve ring in mid-pharynx region. Tail sharply pointed, ratio $c' = 3.6$. Hyaline region short, comprising *ca* 40% of tail length. All mature infective juveniles bearing minute refractive spine in tail tip.

TYPE HOST AND LOCALITY

Bovien (1937) collected this species, together with *S. feltiae*, in larvae of the bibionid fly, *Phyla febrilis* (L.), (Bibionidae: Diptera), collected near the city of Skive in Denmark.

TYPE MATERIAL

Bovien (1937) designated no types for *S. affine*. Poinar (1988) examined Bovien's original slides of *S. affine*, Bovien designating his preserved material of *S. affine* as *Neoaplectana bibio* II. Unfortunately, the slides were mostly broken and the nematodes had dried out, so it was impossible to obtain any meaningful quantitative data from these specimens. However, Poinar (1988) designated a lectotype on a slide containing males of *S. affine* and two paralectotypes, one with males and one with females. These, as well as topotypes of both sexes from the subsequent collection of Poinar and Lindhardt (1971), are deposited in the University of California-Davis Nematode Collection, Davis, CA, USA.

RELATIONSHIPS

Steinernema affine has short infective juveniles with a body length of *ca* 700 μm. The habitus is arcuate to C-shaped. The head is rounded and slightly offset from the rest of body. The lateral fields have six ridges with the submarginal pair most prominent. The tail is sharply pointed with a refractive spine in its tip and has a short hyaline region comprising *ca* 40% of the tail length. The excretory pore is located in the mid-pharynx region. Males have robust spicules with the manubrium rounded and a well developed rostrum. The spicule tip is blunt. A mucron on the tail terminus is absent in both generations. Females lack a distinct postanal swelling but have papilla-like projections on the tail tip.

Bovien (1937) provided only two quantitative characters of *S. affine* and these were the length (1.0-1.62 mm) and body diam. (59-100 μm) of the males (no designation regarding whether these were the first or second generation). Additional descriptive remarks were restricted to qualitative characters which separated *S. affine* from *S. feltiae.* As Bovien (1937) pointed out, *S. affine* could be separated from *S. feltiae* by: *i*) colour of the spicules (grey or colourless in *S. affine* but orange-brown in *S. feltiae*); *ii*) the shape of the spicules (shorter capitulum and more curved calomus in *S. affine*); *iii*) shorter average size of infective stages (693 μm for *S. affine* and 825 μm for *S. feltiae*), although there is some overlap in ranges; and *iv*) the presence of a refractive spine in the tail tip of the infective stages of *S. affine* (but absent in *S. feltiae*). Bovien (1937) also stated that the gubernaculum of *S. affine* was more evenly curved and lacked a proximal knob or hook, although Poinar (1988) observed both knobbed and hooked gubernacula in his study.

Poinar (1988) mentioned that the size of the infective stage, presence of a spine in the infective stage tail, colour and shape of the spicules, and presence of a minute cuticular male tail spine separated *S. affine* from the previously described *S. glaseri*, *S. carpocapsae*, *S. intermedium*, *S. arenarium* and *S. rara.* The presence of a spine in the tail tip of the infective is unusual in steinernematids. In *S. affine*, this structure appears only in the fully mature infective and is minute.

Steinernema affine belongs to the *intermedium*-group which includes *S. intermedium*, *S. beddingi* and *S. sichuanense*. *Steinernema affine* is easily separated from these species by the presence of the above mentioned refractive spine in the infective juvenile tail. Moreover, it differs from *S. beddingi* by its six ridges in lateral field compared to four for *S. beddingi*. For the male, a mucron is absent in both generations of males of *S. affine*, *S. intermedium* (with some exceptions) and *S. beddingi*, whereas it is present in the second generation of *S. sichuanense*. Morphology of the four species in the *S. intermedium*-group, together with the morphometrics of spicules and gubernacula, are quite different based on SEM photographs. For the female, the postanal swelling is absent in the first generation of *S. sichuanense*, whereas *S. affine* and *S. intermedium* have a slight swelling and *S. beddingi* has a conspicuous swelling.

MOLECULAR CHARACTERISATION

Steinernema affine is characterised by sequence lengths (flanked by primers 18S and 26S) of ITS regions, 1066 bp, ITS1 = 306 bp, ITS2 = 350 bp and its composition, A = 0.25797, C = 0.17167, G = 0.21576, T = 0.35460. The sequences of ITS and D2D3 regions are deposited in GenBank under accession numbers AY230159 and AF331899, respectively.

BIONOMICS

This species is easily multiplied under laboratory conditions in larvae of *Galleria mellonella* In field soil samples and laboratory experiments, *S. affine* appeared to be dominant when compared with *S. kraussei*. The *S. affine* population was sustained for longer and only the progeny of *S. affine* emerged from *G. mellonella* cadavers initially infected with both *S. affine* and *S. kraussei* in laboratory experiments. Only in the single species treatment did *S. kraussei* develop a sustained population (Mráček *et al.*, 2005b).

DISTRIBUTION

This species is widely distributed throughout Europe (Sturhan & Lišková, 1999; Sturhan & Mráček, 2000; Mráček *et al.*, 2005a), including the UK and Ireland (Griffin *et al.*, 1991; Hominick *et al.*, 1995; Hominick, 2002). *Steinernema affine* was discovered in the Vancouver area, British Columbia, Canada (Mráček *et al.*, 2005b). *Steinernema affine* prevails in open and semi-open habitats (Hominick *et al.*, 1995; Mráček *et al.*, 2005a).

References

BOVIEN, P. (1937). Some types of association between nematodes and insects. *Videnskabelige Meddelelser fra Dansk Naturhistorisk. Forening* 101, 1-114.

GRIFFIN, C.T., MOORE, J.F. & DOWNES, M.J. (1991). Occurrence of insect-parasitic nematodes (Steinernematidae, Heterorhabditidae) in the Republic of Ireland. *Nematologica* 37, 92-100.

HOMINICK, W.M. (2002). Biogeography. In: Gaugler, R. (Ed.). *Entomopathogenic nematodes*. Wallingford, UK, CABI Publishing, pp. 115-143.

HOMINICK, W.M., REID, A.P. & BRISCOE, B.R. (1995). Prevalence and habitat specificity of steinernematid and heterorhabditid nematodes isolated during soil surveys of the UK and The Netherlands. *Journal of Helminthology* 69, 27-32.

MRÁČEK, Z., BEČVÁŘ, S., KINDLMANN, P. & JERSÁKOVÁ, J. (2005a). Habitat preference for entomopathogenic nematodes, their insect hosts and new faunistic records for the Czech Republic. *Biological Control* 34, 27-37.

MRÁČEK, Z., KINDLMANN, P. & WEBSTER, J.M. (2005b). *Steinernema affine* (Nematoda: Steinernematidae), a new record for North America and its distribution relative to other entomopathogenic nematodes in British Columbia. *Nematology* 7, 495-501.

POINAR JR, G.O. (1988). Redescription of *Neoaplectana affinis* Bovien (Rhabditida: Steinernematidae). *Revue de Nématologie* 11, 143-147.

POINAR JR, G.O. & LINDHARDT, K. (1971). The re-isolation of *Neoaplectana bibionis* Bovien (Nematodea) from Danish bibionids (Diptera) and their possible use as biological control agent. *Revue de Nématologie* 11, 143-147.

PŮŽA, V. (2001). *Comparison of an infective juvenile tail morphology of the family Steinernematidae*. B.Sc. Thesis, University of South Bohemia, 23 pp.

STURHAN, D. & LIŠKOVÁ, M. (1999). Occurrence and distribution of entomopathogenic nematodes in the Slovak Republic. *Nematology* 1, 273-277.

STURHAN, D. & MRÁČEK, Z. (2000). Comparison of the *Galleria* baiting technique and a direct extraction method for recovering *Steinernema* infective-stage juveniles from soil. *Folia Parasitologica* 47, 315-318.

WOUTS, W.M., MRÁČEK, Z., GERDIN, S. & BEDDING, R.A. (1982). *Neoaplectana* Steiner, 1929 a junior synonym of *Steinernema* Travassos, 1927 (Nematoda: Rhabditida). *Systematic Parasitology* 4, 147-154.

Steinernema akhursti[*,**] Qiu, Hu, Zhou, Mei, Nguyen & Pang, 2005
(Figs 36-38)

MEASUREMENTS

See Table 8.

DESCRIPTION

Male (first generation)

Body usually J-shaped when heat-relaxed. Cephalic extremity rounded, continuous with body contour. Six prominent labial papillae, two amphidial apertures and four cephalic papillae present. Stoma shallow, cheilorhabdions and prorhabdions sclerotised, small, sometimes indistinct. Pharynx muscular with cylindrical procorpus, metacorpus slightly swollen; isthmus distinct, surrounded by nerve ring, basal bulb prominent. Excretory pore located anterior to nerve ring at base of metacorpus. Genital system monorchic, reflexed. Spicules paired, brown in colour. Length of spicule head greater than width, blade moderately curved, dorsal lobe and lateral lobe prominent, starting from head and extending to tip of blade. Ventral lobe indistinct. Spicule tip blunt with aperture on ventral side close to tip. Velum well developed, starting from anterior end of ventral lobe and ending at one-third of spicule length from spicule tip. Gubernaculum boat-shaped in lateral view, gradually tapering anteriorly. Anterior part of gubernaculum curved ventrally. Cuneus long, needle-shaped. Eleven pairs and one single precloacal genital papilla distributed as in Figure 37A, B. Tail conoid, usually concave on ventral side. Tail tip with prominent mucron.

Male (second generation)

Similar first generation but most morphometrics such as body, spicule and gubernaculum lengths smaller. Tail mucron 1.5-2 times as long as that in first generation male.

Female (first generation)

Body usually C-shaped on heat relaxation, cuticle smooth or with faint annules. Lateral fields and phasmids absent. Cephalic extremity bluntly

[*] Named after R.J. *Akhurst*, a prominent worker in the field of entomopathogenic nematology.

[**] Description after Qiu *et al.* (2005). Revised by Khuong B. Nguyen.

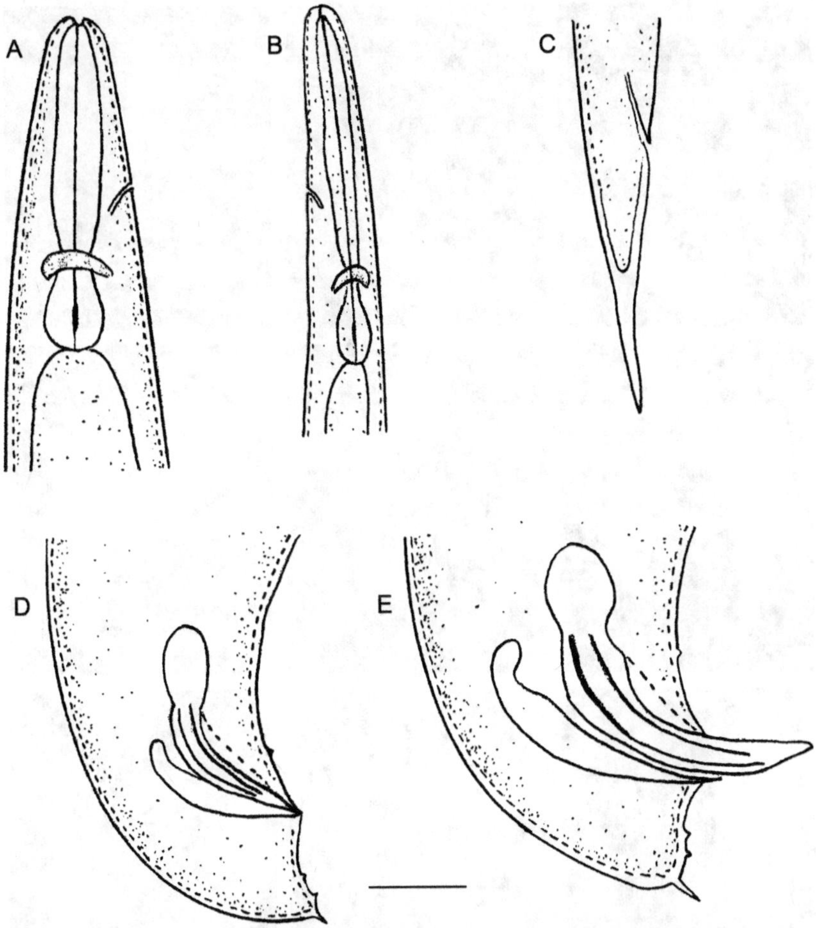

Fig. 36. Steinernema akhursti. *A, D: Anterior and tail regions of first generation male in lateral view; B, C: Anterior region and tail of infective juvenile; E: Tail of second generation male showing long, thin mucron on tail tip. (Scale bar: A = 50 μm; B = 30 μm; C = 25 μm; D = 35 μm; E = 20 μm.) Redrawn after the original authors, courtesy of* Journal of Invertebrate Pathology.

rounded, continuous with body contour. Six conspicuous labial papillae, four cephalic papillae and two amphidial apertures. Stoma shallow. Cheilorhabdions small, well sclerotised. Prorhabdions distinct. Pharynx muscular with cylindrical procorpus, metacorpus swollen, diam. similar to that of basal bulb, isthmus distinct. Excretory pore located anterior to nerve ring, near mid-pharynx. Genital system amphidelphic, reflexed.

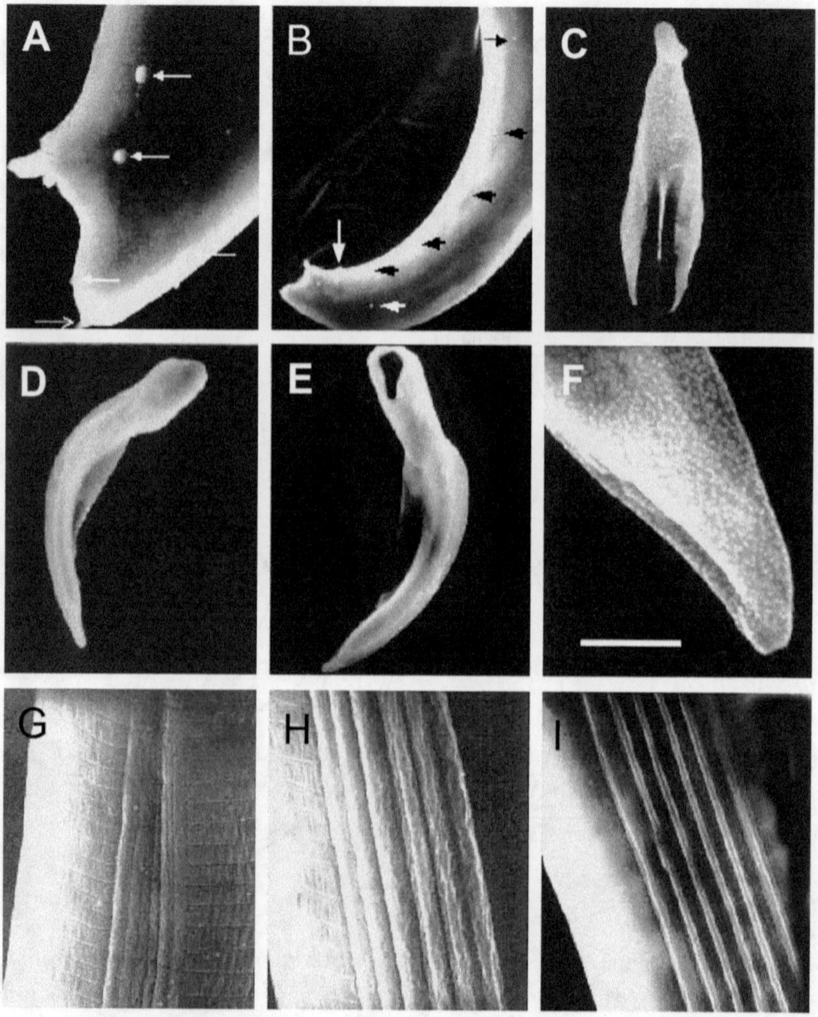

Fig. 37. Steinernema akhursti. *SEM photographs. A, B: Posterior region of first generation males showing arrangement of genital papillae (arrows); C: Gubernaculum, ventral view showing needle-shaped cuneus; D-F: Spicules and spicule tip; G-I: Lateral field pattern. (Scale bar: A = 40 μm; B = 150 μm; C = 30 μm; D = 30 μm; E = 27 μm; F = 4 μm; G = 12 μm; H = 10 μm; I = 12 μm.) After the original authors, courtesy of* Journal of Invertebrate Pathology.

Vulva a transverse slit, symmetrical and slightly protruding. Double-flapped epiptygmata present. Vagina short and muscular. Postanal

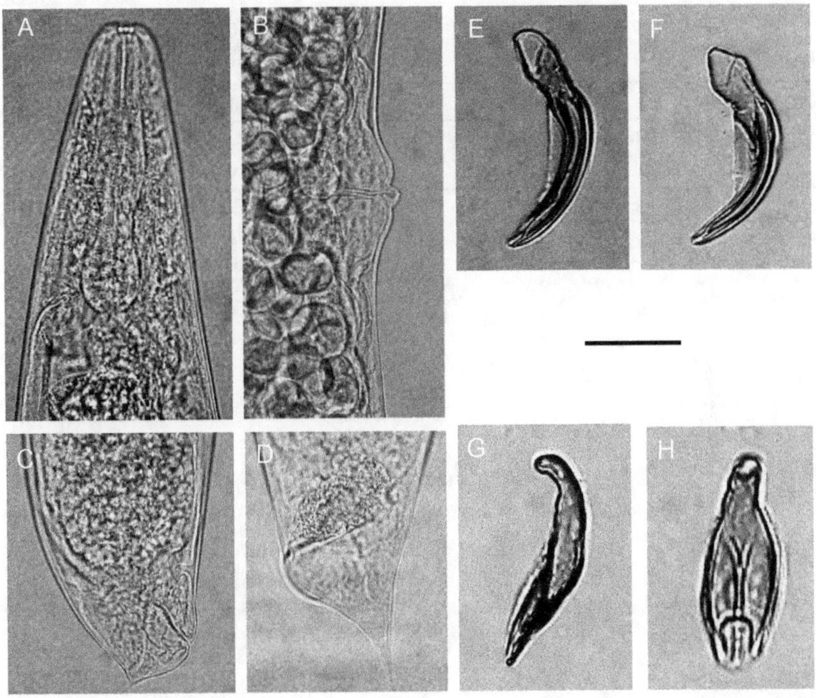

Fig. 38. Steinernema akhursti. *Light micrographs. A-C: Anterior region, vulva and tail region of first generation female; D: Tail of second generation female; E-H: Spicules and gubernacula of first generation male. (Scale bar: A = 90 μm; B-D = 100 μm; E, F = 30 μm; G, H = 25 μm.) After the original authors, courtesy of* Journal of Invertebrate Pathology.

swelling present but not prominent. Tail conoid, shorter than anal body diam. Short, spike-like mucron observed on tail tip of *ca* 80% of specimens examined.

Female (second generation)

Similar to first generation female but smaller. Vulva slightly protuberant. Epiptygmata present. Tail longer than anal body diam., tapering to a pointed terminus. Ventral postanal swelling present.

Infective juvenile

Body elongate and ensheathed in second-stage cuticle immediately after harvesting, but many losing sheath in storage or in soil. Cephalic extremity rounded anteriorly, continuous with body contour. Cephalic region smooth. Four distinct cephalic papillae. Amphidial apertures

Table 8. *Morphometrics of* Steinernema akhursti. *Measurements are in μm and in the form: mean ± SD (range).*

Character	First generation			Second generation		Infective juvenile
	Holotype (male)	Male	Female	Male	Female	
n	–	20	20	20	20	20
L	1625	1589 ± 154 (1350-1925)	7283 ± 926 (5625-9000)	1042 ± 47 (950-1100)	2485 ± 464 (2050-3250)	812 ± 19 (770-835)
a	–	–	–	–	–	24 ± 0.9 (23-26)
b	–	–	–	–	–	6.8 ± 0.16 (6.6-7.2)
c	–	–	–	–	–	11 ± 0.54 (10-12)
c′	–	–	–	–	–	3.5 ± 0.09 (3.3-3.7)
V	–	–	51 ± 1 (48-53)	–	52 ± 1.1 (50-54)	–
Body diam.	135	131 ± 9.7 (115-150)	239 ± 27 (200-270)	67 ± 6.7 (55-78)	139 ± 7.1 (125-150)	33 ± 1.0 (33-35)
EP	100	102 ± 6 (93-113)	126 ± 7 (113-138)	75 ± 4.5 (70-80)	93 ± 11 (80-108)	59 ± 1.5 (55-60)
NR	145	136 ± 10 (120-163)	164 ± 11 (150-175)	101 ± 5.3 (88-105)	128 ± 18 (113-150)	90 ± 4 (83-95)
ES	190	182 ± 10 (168-205)	139 ± 17 (213-258)	138 ± 6.6 (125-150)	181 ± 18 (163-205)	119 ± 2.5 (115-123)
Tail length (T)	32.5	35 ± 3.3 (30-40)	49 ± 7	22 ± 2.6 (18-25)	63 ± 8 (58-83)	73 ± 2.9 (68-75)
Hyaline (H)	–	–	–	–	–	39 ± 1.6 (36-41)
Anal body diam. (ABD)	50	51 ± 5.7 (45-70)	86 ± 11 (68-100)	37 ± 3 (33-43)	48 ± 4.6 (43-55)	20 ± 0.6 (10-20)
Spicule length (SL)	85	90 ± 4.6 (85-100)	–	72 ± 3.9 (65-78)	–	–
Gubernaculum length (GL)	62.5	64 ± 2.7 (58-68)	–	48 ± 1.7 (45-50)	–	–
D% = EP/ES × 100	53	56 ± 2.6 (52-61)	–	55 ± 3.6 (49-60)	–	47 ± 1.4 (45-50)
E% = EP/T × 100	–	–	–	–	–	77 ± 4.5 (73-86)
SW% = SL/ABD × 100	170	180 ± 14 (140-200)	–	200 ± 16 (160-210)	–	–
GS% = GL/SL × 100	74	71 ± 4 (65-77)	–	66 ± 3 (61-70)	–	–
H% = H/T × 100	–	–	–	–	–	52 ± 2.1 (49-56)

EP = distance from anterior end to excretory pore; NR = distance from anterior end to nerve ring; ES = distance from anterior end to end of pharynx.
After the original authors.

slit-like but not prominent. Cuticle marked with transverse striations. Pharynx long and thin. Excretory pore located anterior to nerve ring. Lateral field beginning anteriorly with one line followed by two additional lines to form two ridges. At about excretory pore level each of the two ridges separating into three, making six ridges in all. Six ridges evenly distributed, at least in middle body portion. Six-ridge pattern (*i.e.*, seven lines) extending posteriorly to anus level. After that, ridges merging to two before disappearing. Formula for lateral field = 2, 6, 2. Phasmid not prominent. Tail attenuate and tapering gradually with constriction at level of hyaline region. Hyaline region comprising *ca* 54% of tail length.

TYPE HOST AND LOCALITY

The type host of this nematode in nature is unknown as it was recovered from soil using *Galleria mellonella* larvae as bait. Isolate YNb112 was designated as the type and was recovered from a soil sample collected from grassland in Jiantang town, Zhongdian county, Diqing district, Yunan province, People's Republic of China.

OTHER LOCALITIES

This species was recovered from many soil samples collected from various localities in temperate areas of Yunan province, People's Republic of China.

TYPE MATERIAL

Holotype (male), allotype female, paratype deposited in the State Key Laboratory for Biocontrol, College of Life Sciences, Zhongshan University, Guangzhou 510275, People's Republic of China. Additional paratype specimen deposited in the University of California-Davis Nematode Collection, Davis, CA, USA.

RELATIONSHIPS

Steinernema akhursti is characterised morphologically by the combination of the features of various developmental stages. For the first generation males, they can be recognised by a spicule length of 90 μm; spicule tip blunt with an aperture on the ventral side; gubernaculum with a long and needle-shaped cuneus; and tail conoid with a prominent mucron on the tip and concave ventrally. The second generation males are characterised by a long, thin mucron on the tail tip. For the infective juvenile, the combination of the following characters: body length (812 μm), distance from anterior end to excretory pore (59 μm), tail length (73 μm), E% (77), lateral field with six evenly distributed and identical ridges at the mid-body, and tail with long and slightly constricted hyaline region. The female *S. akhursti* is recognised by a conoid tail with a short mucron, a slightly swollen anal region, and a slightly protuberant and symmetrical vulva with conspicuous double-flapped epiptygmata.

Both molecular and morphometric data showed that *S. akhursti* belongs to the *feltiae*-group and is closely related to *S. feltiae*, *S. oregonense*, *S. kraussei*, *S. weiseri*, *S. sangi*, *S. thanhi* and *S. monticolum*. The shape of the spicule and gubernaculum of the first generation male, the tail and lateral field morphology of the infective juvenile, the tail and vulva shape of the first generation females, and the tail of the second generation males (with a long mucron) can be used to distinguish *S. akhursti* from the others. For example, based on the shape of the first generation male spicule, *S. akhursti*, with well developed velum and an aperture on the ventral side of the tip (Fig. 37D, E) can be easily separated from *S. feltiae* and *S. oregonense* (which lack a velum and terminal aperture), *S. sangi* (head length shorter than width) and *S. thanhi* (velum much thinner or indistinct). With regard to the lateral field morphology, *S. akhursti* has six evenly distributed ridges in the mid-body region whilst *S. feltiae* possesses eight ridges with the submarginal pair indistinct.

Molecular characterisation

The species is characterised genetically by the sequence length of ITS regions of rDNA, 973 bp (A = 0.25282, G = 0.24029, T = 0.31243, C = 0.19424) (accession no. DQ375757), and partial 28S rDNA, 459 bp (A = 0.23529, G = 0.32244, T = 0.24400, C = 0.19825) (accession no. AY177188) and pair wise distances of these sequences (Qiu *et al.*, 2005).

Bionomics

The symbiotic bacterium of *S. akhursti* has been isolated, purified, and preserved in liquid nitrogen. The bacteria initially produced a red pigment on an artificial medium containing corn flour, egg, and oil and then the colour of the medium gradually turned dark brown when the nematodes proliferated. This feature of pigment production was observed for all *S. akhursti* isolates and is unique for all *Steinernema* species observed so far. Preliminary studies indicated that the symbiotic bacteria of *S. akhursti* possessed many characters of *Xenorhabdus* species, such as phase variation, negative to peroxide, *etc.* A phylogenetic tree based on the 16S rDNA sequence showed that it formed a monophyletic clade with, but was clearly distinguished from, *Xenorhabdus nematophila* and *Xenorhabdus japonica* (Qiu *et al.*, 2005).

DISTRIBUTION

About 30 isolates of this species were obtained from various localities in Yunan province. *Steinernema akhursti* isolates were recovered from places with mean annual temperature between 8°C (Zhongshan) and 16°C (Kunming, Dali and Qujing). *Steinernema akhursti* was not found in warm areas, such as Jinghong and Luxi districts, despite extensive entomopathogenic nematode surveys (Qiu *et al.*, 2005).

Reference

QIU, L., HU, X., ZHOU, Y., MEI, S., NGUYEN, K.B. & PANG, Y. (2005). *Steinernema akhursti* sp. n. (Nematoda: Steinernematidae) from Yunan, People's Republic of China. *Journal of Invertebrate Pathology* 90, 151-160.

Steinernema anatoliense[*,**] Hazir, Stock & Keskin, 2003 (Fig. 39)

MEASUREMENTS

See Table 9.

DESCRIPTION

Male (first generation)

Body curved posteriorly, J-shaped when heat-relaxed, larger than second generation. Cuticle, labial region, stoma and pharyngeal region as in female. Single reflexed testis, consisting of germinal growth zone leading to seminal vesicle. *Vas deferens* with inconspicuous walls. Spicules paired, symmetrical, curved, with light yellow colouration. Manubrium rectangular or oval. Calomus distinct, short. Lamina with two internal ribs, rostrum (retinaculum) present. Velum slightly developed, narrow. Spicule tip rounded, with a concavity in ventral view. Gubernaculum *ca* half length of spicules, straight or slightly curved. In ventral view, cuneus short, needle shaped, not reaching end of corpus. Tail conoid, mucronate. Complement of 23 genital papillae (11 pairs and one single) disposed as follows: six precloacal subventral pairs; one pair adcloacal sublateral; single midventral precloacal; one pair postcloacal subventral; two pairs terminal postcloacal, near tail tip; one pair subdorsal postcloacal.

Male (second generation)

Similar to first generation male but smaller. Tail conoid and mucronate, but mucron (average = *ca* 1 μm) shorter than in first generation.

Female (first generation)

Body C-shaped when heat-relaxed, larger than second generation female. Cuticle smooth under light microscopy, but with fine transverse striae visible under SEM. Lateral field and phasmids inconspicuous. Cephalic extremity truncate to slightly rounded, continuous with body. Six amalgamated lips with distinct tips, each bearing single labial

[*] Name derived from *Anatolia*, an ancient name for the country of origin.

[**] Description after Hazir *et al.* (2003b). Revised by David J. Hunt.

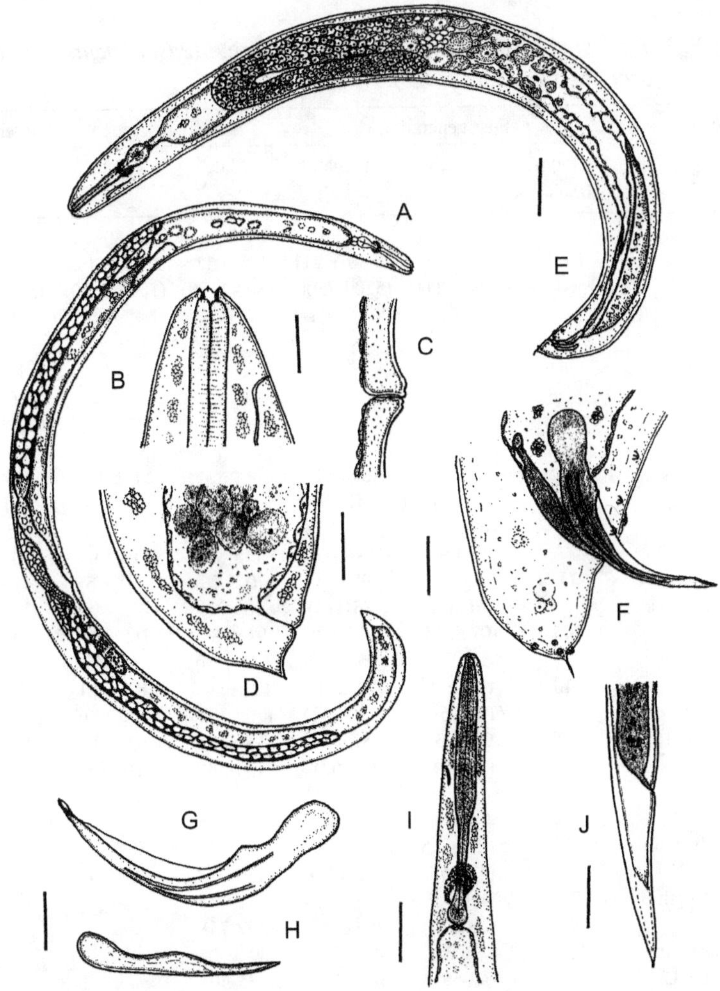

Fig. 39. Steinernema anatoliense. *A-D: Female. A: Entire worm; B: Anterior region of first generation; C: Vulva of second generation; D: Tail region of first generation. E-H: First generation male. E: Entire worm; F: Posterior region, lateral view; G: Spicule, lateral view; H: Gubernaculum, lateral view. I, J: Infective juvenile. I: Anterior region; J: Posterior region. (Scale bars: A = 205 μm; B = 54 μm; C, D = 61 μm; E = 80 μm; F = 16 μm; G = 14 μm; H = 12.5 μm; I = 17.5 μm; J = 17 μm.) Reproduced from* Systematic Parasitology, *55 (2003), p. 214; A new entomopathogenic nematode,* Steinernema anatoliense *n. sp. (Rhabditida: Steinernematidae), from Turkey, by Hazir, S., Stock, S.P. and Keskin, N. Figure 1. With kind permission of Springer Science and Business Media.*

Table 9. *Morphometrics of* Steinernema anatoliense. *Measurements are in μm and in the form: mean ± SD (range).*

Character	First generation			Second generation		Infective juvenile
	Holotype (male)	Male	Female	Male	Female	
n	–	19	19	20	20	20
L	1531	1532 ± 195 (1236-1841)	6830 ± 211 (5500-7500)	995 ± 116 (800-1100)	2235 ± 145 (1500-2850)	545 ± 21 (507-580)
a	–	–	–	–	–	22 ± 2 (19-27)
b	–	–	–	–	–	5.1 ± 0.3 (4.6-5.7)
c	–	–	–	–	–	10.5 ± 0.5 (9.4-11.8)
V	–	–	57 ± 1 (51-59)	–	51 ± 1.5 (49-54)	–
Body diam.	95	109 ± 15 (87-150)	195 ± 24 (187-200)	75 ± 7.5 (55-80)	124 ± 8 (110-135)	24.5 ± 2 (21-28)
EP	71	67 ± 0.5 (61-88)	95 ± 8 (85-113)	70 ± 7.5 (65-93)	130 ± 5 (122-145)	37 ± 1 (36-39)
NR	105	107 ± 15 (87-131)	175 ± 8 (162-190)	91 ± 8.5 (84-120)	155 ± 9 (139-165)	76 ± 3 (71-82)
ES	240	144 ± 8 (129-160)	235 ± 11 (220-241)	125 ± 9.5 (105-154)	182 ± 10 (175-200)	107 ± 6 (97-124)
Tail length (T)	33.5	30 ± 3 (26-34)	57 ± 7 (47-63)	33 ± 4.5 (30-38)	63 ± 4 (60-70)	53 ± 3 (46-58)
Hyaline (H)	–	–	–	–	–	22.5 ± 1 (19-26)
Anal body diam. (ABD)	54	52.5 ± 3.5 (47-60)	88 ± 9 (72-110)	43 ± 3 (36-47)	50 ± 3 (45-55)	12 ± 0.7 (11-14)
Spicule length (SL)	75	74 ± 3.5 (68-84)	–	68 ± 2.5 (57-75)	–	–
Gubernaculum length (GL)	45.5	47 ± 3.5 (42-59)	–	39 ± 2.5 (35-50)	–	–
D% = EP/ES × 100	48	48.5 ± 3.5 (46.5-55)	40.5 ± 4 (38.5-47)	62 ± 4 (56-68)	69 ± 3 (65-72)	35 ± 2.5 (31.5-39)
E% = EP/T × 100	–	–	–	–	–	72 ± 3.9 (64-81.5)
SW% = SL/ABD × 100	170	175 ± 10 (160-190)	–	160 ± 10 (150-180)	–	–
GS% = GL/SL × 100	60	55 ± 4 (40-65)	–	60 ± 10 (50-70)	–	–
H% = H/T × 100	–	–	–	–	–	43 ± 0.8 (41-45)

EP = distance from anterior end to excretory pore; NR = distance from anterior end to nerve ring; ES = distance from anterior end to end of pharynx.

papilla. Four cephalic papillae present. Amphidial apertures small, located posterior to lateral labial papillae. Stoma reduced, short and broad, with inconspicuous sclerotised walls. Cheilorhabdions and telorhabdions vestigial. Pharynx muscular; procorpus cylindrical; metacorpus slightly swollen and non-valvate; indistinct isthmus followed by pyriform basal bulb containing reduced valve. Pharynx set off from intestine. Nerve ring usually surrounding isthmus or anterior part of basal bulb. Excretory pore opening circular, located in anterior third of metacorpus. Ovaries opposed, reflexed in dorsal position; oviduct well developed, glandular spermatheca and uterus in ventral position. Vagina short, with muscular walls. Vulva located near mid-body. Vulval lips symmetrical, slightly protruding, epiptygmata absent. Tail conoid, without mucron, no postanal swelling.

Female (second generation)

Similar to first generation but smaller. Vulval lips slightly protruding and symmetrical, epiptygmata absent. Tail conoid, without mucron. Postanal swelling slightly developed.

Infective juvenile

Body slender, tapering regularly from base of pharynx to anterior end and from anus to terminus. Labial region smooth, oral aperture closed. Lips indistinct, with six labial papillae. Amphidial apertures circular, small. Cuticle with transverse striae. Lateral field distinct with 6-8 longitudinal ridges (*i.e.*, 7-9 incisures) in mid-body region, with central ridges closer to each other than marginal and submarginal ridges. Pharynx long, narrow, with slender isthmus and sub-pyriform basal bulb. Cardia present. Nerve ring located at isthmus level. Excretory pore located in anterior third of pharynx. Anterior portion of intestine with dorsally displaced pouch containing symbiotic bacterium. Lumen of intestine narrow, rectum long, anus distinct. Genital primordium evident. Tail conoid with pointed terminus. Hyaline region extending for *ca* half of tail length.

TYPE HOST AND LOCALITY

Natural host unknown, material isolated by baiting soil with *Galleria mellonella* larvae. Collected from natural pastures, Kars, East Anatolia, Turkey.

TYPE MATERIAL

Holotype male and paratype males, females and juveniles in the University of California-Davis Nematode Collection, Davis, CA, USA. Other paratypes (all stages) at USDA Nematode Collection, Beltsville, MD, USA.

RELATIONSHIPS

Steinernema anatoliense belongs to the *carpocapsae*-group of species. It has been shown to be reproductively isolated from *S. abbasi*, *S. carpocapsae*, *S. scapterisci*, *S. siamkayai* and *S. rarum* (Hazir *et al.*, 2003b). The species is characterised by the following combination of features: infective juveniles with a body length of 545 (507-580) μm and lateral field with 6-8 longitudinal ridges (*i.e.*, 7-9 incisures); first generation male tail bearing a mucron, spicules arcuate and robust with rectangular to oval-shaped manubrium, short calomus and very thin velum, spicule tip rounded with small concavity in ventral view, gubernaculum straight and about half as long as spicule, 23 genital papillae arranged as one midventral precloacal papilla, six precloacal pairs, an adcloacal ventrosublateral pair, one subventral pair postcloacal, two postcloacal pairs at the tail terminus and one subdorsal postcloacal pair. Second generation males have a shorter mucron than the first generation male (*ca* 1 μm) whilst the first generation females have slightly protruding vulval lips and no postanal swelling.

MOLECULAR CHARACTERISATION

The species was originally distinguished from *S. carpocapsae*, *S. rarum*, *S. tami*, *S. abbasi* and *S. siamkayai* by ITS RFLP profile. Additionally, the sequence of D2D3 regions is deposited at GenBank, accession no. AY841761, 12S rRNA, AY944004 and cytochrome c oxidase subunit I (cox1), AY943978.

BIONOMICS

The life-cycle is similar to other species of *Steinernema*, the infective juvenile entering the host through the mouth or spiracles and then penetrating the haemocoel. Once in the haemocoel, the symbiotic bacteria are released by the nematode and kill the host within 36-48 h.

Two amphimictic generations occur within the host. Observations using *Galleria mellonella* larvae as host indicated that the infective juveniles reached the first generation adult stages in 48-72 h at 22°C. Second generation adults appeared in 5-6 days with infective juveniles emerging from the cadaver 7-8 days after infection (Hazir *et al.*, 2003b).

DISTRIBUTION

Known only from the type locality in East Anatolia, Turkey (Hazir *et al.*, 2003a, b).

References

HAZIR, S., KESKIN, N., STOCK, S.P., KAYA, H.K. & ÖZCAN, S. (2003a). Diversity and distribution of entomopathogenic nematodes (Rhabditida; Steinernematidae and Heterorhabditidae) in Turkey. *Biodiversity and Conservation* 12, 375-386.

HAZIR, S., STOCK, S.P. & KESKIN, N. (2003b). A new entomopathogenic nematode, *Steinernema anatoliense* n. sp. (Rhabditida: Steinernematidae), from Turkey. *Systematic Parasitology* 55, 211-220.

Steinernema apuliae[*,**] Triggiani, Mráček & Reid, 2004
(Figs 40-42)

MEASUREMENTS

See Table 10.

DESCRIPTION

Male (first generation)

Body curved ventrally posteriorly, tapering towards labial region, J-shaped in heat-relaxed specimens. Cuticle with very faint transverse striae, lateral fields absent. Cephalic extremity rounded, almost continuous with body. Six lips fused at base, each lip bearing a labial papilla. Outer circle of four cephalic papillae present. Amphidial apertures distinct, located laterally between labial and cephalic circle of papillae. Excretory pore variable in position, rarely slightly anterior, mostly at, and sometimes posterior to, nerve ring (ranging from 75-92% of length from anterior body end to base of pharynx). Stoma short, shallow, cheilorhabdions distinct, sclerotised, followed by sclerotised smaller, sometimes indistinct, prorhabdions. Pharynx muscular with procorpus extending into slightly swollen, non-valvate, metacorpus, a narrower isthmus region surrounded by a nerve ring, and a rounded basal bulb. Cardia rather long, intestine generally with a wide lumen. Distinct, papilliform deirids situated at or slightly posterior to pharyngeal bulb. Well developed and reflexed testis. Paired symmetrical, slightly curved spicules, light brown in colour, distally bluntly pointed. Spicule manubrium distinctly elongated, with length : width ratio of 1.2-1.7 : 1; manubrium less then one-third of spicule length (*ca* 29%); calomus distinct, lamina slightly arched, ventral rostrum absent; velum narrow, indistinct, parallel with lamina, extending from calomus to distal part of lamina. Gubernaculum boat-shaped in lateral view, distally much attenuated, proximal part varying from straight and bluntly pointed to slightly bent. Tail bluntly conoid, usually concave at ventral side, mucron absent. Complement of 23 genital papillae comprising 11 pairs and a single midventral papilla

[*] Named after *Apulia*, a region in southern Italy.

[**] Description after Triggiani *et al.* (2004). Revised by Zdeněk Mráček.

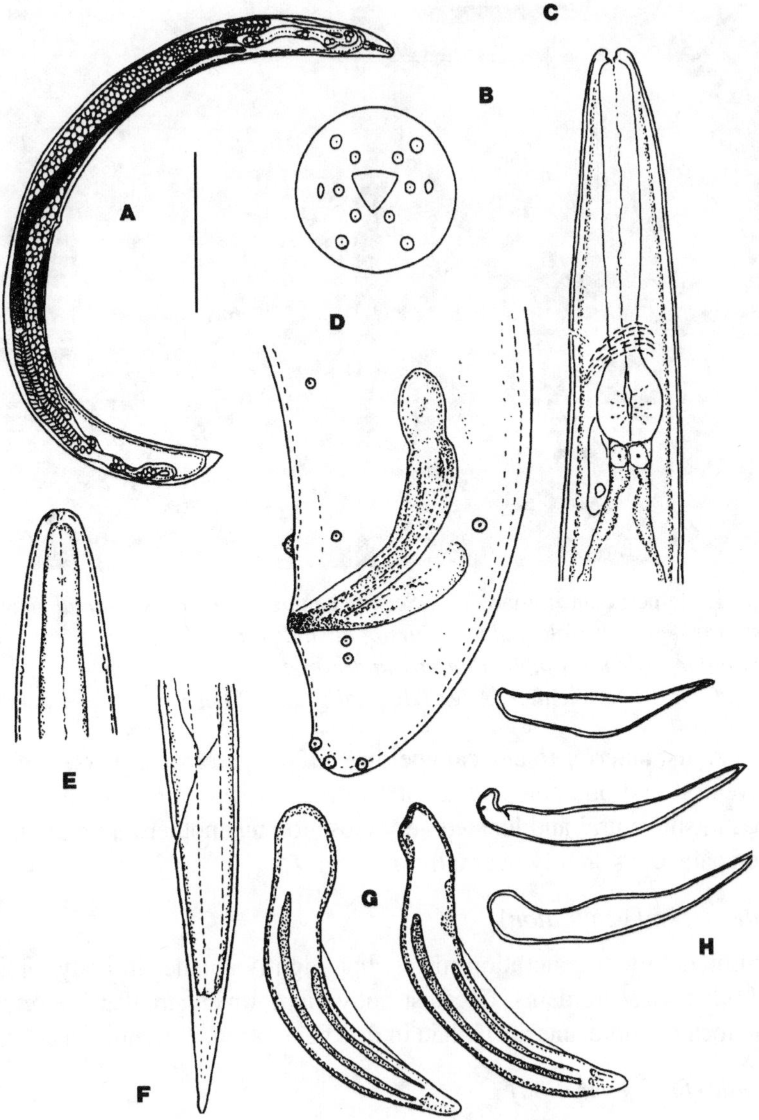

Fig. 40. Steinernema apuliae. *A: First generation female; B: Adult arrangement of labial and cephalic papillae, and amphidial apertures. C, D: Male. C: Pharynx and excretory pore; D: Tail region showing arrangement of genital papillae. E, F: Infective juvenile. E: Rounded head; F: Anus and hyaline portion. G: Spicule variation; H: Gubernaculum variation. (Scale bar: A = 1000 μm; C = 72 μm; D = 38 μm; E = 25 μm; F = 17 μm; G = 31 μm; H = 28 μm.) After Triggiani* et al. *(2004), courtesy of* Zootaxa.

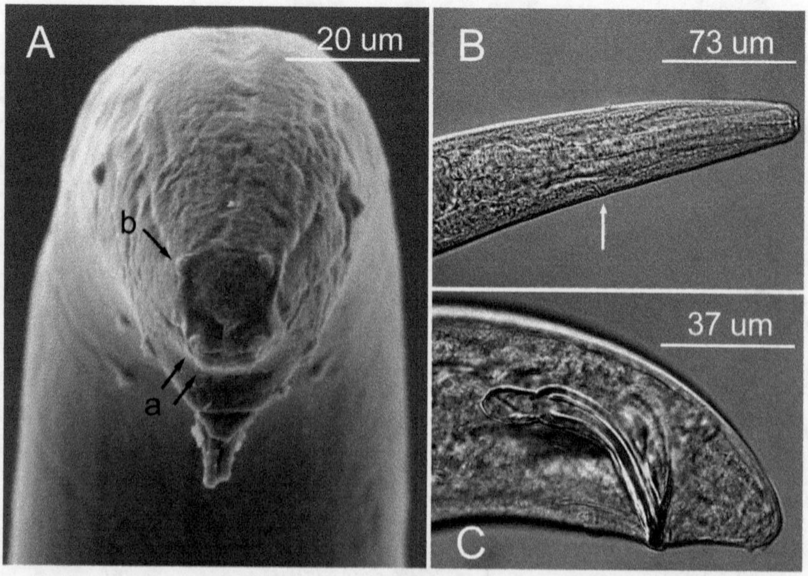

Fig. 41. Steinernema apuliae. *A-C: Second generation male. A: Dorsal view of posterior region of male and two pairs of adcloacal papillae (a, b), three pairs of papillae near tail tip; B: Pharynx and excretory pore; C: Tail region with spicules and gubernaculum. After Triggiani* et al. *(2004), courtesy of* Zootaxa.

located just anterior to cloacal aperture. Of six precloacal pairs, five are subventral and one pair lateral; of five postcloacal pairs, two pairs adcloacal, subventral and located very close to one another and remaining three pairs consistently near tail tip.

Male (second generation)

Similar to first generation males, but slightly smaller in body length and other measurements. Greatest body diam. much smaller. Excretory pore located more anteriorly than in first generation. Tail mucron absent.

Female (first generation)

Body usually C-shaped on heat relaxation. Cuticle only faintly striated. Cephalic extremity bluntly rounded, slightly tapering anteriorly. Morphology of head region and pharynx similar to male. Excretory pore located rarely anterior, mostly at, and sometimes posterior to, nerve ring. Genital tract amphidelphic with reflexed ovaries. Vulva asymmetrical and oblique slit to symmetrical and transverse slit, slightly (3 μm) to distinctly (35 μm) protruding from body contour (depending on age).

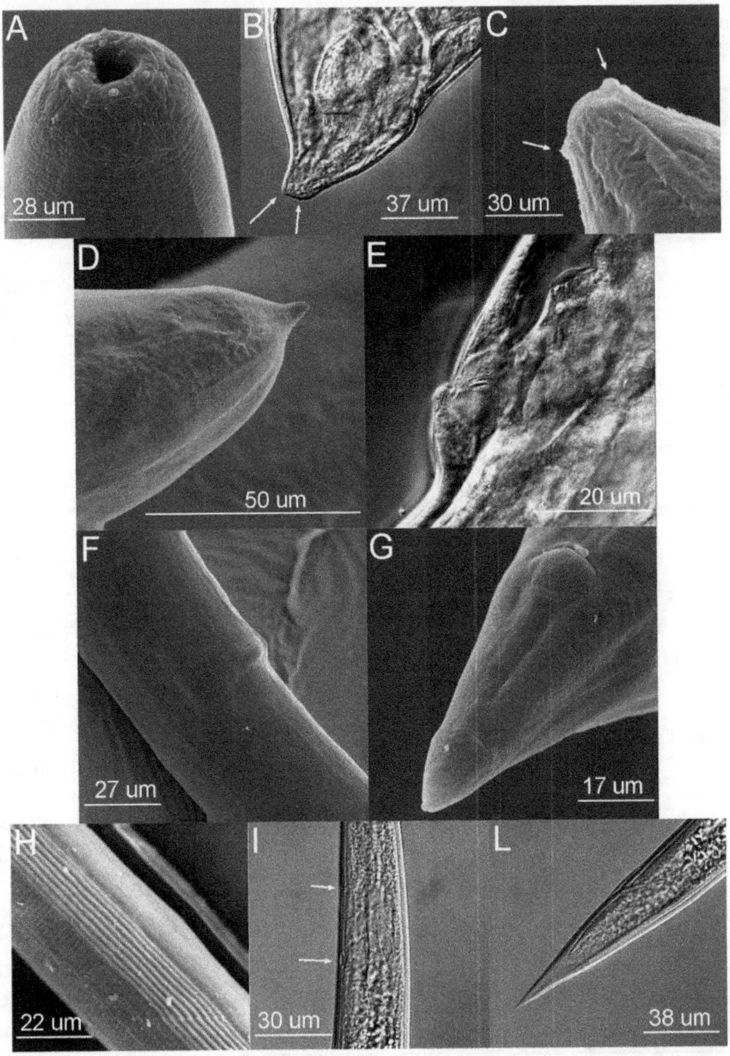

Fig. 42. Steinernema apuliae. *A-D: First generation female. A: Head with six labial and four cephalic papillae; B: Terminal papillae or protuberances; C: Terminal papillae or protuberances (arrows); D: Lateral view of tail with conoid tip and terminal papilla or protuberance. E-G: Second generation female. E: Asymmetrical vulva; F: Asymmetrical vulva; G: Ventral view of tail with an apical protuberance. H-L: Infective juvenile. H: Lateral field with eight identical ridges; I: Pharyngeal region with excretory pore (upper arrow) and hemizonid (lower arrow); L: Tail with hyaline portion. After Triggiani* et al. *(2004), courtesy of* Zootaxa.

Table 10. *Morphometrics of* Steinernema apuliae. *Measurements are in μm and in the form: mean ± SD (range).*

Character	Holotype (IJ)	First generation		Second generation		Infective juvenile
		Male	Female	Male	Female	
n	–	20	15	20	20	19
L	1192	1412 ± 114 (1272-1616)	5086 ± 1700 (2828-9474)	1135 ± 112 (929-1394)	2362 ± 433 (1818-3192)	1064 ± 66 (945-1212)
a	31	14 ± 1.3 (11-16)	19 ± 5.7 (11-33)	15 ± 1.1 (13-17)	16 ± 1.6 (12-19)	29 ± 1.4 (26-31)
b	8.2	8.1 ± 0.7 (7.1-10)	22 ± 6.8 (13-39)	6.1 ± 0.7 (4.9-7.5)	14 ± 2.1 (12-21)	7.3 ± 0.3 (6.8-7.8)
c	16	38 ± 3.8 (31-48)	82 ± 29 (52-173)	31 ± 2.7 (26-37)	46 ± 5.6 (40-64)	15 ± 0.7 (14-17)
c′	2.9	3.8 ± 0.1 (3.3-4.3)	3.0 ± 0.3 (2.4-3.6)	3.7 ± 0.3 (3.1-4.1)	1.3 ± 0.1 (1-1.5)	3.0 ± 0.2 (2.7-3.3)
V	–	–	55 ± 1.8 (50-58)	–	57±1.9 (55-61)	–
Body diam.	38	104 ± 11 (86-124)	271 ± 40 (227-383)	76 ± 4.1 (70-89)	144 ± 25 (102-196)	37 ± 1.9 (34-41)
EP	100	141 ± 8.6 (124-153)	184 ± 15 (141-199)	135 ± 2.9 (113-162)	158 ± 18 (129-187)	95±4.2 (86-102)
NR	119	–	–	–	–	111 ± 5.3 (103-114)
ES	146	175 ± 9.7 (156-188)	229 ± 11 (207-246)	184 ± 9.4 (164-199)	225 ± 24 (196-266)	145 ± 6.4 (137-159)
Tail length (T)	73	38 ± 2.3 (33-41)	62 ± 8.2 (47-74)	36 ± 2.8 (33-45)	61 ± 4.4 (51-67)	71 ± 5.4 (63-80)
Hyaline (H)	33	–	–	–	–	30 ± 2.7 (25-34)
Anal body diam. (ABD)	25	44 ± 2.4 (41-48)	69 ± 6.2 (54-78)	41 ± 3.0 (37-48)	46 ± 9.7 (35-70)	23 ± 1.4 (20-26)
Spicule length (SL)	–	72 ± 5.0 (64-80)	–	66 ± 3.1 (60-72)	–	–
Gubernaculum length (GL)	–	50 ± 2.5 (46-54)	–	47 ± 1.4 (15-49)	–	–
Gubernaculum width	–	9.7 ± 0.4 (9.5-10)	–	8.3 ± 0.8 (6.0-9.5)	–	–

Vagina short, slanted, leading into paired uteri. Rectum narrow, anal aperture distinct. Tail of mature female short and obtuse, with small postanal lobe, almost as long as broad; in young female, tail longer with conoid tip, tip more prominent in older, obtuse tailed, females. Tip bearing one large plus two or three minute protuberances or papillae protruding from body contour. Pygmy forms rarely observed.

Table 10. *(Continued).*

Character	First generation			Second generation		Infective juvenile
	Holotype IJ	Male	Female	Male	Female	
D% = EP/ES × 100	68	81 ± 0.1 (75-92)	80 ± 4.7 (68-87)	74 ± 8.6 (61-90)	70 ± 5.1 (60-80)	66 ± 2.3 (63-70)
E% = EP/T × 100	140	371*	511*	375*	259*	1.3 ± 0.1 (1.2-1.5)
SW% = SL/ABD × 100	–	160 ± 3 (150-180)	–	160 ± 1 (140-180)	–	–
GS% = GL/SL × 100	–	70 ± 1 (70-80)	–	70 ± 3 (70-80)	–	–
H% = H/T × 100	46	–	–	–	–	42 ± 2.6 (38-47)

EP = distance from anterior end to excretory pore; NR = distance from anterior end to nerve ring; ES = distance from anterior end to end of pharynx.

* Recalculated according to Triggiani *et al.* (2004).

Female (second generation)

Similar to first generation female in general morphology, but smaller in dimension. Excretory pore located more anteriorly than in first generation female. Vulva distinctly asymmetrical and in form of oblique slit in all observed specimens. Tail conical, pointed, with short protuberance or papilla-like structures on tip, but not mucronate; tail longer than body diam. at anus. Pygmy forms not observed.

Infective juvenile

Body of heat-relaxed specimens slightly C-shaped to almost straight, slender, slightly tapering towards anterior and posterior ends. Specimens from *Galleria* cultures often enclosed in second-stage cuticle that is indistinctly annulated and 1.5-2 μm thick in mid-body region. Lateral fields at mid-body region with eight equally distributed ridges (*i.e.*, nine lines); all ridges and lines distinct and spaced at approximately even distances. Phasmids weakly developed, posterior to anus, with their apertures located *ca* 40% of tail length at ventral edge of lateral fields. Cephalic extremity finely rounded, mostly slightly off-set from body contour by a depression, 11-13 μm diam. Four cephalic papillae present, amphidial apertures indistinct. Papilliform deirids situated slightly posterior to pharyngeal bulb. Stoma closed. Pharynx with slightly expanded procorpus, narrower isthmus and subpyriform basal bulb with weakly developed valve apparatus. Basal bulb usually slightly displaced towards dorsal side by excretory gland. Lentil-like hemizonid, *ca* 5 μm long, mostly varying in position from beginning of bulb to its midpoint. Excretory pore located posterior to mid-pharynx region (on average *ca* 66% of length from anterior end to base of pharynx), cuticularised, *ca* 1 μm diam., excretory duct extending posteriorly to cardia region. Intestine separated from pharynx by a long cardia, which is followed by a slender bacterial vesicle. Gut filled with numerous fat globules, lumen not visible. Rectum straight, almost as long as anal body diam. Anus 4-5 μm wide, sickle-shaped. Tail relatively broad (low ratio c′), conoid, straight, with terminus finely pointed, tip devoid of refractive inclusion. Hyaline region comprising *ca* 41-42% of tail length.

TYPE HOST AND LOCALITY

The natural host is unknown. The type population was collected by using *Galleria mellonella* larvae as bait from a soil sample picked up from a salt pan near Margherita di Savoia, in the district of Bari, Italy, at an altitude of 30 m a.s.l., in October 1996.

TYPE MATERIAL

Holotype infective juvenile, paratype infective juveniles, males and females of the first and second generations deposited in the Laboratory of Insect Pathology, Institute of Entomology, Czech Academy of Sciences, České Budějovice. Additional paratypes (infective juveniles, males and females) are at the Department of Biology and Chemistry, Section of Entomology and Zoology, University of Bari, Italy; paratype infective juveniles are in the CABI Europe-UK Nematode Collection, Egham, Surrey, UK.

RELATIONSHIPS

Steinernema apuliae is mainly characterised by the following morphological characteristics: male with light brown, slightly curved spicules with longitudinally shaped manubrium and absence of a tail mucron in both generations. Female with a characteristically asymmetrical oblique slit vulva located distinctly posterior to the mid-body. Position of excretory pore is mostly posterior in both sexes. One distinct and two or three minute protuberances or papillae are located on the first generation female tail tip. Infective juveniles have a long body, on average over

1000 μm, the hyaline region is medium long, comprising slightly over 40% of the tail length, the excretory pore is located near the posterior pharynx (58-77%), the labial region is broad, continuous, and rounded, and the lateral fields have eight equally developed and evenly spaced ridges (equivalent to nine lines).

Steinernema glaseri and *S. arenarium* are the only described *Steinernema* species found in Europe that closely resemble *S. apuliae* in morphometrics and other morphological characters. However, females of *S. apuliae* have the vulva in the form of an asymmetrical, oblique slit, whereas in *S. glaseri* and *S. arenarium* it is more symmetrical and transverse. The vulval aperture is slightly more posterior in *S. apuliae* females. Males of *S. glaseri* possess a hooked spicule tip that is never observed in *S. apuliae* or *S. arenarium*. The hyaline tail region of IJ of *S. apuliae* seems to be slightly shorter. The pair of marginal ridges in the IJ lateral field is slightly annulated in *S. glaseri* and smooth in *S. arenarium* and *S. apuliae*. Finally, PCR-RFLP analysis can be used to differentiate these species.

Many species, such as *S. cubanum*, *S. puertoricense*, *S. longicaudum*, *S. karii*, *S. loci* and *S. diaprepesi*, belong to the *glaseri*-group. There is an important zoogeographical barrier for tropical species such as *S. cubanum*, *S. puertoricense*, *S. karii* and *S. loci*. The infective juvenile value D% is very low for *S. karii* (56.6%) and *S. loci* (57%), whereas it is 66% for *S. apuliae*. The ratio c′ for *S. longicaudum* IJ is very high (4.22) due to the tail being over 90 μm long, as it is for *S. puertoricense*. *Steinernema diaprepesi* has a shorter IJ of *ca* 1000 μm, whilst D% is close to 50.

The morphology of males is similar in these species, although females of *S. apuliae* possess a strongly asymmetrical vulva that differs from that of *S. puertoricense*, which has double flapped epiptygmata. *Steinernema cubanum* is morphologically and morphometrically very similar to *S. apuliae*, although the vulva is positioned at *ca* 50% of the body length whereas in *S. apuliae* it is significantly posterior to the mid-body, varying mostly between 55-57% of the body length.

MOLECULAR CHARACTERISATION

The RFLP profiles yielded by digestion of the ITS region with 17 restriction enzymes for *S. apuliae* were compared to a database containing over 50 different *Steinernema* isolates (including 22 described species).

Examination of the RFLP profiles showed *S. apuliae* to be unique. *Steinernema apuliae* can be differentiated from *S. arenarium* by 9/17 restriction enzymes (*Alu*I, *Dde*I, *Hae*III, *Hha*I, *Hind*III, *Hinf*I, *Pvu*II, *Rsa*I and *Sal*I) and from *S. glaseri* by 11/17 enzymes (*Alu*I, *Dde*I, *Hae*III, *Hha*I, *Hind*III, *Hinf*I, *Hpa*II, *Pvu*II, *Rsa*I, *Sal*I and *Sau*3AI).

BIONOMICS

Steinernema apuliae is easily reared on *G. mellonella* larvae in the laboratory at 20-22°C. *Galleria* larvae were killed 2 days after exposure to the IJ; first generation adults appearing 3 days after host death and new IJ starting to emerge after 1 week.

DISTRIBUTION

Steinernema apuliae has been found exclusively in southern Italy in the Apulia and Basilicata regions (Triggiani & Tarasco, 2002). Isolates were mostly obtained from sandy soil with one exception from silty soil. They were localised to coastal regions with scant vegetation and pine forests and where *Thaumetopoea pityocampa* (Lepidoptera, Thaumetopoeidae) was common.

References

TRIGGIANI, O. & TARASCO, E. (2002). Long nematode strains from Italy. In: *COST 850 Biocontrol Symbiosis – WG 'Biotechnology' workshop and meeting, molecular identification and taxonomy of undescribed species of long-dauer juvenile phenotype from Europe and southern Mediterranean regions). Budapest and Debrecen, Hungary, 4-6 April, 2002.*

TRIGGIANI, O., MRÁČEK, Z. & REID, A. (2004). *Steinernema apuliae* sp. n. (Rhabditida: Steinernematidae), a new entomopathogenic nematode from southern Italy. *Zootaxa* 460, 1-12.

*Steinernema arenarium**,** (Artyukhovsky, 1967) Wouts, Mráček, Gerdin & Bedding, 1982

= *Neoaplectana arenaria* Artyukhovsky, 1967
= *Neoaplectana anomali* Kozodoi, 1984
= *Steinernema anomalae* (Kozodoi, 1984) Curran, 1989
(Figs 43-47)

MEASUREMENTS

See Table 11.

DESCRIPTION

Male (first generation)

Body yellowish, curved ventrally when heat-relaxed; tail without mucron. Lateral fields and phasmids not observed. Labial region truncate with or without perioral disc; six labial and four prominent cephalic papillae; amphidial apertures small. Stoma bowl-shaped. Pharynx muscular, metacorpus and basal bulb enlarged. Slightly cuticularised valves discernible in bulb. Cardia 5-7 μm long. Excretory pore 2 μm diam. Deirids 2-3 μm high, 4-6 μm diam., situated 80-90 μm posterior to basal bulb. Spicule head short, comprising *ca* 22% of spicule length, slightly variable in shape; shaft reduced; blade robust, more than 3.5 times as long as head, curvature variable; velum absent. Spicule reddish brown in colour, tip swollen with a prominent dorsal aperture, making tip appear constricted. Gubernaculum much shorter than spicule, anterior part curved ventrally, posterior part forked; corpus broad with two wings. Cuneus short, V-shaped. Complement of genital papillae ranging from 11-14 pairs plus a single midventral precloacal papilla. When 11 pairs of papillae present, distributed as five pairs precloacal and subventral, two pairs adcloacal, one pair lateral and three caudal pairs. When 14 pairs of papillae present there are eight precloacal pairs.

Male (second generation)

Data not available.

* From Latin *arenarius* = pertaining to sand.

** Description after Artyukhovsky *et al.* (1997). Revised by Khuong B. Nguyen.

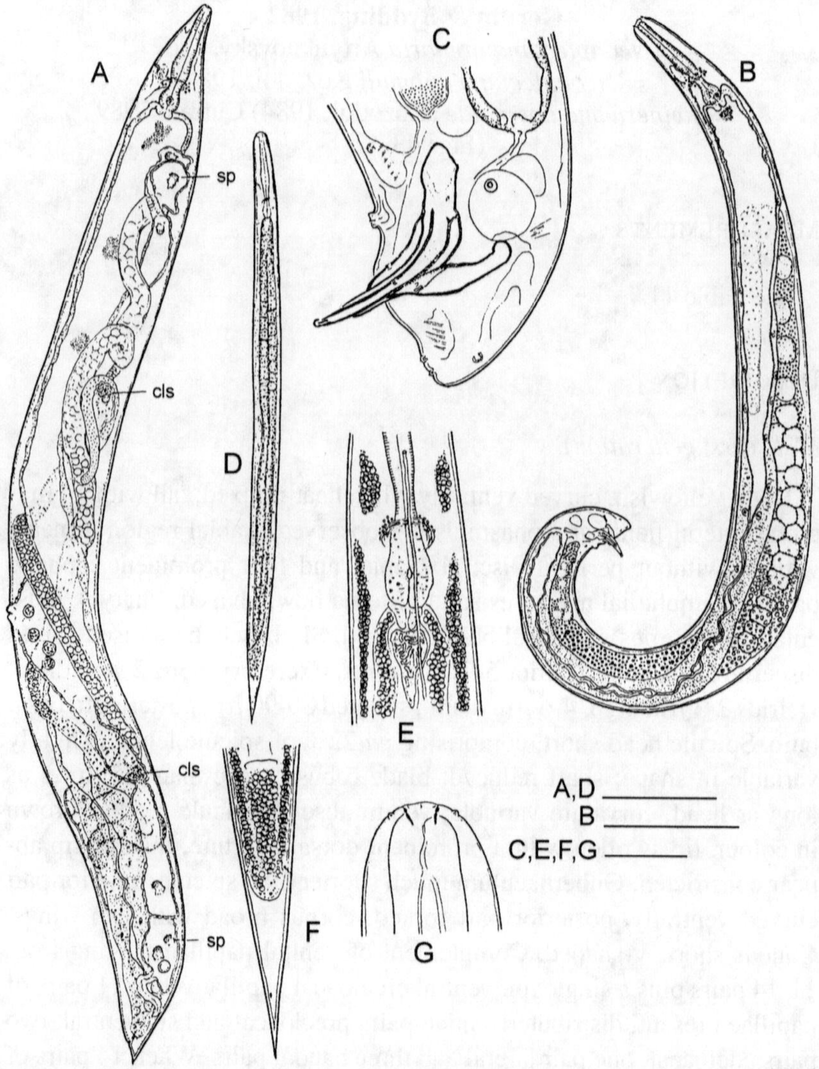

Fig. 43. Steinernema arenarium. *A: Female, sp = sperm, cls = clusters of immature spermatozoa; B: Male; C: Posterior region of male; D: Entire infective juvenile; E: Pharyngeal region showing basal bulb and bacterial vesicle; F: Infective juvenile tail with hyaline portion; G: Male stoma. (Scale bars: A, D = 250 μm; B = 200 μm; C, E, F, G = 25 μm.) After Artyukhovsky* et al. *(1997), courtesy of* Russian Journal of Nematology.

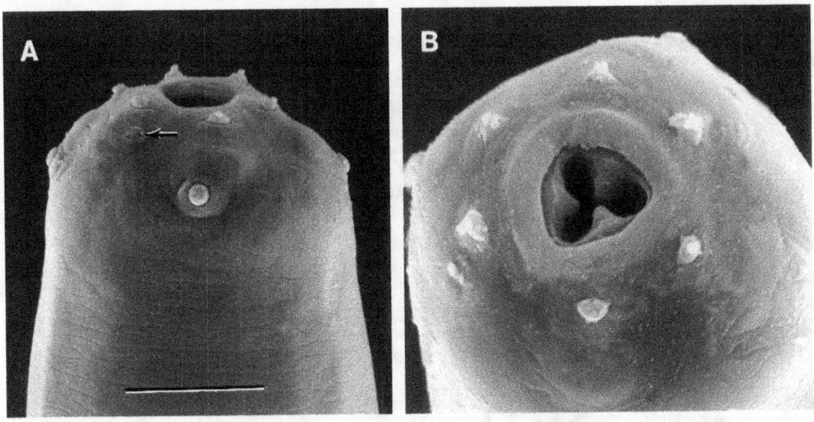

Fig. 44. Steinernema arenarium *male head. A: Sublateral view showing six labial papillae, three of the four cephalic papillae, and one of the two amphidial apertures (arrow); B:* En face *view showing perioral disc and triradiate mouth. (Scale bar: A = 8.8 μm; B = 6 μm.) After Nguyen and Smart (1993), courtesy of* Journal of Nematology.

Female (first generation)

Body with yellowish intestine and colourless genital system. Lateral fields and phasmids not observed. Labial region broadly rounded with six pointed labial papillae and four rounded cephalic papillae. Amphidial apertures present, but small. Stoma with cuticularised ring (cheilorhabdions), 1.5 μm thick, posterior part similar to that of other steinernematids. Pharynx with prominent metacorpus. Valves in basal bulb present. Excretory duct 3 μm diam. anteriorly. Vulva with thickened posterior lip. In some specimens anterior branch of genital system apparently less developed than posterior branch and with few eggs; posterior genital system having more loops and passing through all sectors of the body. Uterus bipartite; part near vulva with walls consisting of swollen cells with 2-3 nuclei, and part near ovary with thinner walls. In young females these parts of the uterus are divided by a constriction. Tail with rounded terminus, *ca* 25% of population having 2-3 μm long spike; two such spikes being observed in one female.

Female (second generation)

Data not available.

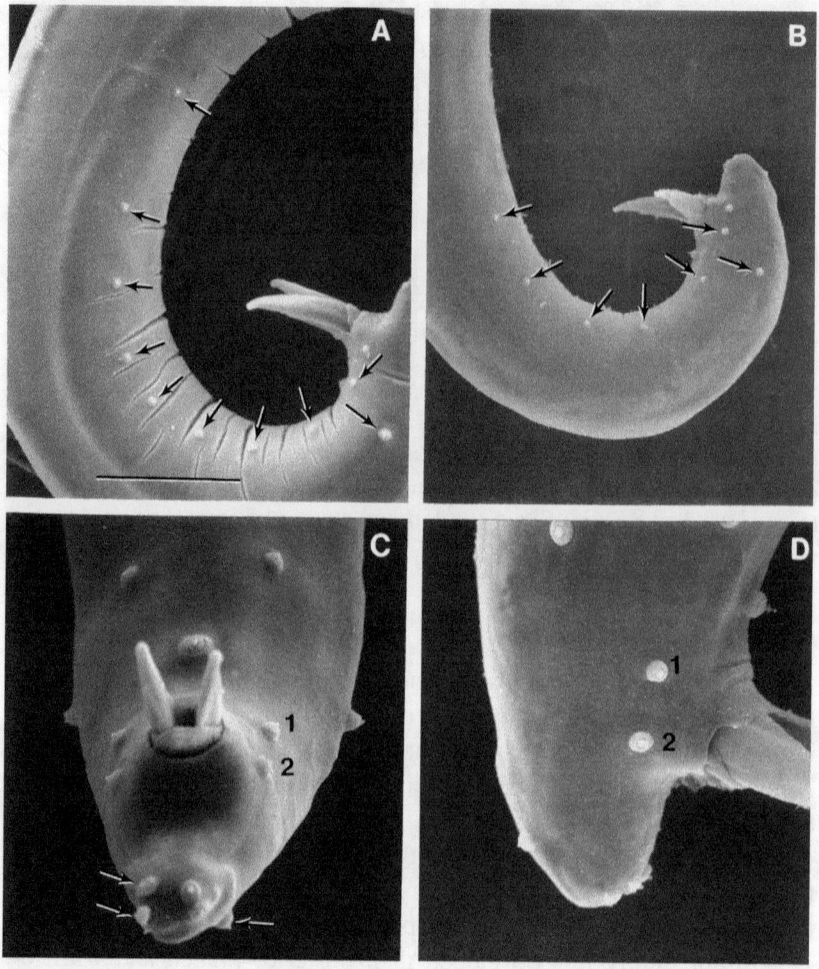

Fig. 45. Steinernema arenarium *male posterior region. A: Lateral view showing nine preanal subventral pairs of papillae, one lateral pair of papillae and single midventral papilla (arrows); B: Lateral view showing seven pairs of preanal papillae (arrows), single midventral preanal papilla, tail lacking mucron, and protruding spicules; C: Ventral view showing seven pairs of papillae (1, 2 = adanal, right arrow = caudal, subdorsal, left arrows = caudal, subterminal) and one single preanal papillae, the two spicule tips and the distal end of gubernaculum; D: Enlargement of end portion of B. (Scale bar: A = 50 μm; B = 60 μm; C, D = 20 μm.) After Nguyen and Smart (1993), courtesy of* Journal of Nematology.

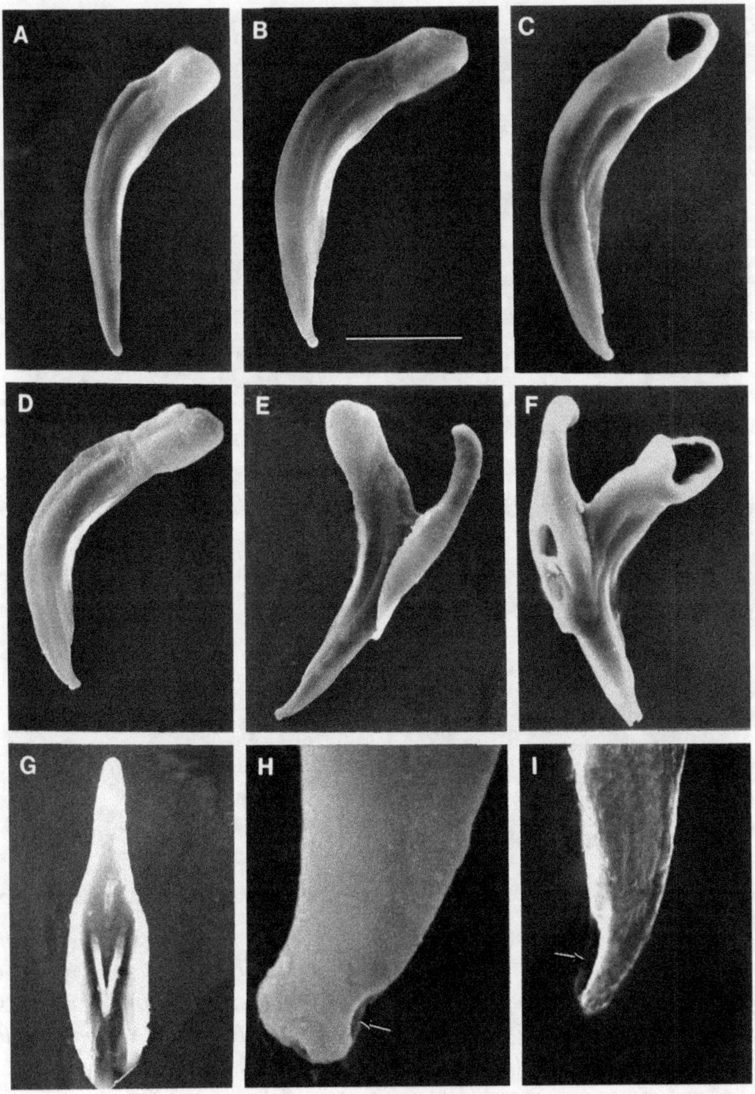

Fig. 46. Steinernema arenarium *spicule and gubernaculum. A-D: Spicule with short spicule head, reduced shaft, and enlarged tip; E, F: Spicules and gubernacula, with gubernaculum in lateral and subdorsal views, respectively; G: Gubernaculum, ventral view. Note V-shaped cuneus. H: Spicule tip with a dorsal aperture (arrow); I: Spicule tip of* S. glaseri *with ventral aperture (arrow), which is longer than that of* S. arenarium. *(Scale bar: A = 30 μm; B, D = 23 μm; C, F = 25 μm; E = 27 μm; G = 20 μm; H = 2.5 μm; I = 7 μm.) After Nguyen and Smart (1993), courtesy of* Journal of Nematology.

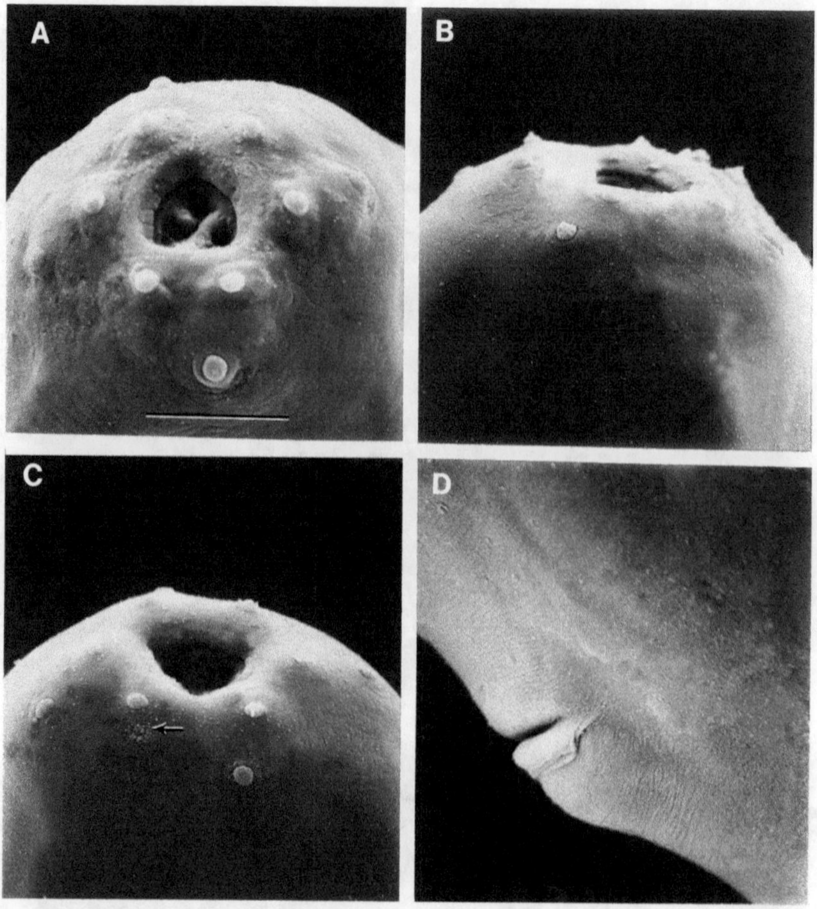

Fig. 47. Steinernema arenarium *female. A: Face view showing six labial papillae and four cephalic papillae; B: Anterior end, sublateral view; C: Sublateral view showing one of the two small amphidial apertures (arrow), six labial papillae and three of four cephalic papillae; D: Vulval region showing a thickened posterior lip. (Scale bar: A = 10 μm; B, D = 12 μm; C = 13.6 μm.) After Nguyen and Smart (1993), courtesy of* Journal of Nematology.

Infective juvenile

Body long and relatively robust. Lateral field with eight ridges (*i.e.*, nine lines). Labial region rounded, occasionally with slightly offset cephalic capsule, 12-14 μm diam. Cephalic extremity mostly smooth, four prominent cephalic papillae; amphidial apertures pronounced.

Table 11. *Morphometrics of* Steinernema arenarium. *Measurements are in μm and in the form: mean ± SD (range).*

Character	First generation		Infective juvenile
	Male	Female	
n	50	50	50
L	1845 ± 236 (1280-2320)	5807 ± 1682 (2600-9720)	1217 ± 105 (930-1580)
a	–	–	33.3 ± 3.5 (25-40)
b	–	–	7.8 ± 0.6 (5.9-9.2)
c	–	–	15.1 ± 1.2 (12.2-17.9)
V	–	50 ± 2.2 (48-60)	–
Body diam.	117 ± 23 (70-165)	279 ± 52 (170-400)	37 ± 2.8 (31-44)
ES	168 ± 15 (125-209)	228 ± 23 (180-280)	156 ± 9.3 (132-187)
Tail length (T)	32 ± 6.6 (14-40)	68 ± 14.3 (40-105)	81 ± 4.5 (65-95)
Spicule length (SL)	76 ± 6.6 (63-93)	–	–
Gubernaculum length (GL)	53 ± 4.1 (45-63)	–	–
D% = EP/ES × 100	78 ± 10 (52-96)	–	–
H% = Hyaline/T × 100	–	–	41 ± 5 (23-57)

EP = distance from anterior end to excretory pore; NR = distance from anterior end to nerve ring; ES = distance from anterior end to end of pharynx. After Artyukhovsky *et al.* (1997).

Labial papillae absent (see Nguyen & Smart, 1993). Tail conical with spherical or irregular border between hypodermal and hyaline region.

TYPE HOST AND LOCALITY

The species was isolated from larvae and pupae of the May beetle, *Melolontha hippocastani* (Coleoptera), in the Usmanski forest, Voronezh Region, Russia.

TYPE MATERIAL

Neotype male deposited in the collection of the Zoological Museum of Moscow State University (slide number Ic-396).

RELATIONSHIPS

Infective juvenile rather long measuring 1217 (930-1580) μm in length and with eight ridges (*i.e.*, nine lines or incisures) in lateral field. Structures resembling bacterial vesicles were seen posterior to the cardia in some infective juveniles, each vesicle appearing as a transparent sac-like body with numerous internal folds. No separate bacterial cells could be distinguished inside the vesicle. In the male the spicule is thick with a ball-shaped tip and the cuneus of the gubernaculum is V-shaped. In the postovarial part of the uterus, large motile spermatozoa can be seen. Immobile cells with a coarse surface (probably spermatocytes) can be seen close to the constriction in both anterior and posterior branches of the young female genital system.

This species belongs to the *glaseri*-group and is similar to *S. glaseri* but can be distinguished from this nematode by its swollen spicule tip, compared to the notched tip in *S. glaseri* and by the V-shaped cuneus *vs* Y-shaped in *S. glaseri*.

MOLECULAR CHARACTERISATION

Steinernema arenarium is characterised by the sequence length of ITS regions, 996 bp (ITS1 = 287 bp, ITS2 = 301 bp), composition of ITS regions of rDNA, A = 0.24498, C = 0.20482, G = 0.26406, T = 0.28614 (accession no. DQ314288), and D2D3 region length, 876 bp and composition, A = 0.23973, C = 0.20890, G = 0.31621, T = 0.23516 (accession no. AF331892).

DISTRIBUTION

Apart from the type locality in Voronezh region, Russia, this species (as *S. anomalae*) has also been recorded from the Rjazan region, Russia (Kozodoi, 1984) and from Italy and Spain (see Hominick, 2002).

References

ARTYUKHOVSKY, A.K. (1967). [*Neoaplectana arenaria* nov. sp. (Steinernematidae, Nematoda) inducing nematode disease in chafers of the Voronezh region.] *Trudy Voroneszhskogo Gosudarstvennogo Zapovednika* 15, 94-100.

ARTYUKHOVSKY, A.K., KOZODOI, E.M., REID, A.P. & SPIRIDONOV, S.E. (1997). Redescription of *Steinernema arenarium* (Artyukhovsky, 1967) topotype from Central Russia and a proposal for *S. anomalae* (Kozodoi, 1984) as a junior synonym. *Russian Journal of Nematology* 5, 31-37.

CURRAN, J. (1989). Chromosome numbers of *Steinernema* and *Heterorhabditis* species. *Revue de Nématologie* 12, 145-148.

HOMINICK, W.M. (2002). Biogeography. In: Gaugler, R. (Ed.). *Entomopathogenic nematodes*. Wallingford, UK, CABI Publishing, pp. 115-143.

KOZODOI, E.M. (1984). A new entomopathogenic nematode *Neoaplectana anomali* sp. n. (Rhabditida, Steinernematidae) and observations on its biology. *Zoologichesky Zhurnal* 63, 1605-1612.

NGUYEN, K.B. & SMART JR, G.C. (1993). Scanning electron microscope studies of *Steinernema anomali* Kozodoi, 1984. *Journal of Nematology* 25, 486-492.

WOUTS, W.M., MRÁČEK, Z., GERDIN, S. & BEDDING, R.A. (1982). *Neoaplectana* Steiner, 1929 a junior synonym of *Steinernema* Travassos, 1927 (Nematoda: Rhabditida). *Systematic Parasitology* 4, 147-154.

Steinernema ashiuense[*,**] Phan, Takemoto & Futai, 2006
(Figs 48, 49)

MEASUREMENTS

See Table 12.

DESCRIPTION

Male (first generation)

Body curved ventrally on heat relaxation; habitus C- or J-shaped. Cuticle appearing smooth under LM, lateral field not observed. Cephalic extremity rounded, slightly offset from body. Six pointed labial papillae and four cephalic papillae visible under SEM. Amphidial apertures inconspicuous. Buccal cavity funnel- or cup-shaped. Stoma shallow. Pharynx muscular; procorpus cylindrical; metacorpus slightly swollen, non-valvate; isthmus distinct; basal bulb pyriform, valve distinct. Nerve ring located just anterior to basal bulb. Cardia prominent and protruding into intestinal lumen. Excretory pore located at middle of pharynx. Excretory duct cuticularised; excretory gland swollen and elongated. Genital system monorchic, reflexed. Spicules paired, slightly yellowish in colour, strongly curved; length/width ratio = 4 (3.5-4.4); head (manubrium) as long as wide. Blade arcuate with straight tip and three well-defined lobes; dorsal lobe well curved, terminating at spicule tip; lateral lobe prominent, usually greater in width anteriorly, terminating at spicule tip; ventral lobe enlarged anteriorly at dorsal and ventral side to form prominent apex and rostrum, terminating posterior to rounded spicule tip. Velum large, not covering spicule tip. Gubernaculum *ca* 40-70% of spicule length, boat-shaped in lateral view, swollen at middle, with or without slightly ventrally curved knob at proximal end; in ventral view, cuneus long, needle-shaped, pointed posteriorly. Wing of corpus expanding laterally. Genital papillae disposed as single midventral precloacal papilla and ten pairs; latter arranged as follows: five pairs subventral precloacal, one pair lateral precloacal; one pair adcloacal; two pairs subventral postcloacal and one pair subdorsal postcloacal. Tail conoid with small mucron usually present. Phasmids inconspicuous.

---

[*] Specific epithet derived from *Ashui*, the name of the type locality.

[**] Description after Phan *et al.* (2006). Revised by David J. Hunt.

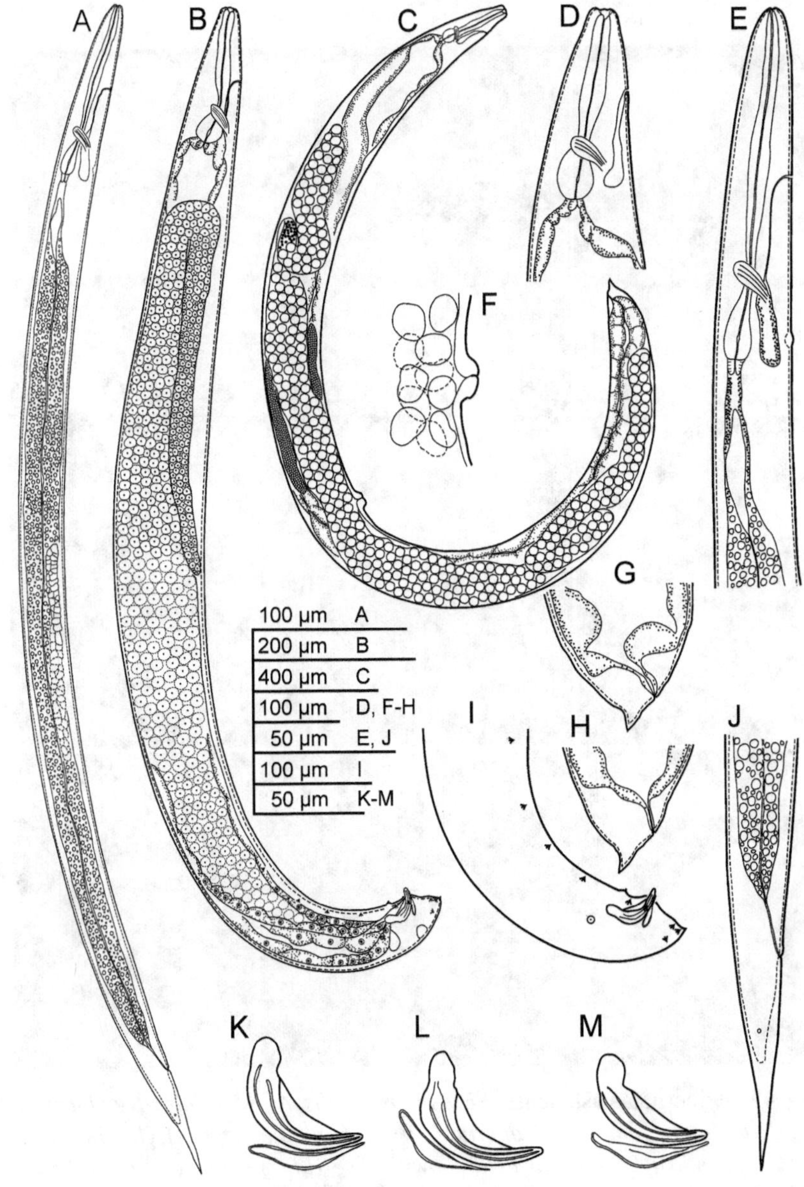

Fig. 48. Steinernema ashiuense. *A, E, J: Infective juvenile. A: Entire view; E: Pharynx region; J: Tail region. B, I, K-M: Male, first generation. B: Entire view; I: Tail region; K-M: Spicule and gubernaculum. C, D, F, G, H: Female, first generation. C: Entire view; D: Pharynx region; F: Vulva region; G, H: Tail region. After Phan* et al. *(2006), courtesy of* Nematology.

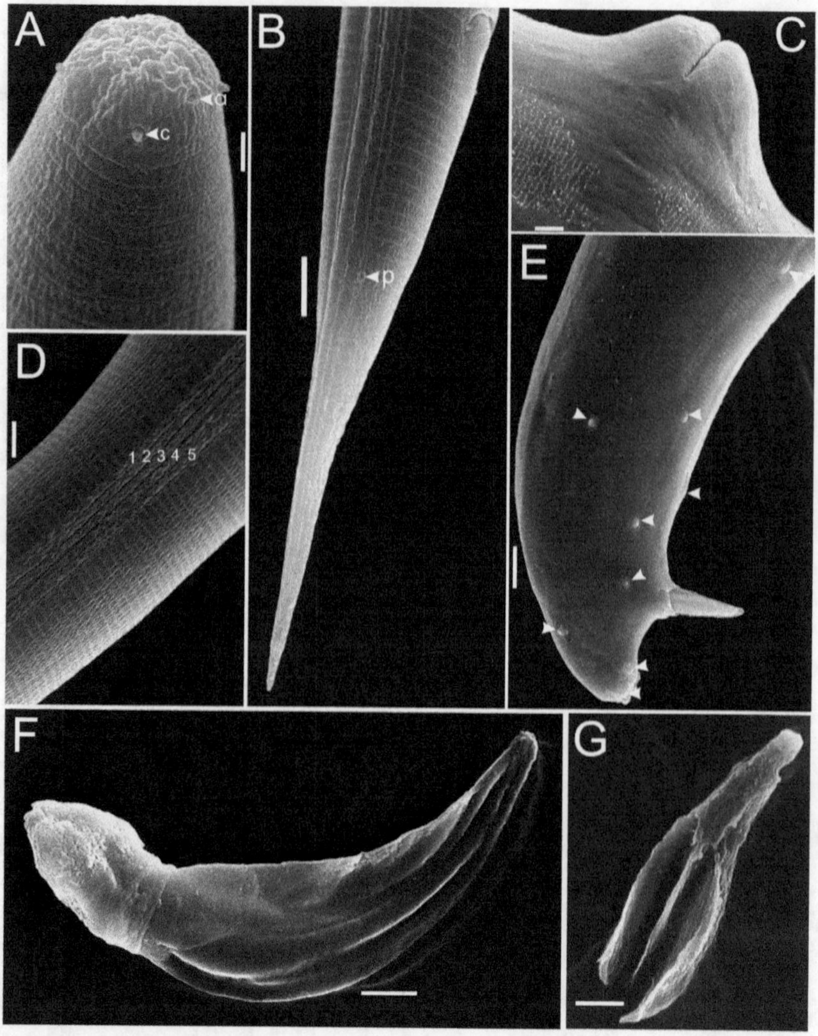

Fig. 49. Steinernema ashiuense. *SEM micrographs. A, B, D: Infective juveniles. A: Head (a = amphidial aperture, c = cephalic papilla); B: Tail view (p = phasmid); D: Lateral field showing five ridges (numbers). Female. C: Vulva in lateral view. E-G: Male. E: Tail in lateral view (arrows indicate genital papillae); F: Spicule; G: Gubernaculum. (Scale bars: A = 1.42 μm; B, F, G = 5 μm; C = 6.66 μm; D = 3.33 μm; E-G = 10 μm.) After Phan* et al. *(2006), courtesy of* Nematology.

Male (second generation)

Similar to first generation but slender, body diam. *ca* half that of first generation. Testis flexure extending more posteriorly. Mucron more distinct, *ca* 2.5-7.5 μm long.

Female (first generation)

Body robust, habitus C-shaped. Cuticle appearing smooth under LM, lateral field not observed. Cephalic extremity broadly rounded. Six pointed labial papillae and four cephalic papillae visible with SEM. Amphidial apertures inconspicuous. Buccal cavity funnel- or cup-shaped. Stoma shallow. Pharynx with cylindrical procorpus; metacorpus slightly swollen and non-valvate; isthmus distinct; basal bulb pyriform, valve visible. Nerve ring located just anterior to basal bulb. Excretory pore usually located at mid-pharynx level. Excretory duct cuticularised and excretory gland swollen. Cardia prominent, protruding into intestinal lumen. Genital system didelphic, amphidelphic, reflexed; filled with eggs. Vulva in form of a median transverse slit, protruding from body, epiptygmata absent. Vagina short, oblique with muscular walls. Tail dome-shaped, shorter than anal body diam., terminal peg present.

Female (second generation)

Similar to first generation but smaller, body length *ca* half that of first generation and one-third of body diam. Eggs arranged in single row. Tail conoid without mucron. Postanal region slightly swollen.

Infective juvenile

Body slender, habitus moderately ventrally curved when heat-relaxed, often enclosed in cuticle of second-stage, tapering regularly from base of pharynx to anterior end and from anus to terminus. Oral aperture and anus closed. Cephalic extremity broad, offset from body; labial papillae not observed; transverse slit-like amphidial apertures situated posterior to labial disc but anterior to cephalic papillae. Pharynx long, narrow; isthmus distinct, surrounded by nerve ring; basal bulb elongate with valve. Cardia prominent. Excretory pore located at mid-pharynx level. Hemizonid distinct, located anterior to base of basal bulb. Bacterial vesicle elongate (21-34 μm long $\times$ 7-10 μm diam.). Genital primordium clear, *ca* 120-160 μm long. Lateral field at mid-body with five equally-developed ridges (*i.e.*, six lines or incisures). Tail conical, straight, often bent towards dorsal side; terminus finely rounded. Hyaline region

Table 12. *Morphometrics of type population of* Steinernema ashiuense. *Measurements are in μm and in the form: mean ± SD (range).*

Character	First generation			Second generation		Infective juvenile
	Holotype (male)	Male	Female	Male	Female	
n	–	20	20	20	20	20
L	1360	1279 ± 99 (1050-1395)	3387 ± 237 (3020-3830)	1037 ± 81 (900-1468)	1554 ± 118 (1350-1760)	768 ± 23 (720-800)
a	12	12 ± 1 (11-14)	16 ± 2 (13-19)	18 ± 1.2 (16-20)	19 ± 2 (16-22)	25 ± 1 (24-27)
b	8	8 ± 1 (7-10)	15 ± 1 (13-18)	7 ± 1 (6-8)	9 ± 1 (8-11)	6 ± 1 (6-7)
c	60	53 ± 4 (45-61)	85 ± 9 (64-105)	47 ± 7 (34-64)	28 ± 3 (24-36)	11 ± 1 (10-12)
c′	–	–	–	–	–	
V	–	–	51 ± 2 (47-54)	–	52 ± 4 (45-60)	–
Body diam.	115	106 ± 14 (80-125)	221 ± 20 (180-260)	58 ± 6 (50-75)	82 ± 12 (70-100)	30 ± 1 (28-33)
EP	88	78 ± 5 (70-88)	75 ± 13 (50-93)	73 ± 5 (63-80)	76 ± 7 (65-100)	55 ± 2 (51-59)
NR	123	118 ± 6 (108-128)	149 ± 11 (133-168)	108 ± 8 (98-123)	119 ± 7 (108-130)	86 ± 3 (77-91)
ES	163	157 ± 8 (143-170)	221 ± 11 (200-238)	150 ± 9 (138-165)	166 ± 9 (148-185)	119 ± 3 (113-128)
Testis reflexion	425	359 ± 39 (290-425)	–	198 ± 34 (110-248)	–	–
Tail length (T)	23	24 ± 2 (20-28)	40 ± 5 (30-50)	22 ± 3 (15-26)	56 ± 5 (45-65)	71 ± 3 (66-76)
Anal body diam. (ABD)	40	40 ± 3 (35-45)	63 ± 10 (45-88)	30 ± 2 (28-33)	–	16 ± 1 (15-18)
Spicule length (SL)	58	59 ± 4 (50-65)	–	50 ± 2 (45-55)	–	–
Spicule width	15	4.0 ± 0.2 (3.5-4.4)	–	11 ± 1 (10-13)	–	–
Gubernaculum length (GL)	25	37 ± 4 (25-43)	–	29 ± 3 (23-35)	–	–
Gubernaculum width	6	7 ± 1 (5-8)	–	5 ± 1 (4-8)	–	–

distinct, comprising *ca* 36% of tail length. Phasmids distinct, located in anterior half of tail.

TYPE HOST AND LOCALITY

The type host of this species in nature is unknown as the nematode was collected from soil by baiting with *Galleria mellonella*. The soil

Table 12. *(Continued).*

Character	First generation			Second generation		Infective juvenile
	Holotype (male)	Male	Female	Male	Female	
D% = EP/ES × 100	54	50 ± 4 (44-56)	34 ± 6 (23-46)	49 ± 3 (44-55)	46 ± 3 (41-54)	46 ± 2 (43-50)
E% = EP/T × 100	–	–	–	–	–	78 ± 4 (70-85)
SW% = SL/ABD × 100	144	149 ± 10 (128-167)	–	164 ± 2 (138-191)	–	–
GS% = GL/SL × 100	43	63 ± 6 (43-73)	–	59 ± 5 (47-68)	–	–
H% = H/T × 100	–	–	–	–	–	36 ± 3 (30-42)

EP = distance from anterior end to excretory pore; NR = distance from anterior end to nerve ring; ES = distance from anterior end to end of pharynx.

sample was collected from a grassland habitat along the bank of the Yura river, near the Office of Ashiu Forest Research Station of Kyoto University, Myama town, Kyoto prefecture, Japan (longitude 135°43′E, latitude 35°18′N, altitude 356 m a.s.l., annual average of temperature 11.7°C, precipitation 2353 mm/year, depth of snow during winter *ca* 1 m, flora transition zone from cool-temperate to warm-temperate). Soil type is dark grey-brown sandy clay silt.

TYPE MATERIAL

Holotype male and paratype male, female and infective juveniles deposited in the nematode collection at the Herbarium and Insect Museum of the National Institute for Agro-Environmental Sciences, Tsukuba, Ibaraki, Japan. Additional paratype slides at the Laboratory of Environmental Mycoscience, Graduate School of Agriculture, Kyoto University, Japan and Institute of Ecology and Biological Resources, Vietnamese Academy of Sciences and Technology, Vietnam.

RELATIONSHIPS

Steinernema ashiuense is characterised by an IJ body length of 768 (720-800) μm, lateral field with five, equally-developed, ridges (six lines or incisures), head smooth, lacking horn-like structures, excretory pore located at mid-pharynx level, hyaline region short, comprising *ca* one-third of tail length. Males have broad, slightly yellowish, spicules

50-65 μm long; spicule length/spicule width = 4 (3.5-4.4); and two subventral and one subdorsal pairs of postcloacal genital papillae.

The lateral field of *S. ashiuense* has five, equally-developed, ridges, a feature that distinguishes it from all species in the *feltiae*-group: *S. kraussei* and *S. oregonense* have five, equally-developed, ridges plus two weak, submarginal, ones (total of seven ridges); *S. feltiae*, *S. robustispiculum* and *S. sangi* have eight ridges with the two submarginal and two central ridges less distinct; *S. akhursti* and *S. jollieti* have six, equally-developed, ridges; and *S. litorale*, *S. silvaticum* and *S. weiseri* have eight, equally-developed, ridges.

Steinernema ashiuense differs from *S. monticolum* by the location of the excretory pore opening in IJ and male (located at mid-pharynx level *vs* at posterior one-third of pharynx), and some characters of the male such as a shorter spicule length of 59 (50-65) *vs* 70 (61-80) μm, a relatively broader spicule with spicule length/spicule width = 4 (3.5-4.4) *vs* 8.7 (8-8.8) and spicule head (manubrium) elongate *vs* round.

Steinernema ashiuense has broad spicules as found in *S. kushidai*, a species that is also found in Japan, but can be separated from the latter by longer IJ body length of 768 (720-800) *vs* 589 (524-662) μm, more posterior excretory pore of 55 (51-59) *vs* 46 (42-50) μm, longer IJ tail length of 71 (66-76) *vs* 50 (44-59) μm, lateral lobe of spicule terminating at spicule tip *vs* not terminating at spicule tip, and cuneus of the gubernacula long, needle-shaped and pointed posteriorly *vs* not well seen (see Adams & Nguyen, 2002).

MOLECULAR CHARACTERISATION

The ITS1 + 5.8S + ITS2 sequence length of *S. ashiuense* is 662 bp which is the shortest amongst 11 species in the *feltiae*-group (Phan *et al.*, 2006). Sequence divergence ranged from 3.0-20%. The sequence of *S. ashiuense* differed from those of other *Steinernema* species in 20-135 substitutions. Maximum parsimony analysis of the alignment of *Steinernema* species in this group revealed 156 parsimony informative characters. In the single unrooted maximum parsimony tree, *S. ashiuense* clustered with *S. robustispiculum* + *S. monticolum* with high bootstrap support.

Maximum parsimony analysis of the alignment of this group with an out group taxon revealed 196 parsimony informative characters. The single MP tree (tree length = 573) and ML tree (DNA model = GTR +

G, – ln likelihood = 3655.07) had a similar topology. *Steinernema ashiuense* clustered with *S. robustispiculum* + *S. monticolum* with high to moderate bootstrap support.

The alignment of the ITS1 + 5.8S + ITS2 of *S. ashiuense* with two closely related species (*S. robustispiculum* and *S. monticolum*) show that the former differs from these species by nine unique substitutions in the ITS1 and eight unique substitutions in the ITS2.

The ITS sequence is deposited in GenBank under accession no. DQ354694.

BIONOMICS

Under laboratory conditions, this nematode infected both *Arge similis* (Argeridae, Hymenoptera) and *Atractomorpha latta* (Pyrgomorphidae, Orthoptera), although neither appeared to be good hosts.

DISTRIBUTION

Known only from the bank of the Yura River near the Ashui Forest Research Station, Kyoto prefecture, Japan.

References

ADAMS, B.J. & NGUYEN, K.B. (2002). Taxonomy and systematics. In: Gaugler, R. (Ed.). *Entomopathogenic nematology*. Wallingford, UK, CABI Publishing, pp. 1-33.

PHAN, L.K., TAKEMOTO, S. & FUTAI, K. (2006). *Steinernema ashiuense* sp. n. (Nematoda: Steinernematidae), a new entomopathogenic nematode from Japan. *Nematology* 8, 681-690.

Steinernema asiaticum[*,**] Anis, Shahina, Reid & Rowe, 2002
(Figs 50, 51)

MEASUREMENTS

See Table 13.

DESCRIPTION

Male (first generation)

Body ventrally curved, J-shaped on heat-relaxation. Cuticle with very faint transverse striae. Lateral field 3.2-4.0 μm wide, bearing four incisures (*i.e.*, three ridges) which are more prominent in posterior part of body. Cephalic extremity truncate to slightly rounded, continuous with body. Labial region bearing six labial papillae, two amphidial apertures and four cephalic papillae. Stoma shallow, triangular at base, cheilorhabdions located posterior to lips, forming a thick, cuticularised ring. Excretory pore located anterior to nerve ring at level of metacorpus. Pharynx muscular, more or less cylindrical, short and broad in relation to body length. Procorpus slightly expanded, extending into a non-valvate metacorpus, slightly narrowing at isthmus level and terminating in rounded basal bulb. Basal bulb with valve plates and three pharyngeal glands. Pharyngo-intestinal valve rounded. Genital system monorchic, testis reflexed, consisting of germinal zone leading to seminal vesicle; distance from base of pharynx to anterior end of testis always greater than length of pharynx. Spicules paired, uniformly curved, symmetrical, golden to dark yellow in colour. Head of spicules large, elongate, rounded. Shaft short/absent, blade thick, tapering gradually to terminus, *ca* four times longer than head. Velum present. Each spicule with two internal ribs of varying length; dorsal rib curved ventrally, ventral rib extended to posterior tip. Posterior end of spicule blunt. Shape of spicule and width of head variable. Gubernaculum boat-shaped in lateral view, ventrally curved, anterior part curved anteriorly, enlarging abruptly and bifurcated to form broad corpus with wings extending posteriorly. Proximal end with or without knob/hook, cuneus needle-shaped, pointed posteriorly. Twenty-three genital papillae comprising 11 pairs and a single, precloacal, midventral papilla. Papillae arranged

[*] Named after *Asia*, the continent where it was found.

[**] Description after Anis *et al.* (2002). Revised by David J. Hunt.

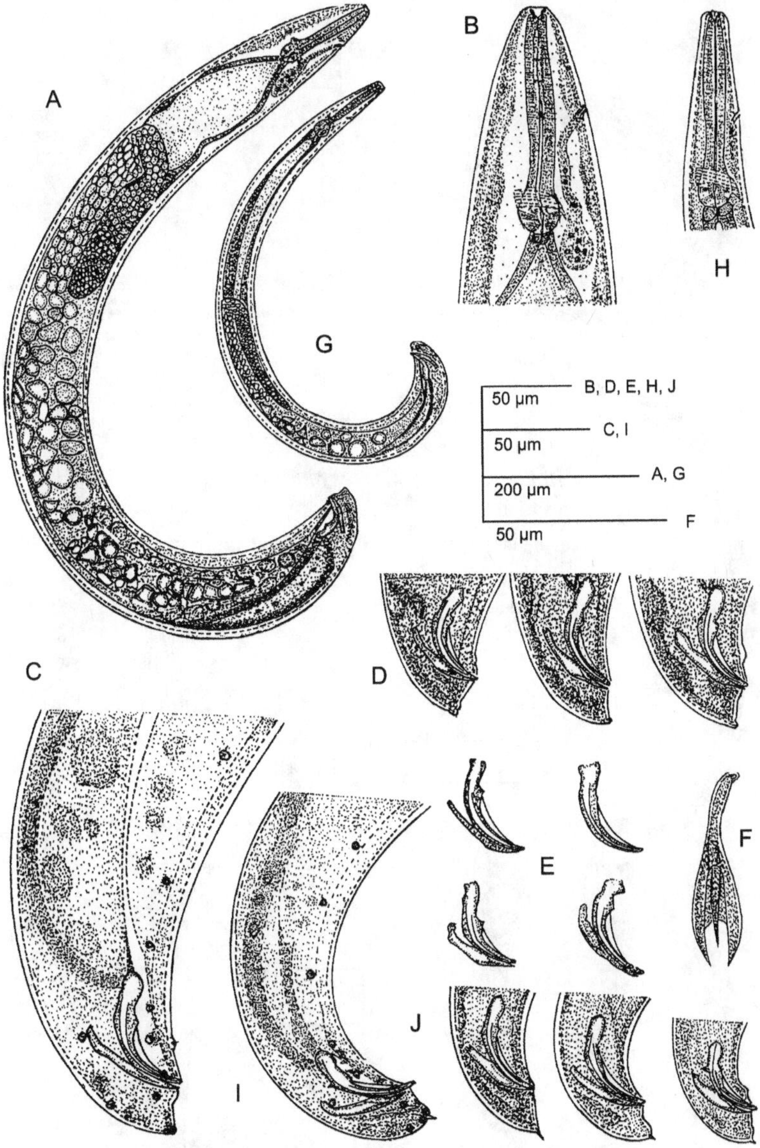

Fig. 50. Steinernema asiaticum. *A-F: First generation male. A: Entire body; B: Pharyngeal region, C: Tail region; D: Variation in tail shape; E: Variation in spicule and gubernaculum shape; F: Gubernaculum, ventral view. G-J: Second generation male. G: Entire body; H: Pharyngeal region; I: Tail region; J: Variation in tail shape. After Anis* et al. *(2002), courtesy of* International Journal of Nematology.

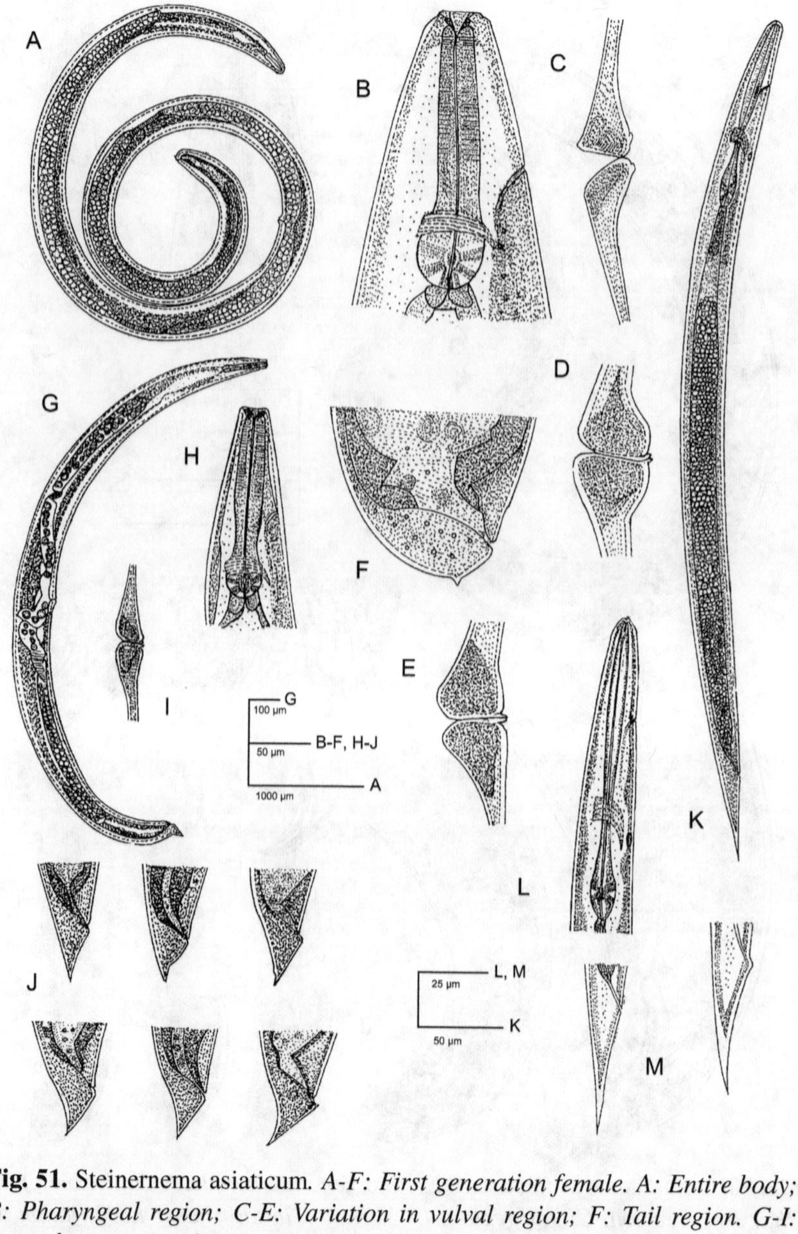

Fig. 51. Steinernema asiaticum. *A-F: First generation female. A: Entire body; B: Pharyngeal region; C-E: Variation in vulval region; F: Tail region. G-I: Second generation female. G: Entire body; H: Pharyngeal region; I: Vulval region. J: Variation in tail shape. K-M: Infective juvenile. K: Entire body; L: Pharyngeal region; M: Variation in tail shape. After Anis* et al. *(2002), courtesy of* International Journal of Nematology.

Table 13. *Morphometrics of* Steinernema asiaticum.* *Measurements are in μm (except where indicated) and in the form: mean ± SD (range).*

Character	First generation			Second generation		Infective juvenile
	Holotype (male)	Male	Female	Male	Female	
n	–	20	20	20	20	20
L (mm)	1.27	1.28 ± 0.15 (0.9-1.6)	9.9 ± 1.0 (8.3-11.7)	0.9 ± 0.09 (0.7-1.1)	1.4 ± 0.28 (1.1-2.2)	0.42 ± 0.02 (0.36-0.45)
a	11.8	12.3 ± 0.9 (10.9-14)	33 ± 2.2 (30-37)	15.3 ± 0.95 (13-16.6)	16 ± 1.36 (14-19)	17.6 ± 0.76 (16-19)
b	10.3	9.6 ± 0.88 (8.0-11.6)	44 ± 4.2 (40-52)	7.6 ± 0.54 (6.5-8.8)	10.4 ± 1.43 (9-15)	4.3 ± 0.76 (4.2-4.6)
c	57	51 ± 5 (40-63)	416 ± 109 (299-566)	43 ± 4.7 (34-56)	32 ± 4.93 (25-42)	10.5 ± 0.82 (9.4-12.5)
V	–	–	50 ± 2.1 (46-52)	–	55 ± 1.65 (52-57)	–
Body diam.	108	104 ± 11.7 (86-130)	298 ± 14.7 (270-317)	61 ± 7.1 (53-77)	91 ± 14.2 (78-126)	23 ± 1.5 (20-25)
EP	56	60 ± 8.6 (48-82)	106 ± 11.9 (90-132)	67 ± 7.0 (52-82)	73 ± 7.6 (62-85)	32 ± 1.3 (28-34)
NR	100	97 ± 6.7 (90-110)	165 ± 6.8 (152-174)	93 ± 5.8 (85-105)	102 ± 9.4 (90-119)	56 ± 2.8 (49-61)
ES	123	133 ± 8.1 (122-148)	224 ± 8.6 (208-240)	121 ± 6.3 (109-134)	138 ± 9.0 (127-153)	96 ± 5.8 (82-104)
Tail length (T)	22	25 ± 2.1 (22-32)	25 ± 8.1 (19-36)	21 ± 1.7 (18-25)	45 ± 4.8 (35-56)	41 ± 5.1 (28-46)
Anal body diam. (ABD)	36	34 ± 3.1 (30-40)	107 ± 23.8 (86-160)	28 ± 2.4 (26-35)	32 ± 3.6 (27-39)	–
Spicule length (SL)	68	68 ± 4.2 (61-74)	–	61 ± 4.2 (53-68)	–	–
Gubernaculum length (GL)	53	53 ± 4.3 (46-62)	–	45 ± 3.8 (38-50)	–	–
D% = EP/ES × 100	45	44 ± 5 (35-57)	47 ± 5 (40-58)	50 ± 5 (45-70)	52 ± 3 (47-61)	32 ± 1 (30-36)
E% = EP/T × 100	250	230 ± 25 (200-310)	460 ± 130 (300-620)	300 ± 40 (240-400)	161 ± 17.9 (139-200)	78 ± 7 (60-90)
SW% = SL/ABD × 100	189	200 ± 15 (161-220)	–	210 ± 14 (180-230)	–	–
GS% = GL/SL × 100	77	76 ± 5 (68-87)	–	74 ± 5 (64-88)	–	–
H% = H/T × 100	–	–	–	–	–	~40

EP = distance from anterior end to excretory pore; NR = distance from anterior end to nerve ring; ES = distance from anterior end to end of pharynx.

* Note that in the original description the table has been set incorrectly and that most morphometric values do not correspond to the specified character. These (and other) errors have been corrected herein.

as six pairs precloacal subventral, posteriormost pair being near cloacal aperture; one precloacal sublateral pair at level of subventral pair 5; one postcloacal subventral pair; one postcloacal subventral pair near tail tip; one postcloacal subdorsal pair; one postcloacal subdorsal pair near tail tip. Tail short, dorsally convex, conoid. Fine mucron visible on tail of some males with SEM.

Male (second generation)

Similar to first generation but smaller and more slender with spicule and gubernaculum shorter and thinner. Tail with well developed mucron.

Female (first generation)

Heat-relaxed specimens usually spiral form. Body cuticle smooth or very faintly annulated. Phasmids inconspicuous. Cephalic extremity truncate to slightly rounded, bearing six labial and four cephalic papillae plus lateral amphidial apertures. Lips fused. Oral aperture subtriangular. Stoma shallow, triangular at base. Cheilorhabdions strongly sclerotised, meso-, meta- and telorhabdions vestigial. Pharynx with cylindrical procorpus, muscular, metacorpus swollen, isthmus distinct, basal bulb, rounded, valvate. Nerve ring usually surrounding anterior region of basal bulb. Cardia valve well developed. Excretory pore located anterior to nerve ring. Genital system amphidelphic, reflexed, often containing many eggs. Genital system amphididelphic, reflexed. Vulva a transverse slit; lips slightly protruding; moderately developed double-flapped epiptygmata present. Tail shorter than anal body diam., not constricted posteriorly, bluntly conical to dome-shaped with short terminal spine.

Female (second generation)

Similar to first generation female but smaller, more slender and with a truncate head. Postanal region constricted, longer, conoid, more tapering and ending in fine, mucronate, process. Tail variable in shape with postanal swelling.

Infective juvenile

Heat-relaxed specimens dying straight to slightly ventrally arcuate. Newly moulted IJ often enclosed in cuticle of J2. Body slender, gradually tapering at both extremities (*i.e.*, from base of pharynx to anterior end and from anus to tail terminus). Cuticle with faint transverse striations. Lateral field distinct, with six ridges or seven incisures. Labial region smooth, continuous, rounded to hemispherical. Excretory pore located in anterior region of pharynx; excretory duct long, usually distinct, extending posteriorly to region of basal bulb. Pharynx with cylindrical procorpus, distinctly narrowed at nerve ring, terminating in an elongate, dorsally displaced, valvate basal bulb. Cardia present. Nerve ring distinct, located anterior to basal bulb. Hemizonid distinct, at level of anterior margin of basal bulb. Bacterial pouch well developed, in anterior part of intestine close to cardia. Gut lumen containing numerous fat globules. Rectum long, narrow; anus distinct. Tail variable in shape, elongate conoid, tapering to a finely pointed terminus. Hyaline region comprising *ca* 40% of tail length. Phasmids inconspicuous.

TYPE HOST AND LOCALITY

Natural host unknown. The species was collected from sandy soil around coconut trees, grass lawns and ornamental plants in the Malir area, Gulsham-e-Maymar, Karachi, Sindh, Pakistan.

TYPE MATERIAL

Numerous paratype males, females (of both generations) and infective juveniles in the CABI Europe-UK Nematode Collection, Egham, Surrey, UK, and other paratype material in the type collection of Rothamsted Research, Harpenden, Herts., UK.

RELATIONSHIPS

Steinernema asiaticum is characterised by the very small infective juvenile, which is 425 (360-450) μm long, EP = 32 (28-34) μm, tail length = 40 (28-46) μm and E% = 78 (60-90). There are six ridges (*i.e.*, seven incisures) in the lateral field and horn-like structures on the cephalic region are absent. The first generation male is characterised by a spicule length of 68 (61-74) μm; D% = 44 (35-57) and tail mucron very fine and only usually visible with SEM. The female has a vulva with moderately developed double-flapped epiptygmata and the bluntly conical to dome shaped tail bears a short terminal spine.

Steinernema asiaticum was distinguished from all other species in the genus by having the smallest infective juvenile body length of 425 (360-450) μm. This places it in the *carpocapsae*-group and it is close to *S. siamkayai* (L = 446 (398-495) μm), *S. tami* (L = 530 (400-600) μm) and *S. carpocapsae* (L = 558 (438-650) μm). It can be further distinguished by having six ridges in the IJ lateral field *vs* 6-8 in *S. siamkayai*, eight in *S. tami* and seven in *S. carpocapsae*, and by E% value of 78 (60-90) *vs* 96 (85-112) in *S. siamkayai* and 60 (54-66) in *S. carpocapsae*. The E% of *S. tami* is similar at 73 (67-86).

It can also be differentiated from *S. siamkayai*, the most similar species, by the disposition of the male genital papillae, spicule head elongate-rounded *vs* rhomboid, spicule of first generation male golden to dark yellow in colour *vs* orange brown, anterior part of gubernaculum longer, curved anteriorly and proximal part with or without a knob- or hook-like terminus *vs* gubernaculum large, slender, arcuate, lacking a proximal knob- or hook-like terminus, D% of 44 (35-57) *vs* 60 (51-68) and tail mucron in first generation male, if present, fine, indistinct, *vs* prominent.

Molecular characterisation

The species was characterised by the RFLP profile of the ITS region. Isolate 211 from Pakistan had an identical RFLP profile to isolate SSL2 from Sri Lanka (Anis *et al.*, 2002).

Bionomics

The species was isolated from sandy soil in Sindh, Pakistan, by baiting with *Galleria mellonella* larvae (Anis *et al.*, 2002). Amarasinghe *et al.* (1994) had previously found the same species along the southern and western coastal areas of Sri Lanka in sandy soils within 5-100 m of the sea. The associated vegetation was coconut with either *Ipomoea* creepers or common grasses. As in Pakistan, no natural hosts were detected.

Tabassum and Shahina (2004) reported a technique for mass-production of *S. asiaticum* on chicken offal whilst Shahina *et al.* (2004) reported on the bacterial associate of the species.

Distribution

The species is known from the type locality in Sindh province, Pakistan (Anis *et al.*, 2002), and also occurs in Sri Lanka (Amarasinghe *et al.*, 1994; Hominick *et al.*, 1997). Amarasinghe *et al.* (1994) reported the species (then undescribed but known to be new from its distinctive RFLP profile) from the southwest coast of Sri Lanka where it was restricted to sandy soils within 5-100 m of the sea. The nematodes tolerated high temperatures (*ca* 30°C), neither monsoon nor drought affecting prevalence. In Sri Lanka, *S. asiaticum* was found together with two other *Steinernema* species and two *Heterorhabditis* species.

References

AMARASINGHE, L.D., HOMINICK, W.M., BRISCOE, B.R. & REID, A.P. (1994). Occurrence and distribution of entomopathogenic nematodes in Sri Lanka. *Journal of Helminthology* 68, 277-286.

ANIS, M., SHAHINA, F., REID, A.P. & ROWE, J. (2002). *Steinernema asiaticum* sp. n. (Rhabditida: Steinernematidae) from Pakistan. *International Journal of Nematology* 12, 220-231.

HOMINICK, W.M., BRISCOE, B., DEL PINO, F.G., HENG, J., HUNT, D.J., KOZODOY, E., MRÁČEK, Z., NGUYEN, K.B., REID, A.P., SPIRIDONOV, S., STOCK, P., STURHAN, D., WATURU, C. & YOSHIDA, M. (1997). Biosystematics of entomopathogenic nematodes: current status, protocols and definitions. *Journal of Helminthology* 71, 271-298.

SHAHINA, F., MANZAR, H. & TABASSUM, K.A. (2004). Symbiotic bacteria *Xenorhabdus* and *Photorhabdus* associated with entomopathogenic nematodes in Pakistan. *Pakistan Journal of Nematology* 22, 117-128.

TABASSUM, K.A. & SHAHINA, F. (2004). *In vitro* mass rearing of different species of entomopathogenic nematodes in monoxenic solid culture. *Pakistan Journal of Nematology* 22, 167-175.

Steinernema backanense[*],[**] Phan, Spiridonov, Subbotin & Moens, 2006
(Fig. 52)

MEASUREMENTS

See Table 14.

DESCRIPTION

Male (first and second generation not distinguished in original description)

Body fusiform, obese, C- or broadly J-shaped in heat-relaxed specimens, tapering towards anterior end. Cuticle annulation 1.3-1.5 μm wide, indistinct throughout body. Anterior extremity rounded. Six labial spike-like papillae and four protruding cephalic papillae present. Cheilostom cuticle connected to a thick, strongly cuticularised, 2 μm long ring of 5 μm diam., forming middle part of stoma walls. Additional thinner cuticular ring situated posterior to thicker one, embedded in pharyngeal tissue. Pharynx corpus cylindrical, without visible collar around stoma, but with prominent metacorpal swelling (mid-corpus diam. = *ca* 16 μm, metacorpus diam. = *ca* 19-20 μm). Basal bulb subspherical. Nerve ring situated on anterior slope of basal bulb in most males. Cardia very flat, only 5 μm long. Deirids 2 μm diam. situated posterior to pharyngo-intestinal junction. Excretory channel strongly cuticularised for 4-5 μm from pore. Intestine with broad lumen up to mid-body but with only narrow, meandering, lumen from there to rectum. Separate spermatocytes visible up to flexure, being replaced with complex rachis structure in growth zone. *Vas deferens* walls not vacuolated, but with spherical inclusions of glandular appearance. Characteristic steinernematid set of genital papillae with a row of usually nine pairs of subventral papillae extending from tail terminus to mid-body region, one large, unpaired, precloacal papilla in midventral position, one pair of papillae in lateral position at manubrium level when spicules retracted, and one subterminal pair of dorsolateral papillae in postcloacal area. Third pair subventral papillae from tail tip located near mid-tail. Pair of postdeirids situated sublaterally anteriorly to first pair of precloacal papillae. Spicules

[*] Named after the type locality in the *Backan* province of Vietnam.
[**] Description after Phan *et al.* (2006). Revised by Zdeněk Mráček.

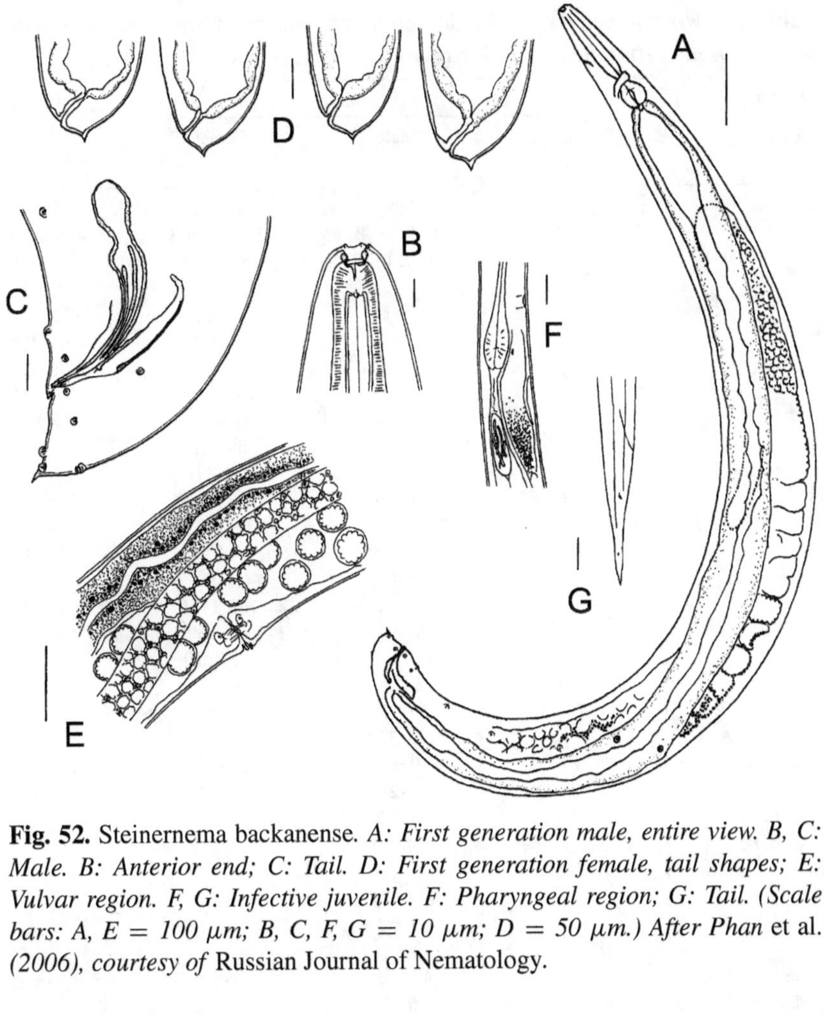

Fig. 52. Steinernema backanense. *A: First generation male, entire view. B, C: Male. B: Anterior end; C: Tail. D: First generation female, tail shapes; E: Vulvar region. F, G: Infective juvenile. F: Pharyngeal region; G: Tail. (Scale bars: A, E = 100 μm; B, C, F, G = 10 μm; D = 50 μm.) After Phan* et al. *(2006), courtesy of* Russian Journal of Nematology.

with elongated, funnel-shaped manubrium, tapering towards distal end, with round extremity and more or less pronounced terminal duct. No bifurcation of outer line of velum and ventral rib of spicule observed near rostrum. All four main 'ribs' of spicule discernible, but ventral most one tightly adpressed to inner one. Spicule body gradually narrowing from rostrum to distal end. Cloacal aperture covered by flap when spicules retracted. Two papilla-like structures on margin of cloacal aperture. Distal half of velum edge strongly cuticularised. Gubernaculum usually straight with ventrally curved proximal end, often with cuticular knobs on inner

Table 14. *Morphometrics of* Steinernema backanense. *Measurements are in μm and in the form: mean ± SD (range).*

Character	First generation			Infective juvenile
	Holotype (male)	Male	Female	
n	–	15	15	25
L	1125	1138 ± 93 (1005-1320)	3184 ± 363 (2610-3990)	438 ± 20 (391-473)
a	11	11 ± 1 (9.8-15)	19 ± 2 (16-21)	22 ± 1 (20-27)
b	8	8.2 ± 0.6 (7.2-9.4)	16 ± 1 (14-19)	4.8 ± 0.2 (4.5-5.2)
c	58	61 ± 8 (51-74)	152 ± 25 (114-194)	10 ± 0.4 (9.8-11)
V	–	–	57 ± 1 (54-59)	–
Body diam.	105	101 ± 14 (75-135)	168 ± 14 (150-195)	20 ± 1 (18-22)
EP	51	56 ± 5 (45-63)	74 ± 11 (60-99)	31 ± 1 (28-33)
NR	81	91 ± 5 (81-99)	135 ± 17 (105-168)	62 ± 2 (58-65)
ES	137	138 ± 7 (123-150)	203 ± 19 (164-236)	91 ± 3 (87-97)
Testis reflexion	284	256 ± 45 (156-330)	–	–
Tail length (T)	17	19 ± 2 (15-21)	28 ± 3 (24-33)	42 ± 2 (39-46)
Anal body diam. (ABD)	26	26 ± 2 (24-29)	65 ± 5 (59-72)	11 ± 1 (10-11)
Spicule length (SL)	59	58 ± 3 (53-62)	–	–
Spicule width	11	10 ± 1 (9-12)	–	–
Gubernaculum length (GL)	47	43 ± 3 (36-48)	–	–
Gubernaculum width	8	6 ± 1 (4.5-7.5)	–	–

surface and prominent, 10-12 μm long, hole. Tail mucron spike-like, 2 ± 0.7 (1.5-3) μm long.

Female (first and second generation not distinguished in original description)

Cuticle annulation 1.5 μm wide, discernible only near vulva lips. Cephalic extremity with six labial and four cephalic papillae. Cuticular

Table 14. *(Continued).*

Character	First generation			Infective juvenile
	Holotype (male)	Male	Female	
D% = EP/ES × 100	41	41 ± 5 (30-48)	37 ± 4 (30-43)	34 ± 2 (30-37)
E% = EP/T × 100	–	–	–	73 ± 4 (64-83)
SW% = SL/ABD × 100	2.1	219 ± 18 (184-256)	–	–
GS% = GL/SL × 100	0.8	74 ± 5 (66-79)	–	–
H% = H/T × 100	–	–	–	42 ± 2 (39-46)

EP = distance from anterior end to excretory pore; NR = distance from anterior end to nerve ring; ES = distance from anterior end to end of pharynx.

ring forming main part of stoma walls 5-6 μm thick and 2.5-3 μm long. Stoma cavity 10 μm diam. Excretory duct strongly cuticularised, entering body at angle less than 30°. Equatorial vulva without discernible epiptygmata in some specimens, and often with unequal vulva lips protruding up to 30-35 μm. Tail end subconical, with spike-like, 7-9 μm long, mucron.

Infective juvenile

Body of heat-relaxed juveniles quite obese, slightly C-shaped, tapering towards anterior end. Cuticular annulation *ca* 1 μm. wide visible throughout body. Lateral field in anterior half of body with four ridges in centre, two pronounced marginal ridges and poorly discernible submarginal lines. Such regular pattern not observed in body posterior half, with ventral gap between central and marginal ridges disappearing. Lateral field 5-6 μm wide at mid-body. Frontal surface of anterior end always regularly round. Labial papillae indistinct, four cephalic papillae and amphidial pouches occasionally visible. Pharynx corpus slender, only slightly widening to metacorpal part. Nerve ring located slightly posterior to corpus-isthmus junction. Hemizonid on isthmus level, flattened. Round deirid body beneath lateral field at hemizonid level. Basal bulb with poorly visible valves. Cardia *ca* 5 μm long, conical, narrowing to posterior end. Bacterial vesicle elliptical with poorly visible walls, *ca* 12 μm long, loosely filled with 10 μm long, curved, bacterial cells. Rectum *ca* two anal body diam. long. Genital primordium 50 μm long.

Phasmids located at one-third of tail length posterior to anus (phasmid-mucron distance *ca* 30 μm), with aperture near ventral edge of ventral marginal line. Lateral field lines disappearing at level of phasmids. Tail with slightly pronounced ventral concavity, regularly conical, without prominent ventral curvature, terminus sharply pointed. Hyaline region comprising slightly more than 40% of tail length.

TYPE HOST AND LOCALITY

Collected by *Galleria mellonella* baiting method in Lungchu (Babe lake) in Backan province, Vietnam (longitude 105°34′E, latitude 22° 26′N, altitude 700 m a.s.l.) from lixisols soil of a primary, closed, broad-leaved seasonal forest situated on the slope of a limestone mountain.

OTHER LOCALITY

Banks of the Thacbac stream in the Tamdao mountains of Vinhphuc province, Vietnam (longitude 105°35′E, latitude 21°27′N, altitude 900 m a.s.l.). The habitat was primary, closed, broad-leaved forest on the slopes of rhyolite mountains.

TYPE MATERIAL

Holotype male and nine paratype males are deposited at Institute of Parasitology of Russian Academy of Sciences, Leninskii prospect 33, Moscow, 119071, Russia. Slides with paratype males, first generation females and infective juveniles deposited in Department of Nematology, Institute of Ecology and Biological Resources, National Center for Science and Technology, Hoang Quoc Viet 18, Nghiado, Caugiay, Hanoi, Vietnam.

RELATIONSHIPS

This species is characterised by males with two papilla-like structures on the cloacal aperture edges, spicules with elongated manubrium and well developed rostrum and a short, spike-like, mucron present in both generations. First generation females have a pronounced, conical, tail mucron and epiptygmata are mostly developed. The body length of the infective juvenile is shorter, on average, than 500 μm, the hyaline region comprises less than half of the tail length and the lateral fields have four ridges (or five lines) in the mid-body region.

Steinernema backanense can be distinguished from all of the other three similar Vietnamese species (*i.e.*, *cumgarense*, *eapokense* and *sasonense*), and also from *S. scapterisci*, *S. siamkayai* and *S. tami* by the short spicules of 58 (53-62) μm. Spicules shorter than 63 μm have sometimes been reported for *S. carpocapsae* (66 (58-77) μm), but *S. backanense* differs in the D% value of the IJ (29-35 *vs* 23-28).

Molecular characterisation

Steinernema backanense is characterised by sequence lengths of ITS regions (ITS1 + 5.8S + ITS2), 736 bp, ITS1 = 202 bp, ITS2 = 307 bp and its composition, A = 0.23641, C = 0.16711, G = 0.21739, T = 0.37908. The sequence of ITS regions is deposited in GenBank (accession no. AY487918).

Remark

More precise description of taxonomically important characters, such as the number of ridges in lateral fields of infective juveniles, first generation female postanal swelling and occurrence of papilla-like mucrons on the female tail tip are missing or unsatisfactorily described in the original paper.

Distribution

The species has only been found in Vietnam (Phan *et al.*, 2006).

Reference

PHAN, K.L., SPIRIDONOV, S.E., SUBBOTIN, S.A. & MOENS, M. (2006). Four new species of *Steinernema* Travassos, 1928 with short infective juveniles from Vietnam. *Russian Journal of Nematology* 14, 11-29.

Steinernema beddingi[*,**] Qiu, Hu, Zhou, Pang & Nguyen, 2005
(Figs 53-55)

MEASUREMENTS

See Table 15.

DESCRIPTION

Male (first generation)

Body C-shaped and curved posteriorly when heat-relaxed. Cuticle smooth. Stoma shallow and broad. Pharynx extending nearly to oral aperture. Cheilorhabdions and prorhabdions sclerotised, small but distinct. Pharynx muscular with a cylindrical procorpus, metacorpus swollen, isthmus well defined, basal bulb enlarged. Nerve ring often surrounding anterior portion of basal bulb. Excretory pore opening located near posterior part of metacorpus. Lateral fields and phasmids not observed. Genital system monorchic, reflexed. Spicules paired, colourless and moderately curved. Spicule length/width ratio = *ca* 4.6. Spicule head shorter than wide with a length/width ratio *ca* 1.2. Lateral lobe prominent, beginning anteriorly at shaft and extending posteriorly to spicule tip. Ventral lobe enlarged anteriorly and ventrally to form a prominent rostrum and ending before, but close to, spicule tip. Dorsal lobe starting from shaft and terminating at a level similar to ventral lobe. Spicule terminus blunt. Velum not observed. Gubernaculum boat-shaped in lateral view. In ventral view, gubernaculum tapering anteriorly and evenly. Corpus fused and cuneus not seen. Tail long and concave ventrally. No mucron-like projection was observed at tip of tail. Twenty-three genital papillae seen with SEM, including five precloacal pairs and one single precloacal papilla, two pairs adcloacal, one pair lateral and three pairs postcloacal papillae (two pairs subterminal and one subdorsal).

---

[*] Patronym in honour of R.A. *Bedding* from the Australian Center for International Agricultural Research (ACIAR).

[**] Description after Qiu *et al.* (2005). Revised by Khuong B. Nguyen.

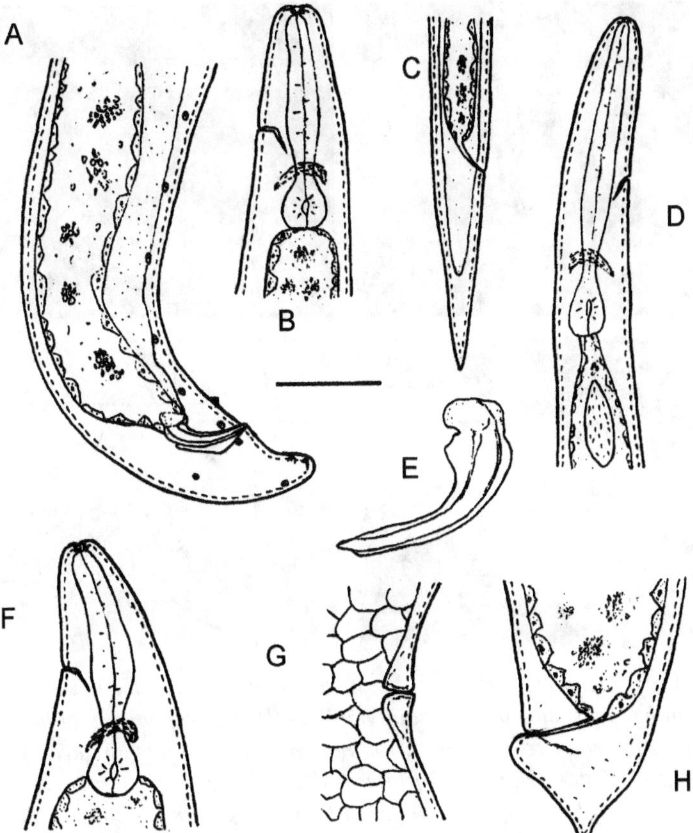

Fig. 53. Steinernema beddingi. *A, B, E: Anterior region (A), tail (B) and spicule (E) of first generation male in lateral view; C, D: Tail (C) and anterior region (D) of infective juvenile. (Scale bar: A = 80 μm; B = 70 μm; C = 27 μm; D = 30 μm; E = 25 μm; F = 125 μm; G, H = 110 μm.) After Qiu* et al. *(2005), courtesy of* Nematology.

Male (second generation)

Morphologically similar to first generation male, including lack of mucron, but most measurements are smaller and excretory pore located more anterior than in the first generation (Table 15).

Female (first generation)

Body C-shaped when heat-relaxed. Cuticle smooth, head rounded and slightly truncate anteriorly. Six lips fused, each with a papilla. Four

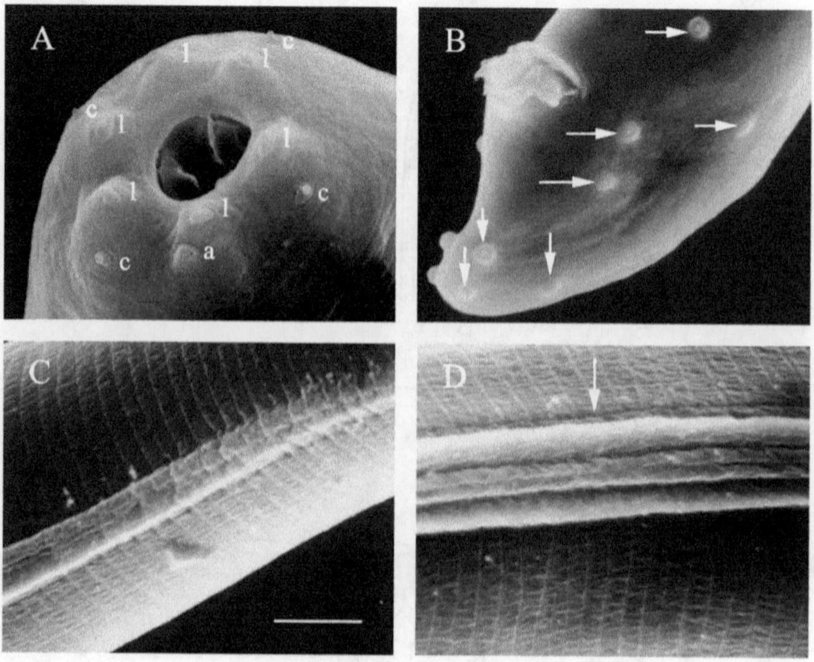

Fig. 54. Steinernema beddingi. *SEM micrographs. A: Face view of first generation female showing six labial papillae (l), four cephalic papillae (c) and amphidial aperture (a); B: Posterior region of male showing seven papillae (arrows); C: Anterior portion of lateral field of infective juvenile; D: Middle portion of lateral field of infective juvenile showing four prominent ridges plus a faint outer line (arrow) occasionally visible in some specimens. (Scale bar: A = 9 μm; B = 20 μm; C = 8 μm; D = 3.5 μm.) After Qiu* et al. *(2005), courtesy of* Nematology.

cephalic papillae occurring further posteriorly on head. Amphidial apertures distinct and located posterior to lateral labial papillae. Morphology of anterior region of female similar to male. Lateral fields and phasmids not observed. Genital system didelphic, reflexed. Vulva a transverse slit, usually slightly protuberant, symmetrical. Vagina short, muscular. Epiptygmata not observed. Tail broad, with wedge-like projection on tip. Postanal swelling distinct.

Female (second generation)

Similar to first generation females but much smaller. Vulva slightly more protuberant than in first generation. Postanal swelling and tail projection as prominent as those of first generation.

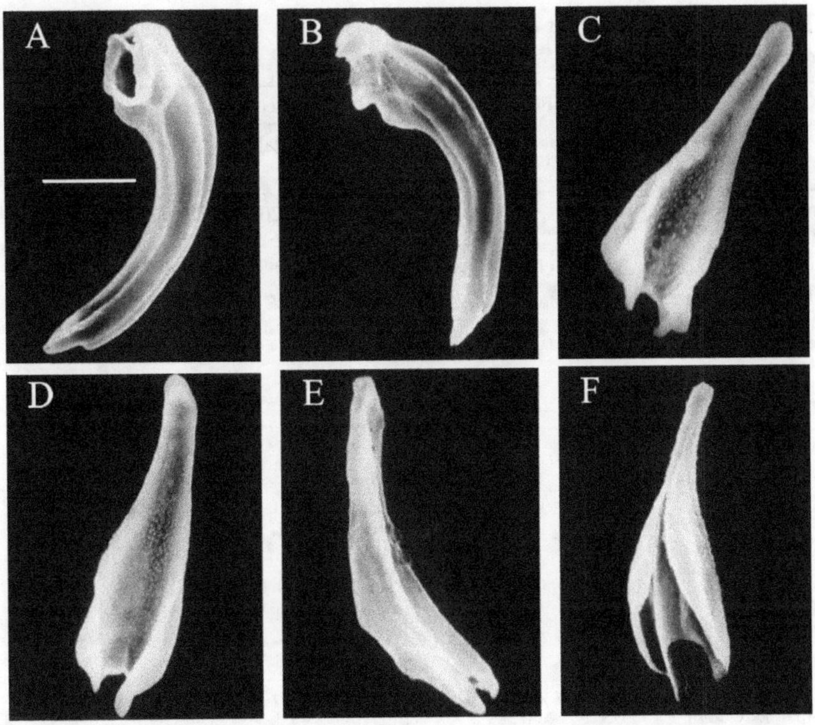

Fig. 55. Steinernema beddingi. *SEM micrographs. A, B: Spicules of first generation males; C-F: Gubernacula of first generation males in ventral (C, D), lateral (E) and dorsal (F) view. (Scale bar: A = 20 μm; B = 22 μm; C-F = 15 μm.) After Qiu* et al. *(2005), courtesy of* Nematology.

Infective juvenile

Body slender, enclosed in a sheath. Cephalic region smooth, rounded, papillae not seen. Amphidial apertures slit-like. Four prominent cephalic papillae. Oral aperture and anus closed. Pharynx long and narrow, isthmus distinct and surrounded by nerve ring. Excretory pore located anterior to nerve ring, near middle of pharynx. Cuticle with prominent transverse striations. Lateral field beginning anteriorly with two slightly raised ridges at a position slightly anterior to excretory pore. Another two ridges appearing between previous two at a position close to base of pharynx and become gradually more prominent at middle portion of body and then fading gradually after level of anus. Middle two ridges slightly smaller than other two and whole lateral field distinctly protruding from body surface. Tail attenuate and tapering gradually with

Table 15. *Morphometrics of* Steinernema beddingi. *Measurements are in μm and in the form: mean ± SD (range).*

Character	First generation			Second generation		Infective juvenile
	Holotype (male)	Male	Female	Male	Female	
n	–	20	20	20	20	20
L	1820	2215 ± 355 (1750-2700)	3199 ± 587 (2200-3950)	945 ± 45 (790-965)	1520 ± 85 (1375-1675)	743 ± 25 (700-790)
a	–	–	–	–	–	25 ± 1.4 (23-27)
b	–	–	–	–	–	6.1 ± 0.26 (5.8-6.6)
c	–	–	–	–	–	9.6 ± 0.45 (8.7-10.7)
V	–	–	49 ± 2 (46-52)	–	49 ± 2 (47-55)	–
EP	100	114 ± 12 (95-138)	117 ± 10 (100-138)	71 ± 5.3 (63-78)	75 ± 4 (70-83)	70 ± 3.1 (64-75)
NR	132	143 ± 11 (125-165)	162 ± 14 (145-200)	127 ± 5.3 (120-137)	133 ± 6 (121-143)	96 ± 7.5 (80-113)
ES	180	195 ± 13 (175-218)	230 ± 17 (200-263)	173 ± 4.6 (163-180)	188 ± 7 (180-205)	125 ± 4.4 (113-130)
Tail length (T)	41	52 ± 6.5 (41-63)	85 ± 12 (68-105)	48 ± 2.2 (45-52)	80 ± 4 (73-88)	77 ± 3.1 (72-83)
Anal body diam. (ABD)	53	66 ± 9.5 (55-82)	85 ± 10 (63-108)	43 ± 1.3 (40-47)	48 ± 2 (43-50)	–
Spicule length (SL)	65	71 ± 4.8 (63-78)	–	67 ± 2.3 (64-70)	–	–
Gubernaculum length (GL)	43	43 ± 3.3 (38-48)	–	42 ± 1.8 (38-46)	–	–
D% = EP/ES × 100	55.6	58 ± 3 (54-63)	–	42 ± 4.5 (38-50)	–	57 ± 2.6 (52-64)
E% = EP/T × 100	–	–	–	–	–	92 ± 5.2 (84-103)
SW% = SL/ABD × 100	126	108 ± 13 (88-132)	–	–	–	–
GS% = GL/SL × 100	66	61 ± 4 (55-66)	–	–	–	–

EP = distance from anterior end to excretory pore; NR = distance from anterior end to nerve ring; ES = distance from anterior end to end of pharynx.
After the original authors.

hyaline region comprising *ca* 45% of tail length. No spine-like structure seen on tail tip. Phasmid inconspicuous. No substantial difference between IJ produced from the artificial medium and from *Galleria* larvae except that former was slightly shorter than latter.

TYPE HOST AND LOCALITY

The type host of this nematode in nature is unknown as it was recovered from soil using *Galleria* larvae as bait. The soil sample was collected from a cabbage field in An-fen-ying town (latitude N24.58, longitude E102.28, altitude 1850 m, mean annual temperature 16°C, precipitation 1000 mm year^{-1}), about 60 km from Kunming, Yunan province, People's Republic of China.

TYPE MATERIAL

Holotype male, allotype female, five paratype first generation males, five paratype females, 15 paratype infective juveniles and other population slides deposited in the State Key Laboratory for Biocontrol, School of Life Sciences, Zhongshan University, Guangzhou 510275, People's Republic of China. One slide that contains about ten paratype IJ and two slides, each with one paratype male and one female, deposited in USDA Nematode Collection, Beltsville, MD, USA. Living infective juveniles are also preserved in liquid nitrogen in the nematode collection of SKLB, Zhongshan University, People's Republic of China.

RELATIONSHIPS

Steinernema beddingi is characterised morphologically by a combination of features of various developmental stages. For IJ, body length averaging 743 (700-790) μm; distance from anterior end to excretory pore 70 (64-75) μm, to the base of pharynx 125 (113-130) μm; tail length 77 (72-83) μm; D and E value 57 (52-64) and 92 (84-103), respectively; lateral field at the mid-body position with only four ridges, which are distinctly raised from body surface; tail attenuated and tapering gradually without a dorsal constriction or a spine on the tail tip. For males, tail has a concavity in the ventral part and lacks a mucron on tip; spicules and gubernaculum are colourless; spicule head very short, dorsal and ventral lobes ending before spicule tip whilst distal end of lateral lobe suddenly narrows both dorsally and ventrally to form a blunt tip; gubernaculum without cuneus. For females, vulva is a symmetrical, transverse slit without epiptygmata, and the tail has a wedge-like structure on the tip and a conspicuous postanal swelling.

Both morphometrical (Table 15) and molecular data show congruently that *S. beddingi* is close to *S. affine* and *S. intermedium*. However, it can be separated from these two species by the morphological characters of various developmental stages. The morphometrics of *S. beddingi* IJ are very close to those of *S. affine* and *S. intermedium* except that the tail is longer, 77 (72-83) μm compared to 64 (53-72) μm for *S. intermedium* and 66 (64-74) μm for *S. affine*. The infective juvenile of *S. beddingi* can also be distinguished from the two described species mentioned above by morphology of the tail and lateral field: *S. beddingi* IJ tail attenuates and tapers gradually with no spine on the tail tip, whereas *S. affine* IJ usually has a spine-like structure on tail tip, and the tail of *S. intermedium* IJ usually has a constriction on the dorsal side. The middle portion of the lateral field of *S. beddingi* has four raised ridges whilst that of the other two species has six ridges. For the first generation males, the spicule and gubernaculum length of *S. beddingi*, 71 (63-78) μm and 43 (38-48) μm, respectively, are shorter than those of *S. intermedium* (91 (84-100) μm and 64 (56-75) μm, respectively). The morphology of spicule and gubernaculum also differs substantially from the two described species; for example, no velum was observed on spicules of *S. beddingi*, whilst it was observed clearly on the spicules of both *S. intermedium* and *S. affine*; no cuneus was observed on *S. beddingi* gubernacula, whilst both *S. affine* and *S. intermedium* have a needle-type cuneus. For the first generation females, *S. beddingi* has a distinct postanal swelling that is absent in the other two species. *Steinernema beddingi* can also be distinguished from *S. affine* and *S. intermedium* genetically by either the ITS or 28S D3 domain rDNA sequence, and biologically by cross tests.

Steinernema beddingi can be distinguished from other described *Steinernema* species that have a similar mean IJ body length (ranging from 650-850 μm), including *S. pakistanense*, *S. ceratophorum*, *S. bicornutum*, *S. feltiae*, *S. thanhi*, *S. sangi*, *S. weiseri* and *S. monticolum* by morphological characters of various stages. For example, it can be distinguished from all nematodes mentioned above by the spicule and gubernaculum morphology of the first generation male (see Fig. 4, Adams & Nguyen, 2001) as well as lateral field morphology and ridge number of IJ (the number of ridges in the mid-body region of the above mentioned nematodes is six to eight whilst *S. beddingi* has four and the whole lateral field is distinctly raised from the body surface). *Steinernema beddingi* can also be separated from *S. pakistanense*, *S. ceratophorum* and *S. bicornutum* by IJ head structure (IJ of these latter three species possess two horn-like structures on the head).

MOLECULAR CHARACTERISATION

Steinernema beddingi is genetically characterised by the sequence length of ITS regions, 1069 bp (ITS1 = 308 bp, ITS2 = 349 bp), and its composition, A = 0.26193, C = 0.17025, G = 0.21890, T = 0.34892, GC% = 38.9 (accession no. AY603397), and partial 28S rDNA, 465 bp, and its composition, A = 0.26451, G = 0.29677, T = 0.27311, C = 0.16559, GC% = 46.2 (accession no. AY603396).

BIONOMICS

Steinernema beddingi reproduced poorly in *Galleria* larvae, usually without second generation adults in the infected *Galleria* larval cadaver, and only a few thousand IJ on average were obtained from a larvae. However, it proliferated well in an artificial medium. IJ produced either from *Galleria* larvae or the artificial medium were highly pathogenic to *Galleria* larvae, with most larvae dying within 24 h when 15 larvae were exposed to about 2000 IJ in a Petri dish lined with two layers of moist filter paper at 23°C. It is not known why *S. beddingi* reproduced poorly in *Galleria* larvae.

The symbiotic bacterium of *S. beddingi* has been isolated and preserved in liquid nitrogen. It produces a yellow pigment on the artificial medium and possesses many characters of *Xenorhabdus* species, such as phase variation, negative to peroxide, *etc*. A phylogenetic tree based on the 16S rDNA sequence showed the symbiotic bacteria of *S. beddingi* claded with, but can be clearly distinguished from, *X. bovienii* (Qiu *et al.*, data unpubl.).

DISTRIBUTION

Known only at the type locality, Kunming, Yunan province, People's Republic of China (Qiu *et al.*, 2005).

References

ADAMS, B.J. & NGUYEN, K.B. (2001). Taxonomy and systematics. In: Gaugler, R. (Ed.). *Entomopathogenic nematology*. Wallingford, UK, CABI Publishing, pp. 1-33.

QIU, L., HU, X., ZHOU, Y., PANG, Y. & NGUYEN, K.B. (2005). *Steinernema beddingi* n. sp. (Nematoda: Steinernematidae), a new entomopathogenic nematode from Yunan, China. *Nematology* 7, 737-749.

*Steinernema bicornutum**,** Tallosi, Peters & Ehlers, 1995
(Figs 56, 57)

Measurements

See Table 16.

Description

Male (first generation)

Cephalic extremity truncate. Face with six lips bearing a single labial papilla on top. Outer circle of four cephalic papillae further back on the head. Stoma short and broad. Pharynx muscular and almost cylindrical, but widening towards base. Metacorpus non-valvate; isthmus much thinner than metacorpus; nerve ring surrounding pharynx between isthmus and basal bulb. Valve ridges of basal bulb indistinct; basal bulb slightly broader or as broad as metacorpus. Genital system monorchic reflexed. Sperm 4-5 μm diam. Row of six pairs of precloacal ventrolateral genital papillae plus a single ventral precloacal papilla. Three pairs papillae located posterior to gubernaculum and another three pairs clustered at tail tip. Spicules paired and light brown in colour. Spicule head typified by ventral projection and with width greater than length. Shaft prominent. Blade well curved anteriorly; posterior part slightly curved or almost straight. In ventral view, gubernaculum tapering anteriorly to a ventrally curved end. Corpus separated posteriorly. Two different types of spicules occur in first generation males. Type I spicules are curved; capitulum distinct and well separated from lamina, proximally broad and flattened; calomus with a small pointed rostrum. From this point a distinct velum extends almost to the proximal end of the lamina. Spicule tip with a transverse incision, similar to the spicule of *S. carpocapsae*. Type II spicules (10-20% of the population) are much less curved; lamina extending to capitulum without distinct demarcation; velum absent or indistinct; spicule tip without incision; colour usually a little darker; proximal hook of gubernaculum usually more pronounced than in Type I males. Tails of first generation males of both types are short, rounded and lack a mucron.

---

* From the Latin, *bi* and *cornus*, referring to the presence of two horn-like structures on the IJ head.

** Description after Tallosi *et al.* (1995). Revised by Khuong B. Nguyen.

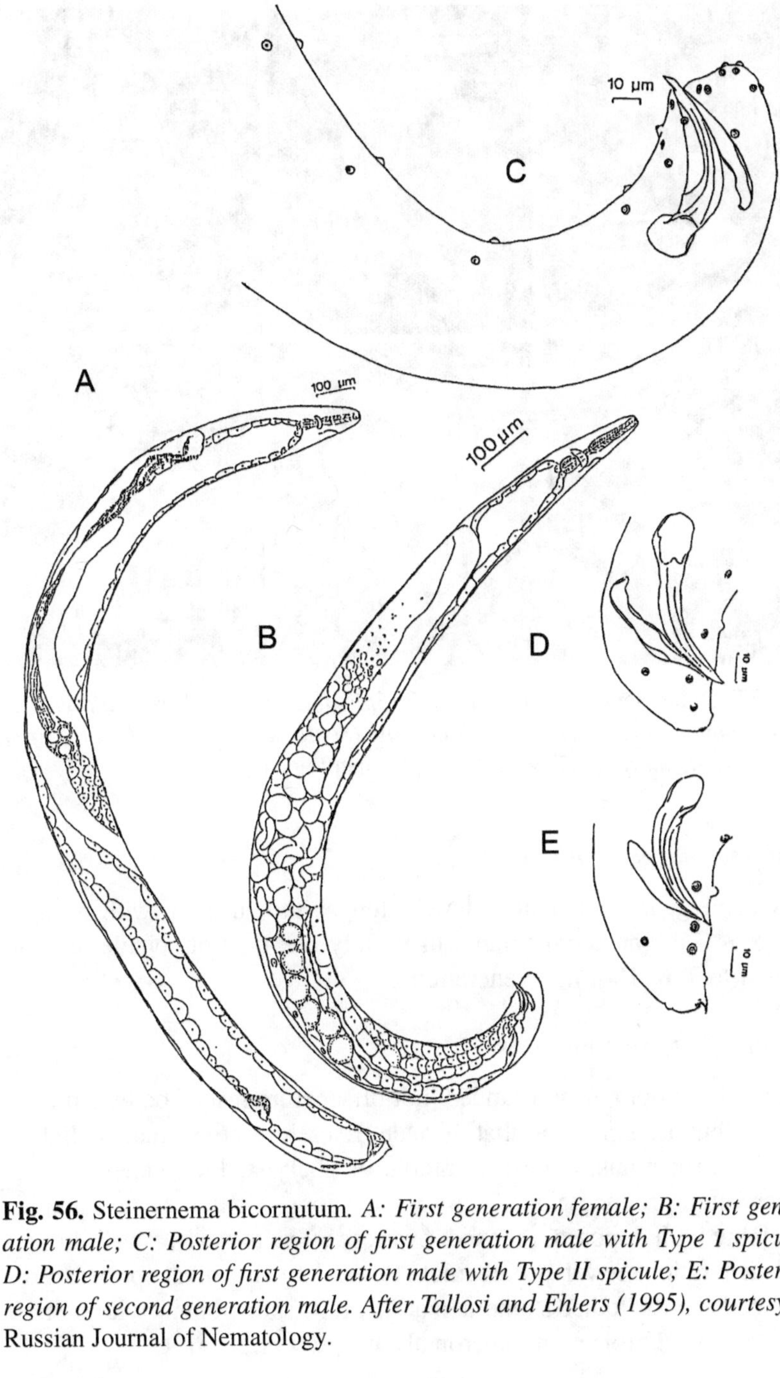

Fig. 56. Steinernema bicornutum. *A: First generation female; B: First generation male; C: Posterior region of first generation male with Type I spicule; D: Posterior region of first generation male with Type II spicule; E: Posterior region of second generation male. After Tallosi and Ehlers (1995), courtesy of* Russian Journal of Nematology.

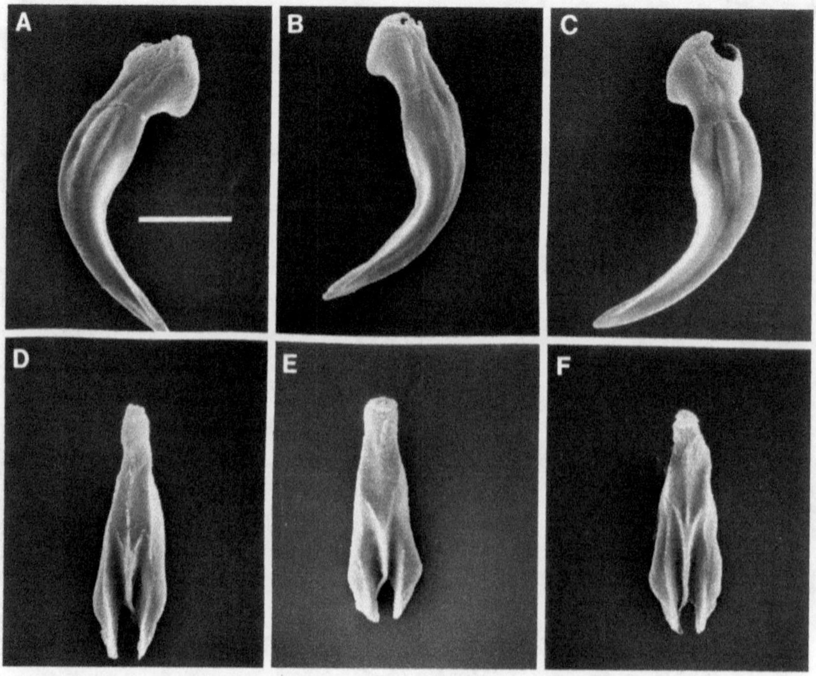

Fig. 57. Steinernema bicornutum. *A-C: Spicules; D-F: Gubernacula, ventral views showing arrowhead-shaped cuneus. (Scale bar: A-F = 20 μm.) After Nguyen and Smart (1997), courtesy of* Nematologica.

Male (second generation)

Second generation males have a longer tail, usually with a small mucron. Gubernaculum spindle-like. Only one type of spicule present, similar to Type II of first generation.

Female (first generation)

Cuticle smooth, head rounded, not offset from rest of body. Anterior part of females similar to that of male. Excretory pore situated slightly anterior to isthmus, almost at middle of pharynx. First generation females often produce giant forms >1 cm long. Genital system amphidelphic with opposed, reflexed, ovaries. Vulva a transverse slit in second generation, never protuberant. Protruding of vulva lips less pronounced than in other steinernematids. Tail, short, blunt, with a short wedge-shaped projection, mucron always present.

Table 16. *Morphometrics of* Steinernema bicornutum. *Measurements are in μm and in the form: mean ± SD (range).*

Character	First generation		Second generation		Infective juvenile
	Male	Female	Male	Female	
n	20	20	20	20	20
L	1353 ± 149 (945-1539)	6329 ± 2068 (2727-6329)	1007 ± 68 (864-1134)	1616 ± 146 (1269-1809)	770 ± 52 (648-873)
a	–	–	–	–	26.5 ± 1.5 (23-29)
b	–	–	–	–	6.2 ± 0.3 (5.6-6.9)
c	–	–	–	–	10.7 ± 0.66 (9.7-12)
V	–	50	–	54	
Body diam.	108 ± 11 (80-128)	243 ± 35 (170-305)	64 ± 2 (63-68)	102 ± 11 (80-123)	–
EP	82 ± 7.8 (67-98)	78 ± 11 (58-95)	71 ± 5 (60-80)	67 ± 4 (60-73)	–
NR	123 ± 8 (108-137)	126 ± 14 (100-158)	101 ± 4.4 (95-110)	104 ± 5 (95-110)	–
ES	156 ± 7 (138-167)	188 ± 13 (163-210)	131 ± 4 (125-140)	146 ± 8 (135-163)	–
Tail length (T)	32 ± 2.5 (25-35)	30 ± 6 (20-40)	30 ± 2.6 (25-35)	62 ± 5 (58-75)	72 ± 5 (63-78)
Spicule length (SL)	65 ± 4.3 (53-70)	–	51 ± 1.5 (48-53)	–	–
Gubernaculum length (GL)	47 ± 3.5 (38-50)	–	33 ± 2.9 (25-38)	–	–
D% = EP/ES × 100	50 ± 3 (50-60)	–	50 ± 3 (50-60)	–	50 ± 3 (40-60)
E% = EP/T × 100	–	–	–	–	80 ± 6 (80-100)
SW% = SL/ABD × 100	–	–	–	–	–
GS% = GL/SL × 100	72	–	65	–	–
H% = Hyaline/T × 100	–	–	–	–	45 (37-55)

EP = distance from anterior end to excretory pore; NR = distance from anterior end to nerve ring; ES = distance from anterior end to end of pharynx.
After the original authors.

Female (second generation)

Similar to first generation female but smaller. Vulva protuberant. Tail longer, always with mucron.

Infective juvenile

Body slender with transverse annulations. Lateral field with eight ridges (*i.e.*, nine lines) of which submarginal ridges are less distinct. Cephalic extremity hemispherical, cephalic region bearing two horn-like structures (which may correspond to laterally protruding labial papillae) located on an oval ring around oral aperture. Infective juvenile generally ensheathed by second-stage juvenile cuticle. Observation of horn-like structures occasionally prevented by presence of second-stage cuticle. Amphidial apertures lateral, located slightly further back from labial region, almost covered by the two horns. Four distinct cephalic papillae arranged medially on head at a 45° angle to horn-like structures. Excretory pore situated halfway between head and bulb. Pharynx and intestine lumen collapsed. Pouch containing cells of symbiotic bacterium, probably *Xenorhabdus* sp., present in anterior portion of intestine. Tail constricted with a pointed end.

TYPE HOST AND LOCALITY

Type host unknown as the species was isolated using *Galleria mellonella* as bait from soil at Strazilovo, Vojvodina, south of Novi Sad, in the Furska Gora Mountains, Serbia (formerly Yugoslavia).

TYPE MATERIAL

Holotype and paratypes deposited in the German Nematode Collection at Biologische Bundesanstalt, Institute for Nematology and Vertebrate Research, Münster, Germany. The strain designation of the type population is YuS-Wo 6.

RELATIONSHIPS

Steinernema bicornutum is characterised by the infective juvenile and male. The infective juvenile is characterised by the presence of horn-like structures in the cephalic region, body length of 770 μm, distance from anterior end to excretory pore = 61 μm, tail length = 72 μm and D% = 50. For the male, spicule and cuneus shape are unique and SW ratio is very high at 222.

Steinernema bicornutum is a member of a small group of species bearing horn-like structure on the IJ head. This character distinguishes the group from all other described of *Steinernema*. Within the *bicornutum*-group, it can be differentiated from all others by the long body length of its infective juveniles, *i.e.*, 770 μm compared to 709 μm or less in the others.

MOLECULAR CHARACTERISATION

The species is characterised genetically by sequence length of ITS regions of rDNA, 1016 bp (ITS1 = 281 bp, ITS2 = 329 bp), and its composition (A = 0.26476, C = 0.17125, G = 0.22637, T = 0.33759) (accession no. AF121048), and the length of D2D3 regions, 870 bp and its composition, A = 0.25402, C = 0.16667, G = 0.30230, T = 0.27701 (accession no. AF331904) (Nguyen *et al.*, 2004).

DISTRIBUTION

Known from the type locality in Serbia (Tallosi *et al.*, 1995) and from the Czech Republic, Denmark, Germany, Slovak Republic, Switzerland, Jamaica and the Canary Islands (see Hominick, 2002).

References

HOMINICK, W.M. (2002). Biogeography. In: Gaugler, R. (Ed.). *Entomopathogenic nematodes*. Wallingford, UK, CABI Publishing, pp. 115-143.

NGUYEN, K.B., TESFAMARIAM, M., GOZEL, U., GAUGLER, R. & ADAMS, B.J. (2004). *Steinernema yirgalemense* n. sp. (Rhabditida: Steinernematidae) from Ethiopia. *Nematology* 6, 839-856.

TALLOSI, B., PETERS, A. & EHLERS, R.-U. (1995). *Steinernema bicornutum* sp. n. (Rhabditida: Steinernematidae) from Vojvodina, Yugoslavia. *Russian Journal of Nematology* 3, 71-80.

Steinernema carpocapsae*[*,] (Weiser, 1955) Wouts, Mráček, Gerdin & Bedding, 1982**

= *Neoaplectana carpocapsae* Weiser, 1955
= *Neoaplectana feltiae sensu* Stanuszek, 1974, *nec* Filipjev, 1934
= *Neoaplectana feltiae pieridarum* Stanuszek, 1974
= *Neoaplectana pieridarum* Stanuszek, 1974
= *Steinernema feltiae pieridarum* (Stanuszek, 1974) Wouts, Mráček, Gerdin & Bedding, 1982
= *Steinernema pieridarum* (Stanuszek, 1974) Wouts, Mráček, Gerdin & Bedding, 1982
= *Neoaplectana carpocapsae pieridarum* Stanuszek, 1974
= *Neoaplectana dutkyi* Turco, Thames & Hopkins, 1971
= *Steinernema dutkyi* (Turco, Thames & Hopkins, 1971) Wouts, Mráček, Gerdin & Bedding, 1982
(Figs 58-60)

MEASUREMENTS

See Table 17.

DESCRIPTION

Male (first generation)

Cuticle smooth, head truncate to slightly rounded, lips fused, two circles of anterior papillae: six inner labial papillae and four outer cephalic papillae. Amphidial apertures small, posterior to labial papillae. Stoma partially collapsed with only an anterior vestibule remaining. Collar lacking, pharyngeal tissue reaching to base of vestibule. Cheilorhabdions represented as lightly sclerotised areas lining the inside of the labial region anterior to pharynx. A small sclerotised area just beneath the cheilorhabdions probably represents the modified prorhabdions. Posterior to these rhabdions, an asymmetrical funnel structure connecting with pharynx lumen at the stoma bottom. Pharynx muscular, anterior portion of procorpus slightly expanded just posterior to stoma, then extending into a slightly enlarged metacorpus, followed by an isthmus and a basal bulb containing a small valve. Nerve ring surrounding isthmus just anterior to basal bulb. Anterior region similar to females. Genital system monorchic, reflexed. Spicules paired, symmetrical, moderately

[*] Named after the genus of the type host, *Carpocapsa pomonella*.

[**] Description after Poinar (1967). Revised by Khuong B. Nguyen.

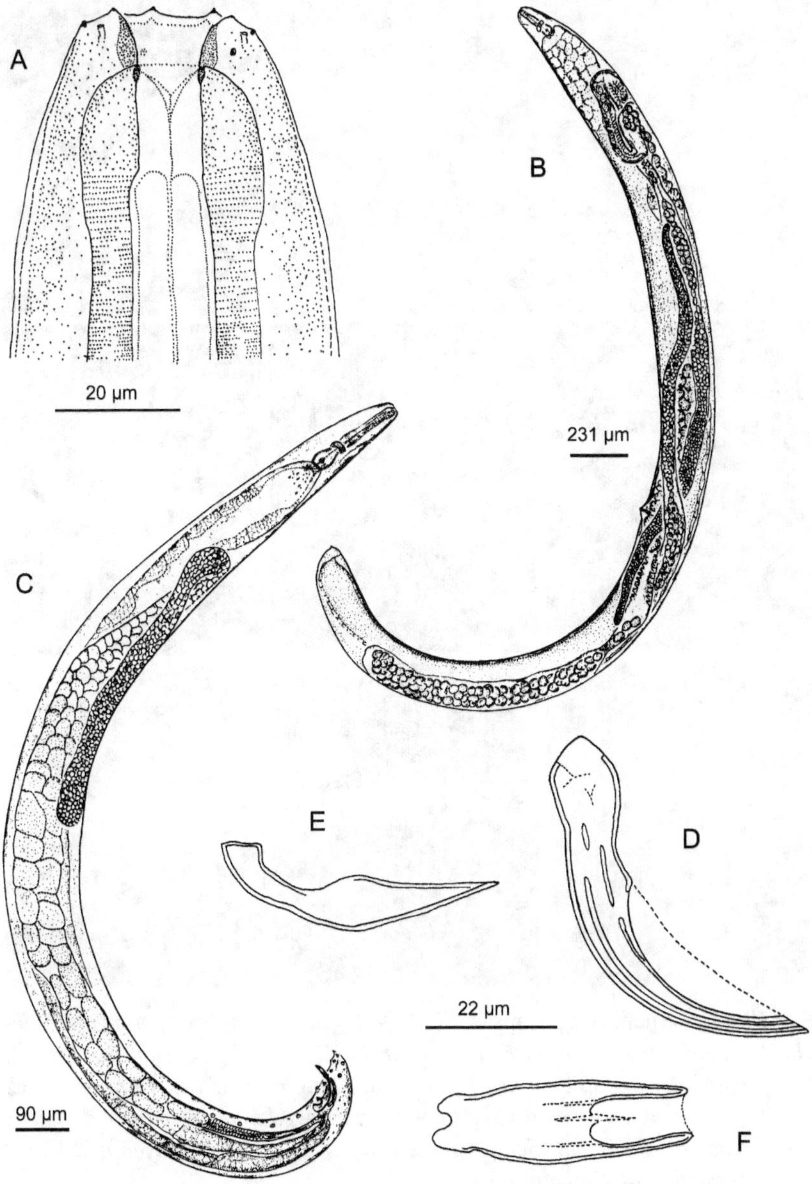

Fig. 58. Steinernema carpocapsae. *A: Female head region; B: Entire female; C: Entire male; D: Spicule, lateral view; E: Gubernaculum, lateral view; F: Gubernaculum, dorsal view. After Poinar (1967), courtesy of* Proceedings of the Helminthological Society of Washington.

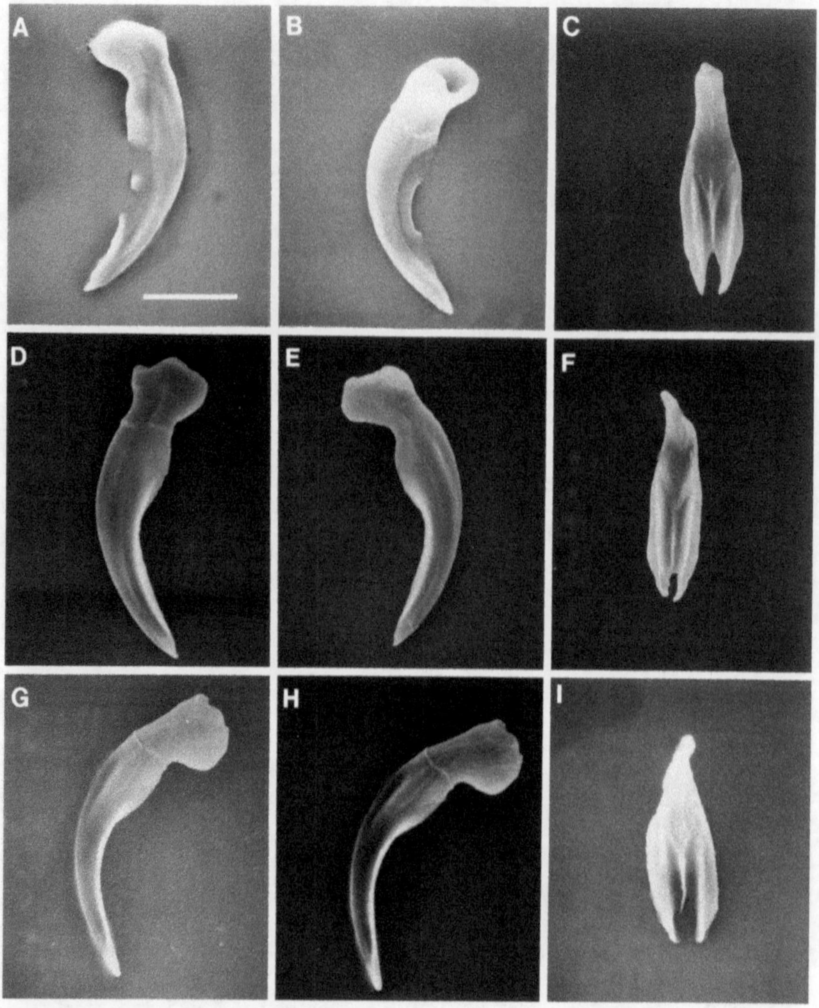

Fig. 59. Steinernema carpocapsae. *SEM photographs of spicules and gubernacula. A-C: Agriotos strain; D-F: Breton strain; G-I: Mexican strain. A, D, E, G, H: Spicule showing spicule head wider than long and ventrally projected; C, F, I: Gubernaculum, ventral view showing narrowed anterior part and cuneus arrowhead- or Y-shaped. (Scale bar: A-I = 20 μm.) After Nguyen and Smart (1997), courtesy of* Nematologica.

or slightly curved. Spicule head wider than long, ventrally projected (see Nguyen & Smart, 1997). Shaft present. Blade moderately curved, tapering to a point or with bluntly pointed terminus. Velum present. In

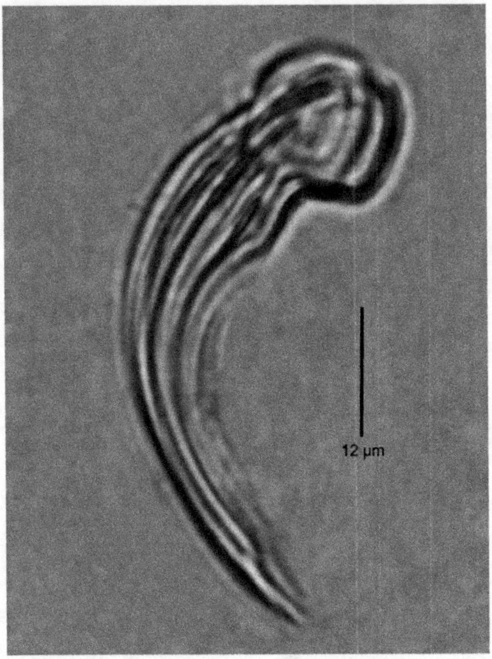

Fig. 60. Steinernema carpocapsae *light microscope photograph. Spicules, lateral view.*

ventral view, gubernaculum tapering anteriorly to form a short narrow part. Cuneus arrowhead-shaped or Y-shaped. Corpus separated posteriorly. Male tail with 23 genital papillae (11 pairs and a single midventral precloacal papilla) arranged as: five pairs subventral precloacal; two pairs cloacal; one pair lateral; and three pairs post cloacal. Tail tip with a mucron. Bursa absent.

Male (second generation)

Smaller than first generation male. Body diam. much smaller, other measurements shorter.

Female (first generation)

Variable in size, some giant forms reaching 10 mm. Anterior region similar to male but much larger. Excretory pore usually located anterior to nerve ring. Lateral fields and phasmids inconspicuous. Genital system amphidelphic, reflexed. Vulva in form of transverse slit, situated on a protuberance. Vagina short, with muscular walls. Tail bluntly conical to dome-shaped with or without a short spine on tip.

Table 17. *Morphometrics of* Steinernema carpocapsae. *Measurements are in μm and in the form: mean (range).*

Character	First generation		Infective juvenile
	Male	Female	
n	25	25	25
L	1450 (1090-1710)	3680 (2800-5160)	558 (438-650)
a	–	–	21 (19-24)
b	–	–	4.4 (4.4.8)
c	–	–	10 (9-11)
V	–	54 (52-56)	–
Body diam.	102 (77-131)	148 (123-185)	25 (20-30)
EP	61 (47-74)	61 (47-74)	38 (30-56)
NR	110 (93-124)	137 (118-161)	85 (76-99)
ES	155 (136-167)	191 (161-217)	120 (103-190)
Testis reflexion	563 (400-808)	–	–
Tail length (T)	30 (23-39)	36 (28-47)	53 (46-61)
Anal body diam. (ABD)	43 (33-55)	–	–
Spicule length (SL)	65 (59-72)	–	–
Spicule width	11 (9-13)	–	–
Gubernaculum length (GL)	47 (39-56)	–	–
Gubernaculum width	5 (4-7)	–	–
D% = EP/ES × 100	39	–	26 (23-28)
E% = EP/T × 100	–	–	60 (54-66)
SW% = SL/ABD × 100	151	–	–
GS% = GL/SL × 100	72	–	–

EP = distance from anterior end to excretory pore; NR = distance from anterior end to nerve ring; ES = distance from anterior end to end of pharynx.
After Poinar (1967, 1990).

Female (second generation)

Second and succeeding generation females in host are smaller in size. Pygmy forms never found.

Infective juvenile

Infective stage, when newly formed, always enclosed in cuticle of second-stage juvenile as a sheath. However, sheath is lost rather easily, even in storage, and thus may not always be present. Oral aperture and anus closed. Pharynx degenerate and thus not clearly visible, but basal bulb elongate and with valve. Tail pointed; lateral fields distinct with six ridges.

TYPE HOST AND LOCALITY

The species was isolated from the codling moth, *Carpocapsa pomonella*, in northeast Bohemia (Holovousy village), former Czechoslovakia.

TYPE MATERIAL

The type material from Czechoslovakia is lost (Mráček, personal communication). Many specimens, under the names *Steinernema carpocapsae*, *Neoaplectana carpocapsae*, or *N. dutkyi*, are deposited in the USDA Nematode Collection, Beltsville, MD, USA.

RELATIONSHIPS

Steinernema carpocapsae is characterised by the short body of the infective juvenile (558 μm) and E% = 60. For the male, the spicule head is wider than long, ventrally projected, and *ca* 66 μm long; ratio spicule length/spicule width *ca* 5.2 (4.7-6.0). The gubernaculum in ventral view is broad, the anterior part being short and the cuneus arrowhead- or Y-shaped.

This species can be differentiated from all other species of the *carpocapsae*-group by body length of the infective juvenile, 558 μm compared to 402 μm for that of *S. cumgarense* and *S. eapokense*, 438 μm for *S. asiaticum*, 446 μm for *S. siamkayai*, 486 μm for *S. sasonense*, 499 μm for *S. rarum*, 510 μm for *S. ritteri* and 530 μm for *S. tami*, and by the shapes of the spicule and gubernaculum (Figs 59, 60). This species can be distinguished from *S. scapterisci*, the closest species, by its shorter spicule measuring 66 (58-77) *vs* 83 (72-92) μm and by the presence of prominent epiptygmata in female *S. scapterisci.*

MOLECULAR CHARACTERISATION

Steinernema carpocapsae is characterised by sequence length of ITS regions, 987 bp (ITS1 = 279 bp, ITS2 = 301 bp) and its composition, A = 0.24417, C = 0.17427, G = 0.23202, T = 0.34954 (accession no. AF121049), and the length of D2D3 regions, 885 bp and its composition, A = 0.26328, C = 0.19322, G = 0.29040, T = 0.25311 (accession no. AF331900).

BIONOMICS

Poinar (1979) reported that 250 species of insects belonging to ten orders can be infected and killed by *S. carpocapsae*. Many studies have been made of this nematode species. These are too numerous to be reported in this monograph and the reader is referred to Poinar (1979) for more information.

DISTRIBUTION

This species is widespread in Europe and North America and has also been found in South America (Argentina, Brazil), China, Korea, Taiwan, Australia and New Zealand (Poinar, 1990; Hominick, 2002).

References

FILIPJEV, I.N. (1934). Miscellanea Nematologica 1. Eine neue Art der Gattung *Neoaplectana* Steiner nebst Bermerkungen über die systematische Stellung der letzteren. *Parazitologichesky* 4, 229-240.

HOMINICK, W.M. (2002). Biogeography. In: Gaugler, R. (Ed.). *Entomopathogenic nematodes*. Wallingford, UK, CABI Publishing, pp. 115-143.

NGUYEN, K.B. & SMART JR, G.C. (1997). Scanning electron microscopic studies of spicules and gubernacula of *Steinernema* spp. (Nemata: Steinernematidae). *Nematologica* 43, 465-480.

POINAR JR, G.O. (1967). Description and taxonomic position of the DD-136 nematode (Steinernematidae: Rhabditoidea) and its relationship to *Neoaplectana carpocapsae* Weiser. *Proceedings of the Helminthological Society of Washington* 34, 199-209.

POINAR JR, G.O. (1979). *Nematodes for biological control of insects*. Boca Raton, FL, USA, CRC Press, 277 pp.

POINAR JR, G.O. (1990). Taxonomy and biology of Steinernematidae and Heterorhabditidae. In: Gaugler, R. & Kaya, H.K. (Eds). *Entomopathogenic nematodes in biological control*. Boca Raton, FL, USA, CRC Press, pp. 23-60.

STANUSZEK, S. (1974). [*Neoaplectana feltiae pieridarum*, n. ecotype (Nematoda: Rhabditoidea, Steinernematidae) – a parasite of *Pieris brassicae* L. and *Mamestra brassicae* L. in Poland, morphology and biology.] *Zeszyty Problemowe Postepow Nauk Rolniczych* 154, 361-393.

TURCO, C.P., THAMES, W.H. & HOPKINS, S.H. (1971). On the taxonomic status and comparative morphology of species of the genus *Neoaplectana* Steiner (Neoaplectanidae: Nematoda). *Proceedings of the Helminthological Society of Washington* 38, 68-79.

WEISER, J. (1955). *Neoaplectana carpocapsae* n. sp. (Angullulinata: Steinernematidae) novy, cizopasnic housenik obalece jableeneho, *Carpocapsa pomonella* L. *Westnik Ceskoslovenske Zoologicke Spolecnosti* 19, 44-52.

WOUTS, W.M., MRÁČEK, Z., GERDIN, S. & BEDDING, R.A. (1982). *Neoaplectana* Steiner, 1929 a junior synonym of *Steinernema* Travassos, 1927 (Nematoda: Rhabditida). *Systematic Parasitology* 4, 147-154.

Steinernema caudatum[*,**] Xu, Wang & Li, 1991
(Fig. 61)

MEASUREMENTS

See Table 18.

DESCRIPTION

Male (first generation)

Cephalic extremity truncate with six labial and four cephalic papillae. Male with 23 genital papillae. Papillae usually distributed as follows: six pairs precloacal and subventral, a single midventral precloacal papilla, two pairs adcloacal, one pair postcloacal and subdorsal, and two pairs postcloacal and subventral. Spicule symmetrical, long and slender, yellowish-brown, curved 45-65°. Spicule head variable in shape, somewhat longer than wide. Shaft indistinct. Spicule tip blunt. Two internal ribs, dorsal rib curved to ventral side at posterior end, ventral rib becoming slender or disappearing slightly anterior to posterior end. Velum present, starting at end of spicule head and terminating at about mid-spicule. Gubernaculum slender and long in lateral view, anterior end curved ventrally, posterior end truncate, rounded, or bifurcate. Ventral view of gubernaculum: bifurcate at end, inner side of two branches parallel, ends of corpus relatively truncate. Membranous structure connecting the two branches. Ventral side of male tail slightly concave, which makes the tail appear slender. Tail terminus truncate, without mucron in first generation male, but pointed in second generation male because of presence of projection at the tail tip.

Female (first generation)

Anterior region as in male. Vulva located slightly posterior to mid-body. Uteri paired. Tail with postanal swelling, posterior part of tail usually slightly constricted. Shape of projections variable, but never pointed or sharp. The tail tapers gradually in second generation females.

[*] From Latin, *cauda* = tail.

[**] Description after Xu *et al.* (1991). Revised by Khuong B. Nguyen.

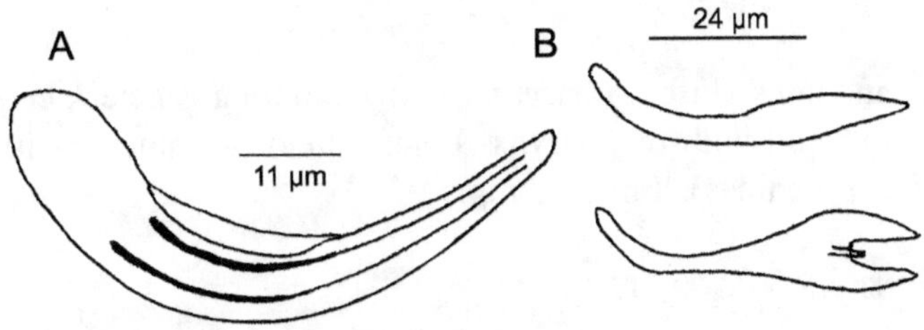

Fig. 61. Steinernema caudatum. *A: Diagrammatic spicule; B: Diagrammatic gubernaculum in lateral (upper) and ventral view (lower). Redrawn from the original description.*

Table 18. *Morphometrics of* Steinernema caudatum. *Measurements are in μm and in the form: mean (range).*

Character	First generation		Second generation		Infective juvenile
	Male	Female	Male	Female	
n	?	?	?	?	?
L	1620	3270	1430	1930	1106 (933-1296)
a	–	–	–	–	30.7
b	–	–	–	–	7.1
c	–	–	–	–	12.6
V	–	52	–	53	–
Body diam.	91	251	65	123	36 (34-41)
EP	116	114	116	109	82 (76-89)
NR	128	158	153	160	109 (104-126)
ES	164	218	199	235	156 (149-175)
Tail length (T)	44	49	48	89	88 (80-100)
Anal body diam. (ABD)	34	74	33	44	–
Spicule length (SL)	75	–	62	–	–
Spicule width	13	–	11	–	–
Gubernaculum length (GL)	52	–	42	–	–
Gubernaculum width	7.6	–	7	–	–
D% = EP/ES × 100	71	–	58	–	–
E% = EP/T × 100	–	–	–	–	94 (87-100)
SW% = SL/ABD × 100	221	–	188	–	–
GS% = GL/SL × 100	69	–	68	–	–

EP = distance from anterior end to excretory pore; NR = distance from anterior end to nerve ring; ES = distance from anterior end to end of pharynx.
After the original authors.

Infective juvenile

Stoma and anus closed. Pharynx narrow and long. Bacterial chamber connected to end bulb of pharynx. Lateral field prominent, with eight ridges (*i.e.*, nine lines). Tail tapering.

TYPE HOST AND LOCALITY

Type host unknown. Nematodes were collected by baiting with *Galleria mellonella* larvae from soil in Shangtou, Guangdong province, People's Republic of China.

TYPE MATERIAL

Holotype, allotype and paratypes deposited in Insect Collection, Department of Plant Protection, Southern China Agricultural University, People's Republic of China. Some paratypes were also deposited in Specimen Collection, Biology Department, Guangzhou Normal University, People's Republic of China.

RELATIONSHIPS

Steinernema caudatum is characterised by the tail length, E% of the infective juvenile and spicule length, D% and SW% of male and can be distinguished from other species in the *glaseri*-group by the above diagnostic characters.

COMMENT

This species is not well described yet appears to be distinct.

DISTRIBUTION

Steinernema caudatum is recorded from Guangdong province, People's Republic of China (Xu *et al.*, 1991; Han, 1994).

References

HAN, R. (1994). Advances in the research of entomopathogenic nematodes *Steinernema* and *Heterorhabditis* in China. *Entomologia Sinica* 1, 346-364.

XU, Z., WANG, G. & LI, X. (1991). A new species of the genus *Steinernema* (Rhabditida: Steinernematidae). *Zoological Research* 12, 17-20.

Steinernema ceratophorum[*,**] Jian, Reid & Hunt, 1997
(Figs 62-64)

MEASUREMENTS

See Table 19.

DESCRIPTION

Adults

Cuticle with fine transverse striae visible under SEM. Lateral field and phasmids inconspicuous. Cephalic extremity truncate to slightly rounded, continuous with body; oral aperture circular. Ten sensory papillae: six labial, four cephalic arranged in usual pattern; six lips, each with one papilla. Labial papillae not evenly distributed, two dorsal and two ventral papillae being located closer together. Stoma partially collapsed; pharyngeal collar absent; pharynx extending near to oral aperture. Cheilorhabdions located beneath lips and composed of thick, cuticularised ring. Posterior to this is another cuticularised ring, presumably prorhabdions; meso-, meta- and telorhabdions are vestigial and presumably occur in collapsed area of stoma. Pharynx typical of family; procorpus cylindrical and muscular; metacorpus slightly swollen, non-valvate; isthmus distinct; basal bulb muscular, with small, distinct, valve. Nerve ring surrounding isthmus and anterior portion of basal bulb. Pharyngo-intestinal valve almost bilaterally symmetrical and projecting into intestine. Excretory pore opening usually anterior to nerve ring; somewhat variable in location.

Male (first generation)

Cuticle, labial region, stoma and pharyngeal region as above. Body curved posteriorly, C- or J-shaped when heat-relaxed. Genital system monorchic, reflexed. Spicules paired, symmetrical, curved, of distinct dark yellow colour; spicule head elongate; head length about 1.5 times or more than head width, somewhat angular in shape; small rostrum or retinaculum present; velum extending nearly to the tip of lamina. Each spicule with two internal ribs of variable length. Gubernaculum

[*] Derived from the Greek *keratos* = a horn; and *phoreus* = a bearer, and referring to the horn-like cephalic protuberances.

[**] Description after Jian *et al.* (1997). Revised by David J. Hunt.

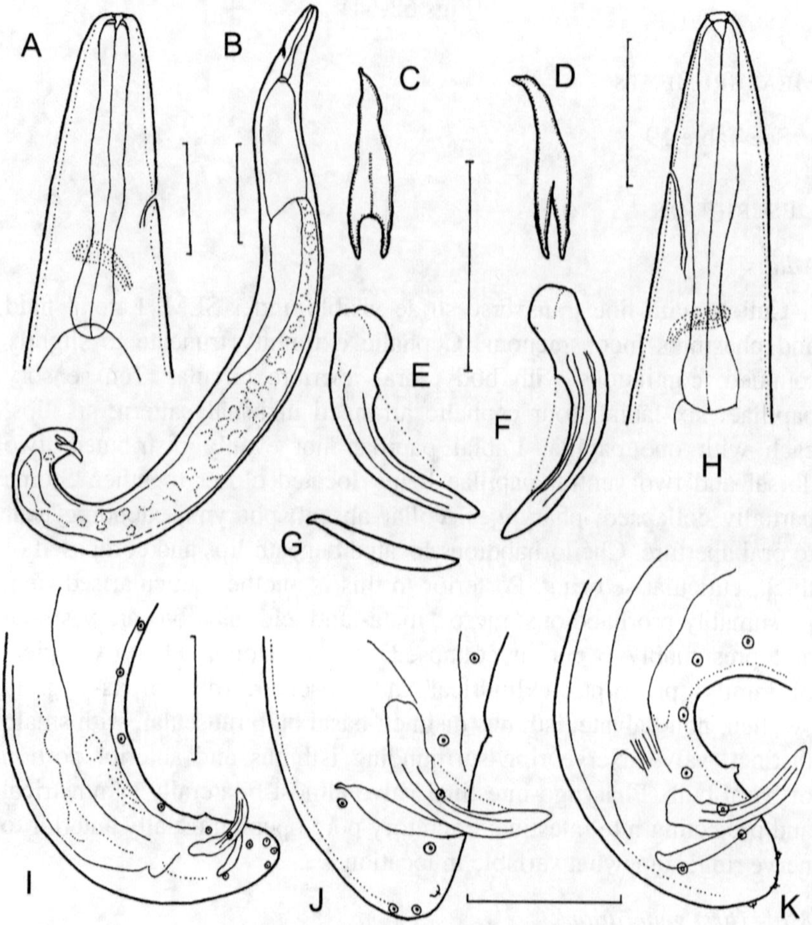

Fig. 62. Steinernema ceratophorum. *Male. A: Head and pharyngeal region of first generation; B: Entire body of first generation; C, D: Ventral view of gubernaculum; E, F: Spicules of first generation; G: Lateral view of gubernaculum; H: Head and pharyngeal region of second generation; I, J: Tail of first generation showing the location of genital papillae; K: Tail of second generation. (Scale bars: A, C-H, J, K = 50 μm; B = 200 μm; I = 100 μm.) Reproduced from* Systematic Parasitology, *37 (1997), p. 119;* Steinernema ceratophorum *n. sp. (Nematoda: Steinernematidae), a new entomopathogenic nematode from north-east China, by Jian, H., Reid, A.P. and Hunt, D.J., Figure 1. With kind permission of Springer Science and Business Media.*

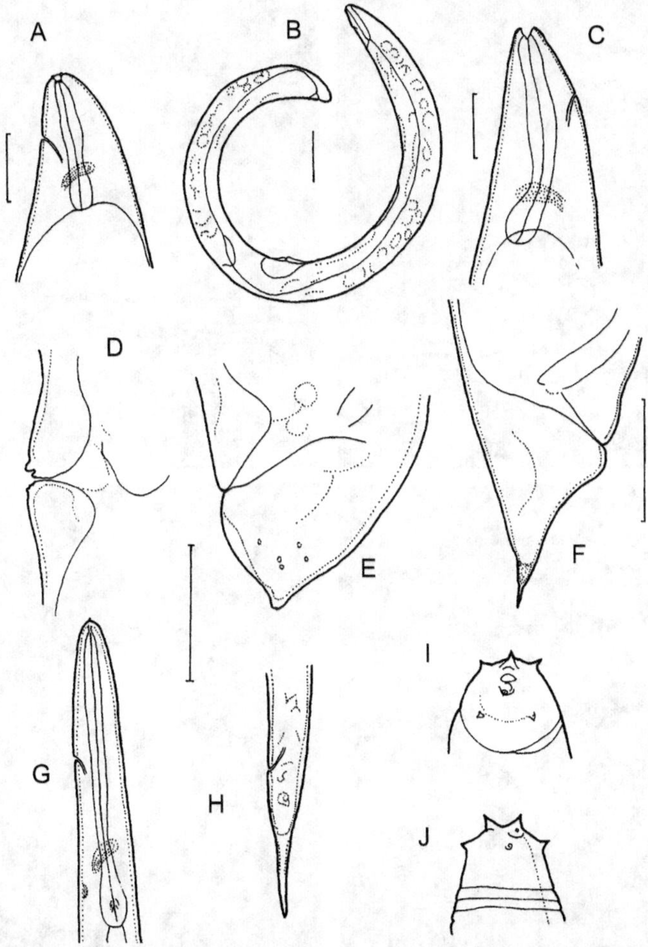

Fig. 63. Steinernema ceratophorum. *A-F: Female. A: Head and pharyngeal region of first generation; B: Entire body of first generation; C: Head and pharyngeal region of second generation; D: Vulva region of first generation; E: Tail of first generation showing small mucronate projection; F: Tail of second generation showing fine mucronate process. G-J: Infective juvenile. G: Lateral view of anterior portion; H: Tail; I, J: Head region showing two horn-like structures in lip region (from SEM picture). (Scale bars: A = 100 μm; B = 200 μm; C-J = 50 μm.) Reproduced from* Systematic Parasitology, *37 (1997), p. 120;* Steinernema ceratophorum *n. sp. (Nematoda: Steinernematidae), a new entomopathogenic nematode from north-east China, by Jian, H., Reid, A.P. and Hunt, D.J., Figure 2. With kind permission of Springer Science and Business Media.*

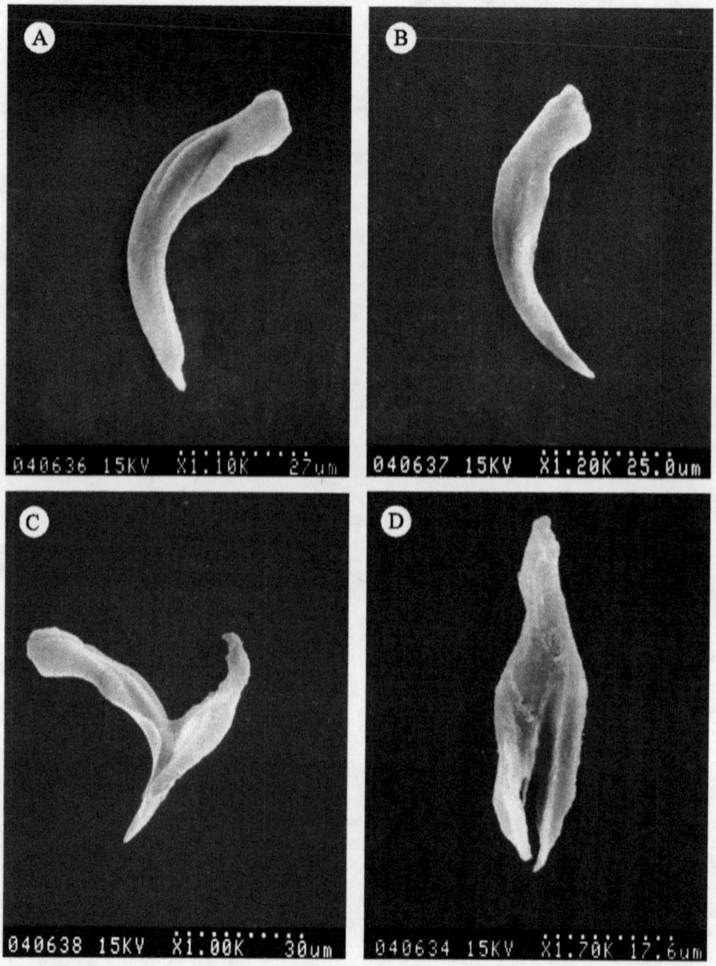

Fig. 64. Steinernema ceratophorum. *SEM micrographs. A, B: Spicule, lateral view; C: Spicule and gubernaculum, lateral view; D: Gubernaculum, ventral view. Courtesy of K.B. Nguyen.*

boat-shaped in lateral view and distally forked in ventral or dorsal view; anterior part thin with or without a proximal hook, ventrally curved, cuneus simple in dorsal view. Bursa absent. Twenty-three genital papillae (11 pairs plus one single) arranged as: six precloacal subventral pairs, most posterior being near cloaca; one single papilla located midventrally just anterior to cloaca; one adcloacal subdorsal pair located near the proximal section of gubernaculum; two postcloacal subventral

Table 19. *Morphometrics of* Steinernema ceratophorum. *Measurements are in μm and in the form: mean ± SD (range).*

Character	First generation		Second generation		Infective juvenile
	Male	Female	Male	Female	
n	35	35	35	35	45
L	1385 ± 134 (1136-1694)	5370 ± 1593 (2718-11899)	835 ± 52 (727-927)	1802 ± 396 (1227-2536)	706 ± 62 (591-800)
a	–	–	–	–	25.9 (23.7-27.9)
c	–	–	–	–	10.6 (8.8-12.9)
c′	0.6 (0.4-0.8)	0.41 (0.19-0.63)	0.6 (0.5-0.9)	1.0 (0.6-1.4)	4.2 (3.3-5.1)
V	–	54.4 (45-71)	–	58.5 (53-65)	–
Body diam.	146 ± 21 (104-185)	249 ± 36 (163-338)	65 ± 6 (54-79)	115 ± 20 (79-151)	27 ± 3 (23-34)
EP	85 ± 11 (50-104)	68 ± 18 (36-101)	64 ± 8 (45-74)	58 ± 8 (41-74)	55 ± 5 (47-70)
NR	123 ± 14 (90-147)	145 ± 12 (124-176)	106 ± 8 (83-120)	114 ± 9 (99-130)	92 ± 6 (79-103)
ES	165 ± 10 (149-190)	193 ± 12 (162-225)	135 ± 5 (122-144)	156 ± 7 (142-171)	123 ± 7 (108-144)
Testis reflexion	393 ± 94 (163-574)	–	125 ± 28 (77-203)	–	–
Tail length (T)	30 ± 4 (23-38)	36 ± 13 (11-65)	23 ± 3 (16-27)	48 ± 10 (25-72)	66 ± 5 (56-74)
Anal body diam. (ABD)	52 ± 5 (45-70)	87 ± 25 (54-160)	37 ± 3 (27-43)	46 ± 9 (34-68)	15 ± 2 (9-18)
Spicule length (SL)	71 ± 7 (54-90)	–	54 ± 4 (45-63)	–	–
Spicule width	9 ± 1 (7-11)	–	7 ± 1 (7-9)	–	–
Gubernaculum length (GL)	40 ± 4 (25-45)	–	31 ± 5 (23-41)	–	–
Gubernaculum width	7 ± 1 (5-9)	–	5 ± 1 (5-7)	–	–
D% = EP/ES × 100	51 ± 7 (33-69)	35 ± 9 (18-51)	47 ± 5 (34-59)	36 ± 5 (25-46)	45 ± 3 (40-56)
E% = EP/T × 100	283	296	278	121	84 ± 6 (74-96)
SW% = SL/ABD × 100	140 ± 20 (100-200)	–	150 ± 20 (120-210)	–	–
GS% = GL/SL × 100	60 ± 10 (40-80)	–	60 ± 10 (40-80)	–	–
H% = H/T × 100	–	–	–	–	~50

EP = distance from anterior end to excretory pore; NR = distance from anterior end to nerve ring; ES = distance from anterior end to end of pharynx.

pairs; and two subdorsal pairs near tail tip. Tail short, rounded conoid and lacking mucron.

Male (second generation)

Smaller than first generation male (average length: 0.9 *vs* 1.38 mm). Tail lacking mucron.

Female (first generation)

Body usually spiral or C-shaped when heat-relaxed. Vulva a transverse slit; vulval lips usually protruding slightly from body surface. Vagina short, leading into paired, amphidelphic genital tracts. Tail shorter than anal body diam, usually conoid in shape with small mucronate projection on tip.

Female (second generation)

Second generation females smaller than first generation females (average length: 1.7 *vs* 5.37 mm). Posterior region tapering evenly to salient posterior anal lip and ending in fine mucronate process *ca* 5.5 μm long.

Infective juvenile

Emerging infective juveniles often enclosed in second-stage cuticle. Body slender, gradually tapering from base of pharynx to anterior end and from anus to terminus. Cuticle with fine transverse striae; lateral field distinct with six or eight longitudinal ridges in mid-body region as seen under light microscope and with SEM. Marginal and submarginal ridges are less distinct than other ridges, and space (groove) between marginal and submarginal ridge is less than that of any two central ridges. Labial region continuous, rounded and hemispherical with pair of lateral horn-like structures near the oral aperture; oral aperture and anus closed; pharynx and intestine collapsed. Excretory pore located in anterior region of pharynx (D% = 45.6). Anterior portion of intestine with pouch containing cells of symbiotic bacteria. Tail straight, conoid, tapering evenly to finely pointed terminus; terminal spine absent. Hyaline region comprising *ca* 50% of tail length.

TYPE HOST AND LOCALITY

Type host unknown as the species was collected by baiting with larvae of *Galleria mellonella*. Collected in September 1994 from a tobacco field near Youn Ling town in Xin Bin county, Liaoning Province, north-east China.

TYPE MATERIAL

Holotype male, allotype female, ten paratype first generation males, five paratype females, 15 paratype infective juveniles and other population slides deposited in the CABI Europe-UK Nematode Collection, Egham, Surrey, UK; 18 paratype first generation males, 26 paratype first generation females, 15 paratype infective juveniles and other slides deposited in the Institute of Biological Control, Chinese Academy of Agricultural Sciences, Beijing 100081, People's Republic of China; seven paratype first generation males, four paratype first generation females, 15 paratype infective juveniles and other population slides deposited in Muséum national d'Histoire naturelle, Paris, France.

RELATIONSHIPS

Steinernema ceratophorum is characterised by the following combination of characters: infective stage juvenile with a body length of 706 (591-800) μm; juvenile labial region bearing a pair of horn-like structures and lateral field with six or eight longitudinal ridges; tail tip of both first and second generation males lacking a cuticular mucron; spicules curved with a distinctive dark yellow colour and *ca* 71 and 54 μm long in first and second generation respectively; 11 pairs of genital papillae plus a single genital papilla located on the anterior cloacal lip of male; first generation female tail conoid with a small mucronate projection; and second generation female tail tapering evenly posterior to salient posterior anal lip and ending in a fine mucronate process *ca* 5.5 μm long.

Steinernema ceratophorum, in having horn-like structures on the infective juvenile labial region, belongs to the *bicornutum*-group (which also includes *S. abbasi*, *S. bicornutum*, *S. pakistanense*, *S. riobrave* and *S. yirgalemense*). It can be distinguished from *S. bicornutum* by the larger spicules of both first and second generation males (71 *vs* 65 μm and 54 *vs* 51 μm), the shorter gubernaculum (40 *vs* 48 μm and 31 *vs* 33 μm, respectively), and mucron absent on second generation male tail *vs* present; from *S. riobrave* by longer IJ body length of 706 (591-800) *vs* 622 (561-701) μm, longer tail of 66 (56-74) *vs* 54 (46-58) μm, lower E% of 84 (74-96) *vs* 105 (93-111) and higher c′ value of 4.2 *vs* 3.3; from

S. abbasi by longer IJ body length of 706 (591-800) *vs* 541 (496-579) μm, EP = 55 (47-70) *vs* 48 (46-51), D% = 45 (40-56) *vs* 53 (51-58); from *S. pakistanense* by slightly longer tail of 66 (56-74) *vs* 58 (53-62) μm, male spicule shape, lower first generation male D% of 51 (33-65) *vs* 60 (50-60) and lower SW% of 140 (100-200) *vs* 180 (100-220); and from *S. yirgalemense* by the longer IJ body of 706 (591-800) *vs* 635 (578-693) μm and by the presence of 11 *vs* 12 pairs of genital papillae in the male.

MOLECULAR CHARACTERISATION

This species was originally diagnosed by its unique RFLP profile of ITS regions. The species is characterised by the sequence length of ITS regions, 992 bp (ITS1 = 244 bp, ITS2 = 342 bp) and its composition, A = 0.26915, C = 0.17137, G = 21976, T = 33972. The ITS region sequence is deposited with GenBank under accession no. AF440765.

BIONOMICS

The life cycle is similar to other steinernematids, that of the D-4-3 isolate being studied in detail by Jian *et al.* (1999). The life cycle in *Galleria* larvae took 232-240 h at 25°C. Infective juveniles had a higher tolerance to higher temperatures than *S. carpocapsae*.

Yan *et al.* (2005) characterised the associated bacterium and suggested that it was a new species as all four isolates examined formed a distinct clade when compared to known species of *Xenorhabdus*. Xiao *et al.* (2005) reported upon the antibiotic activities and properties of metabolites from *Xenorhabdus* sp. CB43, the bacterium isolated from *S. ceratophorum*.

Trials to assess the efficacy of *S. ceratophorum* in controlling various insects were reported by: Jian *et al.* (2002) for the bark beetle *Dendroctonus valens*; Yan *et al.* (2003) for the chrysomelid beetle *Luperomorpha suturalis* Chen; and Gao *et al.* (2004) for females of the tick *Haemaphysalis longicornis* Neumann.

DISTRIBUTION

In addition to the type locality near Youn Ling, Liaoning Province, north-east China, the species was also isolated from two sites in agricultural soils from Mei Hao county, Jining Province, north-east China.

References

GAO, Z.H., YANG, X.L., LIU, J.Z. & JIAN, H. (2004). [Virulence of entomopathogenic nematodes (Steinernematidae and Heterorhabditidae) to female ticks of *Haemaphysalis longicornis* Neumann (Acari: Ixodidae).] *Acta Entomologica Sinica* 47, 20-24.

JIAN, H., REID, A.P. & HUNT, D.J. (1997). *Steinernema ceratophorum* n. sp. (Nematoda: Steinernematidae), a new entomopathogenic nematode from north-east China. *Systematic Parasitology* 37, 115-125.

JIAN, H., YANG, X.F., ZHAO, X.Y. & YANG, H.W. (1999). [Study on the biology of entomopathogenic nematode, *Steinernema ceratophorum*.] *Chinese Journal of Zoology* 34, 2-6.

JIAN, H., YANG, X.F., LIU, X., YANG, H.W. & MIAO, Z.W. (2002). [Infectivity of entomopathogenic nematode *Steinernema ceratophorum* to bark beetle *Dendroctonus valens* larvae.] *Scientia Silvae Sinicae* 38, 157-159.

XIAO, L., YANG, X.F., DAI, L.Y., QIU, D.W., LIU, Z. & YUAN, J.J. (2005). [Antibiotic activities and properties of metabolites from *Xenorhabdus* sp. CB43.] *Journal of Hunan Agricultural University* 31, 412-414.

YAN, X., JIAN, H., LIU, Z., YANG, H., QUANLI, Z. & SHUANGYUE, L. (2003). Evaluation of entomopathogenic nematodes for control of the beetle, *Luperomorpha suturalis* Chen (Col., Chrysomelidae). *Journal of Applied Entomology* 127, 377-382.

YAN, X., XIN, Z.H., PANG, Y. & QIU, L.H. (2005). [Molecular identification of the symbiotic bacteria of *Steinernema ceratophorum*.] *Acta Scientiarum Naturalium Universatis Sunyatseni* 44, 116-118.

Steinernema cubanum[*,**] Mráček, Hernández & Boemare, 1994
(Figs 65-67)

MEASUREMENTS

See Table 20.

DESCRIPTION

Male (first generation)

Body elongate, curved posteriorly when heat-relaxed or fixed in 4% formalin or TAF. Cuticle smooth, slightly annulated when observed with SEM, without any distinct protuberances. Cephalic extremity rounded, slightly offset from rest of body. Three [*sic*] inconspicuous lips and two circles of head papillae: six labial papillae and four cephalic papillae. Amphidial apertures distinct, located laterally between labial and cephalic circle. Excretory pore variable in position, from slightly anterior to nerve ring to beginning of bulb, rarely to mid-bulb. Stoma and pharynx similar to those in female. Genital system monorchic, reflexed. There are 23 genital papillae comprising 11 pairs and a single midventral adcloacal papilla located just anterior to cloacal aperture. Of these 11 pairs, five pairs subventral and precloacal, six pairs postcloacal with three pairs consistently found on tail tip and three other pairs located more laterally, posterior to gubernaculum. Spicules colourless, paired, slightly curved, distally bluntly pointed. Spicule head (manubrium, capitulum) elongated, ratio length to width 1.5 : 1. Shaft (calomus) indistinct, lamina slightly arched, ventral rostrum either scarcely developed or much reduced. Velum narrow, extending from shaft or manubrium (if developed) nearly to tip of lamina. In lateral view, gubernaculum boat-shaped, remarkably attenuated in anterior region, posterior end variable, from straight, bluntly pointed to hook-like. Tail bluntly conoid. Mucron mostly absent, rarely developed as a minute spine in 5% of male specimens.

---

[*] Named after *Cuba*, the country of origin.

[**] Description after Mráček *et al.* (1994). Revised by Zdeněk Mráček.

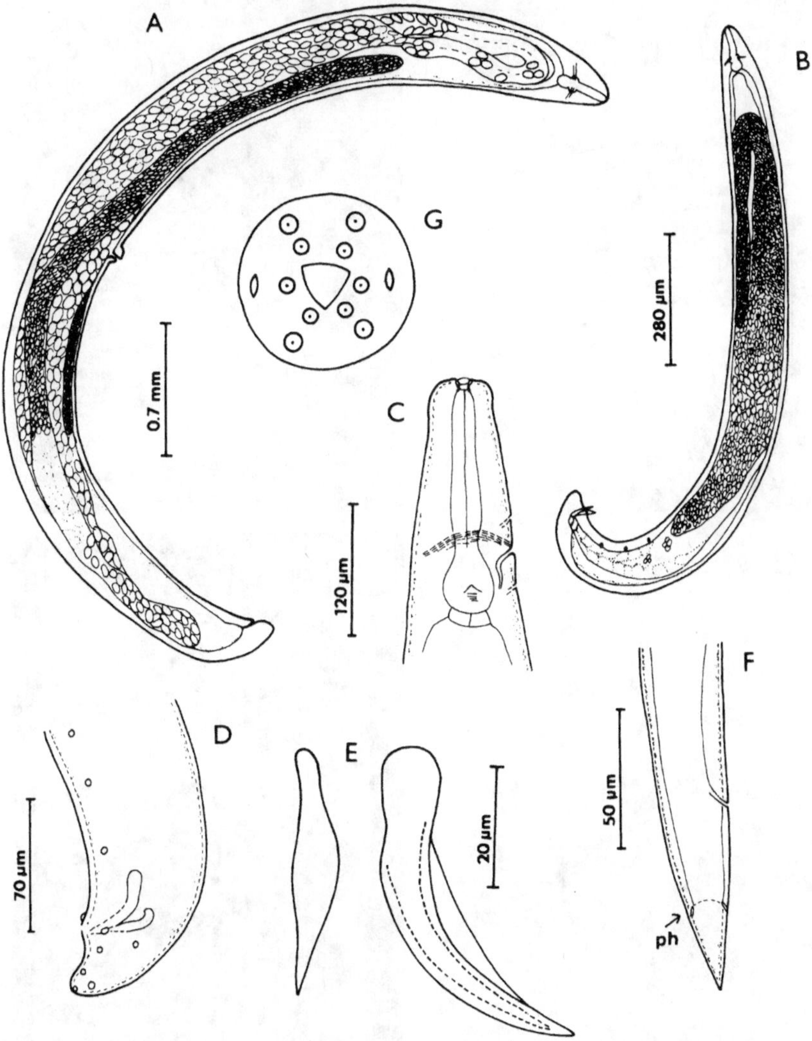

Fig. 65. Steinernema cubanum. *A: Female, first generation; B: Male, first generation; C-E: Male. C: Pharynx showing variability of excretory pore position; D: Tail, arrangement of genital papillae; E: Spicule and gubernaculum. F: Infective juvenile, tail with minute phasmids; G: Arrangement of head papillae and amphidial apertures. After Mráček* et al. *(1994), courtesy of* Journal of Invertebrate Pathology.

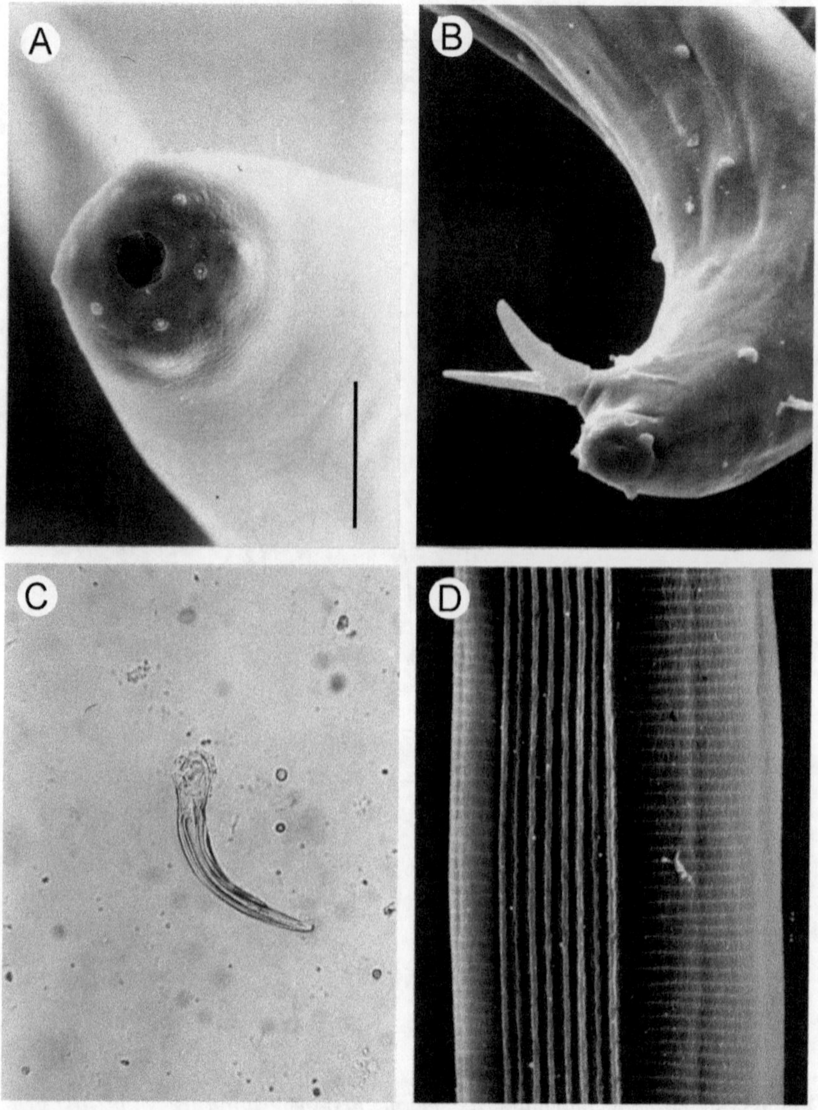

Fig. 66. Steinernema cubanum. *A-C: Male, first generation. A: Head with oral opening and circles of labial and cephalic papillae; B: Tail with spicules showing arrangement of genital papillae; C: Spicule. D: Infective juvenile, lateral field with eight ridges. (Scale bar: A = 18 μm; B = 44 μm; C = 39 μm; D = 16 μm.) After Mráček* et al. *(1994), courtesy of* Journal of Invertebrate Pathology.

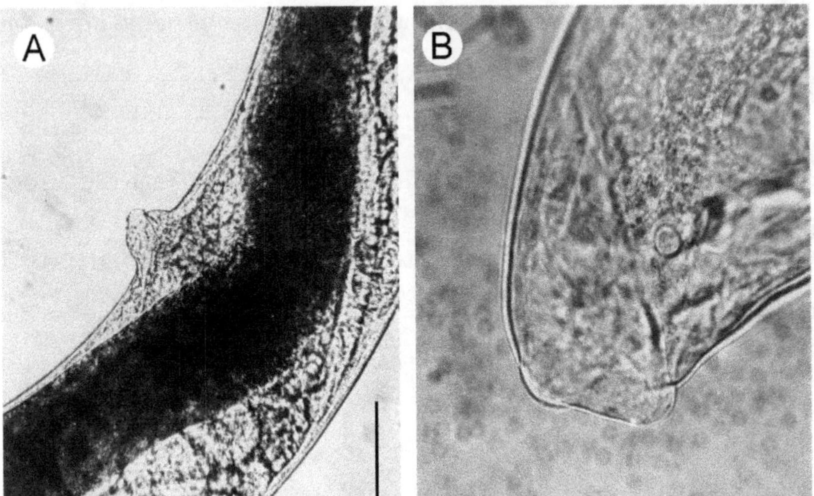

Fig. 67. Steinernema cubanum. *A, B: Female, first generation. A: Asymmetrical vulval opening; B: Blunt tail. C: Female, second generation, pointed tail tip; D: Infective juvenile, tail. (Scale bar: A = 62 μm; B = 46 μm; C = 72 μm; D = 50 μm.) After Mráček* et al. *(1994), courtesy of* Journal of Invertebrate Pathology.

Table 20. *Morphometrics of* Steinernema cubanum. *Measurements are in μm and in the form: mean (range).*

Character	First generation			Second generation		Infective juvenile
	Holotype (male)	Male	Female	Male	Female	
n	–	20	20	20	20	25
L	1333	1400 (1200-1700)	5500 (4000-9000)	1000 (800-1200)	2100 (1500-2700)	1283 (1149-1509)
a	–	15	21	15	15	35
b	7.7	8.9	25	6.6	11	8.6
c	51	39	105	29	38	19
c′	0.8	0.9	0.8	0.9	1.1	2.7
V	–	–	50	–	51	–
Body diam.	–	96 (77-117)	263 (178-391)	65 (53-71)	142 (107-231)	37 (33-46)
EP	148	124 (103-143)	182 (146-215)	113 (100-139)	137 (120-164)	106 (101-114)
NR	–	116 (106-148)	155 (141-164)	106 (94-119)	140 (125-146)	116 (106-130)
ES	174	156 (138-204)	219 (199-225)	150 (132-161)	188 (155-199)	148 (135-159)
Tail length (T)	26	36 (32-42)	52 (46-64)	34 (32-42)	55 (47-63)	67 (61-77)
Hyaline (H)	–	–	–	–	–	31 ± 0.4* (27-33)*
Anal body diam. (ABD)	36	41 (37-45)	62 (46-73)	38 (36-42)	48 (40-54)	25 (24-27)
Spicule length (SL)	49	58 (50-67)	–	52 (50-56)	–	–
Gubernaculum length (GL)	39	39 (37-42)	–	35 (32-40)	–	–
Gubernaculum width	–	6.3 (5.3-7.4)	–	5.5 (5.0-6.1)	–	–
D% = EP/ES × 100	85	70	80	70	70	70
E% = EP/T × 100	57	340	340	320	240	160
SW% = SL/ABD × 100**	136	141	–	137	–	–
GS% = GL/SL × 100**	80	67	–	67	–	–
H% = H/T × 100	–	–	–	–	–	40 ± 0.3 (38-43)*

EP = distance from anterior end to excretory pore; NR = distance from anterior end to nerve ring; ES = distance from anterior end to end of pharynx.

* According to Půža (2001).

** Recalculated from original data.

Male (second generation)

Second generation male similar to first generation in morphology but rather smaller in both body and organ sizes. Mucron absent, if rarely developed it looks like a minute spine.

Female (first generation)

Body elongate, moderately curved when heat-relaxed or fixed in 4% formalin or TAF. Cuticle smooth, slightly annulated when observed with SEM. Cephalic extremity bluntly rounded, slightly tapering anteriorly. Tail of mature female obtuse short, with a characteristic postanal lobe and papilla-like minute projections on end. In young female, tail longer with finger-like tip. Vulva a transverse slit, strongly protruding from the body surface, asymmetrical and situated at mid-body. Oral aperture surrounded by three [*sic*] inconspicuous lips, one circle of six labial papillae and one circle of four cephalic papillae. Amphidial apertures distinct, located laterally between labial and cephalic papillal circles. Stoma short, shallow. Cheilorhabdions prominent, sclerotised. Prorhabdions posterior to cheilorhabdions, smaller, sclerotised. Pharynx muscular with cylindrical corpus, slightly swollen metacorpus, distinct isthmus and well developed basal bulb containing valve. Nerve ring surrounding isthmus. Excretory pore located posteriorly to nerve ring, often at middle region of basal bulb. Genital system amphidelphic with reflexed ovaries. Vagina short, leading into paired uteri. Females oviparous, but later juveniles hatch inside body (= *endotokia matricida*). Rectum narrow, anal aperture distinct.

Female (second generation)

Similar to first generation female in body shape and arrangement of internal organs but smaller. Tail tip pointed, but not mucronate, tail length longer than tail diam. at anus. Excretory pore located more anteriorly than in first generation, close to nerve ring. Pygmy forms observed.

Infective juvenile

Body narrow, elongate, often enclosed in second-stage cuticle, sometimes this cuticle is lost. Oral aperture and anus closed. Labial papillae not developed, cephalic papillae and amphidial apertures distinct. Cuticle smooth, slightly annulated. Lateral fields composed of eight equally spaced and protruding ridges (*i.e.*, nine lines). Alimentary tract collapsed. Excretory pore slightly anterior to nerve ring, at *ca* 70% of pharynx length. Hemizonid at level of basal bulb. Tail pointed, broad with a low c′ ratio, but not attenuated. Hyaline region comprising *ca* 40% of tail length. Minute phasmid observed from dorsal or ventral view.

TYPE HOST AND LOCALITY

Natural host unknown. Collected from citrus plantations, Enrique Troncoso Enterprise, Pinar del Rio province, western Cuba, using the *Galleria* trap method. The plantation was damaged by weevils, *Cylas formicarius*, *Cosmopolites sordidus*, *Pachneus litus* and the moth, *Spodoptera frugiperda*.

TYPE MATERIAL

Holotype (male), allotype (female) deposited in the collection of the Department of Insect Pathology, Institute of Entomology, Czech Academy of Sciences, České Budějovice. Paratypes at the same place and at the Laboratoire de Pathologie Comparée, Université Montpellier, France.

RELATIONSHIPS

This species is characterised by very long, on average *ca* 1300 μm, infective juveniles which have a relatively posterior excretory pore, broad tail and short hyaline region. The lateral field is composed of eight equally-developed ridges. Males lack a tail mucron. Spicules colourless, slightly curved with elongate manubrium, indistinct calomus, and mostly missing rostrum. First generation females possess a transverse slit vulva, strongly protruding from the body surface, asymmetrical and situated at mid-body. Tail obtuse, with postanal swelling and terminal papilla-like projections.

Steinernema cubanum can be separated easily from all species (absence of mucron, oblong spicule, long infective juvenile, eight equally spaced lateral ridges, low ratio c′, *etc.*) apart from those in the *glaseri*-group, such as *S. glaseri*, *S. arenarium*, *S. puertoricense*, *S. apuliae*, *etc.*

It is differentiated from these similar species by the following characteristics. Males of *S. glaseri* possess a hooked spicule tip never observed in other species of this group. Pair of marginal ridges of lateral field are lightly annulated in *S. glaseri* infective juveniles, whereas they are smooth in *S. cubanum*. Position of the excretory pore of adults of the first and partly of the second generation is more posterior to nerve ring than in *S. glaseri*. Infective juveniles of *S. arenarium* differ from *S. cubanum* by its higher ratio c′ (3.8 *vs* 2.7). The tail of adult *S. arenarium* is much longer than that of *S. cubanum*. *Steinernema apuliae* differs from *S. cubanum* by its slightly shorter infective juveniles (1064 *vs* 1283 μm), much larger H% (60 *vs* 40) and by the more posterior vulval aperture (55 *vs* 50%). Geographically, *S. cubanum* is close to *S. puertoricense* but differs by the presence of double flapped epiptygmata in *S. puertoricense*. There is an important zoogeographical barrier for the African *S. karii* and the and Asian *S. loci*. In addition, the infective juvenile ratio D% is very low for *S. karii* (56.6%) and *S. loci* (57%), whereas it is 70% for *S. cubanum*. The ratio c′ for *S. longicaudum* IJ is very high (4.22) due to the tail length of over 90 μm (*vs* c′ = 2.7 for *S. cubanum*). *Steinernema diaprepesi* has shorter infective juveniles *ca* 1000 μm long and D% is close to 50%.

Molecular characterisation

Steinernema cubanum can be differentiated by RFLP profile of ITS regions. Additionally, this species is characterised by sequence length (flanked by primers 18S and 26S) of ITS regions, 988 bp, ITS1 = 279 bp, ITS2 = 304 bp, and its composition, A = 0.24244, C = 0.21550, G = 0.26574, T = 0.27632 (accession no. AY230166), and D2D3 region sequence length, 875 bp, and its composition, A = 0.24114, C = 0.20572, G = 0.31771, T = 0.23543 (accession no. AF331889).

Bionomics

The nematode is adapted to warm temperatures. It is not able to kill and reproduce in *Galleria mellonella* larvae at 10°C or lower (Mráček *et al.*, 1997). In laboratory rearing, the nematode kills *G. mellonella* at temperatures about 20°C and optimum temperature for reproduction is 26°C (data unpubl.).

Distribution

This species is only known from Cuba although its distribution over the island is unknown.

References

MRÁČEK, Z., HERNÁNDEZ, E.A. & BOEMARE, N.E. (1994). *Steinernema cubana* sp. n. (Nematoda: Rhabditida: Steinernematidae) and the preliminary characterization of its associated bacterium. *Journal of Invertebrate Pathology* 64, 123-129.

MRÁČEK, Z., BEČVÁŘ, S., ŘEZÁČ, P., KINDLMANN, P. & WEBSTER, J.M. (1997). Canadian steinernematid (Nematoda) isolates and their cold activity to greater wax moth (*Galleria mellonella*) larvae. *Biological Control* 8, 160-164.

PŮŽA, V. (2001). *Comparison of an infective juvenile tail morphology of the family Steinernematidae*. B.Sc. Thesis, University of South Bohemia, 23 pp.

Steinernema cumgarense[*],[**] Phan, Spiridonov, Subbotin & Moens, 2006
(Fig. 68)

MEASUREMENTS

See Table 21.

DESCRIPTION

Male (first and second generation not distinguished in original description)

Body fusiform, obese, C-shaped in heat-relaxed specimens, tapering towards anterior end. Cuticle annulation 1-1.3 μm wide, indistinct. No lateral fields visible but irregular anastomoses of cuticle annulation present along lateral body. Anterior extremity rounded with flattened area just around oral aperture. Six labial papillae and more prominent cephalic papillae present. Cheilostom cuticle connected to a thick, strongly cuticularised, 2-3 μm long ring 6-6.5 μm diam. forming middle part of stoma. Additional thinner cuticular ring situated posterior to thicker one, embedded into pharyngeal tissue. Pharynx corpus cylindrical, without visible metacorpal swelling. Basal bulb subpyriform. Nerve ring situated on anterior slope of basal bulb in most males. Valves in shape of platelets of thin folded cuticle. Cardia spherical, protruding for 18-22 μm into intestine proventriculus. Deirids 4-5 μm diam., situated posterior to pharyngeal-intestinal junction. Excretory channel strongly cuticularised for 10 μm from pore. Intestinal lumen open up to midbody. Separate spermatocytes poorly distinguishable, even in reflexed testis part, being replaced with complex rachis structure in testis middle, with characteristic patterns of spermatogenesis up to *vas deferens*. Posterior part of intestinal lumen open, forming a wide cavity. Characteristic steinernematid set of genital papillae with a row of usually nine pairs of subventral papillae extending from tail terminus to body middle, single, large, precloacal papilla in midventral position, one pair of papillae in lateral position at manubrium level when spicules retracted, and one pair of papillae in dorsolateral position in postcloacal area. Latter pair

[*] Named after the type locality, *Cu M'Gar*, Vietnam.

[**] Description after Phan *et al.* (2006). Revised by Zdeněk Mráček.

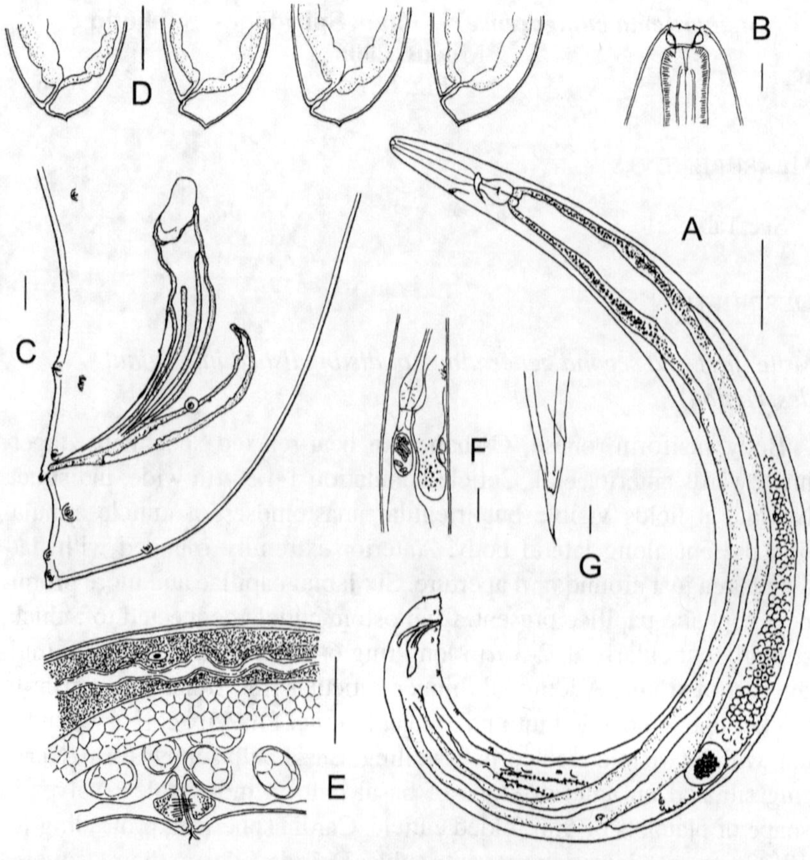

Fig. 68. Steinernema cumgarense. *A: First generation male, entire view. B, C: Male. B: Anterior end; C: Tail. D: First generation female, tail shapes; E: Vulvar region; F, G: Infective juvenile, pharyngeal region. G: Tail. (Scale bars: A, E = 100 μm; B, C, F, G = 10 μm; D = 50 μm.) After Phan* et al. *(2006), courtesy of* Russian Journal of Nematology.

usually closer to equatorial part of tail. Pair of postdeirids situated dorsolaterally at level of anteriormost pair of precloacal papillae. Spicules with elongated, funnel-shaped manubrium, tapering towards distal end, with round extremity and more or less pronounced terminal duct. Bifurcation of outer line or velum and ventral rib of spicule visible near rostrum. All four main ribs of spicule discernible, separated by gaps visible in lateral aspect. Spicule body gradually narrowing from rostrum to distal end. Cloacal aperture covered with flap when spicules retracted. Two

Table 21. *Morphometrics of* Steinernema cumgarense. *Measurements are in μm and in the form: mean ± SD (range).*

Character	First generation			Infective juvenile
	Holotype (male)	Male	Female	
n	–	20	20	25
L	1320	1307 ± 75 (1185-1470)	5178 ± 760 (4275-6045)	402 ± 14 (384-432)
a	15	14 ± 1 (12-15)	23 ± 2 (21-29)	18 ± 1 (17-19)
b	9	9 ± 0.3 (8.9-10)	23 ± 2 (20-26)	4.7 ± 0.2 (4.4-4.9)
c	80	82 ± 7 (69-93)	259 ± 66 (162-354)	11 ± 0.4 (11-12)
V	–	–	54 ± 2 (51-57)	–
Body diam.	90	96 ± 10 (90-120)	222 ± 29 (180-270)	22 ± 1 (21-25)
EP	77	76 ± 7 (62-86)	100 ± 9 (84-113)	35 ± 2 (33-38)
NR	101	102 ± 5 (90-108)	156 v 8 (146-173)	58 ± 2 (54-60)
ES	144	140 ± 6 (128-150)	222 ± 12 (204-240)	86 ± 3 (80-91)
Testis reflexion	396	330 ± 38 (283-396)	–	–
Tail length (T)	17	16 ± 2 (14-20)	21 ± 3 (17-26)	36 ± 2 (34-38)
Anal body diam. (ABD)	29	28 ± 2 (24-32)	64 ± 7 (50-77)	11 ± 1 (10-12)
Spicule length (SL)	75	76 ± 4 (68-81)	–	–
Spicule width	11	12 ± 1.2 (11-15)	–	–
Gubernaculum length (GL)	57	58 ± 5 (48-66)	–	–
Gubernaculum width	8	6 ± 1 (4.5-7.5)	–	–
D% = EP/ES × 100	53	45 ± 5 (35-52)	45 ± 5 (35-52)	41 ± 2 (38-43)
E% = EP/T × 100	–	–	–	99 ± 5 (87-103)
SW% = SL/ABD × 100	2.6 –	277 ± 20 (250-313)	–	–
GS% = GL/SL × 100	0.8	76 ± 4 (71-83)	–	–
H% = H/T × 100	–	–	–	53 ± 3 (48-59)

EP = distance from anterior end to excretory pore; NR = distance from anterior end to nerve ring; ES = distance from anterior end to end of pharynx.

papilla-like structures on cloacal margin. Gubernaculum usually straight with ventrally curved proximal end, often with cuticular knobs on inner surface. Tail mucron spike-like, 1.9 ± 0.6 (1.5-3.0) μm long.

Female (first and second generation not distinguished in original description)

Cuticle annulation 1.5 μm wide, easily discernible near vulva lips. Cephalic extremity with six labial and four cephalic papillae. Cuticular ring forming main part of stoma walls 5-6 μm thick and 7-7.5 μm long. Stoma 10 μm diam. Excretory duct strongly cuticularised, entering body at almost 90° to the longitudinal axis. Vulva equatorial, epiptygmata present. Vulva lips equal in size, protruding up to 20-25 μm. Tail terminus dome-shaped, bearing peg-like, 3-5 μm long, mucron.

Infective juvenile

Body of heat-relaxed juveniles quite obese, moderately C-shaped, tapering towards anterior end. Cuticular annulation *ca* 1 μm wide visible throughout body. Lateral field in anterior half of body with four ridges in centre, two pronounced marginal ridges and poorly discernible submarginal lines. Such regular pattern not observed in posterior half of body, with ventral gap between central and marginal ridges disappearing. Lateral field 8.5-9.5 μm wide at mid-body. Cuticle swollen near anterior end forming 4-5 μm long cephalic capsule. Frontal surface of anterior end always regularly rounded. Labial papillae indistinct, four cephalic papillae and amphidial pouches occasionally visible. Pharynx corpus slender, slightly widening to metacorpal part. Nerve ring always located about corpus-isthmus junction. Hemizonid on isthmus level, often with additional fibres 1-3 μm posterior to main one. Basal bulb with poorly visible valves and large nucleus in dorsal anterior half. Cardia *ca* 10 μm long, narrowing to posterior end. Bacterial vesicle elliptical, 10-13 μm long and 5-6 μm diam. Rectum *ca* two anal body diam. long. Genital primordium 50 μm long. Phasmids at one-third of tail length posterior to anus (phasmid-mucron distance *ca* 25-26 μm), with aperture near ventral edge of ventral marginal line. Lateral field lines disappearing at level of phasmids. Tail with pronounced ventral concavity, ventrally curved with terminus pointed to finely rounded. Hyaline region comprising *ca* half of tail length.

TYPE HOST AND LOCALITY

Natural host unknown as nematodes were obtained by *Galleria mellonella* baiting method. Collected near the village Cu M'Gar in the Daklak province of Vietnam (longitude 108°02′E, latitude 12°29′N, altitude 600 m a.s.l.) from the red basalt soil of highland coffee plantations.

TYPE MATERIAL

Holotype male and seven paratype males deposited at Institute of Parasitology of Russian Academy of Sciences, Leninskii prospect 33, Moscow, 119071, Russia. Slides with paratype males, first generation females and infective juveniles deposited in Department of Nematology, Institute of Ecology and Biological Resources, National Center for Science and Technology, Hoang Quoc Viet 18, Nghiado, Caugiay, Hanoi, Vietnam.

RELATIONSHIPS

Males have two papilla-like structures on the cloacal margin, the spicules have an elongate manubrium and well developed rostrum and there is a short, spike-like, caudal mucron in both generations. The first generation female has a dome-shaped tail and a minute, peg-like, mucron. Vulval epiptygmata are present. The infective juveniles belong to the shortest group with an average body length of 402 μm and the hyaline region comprises *ca* half of the tail length. The lateral fields have four ridges in the mid-body region.

Nematodes of this species have infective juveniles 402 (384-432) μm long and lateral fields with four central ridges and two prominent marginal ridges (at least in the anterior half of the body), but lack horn-like appendages on the cephalic extremity, and thus belong to the *carpocapsae*-group. In IJ body length, this species is similar to *S. eapokense* (L = 402 (370-434) μm) and *S. siamkayai* (L = 446 (308-495) μm), but is shorter than *S. sasonense* (L = 471 (437-516) μm), *S. carpocapsae* (L = 546 (427-640) μm), *S. scapterisci* (L = 572 (517-609) μm) and *S. tami* (L = 530 (400-600) μm).

Steinernema cumgarense is closely related to *S. eapokense* but the IJ can be distinguished by the higher D% of 41 (38-43) *vs* 34 (32-37) and higher E% of 99 (87-103) *vs* 73 (66-80). The IJ of *S. cumgarense* has a relatively thicker body than *S. siamkayai* (a = 17-19 *vs* 19-23). *Steinernema cumgarense* differs from *S. asiaticum* by the length of the spicule and gubernaculum of the first generation male (spicule = 76 (68-81) *vs* 58 (53-62) μm; gubernaculum = 58 (48-66) *vs* 43 (36-48) μm, respectively). The IJ of *S. cumgarense* also differs from *S. sasonense* by the shorter pharynx length of 86 (80-91) *vs* 100 (94-104) μm and shorter tail length of 36 (34-38) *vs* 47 (42-52) μm.

Molecular characterisation

Steinernema cumgarense is characterised by sequence lengths of ITS regions (ITS1+5.8S+ITS2) bp, ITS1 = 269 bp, ITS2 = 306 bp and its composition, A = 0.22877, C = 0.16301, G = 0.21917, T = 0.38901. The sequence of ITS regions is deposited in GenBank (accession no. AY487920).

Comments

More precise description of taxonomically important characters, such as the number of ridges in the infective juvenile lateral field, the form of the first generation female postanal swelling and the occurrence of papilla-like mucrons on the female tail tip are missing or unsatisfactorily described.

Distribution

The species has been found in Vietnam and has been reported from Guangdong, People's Republic of China (Phan *et al.*, 2006).

Reference

Phan, K.L., Spiridonov, S.E., Subbotin, S.A. & Moens, M. (2006). Four new species of *Steinernema* Travassos, 1928 with short infective juveniles from Vietnam. *Russian Journal of Nematology* 14, 11-29.

Steinernema diaprepesi[*],[**] Nguyen & Duncan, 2002
(Figs 69-74)

MEASUREMENTS

See Table 22.

DESCRIPTION

Male (first generation)

Body curved ventrally posteriorly, C-shaped when heat-relaxed. Lateral field present in mid-body with one narrow ridge. Cephalic extremity rounded, usually slightly swollen. Labial region with perioral disc, six labial papillae, two amphidial apertures (usually covered with exudates) and four cephalic papillae. Stoma shallow, cheilorhabdions as small and sclerotised structures at anterior end, sometimes indistinct. Excretory pore near nerve ring, located mostly anterior to basal bulb. Pharynx with cylindrical procorpus, metacorpus absent or slightly swollen, isthmus present, nerve ring around isthmus, basal bulb distinct. Pharyngo-intestinal valve present, usually weak. Genital system monorchic, reflexed. Distance from base of pharynx to anterior end of testis variable. Spicules paired, brown in colour. Head of spicules elongate, length/width ratio from 1.4-2.0 (averaging 1.7 ± 0.2) in some specimens (4/20), twice as long as wide, shaft very short or absent; blade thick, gradually tapering posteriorly, blade terminus blunt with a longitudinal depression/slit in ventral side; velum present. Each spicule with two internal ribs. Gubernaculum boat-shaped in lateral view, anterior part usually with one or two ventral projections. Eleven pairs of genital papillae (occasionally 12) and one single precloacal papilla. Tail conoid, tail terminus without mucron. General morphological characteristics of males collected from *Diaprepes abbreviatus* appear similar to those collected from *Galleria mellonella*, but differing morphometrically. Most measurements of first generation males from *D. abbreviatus* shorter than those collected from *G. mellonella*.

Male (second generation)

Second generation male similar to that of first generation except body, spicule and gubernaculum shorter and thinner, excretory pore much

[*] Named after the genus of the type host, *Diaprepes abbreviatus*.

[**] After Nguyen and Duncan (2002). Revised by Khuong B. Nguyen.

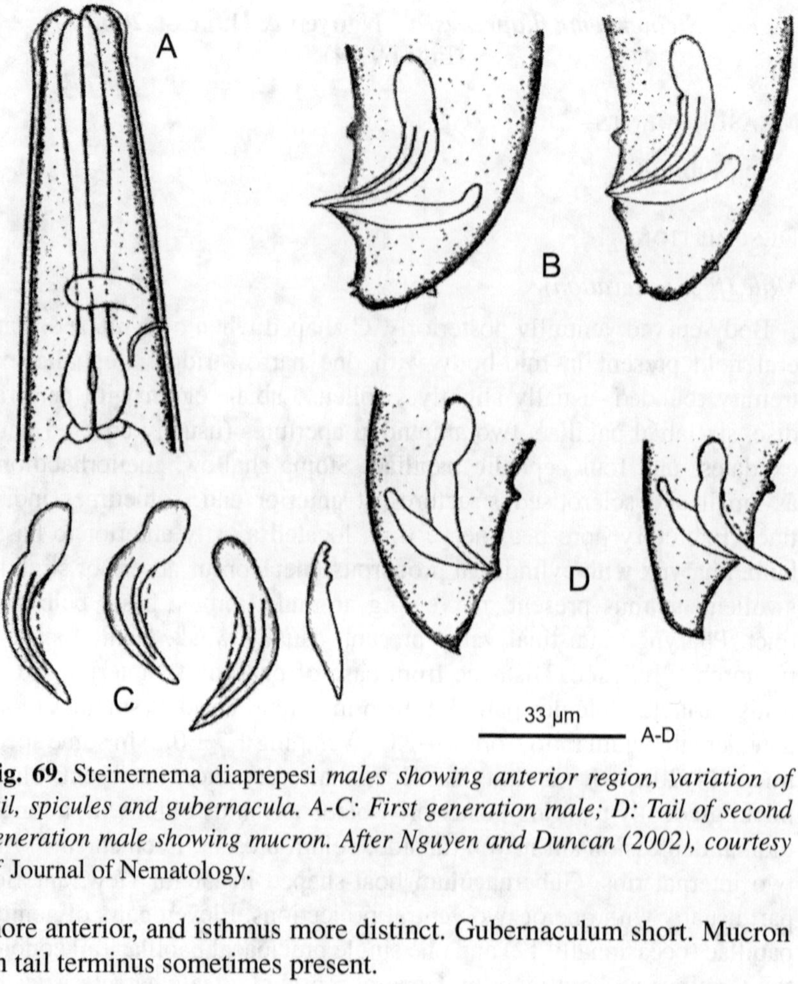

Fig. 69. Steinernema diaprepesi *males showing anterior region, variation of tail, spicules and gubernacula. A-C: First generation male; D: Tail of second generation male showing mucron. After Nguyen and Duncan (2002), courtesy of* Journal of Nematology.

more anterior, and isthmus more distinct. Gubernaculum short. Mucron on tail terminus sometimes present.

Female (first generation)

Body cuticle smooth or with faint annules. Lateral fields and phasmids absent. Cephalic extremity rounded, continuous with body; six labial papillae, four cephalic papillae. Lips indistinct. Amphidial apertures present. Stoma shallow, subtriangular anteriorly, triradiate internally. Cheilorhabdions, well sclerotised but small. Smaller sclerotised structure posterior to cheilorhabdions (presumably the prorhabdions), observed in other species, indistinct in this species. Pharynx with cylindrical procorpus, muscular, metacorpus swollen, isthmus distinct, basal bulb valvate. Nerve ring surrounding isthmus, just anterior to basal bulb.

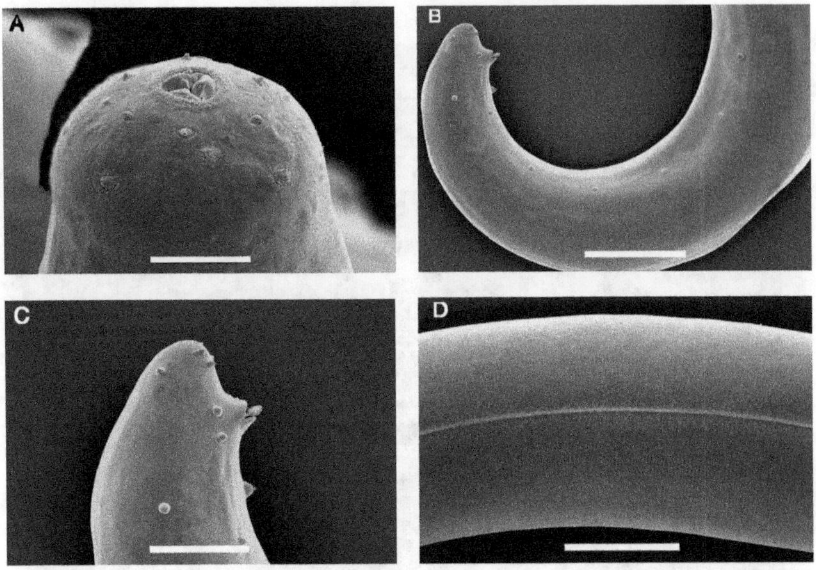

Fig. 70. Steinernema diaprepesi. *SEM of first generation male. A: Slightly swollen head, showing anterior end of stoma with a perioral ring, six labial papillae and two of four cephalic papillae; B, C: Posterior region of male showing number and distribution of genital papillae; D: Lateral field at mid-body with a single ridge. (Scale bars: A = 10 μm; B = 66.7 μm; C, D = 30 μm.) After Nguyen and Duncan (2002), courtesy of* Journal of Nematology.

Pharyngo-intestinal valve present. Excretory pore located just anterior to, or at middle of basal bulb. Genital system amphidelphic, reflexed, often containing many eggs. Vulva, a transverse slit; protruding or not; low double-flapped epiptygmata present. Vagina sclerotised, short. Body diam. greater anterior to vulva than posterior to vulva. Tail shape variable, ventral postanal swelling present, tail shorter than anal body diam. Most females with five papilla-like structures on tail tip. Structures of variable size depending on age of females, usually longer in young females, becoming smaller and shorter as female body enlarges, disappearing in fully mature female.

Female (second generation)

Similar to first generation female but smaller. Vulva less protuberant, epiptygmata usually more prominent than that of first generation

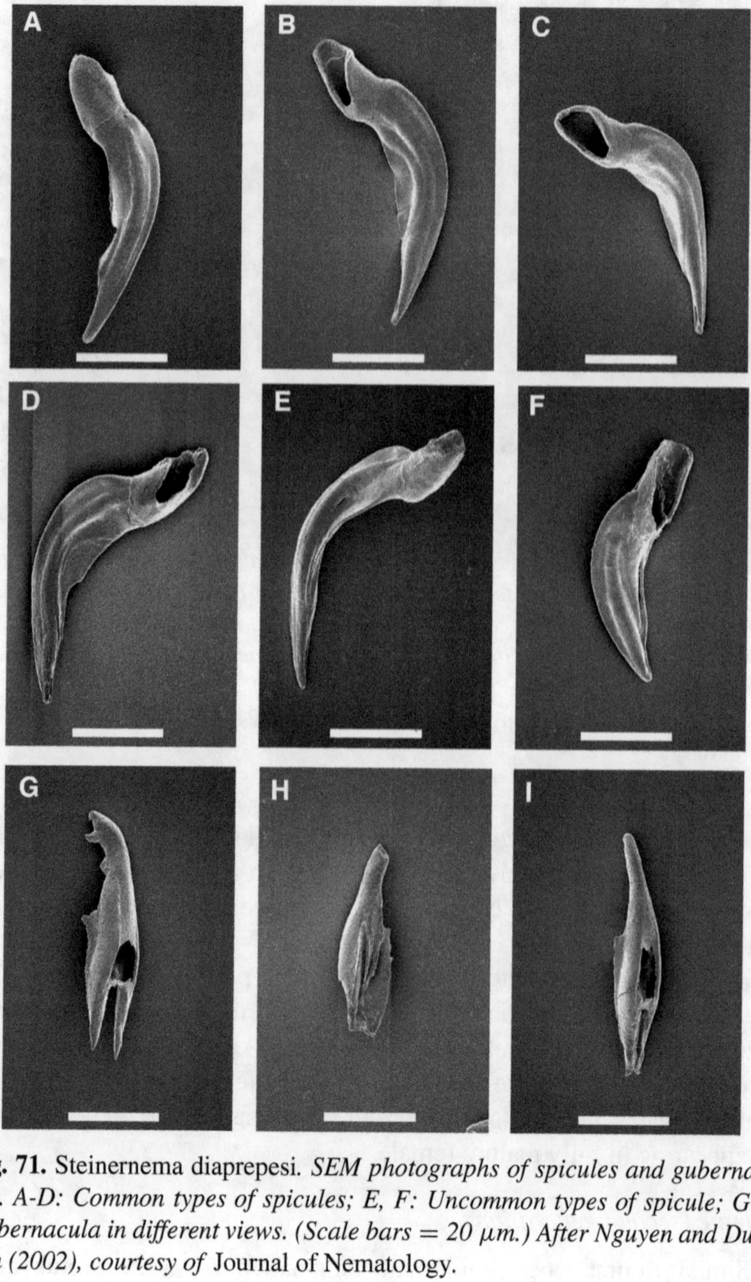

Fig. 71. Steinernema diaprepesi. *SEM photographs of spicules and gubernacula. A-D: Common types of spicules; E, F: Uncommon types of spicule; G-I: Gubernacula in different views. (Scale bars = 20 μm.) After Nguyen and Duncan (2002), courtesy of* Journal of Nematology.

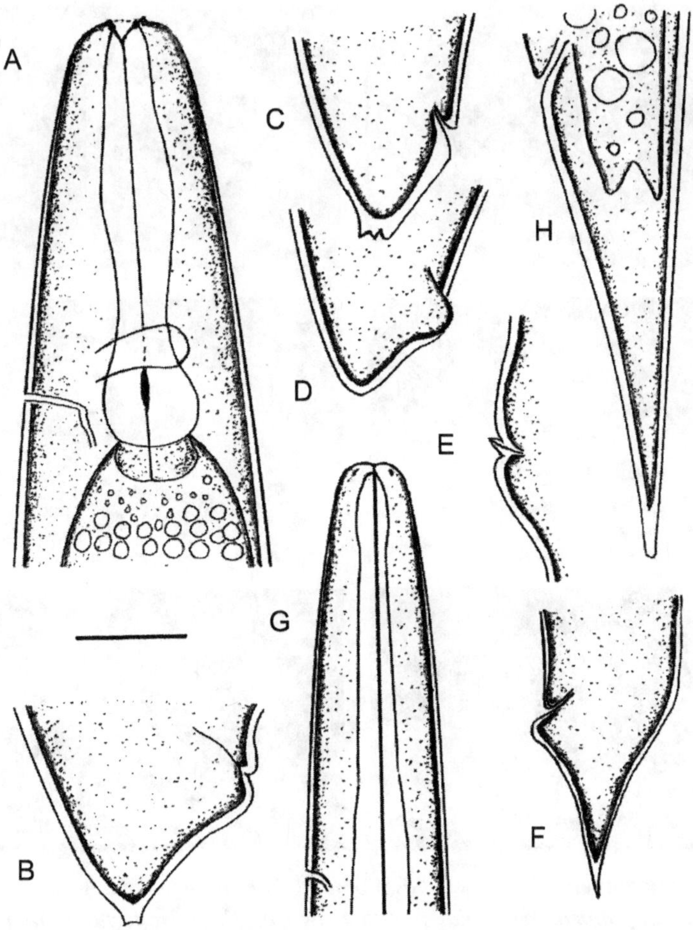

Fig. 72. Steinernema diaprepesi. *A: Anterior region of first generation female; B-D: Variation in tail shape of first generation female; E: Epiptygmata of second generation female; F: Second generation female tail; G, H: Head and tail of infective juvenile. (Scale bar: A-F = 33 μm; G-H = 20 μm.) After Nguyen and Duncan (2002), courtesy of* Journal of Nematology.

females. Tail, tapering to a pointed end, longer than anal body diam.; ventral postanal swelling present.

Infective juvenile

Body elongate. Sheath (= second-stage cuticle) present immediately after harvesting, but many IJ losing their sheath in storage. Cephalic region smooth, continuous, rounded anteriorly. Labial papillae not seen;

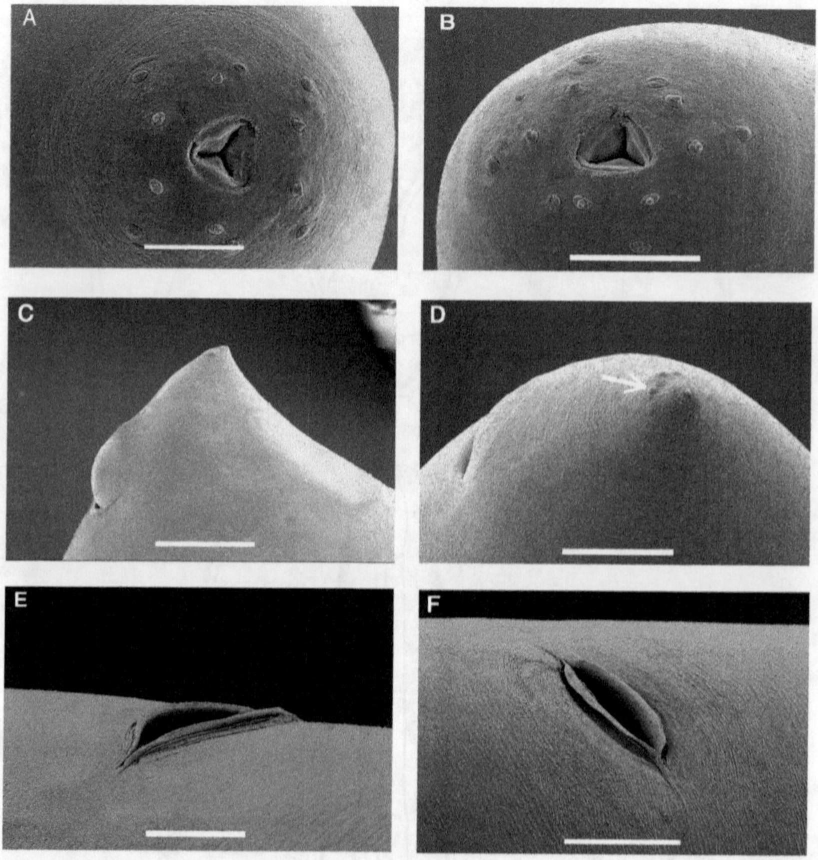

Fig. 73. Steinernema diaprepesi. *SEM of first generation female. A, B: Anterior end showing stoma, six labial papillae, two amphidial apertures, and four cephalic papillae; C, D: Tails showing anus, ventral postanal swelling (C), and tail terminus with papilla-like structures (D, arrow); E, F: Low, double-flapped, epiptygmata. (Scale bars: A, B = 10 μm; C = 20 μm; D = 15 μm; E = 10 μm; F = 15 μm.) After Nguyen and Duncan (2002), courtesy of* Journal of Nematology.

four prominent cephalic papillae. Amphidial apertures slit-shaped, but not prominent. Cuticle marked with prominent transverse striations. Lateral field beginning anteriorly with one line. Two additional lines appearing at annules 14-16 to form two ridges. Near excretory pore level, number of ridges in lateral fields increasing from two to six. An additional central ridge appearing more posteriorly, *ca* a body diam.

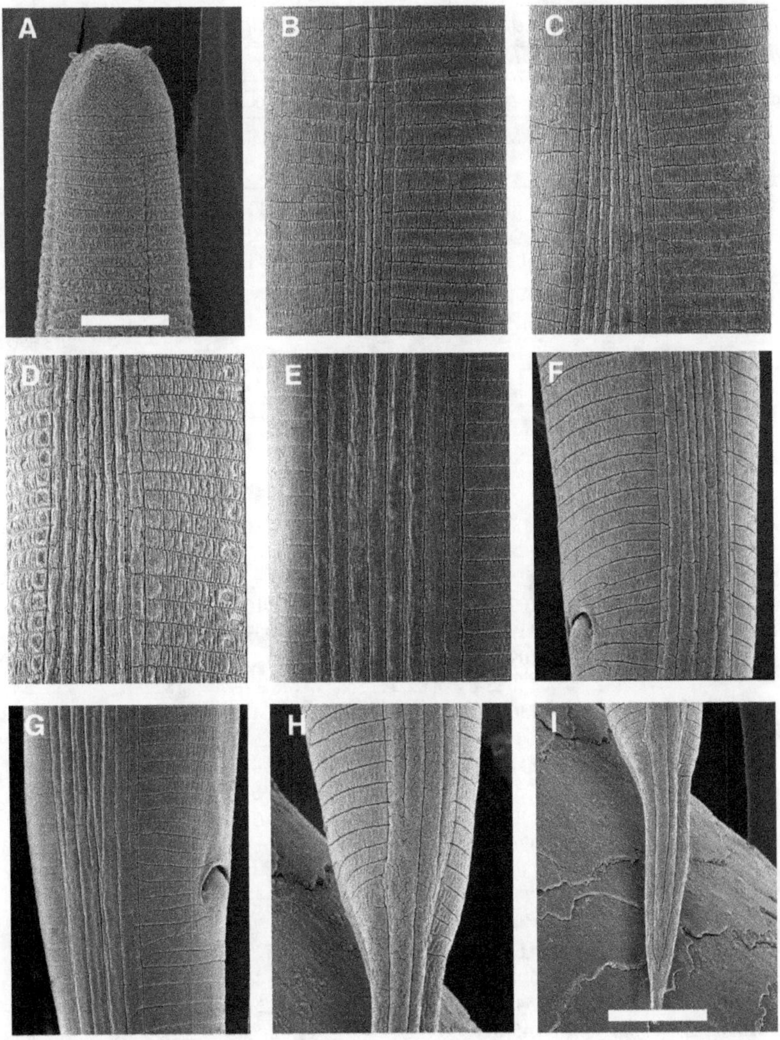

Fig. 74. Steinernema diaprepesi. *SEM of lateral field of infective juvenile. A: Anterior region showing prominent cephalic papillae; B: Lateral field with two ridges anteriorly, six ridges posteriorly; C: Lateral field showing seventh ridge arising in middle of lateral field; D: Lateral field showing middle ridge dividing into two making a total of eight; E: Eight ridges in lateral field; F, G: Left and right lateral fields showing disappearance of two marginal and two central ridges leaving four ridges in all; H: Number of ridges in lateral field reducing to two; I: Tail showing end of lateral field. (Scale bars: A-H = 7.5 μm; I = 12 μm.) After Nguyen and Duncan (2002), courtesy of* Journal of Nematology.

Table 22. *Morphometrics of* Steinernema diaprepesi. *Measurements are in* μm *and in the form: mean* ± *SD (range).*

Character	First generation			Second generation		Infective juvenile
	Holotype (male)	Male	Female	Male	Female	
n	–	20	20	20	20	50
L	1669	1735 ± 156 (1505-2078)	6508 ± 1512 (4061-9878)	1176 ± 117 (1036-1343)	2433 ± 260 (2030-3030)	1002 ± 53 (880-1133)
a	–	–	–	–	–	30 ± 2.6 (23-35)
b	–	–	–	–	–	7.3 ± 0.5 (6.5-8.3)
c	–	–	–	–	–	12.1 ± 0.7 (10.4-13.2)
c′	–	–	–	–	–	3.6
V	–	–	50 ± 2.8 (44-57)	–	54 ± 3 (49-60)	–
Body diam.	127	113 ± 15 (90-145)	94 ± 22 (59-138)	64 ± 7 (54-78)	158 ± 12 (137-180)	–
EP	106	115 ± 9 (100-130)	156 ± 39 (98-214)	83 ± 7 (71-108)	116 ± 15 (94-139)	74 ± 6.4 (66-83)
NR	97	119 ± 6 (109-129)	150 ± 21 (105-179)	110 ± 8 (100-136)	144 ± 12 (120-153)	102 ± 6 (74-109)
ES	132	150 ± 7 (136-162)	200 ± 25 (119-232)	138 ± 12 (129-168)	182 ± 11 (161-211)	138 ± 7.4 (111-152)
Testis reflexion	533	391 ± 68 (241-542)	–	264 ± 16 (187-428)	–	–
Tail length (T)	21	25 ± 3 (20-32)	52 ± 10 (38-68)	24 ± 5 (18-30)	78 ± 6 (67-91)	83 ± 5 (65-91)
Anal body diam. (ABD)	21	42 ± 11 (36-50)	–	34 ± 6 (29-42)	–	23 ± 2.3 (21-27)
Spicule length (SL)	77	79 ± 5 (71-90)	–	69 ± 8 (61-76)	–	–
Spicule width	16.6	16 ± 2 (14-20)	–	12.5 ± 3 (11-15)	–	–
Gubernaculum length (GL)	53	54 ± 5 (45-61)	–	40 ± 6 (30-53)	–	–
Gubernaculum width	10.6	8.7 ± 1.4 (6-12)	–	6.7 ± 1.4 (5-9)	–	–

posterior to excretory pore, thus making a total of seven ridges in lateral field. Central ridge dividing into two near pharynx base, making a total of eight ridges, the maximum number in the lateral field. The portion with eight ridges is the longest part (compared to portions with 2, 6, 7, 4 ridges) of lateral field. At anus level, two marginal and two central ridges disappearing with only four ridges remaining in lateral field. At

Table 22. *(Continued).*

Character	First generation			Second generation		Infective juvenile
	Holotype (male)	Male	Female	Male	Female	
D% = EP/ES × 100	80	80 ± 6 (68-86)	80 ± 19 (46-105)	60 ± 8 (54-67)	64 ± 8.2 (47-74)	54 ± 5 (30-70)
E% = EP/T × 100	–	–	–	–	–	90 ± 6.5 (78-114)
SW% = SL/ABD × 100	202	180 ± 13 (150-200)	–	200 ± 20 (166-233)	–	–
GS% = GL/SL × 100	70	69 ± 10 (59-79)	–	59 ± 7 (46-73)	–	–
H% = Hyaline/T × 100	–	–	–	–	–	57 (50-63)

EP = distance from anterior end to excretory pore; NR = distance from anterior end to nerve ring; ES = distance from anterior end to end of pharynx.
After the original authors.

about mid-tail four ridges in lateral field become two ridges. Near tail terminus, two marginal lines in lateral field converging, and central line disappearing before reaching end of lateral field. Formula for lateral field pattern = 2, 6, 7, 8, 4, 2. Pharynx with thin corpus, basal bulb more or less elongate with visible valve. Tail attenuate, tapering gradually. Hyaline region comprising *ca* 57 (50-63)% of tail length, sometimes, up to 70% (Fig. 72H). Morphometrics of infective juveniles reared from *G. mellonella* and from *D. abbreviatus* differ.

TYPE HOST AND LOCALITY

The nematode was collected from an infested citrus root weevil, *Diaprepes abbreviatus*, in a commercial citrus orchard in Polk County, FL, USA.

TYPE MATERIAL

Holotype (male), allotype and paratype deposited in the USDA Nematode Collection, Beltsville, MD, USA.

RELATIONSHIPS

In the male, spicule length averages 79 (71-90) μm, D% = *ca* 80, ratio SW = *ca* 1.8 and the lateral field has a single, narrow, ridge. In the female, the vulva has short, low double-flapped epiptygmata and the tail terminus usually has five papilla-like structures. In the infective juvenile, body length averages 1002 (880-1133) μm, EP = 74 (66-83) μm, tail length = 83 (65-91) μm, and E% = 89.6 (78-114). The IJ lateral field pattern is variable, the formula for the arrangement of ridges from head to tail being: 2, 6, 7, 8, 4, 2. The portion bearing eight ridges is the longest.

The body length of the infective juvenile of *S. diaprepesi* (1002 μm) is intermediate between that of the *glaseri*-group (*S. arenarium*, *S. cubanum*, *S. glaseri*, *S. longicaudum*, *S. puertoricense*, body length > 1000 μm) and the *feltiae*-group (*S. feltiae*, *S. kraussei*, *S. neocurtillae*, *S. oregonense*, body length > 800 but < 1000 μm). This species can be distinguished from these *Steinernema* species by several morphological characters of the males, females and infective juveniles. Males can be differentiated from other steinernematids by the shape and length of the spicules and gubernaculum (D% = 80 and SW% = 180). Females can be distinguished from other species by the presence of five papilla-like structures on the tail tip (Figs 72, 73) and by the vulva having low, double-flapped, epiptygmata (Figs 72, 73). The body length, distance from anterior end to excretory pore, tail length and E% differentiate the infective juvenile of this species.

MOLECULAR CHARACTERISATION

Steinernema diaprepesi is characterised genetically by sequence lengths of ITS region (1022 bp), ITS1 (301 bp), ITS2 (313 bp), and composition, A = 0.24853, C = 0.21233, G = 0.24951, T = 28963 (accession no. AF122021), and D2D3 region sequence length, 901 bp, and composition, A = 0.23542, C = 0.19654, G = 0.32181, T = 0.24622 (accession no. DQ350609) (Nguyen *et al.*, 2006).

BIONOMICS

Steinernema diaprepesi was isolated from larvae of *D. abbreviatus* that were buried in cages beneath citrus trees. During 2 years of the experiment, indigenous populations of *S. diaprepesi* infected and killed 13-50% of the buried insects within 7 days, with higher rates of parasitism during the summer months compared to spring and autumn months. The nematode was isolated with much lower frequency from the rhizosphere of native plants growing between citrus tree rows than from the rhizosphere of citrus trees. *Steinernema diaprepesi* has been isolated from other orchards located on the central ridge of Florida, which is characterised by deep sandy soils. Attempts to isolate the nematode in heavier-textured soils off of the central ridge have been unsuccessful, although no comprehensive survey has yet been conducted.

DISTRIBUTION

The nematode has been recorded from the USA (Nguyen & Duncan, 2002) and Venezuela (Spiridonov *et al.*, 2005).

References

NGUYEN, K.B. & DUNCAN, L.W. (2002). *Steinernema diaprepesi* n. sp. (Rhabditida: Steinernematidae), a parasite of the citrus weevil *Diaprepes abbreviatus* (L.) (Coleoptera: Curculionidae). *Journal of Nematology* 34, 59-170.

NGUYEN, K.B., MALAN, A.P. & GOZEL, U. (2006). *Steinernema khoisanae* n. sp. (Rhabditida: Steinernematidae), a new entomopathogenic nematode from South Africa. *Nematology* 8, 157-175.

SPIRIDONOV, S.E., REID, A.P., PRODRUCKA, K., SUBBOTIN, S.A. & MOENS, M. (2004). Phylogenetic relationship within the genus *Steinernema* (Nematoda: Rhabditida) as inferred from analyses of the ITS1-5.8S-ITS2 region of rDNA and morphological features. *Nematology* 6, 547-566.

Steinernema eapokense[*,**] Phan, Spiridonov, Subbotin & Moens, 2006
(Fig. 75)

MEASUREMENTS

See Table 23.

DESCRIPTION

Male (first and second generation not distinguished in original description)

Body fusiform, obese, C-shaped in heat-relaxed specimens, tapering towards anterior end. Cuticle annulation 1-1.4 μm wide, poorly distinguishable throughout body. No lateral fields visible but irregular anastomoses of cuticle annulation present along lateral body. Anterior extremity rounded without prominent flattened area around oral aperture. Six labial papillae and four more prominent cephalic papillae present. Cheilostom cuticle connected to a thick, strongly cuticularised, 2 μm long ring of 6-6.5 μm diam., forming middle part of stoma. Additional thinner cuticular ring situated posterior to thicker one, embedded into pharyngeal tissue. Pharynx corpus cylindrical, with visible metacorpal swelling and swelling around basal part of stoma. Basal bulb subspherical. Nerve ring situated on anterior slope of basal bulb in most males. Cardia flat, 8-10 μm long. Deirids 2 μm diam., sitting in centre of 4-5 μm diam. round area of smooth cuticle posterior to level of pharyngo-intestinal junction. Excretory duct strongly cuticularised up to 12-13 μm from pore. Intestinal lumen open throughout body length. Separate spermatocytes poorly distinguishable even in reflexed testis part, being replaced with complex rachis structure in middle region of testis, with characteristic patterns of spermatogenesis up to vacuolated '*vas deferens*'. Characteristic steinernematid set of genital papillae with a row of usually nine pairs of subventral papillae extending from tail terminus to body middle, one unpaired large precloacal papilla in midventral position, one pair of papillae in lateral position at manubrium level when spicules retracted, and one pair of papillae in dorsolateral position in postcloacal area. Latter pair usually closer to equatorial part of tail. Pair

[*] Named after the type locality of *Eapok* in Vietnam.

[**] Description after Phan *et al.* (2006). Revised by Zdeněk Mráček.

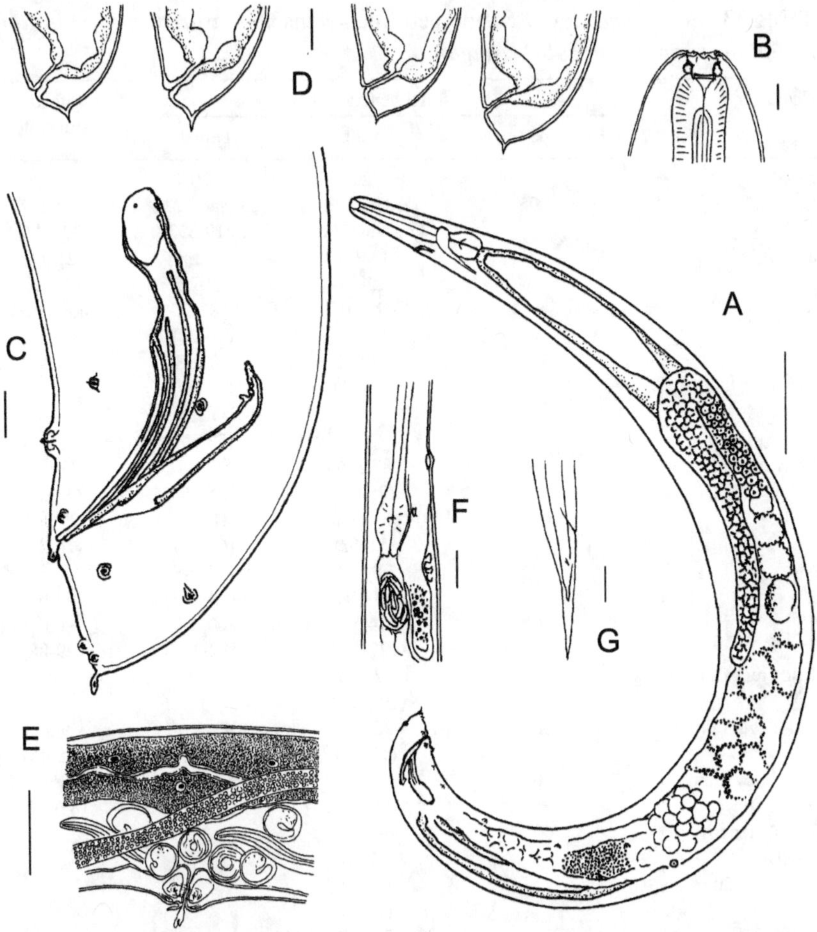

Fig. 75. Steinernema eapokense. *A: First generation male, entire view. B, C: Male. B: Anterior end; C: Tail. D: First generation female, tail shapes; E: Vulvar region. F, G: Infective juvenile. F: Pharyngeal region; G: Tail. (Scale bars: A, E = 100 μm; B, C, F, G = 10 μm; D = 50 μm.) After Phan* et al. *(2006), courtesy of* Russian Journal of Nematology.

of postdeirids situated dorsolaterally at level of anteriormost pair of precloacal papillae. Spicules with elongated, funnel-shaped manubrium, tapering towards distal end, with round extremity and more or less pronounced terminal duct. Bifurcation of outer line of velum and ventral rib of spicule visible near rostrum. All four main 'ribs' of spicule discernible, separated by gaps visible in lateral aspect. Spicule body gradu-

Table 23. *Morphometrics of* Steinernema eapokense. *Measurements are in μm and in the form: mean ± SD (range).*

Character	First generation			Infective juvenile
	Holotype (male)	Male	Female	
n	–	20	20	25
L	1095	1096 ± 50 (1035-1200)	3584 ± 232 (3210-3930)	402 ± 24 (370-434)
a	12	13 ± 1 (12-14)	20 ± 1 (18-22)	21 ± 2 (18-23)
b	8	8.2 ± 0.3 (7.8-8.8)	18 ± 1 (16-21)	4.8 ± 0.3 (4.3-5.3)
c	56	57 ± 7 (49-71)	78 ± 8 (63-90)	10 ± 0.5 (9.6-11)
V	–	–	55 ± 2 (52-57)	–
Body diam.	90	87 ± 6 (75-90)	180 ± 11 (165-195)	19 ± 1 (18-21)
EP	62	57 ± 3 (51-62)	81 ± 9 (69-96)	29 ± 1 (26-31)
NR	98	98 ± 3 (93-104)	143 ± 9 (134-161)	58 ± 3 (54-62)
ES	129	134 ± 5 (127-146)	200 ± 9 (182-209)	84 ± 3 (79-88)
Testis reflexion	218	232 ± 53 (190-288)	–	–
Tail length (T)	20	20 ± 2 (15-23)	47 ± 6 (38-58)	39 ± 3 (35-45)
Anal body diam. (ABD)	30	28 ± 1 (26-30)	58 ± 7 (44-69)	10 ± 0.6 (8.6-11)
Spicule length (SL)	69	69 ± 4 (63-74)	–	–
Spicule width	11	11 ± 1 (9-12)	–	–
Gubernaculum length (GL)	50	50 ± 2 (47-56)	–	–
Gubernaculum width	8	6 ± 1 (5.0-8.0)	–	–
D% = EP/ES × 100	48	43 ± 3 (37-48)	40 ± 4 (33-47)	34 ± 2 (32-37)
E% = EP/T × 100	–	–	–	73 ± 5 (66-80)
SW% = SL/ABD × 100	230	246 ± 19 (221-282)	–	–
GS% = GL/SL × 100	72	73 ± 4 (69-80)	–	–
H% = H/T × 100	–	–	–	53 ± 3 (47-57)

EP = distance from anterior end to excretory pore; NR = distance from anterior end to nerve ring; ES = distance from anterior end to end of pharynx.

ally narrowing from rostrum to distal end. Cloacal aperture covered with flap when spicules retracted. Two papilla-like structures on cloacal margin. Gubernaculum usually straight with ventrally curved proximal end, and prominent dorsal, 10-12 μm long, 'hole'. Tail mucron often cylindrical, 2.9 ± 0.2 (2.2-3.8) μm long with globular inclusion.

Female (first and second generation not distinguished in original description)

Cuticle annulation 1.5 μm wide. Cephalic extremity with six labial and four cephalic papillae. Cuticular ring forming main part of stoma walls 5 μm thick and 5-5.5 μm long. Stoma 10 μm diam. Excretory duct strongly cuticularised, entering body at angle of *ca* 45°. Equatorial vulva with petal-like, 10-12 μm long structures, protruding from aperture. Vulva lips less than 12-15 μm high, equal in size. Tail end dome-shaped, with conical, 10-15 μm long, spike-like mucron.

Infective juvenile

Body of heat-relaxed juveniles quite obese, moderately C-shaped, tapering towards anterior end. Cuticular annulation *ca* 1 μm wide, visible throughout body. Lateral field in anterior half of body with four ridges in centre, two pronounced marginal ridges and poorly discernible submarginal lines. Lateral field 8-9 μm wide at mid-body. Such regular pattern not observed in posterior half of body with ventral gap between central and marginal ridges disappearing. Cuticle swollen near anterior end forming cephalic capsule. Frontal surface of anterior end always regularly rounded. Labial papillae indistinct, four cephalic papillae and amphidial pouches occasionally visible. Pharynx corpus slender, slightly widening to metacorpal part. Nerve ring always around corpus-isthmus junction. Excretory pore encircled by refractive ring of cuticle 1-1.5 μm deep. Hemizonid located at isthmus level, usually protruding from body, 3-4 annules long. Basal bulb with poorly developed valves. Deirids round, subcuticular structures located at hemizonid level. Cardia *ca* 7 μm long, with rounded posterior end. Bacterial vesicle elliptical, 5-7 μm diam. and 10-14 μm long. Postdeirids prominent, situated at one-third of body length from tail end beneath dorsal region of lateral field (appearing just like deirids, *i.e.*, circular subcuticular area with ending or duct approaching cuticle surface). Rectum *ca* 1.5 anal body diam. long. Genital primordium 50 μm long. Phasmids at one-third of tail length posterior to anus (phasmid-mucron distance = *ca* 25-26 μm). Lateral field lines disappearing at phasmid level. Tail with pronounced dorsal concavity, dorsally curved with terminus pointed to finely rounded. Hyaline region comprising approximately half of tail length.

TYPE HOST AND LOCALITY

Type host unknown as nematodes were obtained by the *Galleria mellonella* baiting method. The type isolate was collected in Eapok, Daklak province, Vietnam (longitude 106°04′E, latitude 12°48′N, altitude 600 m a.s.l.), from basalt red soils in coffee plantations on the submontane plateau.

OTHER LOCALITY

Also recorded from a similar habitat to the type near Krongana, Daklak province, Vietnam, at an altitude of 600 m a.s.l.

TYPE MATERIAL

Holotype male and six paratype males deposited at Institute of Parasitology of Russian Academy of Sciences, Leninskii prospect 33, Moscow, 119071, Russia. Slides with paratype males, first generation females and infective juveniles deposited in Department of Nematology, Institute of Ecology and Biological Resources, National Center for Science and Technology, Hoang Quoc Viet 18, Nghiado, Caugiay, Hanoi, Vietnam.

RELATIONSHIPS

The male has two papilla-like structures on the cloacal margin and the spicules have an elongate manubrium and developed rostrum. A short mucron is present in both male generations and is often cylindrical with a globular inclusion. The first generation females possess a pronounced, conical, tail mucron. Infective juveniles belong to the shortest length group with an average body length of 402 μm. The tail tip is dorsally concave and the hemizonid is protuberant in the majority of infective juveniles. The hyaline region comprises *ca* 50% of the tail length and the lateral fields have four ridges in the mid-body region.

The infective juvenile of *S. eapokense* differs from the closely related *S. cumgarense* in lower D% of 34 (32-37) *vs* 41 (38-43) and lower

E% of 73 (66-80) *vs* 99 (87-103). The E% value helps to distinguish *S. eapokense* from *S. siamkayai* (66-80 *vs* 85-112%). The infective juveniles of *S. eapokense* (L = 370-434 μm) are shorter than those of *S. carpocapsae* (438-650 μm), *S. scapterisci* (517-609 μm) and *S. sasonense* (437-516 μm). The spicules of *S. eapokense* first generation males are longer than those of *S. asiaticum* at 69 (63-74) *vs* 58 (53-62) μm, respectively.

Molecular characterisation

Steinernema eapokense is characterised by sequence lengths of ITS regions (ITS1 + 5.8S + ITS2), 723 bp, ITS1 = 262 bp, ITS2 = 306 bp and its composition, A = 0.22544, C = 0.15491, G = 0.22544, T = 0.39419. The sequence of ITS regions is deposited in GenBank (accession no. AY487921).

Comments

More precise description of taxonomically important characters, such as number of ridges in the infective juvenile lateral field, form of first generation female postanal swelling and the occurrence of papilla-like mucrons on the female tail tip are missing and/or unsatisfactorily described.

Distribution

The species is known from two localities in Vietnam (Phan *et al.*, 2006).

Reference

PHAN, K.L., SPIRIDONOV, S.E., SUBBOTIN, S.A. & MOENS, M. (2006). Four new species of *Steinernema* Travassos, 1928 with short infective juveniles from Vietnam. *Russian Journal of Nematology* 14, 11-29.

Steinernema feltiae[*,**] **(Filipjev, 1934) Wouts, Mráček, Gerdin & Bedding, 1982**
= *Neoaplectana feltiae* Filipjev, 1934
= *Neoaplectana bibionis* Bovien, 1937
= *Steinernema bibionis* (Bovien, 1937) Wouts, Mráček, Gerdin & Bedding, 1982
= *Neoaplectana leucaniae* Hoy, 1954
= *Steinernema leucaniae* (Hoy, 1954) Wouts, Mráček, Gerdin & Bedding, 1982
(Figs 76-81)

MEASUREMENTS

See Table 24.

DESCRIPTION

Male (first generation)

Body C-shaped, curved ventrad posteriorly when heat-relaxed. Lateral field indistinct. Cephalic extremity rounded, usually slightly swollen, with perioral disc, six labial papillae, two amphidial apertures (usually covered with exudates) and four cephalic papillae. Stoma shallow, cheilorhabdions small, sclerotised, sometimes indistinct. Excretory pore located near nerve ring, mostly situated anterior to basal bulb. Pharynx with cylindrical procorpus, metacorpus absent or slightly swollen, isthmus present, nerve ring located around isthmus, basal bulb distinct. Pharyngo-intestinal valve present, usually weak. Genital system monorchic, reflexed. Spicules paired, brown in colour. Head of spicules elongate, length/width ratio = 1.7 ± 0.2 (1.4-2.0), in some specimens twice as long as wide, shaft very short or absent; blade thick, gradually tapering posteriorly, rostrum not developed, blade terminus blunt with longitudinal depression/slit ventrally; velum present. Each spicule with two internal ribs. Gubernaculum boat-shaped in lateral view, anterior part usually with one or two ventral projections. Complement of 11 pairs of genital papillae (occasionally 12) and single precloacal midventral papilla present. Precloacal papillae arranged as six lateroventral pairs

[*] Named after *Feltia*, the genus of the type host.

[**] Description modified after Filipjev (1934), Mráček *et al.* (1982), Nguyen *et al.* (2006) and Mráček (unpubl.). Revised by Zdeněk Mráček.

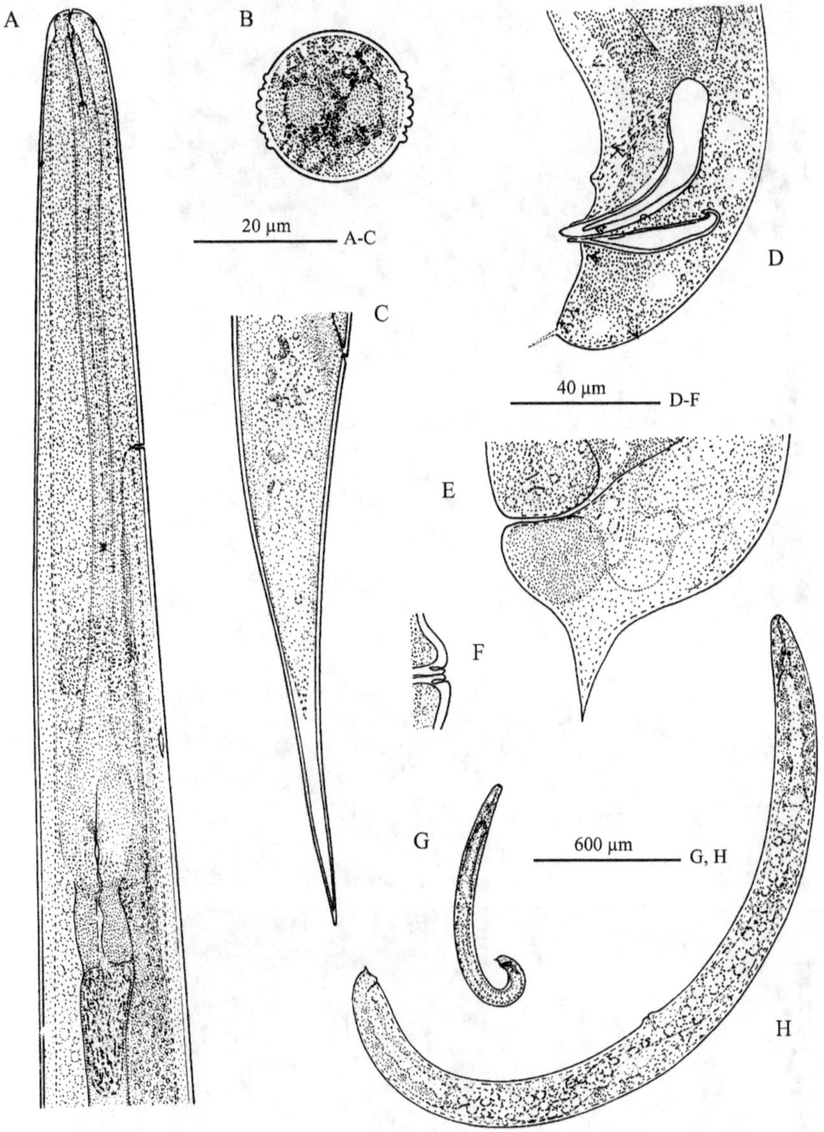

Fig. 76. Steinernema feltiae. *A-C: Infective juvenile. A: Pharyngeal region; B: Cross-section of mid-body showing lateral field with eight ridges; C: Tail region. D: First generation male tail region; E: Female tail region; F: Vulval region; G: Male, entire body; H: Female, entire body. Composite figure after redescription of* S. bibionis *by Wouts (1980), courtesy of* Journal of Nematology.

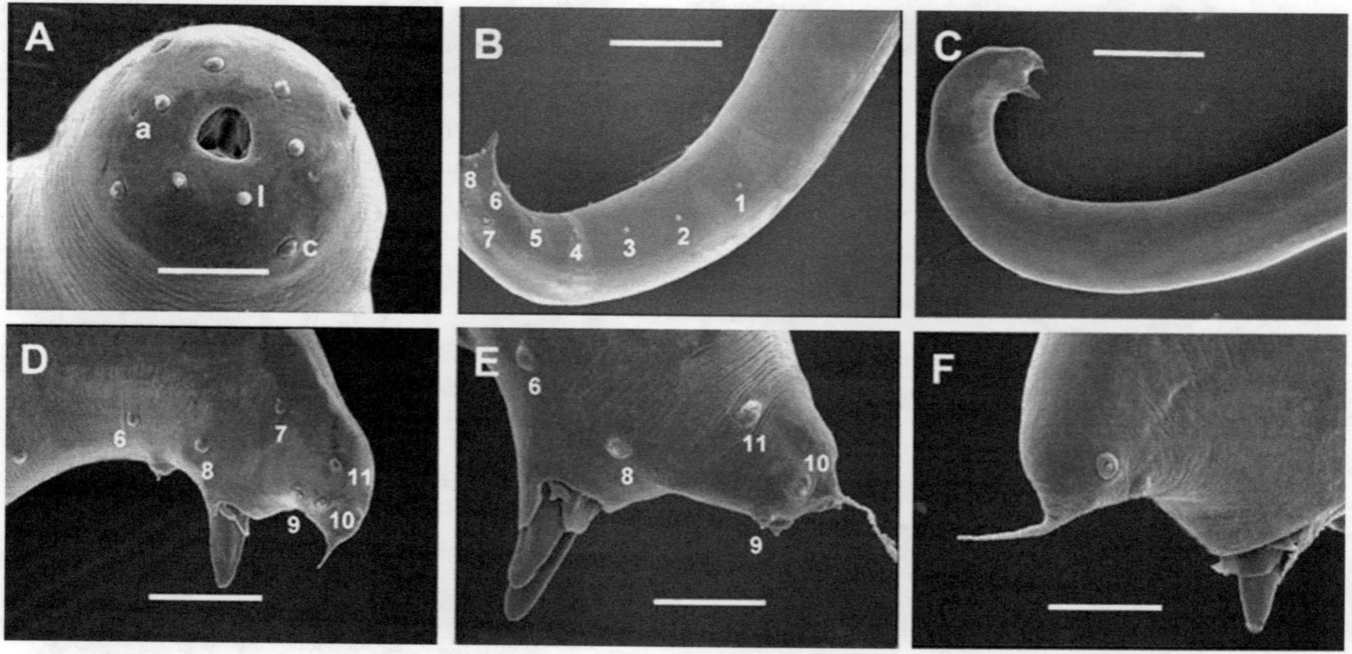

Fig. 77. Steinernema feltiae. *A-F: Male. A: Face view (a = amphidial aperture, c = cephalic papilla, l = labial papilla); B, C: Posterior regions with prominent genital papillae in B and not so prominent in C; D-F: Variation of mucron and tail tip. (Scale bars: A = 7.5 μm; B = 50 μm; C = 86 μm; D = 23 μm; E = 10 μm; F = 9 μm.) After Nguyen* et al. *(2006), courtesy of* Zootaxa.

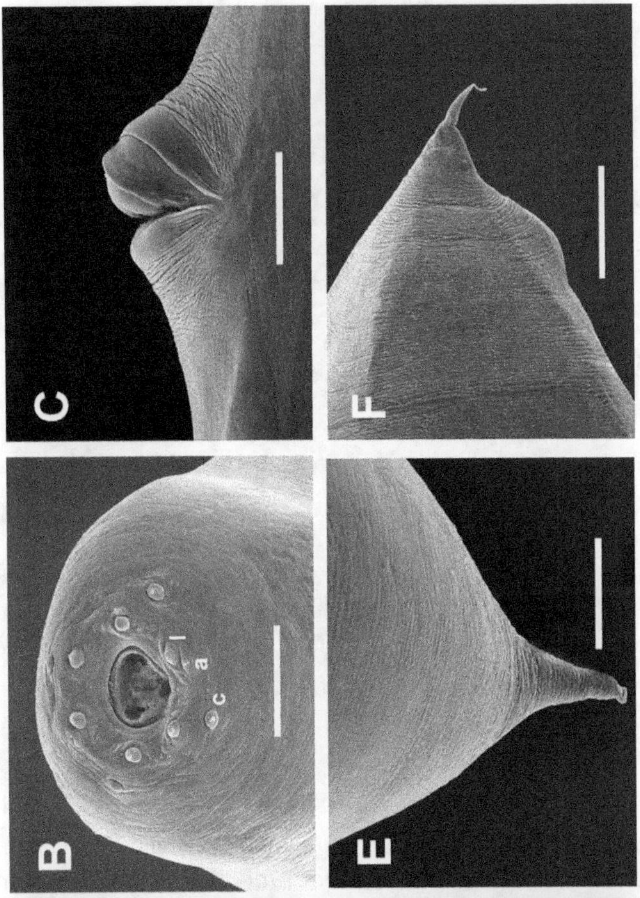

Fig. 78. Steinernema feltiae. *A-F: Female. A, B: Face view (a = amphidial aperture, c = cephalic papilla, l = labial papilla); C: Vulva; D-F: Variations of tail tip. (Scale bars:* A = *7 μm; B = 10 μm; C = 15 μm; D = 12 μm; E = 15 μm; F = 17 μm.) After Nguyen* et al. *(2006), courtesy of* Zootaxa.

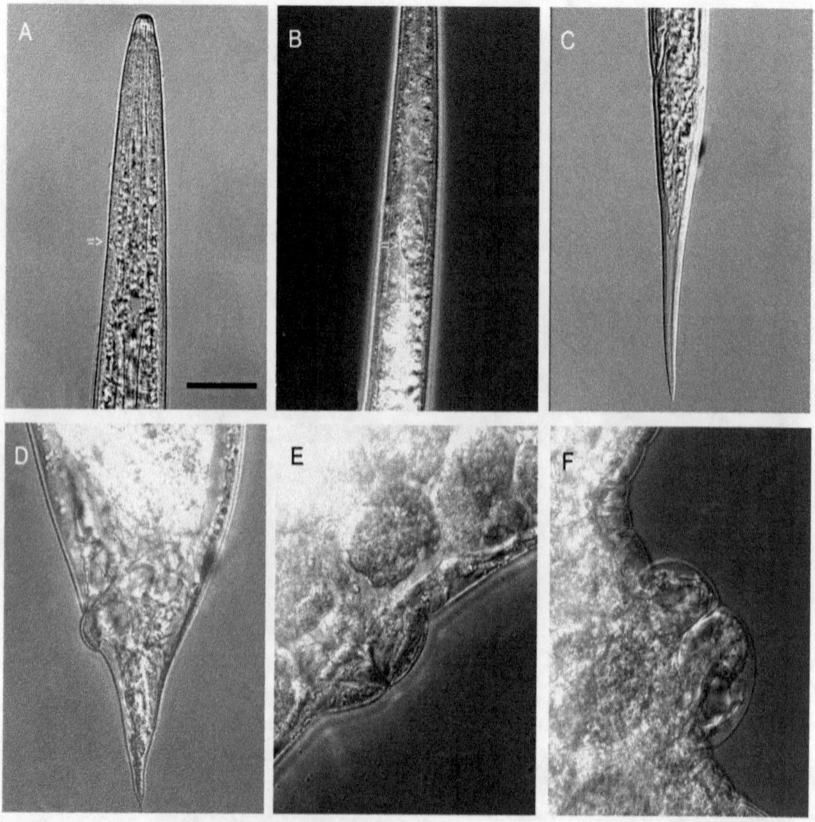

Fig. 79. Steinernema feltiae. *A-C: Infective juvenile. A: Anterior region and excretory pore (arrow); B: Basal bulb (arrow); C: Tail. D-F: Second generation female. D: Tail with postanal swelling; E: Vulva of young specimen; F: Vulva of old specimen. (Scale bar: A-C = 14 μm; D = 32 μm; E, F = 20 μm.) After Nguyen* et al. *(2006), courtesy of* Zootaxa.

plus single midventral precloacal papilla. Postcloacal papillae comprising five subventral pairs. Tail conoid, tail terminus with fine mucron.

Male (second generation)

Similar to that of first generation except body, spicule and gubernaculum shorter and thinner, excretory pore more anterior, and isthmus more distinct. Gubernaculum short. Mucron on tail terminus present and longer than in first generation.

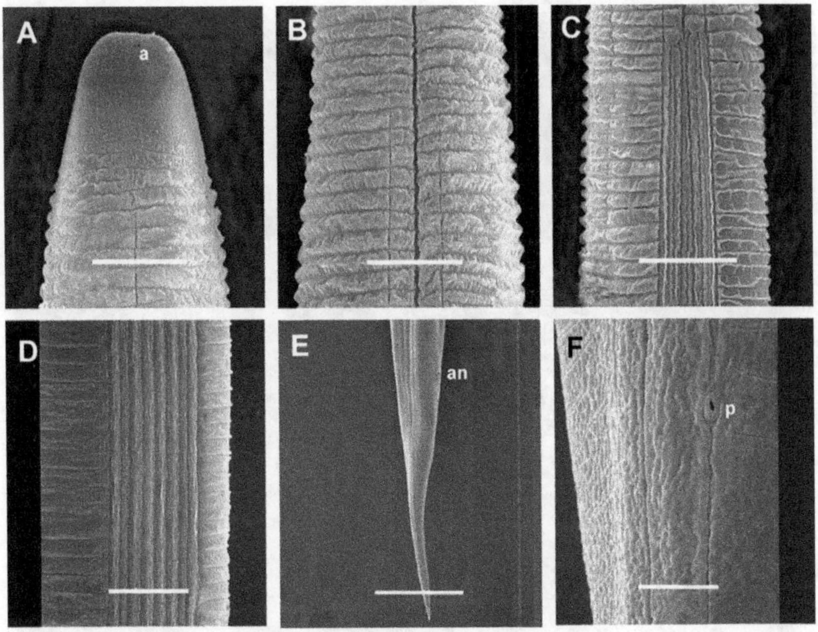

Fig. 80. Steinernema feltiae. *A-F: Infective juvenile. A: Anterior region showing smooth head, amphidial aperture (a) and one line in lateral field; B: Lateral field with two ridges; C: Lateral field with two ridges anteriorly, seven ridges posteriorly; D: Lateral field with eight ridges; E: Lateral field in posterior region showing four ridges near anus (an); F: Lateral field with two ridges and phasmid (p). (Scale bars: A = 4 μm; B = 5 μm; C, D = 8 μm; E = 23 μm; F = 2 μm.) After Nguyen* et al. *(2006), courtesy of* Zootaxa.

Female (first generation)

Body cuticle smooth or with faint annules. Lateral fields and phasmids absent. Cephalic extremity rounded, continuous with body; six labial papillae, four cephalic papillae. Lips indistinct. Amphidial apertures present. Stoma shallow, subtriangular anteriorly, triradiate internally. Cheilorhabdions well sclerotised but small. Smaller, indistinct, sclerotised structure (presumably the prorhabdions) visible posterior to cheilorhabdions. Pharynx with cylindrical procorpus, muscular, metacorpus swollen, isthmus distinct, basal bulb valvate. Nerve ring surrounding isthmus, located just anterior to basal bulb. Pharyngo-intestinal valve present. Excretory pore located just anterior to, or at mid-region of, basal bulb. Genital system amphidelphic, reflexed, often containing

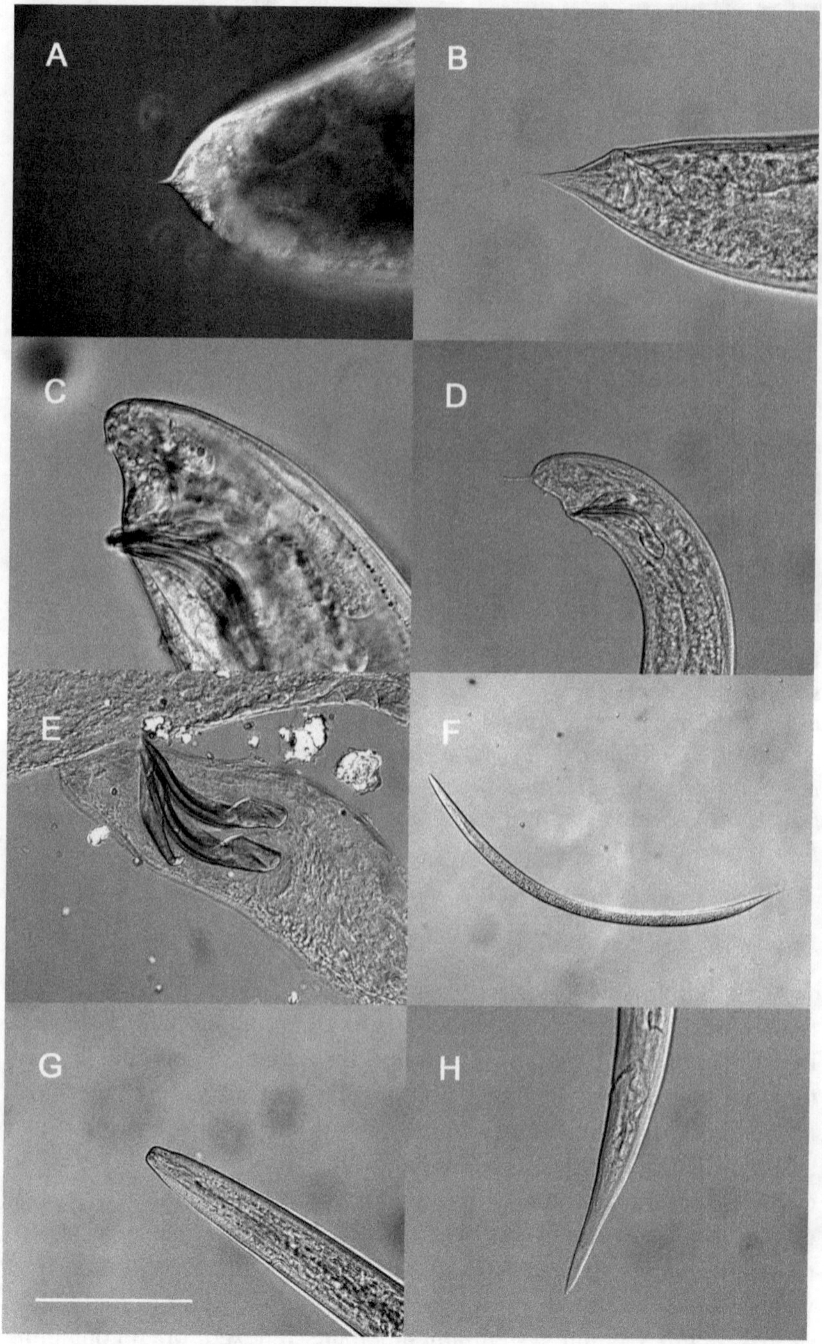
A
B
C
D
E
F
G
H

many eggs. Vulva in form of a transverse slit; usually slightly protuberant; low double-flapped epiptygmata present; vulval aperture posterior to mid-body. Vagina sclerotised, short. Tail shape mostly blunt with conical terminal projection forming a thin mucron, slight ventral postanal swelling present, tail shorter than anal body diam. Females probably lacking papilla-like structures on tail tip.

Female (second generation)

Similar to first generation female but smaller. Vulva less protuberant, epiptygmata usually more prominent than that of first generation. Tail, tapering to a pointed end, mucronate, longer than anal body diam.; slight postanal swelling present.

Infective juvenile

Body elongate and slender, slightly C-shaped. Sheath (second-stage cuticle) present immediately after harvesting, but many infective juveniles losing sheath in storage. Cephalic extremity smooth, continuous, rounded anteriorly. Labial papillae not seen; four prominent cephalic papillae present. Amphidial apertures slit-shaped, not prominent. Cuticle marked with prominent transverse striations. Lateral field beginning anteriorly with one line. Two additional lines appearing at annules 14-16 to form two ridges. Near excretory pore level, number of ridges in lateral fields increasing from two to six. Additional central ridge appearing more posteriorly, *ca* one body diam. posterior to excretory pore, thus making a total of seven ridges in lateral field. Central ridge dividing into two near pharynx base, making a total of eight ridges (= maximum number in lateral field). Submarginal pair less distinct. Portion with eight ridges forming longest part (compared to portions with 2, 6, 7, 4 ridges) of lateral field. At anus level, two marginal and two central ridges disappearing with only four ridges remaining in lateral field. At about mid-tail four ridges in lateral field becoming two ridges.

Fig. 81. Steinernema feltiae. *A, B: Female. A: First generation, tail; B: Second generation, tail. C-E: Male. C: First generation, posterior region with short mucron; D: Second generation, posterior region with a long mucron; E: Spicules and gubernaculum. F-H: Infective juvenile. F: Entire body; G: Anterior region, head and excretory pore; H: Tail with hyaline region. (Scale bar: A, B = 90 μm; C = 50 μm; D = 45 μm; E = 60 μm; F = 340 μm; G, H = 70 μm.) After Z. Mráček.*

Table 24. *Morphometrics of* Steinernema feltiae. *Measurements are in μm and in the form: mean ± SD (range).*

Character		First generation		Second generation		Infective
	Holotype	Male*	Female**	Male*	Female**	juvenile*
n	–	20	20	20	20	20
L	–	1612 ± 88 (1414-1817)	3380 (3095-3774)	1157 ± 100 (929-1313)	1849 (1566-2035)	879 ± 49 (766-928)
a	–	11.5	17 (14-20)	18.4	13 (11-15)	30 ± 1.9 (27-34)
b	–	9.5	14 (12-17)	8,3	10 (10-11)	6.4 ± 0.3 (5.8-6.8)
c	–	41.3	65 (49-88)	36.2	32 (30-36)	10 ± 0.5 (9.4-11)
c′	–	0.8	–	0.8 (0.5-0.9)	– (0.3-0.4)	4.8 ± 0.2 (4.5-5.1)
V	–	–	56 (44-57)	–	58 (55-58)	–
Body diam.	–	140 ± 10 (121-162)	204 (170-254)	63 ± 2.5 (59-67)	148 (106-190)	29 ± 1.9 (26-32)
EP	–	115 ± 3.4 (110-126)	–	87 ± 7.8 (75-99)	–	63 ± 2.3 (58-67)
NR	–	–	–	–	–	113 ± 5.1 (108-117)
ES	–	170 ± 3.4 (164-180)	237 (197-304)	140 ± 4.9 (132-140)	186 (179-197)	136 ± 3.5 (130-143)
Testis reflexion	–	–	–	–	–	–
Tail length (T)	–	39 ± 1.2 (37-43)	52 (39-70)	32 ± 3.8 (27-40)	57 (50-66)	86 ± 2.6 (81-89)
Hyaline (H)	–	–	–	–	–	38 ± 3.2 (32-43)
Anal body diam. (ABD)	–	48 ± 1.7 (43-53)	52 (47-62)	38 ± 3.8 (32-48)	47 (42-52)	18 ± 0.8 (16-19)
Spicule length (SL)	–	66 ± 1.5 (62-68)	–	63 ± 2.8 (57-67)	–	–
Gubernaculum length (GL)	–	52 ± 1.9 (48-56)	–	40 ± 2.7 (38-48)	–	–

Near tail terminus, two marginal lines in lateral field converging, central line disappearing before reaching end of lateral field. Formula for lateral field pattern = 2, 6, 7, 8, 4, 2. Pharynx with narrow corpus, basal bulb more or less elongate with visible valve. Excretory pore located at mid-pharynx level. Tail long, slender (ratio c′ = 4.8), attenuate, tapering gradually. Hyaline region comprising less than 50% of tail length.

Table 24. *(Continued).*

Character	First generation			Second generation		Infective juvenile*
	Holotype	Male*	Female**	Male*	Female**	
D% = EP/ES × 100	–	68 ± 3.1 (64-72)	–	62 ± 6.1 (51-72)	–	46 ± 1.4 (44-50)
E% = EP/T × 100	–	–	–	–	–	74 ± 4 (67-81)
SW% = SL/ABD × 100	–	140 ± 10 (130-150)	–	170 ± 20 (130-200)	–	–
GS% = GL/SL × 100	–	80 ± 3 (70-90)	–	70 ± 10 (60-80)	–	–
H% = H/T × 100	–	–	–	–	–	44 ± 4 (37-51)

EP = distance from anterior end to excretory pore; NR = distance from anterior end to nerve ring; ES = distance from anterior end to end of pharynx.

* According to Nguyen *et al.* (2006).

** According to Mráček *et al.* (1982).

TYPE HOST AND LOCALITY

Locality of topotype material from former Ustinov, today renamed as Izhevsk, Udmurt Republic, Ural region, Russia. This species was originally described from the larvae of *Agrotis* (syn. *Feltia*) *segetum* by Filipjev (1934) and reisolated by Kozodoy *et al.* (1987). Poinar (1989) synonymised *S. bibionis* (originally from Denmark) to the more senior *S. feltiae*.

TYPE MATERIAL

Paratype materials from Dr Kirjanova's collection is deposited at the Institute of Zoology, Russian Academy of Sciences, St Petersburg, Russia.

RELATIONSHIPS

Steinernema feltiae is the most common species characterised by medium length infective juveniles. The excretory pore opening is located approximately at mid-pharynx level. The lateral field has eight lateral ridges with a less distinct submarginal pair in the mid-body portion. The tail is slender with ratio c′ = 4.8 and the hyaline region comprises less than half the tail length. Males possess slightly curved brown spicules with oblong manubrium (lateral view). A mucron is developed in both generations, but in the first generation it is sometimes reduced to an enlarged papilla-like structure. Females have a slightly protruding vulval aperture located posterior to the mid-body and a slight postanal swelling. First generation females have a pointed, mucronate tail and papilla-like projections on the tail tip are probably absent.

Steinernema feltiae belongs to the group of species similar to *S. kraussei*, *S. oregonense*, *S. weiseri*, *S. jollieti* and *S. silvaticum*. Within this assemblage, known as the *feltiae*-group, *S. feltiae* differs from *S. kraussei* by the infective juvenile lateral field, which in the latter has only seven ridges (however, the central, unpaired ridge can be sometimes split into two, less distinct, ridges), whereas in *S. feltiae* there are eight ridges, only the submarginal pair being less prominent. The male differs by the shape of the spicule which is characterised by a rounded manubrium with length to width ratio of 1 : 1 and rostrum present in *S. kraussei* *vs* an oblong manubrium and rostrum absent in *S. feltiae*.

Steinernema jollieti IJ differs by the presence of only six equally developed ridges *vs* eight in *S. feltiae*. Both *S. weiseri* and *S. silvaticum* IJ have a lateral field composed of eight equal ridges but *S. weiseri* possesses a very short hyaline region (H% = 36 *vs* 44 in *S. feltiae*) and males of *S. silvaticum* are characterised by strongly curved spicules, rounded manubrium and presence of rostrum, whereas in *S. feltiae* the spicules are slightly curved, the manubrium is oblong and the rostrum is absent. The closest species, *S. oregonense*, differs mainly by the shorter male mucron and by the first generation females having a tail projection with several mucrons on its tip whereas the tail of *S. feltiae* seems to be smooth.

Molecular characterisation

Steinernema feltiae is characterised by sequence lengths (flanked by primers 18S and 26S) of ITS regions, 980 bp, ITS1 = 275 bp, ITS2 = 298 bp and its composition, A = 0.25816, C = 0.18265, G = 0.23061, T = 0.32857. The sequences of ITS and D2D3 regions are deposited in GenBank under accession no. AF121050 and AF331906, respectively.

BIONOMICS

This species is easily multiplied on *Galleria mellonella* larvae. Several biofactories have reared *S. feltiae* on a large scale for field trials. Laboratory and field experiments have shown *S. feltiae* to be efficacious against a variety of insect species.

DISTRIBUTION

Steinernema feltiae is the most widespread steinernematid species with an almost cosmopolitan distribution. In addition to the type locality in Russia, it is widely distributed throughout Europe (Hominick *et al.*, 1995; Mráček *et al.*, 1999, 2005; Sturhan & Lišková, 1999; Hominick, 2002) and is also found in Asia (Japan) (Yoshida, 2003), North America (Poinar, 1992; Mráček & Webster, 1993), South America (Stock, 1993), Africa (Peters, 1996), Australia and New Zealand (Poinar, 1990; Peters, 1996), *etc.* This species inhabits both open and forested habitats.

References

BOVIEN, P. (1937). Some types of association between nematodes and insects. *Videnskabelige Meddelelser fra Dansk Naturhistorisk Forening, København* 101, 1-114.

FILIPJEV, I.N. (1934). Miscellanea nematodologica, I. Eine neue Art der Gattung *Neoaplectana* Steiner nebst Bemerkungen über die systematische Stellung der letzten. *Parazitologichesky Sbornik* 4, 229-239.

HOMINICK, W.M. (2002). Biogeography. In: Gaugler, R. (Ed.). *Entomopathogenic nematodes*. Wallingford, UK, CABI Publishing, pp. 115-143.

HOMINICK, W.M., REID, A.P. & BRISCOE, B.R. (1995). Prevalence and habitat specificity of steinernematid and heterorhabditid nematodes isolated during surveys of the UK and the Netherlands. *Journal of Helminthology* 69, 27-32.

HOY, J.M. (1954). The biology and host range of *Neoaplectana leucaniae*, a new species of insect-parasitic nematode. *Parasitology* 44, 392-399.

KOZODOY, E.M., VORONOV, D.A. & SPIRIDONOV, S.E. (1987). New data on taxonomic studies of *Neoaplectana feltiae* (Nematoda: Rhabditida). *Zoologichesky Zhurnal* 66, 980-988.

MRÁČEK, Z. & WEBSTER, J. (1993). Survey of Heterorhabditidae and Steinernematidae (Rhabditida, Nematoda) in Western Canada. *Journal of Nematology* 25, 710-717.

MRÁČEK, Z., GUT, J. & GERDIN, S. (1982). *Neoaplectana bibionis* Bovien, 1937, an obligate parasite insects isolated from forest soil in Czechoslovakia. *Folia Parasitologica* 29, 139-145.

MRÁČEK, Z., BEČVÁŘ, S. & KINDLMANN, P. (1999). Survey of entomopathogenic nematodes from the families Steinernematidae and Heterorhabditidae (Nematoda: Rhabditida) in the Czech Republic. *Folia Parasitologica* 46, 145-148.

MRÁČEK, Z., BEČVÁŘ, S., KINDLMANN, P. & JERSÁKOVÁ, J. (2005). Habitat preference for entomopathogenic nematodes, their insect hosts and new faunistic records for the Czech Republic. *Biological Control* 34, 27-37.

PETERS, A. (1996). The natural host range of *Steinernema* and *Heterorhabditis* spp. and their impact on insect populations. *Biocontrol Science and Technology* 6, 389-402

POINAR JR, G.O. (1989). Examination of the neoaplectanid species *feltiae* Filipjev, *carpocapsae* Weiser and *bibionis* Bovien (Nematoda: Steinernematidae). *Revue de Nématologie* 12, 375-377.

POINAR JR, G.O. (1990). Taxonomy and biology of Steinernematidae and Heterorhabditidae. In: Gaugler, R. & Kaya, H.K. (Eds). *Entomopathogenic nematodes in biological control*. Boca Raton, FL, USA, CRC Press, pp. 23-61.

POINAR JR, G.O. (1992). *Steinernema feltiae* (Steinernematidae: Rhabditidae) parasitizing adult fungus gnats (Mycetophylidae: Diptera) in California. *Fundamental and Applied Nematology* 15, 427-430.

STOCK, P. (1993). Description of an Argentinian strain of *Steinernema feltiae* (Filipjev, 1934) (Nematoda: Steinernematidae). *Entomologia Mediterranea* 21, 179-283.

STURHAN, D. & LIŠKOVÁ, M. (1999). Occurrence and distribution of entomopathogenic nematodes in the Slovak Republic. *Nematology* 1, 273-277.

WOUTS, W.M., MRÁČEK, Z., GERDIN, S. & BEDDING, R.A. (1982). *Neoaplectana* Steiner, 1929 a junior synonym of *Steinernema* Travassos, 1927 (Nematoda: Rhabditida). *Systematic Parasitology* 4, 147-154.

YOSHIDA, M. (2003). Intraspecific variation in RFLP patterns and morphological studies on *Steinernema feltiae* and *S. kraussei* from Hokkaido, Japan. *Nematology* 5, 735-746.

Steinernema glaseri[*,**] (Steiner, 1929) Wouts, Mráček, Gerdin & Bedding, 1982

= *Neoaplectana glaseri* Steiner, 1929
(Figs 82-86)

Measurements

See Table 25.

Description

Male (first generation)

Cephalic extremity slightly swollen with perioral disc. Lips not observed. Six prominent labial papillae present plus four larger cephalic papillae. Stoma partially collapsed; cheilorhabdions represented by thick ring of sclerotised material just posterior to lips. Posterior to this another sclerotised ring representing the prorhabdions. Posterior part of stoma forming an asymmetrical funnel with thick anterior end. Pharynx muscular with a cylindrical procorpus followed by a slightly swollen, non-valvate, metacorpus, isthmus, and basal bulb with a valve. Nerve ring surrounding isthmus just anterior to basal bulb. Genital system monorchic, reflexed. Spicule head shorter than wide, comprising *ca* 24 (21-25)% of spicule length; shaft distinct. Blade long, narrow, bearing two ridges. Distal tip of spicules bearing a ventral depression making it appear hook-like. Velum absent. Gubernaculum variable in shape, much shorter than spicule, anterior end curved ventrally, enlarging gradually posteriorly. Two wings of corpus curved upward, cuneus usually forked anteriorly to form Y-shaped structure. Complement of 23 genital papillae (11 pairs plus single midventral precloacal papilla) consistently present with little variation in position. Five of 11 pairs are precloacal, subventral, pair 6 is lateral, pairs 7 and 8 are subventral adcloacal (pair 7 is sometimes precloacal), pairs 9 and 10 are subventral, subterminal, and pair 11 is subdorsal.

Male (second generation)

Smaller than first generation male. Body diam. much smaller.

[*] Named after R.W. *Glaser* who collected the nematode.

[**] After Poinar (1978) and Nguyen and Smart (1995). Revised by Khuong B. Nguyen.

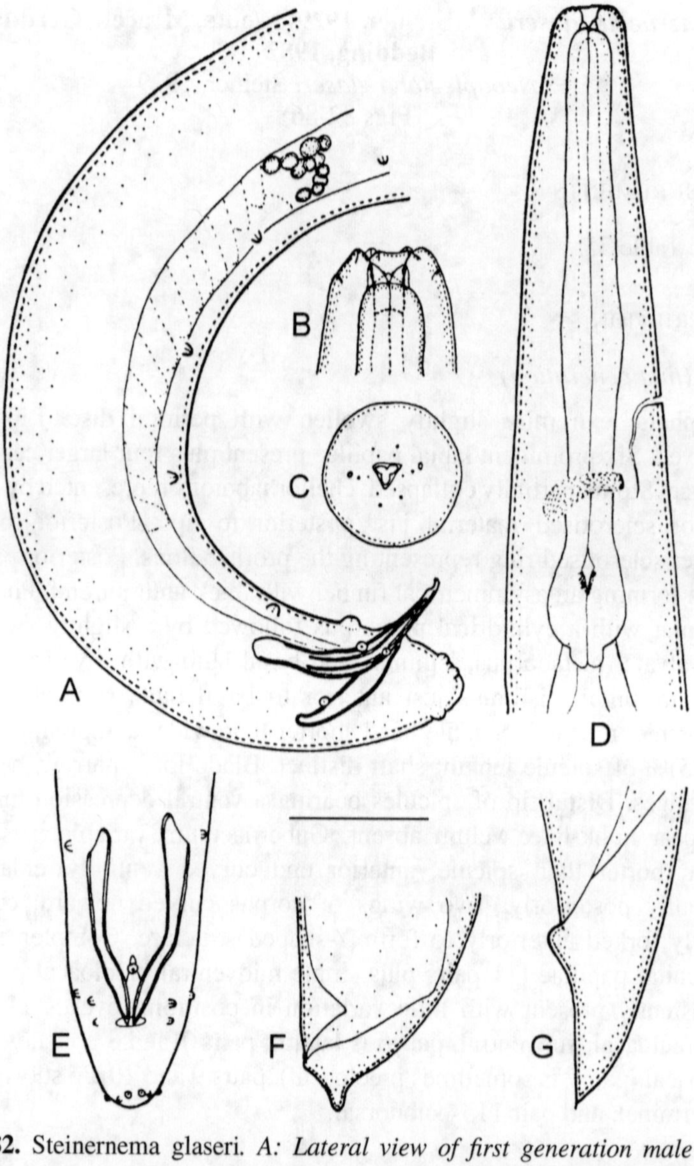

Fig. 82. Steinernema glaseri. *A: Lateral view of first generation male tail; B: Lateral view of head of second generation female; C: Face view of first generation female; D: Lateral view of pharyngeal region of second generation female; E: Ventral view of posterior part of first generation male; F: Lateral view of first generation female tail; G: Lateral view of second generation female tail. (Scale bar: A, C, D, E, G = 53 μm; B = 34 μm; F = 130 μm.) After Poinar (1978), courtesy of* Nematologica.

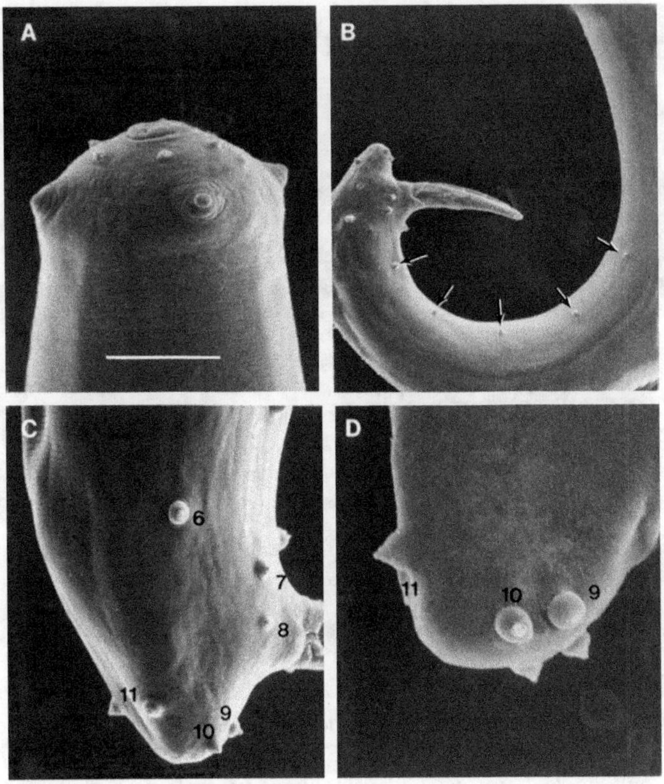

Fig. 83. Steinernema glaseri *male. A: Anterior region showing five of the six labial papillae and three of the four large cephalic papillae; B: Posterior region showing protruding spicules, 11 pairs of genital papillae (precloacal papillae indicated by arrows) and tail terminus without mucron; C: Enlargement of posterior region showing genital papillae pairs 6-11; D: Enlargement of tail terminus showing genital papillae pairs 9-11. (Scale bar: A = 10 μm; B = 50 μm; C = 17.7 μm; D = 8 μm.) After Nguyen and Smart (1995), courtesy of* Nematologica.

Female (first generation)

Cuticle smooth, head slightly rounded. Anterior region similar to male but much larger. Amphidial apertures crescent-shaped, narrow. Excretory pore opening usually anterior to nerve ring. Lateral fields and phasmids inconspicuous. Genital system amphidelphic, reflexed. Vulva a transverse slit, slightly to very protuberant, with or without a thick flap. Vagina short, leading into paired uteri. Eggs deposited initially, but

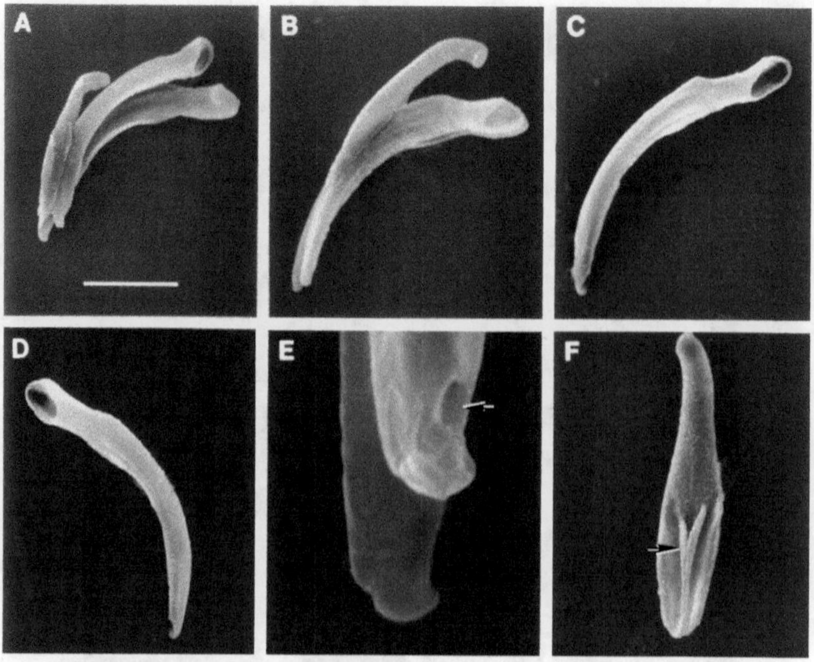

Fig. 84. Steinernema glaseri *spicules and gubernacula. A, B: Variation of spicules and gubernacula; C, D: Spicules showing short head, pronounced shaft, and spicule tip with hooked tip; E: Spicule tip showing ventral aperture (arrow) and hooked tip; F: Gubernaculum, ventral view, showing curved anterior part and Y-shaped cuneus (arrow). (Scale bar: A-D = 25 μm; E = 3.8 μm; F = 17.6 μm.) After Nguyen and Smart (1995), courtesy of* Nematologica.

later juveniles hatching inside females (*endotokia matricida*). Tail with prominent postanal swelling, terminating in rounded projection.

Female (second generation)

Smaller than those of first generation. Tail sometimes terminating in a fine mucron in certain isolates.

Infective juvenile

Cephalic extremity not annulated, labial papillae absent, four cephalic papillae and amphidial apertures pronounced. Oral aperture and anus closed. Pharynx and intestine collapsed. Tail conoid and more or less straight. Hyaline region comprising less than half of tail length. Lateral field pattern beginning anteriorly with one line at third annule, two lateral

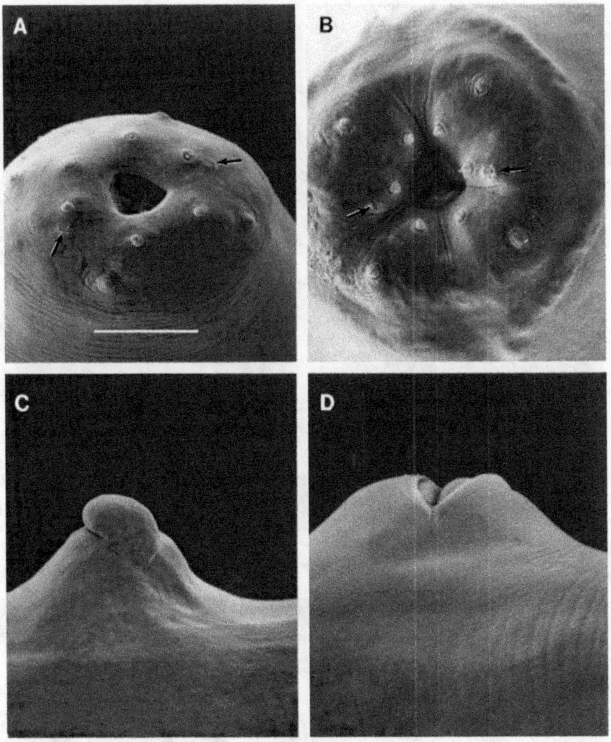

Fig. 85. Steinernema glaseri *female. A: Face view showing six labial papillae, four cephalic papillae, two crescent-shaped amphidial apertures (arrows) and a subtriangular oral aperture; B: Face view of a young female showing amphidial apertures (arrows), six labial papillae interiorly around the stoma, and four cephalic papillae on top of the head; C: Vulva with a thick flap; D: Vulva without a thick flap. (Scale bar: A = 10 μm; B = 13.6 μm; C = 30 μm; D = 8.6 μm.) After Nguyen and Smart (1995), courtesy of* Nematologica.

lines appearing at annules 3-5 to form two ridges. At level of isthmus and basal bulb, number of ridges in lateral field increasing from two to five and lateral field areolated. A short distance posteriorly, central ridge dividing into two, making a total of six ridges in lateral field, areolation becoming less obvious. More posteriorly, two additional lines appearing on either side of lateral field to form two new ridges, making a total of eight ridges, the maximum in lateral field. Near anus level, four central ridges enlarging. Near phasmid level, eight ridges in lateral field becoming two large bands or ridges. Formula of lateral field pattern is 2, 5, 6, 8, 2. Occasionally, eight ridges extending beyond level of

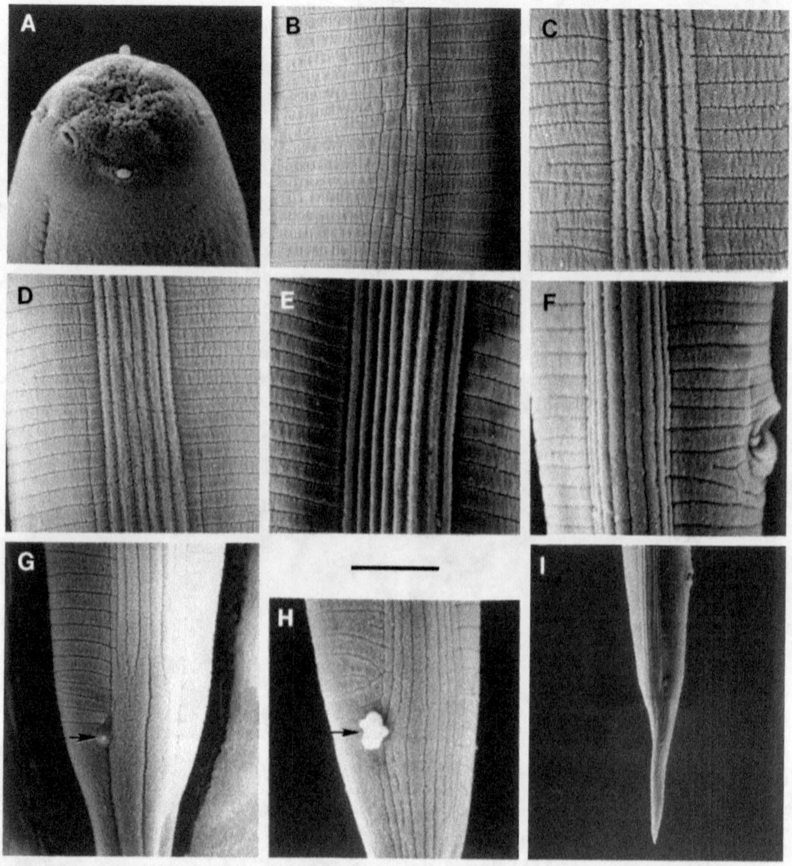

Fig. 86. Steinernema glaseri *infective juveniles. A: Anterior region showing four cephalic papillae, one of the two amphidial apertures and beginning of lateral field; B: Anterior portion of lateral field at point where two ridges become five; C: Lateral field showing position where central ridge splits into two giving a total of six ridges; D: Lateral field at level where central two ridges have not separated fully, and two additional bands are formed exteriorly; E: Lateral field with maximum number of eight ridges; F: Lateral field at anal level showing enlargement of four central ridges; G: Lateral field at phasmid level showing eight ridges becoming two large ridges. The phasmid (arrow), with exudates in central pore, is located ventral to lateral field; H: Lateral field at phasmid level showing less frequent condition where eight ridges are not reduced to two, the phasmid (arrow) is covered with exudates; I: Tail region showing lateral field, anus and phasmid near mid tail. (Scale bar: A, F, H = 6 μm; B, D, E, G = 7.5 μm; C = 5 μm; I = 25 μm.) After Nguyen and Smart (1995), courtesy of* Nematologica.

Table 25. *Morphometrics of* Steinernema glaseri. *Measurements are in μm and in the form: mean (range).*

Character	First generation		Second generation		Infective juvenile
	Male	Female	Male	Female	
n	25	25	25	25	25
L	1700 (1500-1900)	5600 (4000-7900)	1400 (1200-1500)	2100 (1700-2330)	1130 (864-1448)
a	–	–	–	–	29 (26-35)
b	–	–	–	–	7.3 (6.3-7.8)
c	–	–	–	–	14.7 (13.6-15.7)
V	–	51 (48-53)	–	54 (51-56)	–
Body diam.	72 (54-92)	236 (180-300)	55 (54-62)	94 (80-100	29 (25-32)
EP	145 (121-178)	165 (142-180)	115 (105-124)	131 (121-149)	65 (59-69)
NR	132 (99-183)	177 (148-192)	139 (130-146)	159 (149-167)	93 (85-99)
ES	160 (155-187	244 (220-284)	169 (155-180)	214 (205-220)	123 (110-133)
Testis reflexion	176 (84-264)	–	115 (93-139)	–	–
Tail length (T)	30 (28-44)	45 (43-53)	28 (25-31)	58 (56-62)	66 (53-74)
Anal body diam. (ABD)	42 (34-47)	75 (56-93)	37 (34-43)	38 (34-43)	–
Spicule length (SL)	77 (62-90)	–	61 (59-62)	–	–
Spicule width	9 (6-12)	–	7.2 (6.2-9.3)	–	–
Gubernaculum length (GL)	46 (40-50)	–	36 (34-37)	–	–
Gubernaculum width	8 (6-9)	–	6 (60-60)	–	–
D% = EP/ES × 100	91	–	68	–	65 (58-71)
E% = EP/T × 100	–	–	–	–	131 (122-138)
SW% = SL/ABD × 100	183	–	165	–	–
GS% = GL/SL × 100	60	–	59	–	–

EP = distance from anterior end to excretory pore; NR = distance from anterior end to nerve ring; ES = distance from anterior end to end of pharynx.
After Poinar (1978, 1990).

phasmids and gradually reducing in width. Phasmids located near mid-tail, either ventral to lateral fields, or interrupting ventral-most lateral ridges. Phasmids located at almost same level with a central pore; often covered by exudates. No annules observed posterior to phasmids on ventral surface and only a few observed on dorsal surface.

TYPE HOST AND LOCALITY

Found in dead larvae of the Japanese beetle (*Popillia japonica*) collected from Tavistock Golf Course near Haddonfield, NJ, USA (Glaser & Fox, 1930).

TYPE MATERIAL

Lectotype (male) and allolectotype deposited in the USDA Nematode Collection, Beltsville, MD, USA.

RELATIONSHIPS

This species can be distinguished from other species in the *glaseri*-group by characteristics of the male and juvenile. For the male, body diam. is narrower at 72 (54-92) *vs* 188 (184-219) μm for *S. arenarium*, 112 (90-135) μm for *S. guangdongense* and 136 (86-194) μm for *S. longicaudum*. The thin spicule with hook-like tip is typical for this species. The D% of *S. glaseri* (90) is closer to that of *S. arenarium* (93), but much different from that of *S. guangdongense* (70) and *S. longicaudum* (75). The SW% of *S. glaseri* is very different from the above mentioned closely related species at 256 *vs* 210, 175 and 161, respectively. For the infective juvenile, the EP distance is a good diagnostic character at 102 μm *vs* 83, 80 and 82 μm, respectively. E% also can be used for differentiation (131 *vs* 119, 88, 87, respectively).

MOLECULAR CHARACTERISATION

Steinernema glaseri is characterised by the length of ITS regions, 988 bp (ITS1 = 279 bp, ITS2 = 302 bp), its composition, A = 0.236, C = 0.218, G = 0.271, T = 0.275 (accession no. AF122015), and length of its partial 28S region of rDNA, 874 bp, composition, A = 0.240, C = 0.211, G = 0.317, T = 0.232 (accession no. AF331908).

BIONOMICS

The life cycle of *S. glaseri* is similar to that of other steinernematids, but this nematode develops more rapidly in *Galleria mellonella* than others. It takes 3-4 days for *S. glaseri* to develop from infective juveniles to adults compared to *ca* 5 days for *S. carpocapsae* and other species. This species is of tropical origin and survives well between 15-35°C (Kaya, 1990). When infective juveniles of this nematode are placed on the soil surface, most move downwards (Schroeder & Beavers, 1987). When they are placed in the middle of a soil column, more nematodes move downwards than upwards (Georgis & Poinar, 1983). This nematode can survive better in sandy soils than those with loam and clay as the nematodes may expend more energy moving through the smaller pores found in loam and clay soils (Kung *et al.*, 1990). When in soil, this nematode usually moves in various directions to look for hosts. When a host is found, the infective juveniles of *S. glaseri* enter the host through the mouth, spiracles or anus, and then enter the body cavity where they release bacteria and develop as in other steinernematids.

The nematode attacks mainly soil insects, especially those in the order Coleoptera, including the families Chrysomelidae, Curculionidae, Elateridae and Scarabaeidae. The nematode also parasitises some Lepidoptera (Poinar, 1979), Orthoptera (unpubl. data) and maybe other insect Orders.

Bacterial associate

The infective stage of *S. glaseri* carries cells of *Xenorhabdus poinarii* in the intestine. This bacterium was isolated and described by Akhurst and Boemare (1988). It is brown-pigmented but the intensity of pigmentation varies considerably between strains. This bacterium does not adsorb bromothymol blue. Some strains produce antibiotic factors. In this species there are more than two phases and adsorption of neutral red and production of antimicrobial factors are not always associated. All strains grow at 37°C, whilst *X. nematophila* from *S. carpocapsae* does not grow at this temperature.

Biocontrol capability

Steinernema glaseri was the first nematode to be investigated extensively as a biological control agent of insects. Glaser (1932) produced the nematode in large numbers for the first time by an *in vitro* method. At that time, Glaser did not know that bacteria were associated with infective juveniles but fortuitously his method was suitable for the development of bacteria. The nematodes collected from his culture were applied in 73 field plots in New Jersey for control of Japanese beetle. Infected grubs were recovered from 72 out of 73 plots 2 weeks after application (Glaser, 1932; Glaser *et al.*, 1940), the parasitisation of the grub population ranging from 0.3% to 81%. The nematode remained in treated plots for 8.5 years.

It has been shown that *S. glaseri* is very effective as a biological agent of insects. In potted yews, the nematode reduced by more than 90% the populations of Japanese beetle (Wright *et al.*, 1988). Georgis and Hague (1991), who summarised fields trials against Japanese beetle larvae in turf and pasture during the period 1984-1989, reported that the application of *S. glaseri* reduced the insect population as well as, or better than, the standard insecticide (Isofenfos). This nematode also gave the best control of *Adoryphorus couloni* in Australia (Berg *et al.*, 1984, 1987).

DISTRIBUTION

In addition to the type locality in New Jersey, it has also been found in Louisiana, Mississippi, North Carolina (Poinar, 1979), Florida, Texas, Alabama (unpubl.) in the USA, and in Santa Rosa, Brazil (Poinar, 1990). Other records cited by Hominick (2002) are from Argentina, Azores, China, Korea and Spain.

References

AKHURST, R.J. & BOEMARE, N.E. (1988). A numerical taxonomy study of the genus *Xenorhabdus* (Enterobacteriaceae) and proposed elevation of the subspecies *X. nematophilus* to species. *Journal of General Microbiology* 134, 1835-1845.

BERG, G.N., BEDDING, R.A., WILLIAMS, P. & AKHURST, R.J. (1984). Developments in the application of the nematodes for the control of subterranean pasture pests. In: *Proceedings of the 4th Australian Applied Entomology Research Conference, Adelaide, Australia*, pp. 352-356.

BERG, G.N., WILLIAMS, P., BEDDING, R.A. & AKHURST, R.J. (1987). A commercial method of application of entomopathogenic nematodes to pasture for controlling subterranean insect pests. *Plant Protection Quarterly* 2, 174-177.

GEORGIS, R. & HAGUE, N.G.M. (1991). Nematodes as biological insecticides. *Pesticide Outlook* 2, 29-32.

GEORGIS, R. & POINAR JR, G.O. (1983). Effect of soil texture on the distribution of and infectivity of *Neoaplectana glaseri* (Nematoda: Steinernematidae). *Journal of Nematology* 15, 329-332.

GLASER, R.W. (1932). Studies on *Neoaplectana glaseri*, a nematode parasite of Japanese beetle (*Popillia japonica*). New Jersey Department of Agriculture, Circular No. 211.

GLASER, R.W. & FOX, H. (1930). A nematode parasite of Japanese beetle (*Popillia japonica* Newm.). *Science* 70, 16-17.

GLASER, R.W., MCCOY, E.E. & GIRTH, H.B. (1940). The biology and economic importance of a nematode parasitic in insects. *Journal of Parasitology* 26, 479-495.

HOMINICK, W.M. (2002). Biogeography. In: Gaugler, R. (Ed.). *Entomopathogenic nematodes*. Wallingford, UK, CABI Publishing, pp. 115-143.

KAYA, H.K. (1990). Soil ecology. In: Gaugler, R. & Kaya, H.K. (Eds). *Entomopathogenic nematodes in biological control*. Boca Raton, FL, USA, CRC Press, pp. 93-115.

KUNG, S.P., GAUGLER, R. & KAYA, H.K. (1990). Soil type and entomopathogenic nematode persistence. *Journal of Invertebrate Pathology* 55, 401-406.

NGUYEN, K.B. & SMART JR, G.C. (1995). Scanning electron microscope studies of *Steinernema glaseri* (Nematoda: Steinernematidae). *Nematologica* 41, 183-190.

POINAR JR, G.O. (1978). Generation polymorphism in *Neoaplectana glaseri* Steiner (Steinernematidae: Nematoda), redescribed from *Strigoderma arboricola* (Fab.) (Scarabaeidae: Coleoptera) in North Carolina. *Nematologica* 24, 105-114.

POINAR JR, G.O. (1979). *Nematodes for biological control of insects*. Boca Raton, FL, USA, CRC Press, 277 pp.

POINAR JR, G.O. (1990). Taxonomy and biology of Steinernematidae and Heterorhabditidae. In: Gaugler, R. & Kaya, H.K. (Eds). *Entomopathogenic nematodes in biological control*. Boca Raton, FL, USA, CRC Press, pp. 23-60.

SCHROEDER, W.J. & BEAVERS, J.B. (1987). Movement of entomopathogenic nematodes of the families Heterorhabditidae and Steinernematidae in soil. *Journal of Nematology* 19, 257-259.

STEINER, G. (1929). *Neoaplectana glaseri* n. g., n. sp. (Oxyuridae) a new nemic parasite of the Japanese beetle. *Journal of the Washington Academy of Sciences* 19, 436-440.

WRIGHT, P.J., NOONAN, M.J., JACKSON, T.A. & WOUTS, W.M. (1988). Use of nematodes for control of pasture pests in New Zealand. In:

Proceedings of the 5th Australasian Conference on Grassland Invertebrate Ecology, Melbourne, Australia, pp. 82-87.

WOUTS, W.M., MRÁČEK, Z., GERDIN, S. & BEDDING, R.A. (1982). *Neoaplectana* Steiner, 1929 a junior synonym of *Steinernema* Travassos, 1927 (Nematoda: Rhabditida). *Systematic Parasitology* 4, 147-154.

*Steinernema guangdongense**,** Qiu, Fang, Zhou, Pang & Nguyen, 2004
(Figs 87-92)

MEASUREMENTS

See Table 26.

DESCRIPTION

Male (first generation)

Body curved ventrally posteriorly, C-shaped when heat-relaxed. Cephalic extremity rounded, usually slightly swollen, bearing six labial papillae, two amphidial apertures and four prominent cephalic papillae. Stoma shallow. Cheilorhabdions small, sclerotised, sometimes indistinct. Excretory pore located just anterior to basal bulb. Pharynx with cylindrical procorpus, metacorpus slightly swollen, isthmus present, nerve ring around isthmus, basal bulb distinct. Pharyngo-intestinal valve present. Genital system monorchic, reflexed. Distance from base of pharynx to anterior end of testis variable. Spicules paired, brown in colour. Head (manubrium) of spicules with rounded anterior end, almost continuous with shaft. Shaft (calomus) very short or absent, blade (lamina) thick anteriorly, tapering gradually posteriorly, blade terminus blunt, velum present, sometimes very thin. Each spicule with two internal sclerotised ribs. Gubernaculum boat-shaped in lateral view, tapering gradually anteriorly. Genital papillae arranged as 11 pairs plus single midventral precloacal (six pairs precloacal, one adcloacal, one lateral, two postcloacal subterminal and one postcloacal subdorsal). Tail conoid, tail terminus rounded, mucron absent.

Male (second generation)

Similar to that of first generation except body, spicule and gubernaculum shorter and thinner.

Female (first generation)

Body cuticle smooth or with faint annules. Lateral fields and phasmids not observed. Cephalic extremity rounded, continuous with body, bearing six labial and four cephalic papillae. Lips indistinct. Amphidial

* Named after the province of *Guangdong*, People's Republic of China.

** Description after Qiu *et al.* (2004). Revised by Khuong B. Nguyen.

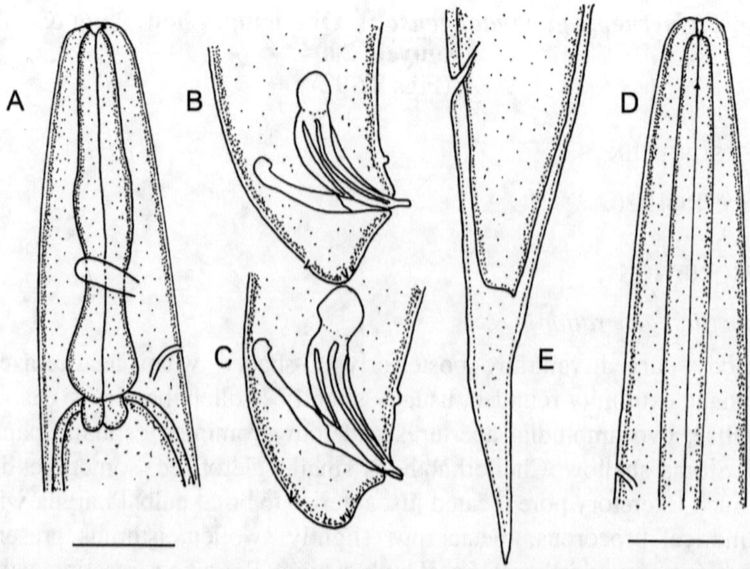

Fig. 87. Steinernema guangdongense, *first generation male and infective juvenile. (Scale bars: A-C = 50 μm; D, E = 20 μm.) After Qiu* et al. *(2004), courtesy of* Zootaxa.

apertures usually inconspicuous even under SEM. Stoma shallow, subtriangular anteriorly; triradiate internally. Cheilorhabdions, well sclerotised but small. Smaller, indistinct, sclerotised structure (presumably prorhabdions) visible posterior to cheilorhabdions. Pharynx with procorpus cylindrical, muscular; metacorpus swollen; isthmus distinct; basal bulb valvate. Nerve ring surrounding isthmus, located just anterior to basal bulb. Pharyngo-intestinal valve present. Excretory pore located near mid-pharynx, anterior to nerve ring. Genital system amphidelphic, reflexed, often containing many eggs. Vulva in form of transverse slit, mostly protuberant, small double-flapped epiptygmata present. Anterior lip larger than posterior. Body diam. immediately anterior to vulva larger than that posterior to vulva. Vagina sclerotised, short. Tail shape variable, with blunt terminus, mucron occasionally present on dorsal side. Ventral postanal swelling present, tail shorter than anal body diam.

Female (second generation)

Similar to first generation female but smaller. Vulva less protuberant, epiptygmata present but less prominent. Tail, tapering to a pointed end, longer than anal body diam.; ventral postanal swelling present.

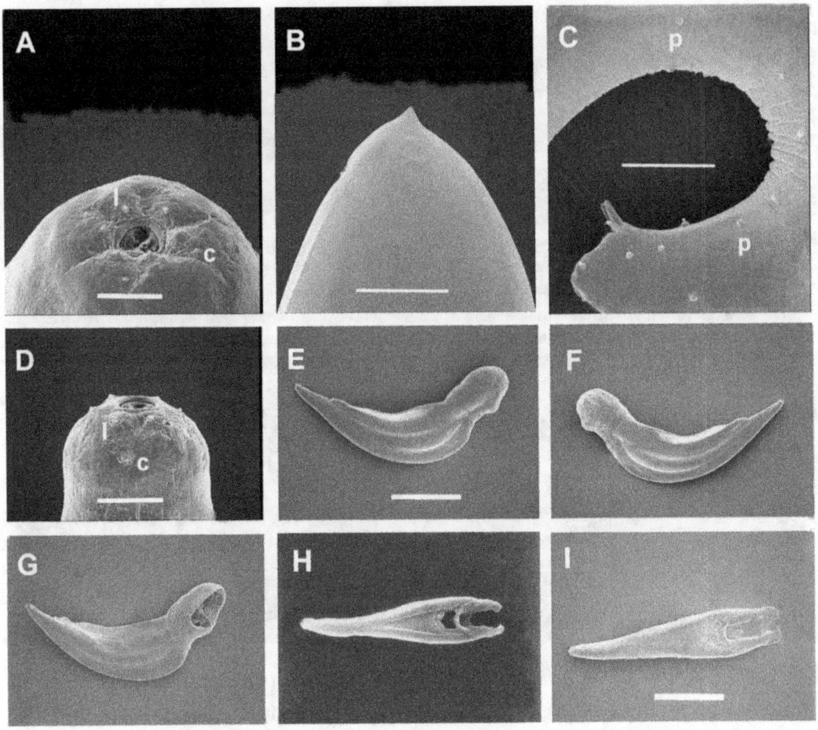

Fig. 88. Steinernema guangdongense. *SEM micrographs. A: Head and tail of a young first generation female showing labial (l) and cephalic (c) papillae; B: Tail of a young female; C: Posterior end of a male showing genital papillae (p); D: Male head showing labial (l) and cephalic papillae (c); E-I: Spicule and gubernaculum. (Scale bars: A, D = 10 μm; B = 42.9 μm; C = 42.5 μm; E-I = 20 μm.) After Qiu* et al. *(2004), courtesy of* Zootaxa.

Infective juvenile

Body elongate, sheath (second-stage cuticle) present immediately after harvesting, but many IJ losing sheath in storage. Labial region smooth, rounded anteriorly, continuous with body. Labial papillae not seen; four prominent cephalic papillae. Amphidial apertures slit-shaped, not prominent, sometimes covered with exudates. Cuticle marked with prominent transverse striations. Lateral field beginning anteriorly with one line at annule 5 or 6. Two additional lines appearing at annules 10-11 to form two ridges. Near excretory pore level, number of ridges in lateral fields increasing from two to seven. Near end of pharynx, central ridge dividing into two, making a total of eight ridges, the maximum number in

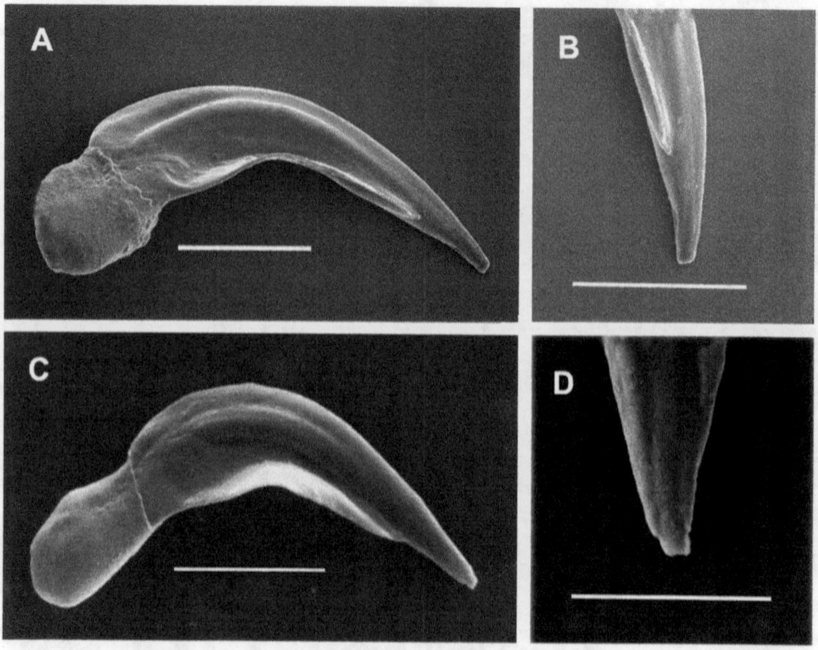

Fig. 89. *Comparative photographs of spicule shape and spicule tips of* Steinernema guangdongense *(A, B) and* S. longicaudum *(C, D). Note: Spicule tip of* S. longicaudum *suddenly narrows to form the tip but not in* S. guangdongense. *(Scale bars: A, B = 20 μm; C, D = 25 μm.) After Qiu* et al. *(2004), courtesy of* Zootaxa.

lateral field. Portion of lateral field with eight ridges is longest compared to portions with 2, 7, 4 ridges. Number of ridges reducing to seven near anus. Seven ridges in lateral field becoming four a few annules posterior to phasmid. Three or four annules posterior to that, four ridges changing to two. Formula of lateral field is 2, 7, 8, 7, 4, 2. Pharynx with narrow corpus, basal bulb more or less elongate with visible valve. Tail attenuate, tapering gradually with constriction on dorsal side. Hyaline region comprising *ca* 57.5 (52-63)% of tail length.

TYPE HOST AND LOCALITY

The type host of this nematode in nature is unknown as it was recovered from soil using *Galleria* larvae as bait. The soil sample was collected in an eucalypt forest planted in Julia town (latitude N20.49,

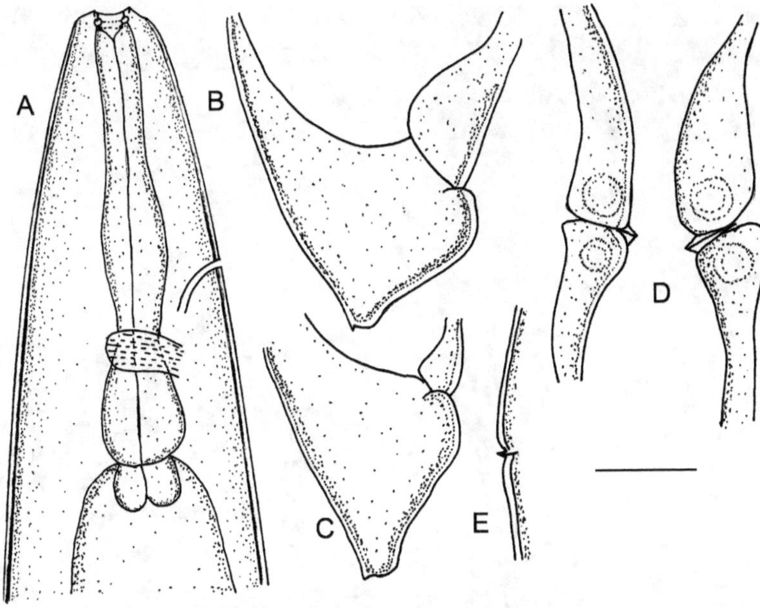

Fig. 90. Steinernema guangdongense *first and second generation females. (Scale bar: A-E = 50 μm.) After Qiu* et al. *(2004), courtesy of* Zootaxa.

longitude E109.58), Lei Zhou, Guangdong province, People's Republic of China.

TYPE MATERIAL

Holotype male, allotype female, ten paratype first generation males, five paratype females, 15 paratype infective juveniles and other mounted material deposited in the State Key Lab for Biocontrol, College of Life Sciences, Zhongshan University, Guangzhou 510275, People's Republic of China. Slides of several males and infective juveniles deposited in the USDA Nematode Collection, Beltsville, MD, USA.

RELATIONSHIPS

Steinernema guangdongense can be recognised by large IJ body diam. of 42 (30-48) μm, distance from anterior end to nerve ring = 102 (88-111) μm; pharynx length = 134 (123-144) μm and a = 25 (22-35). The lateral field pattern is variable, the formula for the arrangement of ridges from head to tail being 2, 7, 8, 7, 4, 2. The species can be

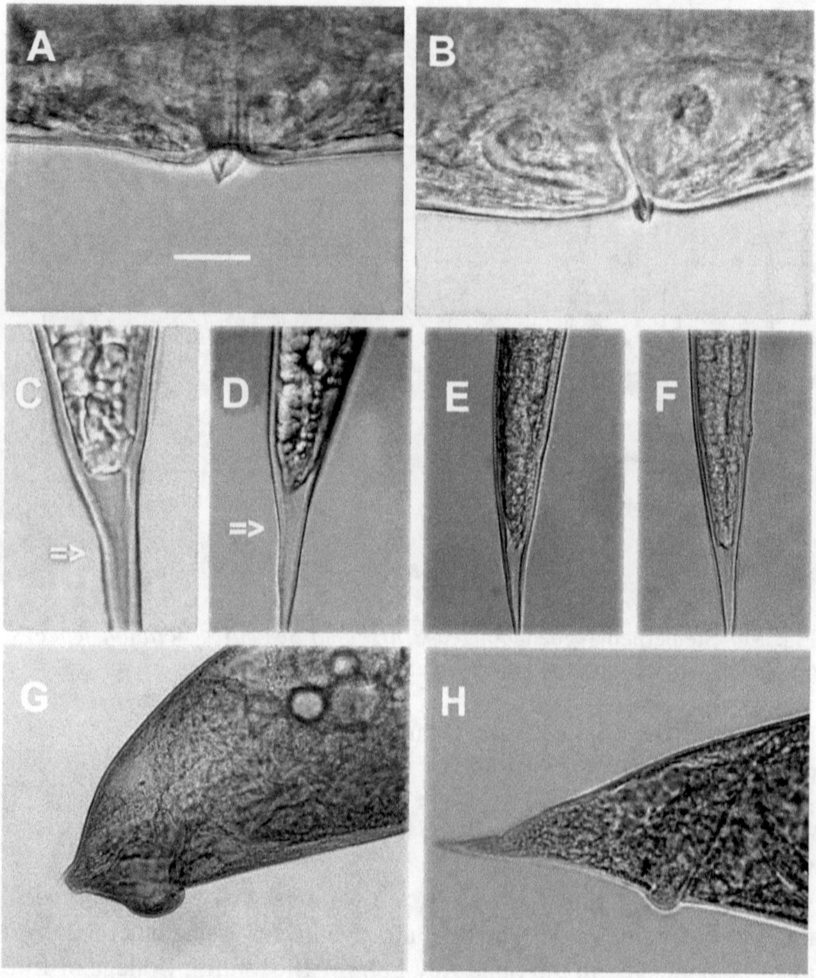

Fig. 91. Steinernema guangdongense. *Light microscope photographs. A, B: Epiptygmata; C, D: Tails of infective juveniles showing dorsal constriction (arrows) as compared to no constriction in* S. longicaudum *in E, F; G: Mature female with prominent post-anal swelling; H: Second generation tail. (Scale bar: A, B = 18 μm; C = 24 μm; D = 22 μm; E, F = 22 μm, G = 44 μm; H = 29 μm.) After Qiu* et al. *(2004), courtesy of* Zootaxa.

further recognised by males with spicule length averaging 86 (80-94) μm, spicule shape, spicule tip and the ratios SW and GS. The female is characterised by the presence of epiptygmata and a prominent postanal swelling.

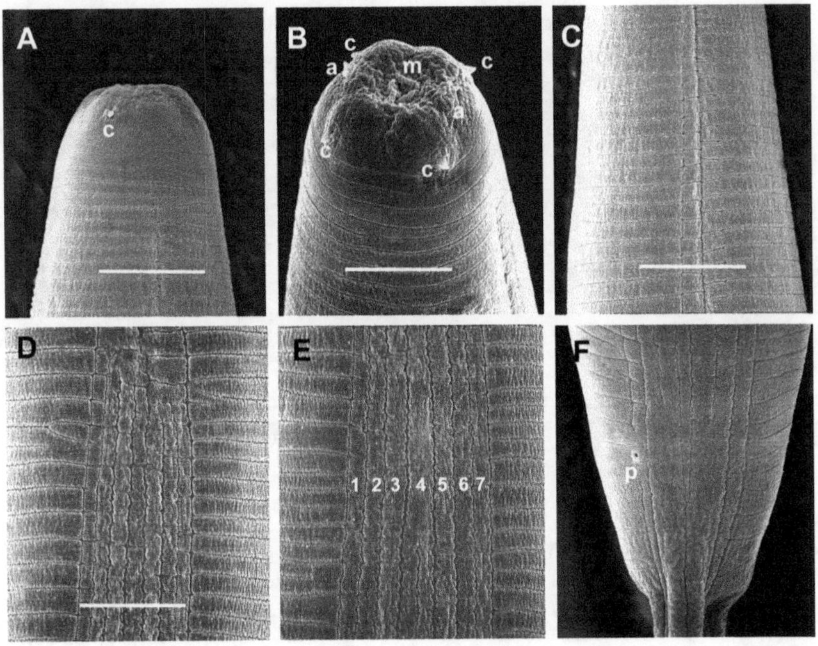

Fig. 92. Steinernema guangdongense. *SEM photographs of infective juvenile. A, B: Anterior region showing single lateral line, closed mouth (m), amphidial apertures (a) and cephalic (c) papillae; C: Lateral field with two ridges (three incisures); D: Lateral field showing change of lateral field pattern from two to seven ridges; E: Lateral field showing seven ridges with middle one (number 4) dividing into two, thereby making eight ridges in all; F: Lateral field showing phasmid (p) and seven ridges changing to four, then two. (Scale bars: A = 6.7 μm; B =* 5 *μm; C = 8.6 μm; D, E, F = 6.7 μm.) After Qiu* et al. *(2004), courtesy of* Zootaxa.

Both morphological and molecular studies showed that the nematode species belongs to the *glaseri*-group, and is closely related to *S. longicaudum. Steinernema guangdongense* can be distinguished from *S. longicaudum* by characters of the infective juveniles (IJ), males and females. For the IJ, although the body length, EP, and E% are similar (1055 *vs* 1063 μm; 80 *vs* 82 μm; 88 *vs* 87%, respectively), the body diam. is larger; the value of NR and the ratio body length/body diam. are smaller (Table 26) and the tail has a dorsal constriction (Fig. 91C, D). For the male, this species has longer spicules that are not strongly curved, shorter spicule head, shaft either not prominent or absent, and spicule tip not suddenly tapered as in *S. longicaudum* (Fig. 89A-D). Also, the ratios

Table 26. *Morphometrics of* Steinernema guangdongense. *Measurements are in µm and in the form: mean ± SD (range).*

Character	First generation			Second generation			Infective juvenile
	Holo-type	Male	Female	Allo-type	Male	Female	
n	–	20	20	–	20	20	20
L	2000	1809 ± 415 (1627-2126)	5050 ± 913 (3800-6800)	6250	1441 ± 43.7 (1350-1500)	276 ± 201 (3130-2500)	1055 ± 49 (98-1145)
a	–	–	–	–	–	–	25 ± 3 (22-35)
b	–	–	–	–	–	–	7.9 ± 0.3 (7.3-8.5)
c	–	–	–	–	–	–	11.6 ± 0.7 (10.2-12.9)
V%	–	–	49.5 ± 0.75 (48-50)	50	–	53 ± 1.5 (51-56)	–
Body diam.	135	112 ± 14 (90-135)	218 ± 29 (175-275)	265	80 ± 3.5 (75-88)	143 ± 8 (130-155)	42 ± 4 (30-48)
EP	123	120 ± 8 (103-132)	158 ± 7 (150-175)	163	119 ± 2.8 (113-123)	119 ± 5.6 (113-130)	80 ± 3.6 (71-85)
NR	130	127 ± 8 (109-139)	163 ± 15 (143-200)	162	136 ± 4.9 (125-143)	165 ± 2.5 (?-168)	102 ± 6 (88-111)
ES	165	162 ± 7 (150-176)	199 ± 16 (183-238)	230	166 ± 5.9 (160-175)	204 ± 4.9 (195-215)	134 ± 5 (123-144)
Tail length (T)	35	31 ± 3 (24-38)	55 ± 8 (43-65)	40	30 ± 1.8 (28-33)	80 ± 3 (75-85)	91 ± 8 (82-103)
Anal body diam. (ABD)	55	50 ± 5 (42-58)	77 ± 15 (63-118)	75	42 ± 3.2 (38-48)	54 ± 2.4 (50-58)	27 ± 2 (24-32)
Spicule length (SL)	85	86 ± 3 (80-94)	–	–	71 ± 3.9 (65-80)	–	–
Gubernaculum length (GU)	70	64 ± 6 (47-73)	–	–	46 ± 2.9 (25-42)	–	–
D% = EP/ES × 100	75	70 ± 17 (67-78)	–	–	71 ± 3 (67-75)	–	59 ± 3 (54-65)
E% = EP/T × 100	–	–	–	–	–	–	88 ± 7 (74-100)
SW% = SP/ABW × 100	155	175 ± 20 (152-216)	–	–	183 ± 16 (158-200)	–	–
GS% = GU/SP × 100	82	75 ± 6 (59-82)	–	–	65 ± 6 (59-74)	–	–
H% = Hyaline/T × 100	–	–	–	–	–	–	57 ± 3 (53-62)

EP = distance from anterior end to excretory pore; NR = distance from anterior end to nerve ring; ES = distance from anterior end to end of pharynx.

SW and GS are smaller. For the female, the presence of small double flapped epiptygmata (Fig. 91A, B), a small projection on the dorsal side of the tail tip and prominent postanal swelling (Figs 90B, C; 91G, H) differentiate this nematode from *S. longicaudum. Steinernema guangdongense* can be distinguished from other related species by morphometrical characters.

MOLECULAR CHARACTERISATION

Steinernema guangdongense is characterised by the length of ITS regions, 989 bp (ITS1 = 285 bp, ITS2 = 293 bp) and its composition, A = 0.241, C = 0.196, G = 0.267, T = 0.296 (accession no. AY170341), and length of its partial 28S region of rDNA, 463bp, and composition, A = 23.8%, C = 19.8%, G = 32.2%, T = 24.2% (accession no. AY169558). More DNA information is available from Qiu *et al.* (2004).

DISTRIBUTION

Steinernema guangdongense was found in a very humid region (latitude N20.49, precipitation = 2000 mm year^{-1}) whereas *S. longicaudum*, the most similar species, was found in a colder and drier region (latitude N39.50, precipitation = 700 mm year^{-1}). The two nematodes may be derived from the same ancestor, differences between the two nematodes being due to their adaptations for survival in different geographic conditions (Qui *et al.*, 2004).

Reference

QIU, L., FANG, Y., ZHOU, Y., PANG, Y. & NGUYEN, K.B. (2004). *Steinernema guangdongense* sp. n. (Nematoda: Steinernematidae), a new entomopathogenic nematode from southern China with a note on *S. serratum* (*nomen nudum*). *Zootaxa* 704, 1-20.

_Steinernema hebeiense_*, Chen, Li, Yan, Spiridonov & Moens, 2006**
(Figs 93, 94)

MEASUREMENTS

See Table 27.

DESCRIPTION

Body J-shaped in heat-relaxed specimens. Cuticle with very faint transverse striae; lateral field indistinct throughout body. Cephalic extremity flattened to slightly rounded anteriorly. SEM *en face* view with six pointed labial papillae and four cephalic papillae. Amphidial apertures visible. Stoma funnel-shaped or cup-shaped. Pharynx muscular; procorpus cylindrical; metacorpus slightly swollen, non-valvate; isthmus indistinct; basal bulb pyriform to round, valve indistinct. Nerve ring surrounding isthmus or just anterior to basal bulb. Cardia prominent and protruding into intestine lumen. Excretory pore usually anterior to nerve ring, located at base of procorpus. Excretory duct cuticularised. Genital system monorchic, reflexed. Spicules paired, yellow-brownish in colour, ventrally curved; spicule head (manubrium) less than one-third of spicule length and longer than wide. Calomus short and indistinct; velum thin, *ca* half of spicule length. Gubernaculum *ca* 80% of spicule length, boat-shaped in lateral view, swollen at middle, with prominent narrow neck; proximal end with knob, occasionally ventrally curved. Gubernaculum wings strongly divided; cuneus pointed. Single midventral precloacal papilla and 11 pairs of genital papillae present and arranged as follows: six pairs subventral precloacal; one pair subdorsal at same level of single midventral precloacal papilla; one pair subventral in cloacal region; two pairs caudal subventral and one pair caudal subdorsal. Tail bluntly conoid, ventrally concave, mucron absent. Phasmids inconspicuous.

Male (second generation)

Similar to first generation male but smaller in body length and other morphometric values except for shorter tail, excretory pore more posteriorly located than in first generation male (D% = 51 *vs* 62), and presence of tail mucron (6 μm long).

---

* Specific epithet derived from the Chinese Province of *Hebe*, the type locality.
** Description after Phan *et al.* (2006). Revised by David J. Hunt.

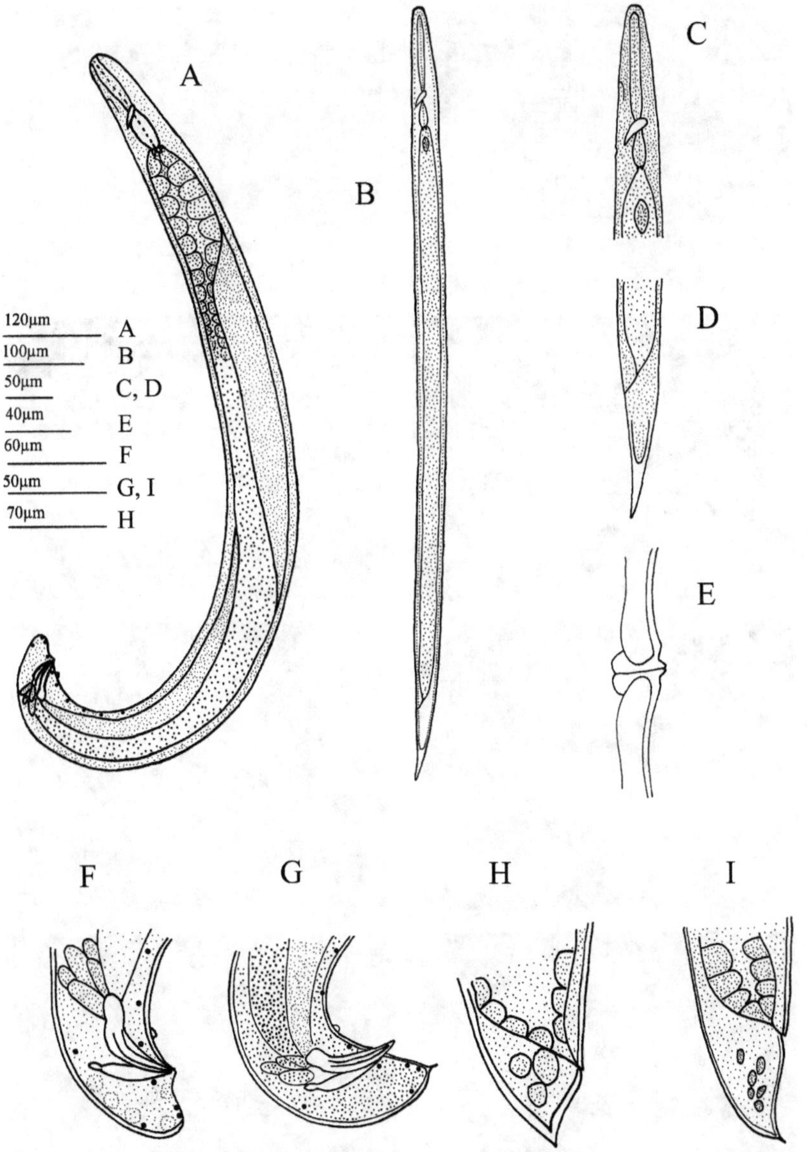

Fig. 93. Steinernema hebeiense. *A: First generation male; B: Infective juveniles, entire view; C: Infective juvenile, anterior region; D: Infective juvenile, tail; E: Female vulva; F: First generation male, posterior end; G: Second generation male, posterior end; H: First generation female, posterior end; I: Second generation female, posterior end. After Chen* et al. *(2006), courtesy of* Nematology.

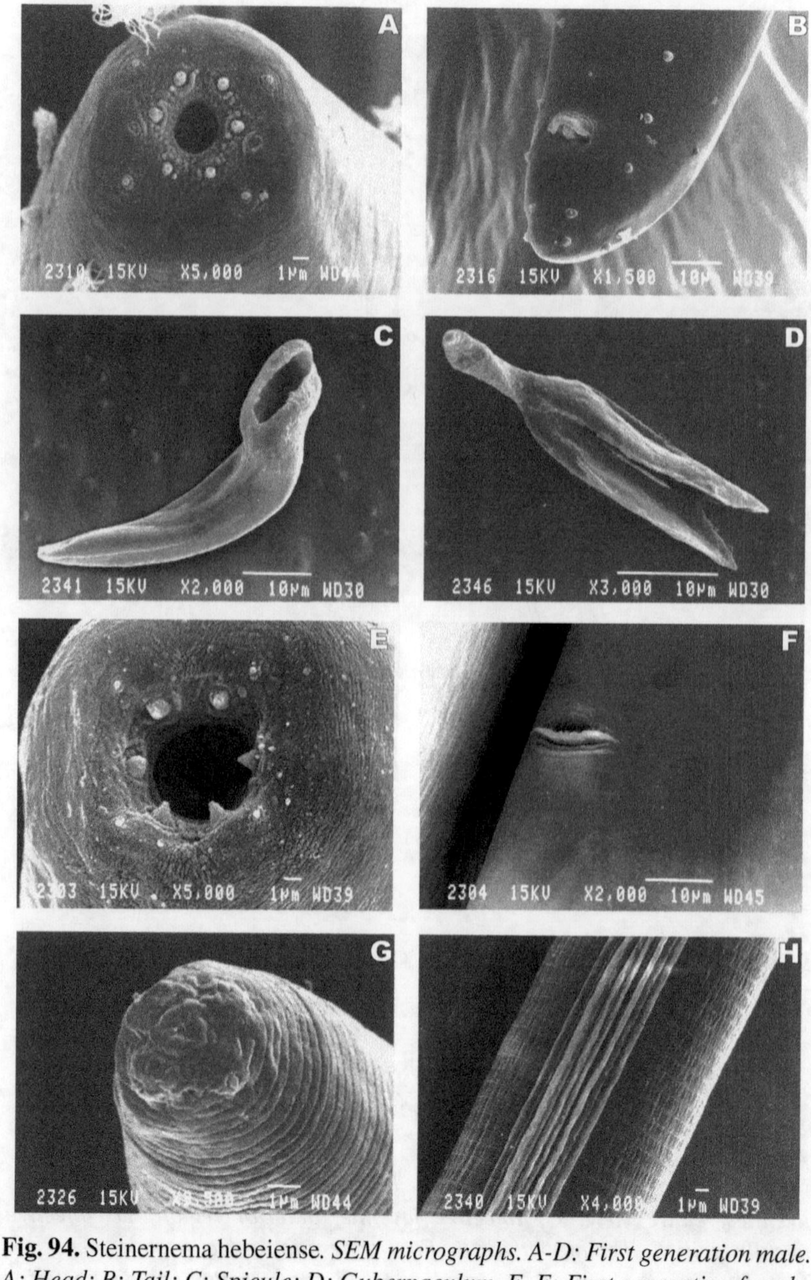

Fig. 94. Steinernema hebeiense. *SEM micrographs. A-D: First generation male. A: Head; B: Tail; C: Spicule; D: Gubernaculum. E, F: First generation female. E: Head; F: Vulva. G, H: Infective juvenile. G: Head; H: Lateral field (mid-body). After Chen* et al. *(2006), courtesy of* Nematology.

Female (first generation)

Body robust, C-shaped when heat-relaxed. Cuticle smooth. Lateral field inconspicuous. Cephalic extremity broadly rounded. SEM *en face* view with six pointed labial papillae and four cephalic papillae. Amphidial apertures inconspicuous. Stoma funnel- or cup-shaped, shallow. Pharynx with cylindrical procorpus; metacorpus slightly swollen and non-valvate; isthmus indistinct; basal bulb pyriform, valve not observed. Excretory pore located between nerve ring and mid-pharynx level. Excretory duct cuticularised, excretory gland swollen. Cardia prominent, protruding into intestine lumen. Didelphic, amphidelphic genital system reflexed and tightly packed with eggs. Vulva in form of transverse slit on a protuberance near middle of body, epiptygmata present. Vagina short, oblique, with muscular walls. Tail pointed, shorter than anal body diam.

Female (second generation)

Similar to first generation female but smaller in body length and other morphometric values, except for tail length. Tail pointed, longer than anal body diam.

Infective juvenile

Body smoothly tapering towards anterior and posterior ends, in heat-relaxed specimens straight or slightly curved ventrally. Oral aperture and anus closed. Cephalic extremity slightly offset from body contour, 6-7 μm diam., truncate. In SEM *en face* view, labial papillae not observed; oval-shaped amphidial apertures situated slightly posterior to labial disc between distinct cephalic papillae. Pharynx long and narrow, isthmus distinct and surrounded by nerve ring, basal bulb elongate or pyriform, valve indistinct in fixed but visible in living specimens. Cardia prominent. Excretory pore located at one-third to mid-pharynx level, cuticularised duct extending to basal bulb. Hemizonid distinct (3.2-3.6 μm long), located at isthmus level. Bacterial pouch round to elongated (11.5-19 μm diam., 12.5-56 μm long), rarely seen in fixed specimens; packed with bacteria, located 26-35 μm posterior to basal bulb. Lateral field with eight ridges (*i.e.*, nine lines), two submarginal ridges reduced at mid-body. Tail conoid without pointed terminus, concave ventrally. Hyaline region comprising *ca* 33-50% of tail length. Phasmid inconspicuous.

Table 27. *Morphometrics of* Steinernema hebeiense. *Measurements are in μm and in the form: mean ± SD (range).*

Character	First generation			Second generation		Infective juvenile
	Holotype (male)	Male	Female	Male	Female	
n	–	20	20	20	20	20
L	1168	1177 ± 117 (1036-1450)	3465 ± 338 (3072-4254)	897 ± 54 (776-981)	1933 ± 272 (1327-2351)	658 ± 29 (610-710)
a	13	14 ± 1.2 (12-17)	21 ± 2.2 (17-25)	17 ± 1.2 (14-19)	18 ± 1.4 (16-21)	26 ± 0.9 (24-28)
b	8.9	9 ± 0.9 (8-11)	24 ± 2.2 (21-29)	9 ± 0.7 (7-10)	16 ± 2.2 (11-19)	6.2 ± 0.25 (5.7-6.7)
c	36	39 ± 5.3 (30-49)	103 ± 17.6 (67-129)	36 ± 3.1 (30-43)	42 ± 7.1 (29-56)	10 ± 0.39 (9.4-11.0)
c′	0.77	0.8 ± 0.1 (0.6-0.9)	0.7 ± 0.13 (0.5-0.9)	0.7 ± 0.06 (0.6-0.9)	1.3 ± 0.12 (1.0-1.5)	–
V	–	–	54 ± 2.1 (50-57)	–	55 ± 2.6 (48-61)	–
Body diam.	88	86 ± 6.3 (74-98)	167 ± 25 (142-245)	54 ± 4.5 (43-63)	107 ± 13 (79-132)	26 ± 1.2 (23-28)
EP	65	64 ± 3.6 (58-73)	65 ± 10 (48-95)	64 ± 4.6 (58-73)	72 ± 5.9 (63-80)	48 ± 2.1 (43-51)
NR	85	84 ± 4.4 (78-93)	104 ± 8.6 (88-123)	73 ± 6.2 (61-83)	84 ± 4.2 (75-93)	78 ± 2.5 (73-83)
ES	131	126 ± 4.0 (118-132)	147 ± 7.2 (133-158)	103 ± 5.9 (90-113)	122 ± 3.8 (115-130)	107 ± 2.8 (100-111)
Tail length (T)	33	30 ± 3.0 (24-35)	35 ± 8.3 (25-50)	25 ± 2.2 (21-30)	46 ± 3.8 (35-54)	66 ± 2.7 (63-71)
Hyaline (H)	–	–	–	–	–	28 ± 2.9 (23-35)
Anal body diam. (ABD)	42	41 ± 3.0 (36-53)	53 ± 5.7 (45-65)	34 ± 1.9 (30-38)	35 ± 1.9 (31-39)	–
Spicule length (SL)	55	57 ± 3.1 (51-63)	–	46 ± 4.1 (39-55)	–	–
Spicule width	12	12 ± 1.6 (10-15)	–	9 ± 0.9 (8-10)	–	–
Gubernaculum length (GL)	47.5	46 ± 3.6 (38-50)	–	36 ± 4.1 (30-48)	–	–
Gubernaculum width	9.5	8.8 ± 1.0 (7-10)	–	8 ± 0.9 (6-10)	–	–

TYPE HOST AND LOCALITY

The natural hosts of *S. hebeiense* are not known as it was isolated by baiting with *Galleria mellonella* from sandy soil in Gu An County, Hebei, People's Republic of China (longitude 116°17′E, latitude 39°26′N, 23 m a.s.l.).

Table 27. *(Continued).*

Character	First generation			Second generation		Infective juvenile
	Holotype (male)	Male	Female	Male	Female	
D% = EP/ES × 100	50	51 ± 3.0 (48-59)	45 ± 6.8 (36-66)	62 ± 6.6 (51-81)	59 ± 4.9 (50-68)	44.6 ± 2.2 (39.5-50)
E% = EP/T × 100	–	–	–	–	–	72 ± 3.5 (65-80)
SW% = SL/ABD × 100	130	140 ± 13 (120-170)	–	140 ± 16 (110-170)	–	–
GS% = GL/SL × 100	90	80 ± 6 (60-90)	–	80 ± 8 (70-100)	–	–
H% = H/T × 100						43 ± 4.6 (32-50)

EP = distance from anterior end to excretory pore; NR = distance from anterior end to nerve ring; ES = distance from anterior end to end of pharynx.

TYPE MATERIAL

Holotype and paratype males, females and infective juveniles deposited at Ghent University, Institute for Zoology, K.L. Ledeganckstraat 35, 9000, Belgium. Slides with paratype first and second generation males, first and second generation females and infective juveniles deposited in the Institute of Plant Protection, Hebei Academy of Agricultural and Forestry Sciences, Baoding, Hebei, People's Republic of China.

RELATIONSHIPS

Infective juveniles of *S. hebeiense* are characterised by a body length of 658 (610-710) μm, distance from the anterior end to excretory pore of 48 (43-51) μm, tail length of 66 (63-71) μm, and E% of 72 (65-80). They have a slightly offset head with a truncate terminus, lateral fields with eight ridges (nine distinct lines) of which the two submarginal ridges are indistinct under the light microscope. Males have light brown, slightly curved, spicules and the manubrium is longer than wide. Spicule length of the first generation male is 57 (51-63) μm and gubernaculum length is 46 (38-50) μm. A mucron is absent in first generation males, but present and 6 (4-10) μm long in males of the second generation. This combination of features resembles those of several species belonging to the *feltiae*-group.

The appearance of the lateral field of *S. hebeiense* infective juveniles is very similar to that of *S. feltiae* which also has the two submarginal ridges reduced. However, infective juveniles of *S. feltiae* have a longer body length of 849 (736-950) *vs* 658 (610-710) μm, a longer distance from the anterior end to the excretory pore of 62 (53-67) *vs* 48 (43-51) μm and a longer tail of 81 (70-92) *vs* 66 (63-71) μm. First generation males of *S. feltiae* have a mucron (*vs* absent) and the mucron of the second generation males is longer (12 *vs* 6 μm). Males of *S. feltiae* have a lower SW value of 1.1 (0.9-1.3) *vs* 1.4 (1.2-1.7) and GS value of 0.59 (0.5-0.6) *vs* 0.8 (0.6-0.9).

Infective juveniles of *S. hebeiense* are similar to those of *S. weiseri* in total body length, but infective juveniles of *S. weiseri* have nine equally developed lines (*i.e.*, eight distinct ridges) in their lateral field, a shorter tail of 55 (46-70) *vs* 66 (63-71) μm and a shorter hyaline region of 16 (12-23) *vs* 28 (23-35) μm. Second generation males of *S. weiseri* have a much shorter mucron (2.5 *vs* 6 μm).

Steinernema hebeiense is also close to *S. oregonense*, *S. kraussei*, *S. jollieti*, *S. sangi* and *S. kushidai*. Infective juveniles of *S. oregonense* can be distinguished from *S. hebeiense* by their longer body length of 980 (820-1110) *vs* 658 (610-710) μm, larger body diam. of 34 (28-38) *vs* 26 (23-28) μm, longer distance from anterior end to excretory pore (66 (60-72) *vs* 48 (43-51) μm) and a higher E% ratio of 100 (90-110) *vs* 72 (65-80).

Infective juveniles of *S. kraussei* can be distinguished from *S. hebeiense* by their longer body length of 951 (797-1102) *vs* 658 (610-710) μm and a longer distance from anterior end to excretory pore of 63 (50-66) *vs* 48 (43-51) μm.

Males of *S. kraussei* are distinguished by a shorter spicule length of 55 (52-57) *vs* 57 (51-63) μm and a lower SW value (1.1 *vs* 1.4 (1.2-1.7)).

Infective juveniles of *S. jollieti* are longer at 711 (625-820) *vs* 658 (610-710) μm and have only six *vs* eight longitudinal ridges in the lateral field.

Steinernema sangi, which belongs to the same clade in the genus, has a lateral field of similar appearance, but can be distinguished from *S. hebeiense* by the longer infective juvenile body of 753 (704-784) *vs* 658 (610-710) μm, very long velum on the spicules, and by the presence of a mucron in the first generation males.

Steinernema kushidai is distinguished from *S. hebeiense* by a shorter IJ tail of 50 (44-59) *vs* 66 (63-71) μm, higher E% value of 92 (84-95) *vs* 72 (65-80) and by the colour of the male spicule (colourless *vs* light brown).

MOLECULAR CHARACTERISATION

In the MP analysis of a total of 924 ITS rDNA characters, 146 characters were constant, 176 parsimony-uninformative and 602 parsimony-informative. According to Chen *et al.* (2006), tree length = 3053, consistency index (CI) = 0.5129, homoplasy index (HI) = 0.4871, retention index (RI) = 0.6039, and g1 = −0.623776.

In all variants of the MP analysis of ITS sequences, *S. hebeiense* constantly formed a single branch with *S. feltiae* and *S. weiseri* being a member of the '*feltiae-kraussei-oregonense*' clade (Spiridonov *et al.*, 2004). Usually, *S. feltiae* was a sister group for *S. hebeiense*, although in some MP cladograms when, for example, gaps were treated as 'fifth base', *S. weiseri* was the sister taxon. The relationships of any pair of these three species were always characterised by a weak (<70%) bootstrap support, unlike the support for the entire subclade of the three species, which ranged from moderate (70-90%) to strong (>90%). MP sequence analysis of the ITS dataset revealed three nucleotide autapomorphies in the ITS rDNA sequence for *S. hebeiense*. The minimal difference in ITS rDNA of *S. hebeiense* with *S. weiseri* was 73 bp or 10% if expressed as mean nucleotide difference adjusted for missing data. Such a level of differences is larger than the known level of intraspecific nucleotide variability in ITS rDNA of steinernematids (*ca* 2.5%; Spiridonov *et al.*, 2004). The difference between the clearly distinguishable species *S. weiseri* and *S. feltiae* and *S. kraussei* was lower at 37 and 64 nucleotides (*i.e.*, 5 and 9%), respectively. Morphological differences between *S. weiseri* and *S. feltiae* further argue against their conspecific status. The MP analysis always revealed a strong bootstrap support for the clade composed by the *feltiae*-group species (with *S. jollieti* in basal position) and *S. monticolum*.

ML analysis of ITS rDNA also revealed close relationships of *S. hebeiense* with *S. feltiae* and *S. weiseri*. The clade including the species of the *feltiae*-group and *S. monticolum* was also strongly supported in ML analysis.

In the MP analysis of a total of 526 of D2D3 LSU rDNA characters, 268 characters were constant, 29 were parsimony-uninformative and 229 were parsimony-informative (tree length = 478, consistency index

(CI) = 0.6799, homoplasy index (HI) = 0.3201, retention index (RI) = 0.8061, and g1 = −2.1984). In all obtained MP, *S. hebeiense* was a sister taxon for *S. weiseri*, although the bootstrap support was always low for this group. The difference between *S. hebeiense* and *S. weiseri* was 13 nucleotides, or 2.6% if expressed as a mean nucleotide difference adjusted to missing data. The difference between the two clearly distinguishable species *S. feltiae* and *S. kraussei* was only nine nucleotides (1.8%). In ML trees the grouping of *S. hebeiense* and *S. weiseri* was moderately or strongly supported.

The ITS and D2D3 sequences of *S. hebeiense* are deposited at GenBank under accession no. DQ105794 and DQ399664, respectively.

Bionomics

Attempts to hybridise *S. hebeiense* with *S. feltiae* and *S. weiseri* failed; adult nematodes developed from juveniles added to the haemolymph drop, but no progeny were observed. Control mating between members of the same species resulted in offspring within 1 week for all three species. The life cycle appears to be similar to other steinernematids. The identity of the associated bacterium is unknown.

Distribution

Known only from the type locality, Gu An County, Hebei, People's Republic of China (Chen *et al.*, 2006).

References

Chen, S.L., Li, X., Yan, A., Spiridonov, S.E. & Moens, M. (2006). A new entomopathogenic nematode, *Steinernema hebeiense* sp. n. (Rhabditida: Steinernematidae), from North China. *Nematology* 8, 563-574.

Spiridonov, S.E., Reid, A.P., Podrucka, K., Subbotin, S.A. & Moens, M. (2004). Phylogenetic relationships within the genus *Steinernema* (Nematoda: Rhabditida) as inferred from analyses of sequence of the ITS1-5.8S-ITS2 region of the rDNA and morphological features. *Nematology* 6, 547-566.

Steinernema hermaphroditum[*],[**] Stock, Griffin & Chaerani, 2004
(Figs 95-97)

MEASUREMENTS

See Table 28.

DESCRIPTION

Hermaphroditic female (first generation)

Body C-shaped when heat-relaxed. Cuticle smooth under light microscopy, but with fine transverse striae visible under SEM. Lateral field and phasmids inconspicuous. Cephalic extremity truncate to slightly rounded, continuous with body. Six lips united but with tips distinct, each bearing one labial papilla. Four cephalic papillae. Amphidial apertures small, located posterior to lateral labial papillae. Stoma reduced (cheilo-, gymno- and stegostom vestigial), short and broad, with inconspicuous sclerotised walls. Pharynx set off from intestine; with cylindrical procorpus; non-valvate and slightly swollen metacorpus, distinct isthmus, and pyriform basal bulb containing reduced valve. Nerve-ring usually surrounding isthmus or anterior part of basal bulb. Excretory pore opening circular, located at posterior third of metacorpus, just anterior to nerve-ring. Ovaries opposed, reflexed in dorsal position; oviduct well developed; glandular spermatheca filled with spermatozoa, uterus in ventral position. Vagina short, with muscular walls. Vulva located in or near middle of body, with protruding and asymmetric vulval lips (anterior lip larger than posterior). Epiptygmata absent. Tail digitate, mucron present in 60% of specimens examined. Postanal swelling present.

Male (first generation)

Cuticle, labial region, stoma and pharyngeal region as in first generation hermaphrodite. Body curved posteriorly, J-shaped when heat-relaxed. Single reflexed testis, consisting of germinal growth zone leading to seminal vesicle. *Vas deferens* with inconspicuous walls. Spicules paired, symmetrical, curved, ochre-brown in colour. Manubrium rectangular, wider than long, calomus short, lamina wide,

[*] The specific epithet reflects the presence of *hermaphrodites* in the first adult generation.

[**] Description after Stock *et al.* (2004). Revised by David J. Hunt.

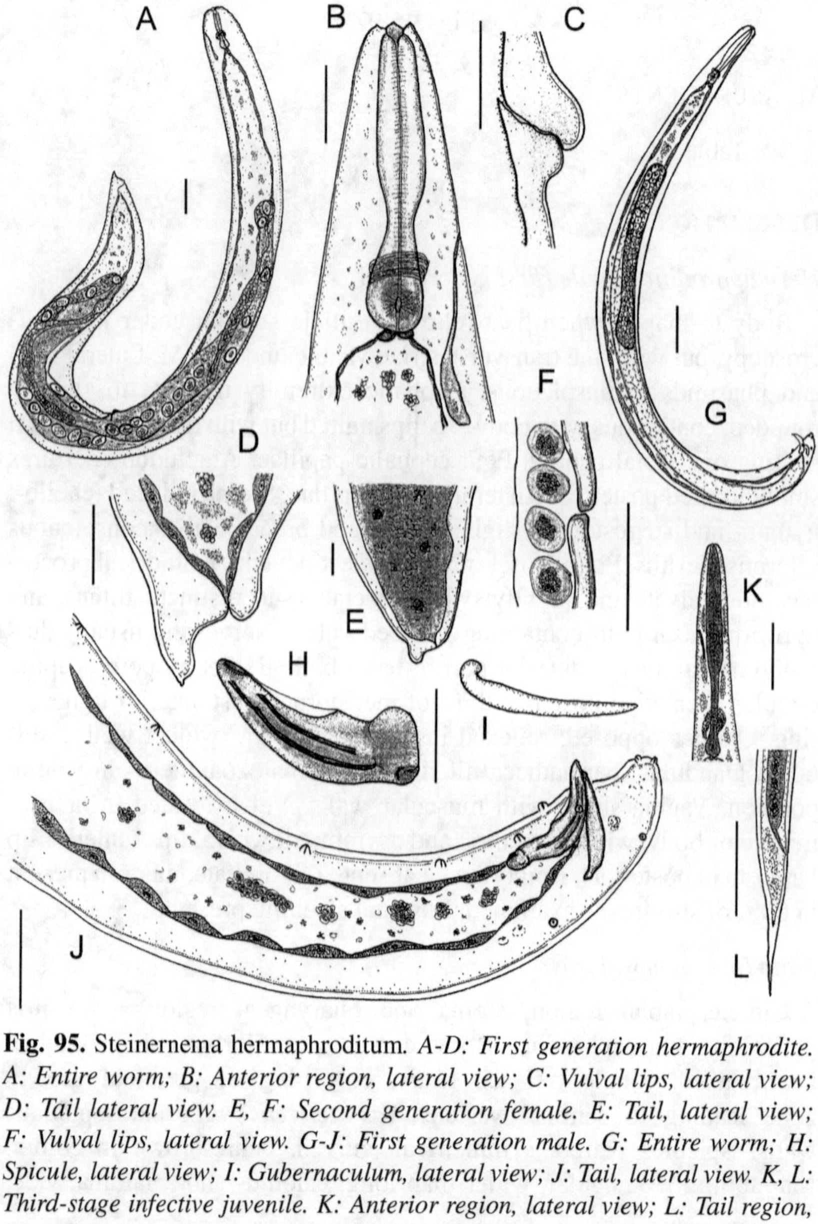

Fig. 95. Steinernema hermaphroditum. *A-D: First generation hermaphrodite. A: Entire worm; B: Anterior region, lateral view; C: Vulval lips, lateral view; D: Tail lateral view. E, F: Second generation female. E: Tail, lateral view; F: Vulval lips, lateral view. G-J: First generation male. G: Entire worm; H: Spicule, lateral view; I: Gubernaculum, lateral view; J: Tail, lateral view. K, L: Third-stage infective juvenile. K: Anterior region, lateral view; L: Tail region, lateral view. (Scale bars: A, G = 100 μm; B-E, J, K, L = 50 μm; F, H, I = 25 μm.) After Stock* et al. *(2004), courtesy of* Nematology.

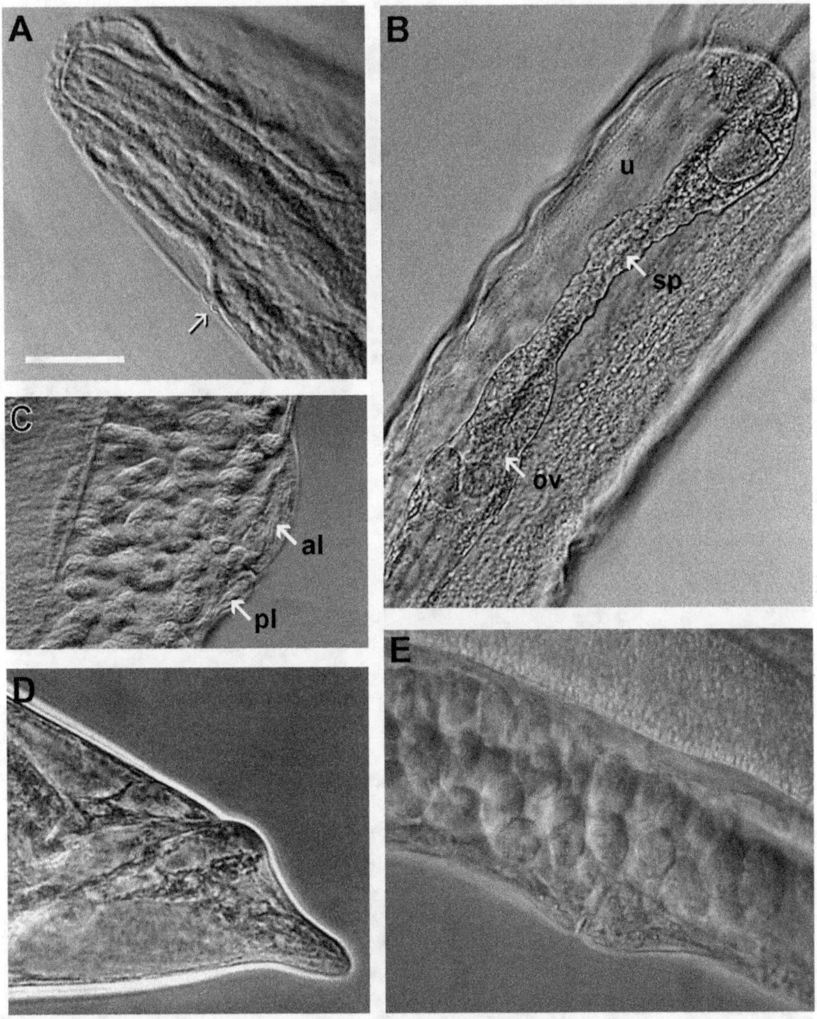

Fig. 96. Steinernema hermaphroditum. *A-D: Hermaphrodite. A: Anterior end showing excretory pore (arrow); B: Anterior part of reproductive tract (ov = ovary; sp = spermatheca; u = uterus); C: Vulva (al = anterior lip; pl = posterior lip); D: Tail, lateral view. E: Second generation female vulva, lateral view. (Scale bar: A = 43 μm; B = 17 μm; C = 10 μm; D = 32 μm; E = 15 μm.) After Stock* et al. *(2004), courtesy of* Nematology.

with two ribs, rostrum present. Velum short, extending from rostrum to distal ventral aperture. Gubernaculum boat-shaped in lateral view, *ca* two-thirds of spicule length. Cuneus long, needle-shaped, pointed

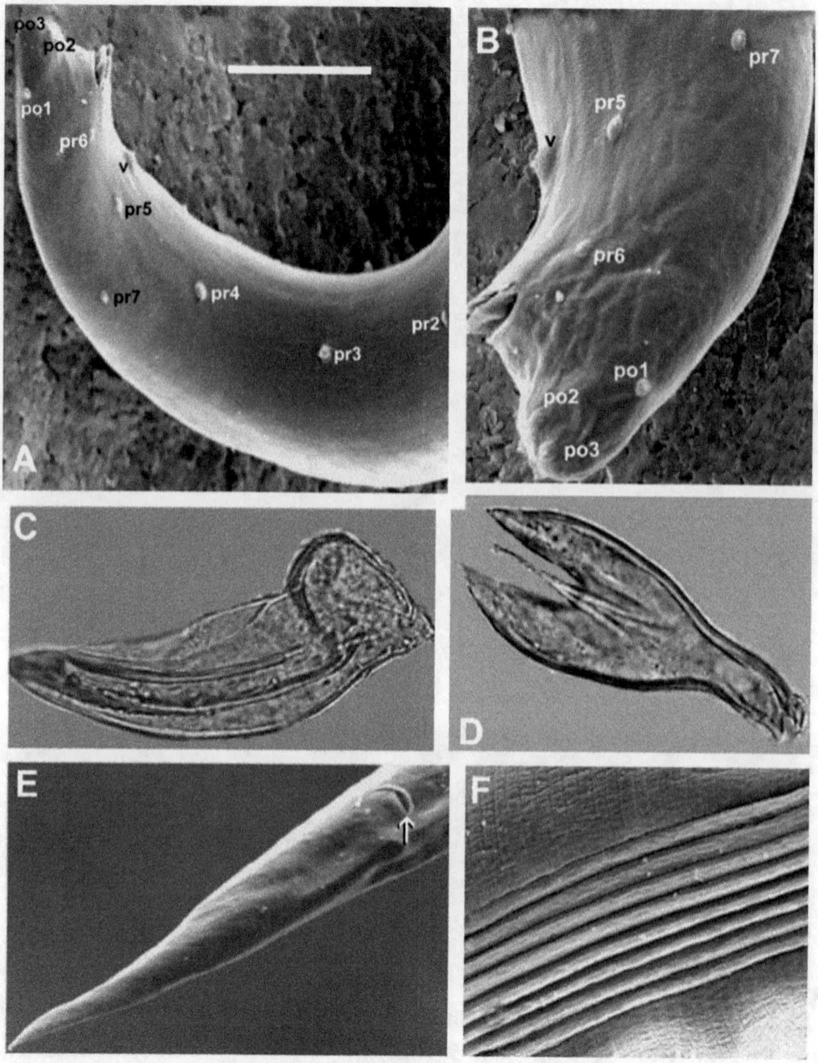

Fig. 97. Steinernema hermaphroditum. *A-D: First generation male. A, B: Posterior end showing distribution of genital papillae (pr = precloacal; v = ventral; po = postcloacal); C: Spicule, lateral view; D: Gubernaculum, dorsal view. E, F: Infective juvenile. E: Tail (ventral view) showing anus (arrow); F: Lateral field pattern at mid-body. (Scale bar: A = 35 μm; B = 33 μm; C = 22 μm; D = 20 μm; E = 23 μm; F = 10 μm.) After Stock* et al. *(2004), courtesy of* Nematology.

Table 28. *Morphometrics of* Steinernema hermaphroditum. *Measurements are in μm and in the form: mean ± SD (range).*

Character	First generation			Second generation		Infective juvenile
	Holotype (male)	Male	Female hermaphrodite	Male	Female	
n	–	19	19	20	20	20
L	1550	1481 ± 19 (1115-1560)	6550 ± 497 (5843-7154)	845 ± 11 (698-975)	5129 ± 125 (3874-5623)	929 ± 99 (700-1000)
a	–	–	–	–	–	29 ± 3.5 (23-35)
b	–	–	–	–	–	6.2 ± 0.5 (5.6-7.1)
c	–	–	–	–	–	10.6 ± 0.7 (9.3-11.7)
V	–	–	55.5 ± 3 (51-61)	–	52 ± 0.5 (50-53)	–
Body diam.	–	–	–	–	–	28.5 (25-32.5)
EP	75	75 ± 5 (65-80)	124 ± 8 (110-134)	53 ± 4 (49-60)	87 ± 7 (79-98)	65 ± 2 (63-68)
NR	117	120 ± 4 (114-125)	147 ± 6 (134-154)	116 ± 4 (100-119)	126 ± 5 (115-133)	103 ± 2 (95-113)
Tail length (T)	33	33.5 ± 2 (28-35)	69 ± 2 (65-72)	21.5 ± 1.5 (19-25)	58 ± 1.5 (50-62)	77 ± 6 (65-83)
Hyaline (H)	–	–	–	–	–	33.4 ± 2 (31-40)
Anal body diam. (ABD)	45	44 ± 1.5 (40-51)	70 ± 2 (66-72)	27 ± 1 (28-35)	57 ± 2 (48-61)	19 ± 2 (16-24)
Spicule length (SL)	66	68 ± 2 (65-70)	–	60 ± 1 (55-64)	–	–
Spicule width	–	–	–	–	–	–
Gubernaculum length (GL)	50	48 ± 1.5 (47-50)	–	41 ± 1.5 (39-44)	–	–
Gubernaculum width	–	–	–	–	–	–
D% = EP/ES × 100	–	48 ± 5 (44-50)	–	40 ± 2 (39-43)	–	50 ± 2.5 (47-55)
E% = EP/T × 100	–	–	–	–	–	85 ± 7 (76-100)
SW% = SL/ABD × 100	133	150 ± 50 (140-160)	–	180 ± 50 (170-190)	–	–
GS% = GL/SL × 100	75	71 ± 1 (70-75)	–	65 ± 2 (60-70)	–	–
H% = H/T × 100	–	–	–	–	–	51 ± 1 (49-53)

EP = distance from anterior end to excretory pore; NR = distance from anterior end to nerve ring; ES = distance from anterior end to end of pharynx.

posteriorly. Twenty-one genital papillae (11 pairs plus single midventral) arranged as follows: six precloacal subventral pairs; single midventral precloacal papilla (located between precloacal pairs 5 and 6); one subdorsal precloacal pair; one postcloacal subdorsal pair; one subventral postcloacal pair; one terminal postcloacal pair.

Male (second generation)

Similar to first generation males, except for body length and diam., position of excretory pore (more anteriorly located in second generation), and size of spicules and gubernaculum.

Amphimictic female (second generation)

Body C-shaped when heat-relaxed. Cuticle, labial region, stoma and pharyngeal region as in first generation hermaphrodite. Paired ovaries, opposed and reflexed in dorsal position; oviduct well developed; spermatheca absent; uterus in ventral position. Vagina short, with muscular walls. Vulva located in or near middle of body, with non-protruding or slightly protruding vulval lips. Epiptygmata absent. Tail round-conoid, without a mucron. Postanal swelling present or absent.

Infective juvenile

Body slender, tapering regularly from base of pharynx to anterior end and from anus to terminus. Labial region smooth; oral aperture closed. Cuticle with transverse striae; lateral field distinct with eight, equidistant, longitudinal ridges (*i.e.*, nine lines or incisures) in mid-body region. Pharynx long, narrow. Nerve-ring located at isthmus level. Excretory pore located *ca* middle of pharynx. Basal bulb valvate. Cardia present. Anterior portion of intestine with dorsally displaced pouch containing symbiotic bacteria. Lumen of intestine narrow, rectum long, anus distinct. Genital primordium evident. Tail conoid with pointed terminus. Hyaline region comprising *ca* half tail length.

TYPE HOST AND LOCALITY

Natural host unknown in nature, material collected by baiting with larvae of *Galleria mellonella*. The type isolate was recovered from scrub and trees along the coastal strip of Kamal, Seram Island, Indonesia (Griffin *et al.*, 2000).

TYPE MATERIAL

Holotype male and paratype hermaphrodite, first generation males, second generation females, second generation males and infective juveniles deposited in the University of California-Davis Nematode Collection, Davis, CA, USA. Other paratypes deposited in the USDA Nematode Collection, Beltsville, MD, USA.

RELATIONSHIPS

Steinernema hermaphroditum is morphologically most similar to a group of species placed in the *feltiae*-group. It is characterised by the presence of hermaphrodites in the first adult generation and by the low proportion of males in both adult generations (average of 1 and 5-6%, respectively (Griffin *et al.*, 2001)). Hermaphrodites are characterised by having asymmetric vulval lips and the presence of a spermatheca filled with sperm. Males are characterised by the number and distribution of the male genital papillae (21 papillae: seven precloacal pairs, three postcloacal pairs and a single midventral papilla), non-mucronate tail and by the morphology of the spicules and gubernaculum. Infective juveniles are distinguished by the lateral field pattern of eight equidistant ridges at the mid-body level and the hyaline region comprising *ca* 50% of the tail length.

MOLECULAR CHARACTERISATION

RFLP analysis of ITS rDNA showed this species has a restriction profile very similar to *S. longicaudum* (isolate CB 2B) (Griffin *et al.*, 2000). These two species differed only in the profiles of two restriction enzymes, *Dde*I and *Hha*I. However, comparison of 28S rDNA sequences revealed these two species differ by 28 bp, providing further evidence for considering *S. hermaphroditum* as a separate species. Moreover, comparison of RFLP profiles of *S. hermaphroditum* with *S. scarabaei* (depicted as a sister taxon in the 28S rDNA sequence phylogenetic analysis), showed that these two species differ in the patterns of eight (*Dde*I, *Hha*I, *Hin*fI, *Hpa*II, *Kpn*I, *Pvu*II, *Sau*3AI and *Xba*I) out of the 17 restriction enzymes use to generate this profile (see Griffin *et al.* (2000) and Stock and Koppenhöfer (2003) for complete RFLP profiles of these two species).

Maximum parsimony (MP) analysis of LSU sequence data yielded 371 parsimony informative characters and produced two most parsimonious trees with a tree length of 1080 steps (CI = 0.60). *Steinernema hermaphroditum* was placed in a clade that included the long-IJ *Steinernema* species known as the *glaseri*-group. This clade comprises *S. glaseri*, *S. arenarium*, *S. puertoricense*, *S. cubanum*, *S. longicaudum*, *S. karii*, *S. scarabaei* and two unidentified *Steinernema* species, T29 from Indonesia and CF VII from California, USA. This species was depicted as a sister taxon of *S. scarabaei*, a species isolated from scarab beetle larvae in New Jersey, USA (Stock & Koppenhöfer, 2003). Evolutionary relationships between these two taxa were strongly supported by bootstrap resampling (100%). Pairwise distance matrix indicated that *S. hermaphroditum* and *S. scarabaei* differed by one character.

Bionomics

Steinernema hermaphroditum was the first self-fertile steinernematid to be reported (Griffin *et al.*, 2000). The presence of sperm in the reproductive tract of unmated first generation females indicated that the mode of reproduction of this species is by hermaphroditism (Griffin *et al.*, 2000). Males of this nematode are present in both adult generations at a low level. Approximately 1-6% of the infective juveniles developed into males, the remainder being hermaphrodites. Observations in *G. mellonella* showed that the infective juveniles were able to reach the adult stage in 48-72 h at 22°C, second generation adults appeared in 4-6 days, and infective juveniles emerged 7-9 days after infection.

Attempts to cross-hybridise *S. hermaphroditum* with *S. scarabaei*, *S. longicaudum* and *Steinernema* sp. (T29 isolate) yielded no progeny. Control crosses using individuals of the same species produced viable offspring. Only 60% (12 out of 20) of the drops with couples of *S. hermaphroditum* produced progeny.

Distribution

In addition to Seram Island, Indonesia, the type locality, the species has also been found in banana/herbs and grass/herbs/coconut habitats in Waai, Ambon Island, Indonesia (Griffin *et al.*, 2000; Stock *et al.*, 2004).

References

GRIFFIN, C.T., CHAERANI, R., FALLON, D., REID, A.P. & DOWNES, M.J. (2000). Occurrence and distribution of the entomopathogenic nematodes *Steinernema* spp. and *Heterorhabditis indica* in Indonesia. *Journal of Helminthology* 74, 143-150.

STOCK, S.P. & KOPPENHÖFER, A.M. (2003). *Steinernema scarabaei* n. sp. (Rhabditida: Steinernematidae), a natural pathogen of scarab beetle larvae (Coleoptera: Scarabaeidae) from New Jersey, USA. *Nematology* 5, 191-204.

STOCK, S.P., GRIFFIN, C.T. & CHAERANI, R. (2004). Morphological and molecular characterisation of *Steinernema hermaphroditum* n. sp. (Nematoda: Steinernematidae), an entomopathogenic nematode from Indonesia, and its phylogenetic relationships with other members of the genus. *Nematology* 6, 401-412.

Steinernema intermedium[*,] (Poinar, 1986) Mamiya, 1988**
= *Neoaplectana intermedia* Poinar, 1986
(Figs 98-101)

MEASUREMENTS

See Table 29.

DESCRIPTION

Male (first generation)

Cuticle smooth, head rounded, not offset from body; six lips united at base but distinct at tips. Each lip bearing a terminal papilla. Four cephalic papillae located more posteriorly on head, located in sub-medial positions. Amphidial apertures distinct, located posterior to lateral papillae in same circle as cephalic papillae. Stoma partially collapsed; pharyngeal collar absent, pharynx extending nearly to oral aperture. Cheilorhabdions represented by thick ring of sclerotised material just posterior to lips. Further posterior is another sclerotised ring representing prorhabdions. Meso-, meta- and telorhabdions vestigial. Pharynx muscular with a cylindrical procorpus followed by a slightly swollen, non-valvate, metacorpus, isthmus and basal bulb containing a modified valve lined with refractive ridges. Nerve ring often surrounding anterior portion of basal bulb. Excretory pore opening variable in location, but in general posteriorly located, particularly in first generation males. Lateral fields and phasmids inconspicuous. Genital system monorchic, reflexed. Seminal vesicle with secondary spermatocytes dividing into spermatids and spermatids shedding their residual bodies. Spicules paired, strongly curved, with blunt tips. Upper section of lamina may protrude slightly over lower portion giving a bifid appearance. Capitulum broad and blunt, rarely pointed; ventral arch usually prominent on calomus yet can be reduced or almost absent. From this arch, velum extending obliquely nearly to tip of lamina. Gubernaculum boat-shaped in lateral view, with proximal portion curved upwards and enclosing spicule shafts. Gubernaculum forked in ventral view with membranous structure running from one side of fork to other. Complement of 23 genital papillae (11 pairs

[*] Named after length of infective juvenile which was *intermediate* between those of *S. carpocapsae* and *S. feltiae*.

[**] Description after Poinar (1986). Revised by Zdeněk Mráček.

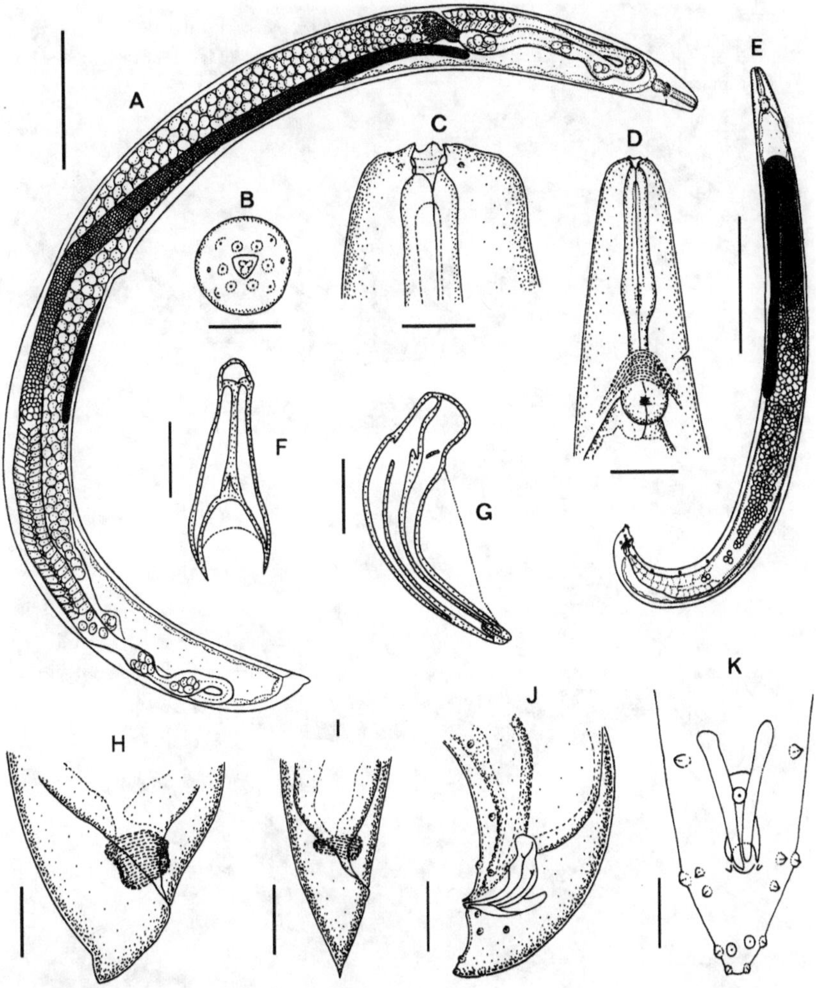

Fig. 98. Steinernema intermedium. *A, I: Second generation female. A: Entire body; I: Tail. C, H: First generation female. C: Head; H: Tail. B, D, E-G, J, K: Male. B: Anterior end,* en face *view; D: Pharyngeal region; E: Entire body; F: Gubernaculum (ventral view); G: Spicule (lateral view); J: Tail (lateral view); K: Tail (ventral view). (Scale bars: A, E = 400 μm; B, C, F, G = 20 μm; D, H, I, J = 50 μm; K = 30 μm.) After Poinar (1985), courtesy of* Revue de Nématologie.

plus single midventral precloacal) consistently present but varying somewhat in position. Three pairs consistently found on tail tip, three pairs in region posterior to gubernaculum. Anterior to cloaca is a row of five sub-

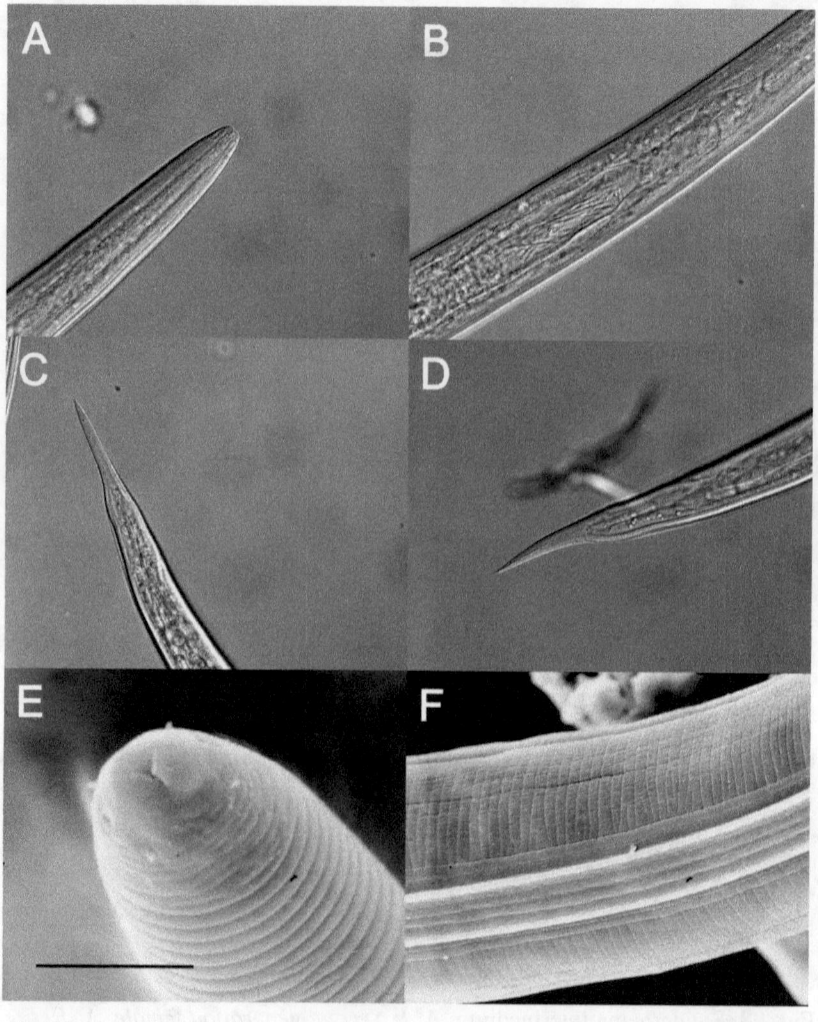

Fig. 99. Steinernema intermedium. *A-F: Infective juvenile. A: Rounded head; B: Pharynx and bacterial vesicle; C: Tail with hyaline layer; D: Tail with spine-like structure; E: Head with cephalic papillae and amphidial apertures; F: Lateral field ridges. (Scale bar: A-D = 50 μm; E, F = 20 μm.) After Z. Mráček.*

ventral pairs and a single midventral papilla just anterior to cloacal aperture that is usually quite consistent in position. Ventral portion of tail usually concave (fairly consistent although degree of concavity varies somewhat). Thin flap of sclerotised cuticle occurring over cloaca and serving to close cloacal aperture when spicules are retracted. Cloacal aperture slit-like with pair of small lateral papilla-like structures on each side.

Male (second generation)

Second generation adults smaller than those of first generation. Papilla-like mucron rarely developed.

Female (first generation)

Amphidelphic with opposed reflexed ovaries. Vulva in form of a transverse slit, usually protruding slightly from body surface. Vagina short, leading into paired uteri. Eggs deposited initially, but later juveniles hatching inside females (*endotokia matricida*). Tail broad with rounded, wedge-like, projection on tip bearing two to four minute papilla-like protuberances. Slight or no postanal swelling. Pygmy forms occurring in some instances.

Female (second generation)

Tail straight and pointed, sometimes with slight postanal ventral swelling. Tail tip mucronate.

Infective juvenile

Cephalic extremity finely rounded. Oral aperture and anus closed, pharynx and intestine collapsed. Tail constricted, especially dorsally. Lateral field composed of six ridges (*i.e.*, seven incisures) in mid-region of body. Submarginal ridges most distinct. Excretory pore located at mid-pharynx. Anterior portion of intestine with pouch containing symbiotic bacteria. Tail slender ($c' = 4$) with characteristic dorsal constriction. Hyaline region comprising *ca* 50% of tail length. No refractile spine in tail.

TYPE HOST AND LOCALITY

Type host not known as the species was recovered by baiting soil from Charleston, SC, USA, with *Galleria mellonella* larvae.

TYPE MATERIAL

Holotype (male) and allotype (female) deposited in the University of California-Davis Nematode Collection, Davis, CA, USA. Paratypes

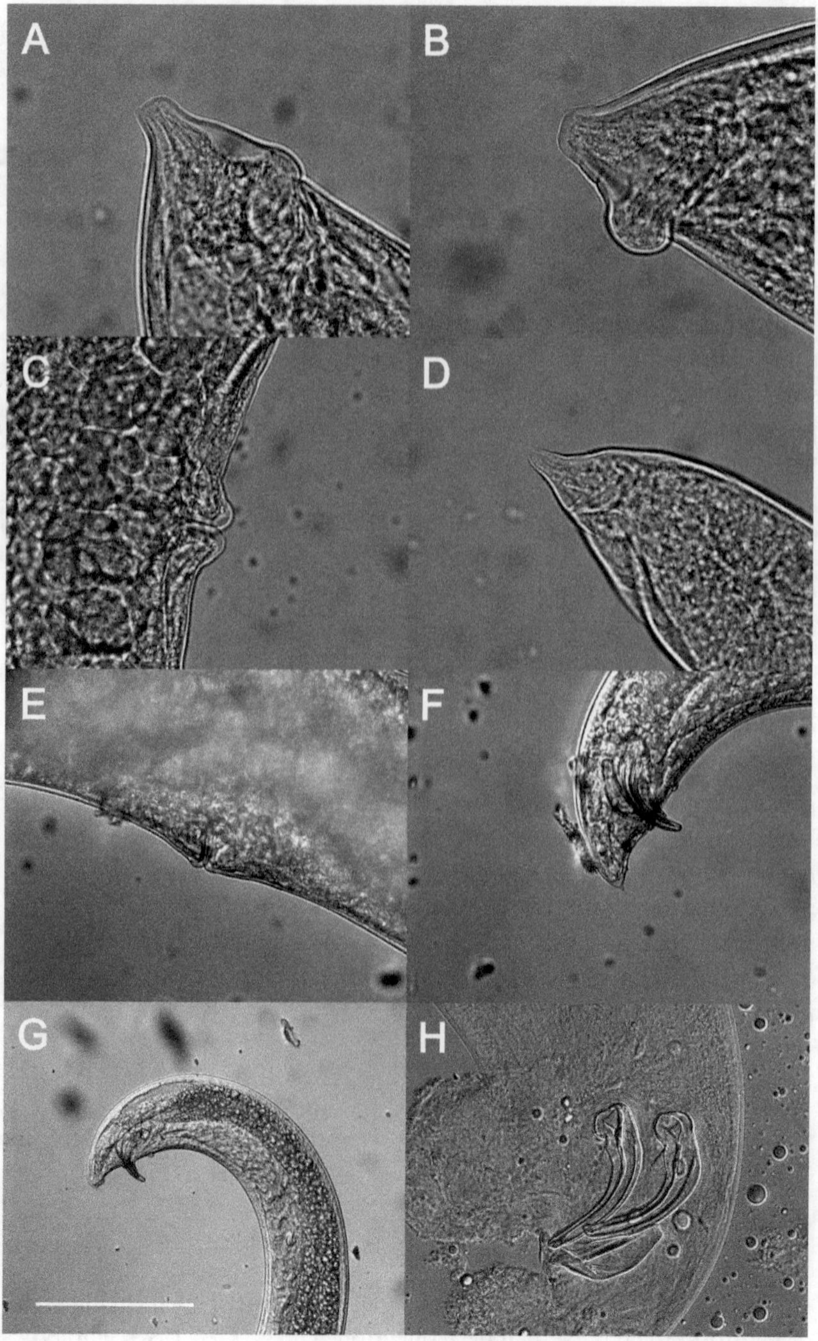
A
B
C
D
E
F
G
H

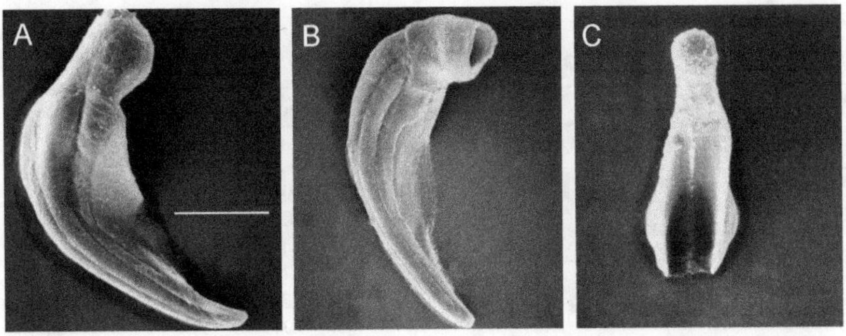

Fig. 101. Steinernema intermedium. *A-C: Male. A, B: Spicule shape; C: Gubernaculum, ventral view. (Scale bar: A-C = 25 μm.) Courtesy of K.B. Nguyen.*

deposited at the Laboratoire des Vers, Muséum national d'Histoire naturelle, Paris, France. The type nematode strain is labelled as SC (South Carolina).

RELATIONSHIPS

Generally, this species, a member of the *intermedium*-group, is characterised in the male by the strongly curved and robust spicules with a distinct rostrum. The tip of the spicule is broad and blunt. The mucron is usually absent in both generation but, if developed in the second generation, then papilla-like and reduced. The infective juvenile is characterised by the lateral field which comprises six ridges with the sublateral pair most prominent, and the characteristic tail constriction.

Poinar (1986) stated that D% was fairly constant in both generations of females (51-61 *vs* 56-66) but in the males there was no overlap (67-80 and 54-63). This latter character is due to the extremely posterior position of the excretory pore in the larger first generation males. According to Poinar (1986) *S. intermedium* can be separated from *S. carpocapsae*, *S. feltiae* and *S. glaseri* by several characters. One of

Fig. 100. Steinernema intermedium. *A-C: First generation female. A: Tail tip with papilla-like structure; B: Tail tip with two papilla-like structures; C: Vulva. D, E: Second generation female. D: Filamentous mucronate tail tip; E: Vulva. F-H: Second generation male. F: Mucronate tail tip; G: Papillate tail tip; H: Spicules and gubernaculum. (Scale bar: A-F = 100 μm; G = 190 μm; H = 70 μm.) After Z. Mráček.*

Table 29. *Morphometrics of* Steinernema intermedium. *Measurements are in μm and in the form: mean (range).*

Character	First generation		Second generation		Infective juvenile
	Male	Female	Male	Female	
n	20	20	20	20	25
L	2400 (1600-3000)	5200 (3300-6400)	1200 (1100-1300)	2200 (1900-2600)	680 (608-800)
a	14	22	17	20	24
b	13	24	7.3	12	5.6
c	44	73	29	35	11
c′	0.7	0.7	0.8	1.2	4
V	–	53 (50-57)	–	55 (52-57)	–
Body diam.	168 (113-207)	236 (170-277)	71 (63-95)	112 (95-126)	28 (25-35)
EP	137 (114-155)	132 (108-143)	97 (85-108)	117 (105-127)	65 (61-69)
NR	146 (120-168)	158 (101-181)	128 (117-136)	146 (133-159)	92 (85-96)
ES	190 (155-209)	218 (193-244)	164 (158-171)	189 (174-203)	121 (110-131)
Testis reflexion	625 (283-1008)	–	225 (183-290)	–	–
Tail length (T)	54 (45-59)	71 (60-89)	41 (35-45)	62 (57-70)	64 (53-72)
Hyaline (H)	–	–	–	–	32 ± 0.6* (27-37)*
Anal body diam. (ABD)	77 (53-88)	101 (73-127)	49 (43-56)	51 (41-63)	16 (13-18)
Spicule length (SL)	93 (80-106)	–	65 (59-69)	–	–
Spicule width	19 (13-26)	–	13 (10-19)	–	–
Gubernaculum length (GL)	62 (48-96)	–	36 (34-40)	–	–
Gubernaculum width	11 (8.0-17)	–	7.3 (5.6-8.0)	–	–
D% = EP/ES × 100	72 (67-80)	60 (51-61)	59 (54-63)	62 (56-66)	54
E% = EP/T × 100**	253	190	237	190	100
SW% = SL/ABD × 100**	121	–	133	–	–
GS% = GL/SL × 100	66	–	55	–	–
H% = H/T × 100**	–	–	–	–	48 ± 0.4* (45-51)

EP = distance from anterior end to excretory pore; NR = distance from anterior end to nerve ring; ES = distance from anterior end to end of pharynx.

* According to Půža (2001).

** Recalculated according to Poinar (1986).

these is the length of the infective juveniles which range from 600-800 μm. The lower end of this range does overlap with the upper end of *S. carpocapsae* (438-650 μm) and the upper end of the range overlaps with the lower range of *S. feltiae* (700-1000 μm), but on average the infective juveniles are larger than those of *S. carpocapsae*. The infective juveniles of *S. glaseri* range from 864-1448 μm. Also, the tail of the infective stage of *S. intermedium* is usually constricted, especially on the dorsal side. Such a constriction is lacking in infective juveniles of *S. feltiae* and, if one is present at all, it occurs on the ventral surface in both *S. carpocapsae* and *S. glaseri*.

The male tail of *S. intermedium* lacks a mucron of any type, thus separating it from both *S. carpocapsae* and *S. feltiae*. The spicules are generally more curved (a line running parallel with the calomus and lamina forms an angle of 70-90°) than those of *S. carpocapsae*, *S. feltiae* and *S. glaseri*. The colourless spicules separates this species from the yellow spicules of *S. feltiae* and their blunt tips separate them from the hooked spicule tips of *S. glaseri* and the pointed spicules of *S. carpocapsae*. The gubernaculum of *S. intermedium* is unusual as the distal portion is bent so as to enclose the distal portion of the spicules. The male tail of *S. intermedium* has a ventral concavity which is either not present or as distinct in the other three steinernematids.

Steinernema intermedium can be distinguished from *S. affine* by the bluntly tipped spicules, more posterior position of the excretory pore and lack of refractive spine in the tail of the infective stages, a character Bovien (1937) claimed was constantly present in *S. affine*.

Steinernema intermedium differs from *S. beddingi* by having six ridges in the lateral field compared to four for *S. beddingi*. In the male a mucron is absent in both generations of *S. intermedium* whereas it is present in the second generation of *S. sichuanense*.

Morphology and morphometrics of spicules and gubernacula of the four species in the *intermedium*-group are, based on SEM photographs, quite different.

MOLECULAR CHARACTERISATION

Steinernema intermedium is characterised by sequence lengths (flanked by primers 18S and 26S) of ITS regions, 1061 bp, ITS1 = 302 bp, ITS2 = 349 bp and its composition, A = 0.25636, C = 0.16682,

G = 0.21961, T = 0.35721. The sequences of ITS and D2D3 regions are deposited in GenBank under accession no. AF122016 and AF331909, respectively.

NOTE

Spiridonov *et al.* (2004) suggested that this species may in fact be a complex of species, '*S. intermedium*'-like isolates from Europe being distinct from the topotype material from the USA and representing three, as yet undescribed, species.

BIONOMICS

Life history studies indicates that *S. intermedium* has a biology comparable to that of other known species of *Steinernema*. Infective stage juveniles enter the haemocoel of insects, liberate their associated bacteria, complete one or two generations, depending on the initial infection rate and then emerge from the cadaver as infective juveniles.

DISTRIBUTION

The species was described from South Carolina, USA, but is not frequently reported from North America. There are many records from Europe (but see the Note above and also Spiridonov *et al.*, 2004), this species apparently showing a preference for tree habitats (Mráček *et al.*, 1999, 2005; Sturhan & Mráček, 2000; Hominick, 2002). There may be a record of this species from Siberia (Ivanova & Spiridonov, 2003).

References

BOVIEN, P. (1937). Some types of association between nematodes and insects. *Videnskabelige Meddelelser fra Dansk Naturhistorisk. Forening* 101, 1-114.

HOMINICK, W.M. (2002). Biogeography. In: Gaugler, R. (Ed.). *Entomopathogenic nematodes.* Wallingford, UK, CABI Publishing, pp. 115-143.

IVANOVA, E.S. & SPIRIDONOV, S.E. (2003). The *intermedium-affine* group of *Steinernema* (Rhabditida, Steinernematidae): morphological and molecular characters. *Russian Journal of Nematology* 11, 138. [Abstr.]

MAMIYA, Y. (1988). *Steinernema kushidai* n. sp. (Nematoda: Steinernematidae) associated with scarabaeid beetle larvae from Shizuoka, Japan. *Applied Entomology and Zoology* 23, 313-320.

MRÁČEK, Z., BEČVÁŘ, S. & KINDLMANN, P. (1999). Survey of entomopathogenic nematodes from the families Steinernematidae and Heterorhabditidae (Nematoda: Rhabditida) in the Czech Republic. *Folia Parasitologica* 46, 145-148.

MRÁČEK, Z., BEČVÁŘ, S., KINDLMANN, P. & JERSÁKOVÁ, J. (2005). Habitat preference for entomopathogenic nematodes, their insect hosts and new faunistic records for the Czech Republic. *Biological Control* 34, 27-37.

POINAR JR, G.O. (1986). *Neoaplectana intermedia* n. sp. (Steinernematidae: Nematoda) from South Carolina. *Revue de Nématologie* 8 (1985), 321-327.

PŮŽA, V. (2001). *Comparison of an infective juvenile tail morphology of the family Steinernematidae*. B.Sc. Thesis, University of South Bohemia, České Budějovice, Czech Republic, 23 pp.

SPIRIDONOV, S.E., REID, A.P., PODRUCKA, K., SUBBOTIN, S.A. & MOENS, M. (2004). Phylogenetic relationships within the genus *Steinernema* (Nematoda: Rhabditida), as inferred from analyses of sequences of the ITS-5.8S-ITS2 region of rDNA and morphological features. *Nematology* 6, 547-566.

STURHAN, D. & MRÁČEK, Z. (2000). Comparison of the *Galleria* baiting technique and a direct extraction method for recovering *Steinernema* infective-stage juveniles from soil. *Folia Parasitologica* 47, 315-318.

Steinernema jollieti[*,**] Spiridonov, Krasomil-Osterfeld & Moens, 2004
(Figs 102, 103)

MEASUREMENTS

See Table 30.

DESCRIPTION

Male (first generation)

Body quite slender, usually J-shaped in heat-relaxed specimens or with posterior end coiled. Cuticle annulation indistinguishable throughout body. Labial region rounded. Six labial papillae present, appearing as small (<0.5 μm high) pointed spikes around oral aperture. Circle of four, more protuberant, cephalic papillae (*ca* 1 μm high) located posterior to labial papillae. Cheilostom cuticle connected to a 3 μm broad and 3.5 μm long strongly cuticularised ring forming middle part of stoma. Pharynx lumen tightly closed up to 7 μm from stoma base. Pharynx corpus widening in diam. from 12-13 μm near base of stoma up to 19-20 μm anterior to isthmus. Isthmus 12-14 μm diam. Basal bulb subpyriform; valves appearing as thin folded platelets. Cardia protruding up to 15 μm into intestine proventriculus, often asymmetrical. Deirids located posterior to level of pharyngo-intestinal junction. No cuticle elevation around excretory pore. Anterior part of intestine with wider lumen and pronounced microvillar layer. Genital system monorchic, reflexed. Testis flexure 250-400 μm posterior to basal bulb. Intestinal lumen collapsed at mid-body but discernible near cloaca. Rectal glandular cells present around spicule muscles. Typical steinernematid set of genital papillae with a row of usually nine pairs of subventral papillae extending from tail terminus to mid-body, one large, unpaired, precloacal papilla in midventral position, one pair of papillae in lateral position at manubrium level when spicules retracted, and one pair of papillae in dorsolateral position in postcloacal area. Latter pair usually positioned in mid-part of tail, being closer to tail terminus only in a few of males examined. Posteriormost pair of subventral papillae close to tail tip, with neighbouring pair

---

* Named after Louis *Jolliet*, the first European explorer of this region of the American Midwest.

** Description after Spiridonov *et al.* (2004). Revised by Zdeněk Mráček.

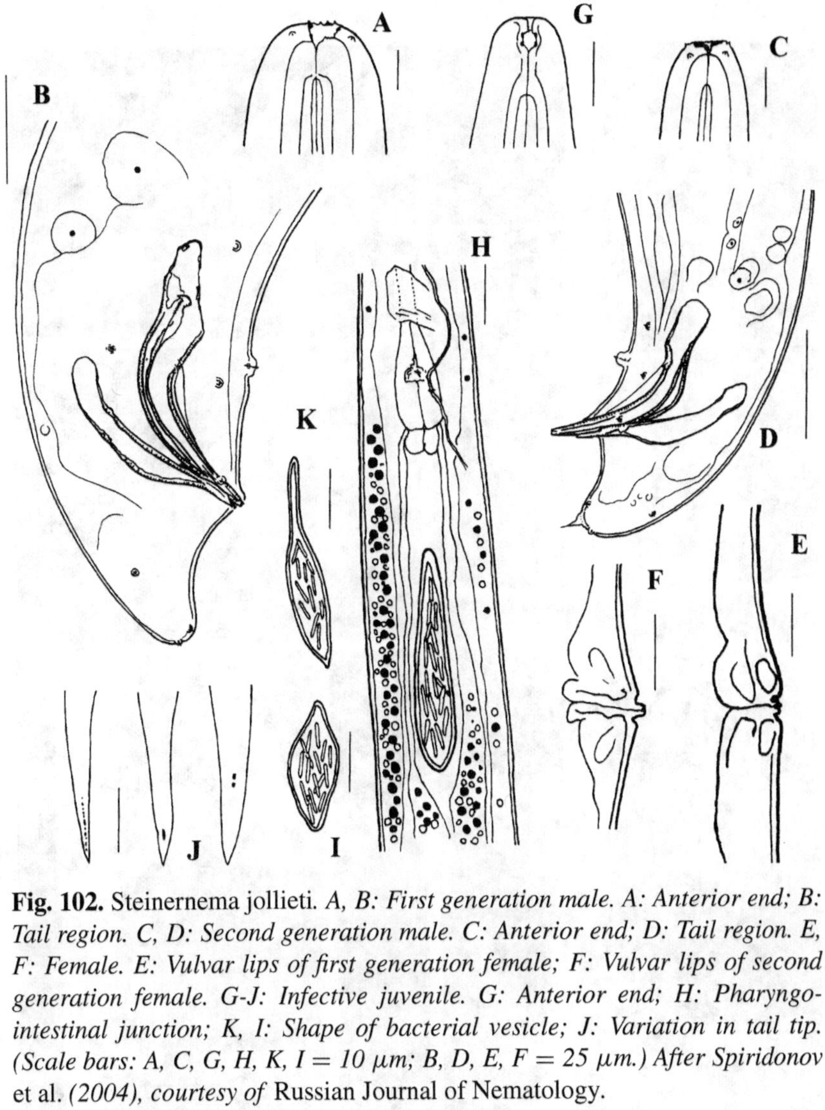

Fig. 102. Steinernema jollieti. *A, B: First generation male. A: Anterior end; B: Tail region. C, D: Second generation male. C: Anterior end; D: Tail region. E, F: Female. E: Vulvar lips of first generation female; F: Vulvar lips of second generation female. G-J: Infective juvenile. G: Anterior end; H: Pharyngo-intestinal junction; K, I: Shape of bacterial vesicle; J: Variation in tail tip. (Scale bars: A, C, G, H, K, I = 10 μm; B, D, E, F = 25 μm.) After Spiridonov* et al. *(2004), courtesy of* Russian Journal of Nematology.

situated 7-8 μm anteriorly. Next pair situated at cloaca level. In two cases ten pairs of subventral papillae were observed with eight in precloacal position. Spicules slightly yellowish in colour, moderately to strongly curved, tapering towards distal end, with rounded tips. Manubrium more or less elongated. Velum indistinct, usually expanding over one-quarter

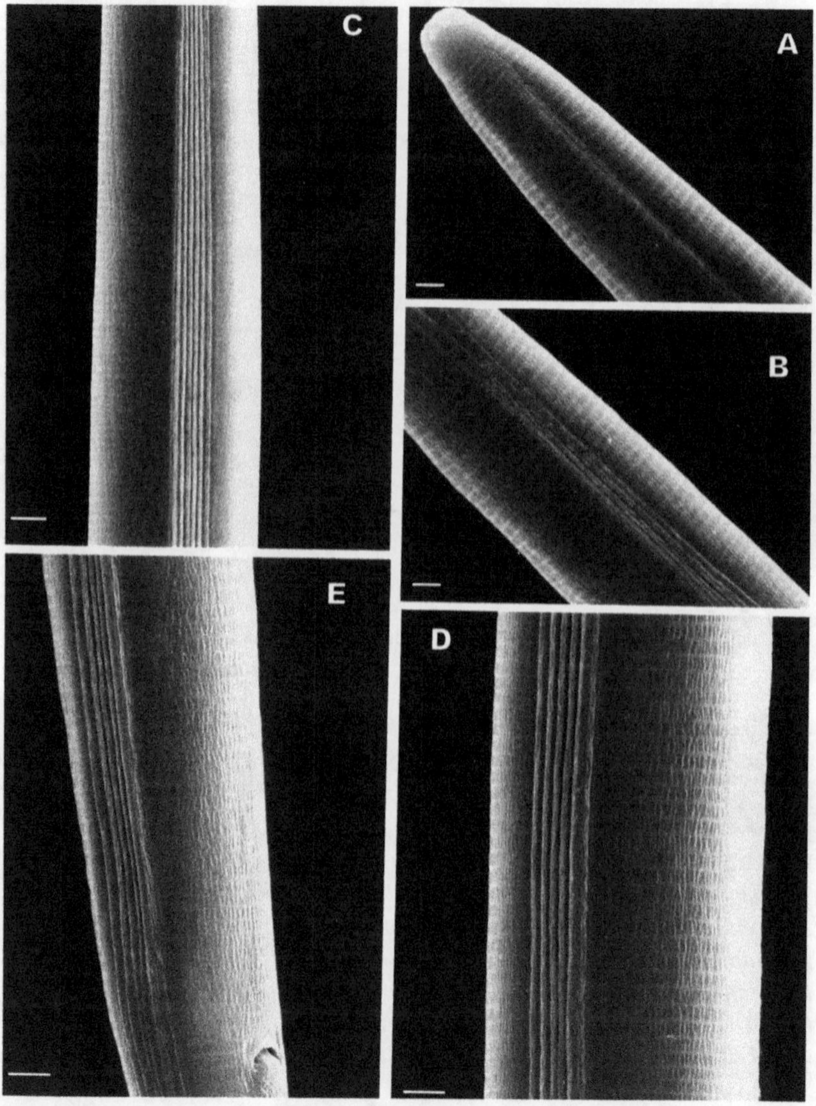

Fig. 103. Steinernema jollieti. *A-E: Infective juvenile, SEM morphology. A: Anterior end; B: Beginning of prominent ridges in lateral field; C: Lateral field appearance in anterior third of body; D: Lateral field at mid-body; E: Lateral field near tail. (Scale bars = 5 μm.) After Spiridonov* et al. *(2004), courtesy of* Russian Journal of Nematology.

Table 30. *Morphometrics of type population of* Steinernema abbasi. *Measurements are in μm and in the form: mean ± SD (range).*

Character	First generation			Second generation		Infective juvenile
	Holotype (male)	Male	Female	Male	Female	
n	–	20	20	20	20	25
L	1653	1662 ± 203 (1296-1952)	5148 ± 662 (3746-6030)	1000 ± 75 (891-1134)	1509 ± 126 (1332-1692)	711 ± 59 (625-820)
a	15	15 ± 1.9 (12-19)	20 ± 2.7 (15-24)	18 ± 1.7 (15-21)	15 ± 2.2 (13-19)	31 ± 2.1 (25-34)
b	11	11 ± 1.5 (8.1-14)	26 ± 3.8 (19-31)	7.9 ± 0.6 (7.1-8.9)	10 ± 0.8 (8.3-11)	5.7 ± 0.3 (4.9-6.4)
c	47	51 ± 9.0 (53-86)	128 - 38 (72-185)	40 ± 3.7 (35-46)	31 ± 5.2 (26-43)	11 ± 0.7 (9.0-12)
c′*	0.8	0.8	–	0.8	–	4.5
V	–	–	51 ± 3.8 (44-56)	–	54 ± 2.8 (52-59)	–
Body diam.	108	115 ± 13 (98-135)	259 ± 24 (219-298)	57 ± 3.7 (50-63)	10 3±17 (85-130)	23 ± 1.9 (20-28)
EP	106	98 ± 8 (83-110)	111 ± 13 (96-136)	79 ± 6.8 (70-93)	64 ± 3.3 (60-68)	60 ± 3.1 (53-65)
ES	151	156±16 (110-168)	214 ± 33 (184-310)	126 ± 6.5 (113-135)	155 ± 7.0 (145-165)	123 ± 5.8 (115-135)
Tail length (T)	34	33 ± 4.2 (24-38)	43 ± 8.0 (31-55)	25 ± 2.4 (20-30)	49 ± 7.3 (38-61)	68 ± 3.7 (60-73)
Hyaline (H)	–	–	–	–	–	37 ± 3.5 (28-43)
Anal body diam. (ABD)	44	44 ± 2.9 (40-50)	–	31 ± 4.9 (24-38)	–	15 ± 1.2 (13-18)
Spicule length (SL)	68	64 ± 5.4 (55-70)	–	58 ± 5.6 (48-63)	–	–
Gubernaculum length (GL)	53	54 ± 4.8 (45-60)	–	42 ± 3.4 (35-45)	–	–
D% = EP/ES × 100	70	64 ± 8.9 (53-83)	52	63 ± 5.2 (53-71)	41	48 ± 1.5 (46-50)
E% = EP/T × 100*	302	296	258	316	130	88
SW% = SL/ABD × 100*	155	145	–	187	–	–
GS% = GL/SL × 100*	78	84	–	72	–	–
H% = H/T × 100	–	–	–	–	–	55 ± 3.3 (46-60)

EP = distance from anterior end to excretory pore; NR = distance from anterior end to nerve ring; ES = distance from anterior end to end of pharynx.

* Recalculated according to Spiridonov *et al.* (2004).

to one-third of spicule length. Gubernaculum boat-shaped with narrowed and flattened distal end and broad inner space in middle.

Male (second generation)

Body quite slender, always J-shaped in heat-relaxed specimens. Cheilostom cuticle connected to a 6-7 μm diam. cuticularised ring forming buccal cavity. In some specimens buccal cavity widely opened. Pharynx lumen collapsed up to 5-6 μm from stoma base. Pharynx corpus widening from 10-12 μm diam. near stoma base up to 16-17 μm anterior to isthmus. Isthmus 11-12 μm diam. Cuticle slightly elevated around excretory pore. Intestinal lumen collapsed at about mid-body level. Testis flexure 165-240 μm posterior to basal bulb. Spicules with velum longer and manubrium more elongate than in first generation males. Genital papillae as in first generation males. Posteriormost pair of subventral papillae located posterior to mucron origin, with next subventral pair situated 5 μm anteriorly. Prominent, 5-8 μm long, conical mucron.

Female (first generation)

Body C-shaped to semicircular in heat-relaxed specimens. Cuticle annulation not discernible. Stoma 8 μm diam. and 6 μm deep, with cylinder-like lumen and strongly cuticularised walls. Excretory pore opening with slightly protruding margins. Vulva aperture on body swelling with vulval lips lower than maximal protrusions of these swellings. Genital tract amphidelphic with reflexed ovaries. Short chains of 3-8 spermatozoa, 6-8 μm diam., in distal part of uterus lumen. Anus lacking postanal swelling. Tail conical, with spike-like terminal mucron.

Female (second generation)

Buccal cavity funnel-shaped, widening from 3-4 μm at base up to 10-11 μm anteriorly. Vulval lips prominently protuberant.

Infective juvenile

Body of heat-relaxed juvenile straight or slightly curved ventrally, tapering towards anterior and posterior ends. Cuticular annulation (*ca* 1.4-1.9 μm wide at mid-body) visible throughout body except at labial region and tail terminus. Cephalic extremity rounded, continuous with body. Labial papillae indistinct, four cephalic papillae and amphidial pouches occasionally visible. Lateral field with six well developed ridges (*i.e.*, seven incisures) in mid-body. Spaces between ridges equal in anterior body half and near mid-body, gaps wider between marginal and submarginal ridges in tail region. Deirids located at basal bulb level, with prominent hemizonid visible as 4 μm long cuticular depression just anterior to deirids. Outer body cuticle over hemizonid slightly swollen or impressed. Pharynx corpus slender, spindle-shaped with only minor widening in middle. Excretory pore located at level of mid-pharynx, encircled by refractive ring containing body cuticle. Strongly cuticularised 2-3 μm long tube linking excretory duct with excretory channel usually situated at bulb-isthmus junction. Cardia 8-9 11 μm long, with rounded posterior end. Bacterial vesicle with well developed 1 μm thick walls. Intestine with lumen visible only around vesicle. Rectum *ca* one anal body diam. long. Genital primordium 130-180 μm long. Phasmid opening near mid-tail. Tail conical, straight or slightly bent to ventral side. Tail tip pointed, usually with spike-like mucron. Refractive inclusion discernible inside tail tip in majority of specimens. Tail hypodermis ending with rounded protuberance, rarely bearing one or two pointed projections.

TYPE HOST AND LOCALITY

The natural host is unknown, the nematode being collected from a soil sample taken close to the bed of a stream in the Bush-Augusta State Park region near St Louis, USA. The vegetation consisted of several North American species of oaks with shrubs as undergrowth. The sample was collected from heaps of soil displaced by the stream and alluvial sediments.

TYPE MATERIAL

Holotype male and paratype males, females and juveniles deposited in the Institute of Parasitology, Russian Academy of Sciences, Moscow, Russia. Additional paratypes deposited in the Natural History Museum, London, UK, Museum national d'Histoire naturelle in Paris, France, and Museum voor Dierkunde, Ghent University, Belgium. The type isolate is presently maintained on *G. mellonella* in the Institute of Parasitology, Russian Academy of Sciences, Moscow.

RELATIONSHIPS

Steinernema jollieti is characterised by the following characters: body of the infective stage juvenile slightly C-shaped or straight and of medium length (625-820 μm), lateral field with six equally spaced and developed ridges at mid-body, cephalic region rather flat and broad, excretory pore at level of mid-pharynx and hyaline region comprising *ca* half of total tail length. Males have yellowish spicules (55-70 μm long in the first generation, 48-63 μm in the second generation), with elongate manubria and a boat-shaped gubernaculum (45-60 μm long in the first generation, 35-45 μm in the second generation). Only two pairs of subventral and one pair of subdorsal papillae are situated in the postcloacal area of the male tail and only second generation males have a mucronate tail.

Phylogenetic analysis of a broad set of steinernematid species and populations placed *S. jollieti* in a clade composed of *S. feltiae*, *S. kraussei*, *S. oregonense*, *S. weiseri* and *S. silvaticum*. Spermatozoa of *S. jollieti* are amoeboid cells 6-8 μm diam. forming short chains of 5-8 cells. Such a structure is also found in *S. feltiae* and *S. kraussei*. The length of both spicule and gubernaculum of *S. jollieti* is very similar to those of *S. feltiae* and *S. weiseri*.

Within the clade, *S. jollieti* can be distinguished from the other species by the structure of the lateral field in infective juveniles. The presence of only six equally-developed ridges separated by gaps of similar width near the middle of the body has not been reported for any other species. *Steinernema kraussei* is characterised by the presence of three ridges in the central part of the lateral field, two well developed marginal ridges, and two less developed submarginal ridges (total number of ridges = seven). Both *S. weiseri* and *S. silvaticum* have a lateral field composed of eight equally spaced ridges. *Steinernema feltiae* is characterised by the presence of four ridges in the centre of the lateral field, two well developed marginal ones and two less developed submarginal ridges (total number of ridges = eight). In some aspects, the gaps between ridges of lateral field of *S. jollieti* do not look equal, but with submarginal ridges being absent, the general pattern is still very different from that of *S. feltiae*. A lateral field pattern similar to that of *S. feltiae* was reported for *S. sangi* described from Vietnam. This latter species can also be distinguished from *S. jollieti* by the presence of submarginal ridges. In the original description of *S. oregonense* the presence of six or eight ridges in the lateral field is mentioned. Examination of some populations identified as *S. oregonense via* sequencing of the ITS revealed a lateral field pattern similar to that of *S. kraussei*, *i.e.*, five well developed ridges and two, weak, submarginal ones (Spiridonov *et al.*, 2004). Additional studies are needed to supplement the original description of *S. oregonense* using specimens from the type isolate, but the lateral fields of *S. jollieti* do not have eight ridges at any point along the infective juvenile body.

The phylogenetic analysis of ITS sequences clusters *S. jollieti* closely to *S. weiseri* and *S. feltiae* Infective juveniles of these three species are quite similar in their general morphometrics. However, *S. feltiae* infective juveniles are usually longer than those of *S. jollieti* and have longer tails. The hyaline region of the tail is shorter in *S. weiseri* and the H% value is correspondingly lower.

The distribution of the genital papillae of *S. jollieti* is also very specific. Only two terminal pairs of subventral papillae and one pair of subdorsal papillae are present in the postcloacal region in this species. All other steinernematids of this group are characterised by the presence of at least one pair of subventral papillae posterior to cloacal level.

The maximum parsimony phylogenetic tree obtained for a clade of species related to *S. jollieti* was in good concordance. Bootstrap support for nodes linking *S. jollieti* with the closely related species (*S. feltiae*, *S. weiseri*) was always weak. *Steinernema jollieti* differs from the closest steinernematid *S. weiseri* in 5.7% of the nucleotides. The reported level of differences between the two closest species (*S. feltiae* and *S. weiseri*) was lower at 4.2-5.3%.

MOLECULAR CHARACTERISATION

Maximum parsimony phylogenetic analysis of a set of steinernematid species consistently placed this species as a sister taxon to the clade composed of *S. feltiae*, *S. kraussei*, *S. oregonense*, *S. weiseri* and *S. silvaticum*. Maximum parsimony analysis based on Clustal X generated alignments of *S. jollieti* placed this species in a clade with *S. feltiae* and *S. weiseri*. Maximum parsimony analysis of distances between the studied sequences supported close relationships between *S. jollieti* and *S. weiseri*. Observed level of nucleotide differences between these two species was equal to, or greater than, that between morphologically distinct species such as *S. feltiae* and *S. weiseri* or *S. kraussei* and *S. oregonense*. The accession number of ITS regions of *S. jollieti* is AY171265.

BIONOMICS

Development of *S. jollieti* was observed in *Galleria mellonella* caterpillars inoculated with 20-50 infective juveniles in Petri dishes lined with wet filter paper and kept at 20-24°C. First generation males and females were abundant on days 4-5 after inoculation, and second generation individuals were found inside insects on days 9-10. The exit of infective juveniles from the cadavers was recorded 13-17 days after inoculation.

DISTRIBUTION

At present, *S. jollieti* is known only from the type locality in the USA (Spiridonov *et al.*, 2004).

Reference

SPIRIDONOV, S., KRASOMIL-OSTERFELD, K. & MOENS, M. (2004). *Steinernema jollieti* sp. n. (Rhabditida: Steinernematidae), a new entomopathogenic nematode from the American Midwest. *Russian Journal of Nematology* 12, 85-95.

Steinernema karii[*,**] Waturu, Hunt & Reid, 1997

(Fig. 104)

MEASUREMENTS

See Table 31.

DESCRIPTION

Male (first generation)

Heat relaxed specimens J-shaped. Cuticle with very faint transverse striae and narrow lateral field 3-4 μm wide, more prominent in posterior region of body, bearing four incisures (*i.e.*, three ridges). Cephalic extremity rounded with six lips fused at base, each lip bearing terminal papilla. Outer circle of four distinct cephalic papillae present with prominent amphidial apertures located laterally. Oral aperture round. Stoma *ca* 7 μm deep and 10 μm diam. Pharynx more or less cylindrical with slightly wider metacorpal region, indistinct isthmus and rounded basal bulb. Cardia well developed. Excretory pore located at *ca* two-thirds of pharynx length. Nerve ring crossing isthmus at slightly more than two-thirds of pharynx length. Genital system monorchic, reflexed, reflexion *ca* 344 (158-490) μm long. Spicules paired, 83 (73-91) μm long, strongly curved and light golden-brown in colour. Capitulum *ca* twice as long as wide, subangular and offset from lamina. Lamina bearing ridges running to spicule tip and with thin velum. Gubernaculum 57 (42-64) μm long, arcuate in lateral view and distally bifurcate in ventral view, proximal end slightly recurved and hook-like. Cuneus present. Twenty-three genital papillae present arranged as follows: six subventral precloacal pairs, a single precloacal midventral papilla; one sublateral precloacal pair usually located at level of spicule neck; one subventral adcloacal pair; two subventral postcloacal pairs and a subdorsal postcloacal pair at about level of, or slightly more anterior than, anteriormost postcloacal subventral papillae. Tail short, dorsally convex conoid, terminus bluntly rounded; mucron absent. Phasmid indistinct, located between last two pairs of caudal papillae.

---

[*] Specific epithet derived from the initials of the Kenya Agricultural Research Institute (*KARI*).

[**] Description after Waturu *et al.* (1997). Revised by David J. Hunt.

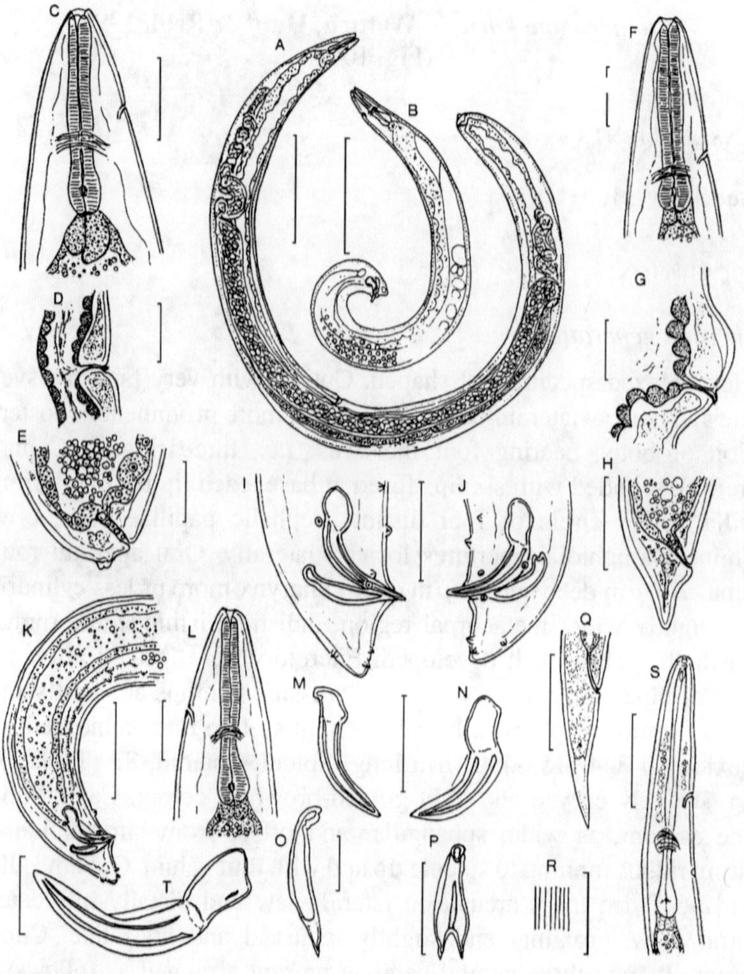

Fig. 104. Steinernema karii. *A: Entire female, first generation; B: Entire male, first generation; C: Pharynx, first generation female; D: Vulval region, first generation female; E: Tail, first generation female; F: Pharynx, second generation female; G: Vulva, second generation female; H: Tail, second generation female; I, K: Tail region, first generation male; J: Tail region, second generation male; L: Pharynx, first generation male; M, N: Spicules, first generation male; O: Gubernaculum, lateral view, first generation male; P: Gubernaculum, ventral view, first generation male; Q: Tail region, infective juvenile; R: Lateral field, infective juvenile; S: Pharynx, infective juvenile; T: Spicule, second generation male. (Scale bars: A = 750 μm; B = 500 μm; C, H, K = 100 μm; D-G, I, J, L-N, P, Q, S = 50 μm; R = 30 μm; O, T = 20 μm.) After Waturu* et al. *(1997), courtesy of* International Journal of Nematology.

Male (second generation)

Similar to first generation in many respects, but slightly smaller (1.1-1.4 *vs* 1.4-2.4 mm). Cuticle with very faint transverse striae and narrow lateral field 3-4 μm wide, more prominent in posterior region of body, bearing four incisures (*i.e.*, three ridges). Spicules and gubernaculum smaller than first generation, being 71 (62-78) μm and 49 (42-58) μm long, respectively. Capitulum *ca* twice as long as wide. Testis reflexion shorter on average, being *ca* 159 (103-217) μm long. Subdorsal pair of postcloacal papillae usually located anteriad to two pairs of subventral papillae near tail tip and precloacal sublateral pair of papillae located nearer to cloaca, *i.e.*, more posterior than in first generation. Occasionally, one of postcloacal subventral papillae located more ventrolaterally. Tail short, dorsally convex conoid with a bluntly rounded terminus lacking a mucron. Phasmids obscure, located in region of last two pairs of caudal papillae.

Female (first generation)

Body usually dying in a closed C-shape on heat relaxation and anteriorly truncate to rounded. Cervical region often slightly contracted in fixed material; cuticle smooth with no lateral lines visible. Labial region with six lips, the lips having fused bases but separate tips, each bearing a minute papilla. Outer circle of four cephalic papillae and lateral amphidial apertures present. Stoma simple, *ca* 9 μm deep and 14 μm diam., tapering into pharynx and lightly cuticularised. Pharynx more or less cylindrical with a barely demarcated basal bulb. Isthmus short and indistinct. Cardia well developed. Excretory pore and distal region of duct cuticularised; located approximately halfway along pharynx with nerve ring at *ca* two-thirds or more of pharynx length. Excretory pore may be apparently located further anterior due to contraction of cervical region in fixed material. Intestine may be displaced anteriorly so as to envelop basal bulb and posterior region of pharynx. Intestine voluminous; lined by large, irregular, nucleolate cells with granular contents, gut lumen filled with numerous fat globules of varying dimension. Rectum short, intestine usually extending around and beyond it into tail region. Rectal lumen usually closed, but may contain a few globules similar in appearance to those in gut lumen. Vulva median in position and slightly raised with double, closely adpressed, epiptygmata. Vagina short, muscular; genital tracts amphididelphic,

Table 31. *Morphometrics of* Steinernema karii. *Measurements are in μm (except for male and female lengths in mm) and in the form: mean±SD (range).*

Character	First generation adults			Second generation adults		Infective juvenile	
	Holotype (male)	Male	Female	Male	Female	7 months old	Freshly emerged
n	–	20	20	25	20	20	28
L	1.97	1.9 ± 0.3 (1.4-2.4)	5.9 ± 2.1 (2.6-10)	1.3 ± 0.1 (1.1-1.4)	2.1 ± 0.3 (1.6-2.6)	904 ± 37 (832-974)	932 ± 24 (876-982)
a	14.7	13.7 (11.7-17.1)	19.2 (13.7-25.6)	15.6 (14.6-17.3)	11.8 (10.0-15.7)	30.6	28.5
b	11.7	11.3 (9.0-13.3)	25.4 (12.7-40.6)	8.8 (7.8-9.8)	11.9 (10.2-14.3)	6.8	
c	43.8	50 (42-70)	108.5 (52-173)	38.2 (32.7-46.0)	40.8 (32-53)	11.9	12.6
c′	0.8	0.7 (0.4-0.9)	0.6 (0.4-0.9)	0.8 (0.7-0.9)	0.9 (0.8-1.1)	3.8	3.4
V	–	–	48 (43-51)	–	53 (51-56)	–	–
Body diam.	134	136 ± 17 (107-166)	298 ± 49 (193-390)	88 ± 3 (76-88)	176 ± 20 (137-215)	30 ± 1.7 (27-35)	33 ± 1.4 (31-35)
EP	125	108 ± 14 (86-138)	86 ± 20 (50-126)	117 ± 9 (96-133)	108 ± 8 (92-119)	76 ± 3.2 (66-82)	74 ± 3 (68-80)
NR	–	–	–	–	–	105 ± 5 (97-112)	–
ES	169	164 ± 7 (146-187)	227 ± 19 (201-269)	149 ± 7 (139-170)	171 ± 13 (139-197)	134 ± 6.8 (122-147)	–
Testis reflexion	–	344 (158-490)	–	–	–	–	–
Tail length (T)	45	37 ± 7.5 (22-48)	55 ± 9 (41-76)	34 ± 2.5 (29-39)	51 ± 4 (43-61)	76 ± 5.8 (66-88)	74 ± 4.5 (64-80)
Hyaline (H)	–	–	–	–	–	–	–
Anal body diam. (ABD)	59	55 ± 5 (43-66)	104 ± 29 (62-168)	44 ± 2 (41-49)	54 ± 6 (43-62)	20 ± 1.6 (19-25)	22 ± 0.6 (21-23)
Spicule length (SL)	80	83 ± 4 (73-91)	–	71 ± 3 (62-78)	–	–	–
Gubernaculum length (GL)	53	57 ± 6 (42-64)	–	49 ± 4 (42-58)	–	–	–
Gubernaculum width	–	–	–	–	–	–	–

filling much of body and highly convoluted. Developing oocytes in multiple rows, uteri containing numerous ovoid eggs. Tail very short, initially dorsally convex-conoid with a ventral digitate-like peg and then becoming short and broadly rounded with a blunt, more or less offset mamillate terminus, the appearance depending greatly on obesity of female.

Table 31. *(Continued).*

Character	First generation adults			Second generation adults		Infective juvenile	
	Holotype (male)	Male	Female	Male	Female	7 months old	Freshly emerged
D% = EP/ES × 100	66	66 (53-78)	37 (24-46)	79 (66-89)	62 (57-69)	56.6	–
E% = EP/T × 100	278	240	–	344	–	100	96
SW% = SL/ABD × 100	136	151	–	161	–	–	–
GS% = GL/SL × 100	66	70 ± 1 (50-80)	–	70 ± 1 (60-80)	–	–	–
H% = H/T × 100	–	–	–	–	–	–	–

EP = distance from anterior end to excretory pore; NR = distance from anterior end to nerve ring; ES = distance from anterior end to end of pharynx.

Female (second generation)

Similar in general respects to first generation female, but smaller in size (1.6-2.6 mm), more slender and with fewer eggs. Intestine less strongly developed and not extending anteriorly around pharyngeal base. Vulval region in younger females, particularly anterior lip, prominently salient, epiptygmata being less obvious as a consequence and directed more posteriorly. Tail proportionately longer, conoid and more tapering to a more or less axial digitate terminus. Strongly offset mamillate tip seen in first generation female not developed as worms do not become obese enough to modify tail contour. Posterior anal lip markedly salient.

Infective juvenile

Heat-relaxed specimens dying straight to slightly ventrally arcuate. Body smoothly tapering towards anterior and posterior extremities; initially enclosed in J2 cuticle which is usually lost with age. Cephalic region smoothly rounded and not set off from body contour. Cuticle marked by transverse striae *ca* 2.1 μm wide at mid-body. Lateral fields at mid-body region marked by eight distinct lines (*i.e.*, seven ridges), submarginal ridges narrower than others. Labial region bearing four distinct cephalic papillae and prominent lateral amphidial apertures. Pharynx narrow with a slender isthmus leading to a subpyriform basal bulb with valve apparatus visible. Basal bulb usually displaced dorsad. Hemizonid distinct, located *ca* level with anterior margin of basal bulb. Excretory pore cuticularised, situated *ca* half way between anterior end and base of pharynx with a long, cuticularised, duct extending posteriorly to region of basal bulb. Intestine separated from pharynx by a cardia, gut lumen containing numerous fat globules. Bacterial pouch well developed, located in anterior part of intestine close to cardia. Tail conoid, tapering to a finely pointed terminus lacking a terminal cuticular spine. Hyaline region comprising *ca* 50% of tail length. Phasmids distinct, located nearly halfway down tail, but anterior to point where hyaline region commences.

TYPE HOST AND LOCALITY

The natural host is unknown as *Galleria mellonella* larvae were used as a trap insect to isolate *S. karii* from soil. The type locality is a field with a clay loam soil with acid-humic top planted to bananas and carrots at Kibingo, Kirinyaga district, Central Province, Kenya. It is situated in the Upper midland agroecological zone at an altitude of 1400-1600 m a.s.l.

TYPE MATERIAL

Holotype male (first generation) and numerous paratypes (males, second generation, female, first and second generation and infective juveniles) deposited in the CABI Europe-UK Nematode Collection, Egham, Surrey, UK. Three first generation males, five second generation males, three first generation females, five second generation females and ten infective juveniles were deposited at the Muséum national d'Histoire naturelle, Paris, France, and the University of California-Davis Nematode Collection, Davis, CA, USA.

RELATIONSHIPS

Steinernema karii is characterised by a combination of morphological characters. These characters include: IJ length of 904 ± 37 (832-974) μm; distance from anterior end to excretory pore in IJ of 76 ± 6 (66-82) μm; IJ tail length of 76 ± 6 (66-88) μm; IJ lateral field with eight lines (*i.e.*, seven ridges); absence of a terminal spine on IJ tail; male first generation spicule length of 83 ± 4 (73-91) μm; lack of a mucron on tail tip of males; spicule shape. *Steinernema karii* shows strong RFLP affinities with members of the long-tailed group of infective juveniles (Reid *et al.*, 1997), although the IJ of *S. karii* is shorter than would normally be expected for this group. It is morphologically similar to:

S. arenarium, *S. neocurtillae*, *S. caudatum*, *S. cubanum*, *S. feltiae* and *S. glaseri*.

Steinernema karii can be distinguished from: *S. feltiae* in longer IJ body length (904 *vs* 849 μm), in lacking a mucron on the male tail and in the more posterior position of the IJ excretory pore (76 *vs* 62 μm); from *S. neocurtillae* in the much more posterior position of the IJ excretory pore (76 *vs* 18 μm) and the absence of a mucron on the male tail; from *S. cubanum* in shorter IJ body length (904 *vs* 1283 μm), longer IJ tail (76 *vs* 67 μm), more anterior position of the IJ excretory pore (76 *vs* 106 μm); from *S. glaseri* in shorter IJ body length (904 *vs* 1130 μm), more anterior position of the IJ excretory pore (76 *vs* 102 μm) and lacking notched spicule tips; from *S. caudatum* in smaller IJ (904 *vs* 1106 μm) and slightly shorter tail (76 *vs* 82 μm); from *S. arenarium* in smaller mean length of IJ (904 *vs* 1034 or 1217 μm), more anterior IJ excretory pore (76 *vs* 83 μm) and spicule lacking a swollen tip.

MOLECULAR CHARACTERISATION

The RFLP analysis of the ITS regions of the rDNA produced restriction fragment bands unique to *S. karii* when compared to a library of 33 *Steinernema* species. This species was readily distinguishable from all other described species except for *S. neocurtillae*, *S. rarum*, *S. ritteri* and *S. caudatum*, all of which lacked RFLP profiles for comparison. *Steinernema karii* shares many RFLPs with *S. arenarium* (10/17 restriction enzymes yield identical profiles). Although seven of these restriction enzymes give an undigested product, indicating no recognition site for these enzymes within the ITS region for these two species, the absence of a *Hin*fI is highly diagnostic as this restriction enzyme commonly cuts the ITS region of steinernematid species. Additionally the *Hpa*II and *Sau*3AI profiles are identical for the two species and there are common bands found in some of the other enzyme profiles such as *Alu*I, *Ma*I, *Pm*II, *Sal*I and *Rsa*I.

Phylogeny reconstruction based on band sharing indicates that *S. karii* and *S. arenarium* are indeed related (Reid *et al.*, 1997). *Steinernema glaseri* and *S. cubanum* display more polymorphisms although these two species also fall into the same group.

The ITS region sequence is deposited at GenBank under accession no. AY230173.

BIONOMICS

Similar to other steinernematids. Mahar *et al.* (2004) compared the pathogenicity of *S. karii*, *S. abbasi*, *S. carpocapsae* and *S. riobrave* to larvae of the diamondback moth, *Plutella xylostella*, at 25°C and 30°C. *Steinernema carpocapsae* was the only species to do better at the lower temperature.

DISTRIBUTION

The species is reported to be widespread throughout central Kenya (see Waturu *et al.*, 1997).

References

MAHAR, A.N., MUNIR, M. & LAGHARI, K.B. (2004). Production and pathogenicity of four steinernematids in diamondback moth, *Plutella xylostella* L. (Lepidoptera: Yponomeutidae). *Asian Journal of Plant Sciences* 3, 484-488.

REID, A.P., HOMINICK, W.M. & BRISCOE, B.R. (1997). Molecular taxonomy and phylogeny of entomopathogenic nematode species (Rhabditida: Steinernematidae) by RFLP analysis of the ITS region of the ribosomal DNA repeat unit. *Systematic Parasitology* 37, 187-193.

WATURU, C.N., HUNT, D.J. & REID, A.P. (1997). *Steinernema karii* sp. n. (Nematoda: Steinernematidae), a new entomopathogenic nematode from Kenya. *International Journal of Nematology* 7, 68-75.

Steinernema khoisanae[*,**] Nguyen, Malan & Gozel, 2006
(Figs 105-110)

MEASUREMENTS

See Table 32.

DESCRIPTION

Male (first generation)

Body slender, C-shaped to strongly curved posteriorly, when heat relaxed. Cephalic extremity rounded, gently tapering anteriorly, slightly swollen in some specimens. Face view with six labial papillae, two amphidial apertures and four cephalic papillae. Stoma shallow and narrow, usually with pronounced cheilorhabdions. Excretory pore situated posterior of nerve ring, located mostly in vicinity of, but sometimes posterior to, basal bulb. Pharynx with cylindrical procorpus, metacorpus slightly swollen, isthmus present. Nerve ring located in anterior part of isthmus, basal bulb distinct. Pharyngo-intestinal valve present. Lateral field with one narrow ridge present in mid-body. Spicules paired, light brown in colour. Head (manubrium) of spicules, somewhat elongate (1.28-1.56); shaft (calomus) very short or absent; blade (lamina) thick, tapering slightly posteriorly; blade terminus blunt; velum present. Each spicule with two internal ribs. Gubernaculum boat-shaped in lateral view, cuneus arrowhead-shaped. Usually a single precloacal midventral papilla and 11 or 12 pairs (66%) of genital papillae arranged in normal position for *Steinernema* (*i.e.*, six or seven pairs precloacal subventral, one pair adcloacal, one pair lateral, two pairs subterminal and one pair subdorsal). In one male (of 12 nematodes for SEM observation), one single papilla and 14 pairs were observed. Number of caudal papillae constant (three pairs), but number of precloacal subventral pairs variable. Tail conoid, mucron absent.

Male (second generation)

Second generation male similar to that of first generation except body length shorter and body diam. less.

[*] Named after the *Khoisan*, the indigenous people living in the region of South Africa from where the nematode originated.

[**] Description after Nguyen *et al.* (2006). Revised by Khuong B. Nguyen.

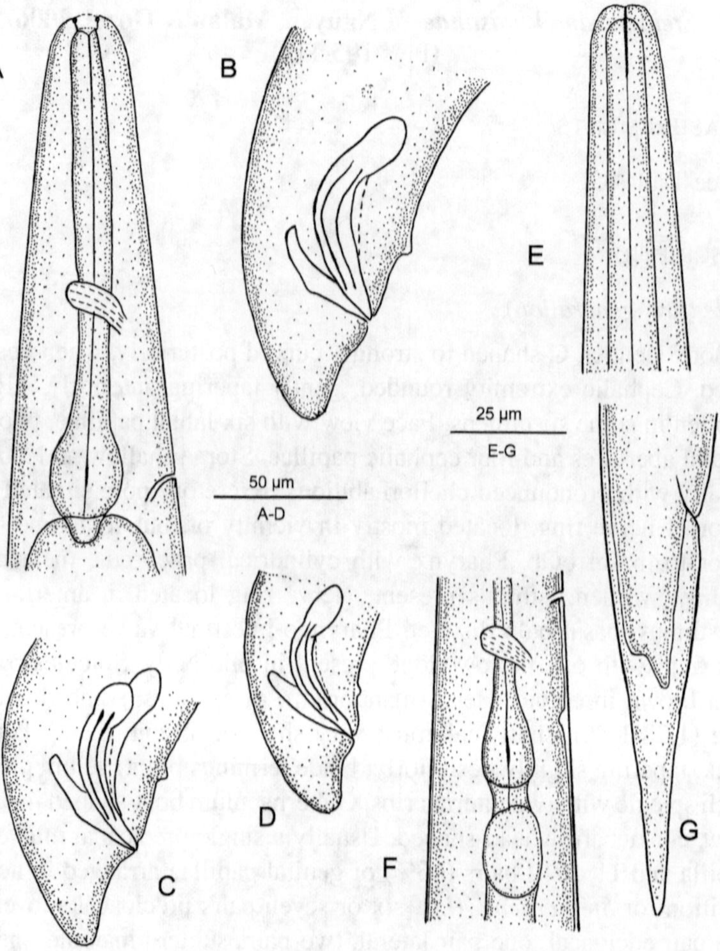

Fig. 105. Steinernema khoisanae *male. A: Anterior region; B, C: Variation of first generation male tail; D: Second generation male tail. E-G: Infective juvenile. E: Anterior region; F: Posterior part of pharynx showing excretory pore, nerve ring, hemizonion, basal bulb and bacterial chamber. After Nguyen* et al. *(2006), courtesy of* Nematology.

Female (first generation)

Body C-shaped or strongly spiralled when heat-relaxed. Cephalic extremity rounded and continuous with body. Body cuticle smooth or with faint annules. Lateral fields with one line and present on all females observed. Phasmids inconspicuous. Cheilorhabdions prominent,

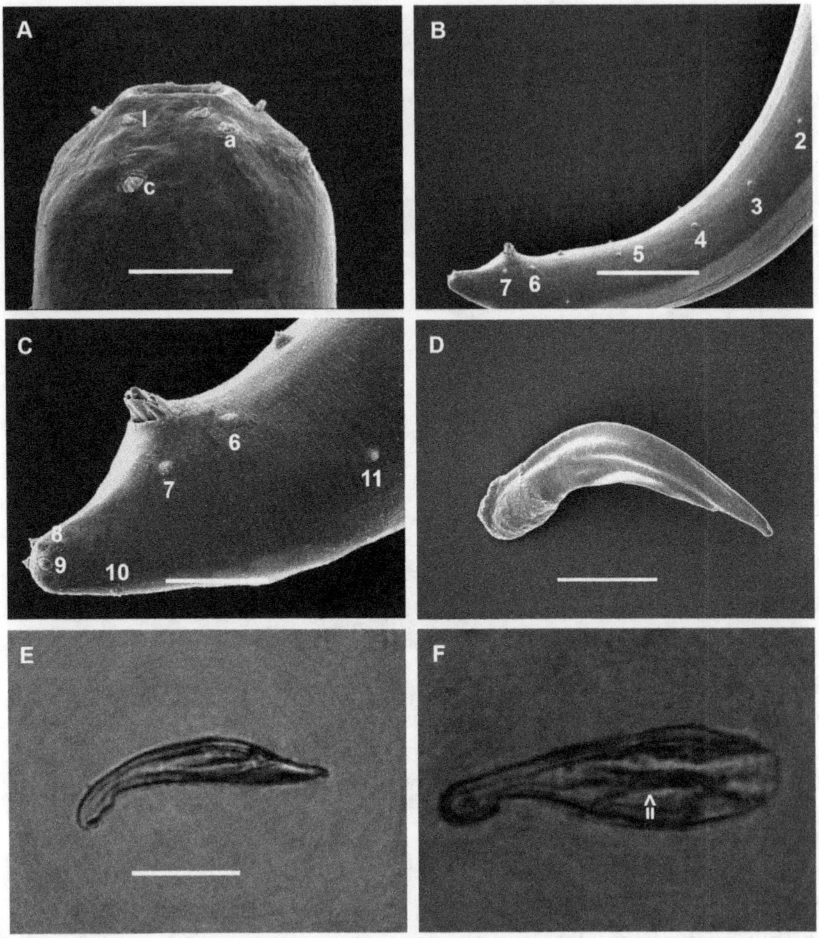

Fig. 106. Steinernema khoisanae. *SEM of first generation male. A: Male head showing labial papillae (l), amphidial aperture (a) and cephalic papillae (c); B: Posterior region showing genital papillae from pairs 2 to 7; C: Pairs 6 to 11; D: Spicule. E, F: Gubernaculum. E: Lateral view; F: Ventral view showing cuneus (arrow). (Scale bars: A = 6.7 μm; B = 60 μm; C = 20 μm; D = 23 μm; E = 23 μm; F = 18 μm.) After Nguyen* et al. *(2006), courtesy of* Nematology.

well sclerotised. Another smaller sclerotised structure present, posterior to cheilorhabdions, presumably prorhabdions. Pharynx with procorpus cylindrical, muscular; metacorpus swollen; isthmus distinct; basal bulb enlarged, valvate. Nerve ring surrounding isthmus, just anterior to basal bulb. Pharyngo-intestinal valve prominent. Excretory pore position

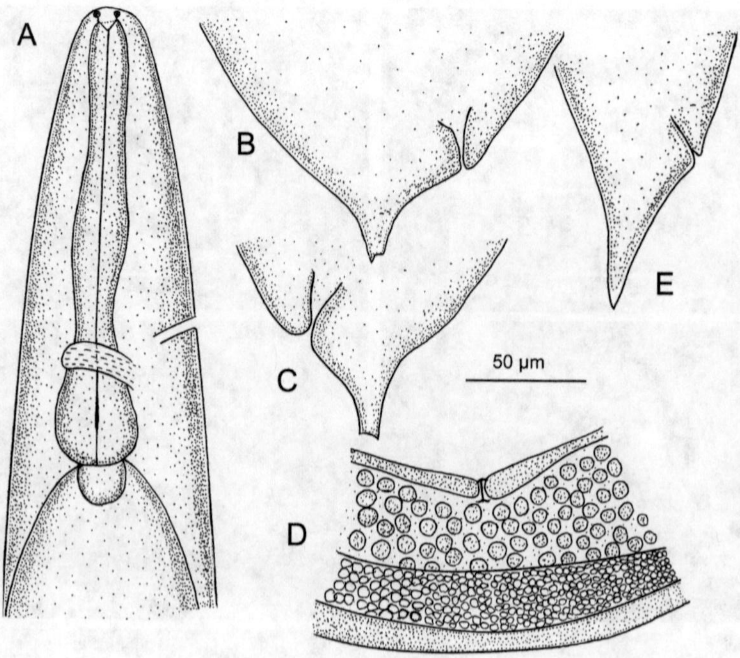

Fig. 107. Steinernema khoisanae *female. A: Anterior region of second generation female; B, C: Tail of first generation females showing posterior projection on tail tip; D: Vulva region of first generation female; E: Second generation tail. After Nguyen* et al. *(2006), courtesy of* Nematology.

variable, from near mid-pharynx to mid-basal bulb. Vulva in form of a transverse slit, not protruding from body surface, without epiptygmata. Tail digitate, with postanal swelling in most females, more pronounced in fully mature females. Tail shorter than anal body diam., tapering to a blunt end. Two projections present at tail end with dorsal projection more pronounced.

Female (second generation)

Body an open C when heat-relaxed. Similar to first generation female but smaller. Body diam. greater anterior to vulva than posterior. Vulva on asymmetrical protuberance and situated at mid-body. Postanal swelling present. Tail tapering gently to a sharp point.

Infective juvenile

Body elongate. Sheath (second-stage cuticle) present immediately after harvesting, but many losing sheath in storage. Exsheathed juvenile

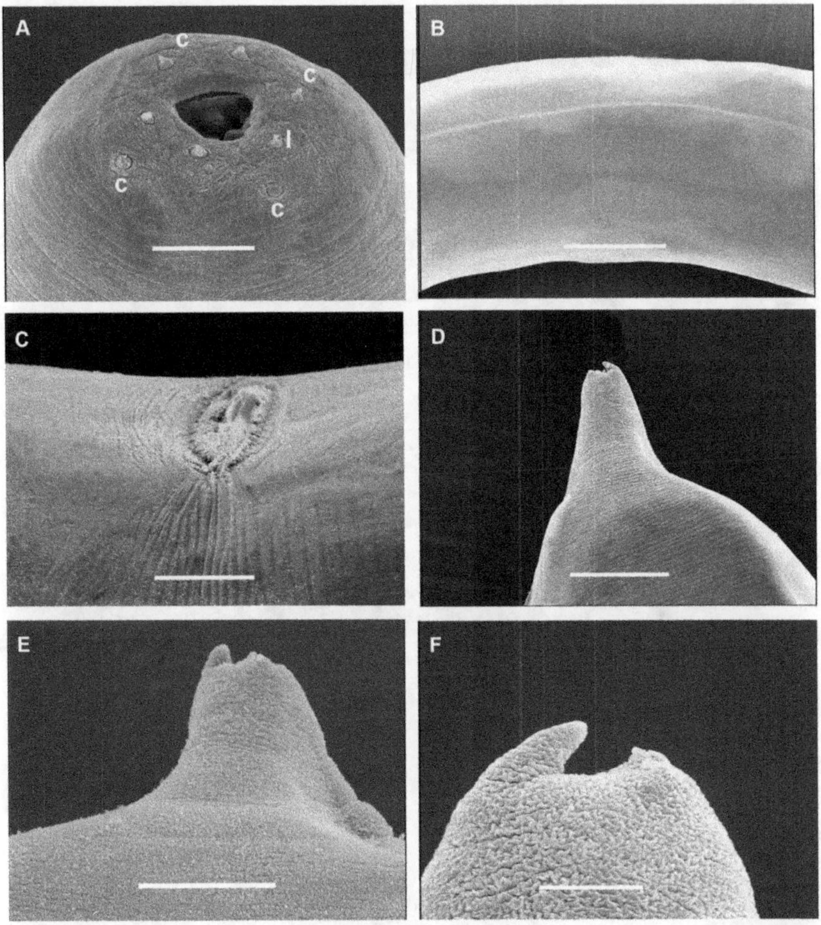

Fig. 108. Steinernema khoisanae. *SEM of female. A: First generation female face view showing mouth, six labial papillae (l) and four cephalic papillae (c); B: First generation female showing lateral field with one line; C: First generation female vulva; D: Second generation female tail with posterior projection at tip; E: Tail tip of first generation female showing posterior projections; F: Higher magnification of tail tip of first generation female. (Scale bars: A = 7.5 μm; B = 75 μm; C = 8.6 μm; D = 15 μm; E = 6.7 μm; F = 3.0 μm.) After Nguyen* et al. *(2006), courtesy of* Nematology.

with four cephalic papillae. Labial region smooth, continuous with body. Amphidial apertures prominent. Cuticle marked with prominent transverse striations. Excretory pore 0.5-1 body diam. anterior to nerve ring. Lateral field beginning anteriorly with one line at first annule.

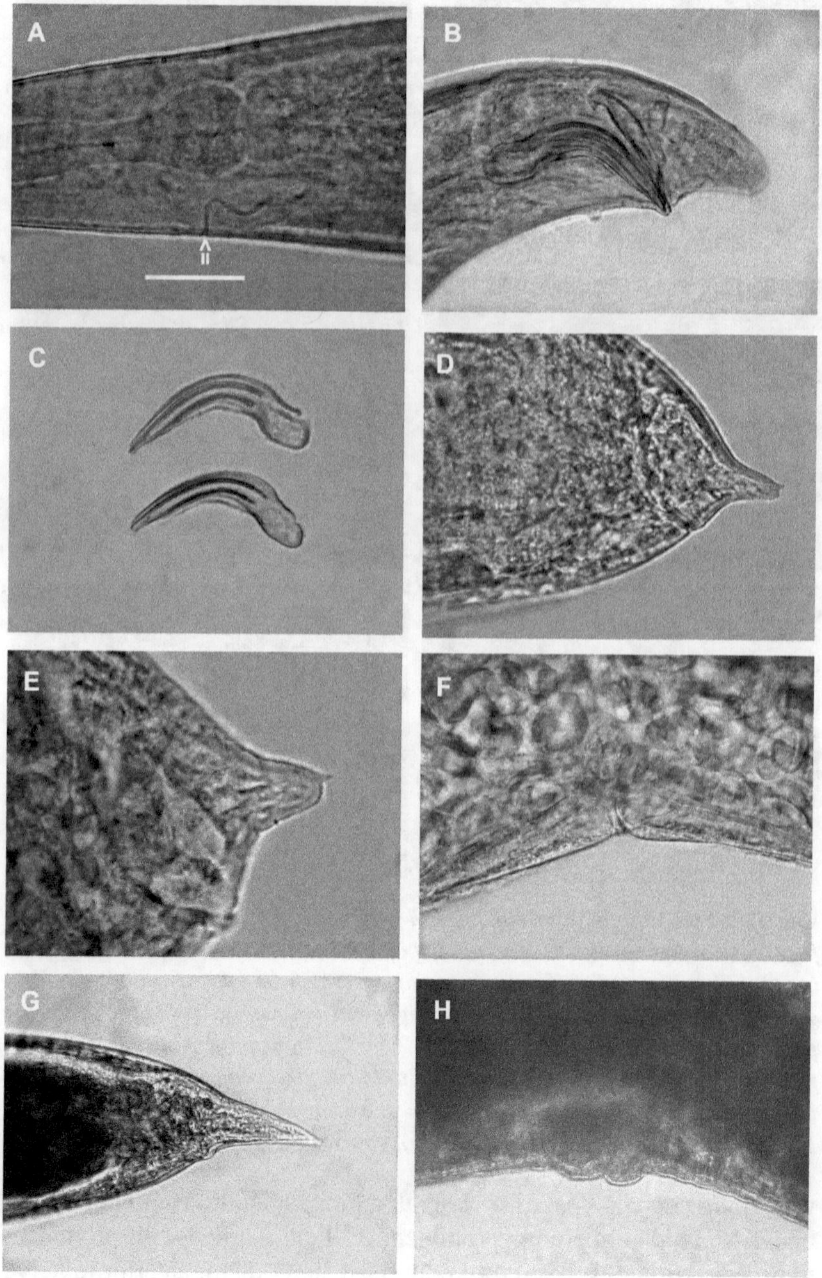
A
B
C
D
E
F
G
H

At annules 8-15, two additional lines appearing forming two ridges. Near excretory pore level, number of ridges in lateral fields increasing from two to seven. Near end of pharynx, central ridge dividing into two, making a total of eight, the maximum number in lateral field. Posterior to mid-body, number of ridges in lateral field becoming six. In posterior third of body, two marginal ridges gradually disappearing, only four ridges remaining in lateral field a short distance anterior to anus. Near mid-tail, four ridges in lateral field reducing to two. With above arrangement, formula of lateral field is 2, 7, 8, 6, 4, 2. Pharynx with narrow corpus, metacorpus slightly swollen, isthmus present, nerve ring usually situated at mid-isthmus level; basal bulb elongate with visible valve. Cardia present. Bacterial pouch located just posterior to cardia, 10.1 (9.3-10.5) μm diam. × 15.2 (15-17) μm long, containing bacterial cells, variable in shape, mostly oval. Genital primordium prominent, 156 (130-207) μm long. Hemizonion very distinct, near mid-basal bulb, *ca* 130 (118-141) μm from anterior end for isolate SF80, and 121 (113-130) μm for isolate SF87. Tail four anal body diam. long, attenuate. Phasmid visible or not; when visible, very small, located near mid-tail. Hyaline region comprising 58 (54-61)% of tail length.

TYPE HOST AND LOCALITY

Natural host unknown as the species was isolated by baiting with *Galleria mellonella* larvae. The type isolate (SF80) was collected from a soil sample from around bluegrass on a road side at latitude 33°12′.33S, longitude 19°06′.57E in the Cape Province, South Africa.

Fig. 109. Steinernema khoisanae. *Light micrographs. A-F: First generation stages. G, H: Second generation female. A: Basal bulb region of male showing excretory pore (arrow) and duct near junction of pharynx and intestine; B: Posterior region of male showing spicules and gubernaculum; C: Two spicules isolated from single male showing different shapes of spicule head; D: Female posterior region showing tail shape and posterior projection; E: Higher magnification of (D) showing posterior projection more clearly; F: Non-protuberant vulva. G: Posterior region; H: Vulva region showing protuberant vulva. (Scale bar: A = 17 μm; B = 38.6 μm; C = 40 μm; D = 82 μm; E = 48 μm; F = 82 μm; G, H = 56 μm.) After Nguyen* et al. *(2006), courtesy of* Nematology.

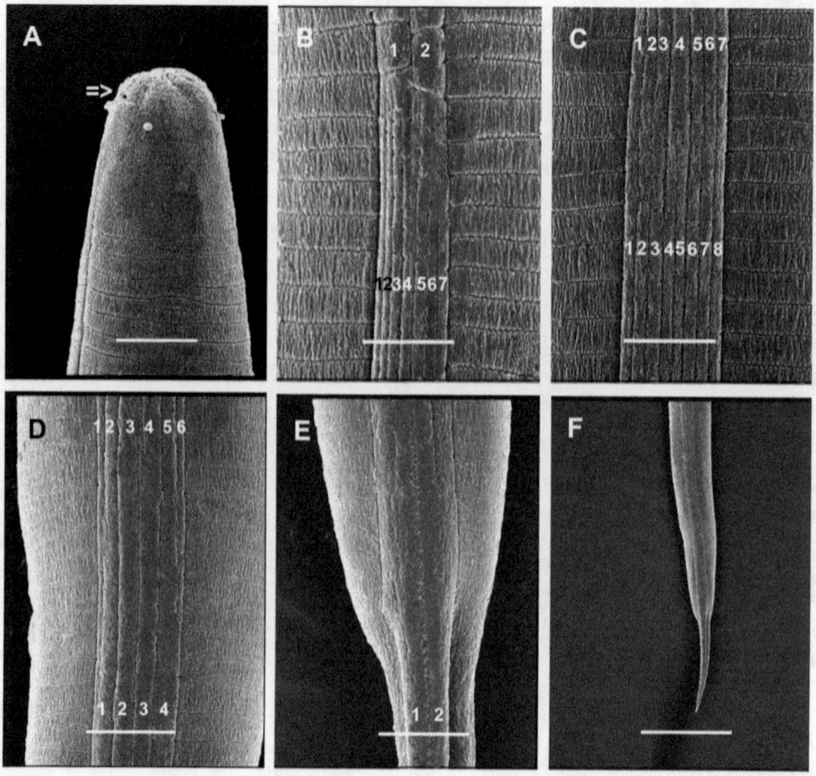

Fig. 110. Steinernema khoisanae. *SEM of infective juvenile. A: Anterior region showing one of two amphidial apertures (arrow), three of four cephalic papillae and single line in lateral field; B: Lateral field showing change from two to seven ridges; C: Lateral field showing change from seven to eight ridges; D: Lateral field showing change from six to four ridges; E: Lateral field in posterior region showing change from four to two ridges; F: Posterior region. (Scale bars: A, D, E = 5 μm; B = 3.8 μm; C = 4.3 μm; F = 37.5 μm.) After Nguyen* et al. *(2006), courtesy of* Nematology.

Table 32. *Morphometrics of* Steinernema khoisanae. *Measurements are in μm and in the form: mean ± SD (range).*

Character	First generation		Second generation		Infective juvenile
	Male	Female	Male	Female	
n	20	20	20	20	25
L	1809 ± 207 (1428-2248)	7213 ± 1155 (4927-9462)	1410 ± 85 (1283-1586)	2300 ± 145 (2021-2593)	1062±63 (904-1159)

Table 32. *(Continued).*

Character	First generation		Second generation		Infective juvenile
	Male	Female	Male	Female	
a	–	–	–	–	35 ± 2 (32-37)
b	–	–	–	–	7.6 ± 0.5 (6.6-8.4)
c	–	–	–	–	12.5 ± 0.8 (10.6-13.8)
c′	–	–	–	–	4.25
V	–	52 ± 1.6 (50-56)	–	51 ± 0.8 (50-53)	–
Body diam.	180 ± 10.4 (90-132)	294 ± 30 (241-352)	71 ± 3 (63-75)	140 ± 11 (122-158)	31 ± 2 (27-34)
EP	146 ± 12 (122-168)	190 ± 24 (146-233)	144 ± 9 (125-158)	133 ± 7 (122-145)	95 ± 5 (87-101)
NR	98 ± 5 (90-110)	134 ± 16 (105-163)	106 ± 7 (83-107)	94 ± 11 (77-112)	109 ± 5 (101-118)
ES	180 ± 6 (163-183)	231 ± 16 (182-245)	179 ± 6 (158-183)	195 ± 7 (185-203)	140 ± 7 (130-155)
Testis reflexion	520 ± 57 (310-586)	–	210 ± 26 (133-263)	–	–
Tail length (T)	36 ± 3 (30-40)	42 ± 8 (25-53)	38 ± 7 (30-48)	65 ± 6 (55-80)	85 ± 5 (76-97)
Anal body diam. (ABD)	40 ± 3 (33-45)	80 ± 18 (58-120)	35 ± 2 (32-38)	48 ± 6 (38-58)	20 ± 2 (15-25)
Spicule length (SL)	85 ± 5 (70-88)	–	74 ± 5 (62-85)	–	–
Spicule width	12 ± 2 (8-17)	–	11 ± 1 (8-13)	–	–
Gubernaculum length (GL)	56 ± 5 (50-63)	–	45 ± 2.7 (40-50)	–	–
Gubernaculum width	7.6 ± 1.2 (5.0-10.0)	–	6.9 ± 0.8 (5.0-8.3)	–	–
D% = EP/ES × 100	85 ± 7 (71-99)	80 ± 14 (57-102)	85 ± 5 (75-91)	68 ± 3 (61-75)	68 ± 4 (60-73)
E% = EP/T × 100	–	–	–	–	112 ± 7 (95-128)
SW% = SL/ABD × 100	203 ± 15 (167-227)	–	189 ± 15 (174-230)	–	–
GS% = GL/SL × 100	70 ± 4 (60-80)	–	60 ± 10 (50-70)	–	–
H% = Hyaline/T × 100	–	–	–	–	54 ± 2.5 (49-58)

EP = distance from anterior end to excretory pore; NR = distance from anterior end to nerve ring; ES = distance from anterior end to end of pharynx.
After the original authors.

TYPE MATERIAL

Holotype male first generation deposited in the USDA Nematode Collection, Beltsville, MD, USA. Allotype (female, first generation) and many paratype males and females of the first generation and several infective juveniles in 4% formalin deposited in the same collection. Several males, females and infective juveniles in formalin deposited in the National Collection of Nematodes, Biosystematics Division, Plant Protection Research Institute, Agricultural Research Council, Pretoria, South Africa.

RELATIONSHIPS

Steinernema khoisanae isolates are characterised by morphometrics of the infective juvenile with body length of 1076 μm, narrow body diam. of 33 μm, distance from anterior end to the excretory pore of 94 μm, 85 μm long tail, a ratio = 33, D% = 68, H% = 57 and E% = 111. The lateral field pattern is 2, 7, 8, 6, 4, 2. The male of the first generation can be recognised by the spicule and the gubernaculum shape, position of the excretory pore near the end of the pharynx, D% = 88, and SW% = 199. The female can be recognised by the non-protuberant vulva and tail with prominent mucron (Figs 108C-F, 109D-F).

Steinernema khoisanae is closely related to species in the *glaseri*-group, which includes *S. apuliae*, *S. arenarium*, *S. cubanum*, *S. diaprepesi*, *S. glaseri*, *S. guangdongense*, *S. longicaudum* and *S. puertoricense.*

The infective juvenile of *S. khoisanae* differs from *S. cubanum* in body length of 1283 (1149-1508) μm *vs* 1075 (904-1214) μm for *S. khoisanae.* It differs from *S. puertoricense* in the body diam. of 51 (47-54) μm *vs* 33 (27-39) μm for *S. khoisanae.* From *S. glaseri*, this species differs in the position of the nerve ring and the length of the pharynx 120 (112-128) μm and 162 (158-168) μm, respectively, *vs* 108 (93-116) μm and 138 (115-147) μm for *S. khoisanae*. From *S. arenarium*, *S. guangdongense*, *S. diaprepesi* and *S. longicaudum*, this species differs in the position of the excretory pore, 83 (76-86) μm, 80 (71-83) μm, 74 (66-83) μm and 82 (76-89) μm, respectively, in these four species *vs* 93 (84-100) μm in *S. khoisanae*.

The first generation male of this species differs from all similar species by the position of the excretory pore, which is near the end of pharynx with the D% = 88, higher than all related species, shape of spicule (see Nguyen & Smart, 1997, Fig. 2). The arrowhead-shape of the cuneus (Fig. 106F) of *S. khoisanae* differs from *S. cubanum*, *S. glaseri*, *S. puertoricense*, *S. diaprepesi*, *S. longicaudum* (the cuneus of these species are Y-shaped) and *S. arenarium* (cuneus V-shaped).

The first generation female of *S. khoisanae* differs from *S. puertoricense*, *S. diaprepesi* and *S. guangdongense* in the absence of epiptygmata. It also differs from *S. puertoricense*, *S. longicaudum* and *S. apuliae* in the absence of a protruding vulva, the vulva of *S. khoisanae* being flush with the body contour or slightly shrunken. The lateral fields have one line (Fig. 108B) in this species while they are absent in *S. aciari*, *S. diaprepesi*, *S. guangdongense*, *S. glaseri* and *S. karii*. Unfortunately, the lateral field of female *S. khoisanae* cannot be compared with that of *S. cubanum*, *S. puertoricense* and *S. longicaudum*, as the character was not reported in the latter species.

MOLECULAR CHARACTERISATION

Steinernema khoisanae is characterised molecularly by sequence lengths and compositions of ITS and D2D3 regions (Nguyen *et al.*, 2006): *i*) ITS region lengths, 1051 base pairs (bp), ITS1 = 314 bp, ITS2 = 330 bp and its composition, A = 0.25309, C = 0.19886, G = 0.24263, T = 0.30542 (accession no. DQ314287); *ii*) D2/D3 region length, 876 bp, its composition, A = 0.24138, C = 0.19397, G = 0.31466, T = 0.25000 (accession no. DQ314289). More DNA information is available in Nguyen *et al.* (2006).

DISTRIBUTION

Known only from two localities, Cape Province, South Africa (Nguyen *et al.*, 2006).

References

NGUYEN, K.B. & SMART JR, G.C. (1997). Scanning electron microscopic studies of spicules and gubernacula for *Steinernema* spp. (Nemata: Steinernematidae). *Nematologica* 43, 465-480.

NGUYEN, K.B., MALAN, A.P. & GOZEL, U. (2006). *Steinernema khoisanae* n. sp. (Rhabditida: Steinernematidae), a new entomopathogenic nematode from South Africa. *Nematology* 8, 157-175.

Steinernema kraussei [*,**] **(Steiner, 1923) Travassos, 1927**

= *Aplectana kraussei* Steiner, 1923
= *Steineria kraussei* (Steiner, 1923) Travassos, 1927
= *Oxysomatium kraussei* (Steiner, 1923) Skrjabin, Shikhobalova & Mozgovoi, 1951
(Figs 111-114)

MEASUREMENTS

See Table 33.

DESCRIPTION

Male (first generation)

Cephalic extremity truncate, slightly offset from rest of body. Lips fused with six labial and four cephalic papillae. Pore-like amphidial apertures located laterally in cephalic papillae circle. Cuticle smooth with slight annulation visible under SEM. Oral aperture triangular, stoma represented by sclerotised cheilorhabdions and posteriorly by another ring of smaller prorhabdions. Pharynx muscular with a characteristic rhabditoid cylindrical corpus, subdivided into procorpus and slightly enlarged metacorpus. Isthmus narrow and distinct; basal bulb valvate. Excretory pore always located anterior to nerve ring in metacorpus region. Genital system monorchic, reflexed. Spicules paired, colourless, symmetrical with a fine arch, pointed tips. Manubrium variable in shape with ratio length to width = 1 : 1, but generally square in lateral view. Calomus well developed. At inner base of lamina, retinaculum often present. Two ridges arising in distal part of manubrium and continuing throughout lamina. Gubernaculum posterior end variable from hook-like to rounded; anterior portion remarkably attenuated. Genital papillae comprising 11 or 12 pairs plus single midventral papilla just anterior to cloacal aperture. Of these, five or six pairs precloacal situated subventrally in two rows, three pairs adcloacal in lateral to dorsolateral position, and three pairs postcloacal at tail tip. Tail region curved ventrally; tip rounded with fine mucron.

Male (second generation)

Only size differences between first and second generation.

[*] Named after Dr *Krausse*, the person who found the original isolate.
[**] Description after Mráček (1994). Revised by Zdeněk Mráček.

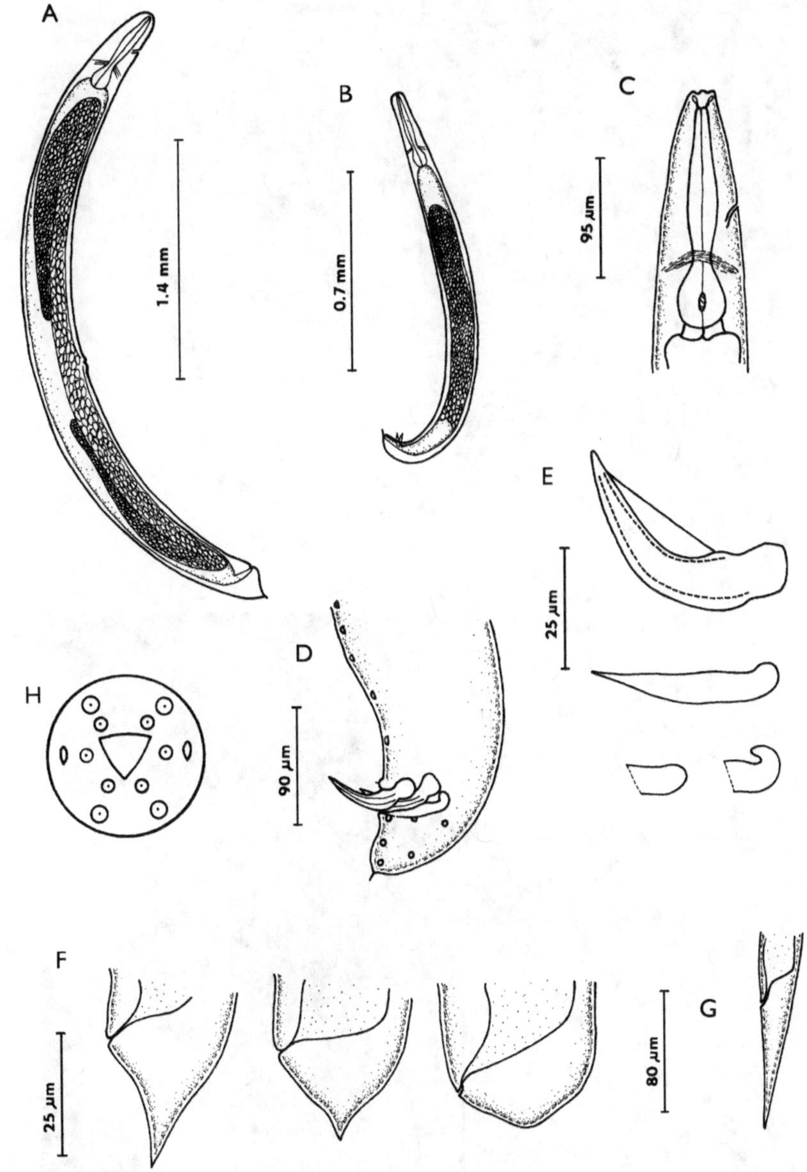

Fig. 111. Steinernema kraussei. *A, F: Female. A: Entire body; F: Tail variability from young (left) to old (right). B-E: Male. B: Entire body; C: Pharynx and excretory pore; D: Tail; E: Spicule and gubernaculum variability. G: Infective juvenile tail; H: Adult arrangement of head papillae and amphidial apertures. After Mráček (1994), courtesy of* Folia Parasitologica.

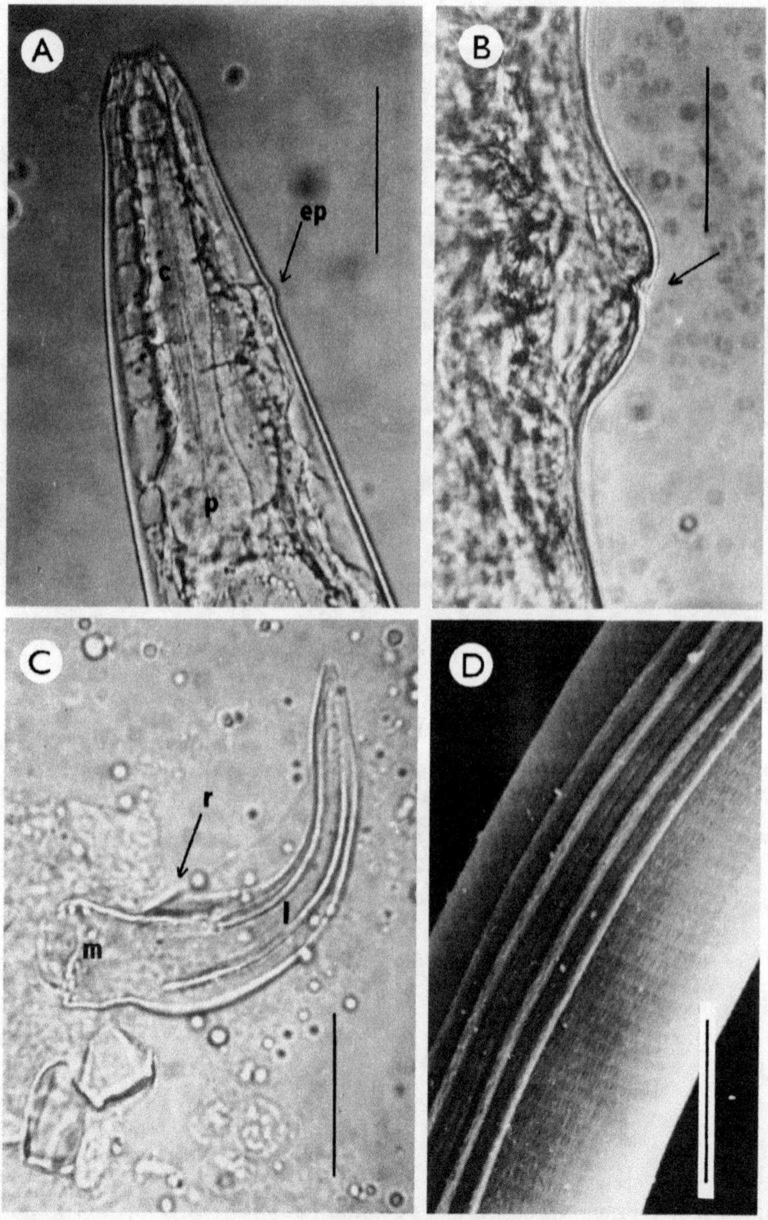

Fig. 112. Steinernema kraussei. *A, C: Male. A: Pharynx and excretory pore (ep); C: Spicule. B: First generation female, vulva (arrow); D: Infective juvenile, lateral field. (Scale bars: A, C = 50 μm; B = 20 μm; D = 15 μm.) After Mráček (1994), courtesy of* Folia Parasitologica.

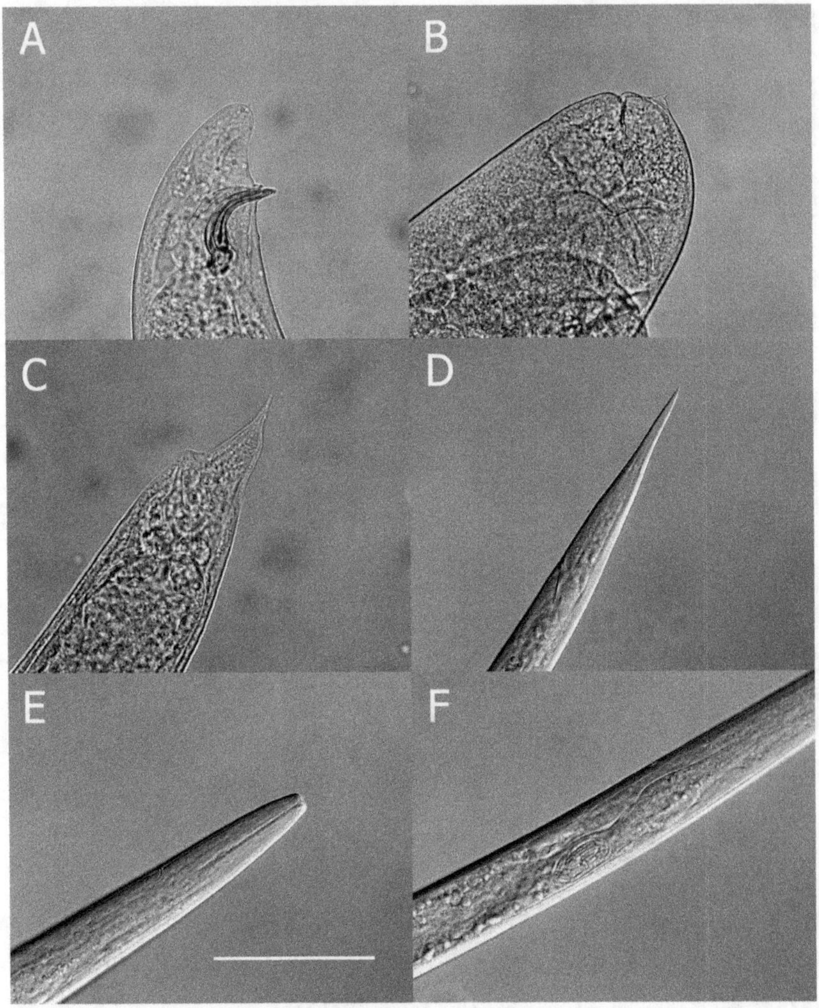

Fig. 113. Steinernema kraussei. *A: First generation male, tail with short mucron; B: First generation female, tail with a projection; C: Second generation female, tail. D-F: Infective juvenile. D: Tail with hyaline region; E: Head region with excretory pore; F: Pharynx with bacterial pouch. (Scale bar: A = 70 μm; B = 180 μm; C = 90 μm; D-F = 60 μm.) After Z. Mráček.*

Female (first generation)

Cephalic extremity rounded, not offset from rest of body. Cuticle smooth and slightly annulated. Anterior region similar to that of

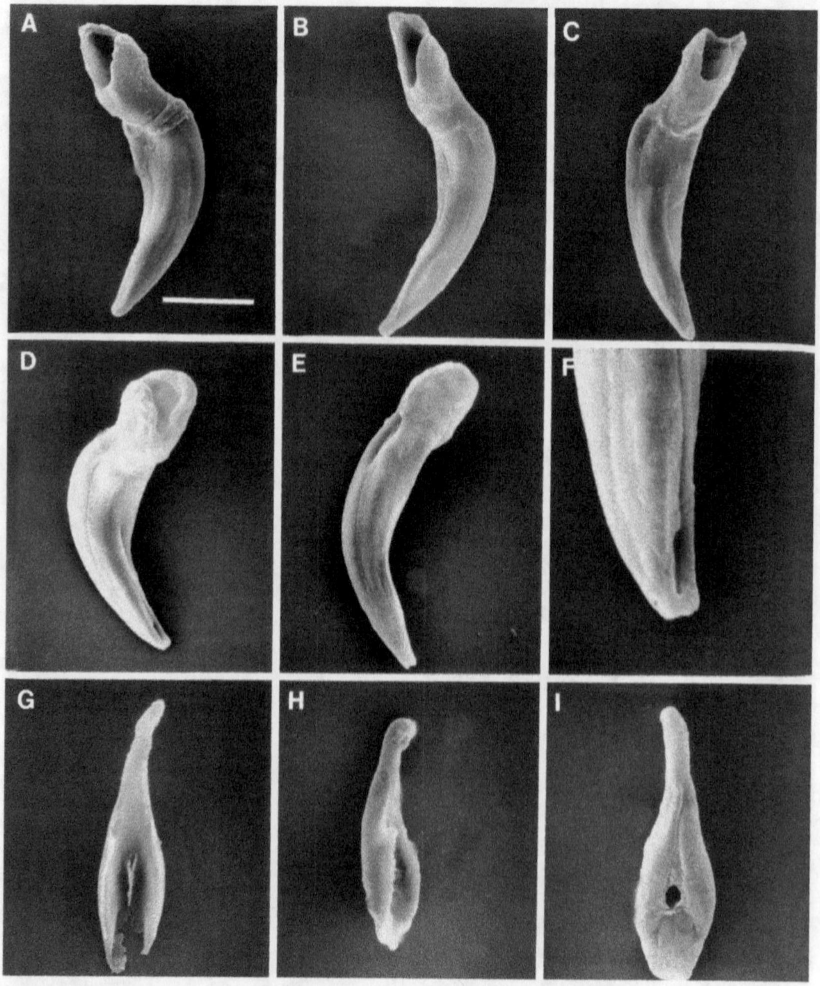

Fig. 114. Steinernema kraussei. *A-I: Male SEM. A-F: Spicules (r = rostrum, l = lamina, m = manubrium); G-I: Gubernacula. (Scale bar: A-F = 15 μm; G-I = 11 μm.) Courtesy of K.B. Nguyen.*

male. Excretory pore located more anteriorly than in male. Vulva slightly or moderately elevated, situated just posterior to mid-body. Genital system paired, reflexed. Tail pointed in young females, lacking a postanal swelling. Older females have blunt tail with a short, spine-like, terminal projection, rarely rounded without any projection.

Table 33. *Morphometrics of* Steinernema kraussei. *Measurements are in μm and in the form: mean (range).*

Character	First generation			Second generation		Infective juvenile
	Holotype (female*)	Male	Female	Male	Female	
n	–	20	20	20	20	25
L	–	1400 (1200-1600)	4200 (2500-5400)	900 (700-110)	1700 (1300-2100)	950 (797-1102)
a	–	11	17	16	17	29
b	–	9	22	7	11	7.1
c	–	37	88	27	29	12
c′	–	0.9**	–	0.9**	–	3.9**
V	–	–	54	–	55	–
Body diam.	–	128 (110-144)	240 (153-288)	57 (51-68)	102 (59-127)	33 (30-36)
EP	–	81 (73-99)	87 (66-99)	71 (66-82)	83 (63-89)	63 (56-66)
NR	–	105 (95-122)	137 (127-146)	98 (89-116)	120 (112-133)	105 (99-111)
ES	–	153 (137-178)	192 (178-205)	135 (121-145)	162 (132-176)	134 (119-145)
Tail length (T)	–	39 (36-44)	48 (33-59)	33 (27-35)	60 (53-69)	81 (72-89)
Hyaline (H)	–	–	–	–	–	31 ± 0.5*** (27-35)
Anal body diam. (ABD)	–	45 (39-50)	–	37 (31-43)	–	20 (19-22)
Spicule length (SL)	–	19 (42-53)	–	44 (40-50)	–	–
Gubernaculum length (GL)	–	33 (29-37)	–	30 (21-35)	–	–
Gubernaculum width	–	7.2 (6.6-8.0)	–	6.8 (5.3-8.0)	–	–
D% = EP/ES × 100	–	53	45	52	51	47
E% = EP/T × 100	–	210	182	215	139	80
SW% = SL/ABD × 100	–	110	–	119	–	–
GS% = GL/SL × 100	–	68	–	68	–	–
H% = H/T × 100	–	–	–	–	–	38 ± 0.4*** (35-40)

EP = distance from anterior end to excretory pore; NR = distance from anterior end to nerve ring; ES = distance from anterior end to end of pharynx.

* Slide destroyed.

** Recalculated according to Mráček (1994).

*** According to Půža (2001).

Female (second generation)

Size difference between the first and second generation; tail always pointed, rarely with postanal swelling.

Infective juvenile

Body thin, cuticle sheath sometimes present. Cephalic extremity slightly offset from rest of body with flattened inconspicuous lips and four cephalic papillae, labial papillae indistinct. Oral aperture closed, pharynx collapsed, basal bulb more elongate than in adults. Excretory pore located far anterior to base of pharynx, at *ca* 46% length of pharynx. Lateral fields with eight ridges, but central and submarginal pair less distinct. Ridges of central pair tending to merge along mid-region of lateral field. Tail pointed, lacking any constriction or spine-like structure. Hyaline region comprising slightly less than 40% of tail length.

TYPE HOST AND LOCALITY

Larvae of the web-spinning sawfly, *Cephalcia abietis*, parasitised with nematodes were collected by Dr Krausse from an outbreak site in Westphalen in the Egge Mountains in Warburg county, near Neuenherrsee, Germany (see Mráček *et al.*, 1991).

NOTE

A major outbreak of sawfly had persisted for several years in this region and Krausse was appointed by the German government to take control. Mráček (1977) redescribed the new South Bohemian isolate of *S. kraussei* and completed Steiner's old description. In March 1991, Mráček visited the original locality in Germany and collected soil samples (Mráček *et al.*, 1991). The precise location of the original type locality was identified, with the help of the local Forest Region employees, as areas Jages 96, 97 and 98 in a spruce forest of an age typical for sawfly infestations (*i.e.*, 60-80 years old). Interestingly, there has not been an outbreak of sawfly in this locality for many years, but another eligible host, wireworms (Elateridae) were recovered from Jages 96-98.

TYPE MATERIAL

Three type specimens (holotype and two paratypes) are in the USDA Nematode Collection, Beltsville, MD, USA. Unfortunately, all represent

females and they are somewhat poorly preserved. Topotype specimens, fixed either in 4% formalin or mounted on slides, are deposited in the Department of Insect Pathology, Institute of Entomology, Biology Centre, Academy of Sciences of the Czech Republic, České Budějovice, Czech Republic.

RELATIONSHIPS

Steinernema kraussei is the type of the genus *Steinernema* Travassos, 1927. The excretory pore opening of the adults is situated far anterior to the nerve ring, close to the middle region of the pharynx. The tail of young females is pointed, although in old females it is more rounded. The male tail bears a fine mucron in both generations, sometimes reduced in the first generation. The spicules of *S. kraussei* have a moderately curved lamina usually bearing a rostrum. The spicule tip is pointed and lacks a hook-like structure. The length to width ratio of the manubrium is 1 : 1. The rostrum is usually developed. Infective juveniles of *S. kraussei* belong to the *feltiae*-group of species, a group characterised by a medium length body of 700-1000 μm. The lateral fields are formed by eight ridges, although the submarginal and central pair are less prominent. The hyaline region comprises at most 40% of the tail length.

Steinernema kraussei must be compared with the second-oldest species, *S. glaseri*. *Steinernema kraussei* differs from most species by the IJ lateral fields bearing eight ridges with the submarginal and central pair less prominent. Such a ridge pattern has not been found in any Holarctic species. *Steinernema glaseri*, by comparison, possesses eight identically prominent ridges. The IJ body length is clearly greater than 1000 μm in *S. glaseri* but is mostly shorter than 1000 μm in *S. kraussei*. The IJ tail of *S. kraussei* is much more slender with average ratio $c' = 4.1$. Females of *S. kraussei* possess a less prominent vulva and do not develop a distinct postanal swelling as in *S. glaseri*. *Steinernema glaseri* males possess less curved spicules with a notched spicule tip *vs* pointed in *S. kraussei*. A rostrum is never presents in *S. glaseri* but is usually well developed in *S. kraussei*. The manubrium is oblong (lateral view) with a length to width ratio of 1.5-2.0 : 1 in *S. glaseri* but is rounded with a ratio of 1 : 1 in *S. kraussei*.

MOLECULAR CHARACTERISATION

Steinernema kraussei is characterised by sequence lengths (flanked by primers 18S and 26S) of ITS regions, 988 bp, ITS1 = 267 bp, ITS2 = 323 bp and its composition, A = 0.25810, C = 0.18522, G = 0.22976, T = 0.32692. The sequences of ITS and D2D3 regions are deposited in GenBank under accession no. AY230176 and AF331896, respectively.

BIONOMICS

This species is prevalent in woodland habitats, both coniferous and deciduous (Mráček *et al.*, 2005). It is rarely found in open habitats with the exception of alpine altitudes (Steiner, 1996; Shishinova *et al.*, 1997).

DISTRIBUTION

The species is widely distributed in the Holarctic region. It is common in Europe (Mráček *et al.*, 1999; Sturhan & Lišková, 1999; Sturhan & Mráček, 2000; Hominick, 2002) and has been found in the Far East (Yoshida, 2003). In North America it is recorded from British Columbia, Canada (Mráček & Webster, 1993; Stock *et al.*, 2000) and from the USA (Stock *et al.*, 2000).

References

HOMINICK, W.M. (2002). Biogeography. In: Gaugler, R. (Ed.). *Entomopathogenic nematodes.* Wallingford, UK, CABI Publishing, pp. 115-143.

MRÁČEK, Z. (1977). *Steinernema kraussei*, a parasite of the body cavity of the sawfly, *Cephalcia abietis*, in Czechoslovakia. *Journal of Invertebrate Pathology* 30, 87-94.

MRÁČEK, Z. (1994). *Steinernema kraussei* (Steiner, 1923) (Nematoda: Rhabditida: Steinernematidae), re-description of its topotype from Westphalen. *Folia Parasitologica* 41, 59-64.

MRÁČEK, Z. & WEBSTER, J. (1993). Survey of Heterorhabditidae and Steinernematidae (Rhabditida, Nematoda) in Western Canada. *Journal of Nematology* 25, 710-717.

MRÁČEK, Z., WEISER, J., BUREŠ, M. & KAHOUNOVÁ, L. (1991). *Steinernema kraussei* (Steiner, 1923) (Nematoda: Rhabditida); rediscovery of its type locality in Germany. *Folia Parasitologica* 39, 181-182.

MRÁČEK, Z., BEČVÁŘ, S. & KINDLMANN, P. (1999). Survey of entomopathogenic nematodes from the families Steinernematidae and Heterorhabditidae (Nematoda: Rhabditida) in the Czech Republic. *Folia Parasitologica* 46, 145-148.

MRÁČEK, Z., BEČVÁŘ, S., KINDLMANN, P. & JERSÁKOVÁ, J. (2005). Habitat preference for entomopathogenic nematodes, their insect hosts and new faunistic records for the Czech Republic. *Biological Control* 34, 27-37.

PŮŽA, V. (2001). *Comparison of an infective juvenile tail morphology of the family Steinernematidae*. B.Sc. Thesis, University of South Bohemia, České Budějovice, Czech Republic, 23 pp.

SHISHINIOVA, M., BUDUROVA, I. & GRADINAROV, D. (1997). Contribution to fauna of entomopathogenic nematodes (Rhabditida: Steinernematidae, Heterorhabditidae) from Bulgaria. *Biotechnology and Biotechnology Equipment* 11, 45-52.

SKRJABIN, K.I., SHIKHOBALOVA, N.P. & MOZOGOVOI, A.A. (1951). [*Descriptive catalogue of parasitic nematodes, volume 2. Oxyurata and Ascaridata.*] Moscow, 631 pp.

STEINER, G. (1923). *Aplectana kraussei* n. sp., eine in der Blattwespe *Lyda* sp. parasitierende Nematodenform, nebst Bemerkungen über das Seitenorgan der parasitischen Nematoden. *Zentralblatt fur Bakteriologie Parasitenkunde Infektionskrankheiten und Hygiene* 59, 14-18.

STEINER, W.A. (1996). Distribution of entomopathogenic nematodes in the Swiss Alps. *Revue Suisse de Zoologie* 103, 439-452.

STOCK, P., MRÁČEK, Z. & WEBSTER, J.M. (2000). Morphological variation among allopatric populations of *Steinernema kraussei* (Steiner, 1923) (Rhabditida: Steinernematidae). *Nematology* 2, 143-152.

STURHAN, D. & LIŠKOVÁ, M. (1999). Occurrence and distribution of entomopathogenic nematodes in the Slovak Republic. *Nematology* 1, 273-277.

STURHAN, D. & MRÁČEK, Z. (2000). Comparison of the *Galleria* baiting technique and a direct extraction method for recovering *Steinernema* infective-stage juveniles from soil. *Folia Parasitologica* 47, 315-318.

TRAVASSOS, L. (1927). Sobre o genero *Oxysomatium*. *Boletim Biologico, São Paulo* 5, 20-21.

YOSHIDA, M. (2003). Intraspecific variation in RFLP patterns and morphological studies on *Steinernema feltiae* and *S. kraussei* from Hokkaido, Japan. *Nematology* 5, 735-746.

Steinernema kushidai[*,**] Mamiya, 1988
(Figs 115-118)

Measurements

See Table 34.

Description

Male (first generation)

Cuticle smooth, lateral fields inconspicuous. Cephalic extremity truncate or slightly rounded. Six lips, fused, but tips distinct with terminal papilla. Four cephalic papillae present. Oral aperture triangular. Stoma short and broad with non-sclerotised walls. Cheilo- and telorhabdions vestigial. Pharynx muscular, procorpus cylindrical, metacorpus slightly swollen, non-valvate, indistinct isthmus followed by pyriform basal bulb. Genital system monorchic, reflexed. Spicules paired, strongly curved, spicule head truncate or rounded. Shaft very short, or indistinct. Blade wide anteriorly, tapering posteriorly to blunt tip. Spicule tip usually slightly curved or with notch dorsally. Velum prominent but short, not covering spicule tip. In ventral view, gubernaculum tapering anteriorly to form a neck, then enlarging slightly to a ventrally curved terminus. In lateral view, bow-shaped, constricted in posterior part to rounded proximal portion. Complement of genital papillae comprising two single, midventral, precloacal papillae plus 12 pairs located as follows: one midventral papilla located just anterior to cloacal aperture, the other far anterior to cloacal aperture and between pairs 1 and 2 of the precloacal series; eight precloacal subventral pairs; one pair dorsolateral at cloacal aperture level; three postcloacal pairs of which two subterminal subventral and one pair subdorsal. Tail conical, tapering to rounded terminus.

Male (second generation)

Similar to that of first generation except body, spicule and gubernaculum shorter and thinner. Tail with minute spine.

Female (first generation)

Cuticle smooth, lateral fields inconspicuous. Body shape and anterior region typical for genus. Cephalic extremity, stoma similar to male but

[*] Named after T. *Kushida* who collected the nematode.

[**] Description after Mamiya (1988). Revised by Khuong B. Nguyen.

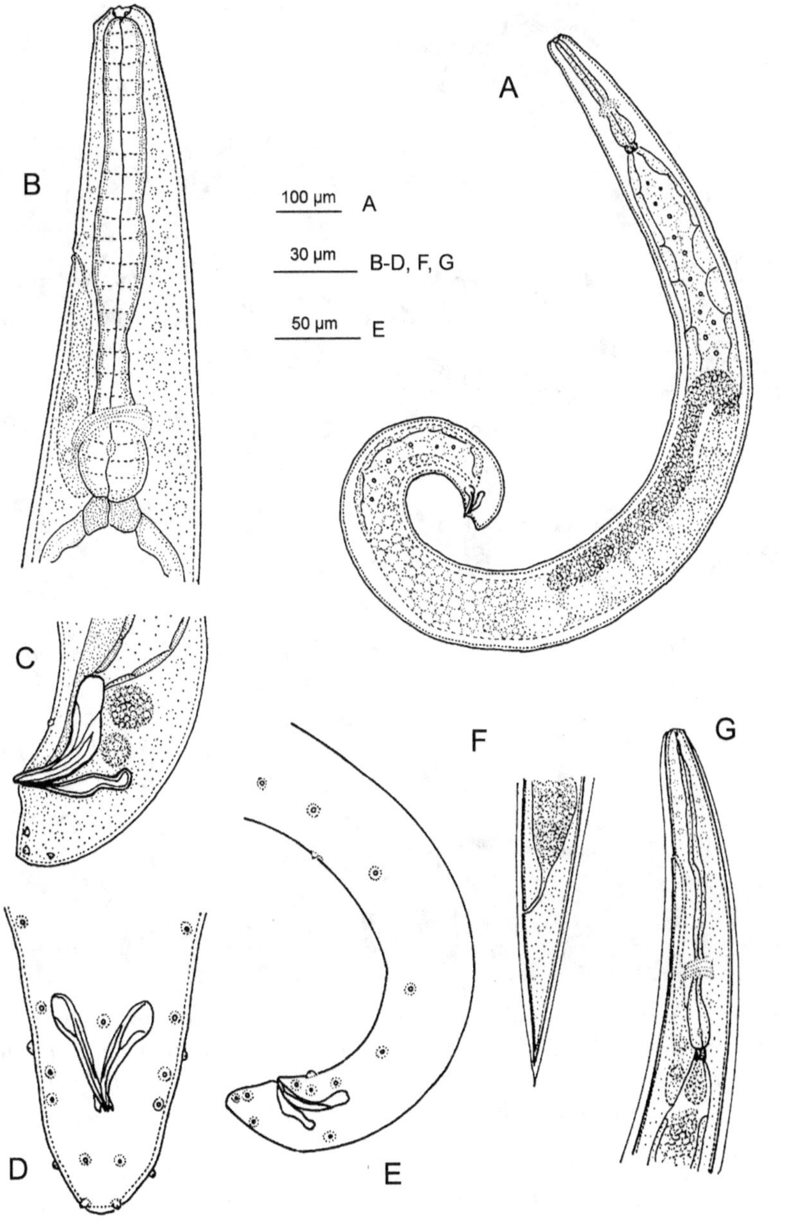

Fig. 115. Steinernema kushidai, *first generation males (A-E) and infective juvenile (F, G). A: Entire body; B: Pharyngeal region; C: Tail, lateral view; D: Tail, ventral view; E: Posterior region; F: Tail; G: Anterior region. After Mamiya (1988), courtesy of* Applied Entomology and Zoology.

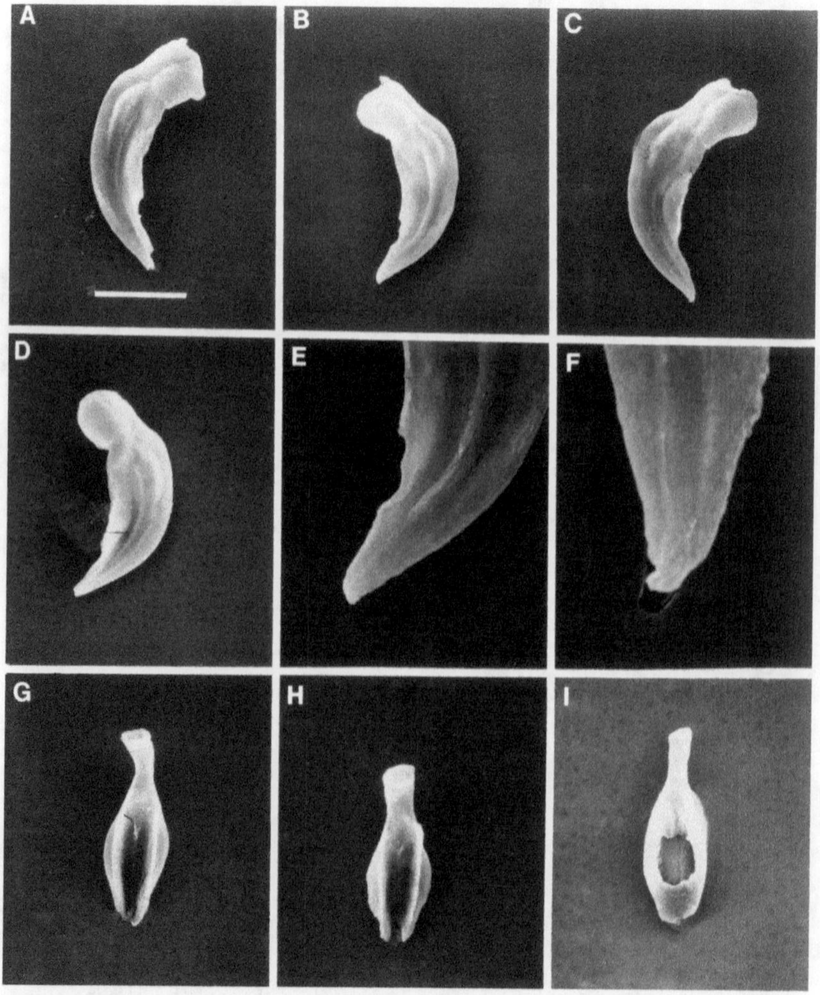

Fig. 116. Steinernema kushidai, *SEM photographs of spicules and gubernacula. A-D: Spicules; E, F: Spicule tips; G-I: Gubernacula, ventral and dorsal views. (Scale bar: A-D, G-I = 20 μm; E = 7.5 μm; F = 3.0 μm.) After Nguyen and Smart (1997), courtesy of* Nematologica.

much larger. Pharynx muscular. Cardia well developed. Excretory pore opening circular, excretory glands large extending to anterior portion of intestine dorsally. Nerve ring located just anterior to basal bulb. Genital system amphidelphic, reflexed. Vulva in form of a transverse slit, protruding slightly from body surface. Tail tapering rapidly to pointed terminus, shorter than diam. at anus.

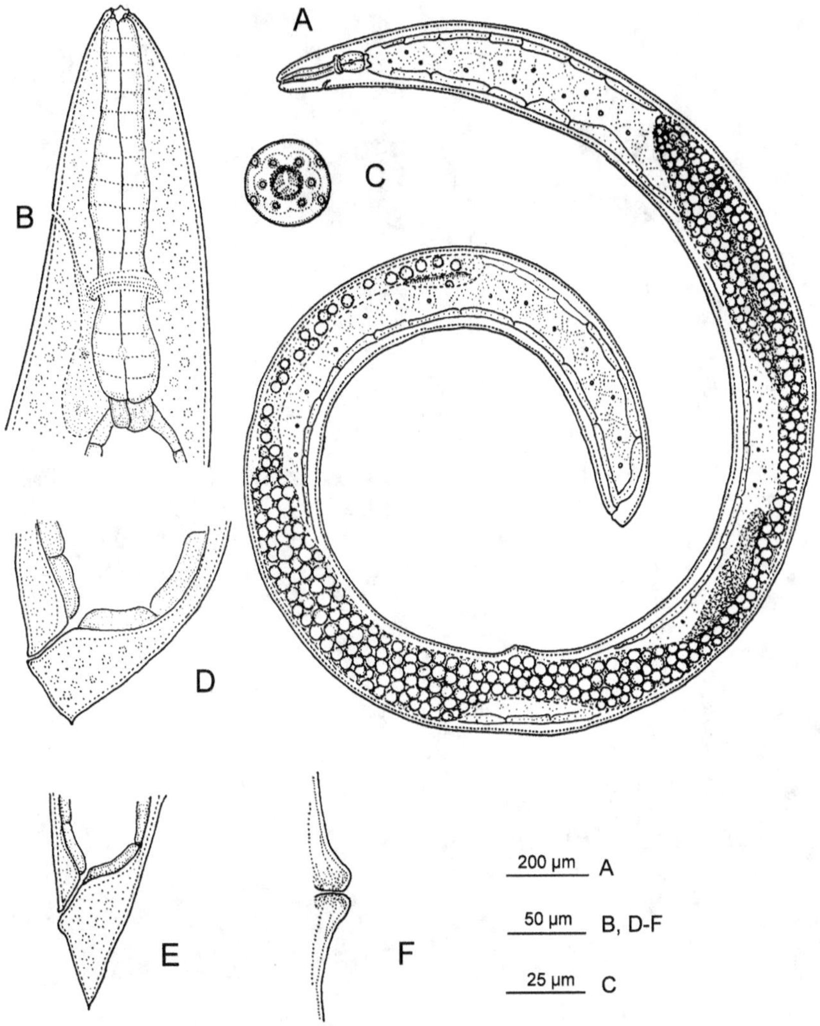

Fig. 117. Steinernema kushidai*; first (A-D) and second (E) generation females. A: Entire body of female; B: Pharyngeal region; C: Face view; D, E: Tail; F: Vulva. After Mamiya (1988), courtesy of* Applied Entomology and Zoology.

Female (second generation)

Similar to first generation females, except for following characters: smaller size, more slender body, tail straight and longer than diam. at anus, minute spine present in tail tip.

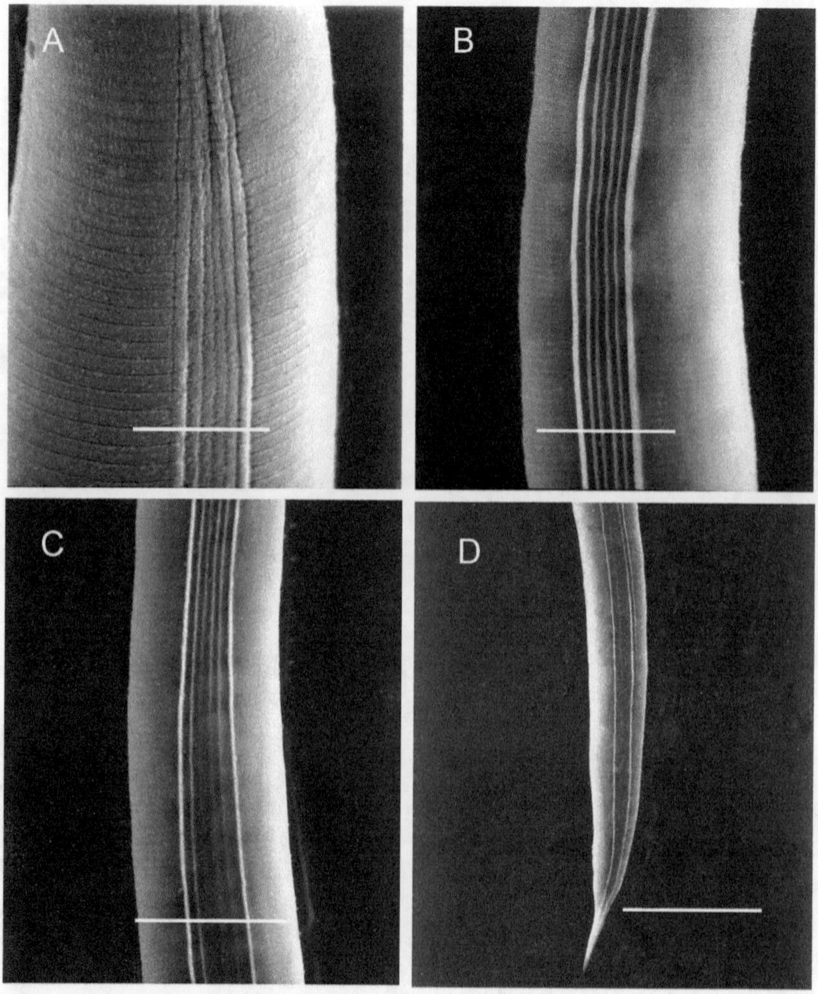

Fig. 118. Steinernema kushidai, *SEM of lateral field of infective juvenile. A: Lateral field with two ridges anteriorly and six ridges posteriorly; B: Lateral field near mid-body with six ridges; C: Lateral field with six ridges anteriorly and four internal ridges disappearing gradually posteriorly; D: Posterior part of lateral field showing two marginal ridges and a large central band. (Scale bars: A = 6 μm; B = 13.6 μm; C = 20.1 μm; D = 38 μm.) Courtesy of K.B. Nguyen.*

Infective juvenile

Body slender, gradually tapering posteriorly, generally ensheathed in cuticle of second-stage juveniles. Labial region not set off. Lateral field

Table 34. *Morphometrics of* Steinernema kushidai. *Measurements are in μm and in the form: mean ± SD (range).*

Character	First generation			Second generation		Infective juvenile
	Holotype (male)	Male	Female	Male	Female	
n	–	20	20	20	20	50
L	1500	1400 ± 16 (1200-1900)	3500 ± 79 (2100-4700)	1200 ± 9 (1000-1300)	2300 ± 15 (2100-2700)	589 ± 39.8 (424-662)
a	–	–	–	–	–	22.5 ± 1.6 (19.3-25.2)
b	–	–	–	–	–	5.3 ± 0.3 (4.9-5.9)
c	–	–	–	–	–	11.7 ± 0.6 (9.9-12.9)
V	–	–	56 ± 1.4 (54-59)	–	56 ± 2.4 (50-60)	–
Body diam.	95	97 ± 19 (75-156)	175 ± 18.5 (153-210)	58 ± 3.3 (54-63)	95 ± 6.7 (84-111)	26 ± 2.5 (22-31)
EP	90	84 ± 7.9 (71-105)	91 ± 6.6 (78-105)	74 ± 3.1 (66-78)	91 ± 11.4 (75-105)	46 ± 2 (42-50)
NR	132	129 ± 5.3 (120-137)	124 ± 9 (111-144)	104 ± 4 (96-111)	136 ± 9.3 (111-150)	76 ± 3.5 (70-84)
ES	168	167 ± 7.9 (156-189)	227 ± 15.2 (204-255)	161 ± 6.8 (147-177)	199 ± 9.7 (174-216)	111 ± 4.1 (106-120)
Tail length (T)	32	33 ± 3 (30-40)	38 ± 4.4 (30-45)	30 ± 6.8 (27-39)	48 ± 8.9 (36-72)	50 ± 3.4 (44-59)
Anal body diam. (ABD)	38	42 ± 5 (36-54)	64 ± 7.2 (54-84)	36 ± 7.4 (36-45)	41 ± 3.8 (36-48)	–
Spicule length (SL)	62	63 ± 5.5 (48-72)	–	58 ± 4 (51-63)	–	–
Gubernaculum length (GL)	47	44 ± 4.9 (39-60)	–	40 ± 3.3 (30-45)	–	–
D% = EP/ES × 100	54	51 ± 4.8 (42-59)	40 ± 2.2 (37-46)	46 ± 3.9 (41-56)	–	41 ± 1.4 (38-44)
E% = EP/T × 100	–	–	–	–	–	92
SW% = SL/ABD × 100	–	150	–	161	–	–
GS% = GL/SL × 100	–	70	–	69	–	–

EP = distance from anterior end to excretory pore; NR = distance from anterior end to nerve ring; ES = distance from anterior end to end of pharynx.
After the original author.

beginning with two ridges. Near excretory pore, number of ridges in lateral field increasing from two to six with marginal ridges prominent. In posterior half of body, two marginal ridges remaining but four internal ridges gradually disappearing posteriorly. In posterior third of body, lateral field with two external ridges and a large internal band. With above description, formula of lateral field = 2, 6, 2. Oral aperture closed. Pharynx long, narrow, terminating in basal bulb, weak and less prominent than in adults. Excretory pore at level of mid-pharynx. Excretory glands displacing basal bulb and anterior end of intestine. Nerve ring distinct, located anterior to basal bulb. Hemizonid at level of nerve ring. Cardia distinct. Bacterial pouch visible in anterior portion of intestine. Intestine dense, dark in appearance. Tail conical with pointed terminus. Small phasmid present in mid-tail region.

TYPE HOST AND LOCALITY

Type host unknown as nematodes were recovered using scarabaeid beetle larvae (*Anomala cuprea*) as bait from soil collected at Hamakita, Shizuoka Prefecture, Japan.

TYPE MATERIAL

Holotype (male) and allotype (female) deposited in the nematology collection at Forestry and Forest Products Research Institute, Ibaraki, Japan. Paratypes deposited at the Herbarium and Insect Museum of the National Institute of Agro-Environmental Sciences, Tsukuba, Ibaraki, Japan; USDA Nematode Collection, Beltsville, MD, USA; University of California-Davis Nematode Collection, Davis, CA, USA; and the CABI Europe-UK Nematode Collection, Egham, Surrey, UK.

RELATIONSHIPS

Steinernema kushidai is closely related to *S. carpocapsae* and *S. intermedium*. It can be separated from the latter two species by the location of the excretory pore of all stages, shape of spicules and gubernaculum of adult males, and the body length of infective juveniles. The body length of infective juveniles of *S. kushidai* is distinctly smaller than *S. intermedium* and slightly longer than *S. carpocapsae*. The excretory pore of infective juveniles of *S. kushidai* (D% = 38-44) is located more anteriorly than in *S. intermedium* (D% = 48-58) but more posteriorly than in *S. carpocapsae* (D% = 23-28). The gubernaculum of the first and second generation males of *S. kushidai* is recognisably different from that of *S. carpocapsae* and *S. intermedium*. In lateral view, the rounded proximal portion is separated from the other part by a constriction, giving a shape clearly distinct from that of the other two species. The gubernaculum of *S. carpocapsae* is narrow with an indistinctly constricted proximal portion, and that of *S. intermedium* is typically bow-shaped in lateral view. The spicules of *S. kushidai* are colourless and have a smoothly rounded manubrium, while those of *S. carpocapsae* are yellow and have an angular manubrium. The bacterial pouch in infective juveniles of *S. kushidai* is smaller than that of *S. intermedium* (27-48 μm). The upper range of pouch length in *S. kushidai* does not overlap with the lower end of that in *S. intermedium.* The body length of infective juveniles of *S. kushidai* is also smaller than in *S. kraussei*, *S. feltiae* and *S. glaseri*, but longer than in *S. rarum.*

MOLECULAR CHARACTERISATION

Steinernema kushidai is characterised by the sequence length of ITS regions, 1032 bp, including primers 18S and 28S, and its composition, A = 0.24418, C = 0.19864, G = 0.24709, T = 0.31001 (accession no. AB243440), and D2D3 region sequence length, 867 bp and its composition, A = 0.24683, C = 0.18454, G = 0.30334, T = 0.26528 (accession no. AF331897).

BIONOMICS

Infective juveniles of *S. kushidai* were capable of invading and killing larvae of the scarabaeids *Heptophylla picea* and *Anomala cuprea* and caused 100% mortality for both insect species. Average time to host death after exposure was 5.0 days for 20 larvae of *H. picea* and 5.2 days for 14 larvae of *A. cuprea.* The infective juveniles started to develop in the host just after host death, irrespective of the time required to kill. The life cycle of *S. kushidai* is similar in many respects to that of other steinernematid nematodes. Exsheathment of infective juveniles after entering the host occurred somewhat later than in other steinernematids. Subsequent host death was delayed more than 3 days in comparison with that caused by other nematodes. Exsheathed juveniles continued to develop into large first generation adults, mated, and deposited many eggs. Eggs and hatched juveniles appeared within 3 days of host death and occasionally developed to second generation adults during the following 2-4 days. In most cases, the first generation adults produced infective juveniles which appeared in the insect cadavers 5 days after the host's death. At this time, parasitic juveniles, probably second-stage juveniles of almost uniform size, outnumbered infective juveniles in the cadavers. The proportion of infective juveniles to total number of juveniles gradually increased and reached nearly 100% 9 days after host death. Even when the second generation adults were produced, most infective juveniles seemed to develop from eggs deposited by the first generation adult females, as the second generation adults were few, if any, and they existed concurrently with infective juveniles. Sometimes juveniles hatched in the body of first generation adults and developed to infective juveniles within the body. Average number of infective juveniles produced per cadaver of *H. picea* was *ca* 50 800 (Mamiya, 1988).

DISTRIBUTION

Known only from the type locality, Hamakita, Shizuoka Prefecture, Japan (Mamiya, 1988).

Reference

MAMIYA, Y. (1988). *Steinernema kushidai* n. sp. (Nematoda: Steinernematidae) associated with scarabaeid beetle larvae from Shizuoca, Japan. *Applied Entomology and Zoology* 23, 313-320.

Steinernema leizhouense[*,**] Nguyen, Qiu, Zhou & Pang, 2006
(Figs 119-124)

Measurements

See Table 35.

Description

Male (first generation)

Body curved ventrally posteriorly. Cephalic extremity rounded, usually slightly swollen. Labial region with six labial papillae, two amphidial apertures, four prominent cephalic papillae. Stoma shallow. Under light microscope, cheilorhabdions appearing as small sclerotised structures at anterior end, sometimes indistinct. Under SEM, cheilorhabdions appearing as subtriangular ring surrounding stoma. Excretory pore located between mid-pharynx and nerve ring. Anterior part of excretory duct well sclerotised. Pharynx with cylindrical procorpus, metacorpus slightly swollen, isthmus present, nerve ring around isthmus (just anterior to basal bulb), basal bulb distinct. Pharyngo-intestinal valve present. Genital system monorchic, reflexed. Distance from base of pharynx to anterior end of testis variable. Spicules paired, brown in colour. Head (manubrium) of spicules enlarged anteriorly, with rounded anterior end, longer than wide (length/width = 1.5-2.0). Shaft (calomus) very short, mostly absent. Blade (lamina) thin, slightly curved, tapering gradually posteriorly, blade terminus blunt, velum very small or absent. Each spicule with two internal sclerotised ribs. Gubernaculum boat-shaped in lateral view, tapering gradually anteriorly. Anterior end usually ventrally curved. Complement of genital papillae disposed as 11 pairs plus single midventral precloacal papilla and distributed as in Figure 120E, F. Tail conoid, curved ventrally; tail terminus rounded, lacking mucron.

Male (second generation)

Similar to that of first generation except body, spicule and gubernaculum shorter and thinner, posterior end spiral-shaped.

[*] Named after the type locality *Leizhou*, People's Republic of China.

[**] After Nguyen *et al.* (2006). Revised by Khuong B. Nguyen.

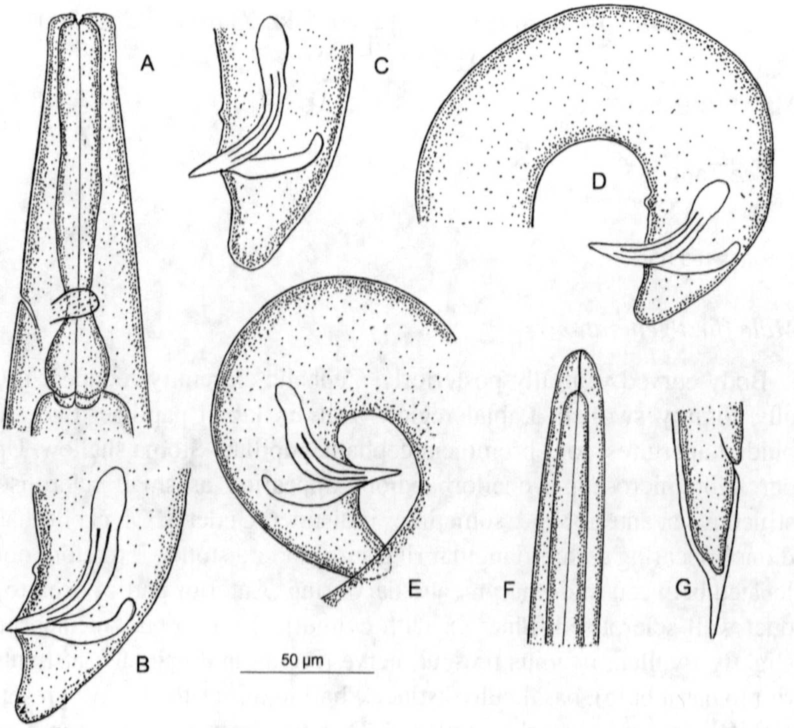

Fig. 119. Steinernema leizhouense, *male and infective juvenile. A: Anterior region of first generation male; B, C: Variations of first generation male tails and spicules and gubernacula; D, E: Variation of second generation male tails, spicules and gubernacula; F, G: Anterior and posterior region of infective juvenile. After Nguyen* et al. *(2006), courtesy of* Russian Journal of Nematology.

Female (first generation)

Body cuticle smooth or with faint annules. Lateral fields and phasmids not observed. Cephalic extremity rounded or flattened anteriorly, continuous with body; six labial and four cephalic papillae present. Lips indistinct. Amphidial apertures usually inconspicuous even under SEM. Stoma shallow. Cheilorhabdions well sclerotised but small. Smaller, indistinct, sclerotised structure (presumably the prorhabdions) located posterior to cheilorhabdions. Pharynx with cylindrical, muscular, procorpus; somewhat swollen metacorpus; distinct isthmus; valvate basal bulb. Nerve ring surrounding isthmus, variable in position. Pharyngo-intestinal valve present. Excretory pore located anterior to mid-pharynx

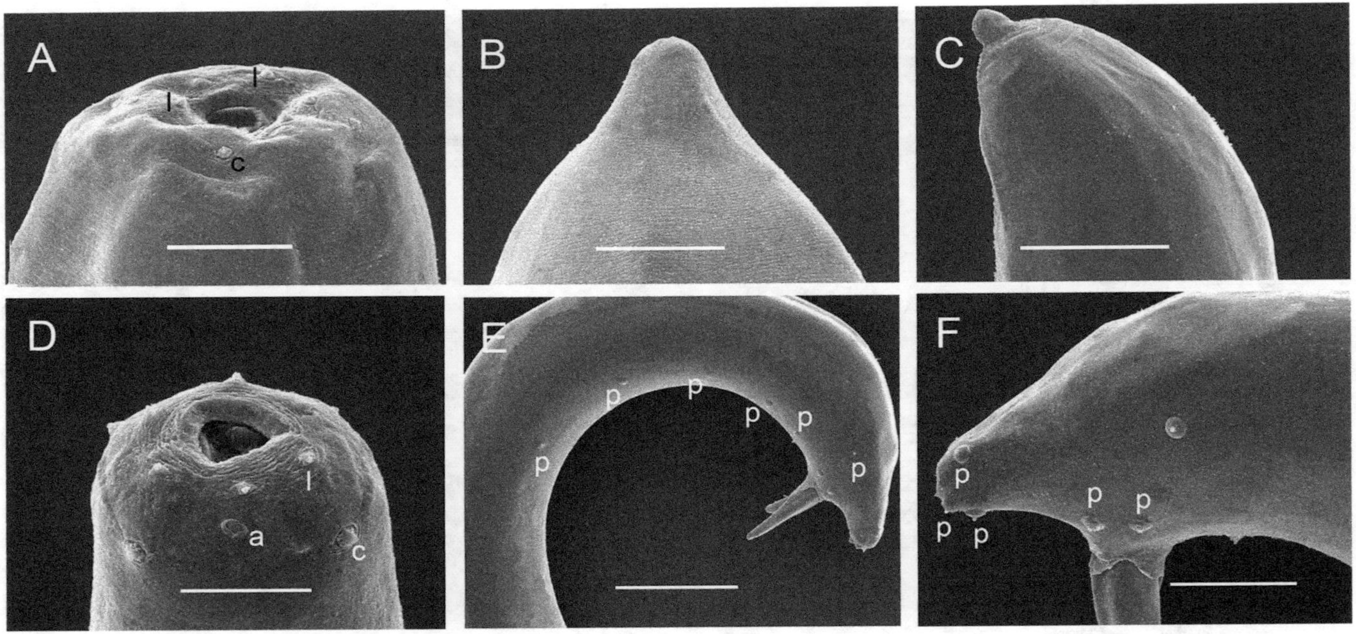

Fig. 120. Steinernema leizhouense. *SEM of first generation female and male. A: Anterior region of female (c = cephalic papilla, l = labial papilla); B, C: First generation female tail; D: Face of first generation male (a = amphidial aperture, c = cephalic papilla, l = labial papilla); E, F: First generation male tails showing distribution of genital papillae (p). (Scale bars: A = 12 μm; B = 16.7 μm; C = 60.0 μm; D = 6.0 μm; E = 37.5 μm; F = 16.7 μm.) After Nguyen* et al. *(2006), courtesy of* Russian Journal of Nematology.

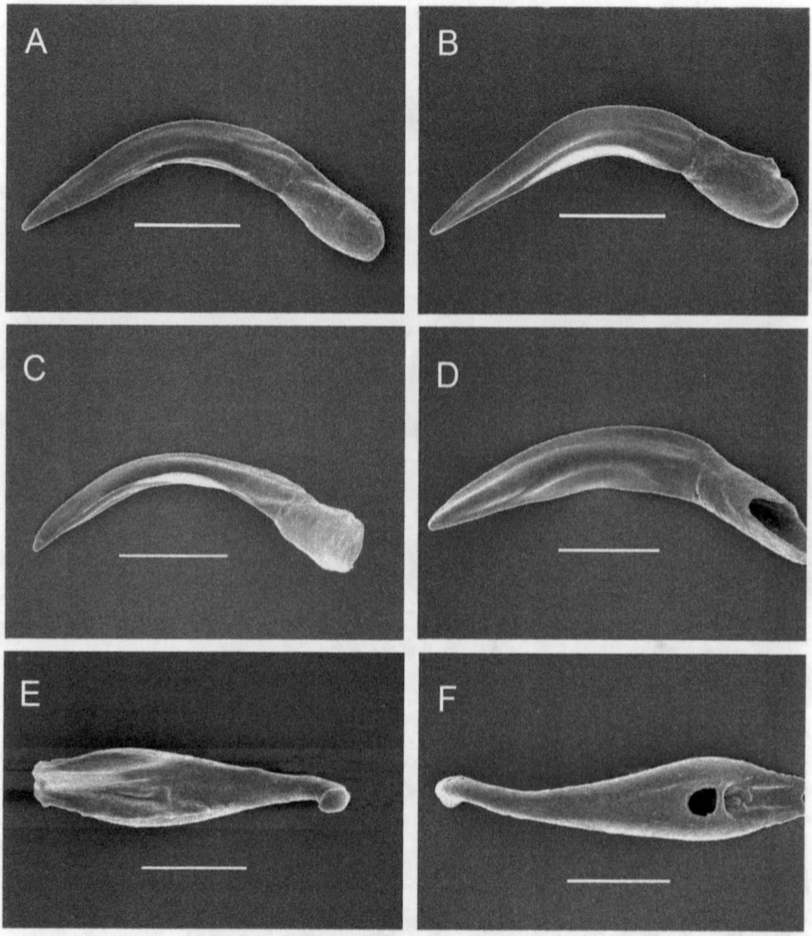

Fig. 121. Steinernema leizhouense. *SEM of spicule and gubernaculum of first generation males. A-D: Variations of spicule shape; E, F: Gubernacula showing ventral and dorsal views. (Scale bars: A, B = 23.1 μm; C, D = 20.0 μm; E, F = 15.0 μm.) After Nguyen* et al. *(2006), courtesy of* Russian Journal of Nematology.

and to nerve ring, excretory duct sclerotised anteriorly. Genital system amphidelphic, reflexed, often containing many eggs. Vulva in form of a transverse slit; mostly protuberant; small double-flapped epiptygmata occasionally visible; posterior lip larger than anterior (75% of specimens). Body diam. immediately posterior to vulva usually larger than that anterior. Vagina sclerotised. Tail shape variable, nipple-shaped, with

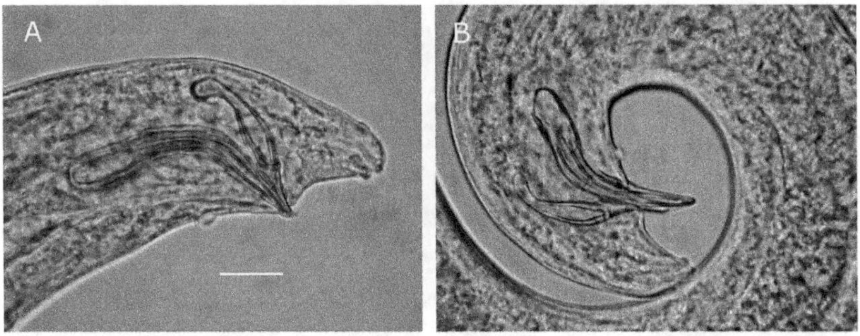

Fig. 122. Steinernema leizhouense *male tails. A: First generation male tail; B: Second generation male tail. (Scale bar: A = 20 μm; B = 18.5 μm.) After Nguyen* et al. *(2006), courtesy of* Russian Journal of Nematology.

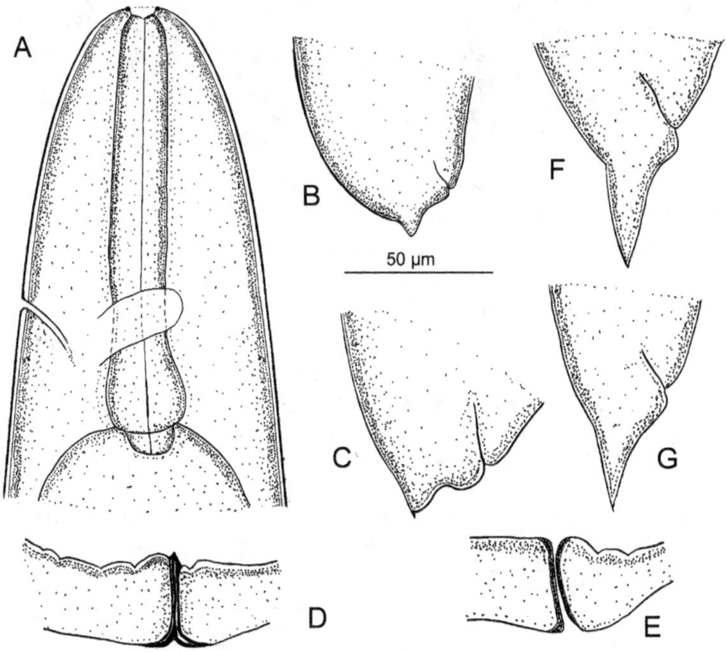

Fig. 123. Steinernema leizhouense *female. A-C: First generation female anterior region and variations in tail shape; D, E: Variations of vulva shape; F, G: Variation of tail shape of second generation females. After Nguyen* et al. *(2006), courtesy of* Russian Journal of Nematology.

blunt terminus, occasionally (*ca* 10% of specimens) a mucron-like structure present on dorsal side of tail terminus (seen more often on young

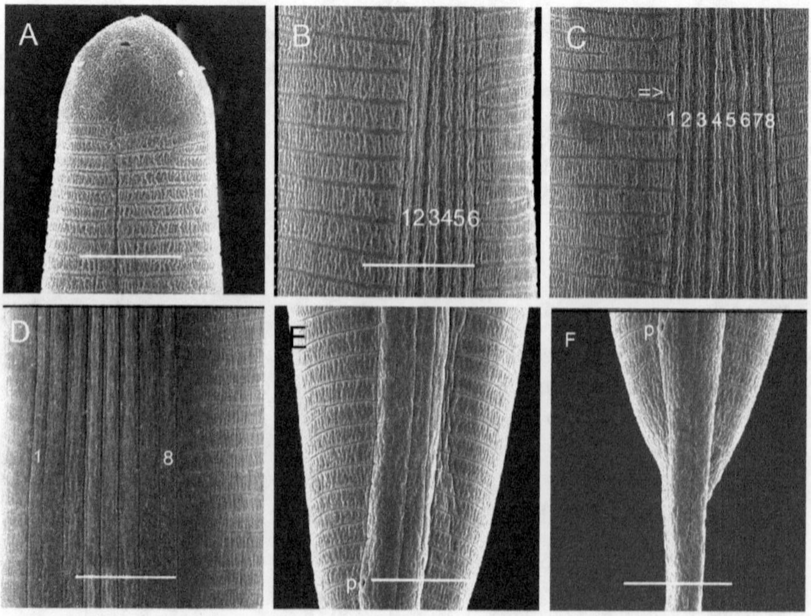

Fig. 124. Steinernema leizhouense. *SEM of lateral field pattern of infective juvenile. A: Anterior region showing smooth head, amphidial aperture and single line in lateral field; B: Lateral field showing number of ridges changing from two to six; C: Lateral field showing a new line appearing (arrow), and middle ridge dividing into two, making eight ridges in all; D: Lateral field with eight ridges; E: Lateral field near anus showing four ridges and phasmid (p); F: Posterior region showing tail shape, lateral field with two ridges and phasmid. (Scale bars: A = 6.0 μm; B, C, E, F = 6.7 μm; D = 7.5 μm.) After Nguyen* et al. *(2006), courtesy of* Russian Journal of Nematology.

Table 35. *Morphometrics of* Steinernema leizhouense. *Measurements are in μm and in the form: mean ± SD (range).*

Character	First generation			Second generation		Infective juvenile
	Holotype (male)	Male	Female	Male	Female	
n	–	20	20	20	20	25
L	1480	1525 ± 197 (1225-1950)	4347 ± 1577 (2500-8900)	1232 ± 101 (1090-1439)	2265 ± 403 (91692-3421)	983 ± 35 (900-1050)
a	–	–	–	–	–	28 ± 4.8 (23-38)
b	–	–	–	–	–	7.5 ± 0.7 (6.8-8.9)

Table 35. *(Continued).*

Character	First generation			Second generation		Infective juvenile
	Holotype (male)	Male	Female	Male	Female	
c	–	–	–	–	–	13.9 ± 2 (11.0-16.8)
V	–	–	52.5	–	52.5 (49-57)	–
Body diam.	105	104 ± 27 (75-143)	219 ± 30 (170-285)	66 ± 6.5 (55-79)	162 ± 22 (114-211)	38 ± 2.3 (35-43)
EP	125	135 ± 11 (118-158)	123 ± 10 (105-150)	94 ± 14 (59-120)	90 ± 16 (68-118)	83 ± 2.6 (80-88)
NR	140	148 ± 15 (125-173)	185 ± 18 (145-218)	116 ± 14 (82-139)	150 ± 24 (118-203)	101 ± 4.1 (95-108)
ES	170	177 ± 12 (148-195)	256 ± 18 (218-288)	161 ± 16 (142-187)	221 ± 33 (174-312)	133 ± 5.7 (125-143)
Testis reflexion	–	–	–	209 ± 60 (164-307)	–	–
Tail length (T)	37	35 ± 5.6 (28-50)	60 ± 9 (50-75)	34 ± 5 (27-48)	59 ± 7 (50-70)	66 ± 4.9 (58-75)
Anal body diam. (ABD)	43	40 ± 4.9 (33-48)	63 ± 11 (53-100)	39 ± 4.5 (33-47)	49 ± 6 (33-56)	26 ± 1.3 (58-75)
Spicule length (SL)	83	82 ± 4.9 (75-93)	–	67 ± 15 (61-79)	–	–
Spicule width	–	–	–	10 ± 1.5 (9-13.6)	–	–
Gubernaculum length (GL)	54	55±3.3 (50-60)	–	48 ± 3.6 (42-55)	–	–
Gubernaculum width	–	–	–	7.9 ± 0.8 (7.6-10.6)	–	–
D% = EP/ES × 100	74	75 ± 5 (67-83)	–	59 ± 7 (43-71)	–	59 ± 3.4 (54-63)
E% = EP/T × 100	–	–	–	–	–	120 ± 12.6 (105-140)
SW% = SL/ABD × 100	193	213 ± 0.23 (177-251)	–	180 ± 2 (150-230)	–	–
GS% = GL/SL × 100	65	67 ± 4 (60-74)	–	59 ± 7 (43-71)	–	–
H% = Hyaline/T × 100	–	–	–	–	–	39 ± 3.4 (33-45)

EP = distance from anterior end to excretory pore; NR = distance from anterior end to nerve ring; ES = distance from anterior end to end of pharynx.
After the original authors.

females). Ventral postanal swelling present but small; tail shorter than anal body diam. Lateral fields and phasmid not observed.

Female (second generation)

Similar to first generation female but smaller. Vulva not as protuberant, epiptygmata absent. Tail tapering to a pointed end, longer than anal body diam.; ventral postanal swelling present.

Infective juvenile

Body elongate. Sheath (= second-stage cuticle) present immediately after harvesting, but often lost in storage. Cephalic region not striated, rounded or slightly angular anteriorly, continuous with body. Labial papillae not seen; four prominent cephalic papillae present. Amphidial apertures slit-like, prominent under SEM. Cuticle marked with prominent transverse striations. Both hemizonid (anterior to excretory pore) and hemizonion (posterior to excretory pore) present. Deirid not observed under either light microscope or SEM. Lateral field beginning anteriorly with one line at first annule. Two additional lines appearing at annules 7-8 forming two ridges. Near excretory pore level, number of ridges in lateral fields increasing from two to six. Near end of pharynx, one marginal line appearing and central ridge dividing into two, making a total of eight ridges, the maximum number in lateral field. Portion of lateral field with eight ridges is longest part (compared to portions with 2, 6, 4 ridges). Near anus, number of ridges reducing to four. Some annules anterior to phasmid, four ridges in lateral field becoming two. With above description, formula of lateral field is 2, 6, 8, 4, 2. Pharynx with narrow corpus, basal bulb more or less elongate with visible valve. Tail attenuate, tapering abruptly. Phasmid present near mid-tail, close to ventral marginal line of lateral field. Hyaline region short, comprising 38 (33-45)% of tail length.

TYPE HOST AND LOCALITY

The type host of this nematode in nature is unknown as it was recovered from soil using *Galleria* larvae as bait. The soil sample was collected in a eucalypt plantation in Tangjia town (20.8°N, 109.8°E), Leizhou, Guangdong province, People's Republic of China.

TYPE MATERIAL

Holotype male, allotype female, ten paratype first generation males, five paratype females, 15 paratype infective juveniles and other population slides deposited in the State Key Lab for Biocontrol, College of Life

Sciences, Zhongshan University, Guangzhou 510275, People's Republic of China. Living infective juveniles are also preserved in liquid nitrogen in the nematode collection of SKLB, Zhongshan University, People's Republic of China. Several paratype specimens deposited in the USDA Nematode Collection, Beltsville, MD, USA.

RELATIONSHIPS

Steinernema leizhouense can be recognised by the large IJ body diam. of 38 (35-43) μm, distance from anterior end to nerve ring = 101 (95-108) μm; pharynx length = 133 (125-143) μm, and a = 28 (23-38). The lateral field pattern is variable, the formula for the arrangement of ridges from head to tail being 2, 6, 8, 4, 2 (Fig. 124). The hyaline region comprises 38 (33-45)% of tail length. This species can be recognised further by male spicule shape, spicule length averaging 82 (75-93) μm, spicule head longer than wide, spicule blade thin, velum present but narrow (Fig. 121) and the ratios SW = 210 and GS = 50. The female is characterised by the nipple-shaped tail terminus (Fig. 120).

Steinernema leizhouense can be distinguished from the closest species, *S. loci* and *S. aciari*, by characteristics of the infective juveniles, males and females. Compared to *S. loci*, IJ of both nematodes have almost the same length but those of *S. leizhouense* have a shorter pharynx of 133 (125-143) *vs* 141 (126-155) μm in *S. loci*, shorter tail, 66 (58-75) *vs* 75 (66-83) μm, and higher E% of 120 (105-140) *vs* 107 (94-120). Males of *S. leizhouense* have a differently shaped (Figs 119, 120 compared to Fig. 2C in Phan *et al.* (2001)) and longer spicule, 82 (75-93) *vs* 71 (60-80) μm, longer gubernaculum, 55 (50-60) *vs* 46 (40-52) μm, and higher SW%, 213 (177-251) *vs* 190 (170-210) (Table 35). Compared to *S. aciari*, IJ of *S. leizhouense* have a shorter body length, 983 (900-1050) *vs* 1113 (975-1250) μm. Because of the shorter body length, all other measurements are smaller, especially tail length, 66 (58-75) *vs* 78 (68-88) μm. Males of *S. leizhouense* have a differently shaped spicule and higher SW% of 213 (177-251) *vs* 204 (180-240).

MOLECULAR CHARACTERISATION

The species is characterised genetically by sequence length and composition of ITS regions (1032 bp, ITS1 = 300 bp, ITS2 = 325 bp and A = 0.252, C = 0.196, G = 0.250 and T = 0.301) (accession no. AY170340), and a partial 28S rDNA sequence (446 bp, A = 0.238,

C = 0.189, G = 0.318, T = 0.255) (accession no. AY169557). The phylogenetic tree of ITS regions showed that *S. leizhouense* and *S. loci* formed a monophyletic group well supported by bootstrap analysis. Additionally, the alignment of ITS regions (not shown) of seven nematode species of the same clade (*S. aciari*, *S. diaprepesi*, *S. glaseri*, *S. guangdongense*, *S. leizhouense*, *S. loci* and *S. longicaudum*) shows that *S. leizhouense* has 28 diagnostic character states (= number of characters in the same column of the alignment present in one sequence, but not in others) and differs from its sister taxa, *S. loci* at 155, and *S. aciari* at 136 total characters of the ITS sequence. Tracing process from MacClade, using a phylogenetic tree of ITS regions of *glaseri*-group with *S. intermedium* as an outgroup, showed that the number of autapomorphies that changed unambiguously on branches was seven for *S. leizhouense*.

The species can also be differentiated from other closely related nematodes by characteristics of its partial 28S sequence. The phylogenetic tree of the partial sequence of 28S shows that *S. leizhouense* and *S. aciari* comprise a monophyletic clade which is sister to the *glaseri*-group. This clade is well supported by bootstrap analysis (Nguyen *et al.*, 2006).

DISTRIBUTION

Steinernema leizhouense is known only from the type locality in Guangdong province, People's Republic of China (Nguyen *et al.*, 2006).

References

NGUYEN, K.B., QIU, L., ZHOU, Y. & PANG, Y. (2006). *Steinernema leizhouense* sp. n. (Nematoda: Steinernematidae), a new entomopathogenic nematode from southern China. *Russian Journal of Nematology* 14, 101-118.

PHAN, K.L., NGUYEN, N.C. & MOENS, M. (2001). *Steinernema loci* sp. n. and *Steinernema thanhi* sp. n. (Rhabditida: Steinernematidae) from Vietnam. *Nematology* 3, 503-514.

Steinernema litorale[*,**] Yoshida, 2004
(Figs 125-129)

MEASUREMENTS

See Table 36.

DESCRIPTION

Male (first generation)

Body J-shaped when heat-relaxed, much smaller and more slender than female. Cuticle, cephalic region, stoma and pharyngeal region similar to female. Excretory pore located between posterior part of metacorpus and nerve ring. Genital system monorchic, reflexed. Spicules paired, heavy, moderately arcuate; manubrium usually elongate, 1.9 (1.5-2.5) times as long as wide, and 28 (24-37)% as long as spicule. Manubrium variable in shape from oval to somewhat angular, rectangular; shaft indistinct or reduced; rostrum usually reduced or absent, sometimes slightly more prominent; velum occasionally absent, usually extending obliquely from indistinct rostrum or proximal part of lamina to distal part of lamina; lamina with two internal ribs, proximal part of lamina often wider than manubrium, slightly tapering to blunt terminus. Gubernaculum variable in form, bow-shaped, narrow distally, usually with constriction between proximal portion and bow-shaped body, proximal portion straight or ventrally directed in lateral view, variable in shape (elongate plate-like, swollen, triangular, hook- or knob-like). Spicules and gubernaculum usually orange-brown to yellow in colour. Complement of genital papillae comprising 11 pairs plus a single unpaired mid-ventral papilla. Eight pairs of papillae located in subventral position, four pairs of which are anterior to retracted spicules, two pairs within range of spicules and two pairs posterior to cloacal aperture; a subdorsal/dorsosublateral pair located adcloacally, another subdorsal pair just anterior to tail terminus and another subdorsal pair slightly more anterior. Tail conical, with postcloacal ventral depression, tapering to rounded terminus with minute spine; terminal spine rarely absent.

[*] From Latin, *litoralis*, referring to the coastal habitat from where the species was recovered.

[**] Description after Yoshida (2004). Revised by Zdeněk Mráček.

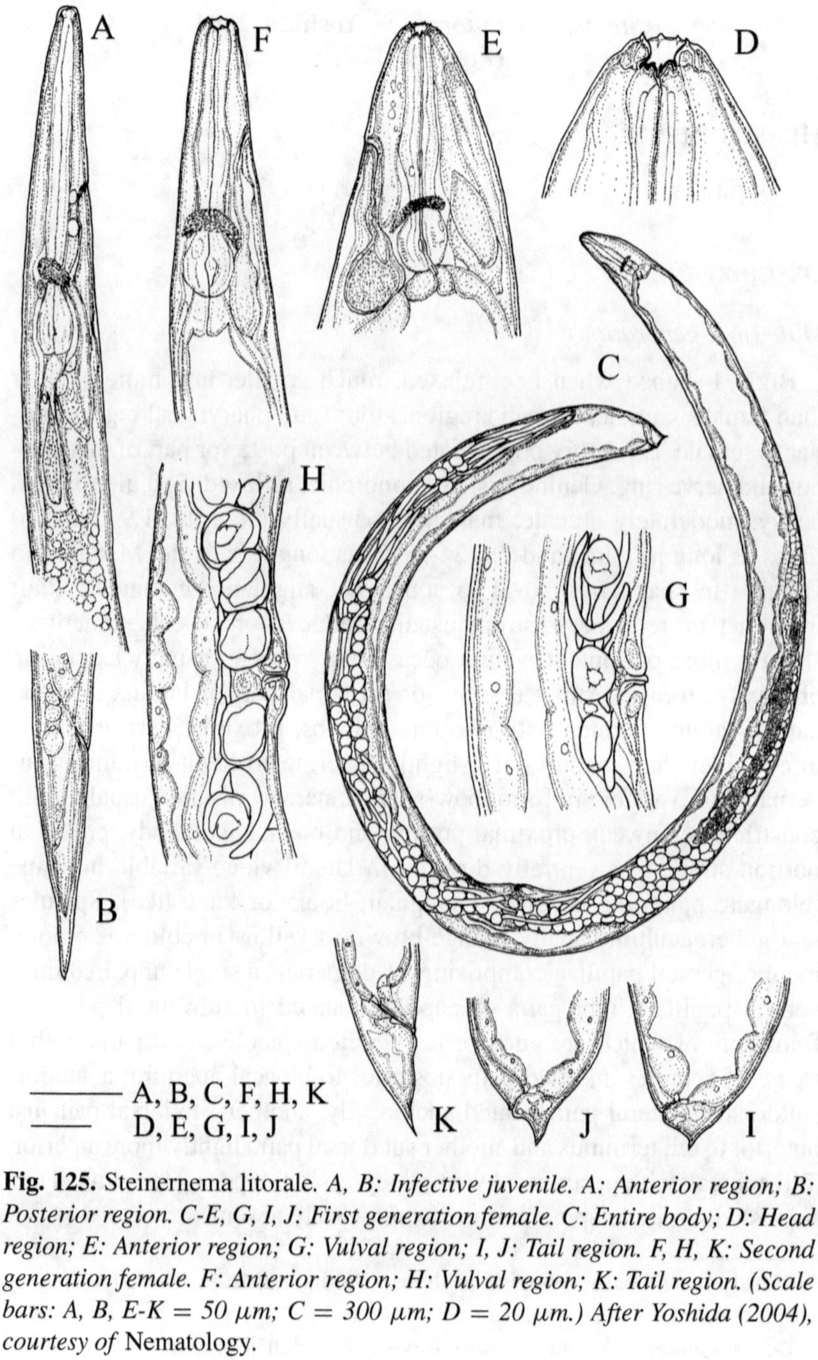

Fig. 125. Steinernema litorale. *A, B: Infective juvenile. A: Anterior region; B: Posterior region. C-E, G, I, J: First generation female. C: Entire body; D: Head region; E: Anterior region; G: Vulval region; I, J: Tail region. F, H, K: Second generation female. F: Anterior region; H: Vulval region; K: Tail region. (Scale bars: A, B, E-K = 50 μm; C = 300 μm; D = 20 μm.) After Yoshida (2004), courtesy of* Nematology.

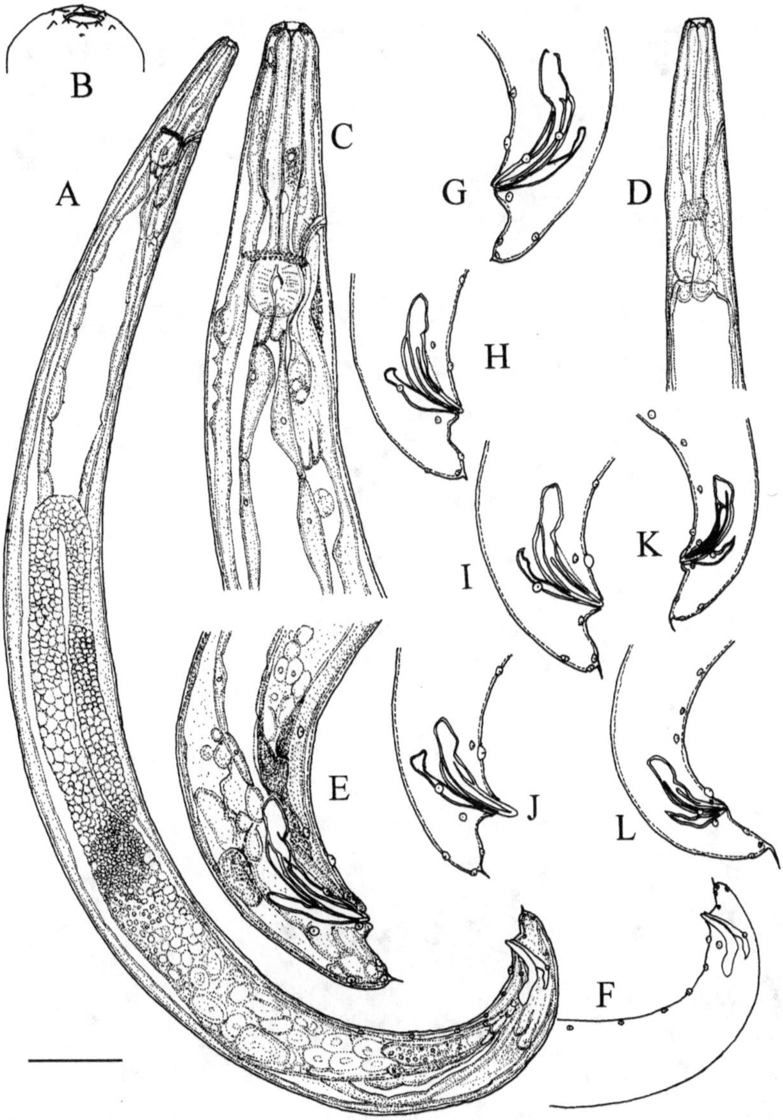

Fig. 126. Steinernema litorale. *A-C, E-J: First generation male. A: Entire body; B: Head region; C: Anterior region; E: Posterior region; F: Posterior region showing arrangement of genital papillae; G-J: Tail region showing variation of typical shaped spicule with various gubernaculum shapes. D, K, L: Second generation male. D: Anterior region; K, L: Tail region showing variation of spicule and gubernaculum. (Scale bar: A, F = 100 μm; B = 20 μm; C-E, G-L = 50 μm.) After Yoshida (2004), courtesy of* Nematology.

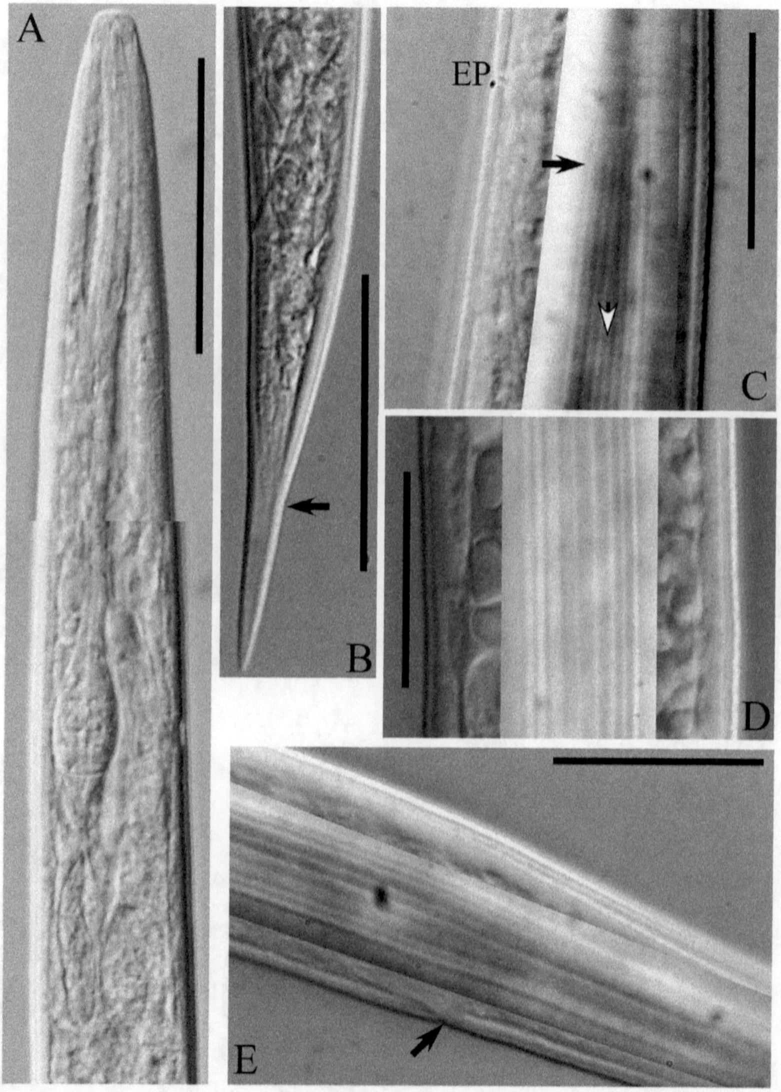

Fig. 127. Steinernema litorale. *A-E: Infective juvenile. A: Anterior region, from head to bacterial vesicle; B: Posterior region; C: Lateral field at anterior region, from anterior part of excretory pore (EP) to upper part of nerve ring region (solid arrow indicates a region where eight lines are discernible and open arrow indicates ninth line appearing in vicinity of posterior part of median bulb); D: Lateral field at mid-body; E: Lateral field at anal (arrow) region. (Scale bars: A, B = 50 μm; C-E = 20 μm.) After Yoshida (2004), courtesy of* Nematology.

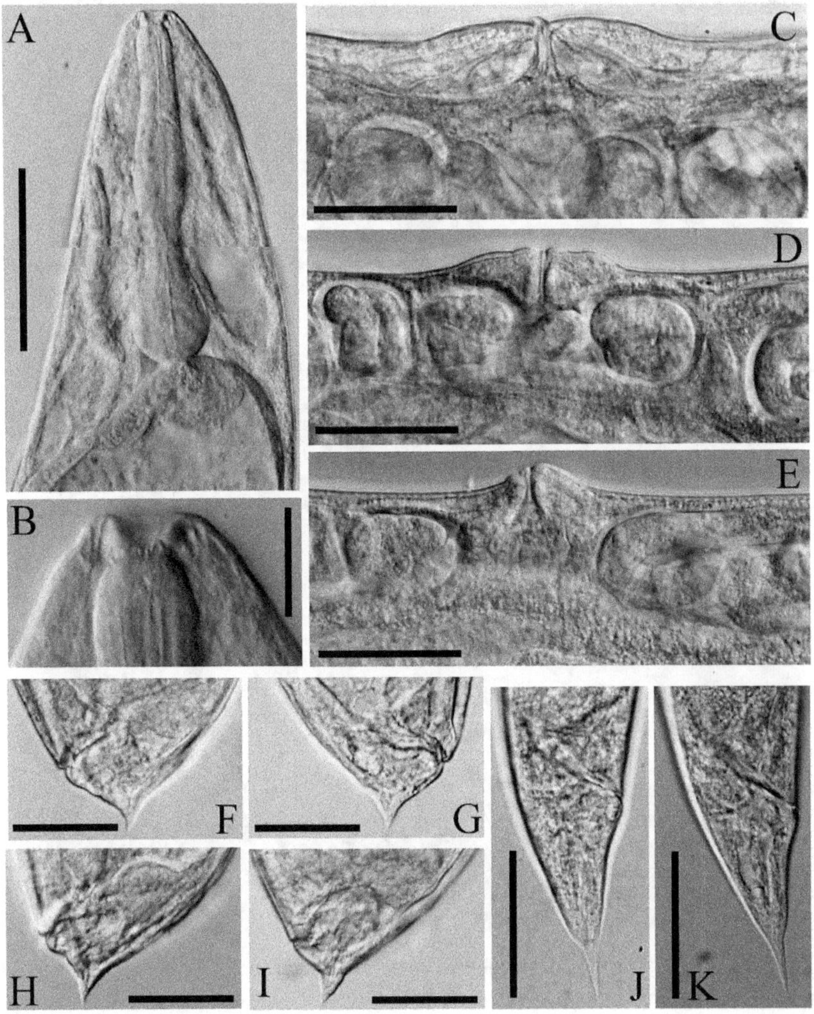

Fig. 128. Steinernema litorale. *A-C, F-I: First generation female. A: Anterior region; B: Head region showing tooth-like structure at bottom of stoma; C: Vulval region showing double flapped epiptygmata; F-I: Tail region showing variation of tail spine. D, E, J, K: Second generation female. D, E: Vulval region showing double flapped epiptygmata; J, K: Tail region. (Scale bars: A = 100 μm; B = 20 μm; C-K = 50 μm.) After Yoshida (2004), courtesy of* Nematology.

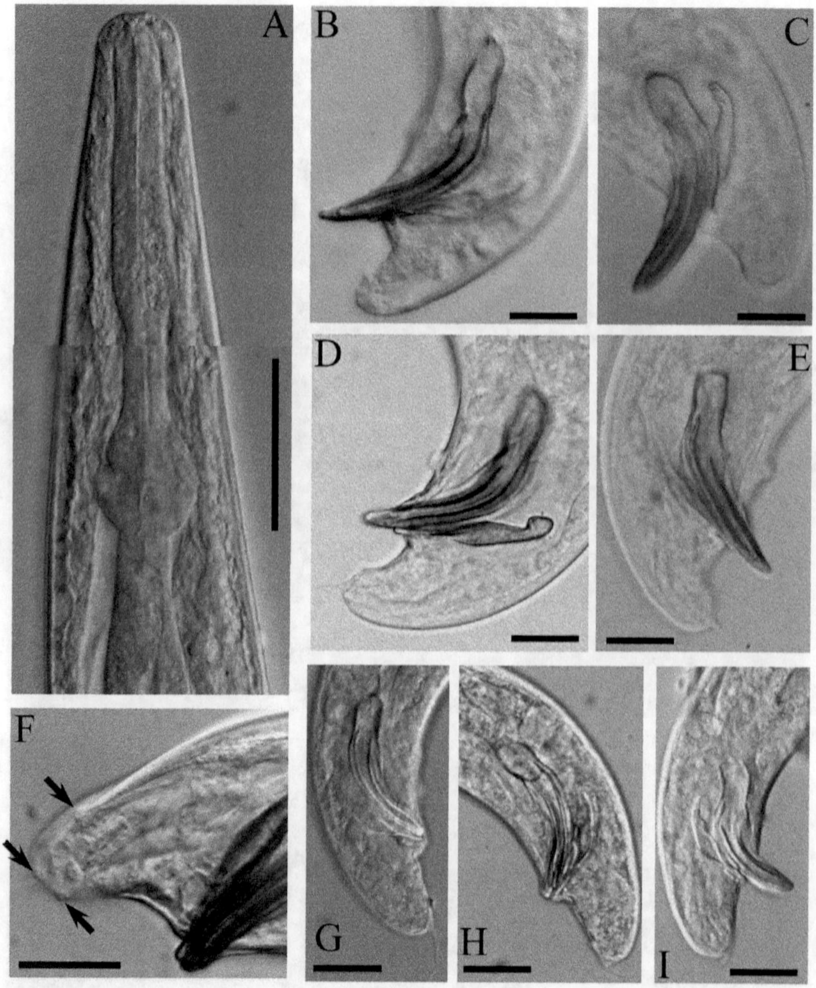

Fig. 129. Steinernema litorale. *A-F: First generation male. A: Anterior region; B-E: Tail region showing variation of spicule; F: Distal part of tail region showing two subdorsal papillae and a subventral papilla (arrows). G-I: Second generation male. G-I: Tail region showing variation of spicule. (Scale bars: A = 50 μm; B-I = 20 μm.) After Yoshida (2004), courtesy of* Nematology.

Male (second generation)

Generally similar to first generation male, but body smaller. Excretory pore located in metacorpus region. Spicules paired, moderately arcuate, smaller than in first generation; manubrium usually elongated, 1.5 (0.7-

2.3) times as long as wide, and 27 (17-38)% as long as spicule. Tail conical, usually without postcloacal ventral depression, tapering to rounded terminus with minute spine, longer than in first generation; terminal spine rarely absent.

Female (first generation)

Body long and robust, spiral to C-shaped when heat-relaxed. Cuticle smooth, lateral fields and phasmids inconspicuous. Cephalic extremity truncate to rounded, continuous with body contour. Six lips, each with terminal papilla; four cephalic papillae located further posterior on head in submedian position. Amphidial apertures often obscure, small, pore-like, located laterally between circlets of labial and cephalic papillae. Stoma short, broad, with sclerotised walls, minute tooth-like structure present at bottom of stoma. Pharynx muscular with cylindrical procorpus, slightly swollen, non-valvate, metacorpus, distinct isthmus, and pyriform basal bulb with distinct valve. Cardia well developed, projecting into intestine. Excretory pore circular, located in anterior region of metacorpus. Excretory glands large, extending to anterior portion of intestine. Nerve ring surrounding anterior portion of basal bulb. Genital tract amphididelphic with opposed, reflexed, ovaries. Vulva in form of a transverse slit. Vulval lips slightly protuberant, with double flapped epiptygmata. Eggs either deposited or juveniles hatch inside female body. Tail conoid, shorter than anal body diam., usually with postanal swelling, abruptly tapering to rounded terminus with mucronate, pointed, projection on tip.

Female (second generation)

Generally similar to first generation female, but much smaller in body length and diam. Body moderately curved ventrally to C-shaped when relaxed by heat. Vulval lips sometimes protuberant, with double flapped epiptygmata. Tail conoid, tapering gradually with slight postanal protuberance, distal half often slightly swollen, terminating with an offset spine, and longer than anal body diam.

Infective juvenile

Body straight to slightly curved when heat-relaxed, gradually tapering towards anterior and posterior ends. Body diam. at excretory pore, labial region and anus equivalent to 68 (62-73), 26 (22-31) and 60 (55-69)%, respectively, of maximum body diam. Lateral fields with eight lines at excretory pore level, a ninth additional central line appearing at *ca*

Table 36. *Morphometrics of* Steinernema litorale. *Measurements are in μm and in the form: mean $\pm$ SD (range).*

Character	Holotype (IJ)	First generation		Second generation		Infective juvenile
		Male	Female	Male	Female	
n	–	25	25	25	25	25
L	908	1360 ± 84 (1230-1514)	4462 ± 325 (3930-5048)	718 ± 46 (595-803)	1316 ± 84 (1162-1530)	909 ± 42 (834-988)
a	28	14 ± 0.9 (12-16)	23 ± 1.4 (21-26)	16 ± 1.2 (13-17)	17 ± 1.2 (15-21)	29.5 ± 0.8 (27-31)
b	7.4	9.3 ± 0.5 (8.3-10)	23 ± 1.5 (20-26)	5.5 ± 0.4 (4.3-6.3)	8.5 ± 0.6 (7.5-9.5)	7.3 ± 0.3 (6.7-7.9)
c	12	41 ± 5.6 (33-56)	117 ± 21 (78-157)	25 ± 2.3 (20-29)	20 ± 1.5 (18-23)	11 ± 0.5 (10-12)
c′	4.1	0.8 ± 0.1 (0.6-0.9)	0.6 ± 0.1 (0.5-0.9)	0.9 ± 0.1 (0.7-1.2)	1.7 ± 0.1 (1.5-2.0)	4.5 ± 0.4 (3.8-5.4)
V	–	–	56 ± 2.2 (50-60)	–	53 ± 1.2 (50-55)	–
Body diam.	32	96 ± 7.9 (82-111)	191 ± 9.1 (175-215)	47 ± 2.1 (43-49)	78 ± 5.4 (67-86)	31 ± 1.4 (28-33)
EP	62	96 ± 7.0 (77-107)	88 ± 12 (65-105)	53 ± 6.2 (44-75)	60 ± 2.8 (57-67)	61 ± 3.3 (54-69)
NR	96	114 ± 11 (94-128)	146 ± 9.1 (130-165)	92 ± 5.7 (79-109)	104 ± 4.0 (99-114)	96 ± 4.0 (89-104)
ES	124	147 ± 7.6 (133-163)	196 ± 8.2 (185-213)	132 ± 8.0 (119-146)	155 ± 6.3 (143-170)	125 ± 4.6 (114-133)
Testis reflexion	–	290 ± 41 (212-380)	–	148 ± 26 (84-183)	–	–
Tail length (T)	79	34 ± 3.7 (26-41)	39 ± 7.6 (25-60)	29 ± 3.2 (25-38)	66 ± 3.9 (57-74)	83 ± 4.5 (72-91)
Hyaline (H)	30	–	–	–	–	28 ± 3.6 (22-37)
Anal body diam. (ABD)	19	43 ± 3.0 (37-49)	62 ± 4.7 (55-75)	31 ± 2.2 (27-37)	38 ± 2.7 (33-44)	19 ± 1.5 (16-22)
Spicule length (SL)	–	75 ± 4.8 (67-89)	–	48 ± 4.0 (41-58)	–	–
Spicule width	–	12 ± 1.1 (10-14)	–	10 ± 1.1 (8.1-12)	–	–
Gubernaculum length (GL)	–	21 ± 1.9 (18-25)	–	13 ± 2.8 (8.1-19)	–	–
Gubernaculum width	–	7.0 ± 0.8 (5.1-8.1)	–	6.0 ± 0.8 (3.4-6.8)	–	–

posterior region of median bulb or anterior region of isthmus. Nine distinct lines (*i.e.*, eight ridges, with four inner ridges wider than two outer ridges at mid-body) from nerve ring region to rectum region. Cephalic extremity dome-like, sometimes slightly set off from body

Table 36. *(Continued).*

Character	Holotype (IJ)	First generation		Second generation		Infective juvenile
		Male	Female	Male	Female	
D% = EP/ES × 100	50	65 ± 4.4 (55-72)	45 ± 6.5 (33-57)	40 ± 4.5 (34-56)	39 ± 5.5 (35-44)	49 ± 2.9 (44-56)
E% = EP/T × 100	78	286 ± 40 (231-382)	230 ± 45 (171-390)	186 ± 27 (139-277)	91 ± 8.0 (77-113)	73 ± 5 (68-84)
SW% = SL/ABD × 100	–	174 ± 13 (154-200)	–	159 ± 19 (124-194)	–	–
GS% = GL/SL × 100	–	71 ± 4 (62-81)	–	65 ± 8 (50-79)	–	–
H% = H/T × 100*	38	–	–	–	–	33

EP = distance from anterior end to excretory pore; NR = distance from anterior end to nerve ring; ES = distance from anterior end to end of pharynx.

* Recalculated according to Yoshida (2004).

contour, truncate. Pharynx narrow, with visible lumen in anterior part, median bulb indistinct or absent, basal bulb slender, pyriform, with valves, cardia distinct. Excretory pore situated anterior to middle of pharynx, with wide duct; excretory glands on ventral side in region of pharyngo-intestinal junction, extending up to region of bacterial vesicle. Nerve ring located *ca* one body diam. anterior to basal bulb. Hemizonid in region of basal bulb, located *ca* 90 (83-97)% of pharynx length from apical extremity. Bacterial vesicle in anterior region of intestine, elongated oval or drop shaped, with thin walls. Lumen of intestine indistinct. Rectum straight, *ca* one anal body diam. long, or less. Tail elongate conoid, almost straight on ventral side, tapering gradually on dorsal side, sometimes with dorsal constriction at beginning of hyaline region, 3.8-5.4 times as long as anal body diam. Hyaline region not clearly demarcated, extending for *ca* 34 (27-44)% of total tail length. Tail terminus pointed, refractile spine absent.

Type host and locality

Natural host unknown. The type isolate (code: AiAt199) was obtained in July 1993 by using a bait insect, *Galleria mellonella*, from a soil sample collected on 17 March 1993 at the type locality. The type locality is a pine-dominated coastal forest on sandy soil at Cape Irago-zaki, Atsumi (Atsumi Peninsula), Mikawa-bay Quasi-National Park, Aichi Prefecture, Honshu Island, Japan.

TYPE MATERIAL

Holotype infective juvenile, 25 paratype first generation males, 25 paratype first generation females and 25 paratype infective juveniles deposited in the nematode collection at the Herbarium and Insect Museum of the National Institute for Agro-Environmental Sciences, Tsukuba, Ibaraki, Japan. Five paratype first generation males, five paratype first generation females and ten paratype infective juveniles deposited in the University of California-Davis Nematode Collection, Davis, CA, USA, and in the CABI Europe-UK Nematode Collection, Egham, Surrey, UK.

RELATIONSHIPS

Steinernema litorale is characterised by the following morphological features. Infective juvenile: body length $= 909 \pm 42$ (834-988) μm, distance from head to excretory pore $= 61 \pm 3.3$ (54-69) μm, tail length $= 83 \pm 4.5$ (72-91) μm, E% $= 73 \pm 5$ (68-84), and lateral field with nine distinct lines (eight ridges, central four ridges wider than outer two ridges in both side). Male: presence of a tail mucron in both male generations, spicule shape (moderately arcuate, heavy, spicule with elongated manubrium), spicule length $= 75 \pm 4.8$ (67-89) μm and gubernaculum length $= 53 \pm 4$ (44-64) μm in first generation male. Female: the presence of a double flapped epiptygmata and a conoid tail with a pointed tip in both female generations, and D% $= 45.2 \pm 6.5$ (33.3-56.8) in the first generation female.

The range of body length (834-988 μm) of the infective juveniles of *S. litorale* is similar to *S. feltiae*, *S. thanhi*, *S. karii*, *S. scarabaei*, *S. kraussei*, *S. oregonense*, *S. loci* and *S. diaprepesi*, but the species can, be distinguished clearly from all of these species, with the exception of *S. feltiae*, by morphological characters.

Steinernema litorale can be distinguished from *S. diaprepesi* by: the shorter distance from the head to the excretory pore of 61 (54-69) *vs* 74 (66-83) μm and lower E% of 73 (63-84) *vs* 90 (78-114) in the infective juvenile; the tail tip structure (mucron present *vs* mucron absent) and lower D% of 65.2 (55.0-72.4) *vs* 80 (68-86) in the first generation male; and by the tail structure (conoid tail terminating with a pointed projection *vs* tail tip with five papilla-like structures) and lower D% of 45.2 (33.3-56.8) *vs* 80 (46-105) in the first generation female.

It can be distinguished from *S. loci* by: the shorter distance from the head to the excretory pore of 61 (54-69) *vs* 80 (71-86) μm and lower

E% of 73 (63-84) *vs* 107 (94-120) in the infective juvenile; the tail tip structure (mucron present *vs* absent) in the first generation male; and the vulval lip structure (double flapped epiptygmata present *vs* absent) and lower D% of 45.2 (33.3-56.8) *vs* 58 (44-77) in the first generation female.

Steinernema litorale is distinguished from *S. oregonense* by: the longer tail length of 83 (72-91) *vs* 70 (64-78) μm and lower E% of 73 (63-84) *vs* 100 (90-110) in the infective juvenile; the tail tip structure (mucron present *vs* absent) and spicule shape (elongated manubrium *vs* broad and slightly angular manubrium) in the first generation male; and by the vulval structure (double flapped epiptygmata present *vs* absent) and the tail tip form (pointed projection present *vs* rounded wedge-shaped projection) in the first generation female.

From *S. scarabaei* it differs by shorter distance between the head and the excretory pore of 61 (54-69) *vs* 77 (72-81.5) μm, lower D% of 48.5 (44.0-56.0) *vs* 60 (50-75); and lower E% of 73 (63-84) *vs* 100 (90-110) in the infective juvenile; spicule shape (moderately curved, elongated manubrium and the absent or indistinct rostrum *vs* curved, rhomboid, manubrium and rostrum present) in the first generation male; tail tip structure (mucron present *vs* absent) in the second generation male; and by the vulval structure (double flapped epiptygmata present *vs* absent) and the tail tip form (pointed projection present *vs* with a ventrally bifurcate mucron) in the first generation female.

Steinernema litorale can be distinguished from *S. karii* by: the shorter head to excretory pore distance of 61 (54-69) *vs* 76 (66-82) μm, E% = 73 (63-84) *vs* 100 and lateral field structure (nine *vs* eight distinct lines) in the infective juvenile; the tail tip form (mucron present *vs* absent) and spicule shape (moderately curved *vs* strongly curved) in the first generation male; and by the tail tip form (pointed projection *vs* ventral, digitate-like, peg) in the first generation female.

Steinernema litorale can be distinguished from *S. thanhi* by: shorter head to excretory pore distance of 61 (54-69) *vs* 75 (68-84) μm, the longer tail length of 83 (72-91) *vs* 63 (52-72) μm and lower E% value of 73 (63-84) *vs* 119 (101-138) in the infective juvenile; the tail tip form (mucron present *vs* absent) in the first generation male; and by the vulval structure (double flapped epiptygmata present *vs* absent), tail tip structure (pointed projection *vs* with a terminal peg) and lower D% value of 45.2 (33.3-56.8) *vs* 66 (58-84) in the first generation female.

Steinernema litorale can be distinguished from *S. kraussei* by: the lateral field structure (nine distinct lines *vs* eight lines, the submarginal lines being indistinct or absent) in the infective juvenile; the spicule shape (moderately curved, elongated manubrium and rostrum absent or indistinct *vs* well curved, square shaped manubrium and rostrum present), the longer spicule of 75 (67-89) *vs* 62.6 (54-70) μm in the Japanese isolate; the longer gubernaculum of 53 (44-64) *vs* 41.8 (30-48) μm in the Japanese isolate and higher SW% value of 174 (154-200) *vs* 134 (110-150) in the Japanese isolate in the first generation male; and by the vulval structure (double flapped epiptygmata present *vs* absent) in the first generation female.

It resembles the Japanese isolates of *S. feltiae* in many morphological characters. The only diagnostic character distinguishing *S. litorale* from *S. feltiae* is the lateral field (nine distinct lines *vs* nine lines, the submarginal lines being indistinct or absent) in the infective juvenile. The infective juvenile is also slightly broader than the Japanese isolates of *S. feltiae*, *i.e.*, 31 (28-33) *vs* 24.5-26.8 (23-30) μm.

This species can be readily distinguished from *S. weiseri*, which is closely related to *S. feltiae* in shape and RFLP profile, by the following characters: body length = 909 (834-988) *vs* 740 (586-828) μm; body diam. = 31 (28-33) *vs* 25 (24-29) μm); tail length = 83 (72-91) *vs* 60 (49-68) μm; lateral field (nine distinct lines, central four ridges wider than outer two ridges in both side at mid-body *vs* nine distinct lines spaced at approximately equal distances at mid-body) in the infective juvenile; higher D% value of 65.2 (55.0-72.4) *vs* 49 (39-60) in the first generation male; and the average mucron length of 10.3 (0-16.5) *vs* 2.5 (0-6.4) μm in the second generation male.

The morphological study and RFLP analysis indicated that there were many phenotypic similarities between *S. litorale* and *S. feltiae*. The biological study (cross-hybridisation tests) indicated the possibility that females of *S. feltiae* hybridised with the males of *S. litorale*. However, progeny were not generated in the reverse cross when *S. litorale* females were mated with *S. feltiae* males.

Molecular characterisation

The sequence analysis of ITS region of rDNA suggested the close relationship between this species and *S. feltiae* and that of Cytochrome oxidase I supported this species as being distinct from *S. feltiae*. In the field, both of these species were detected from the same sampling site in Hokkaido and habitat preference of the two species in Hokkaido are similar, *i.e.*, coastal grassland and sandy soil. Consequently, *S. litorale* can be considered morphologically, molecularly, biologically and ecologically as being closely related to *S. feltiae*.

The RFLP profiles of ITS region in rDNA produced by 17 restriction enzymes (Reid *et al.*, 1997) molecularly characterise *S. litorale*. In particular, the RFLP pattern of *Hin*fI separates *S. litorale* from all other described species for which RFLP data are available. The ITS region sequence is deposited at GenBank (accession no. AB243441). Its sequence length between primers 18S and 26S is 961 bp (ITS1 = 264 bp; ITS2 = 290 bp).

DISTRIBUTION

Steinernema litorale is widely distributed in the eastern part of Japan from the Atsumi Peninsula, Aichi Prefecture, Hokkaido. Most of the isolates were detected from sandy coastal soils, although two isolates were found inland, one in an apple orchard in a mountainous region, and the other from a grass site in a garden in a residential area. The inland isolates were collected from sites with intense human impact, while the coastal sites were 'buffer zones' (grassland or shelter belts) between the sandy beaches and those regions with a more intense human impact. Moreover, over 250 inland sites in Hokkaido and Honshu were checked for the presence of entomopathogenic nematodes and only the two sites mentioned above were positive for this species. As the species was recorded from both grassland and woodland, the associated vegetation has not yet been clarified (Yoshida, 2004).

References

REID, A.P., HOMINICK, W.H. & BRISCOE, B.R. (1997). Molecular taxonomy and phylogeny of entomopathogenic nematode species (Rhabditida: Steinernematidae) by RFLP analysis of the ITS region of rDNA repeat unit. *Systematic Parasitology* 37, 187-193.

YOSHIDA, M. (2004). *Steinernema litorale* n. sp. (Rhabditida: Steinernematidae), a new entomopathogenic nematode from Japan. *Nematology* 6, 819-838.

Steinernema loci[*,**] Phan, Nguyen & Moens, 2001

(Figs 130, 131)

MEASUREMENTS

See Table 37.

DESCRIPTION

Male (first generation)

Body curved ventrally or C-shaped when heat-relaxed. Cuticle smooth. Cephalic extremity rounded, slightly offset from body. SEM face view with six pointed labial papillae and four cephalic papillae. Amphidial apertures inconspicuous. Stoma funnel- or cup-shaped, shallow. Pharynx muscular: procorpus cylindrical; metacorpus slightly swollen, non-valvate; isthmus indistinct; basal bulb pyriform, valve not observed. Nerve ring surrounding isthmus or just anterior to basal bulb. Cardia prominent, protruding into intestine lumen. Excretory pore located between nerve ring and middle of pharynx. Excretory duct cuticularised, excretory gland swollen and elongate. Genital system monorchic, reflexed. Spicules paired, yellow-brownish in colour and ventrally curved; spicule head (manubrium) *ca* 33% of spicule length long and longer than wide. Velum thin. Gubernaculum *ca* 65% of spicule length, boat-shaped, swollen at middle and proximal end with ventrally curved knob. Single midventral precloacal papilla and 11 pairs of genital papillae arranged as follows: five pairs subventrally precloacal; one pair subdorsal at same level of single ventral precloacal papillae; two pairs subventral in cloacal region; two pairs caudal, subventral and one pair caudal, subdorsal. Tail conoid, mucron absent. Phasmids inconspicuous.

Male (second generation)

Data not available.

Female (first generation)

Body robust, spiral-shaped or C-shaped when heat-relaxed. Cuticle smooth. Cephalic extremity broadly rounded. SEM face view with six pointed labial papillae and four cephalic papillae. Amphidial apertures

---

[*] Named after Phan ke *Loc*, father of the first author.

[**] After Phan *et al.* (2001). Revised by Khuong B. Nguyen.

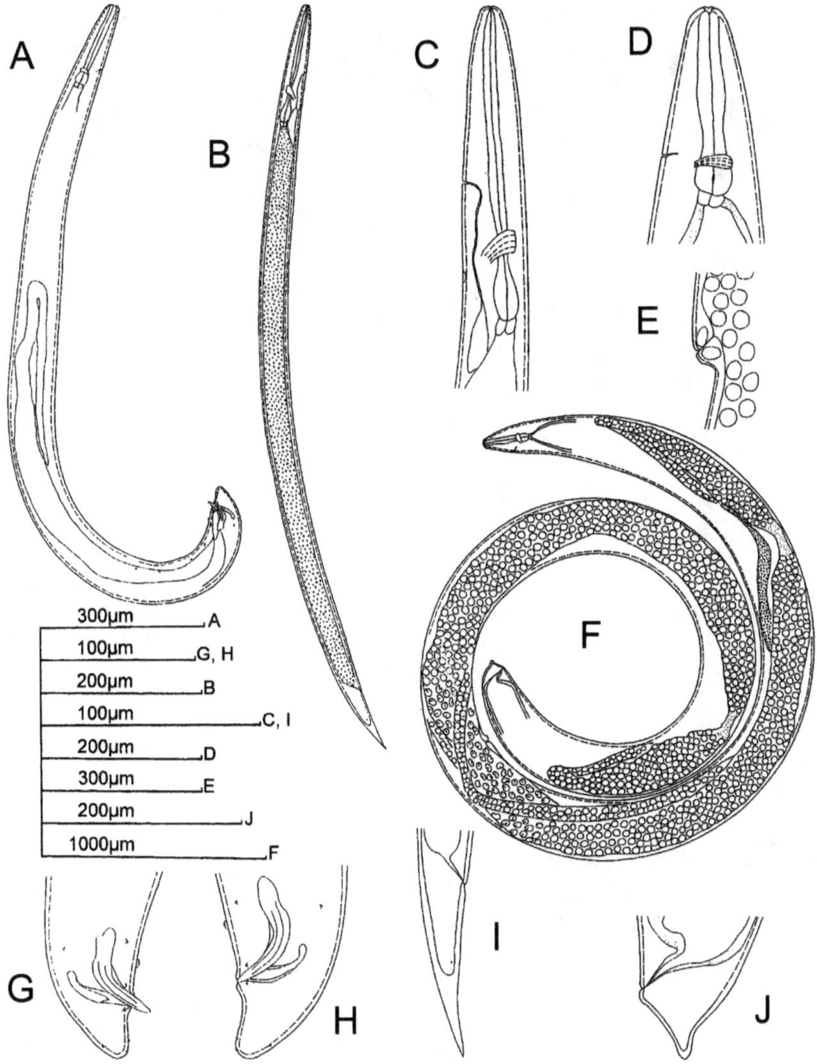

Fig. 130. Steinernema loci. *A, G, H: First generation male. A: Entire male; G, H: Tail in lateral view. D, E, F, J: First generation female. D: Pharyngeal region; E: Vulva region; F: Entire female; J: Tail, lateral view. B, C, I: Third-stage infective juvenile. B: Entire body; C: Pharyngeal region; I: Tail, lateral view. After Phan* et al. *(2001), courtesy of* Nematology.

inconspicuous. Stoma funnel- or cup-shaped, shallow. Pharynx with cylindrical procorpus; metacorpus slightly swollen and non-valvate;

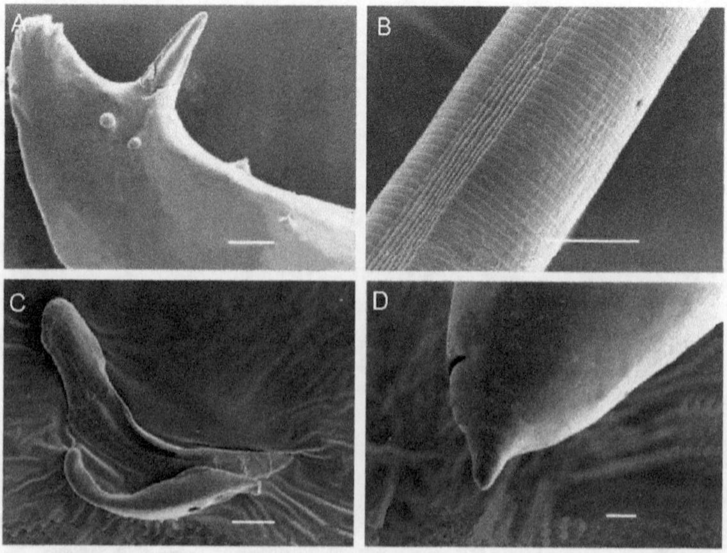

Fig. 131. Steinernema loci *SEM micrographs. A: Posterior region of first generation male showing posterior papillae; B: Lateral field of IJ showing seven ridges in anterior region; C: Spicule and gubernaculum; D: Female tail. (Scale bar = 10 μm.) Modified after Phan* et al. *(2001), courtesy of* Nematology.

Table 37. *Morphometrics of* Steinernema loci. *Measurements are in μm and in the form: mean ± SD (range).*

Character	First generation			Infective juvenile
	Holotype (male)	Male	Female	
n	–	20	20	50
L	1350	1397 ± 83 (1128-1500)	5426 ± 1541 (3630-9630)	985 ± 45 (896-1072)
a	15	–	–	27 ± 2 (24-30)
b	8.9	–	–	7 ± 1 (6-8)
c	–	–	–	13 ± 1 (11-15)
c′	–	–	–	3.6
V	–	–	53 ± 2 (51-58)	–
Body diam.	91	85 ± 10 (60-100)	287 ± 55 (210-436)	37 ± 3 (30-45)

Table 37. *(Continued).*

Character	First generation			Infective juvenile
	Holotype (male)	Male	Female	
EP	116	121 ± 6 (106-133)	141 ± 18 (116-190)	80 ± 3 (71-86)
NR	–	138 ± 7 (124-148)	158 ± 13 (136-186)	102 ± 5 (89-111)
ES	152	165 ± 6 (152-175)	243 ± 17 (220-275)	141 ± 6 (126-155)
Testis reflexion	–	294 ± 82 (80-384)	–	–
Tail length (T)	30	27 ± 3 (24-33)	58 ± 8 (46-76)	75 ± 4 (66-83)
Anal body diam. (ABD)	39	38 ± 2 (32-42)	80 ± 11 (64-108)	21 ± 2 (18-27)
Spicule length (SL)	72	71 ± 4 (60-80)	–	–
Spicule width	–	13 ± 1 (11-15)	–	–
Gubernaculum length (GL)	45	46 ± 3 (40-52)	–	–
Gubernaculum width	–	9 ± 1 (8-10)	–	–
D% = EP/ES × 100	80	73 ± 5 (61-80)	–	57 ± 3 (52-63)
E% = EP/T × 100	–	–	–	107 ± 7 (94-120)
SW% = SL/ABD × 100	180	190 ± 10 (170-210)	–	–
GS% = GL/SL × 100	60	70 ± 10 (50-80)	–	–
H% = Hyaline/T × 100	–	–	–	46 ± 3 (40-52)

EP = distance from anterior end to excretory pore; NR = distance from anterior end to nerve ring; ES = distance from anterior end to end of pharynx.
After the original authors.

isthmus indistinct; basal bulb pyriform, valve not observed. Excretory pore located between nerve ring and middle of pharynx. Excretory duct cuticularised and excretory gland swollen. Cardia prominent, protruding into intestine lumen. Didelphic, amphidelphic, genital system reflexed and tightly filled with eggs. Vulva median, in form of a transverse slit on a protuberance, no epiptygmata. Vagina short, oblique, with muscular walls. Tail shorter than anal body diam., terminal peg present.

Female (second generation)

Data not available.

Infective juvenile

Often enclosed in cuticle of second-stage. Body long, slender. Labial region smooth, slightly offset from body. Oral aperture and anus closed. In SEM face view, labial papillae not observed; pore-like amphidial apertures situated posterior to labial disc just anterior to cephalic papillae. Pharynx long and narrow, isthmus distinct and surrounded by nerve ring, basal bulb elongate, with valve. Cardia prominent. Excretory pore located between nerve ring and middle of pharynx. Hemizonid at isthmus level, distinct. Bacterial pouch rarely seen but packed with bacteria. Lateral field with two, six, seven and eight ridges anteriorly, then reducing to four and two. With above description, the formula of lateral field is 2, 6, 7, 8, 4, 2. Tail conoid, mucron absent. Hyaline region pronounced, comprising *ca* 46% of tail length. Phasmids distinct, located in anterior half of tail.

TYPE HOST AND LOCALITY

Natural host unknown. Collected from sandy soil on a marine beach in Thachkhe, Thachha, Hatinh, Vietnam.

TYPE MATERIAL

Holotype at Ghent University, Institute of Zoology, K.L. Ledeganckstraat 35, 9000, Belgium.

RELATIONSHIPS

Steinernema loci is characterised by the following characters: infective juveniles have a rounded and broad cephalic extremity which is smooth and slightly offset from the body and the lateral field formula is 2, 6, 7, 8, 4, 2. The male spicule head (manubrium) is longer than wide.

Steinernema loci is morphologically similar to *S. arenarium* and *S. glaseri*. It can be distinguished from similar species on the basis of IJ characters. From *S. glaseri* by a shorter body length of 985 (896-1072) *vs* 1130 (864-1448) μm and a more anterior position of the excretory pore at 80 (71-86) *vs* 102 (87-110) μm; from *S. arenarium* by a shorter body length of 985 (896-1072) *vs* 1034 (724-1408) μm, and a more anterior excretory pore at 80 (71-86) *vs* 83 (76-86) μm from the cephalic extremity.

The morphometrics of *S. loci* are close to those of *S. longicaudum*, *S. oregonense*, *S. kraussei* and *S. karii*. The species can be separated from *S. longicaudum* by a shorter IJ body length of 985 (896-1072) *vs* 1063 μm, and a shorter IJ tail length of 75 (66-83) *vs* 95 μm; from *S. oregonense* by a more posterior IJ excretory pore at 80 (71-86) *vs* 66 (60-72) μm, longer IJ pharynx of 141 (126-155) *vs* 132 (116-148) μm and longer IJ tail of 75 (66-83) *vs* 70 (64-78) μm; from *S. kraussei* by a more posterior IJ excretory pore at 80 (71-86) *vs* 60 (56-66) μm; and from *S. karii* by a more posterior IJ excretory pore at 80 (71-86) *vs* 74 (68-80) μm from the cephalic extremity.

Molecular characterisation

Steinernema loci is genetically characterised by the length of ITS regions (792 bp) and its composition, A = 0.23737, C = 0.18434, G = 0.24621, T = 0.33207 (accession no. AY355443).

Distribution

The species is known only from the type locality in Vietnam (Phan *et al.*, 2001).

Reference

Phan, K.L., Nguyen, N.C. & Moens, M. (2001). *Steinernema loci* sp. n. and *Steinernema thanhi* sp. n. (Rhabditida: Steinernematidae) from Vietnam. *Nematology* 3, 503-514.

Steinernema longicaudum[*,**] Shen & Wang, 1992
= *S. serratum* Liu, 1992 (= *nomen nudum*)
(Figs 132, 133)

MEASUREMENTS

See Table 38.

DESCRIPTION

Male (first generation)

Body curved posteriorly, C- or J-shaped when heat-relaxed. Cuticle finely striated under SEM, lateral field and phasmid inconspicuous. Cephalic extremity truncate or slightly rounded. Oral aperture circular or subtriangular. Six labial and four prominent cephalic papillae present. Stoma shallow. Pharynx extending close to oral aperture. Pharynx typical of genus. Cardia present, projecting into intestine. Genital system monorchic, reflexed. Under SEM, spicules well curved, relatively thick. Ratio spicule length/spicule width averaging 5.1 (4.3-5.5). Spicule head narrowing posteriorly to form a distinct shaft. Blade well curved, bearing two depressions laterally dividing blade into three lobes. Lateral lobe narrowing anteriorly to a point and terminating before reaching shaft, disappearing posteriorly before reaching blade terminus. Ventral lobe enlarging anteriorly and terminating posteriorly at same point as velum. Dorsal lobe usually very uniform in width. Near posterior end, blade suddenly reduced in width to form a tip with flattened terminus. Velum present, terminating a distance from spicule tip. In lateral view, gubernaculum variable in shape, mostly boat-shaped, curved ventrally, tapering to both ends. In ventral view, cuneus Y-shaped, not reaching posterior end of corpus. Corpus separated posteriorly. Complement of 23 genital papillae comprising 11 pairs plus single, precloacal, midventral papilla. Tail rounded, mucron absent.

Male (second generation)

Similar to first generation but smaller. Spicules as in first generation but capitulum more elongate.

---

[*] From Latin *longus* = long, *cauda* = tail.

[**] Description after Shen and Wang (1992) and Stock *et al.* (2001). Revised by Khuong B. Nguyen.

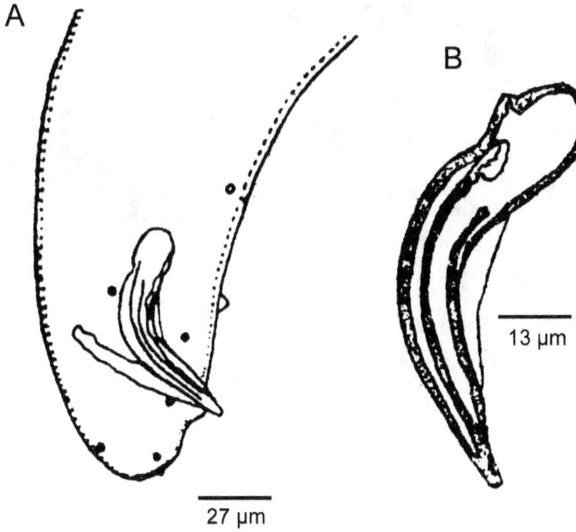

Fig. 132. Steinernema longicaudum. *A: Posterior region of first generation male; B: Spicule of first generation male. Modified after Shen and Wang (1992).*

Female (first generation)

Body cuticle smooth. Cephalic extremity truncate. Lips fused, triangular, with six labial and four cephalic papillae. Basal bulb of pharynx swollen, with vestigial valve. Nerve ring located anterior to basal bulb in isthmus region. Excretory pore located anterior to basal bulb, usually anterior to nerve ring. Vulva near mid-body, in form of a transverse slit, located on a protuberance, epiptygmata absent. Genital system didelphic, usually reflexed. Vagina short, initially angled slightly anteriad. Eggs deposited externally but later developing and juveniles hatching in reproductive system. Tail longer than anal body diam., rounded with one to several small projections at terminus.

Female (second generation)

Similar to first generation female but smaller. Excretory pore located anterior to nerve ring. Genital system mostly reflexed. Tail evenly conoid to a pointed terminus, longer than anal body diam.

Infective juvenile

Usually enclosed in a sheath (= second-stage juvenile cuticle) often lost in storage. Body slender, gradually tapering to both ends. Cephalic extremity rounded. Cuticle finely annulated. Lateral field

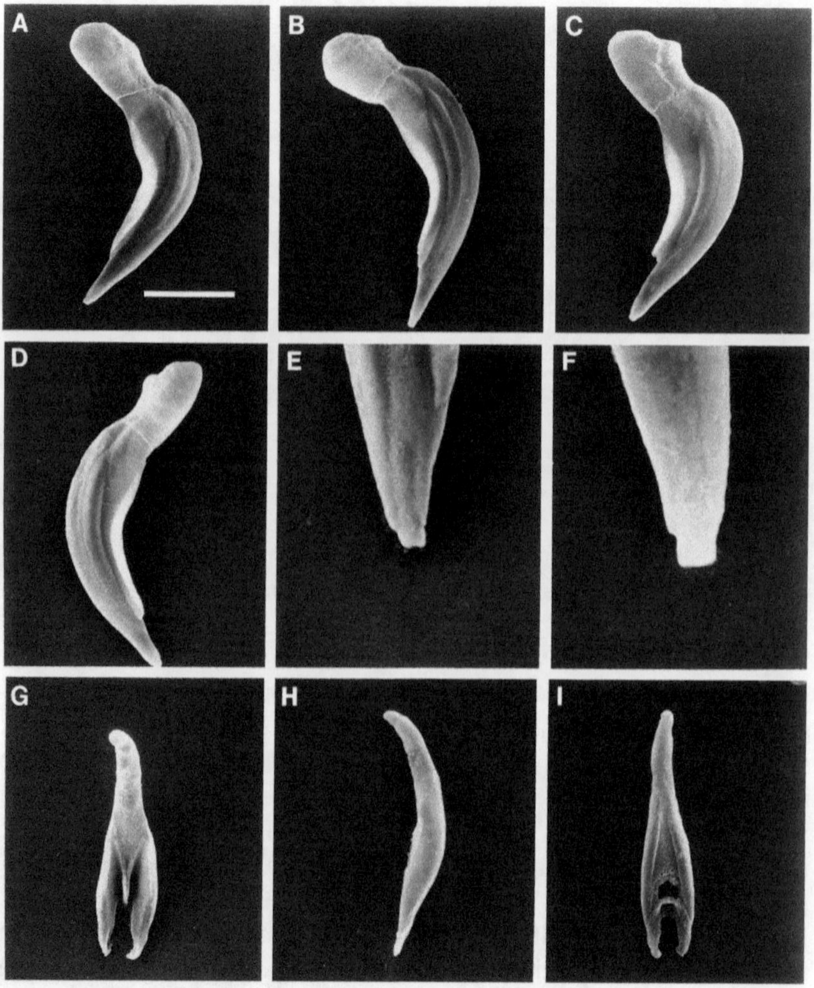

Fig. 133. Steinernema longicaudum. *SEM photographs of spicules and gubernacula of first generation male. A-D: Spicule showing spicule head, blade with three lobes and velum; E, F: Spicule blade narrowing suddenly to form spicule tip; G: Gubernaculum, ventral view showing Y-shaped cuneus; H, I: Gubernacula, lateral and dorsal views. (Scale bar: A-D, G-I = 25 μm; E = 7.5 μm; F = 5 μm.) Courtesy of K.B. Nguyen.*

with eight evenly spaced ridges (*i.e.*, nine lines) at mid-body. Oral aperture and anus closed. Pharynx degenerate. Internal structure of basal bulb not as clear as in adults. Excretory pore located in postmedian region of pharynx. Excretory duct cuticularised. Hemizonid and smaller

Table 38. *Morphometrics of* Steinernema longicaudum. *Measurements are in* μm *and in the form: mean* $\pm$ *SD (range).*

Character		First generation		Second generation		Infective juvenile
	Holotype (female)	Male	Female	Male	Female	
n	–	23	22	23	22	28
L	–	1788 ± 309 (1412-2733)	6935 ± 1977 (3188-10999)	826 ± 36 (774-911)	2358 ± 293 (1847-2986)	1043 ± 73 (929-1170)
a	–	–	–	–	–	28 ± 1.4 (26-31)
b	–	–	–	–	–	7.4 ± 0.5 (6.3-8.1)
c	–	–	–	–	–	11.1 ± 0.4 (10.5-12.1)
c′	–	–	–	–	–	4.1
V	–	–	51 (44-57)	–	52 ± 2.2 (48-56)	–
Body diam.	–	136 ± 27 (86-194)	347 ± 65 (227-518)	82 ± 3.6 (77-90)	144 ± 19 (113-214)	37 ± 1.4 (34-40)
EP	–	127 ± 120 (79-162)	129 ± 23 (83-169)	114 ± 6.2 (101-124)	90 ± 12 (63-124)	82 ± 4 (74-92)
NR	–	151 ± 16 (120-176)	176 ± 23 (140-221)	144 ± 6.3 (131-156)	136 ± 5.7 (126-149)	111 ± 6.9 (98-129)
ES	–	165 ± 23 (79-192)	224 ± 19 (176-268)	164 ± 4.3 (153-171)	177 ± 4.6 (169-183)	142 ± 4.8 (134-150)
Tail length (T)	–	30 ± 5.8 (20-43)	57 ± 24 (26-130)	33 ± 3.5 (25-41)	50 ± 8.7 (18-65)	94 ± 6, 9 (79-105)
Anal body diam. (ABD)	–	58 ± 9.9 (43-74)	97 ± 39 (54-216)	48 ± 4 (43-59)	40-4.6 (27-50)	23 ± 1.3 (21-27)
Spicule length (SL)	–	91 ± 3.8 (72-108)	–	75 ± 3.4 (56-86)	–	–
Gubernaculum length (GL)	–	60 ± 3.4 (54-65)	–	37 ± 3.9 (30-45)	–	–
D% = EP/ES × 100	–	75 ± 2.2 (56-92)	–	70 ± 2 (63-75)	–	57 ± 2.4 (52-63)
E% = EP/T × 100	–	–	–	–	–	87 ± 7 (76-104)
SW% = SL/ABD × 100	–	161 ± 27 (116-225)	–	158 ± 19 (114-200)	–	–
GS% = GL/SL × 100	–	66 ± 8 (56-88)	–	50 ± 7 (39-68)	–	–

EP = distance from anterior end to excretory pore; NR = distance from anterior end to nerve ring; ES = distance from anterior end to end of pharynx.
After Stock *et al.* (2001).

hemizonion present just anterior to basal bulb. Anterior region of intestine with bacterial pouch. Tail long, averaging 95 μm long, straight or slightly curved posteriad with a pointed terminus. Hyaline region prominent, comprising *ca* half of tail length. Phasmids located anterior to mid-tail at *ca* 10-20% of tail length.

TYPE HOST AND LOCALITY

Natural host unknown. The soil sample from which the species was isolated was collected from an orchard in June 1985 in Laiyang, Shandong, People's Republic of China (37°N latitude, 121°E longitude).

TYPE MATERIAL

Holotype (first generation female), allotype (first generation male) deposited in the Insect Collection, Department of Plant Protection, Laiyang Agriculture College, Southern China Agricultural University, People's Republic of China.

RELATIONSHIPS

This nematode is mainly characterised by features of the infective juvenile, such as the long tail of 95 (78-110) μm, body length, distance between excretory pore and anterior end, distance between nerve ring and anterior end, and length of pharynx. Other characteristics are: first generation female with one to several small projections on tail tip; male spicules broad, moderately curved with the tip lacking a hook-like structure. Near the posterior end, the spicule blade is suddenly reduced in width to form a tip with flattened terminus.

This species is close to *S. arenarium*, *S. cubanum*, *S. glaseri* and *S. guangdongense*, but can be distinguished from these species by the unique tip of the spicule (blade suddenly reduced in width to form a tip with flattened terminus *vs* ball-shaped tip in *S. arenarium*, hook-shaped tip in *S. glaseri*, and smooth tip in *S. cubanum* and *S. guangdongense*). The values of c and E% of the infective juvenile are smaller than those of *S. glaseri* and there is no overlap between these two species.

MOLECULAR CHARACTERISATION

Steinernema longicaudum is characterised molecularly by sequences of ITS and D2D3 regions: *i*) length of ITS regions 956 base pairs (bp), ITS1 = 257 bp, ITS2 = 292 bp and its composition, A = 0.251,

C = 0.196, G = 0.258, T = 0.295 (accession no. AY170338); and *ii*) length of D2/D3 region 462 bp and its composition, A = 0.234, C = 0.201, G = 0.325, T = 0.240 (accession no. AY169553).

DISTRIBUTION

Known from the type locality in People's Republic of China (Shen & Wang, 1992) and also from turf grass in Gonju, Chungnam province, Korea, and from lodgepole pine forest, Sierra Nevada mountains, California, USA (see Stock *et al.*, 1999, 2001). Hominick (2002) cited an unpublished record from Australia.

References

HOMINICK, W.M. (2002). Biogeography. In: Gaugler, R. (Ed.). *Entomopathogenic nematodes*. Wallingford, UK, CABI Publishing, pp. 115-143.

LIU, J. (1992). Taxonomic study of the genus: *Steinernema* Travassos and *Heterorhabditis* Poinar. In: *Proceedings of the XIX International Congress of Entomology, 28 June-4 July 1992, Beijing, China*, p. 318.

SHEN, C.P. & WANG, G.H. (1992). Description and studies of an entomopathogenic nematode: *Steinernema longicaudum* sp. nov. In: *Proceedings of the first national academy symposium of young and middle aged science and technology workers on plant protection, Beijing, China*. Chinese Science and Technology Press, pp. 220-231.

STOCK, S.P., PRYOR, B.M. & KAYA, H.K. (1999). Distribution of entomopathogenic nematodes (Steinernematidae and Heterorhabditidae) in natural habitats in California. *Biodiversity and Conservation* 8, 535-549.

STOCK, S.P., HENG, J., HUNT, D.J., REID, A.P., SHEN, X. & CHOO, H.Y. (2001). Redescription of *Steinernema longicaudum* Shen & Wang (Nematoda: Steinernematidae); geographic distribution and phenotypic variation between allopatric populations. *Journal of Helminthology* 75, 81-92.

Steinernema monticolum[*,**] Stock, Choo & Kaya, 1997
(Figs 134, 135)

MEASUREMENTS

See Table 39.

DESCRIPTION

Male (first generation)

Cuticle smooth. Cephalic extremity truncate to slightly rounded. Lips six, united but tips distinct, each bearing labial papilla. Four cephalic papillae. Stoma short, broad, with inconspicuous sclerotised walls. Cheilorhabdions and telorhabdions vestigial. Pharynx muscular, procorpus cylindrical, metacorpus slightly swollen, non-valvate; indistinct isthmus followed by a pyriform basal bulb containing a reduced valve. Pharynx set off from intestine. Nerve ring usually surrounding isthmus or anterior part of basal bulb. Excretory pore opening circular, located at posterior third of metacorpus, just anterior to nerve ring. Genital system monorchic, reflexed, consisting of germinal growth zone leading to seminal vesicle. *Vas deferens* with inconspicuous walls. Complement of 21-23 genital papillae arranged as 10-11 pairs plus a single midventral, precloacal papilla. Papillae disposed as six subventral precloacal pairs (pair 6 may be adcloacal), one pair lateral precloacal, one pair lateral postcloacal, two pairs ventral postcloacal and one pair subterminal (but occasionally absent). Spicules brownish-orange in colour, paired and curved. Manubrium rounded, robust, slightly separated from calomus. Lamina with two internal ribs, velum present, extending from base of manubrium to proximal two-thirds of lamina. Gubernaculum arcuate, of same colour as spicules, large, from half to two-thirds of spicule length in size. Anterior end of gubernaculum (capitulum) curved ventrally, posterior end forked. Cuneus long. Corpus broad, large, bifurcate. Tail conoid, mucronate.

Male (second generation)

Similar to first generation, but smaller. Tail mucron 7 (5-9) μm long.

[*] The species name reflects the *montane* nature of the geographic area from which it was isolated.

[**] Description after Stock *et al.* (1997). Revised by David J. Hunt.

Female (first generation)

Ovaries opposed, reflexed in dorsal position; oviduct well developed, glandular spermatheca and uterus in ventral position. Vagina short, with muscular walls. Vulval lips not protruding. Vulval aperture located near mid-body. Tail conoid, blunt, with mucron.

Female (second generation)

Similar to first generation, but smaller.

Infective juvenile

Body slender, tapering regularly from base of pharynx to anterior end and from anus to terminus. Cuticle with transverse striations. Cephalic extremity smooth, oral aperture closed. Pharynx long and narrow. Nerve ring located at level of isthmus. Excretory pore located in anterior third of pharynx. Basal bulb valvate. Cardia present. Anterior portion of intestine with dorsally displaced pouch containing symbiotic bacterium. Lumen of intestine narrow, rectum long; anus distinct. Genital primordium evident. Tail conoid, terminus pointed.

TYPE HOST AND LOCALITY

Natural host unknown in nature. Material isolated from bait insect (*Galleria mellonella*) in sandy-silt soil under larch (*Larix leptolepis*, Pinaceae), Mount Jiri, Sancheong, Gyeongnam Province, Republic of Korea.

TYPE MATERIAL

Holotype and paratype males, females and infective juveniles in the University of California-Davis Nematode Collection, Davis, CA, USA.

RELATIONSHIPS

All stages of *S. monticolum* can be distinguished from other members of the genus by the location of the excretory pore, which is at the level of the posterior third of the metacorpus. The tail of the first and second generation male has a large mucron. In the male, D% = 55 (49-61). Other diagnostic characters are the shape of the spicules and gubernaculum, the arrangement of the genital papillae, and the tail length. In the infective juvenile, the average body length = 706 μm,

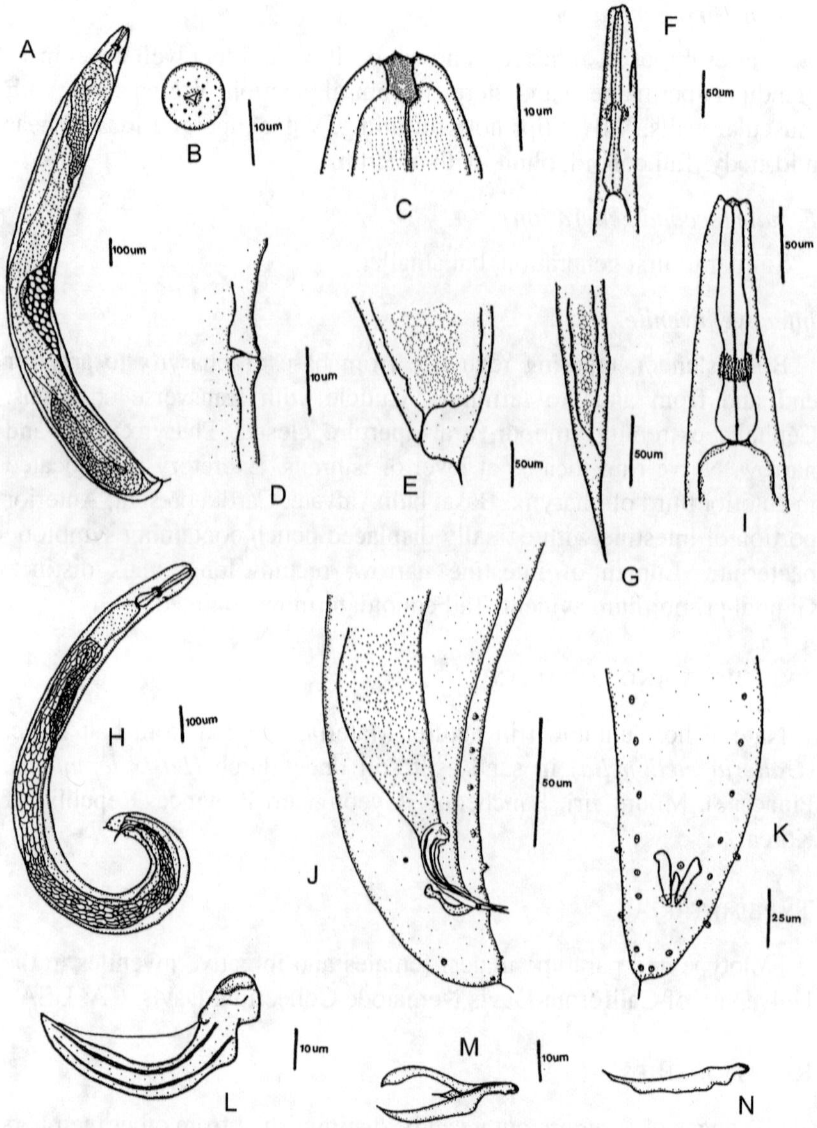

Fig. 134. Steinernema monticolum. *A: First generation female, entire body; B:* En face *view of anterior end of first generation female; C: Anterior end of first generation female; D: Vulva of first generation female; E: Tail of first generation female; F: Anterior end of third-stage infective juvenile; G: Tail of third-stage infective juvenile; H: First generation male, entire body; I: Anterior end of first generation male; J: Tail of first generation male, lateral view.*

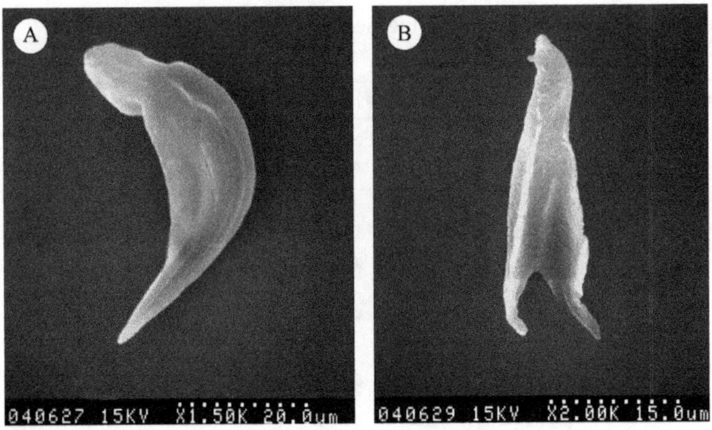

Fig. 135. Steinernema monticolum. *SEM micrographs. A: Spicule, lateral view; B: Gubernaculum, ventral view. Courtesy of K.B. Nguyen.*

the average body diam. = 37 (32-46) μm and ratio a = 19 (14-22). The original description failed to mention the characteristics of the IJ lateral field.

Molecular characterisation

The sequence of the ITS regions is deposited at GenBank under accession no. AF122017. This sequence is characterised by its length (flanked by the primers 18S and 26S), 916 bp, ITS1 = 264 bp, ITS2 = 245 bp, and its composition, A = 0.27402, C = 0.17358, G = 0.23908, T = 0.31332.

Bionomics

The life cycle of *S. monticolum* is similar to other *Steinernema* species. The infective juveniles enter the insect host through the mouth and spiracles (entrance through the anus was not observed) and then penetrate into the haemocoel. Once in the haemocoel, the infective juveniles release their symbiotic bacteria, thereby killing the insect within 36-48 h. Two amphimictic generations occur inside the host,

Fig. 134. *(Continued). K: Tail of first generation male, ventral view; L: Spicule, lateral view; M: Gubernaculum, ventrolateral view; N: Gubernaculum, ventral view. After Stock* et al. *(1997), courtesy of* Nematologica.

Table 39. *Morphometrics of* Steinernema monticolum. *Measurements are in μm and in the form: mean (range).*

Character	First generation			Second generation		Infective juvenile
	Holotype (male)	Male	Female	Male	Female	
n	–	20	20	20	20	?
L	–	1600 (1360-1840)	4410 (3880-6250)	1310 (1269-1355)	2810 (2300-3240)	706 (612-821)
a	–	–	–	–	–	19 (14-22)
b	–	–	–	–	–	5.7 (5.0-6.4)
c	–	–	–	–	–	9.3 (7.6-11.1)
c′	–	–	–	–	–	4.3
V	–	–	52 (50-55)	–	55 (52-59)	–
Body diam.	–	160 (117-206)	287 (250-365)	85 (75-99)	160 (137-179)	37 (32-46)
EP	–	94 (75-113)	91 (75-100)	92 (84-100)	101 (93-112)	58 (54-62)
NR	–	116 (94-128)	130 (97-156)	110 (95-117)	135 (121-157)	88 (81-93)
ES	–	166 (152-184)	219 (189-236)	152 (137-160)	195 (180-205)	124 (120-131)
Testis reflexion	–	370 (334-398)	–	–	–	–
Tail length (T)	–	38 (31-41)	65 (42-81)	38 (33-42)	80 (70-100)	77 (71-95)
Anal body diam. (ABD)	–	49 (38-67)	84 (59-107)	43 (40-46)	57 (49-69)	18 (14-23)
Spicule length (SL)	–	70 (61-80)	–	61 (54-68)	–	–
Spicule width	–	8 (7-10)	–	6 (5-7)	–	–
Gubernaculum length (GL)	–	45 (35-54)	–	29 (25-34)	–	–
Gubernaculum width	–	3 (2.5-5.0)	–	3 (2.5-4.0)	–	–

progeny of the second generation emerging from the insect cadaver as infective juveniles. Observations done on blood drops of *G. mellonella* showed that the infective juveniles are able to reach the adult stage in 24-30 h at 25°C, second generation adults appear in 5-6 days, and infective juveniles emerge 8-10 days after infection.

Table 39. *(Continued).*

Character	First generation			Second generation		Infective juvenile
	Holotype (male)	Male	Female	Male	Female	
D% = EP/ES × 100	–	55 (49-61)	–	61 (60-63)	–	47 (44-50)
E% = EP/T × 100	–	247	140	242	126	76 (63-86)
SW% = SL/ABD × 100	–	140 (120-150)	–	135 (130-140)	–	–
GS% = GL/SL × 100	–	60 (50-70)	–	48 (46-53)	–	–
H% = H/T × 100	–	–	–	–	–	–

EP = distance from anterior end to excretory pore; NR = distance from anterior end to nerve ring; ES = distance from anterior end to end of pharynx.

The infective juveniles of *S. monticolum* carry the cells of their symbiotic bacterium in the ventricular portion of the intestine. The bacterium was isolated from a drop of haemolymph of a nematode-infected *G. mellonella* larva and, like other *Xenorhabdus* species, Phase 1 was characterised by the adsorption of neutral red and bromothymol blue from MacConkey agar and NBTA, respectively.

Steinernema monticolum is a cold-adapted species that was originally isolated from soil in a mountain forest on Mount Jiri, Republic of Korea. Koppenhöfer *et al.* (2000) described the ecological characteristics of the Jiri isolate of this nematode and found that it was best adapted to lepidopteran larvae hosts. *Steinernema monticolum* exhibited an intermediate foraging strategy and because of its adaptation to cold temperatures, Koppenhöfer *et al.* (2000) suggested that it showed most promise against pests, especially Lepidoptera, that hibernate in the upper soil or litter layer.

Campbell and Kaya (2000) showed that infective juveniles of *S. monticolum* did not respond to air movement or host associated volatile cues.

Kang *et al.* (2004) assessed the Jiri isolate of *S. monticolum* against the Japanese lawn grass cutworm, *Spodoptera depravata*, in Korea.

Distribution

Known from the type locality at Mount Jiri, Republic of Korea (Stock *et al.*, 1997).

References

CAMPBELL, J.F. & KAYA, H.K. (2000). Influence of insect associated cues on the jumping behaviour of entomopathogenic nematodes (*Steinernema* spp.). *Behaviour* 137, 591-609.

KANG, Y.J., LEE, D.W., CHOO, H.Y., LEE, S.M., KWEON, T.W. & SHIN, H.K. (2004). [Biological control of *Spodoptera depravata* (Butler) (Lepidoptera: Noctuidae) using entomopathogenic nematodes.] *Korean Journal of Applied Entomology* 43, 61-70.

KOPPENHÖFER, A.M., GANGULY, S. & KAYA, H.K. (2000). Ecological characterisation of *Steinernema monticolum*, a cold-adapted entomopathogenic nematode from Korea. *Nematology* 2, 407-416.

STOCK, S.P., CHOO, H.Y. & KAYA, H.K. (1997). An entomopathogenic nematode, *Steinernema monticolum* sp. n. (Rhabditida: Steinernematidae) from Korea with a key to other species. *Nematologica* 43, 15-29.

Steinernema neocurtillae[*,**] Nguyen & Smart, 1992
(Figs 136-140)

MEASUREMENTS

See Table 40.

DESCRIPTION

Male (first generation)

Body curved posteriorly, C-shaped when heat-relaxed. Cephalic extremity rounded, labial papillae not seen in lateral view. Stoma very shallow to almost absent, usually with a pronounced sclerotised structure at anterior end. Distance from anterior end to excretory pore extremely short, less than body diam. at excretory pore. Pharynx almost cylindrical anteriorly, isthmus indistinct, basal bulb slightly swollen. Genital system monorchic, reflexed. About 60% of males with normal testis, 40% with reduced or collapsed testis. In males with normal testis, distance from base of pharynx to anterior end of testis always shorter than distance from anterior end of nematode to base of pharynx. Spicules paired, brown in colour. Head (manubrium) of spicules elongate, in some, twice as long as wide, about one-third spicule length long; shaft (calomus) very short or absent; blade (lamina) thick, tapering slightly posteriorly, about twice as long as head; blade terminus blunt with depression on ventral side; head/blade angle 132° (120-141); velum rarely seen. Each spicule with two internal ribs. Gubernaculum long, 89 (82-93)% of spicule length long, boat-shaped, anterior part long, ventrally curved. Complement of 11 pairs genital papillae plus single midventral precloacal distributed as usual for genus. Tail conoid; tail terminus of 70% of males with short mucron.

Male (second generation)

Second generation male similar to that of first generation except smaller, thinner, and distance from head to testis reflexion much longer. Collapsed testis (as found in first generation) very rare. Spicules and

[*] Named after the generic name of *Neocurtilla hexadactylla*, the type host.

[**] After Nguyen and Smart (1992). Revised by Khuong B. Nguyen.

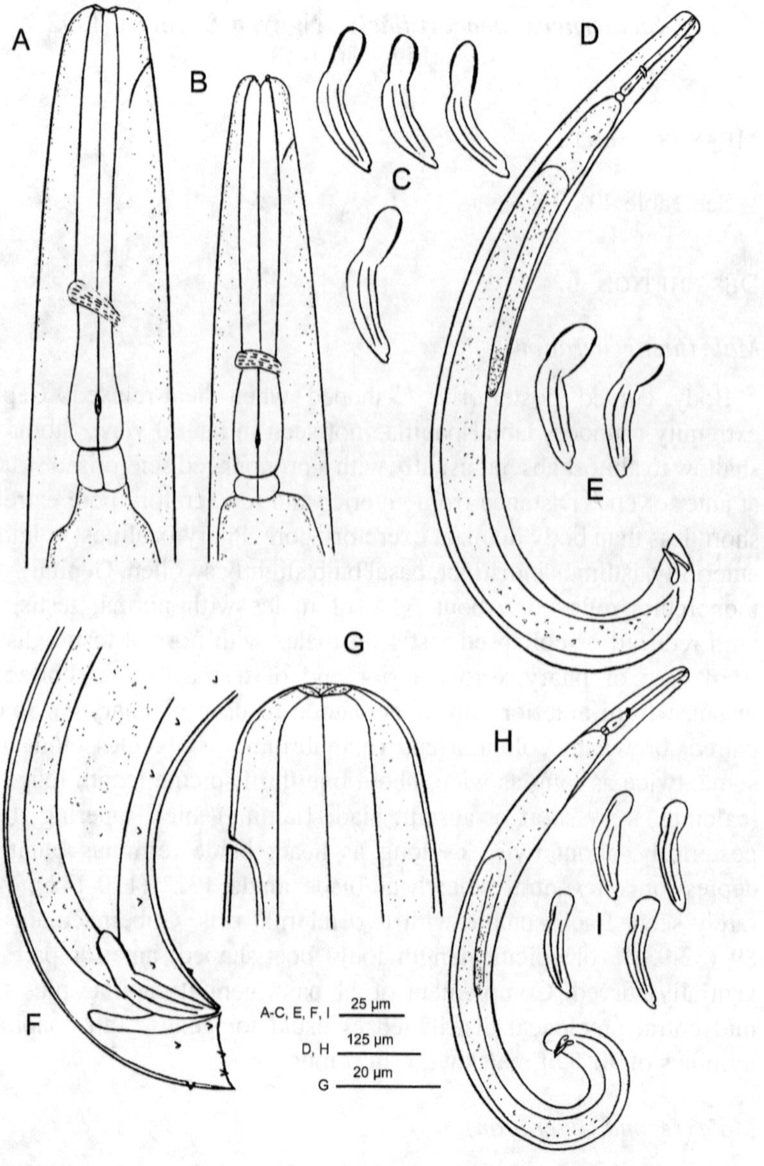

Fig. 136. Steinernema neocurtillae *male. A, C-G: First generation male. A: Anterior region; C, E: Variation in spicule form; D: Entire body; F: Posterior region showing 23 papillae arranged as 11 pairs and single midventral papilla; G: Enlargement of anterior end of A. B, H, I: Second generation male. B: Anterior region; H: Entire body; I: Variation in spicule. After Nguyen and Smart (1992), courtesy of* Journal of Nematology.

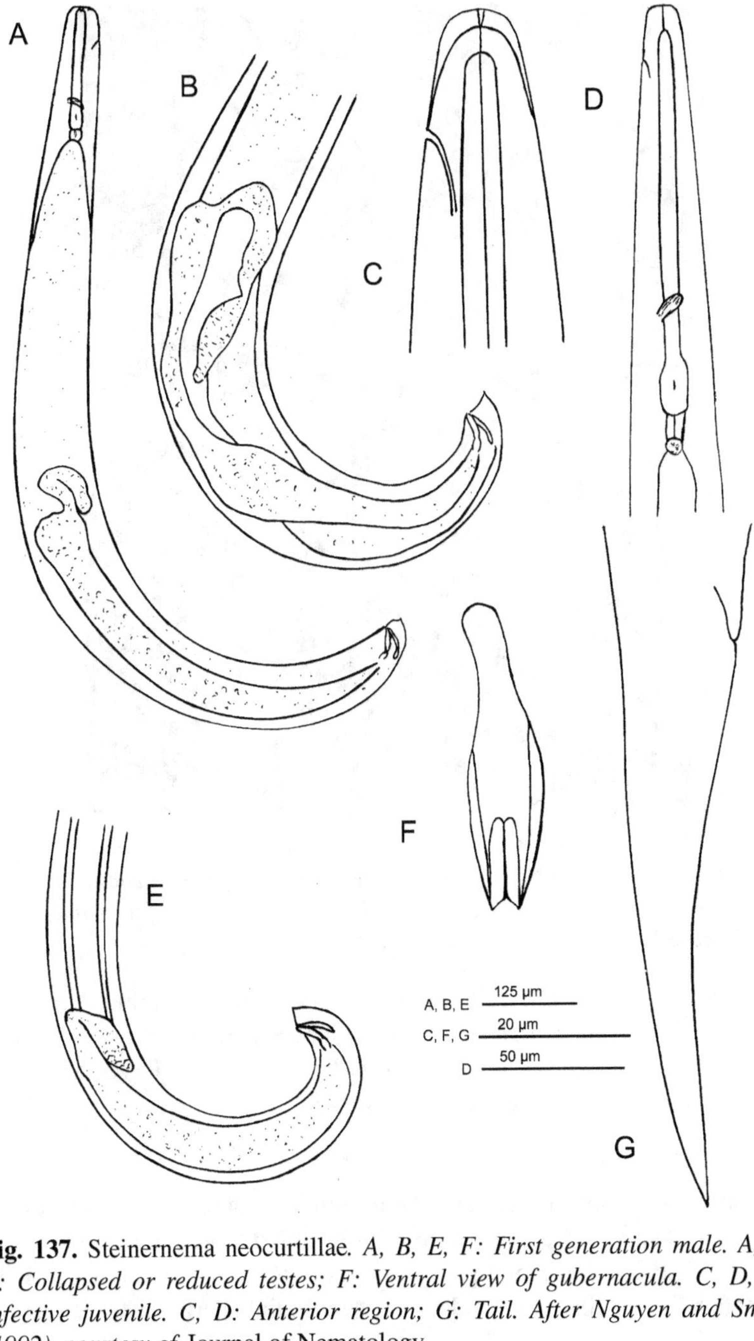

Fig. 137. Steinernema neocurtillae. *A, B, E, F: First generation male. A, B, E: Collapsed or reduced testes; F: Ventral view of gubernacula. C, D, G: Infective juvenile. C, D: Anterior region; G: Tail. After Nguyen and Smart (1992), courtesy of* Journal of Nematology.

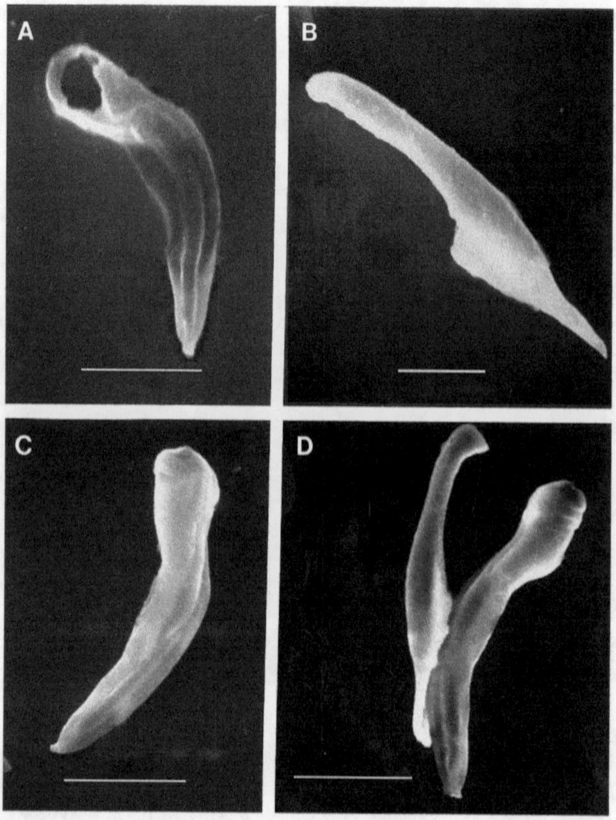

Fig. 138. Steinernema neocurtillae. *A, B: Spicule and gubernaculum of first generation male; C, D: Spicules and gubernacula of second generation male. (Scale bars: A = 20 μm; B = 10 μm; C, D = 15 μm.) After Nguyen and Smart (1992), courtesy of* Journal of Nematology.

gubernaculum shorter and thinner than those of first generation. Spicule with head/blade angle 151° (143-160). Gubernaculum almost as long as spicules, gubernaculum length divided by spicule length (G/S) = 95 (92-100)%. Mucron on tail terminus very rare.

Female (first generation)

Body cuticle smooth or with faint annules. Lateral fields and phasmids not observed. Cephalic extremity rounded, continuous with body; perioral disc present; labial papillae six, obscure, not seen in side view; four cephalic papillae. Lips indistinct. Amphidial apertures not observed. Stoma shallow, posterior part of stoma subtriangular. In lateral view,

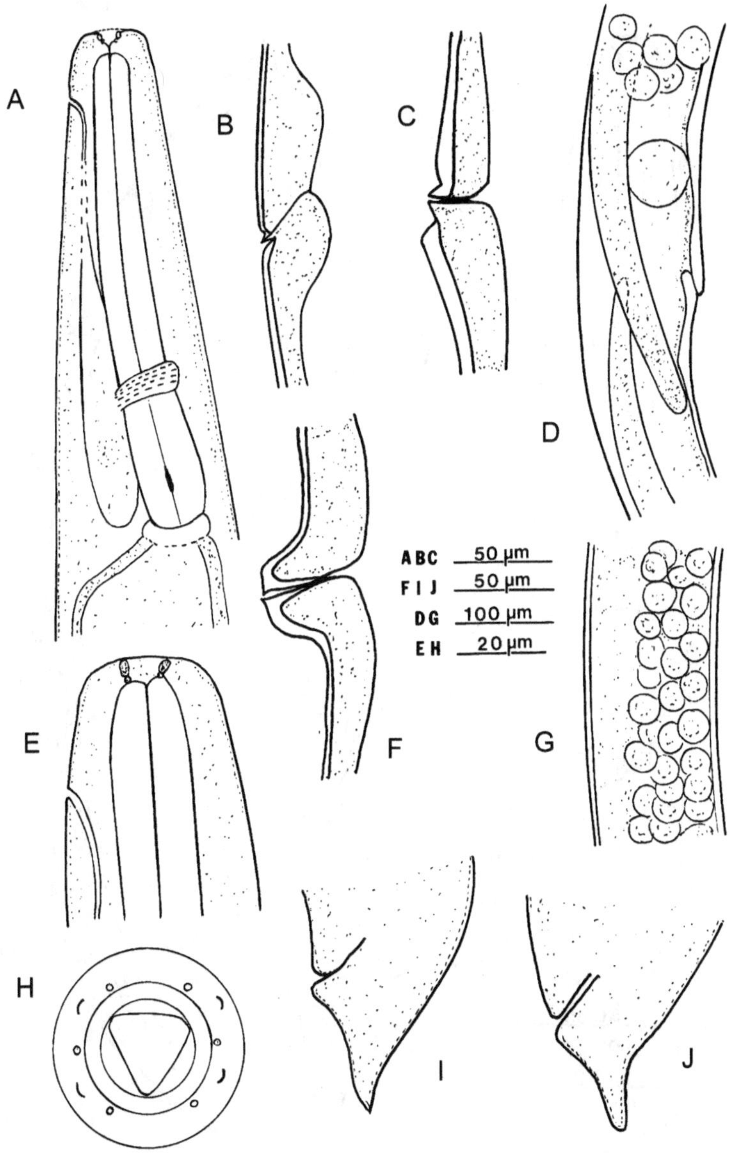

Fig. 139. Steinernema neocurtillae *first generation female. A: Anterior region; B, C, F: Variation of vulval region with variation in form of epiptygmata; D: Vulval region of young female showing short terminal end of reflexed ovaries; E: Enlargement of anterior end of A; G: Portion of body with eggs; H: Face view; I, J: Variation in female tails. After Nguyen and Smart (1992), courtesy of* Journal of Nematology.

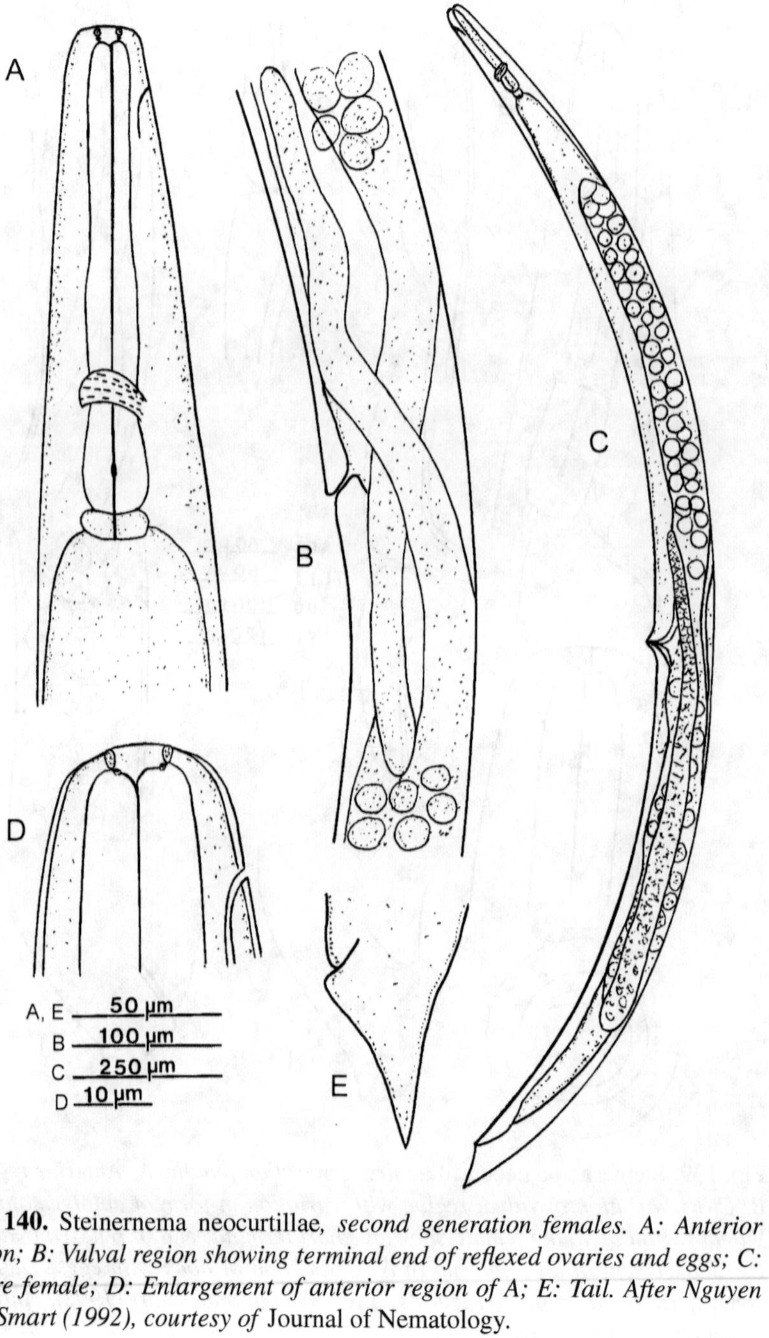

Fig. 140. Steinernema neocurtillae, *second generation females. A: Anterior region; B: Vulval region showing terminal end of reflexed ovaries and eggs; C: Entire female; D: Enlargement of anterior region of A; E: Tail. After Nguyen and Smart (1992), courtesy of* Journal of Nematology.

Table 40. *Morphometrics of* Steinernema neocurtillae. *Measurements are in μm and in the form: mean ± SD (range).*

Character	First generation			Second generation		Infective juvenile
	Holotype (male)	Male	Female	Male	Female	
n	–	20	20	20	20	50
L	1343	1593 ± 231 (1283-1988)	4714 ± 838 (3531-6250)	1115 ± 60 (1042-1217)	1934 ± 288 (1482-2718)	885 ± 44 (741-988)
a	–	–	–	–	–	26 ± 2 (22-29)
b	–	–	–	–	–	6 ± 0.3 (5-7)
c	–	–	–	–	–	11 ± 0.8 (9-14)
V	–	–	50 ± 3 (46-56)	–	53 ± 1.6 (49-55)	–
Body diam.	103	111 ± 20 (77-144)	187 ± 14 (159-215)	56 ± 4 (48-61)	99 ± 13 (78-120)	34 ± 2.7 (28-42)
EP	27	31 ± 5 (19-34)	37 ± 8 (23-53)	26 ± 4 (20-34)	29 ± 4 (20-37)	18 ± 1.7 (14-22)
NR	113	123 ± 15 (108-158)	189 ± 20 (153-231)	112 ± 8 (94-123)	148 ± 11 (127-163)	107 ± 4 (100-119)
ES		165 ± 29 (125-231)	254 ± 31 (181-315)	153 ± 7 (136-161)	197 ± 15 (166-217)	144 ± 7 (130-159)
Testis reflexion		360 ± 90 (279-512)	–	209 ± 21 (180-250)	–	–
Tail length (T)	27	35 ± 6 (19-39)	61 ± 8 (47-78)	18 ± 2.3 (14-22)	67 ± 5 (56-73)	80 ± 5 (64-97)
Anal body diam. (ABD)	39	41 ± 4 (39-50)	65 ± 13 (50-100)	32 ± 1.7 (29-34)	35 ± 4 (30-45)	–
Spicule length (SL)	62	59 ± 4 (52-64)	–	44 ± 1.5 (42-46)	–	–
Spicule width	14	12 ± 1.7 (9-14)	–	8.4 ± 1 (6-9)	–	–
Gubernaculum length (GL)	51	52 ± 4 (44-59)	–	42 ± 1 (41-44)	–	–
Gubernaculum width	7.8	8 ± 1 (6-9)	–	5.5 ± 0.7 (4-6)	–	–
D% = EP/ES × 100	–	19 ± 4 (13-26)	–	17 ± 2 (14-23)	–	12 ± 1 (10-15)
E% = EP/T × 100	–	–	–	–	–	23 ± 2 (18-30)
SW% = SL/ABD × 100	–	143 ± 14 (118-164)	–	140 ± 8 (124-151)	–	–
GS% = GL/SL × 100	–	89 ± 4 (81-93)	–	95 ± 2 (91-100)	–	–

EP = distance from anterior end to excretory pore; NR = distance from anterior end to nerve ring; ES = distance from anterior end to end of pharynx.
After the original authors.

cheilorhabdions prominent, well sclerotised; smaller sclerotisation, presumably prorhabdions, present posterior to cheilorhabdions. Pharynx with cylindrical, muscular, procorpus; metacorpus slightly swollen; isthmus indistinct; basal bulb slightly enlarged. Nerve ring usually obscure, surrounding isthmus. Pronounced pharyngo-intestinal valve present. Excretory pore located far anteriad; distance from anterior end to excretory pore usually shorter than body diam. at excretory pore. Excretory cell very large. Genital system amphidelphic, reflexed, often containing many eggs. Vulva in form of a transverse slit, protuberant in females with juveniles in body, but not so on young females; double-flapped epiptygmata present. Vagina sclerotised, short, *ca* 25% body diam. long, perpendicular to body in mature females, but directed posteriorly in young females. Body diam. greater anterior to vulva than posterior. Tail shape variable, ventral postanal swelling present or absent, tail usually shorter than anal body diam. About 25% of population with tail longer than anal body diam.

Female (giant form of first generation)

In addition to normal females of first generation, a giant form of female also occurs. Giant form characterised by twisted body and two to three times longer than normal females of first generation (10 313-13 000 *vs* 3531-6250 μm). Other structures as described for normal females.

Female (second generation)

Similar to first generation but smaller, median bulb of pharynx slightly more enlarged. Ovaries containing at most three rows of eggs. Tail tapering to a point, longer than anal body diam.; ventral postanal swelling present.

Infective juvenile

Body thin, elongate. Sheath (= second-stage cuticle) present but sometimes lost. Labial region continuous. Oral aperture narrow. Under SEM, labial papillae not seen but four prominent cephalic papillae and amphidial apertures present. Distance from anterior end to excretory pore extremely short, averaging 18 μm. D% = 12 and E% = 23. Pharynx with narrow anterior section, basal bulb more or less elongate with visible valve. Lateral field with seven incisures forming six bands or ridges, sometimes eight incisures (seven bands) present. Tail attenuate, tapering gradually.

Type host and locality

Haemocoel of the northern mole cricket, *Neocurtilla hexadactylla* (Orthoptera: Scapteriscidae), collected near La Crosse (Alachua County), FL, USA.

Type material

Holotype male, slide number NT-460t, deposited in the USDA Nematode Collection, Beltsville, MD, USA.

Note

The original species epithet of *neocurtillis* was changed to *neocurtillae* by Hominick *et al.* (1997) in order to conform with nomenclatural rules. The original authority remains unchanged.

Relationships

The male body is small and the distance from the anterior end to the excretory pore is less than the body diam. at that level and equivalent to only 19% of the pharynx length. The gubernaculum length is always more than 75% of the spicule length, being 89 (82-92)% of spicule length in the first generation, and 95 (92-100)% in the second generation. The female labial papillae were not seen under light microscope but double flapped vulval epiptygmata were present. The body of the infective juvenile is 885 μm in length and D% and E% are both very low, at 12 and 23, respectively.

Steinernema neocurtillae can be distinguished from all other species of the genus by the excretory pore of all stages being located very close to the anterior end. In the IJ this distance is 18 (14-22) μm, compared to 38 (30-110) μm or greater in all other species, resulting in the low D% and E% of 12 (10-15) and 23 (18-30), respectively, compared to D% = 26 or greater and E% = 60 or greater for all other species. This species also differs by the ratio of the gubernaculum length divided by spicule length, which is greater than 75% (89% in the first generation and 95% in the second generation) compared to less than 75% for all other species except for *S. scapterisci.*

Molecular characterisation

Steinernema neocurtillae is characterised by the length of ITS regions (flanked by the primers 18S and 26S), 967 bp and its composition, A = 0.27715, C = 0.14995, G = 0.19752 and T = 0.37538. The sequence of the ITS regions is deposited at GenBank under accession no. AF122018.

Bionomics

So far the only known host for this nematode is the northern mole cricket. It was difficult to produce the nematode *in vivo* in the laboratory. Infective juveniles placed in soil in Petri dishes infected and killed *N. hexadactylla* but few IJ were produced in the cadavers. On the other hand, infected mole crickets collected from the field produced large numbers of IJ.

Distribution

Steinernema neocurtillae has been found in Florida, USA (Nguyen & Smart, 1992) and possibly exists all over the country where the nematode host, the northern mole cricket, *Neocurtilla hexadactylla*, occurs. To date, this mole cricket is the only known host.

References

HOMINICK, W.M., BRISCOE, B.R., DEL PINO, F.G., HENG, J., HUNT, D.J., KOZODOY, E., MRÁČEK, Z., NGUYEN, K.B., REID, A.P., SPIRIDONOV, S., STURHAN, D., WATURU, C. & YOSHIDA, M. (1997). Biosystematics of entomopathogenic nematodes: current status, protocols and definitions. *Journal of Helminthology* 71, 271-298.

NGUYEN, K.B. & SMART JR, G.C. (1992). *Steinernema neocurtillis* n. sp. (Rhabditida: Steinernematidae) and a key to species of the genus *Steinernema*. *Journal of Nematology* 24, 463-477.

Steinernema oregonense[*,**] Liu & Berry, 1996
(Figs 141-143)

MEASUREMENTS

See Table 41.

DESCRIPTION

Male (first generation)

Cuticle, lip, stoma and pharyngeal region similar to first generation female. Body much smaller than female. Genital system monorchic, reflexed, consisting of germinal growth zone leading into seminal vesicle. Single *vas deferens* with inconspicuous muscular walls. Spicules paired, brown/yellow in colour, with ridges and thin velum, usually moderately curved. Capitulum variable in shape, usually broad, slightly angular; lamina bearing two ridges on edges and surface. Distal tip of spicule moderately blunt. Gubernaculum boat-shaped in lateral view, ventrally curved, with a proximal knob. Complement of 23 genital papillae present, variable in position, consisting of nine ventrolateral pairs, two lateral pairs plus a midventral papilla located just anterior to cloacal aperture. Six pairs located precloacally, one pair in region of gubernaculum, four pairs located postcloacal. Cloacal aperture slit-like. Tail conoid, ventral portion usually concave, tip blunt, mucron absent.

Male (second generation)

Similar to first generation but differing by following characters: smaller size, slender body, tail usually bearing a minute, filamentous, mucron.

Female (first generation)

Cuticle smooth, lateral lines and phasmids indistinct. Cephalic extremity rounded or slightly truncate, not offset. Six lips united at base but distinct at tips, each lip bearing one papilla. Four cephalic papillae located further posterior on head in submedial position as an outer circle surrounding six inner labial papillae. Amphidial apertures small, located posterior to lateral labial papillae. Oral aperture triangular, stoma short,

[*] Named after the state of *Oregon*.

[**] Description after Liu and Berry (1996). Revised by Zdeněk Mráček.

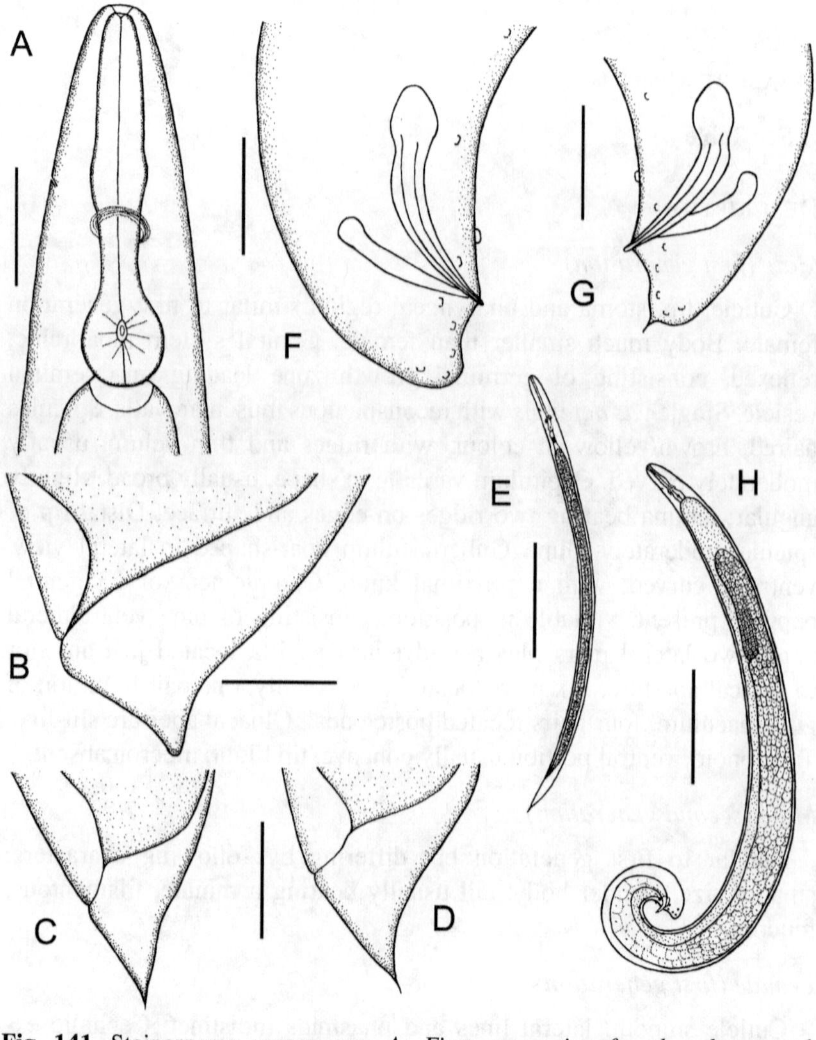

Fig. 141. Steinernema oregonense. *A: First generation female, pharyngeal region; B: First generation female, tail; C, D: Second generation female, tails; E: Infective juvenile, entire body; F: First generation male tail with spicule and gubernacula; G: Second generation male tail with spicule and gubernacula; H: First generation male, entire body. (Scale bars: A = 65 μm; B, G = 35 μm; C, D = 50 μm; E = 255 μm; F = 40 μm; H = 295 μm.) After Liu and Berry (1996), courtesy of* Fundamental and Applied Nematology.

broad, with non-sclerotised walls and some minute, tooth-like, structures posterior to stoma base. Cheilo- and telorhabdions vestigial. Pharynx muscular, extending nearly to oral aperture, procorpus cylindrical, metacorpus slightly swollen, non-valvate, indistinct isthmus followed by pyriform basal bulb containing valve lined with refractive ridges. Nerve ring surrounding anterior portion of basal bulb. Excretory pore opening circular, located anterior to nerve ring. Genital system didelphic, amphidelphic with opposed reflexed ovaries lying in dorsal position. Vulva in form of a transverse slit, usually slightly to moderately protuberant. Vagina short, leading into paired uteri. Tail usually broad, shorter than anal diam., slight or indistinct postanal swelling present. Tail tip with a round, wedge-shaped, projection and two or three mucrons or projections.

Female (second generation)

Similar to first generation but smaller and more slender. Tail straight without a postanal swelling, longer than anal body diam., tapering or pointed tip and bearing a mucron.

Infective juvenile

Often enclosed in cuticle of second-stage. Body slender, tapering regularly from base of pharynx to anterior end and from anus to terminus. Lateral fields distinct, apparently with seven ridges (*i.e.*, eight lines or incisures). Of these ridges, submarginal pair less distinct (Spiridonov *et al.*, 2004). Labial region smooth, usually continuous. Oral aperture and anus closed. Pharynx long, narrow, basal bulb weak and less prominent than in adults. Intestine collapsed. Nerve ring distinct. Hemizonid located at level of basal bulb. Anterior portion of intestine with pouch containing symbiotic bacteria. Tail conical with pointed terminus. Hyaline region comprising *ca* half of tail length.

TYPE HOST AND LOCALITY

Natural host not known as the species was recovered, using *Galleria mellonella* (L.) larvae as bait, from a soil sample collected in a grass field near Grants Pass, OR, USA.

TYPE MATERIAL

Holotype (male, first generation) and allotype (female, first generation) deposited in the University of California-Davis Nematode Collec-

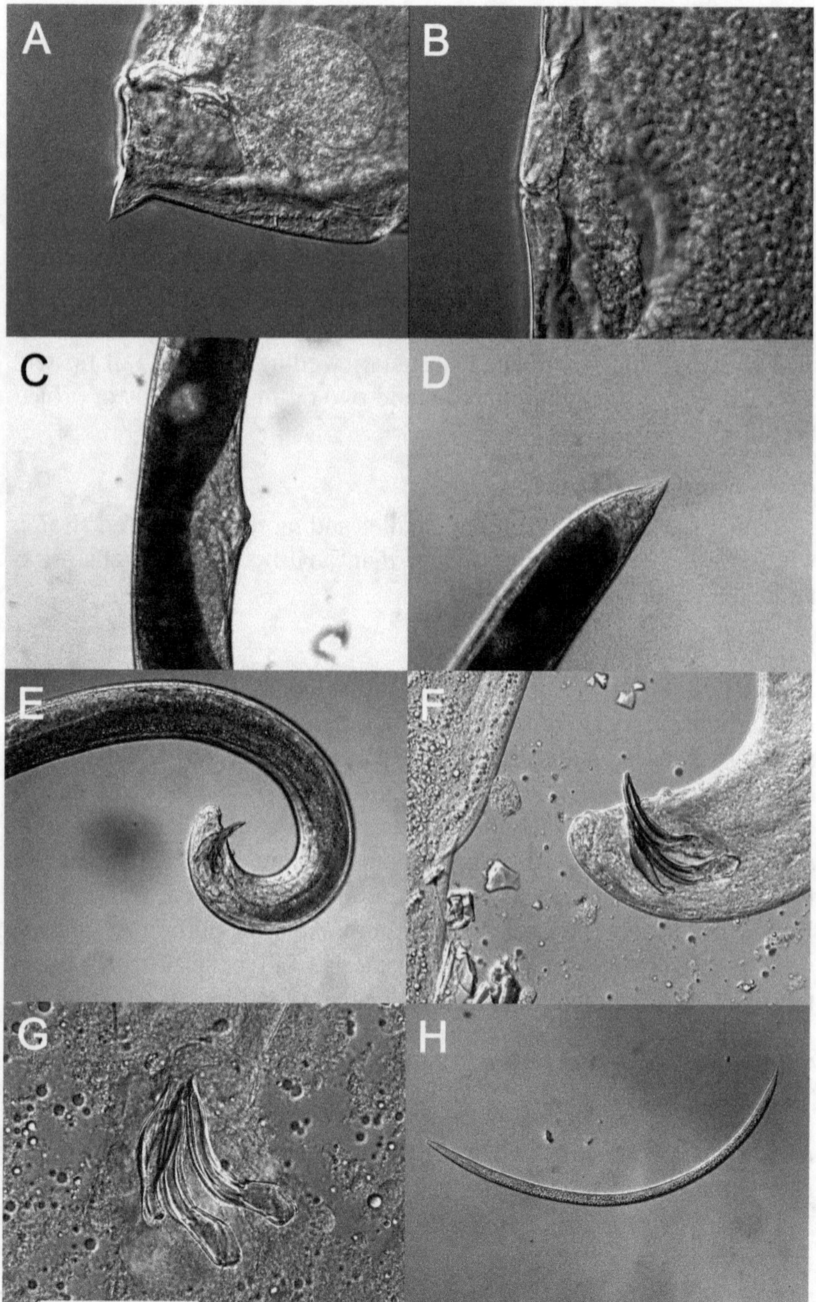
A
B
C
D
E
F
G
H

tion, Davis, CA, USA. Paratypes deposited in the USDA Nematode Collection, Beltsville, MD, USA.

NOTE

This species was named *S. oregonensis* by the original authors but was changed to *S. oregonense* by Hominick *et al.* (1997) in order to comply with International Code of Zoological Nomenclature (ICZN) rules. The original authority is unchanged.

RELATIONSHIPS

This species is characterised by the relatively long infective juveniles of *ca* 1000 μm and other morphological/morphometric features of this stage. The IJ excretory pore is located at mid-pharynx level. The tail is slender and long and has a high c′ ratio of 4.7. The hyaline region comprises *ca* half of the tail length. The lateral field has seven ridges, the submarginal pair being less distinct. Males possess a minute mucron that is absent or reduced in the first generation. The spicules are slightly curved, brown/yellow in colour with an oblong manubrium (lateral view), moderately blunt tip and lack a rostrum. Females of both generations mostly lack a postanal swelling. The tail in the first generation female has two or three terminal projections.

The average length (977 μm), ratio D% (50) and ratio E% (100) of the infective juveniles, the absence (first generation male) and the presence (second generation male) of a tail mucron and the moderately curved spicules, together with RAPD fragments and cross-breeding results, were used to separate *S. oregonense* from all previously described species of *Steinernema.*

The length range of infective juveniles of *S. oregonense* is different from those of *S. rarum*, *S. ritteri*, *S. carpocapsae*, *S. scapterisci*, *S. kushidai*, *S. riobrave*, *S. intermedium* and *S. cubanum*. The length range of *S. oregonense* infective juveniles somewhat overlaps those of

Fig. 142. Steinernema oregonense. *A, B: First generation female. A: Tail with mucrons; B: Vulva with small epiptygmata. C, D: Second generation female. C: Vulva; D: Tail, posterior region with short mucron. E, G: Second generation male. E: Posterior portion of body; F: Tail with short mucron; G: Spicules and gubernaculum. H: Infective juvenile, entire body. (Scale bar: A, B = 60 μm; C, D = 250 μm; E = 150 μm; F = 50 μm; G = 55 μm; H = 380 μm.) After Z. Mráček.*

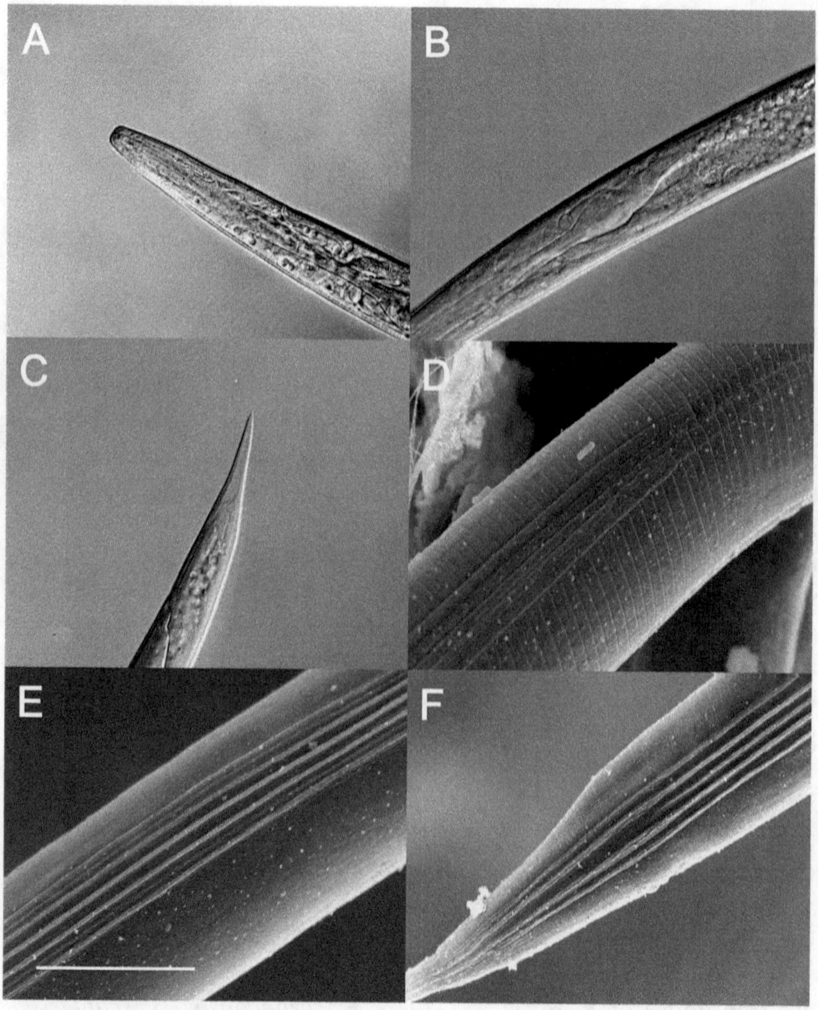

Fig. 143. Steinernema oregonense. *A-F: Infective juvenile. A: Head and excretory pore; B: Pharynx and bacterial vesicle; C: Tail with hyaline portion; D: Lateral field in neck portion; E: Lateral field in mid-body with seven ridges; F: Lateral field in tail portion. (Scale bar: A-C = 70 μm; D-F = 25 μm.) After Z. Mráček.*

S. affine, *S. arenarium*, *S. feltiae*, *S. glaseri*, *S. neocurtillae* and *S. puertoricense*. However, E% of the infective juveniles of *S. oregonense* does not overlap with those of *S. feltiae*, *S. neocurtillae* or *S. glaseri*. The ratio D% separates *S. oregonense* from *S. puertoricense*.

Males of *S. oregonense* can be separated from those of other *Steinernema* species by the absence (in first generation) and the presence (second generation) of a tail mucron and by the moderately curved spicules. The absence of a tail mucron in first generation males separates *S. oregonense* from both *S. affine* and *S. feltiae*. The moderately curved spicules separates *S. oregonense* from *S. intermedium* and *S. riobrave* where the spicules are usually strongly curved (a line running parallel with the calomus and lamina forms an angle of 70-90°). *Steinernema oregonense* is closely related to *S. arenarium*, *S. cubanum* and *S. glaseri* in general morphological characters. The blunt tip spicules separate *S. oregonense* from *S. glaseri* (notched tips) and *S. arenarium* (swollen tips). The presence of a tail mucron in second generation males separates *S. oregonense* from *S. cubanum*. The males of *S. oregonense* have a ventrally concave tail, which differs from that of *S. arenarium* and *S. glaseri*. *Steinernema oregonense* is reproductively isolated from *S. arenarium*, *S. feltiae* and *S. glaseri*.

This species also resembles *S. feltiae*, *S. weiseri* and *S. jollieti*. It differs from these species by the length of the male mucron which is much shorter in *S. oregonense* than all of these species except for *S. jollieti*. The infective juveniles are longer than those of *S. feltiae*, *S. weiseri* and *S. jollieti*. The lateral field in *S. oregonense* differs from *S. feltiae* (eight ridges), *S. weiseri* (eight ridges) and *S. jollieti* (six ridges) in having seven ridges with the submarginal pair less distinct. However, this character should be clarified more precisely. The presence of two or three projections on the first generation female tail seems to be specific for this species.

Molecular characterisation

Steinernema oregonense is characterised by sequence lengths (flanked by primers 18S and 26S) of ITS regions, 973 bp, ITS1 = 267 bp, ITS2 = 298 bp and its composition, A = 0.25385, C = 0.19322, G = 0.23535, T = 0.31757. The sequence of ITS regions is deposited in GenBank under accession no. AF122019.

Bionomics

Steinernema oregonense has a life cycle comparable to that of other described species of *Steinernema*. The infective juveniles enter the haemocoel of insects and liberate their associated bacteria. At room

Table 41. *Morphometrics of* Steinernema oregonense. *Measurements are in μm and in the form: mean ± SD (range).*

Character	First generation			Second generation		Infective juvenile
	Holotype (male)	Male	Female	Male	Female	
n	–	20	20	20	20	25
L	1700	1680 ± 0.08 (1560-1820)	5200 ± 0.3 (4400-6200)	1290 ± 0.05 (1220-1350)	2400 ± 0.2 (1900-2700)	980 ± 0.11 (820-1110)
a	–	–	–	–	–	30 ± 5.2 (24-37)
b	–	–	–	–	–	7.6 ± 0.6 (6.0-8.0)
c	–	–	–	–	–	14 ± 1.6 (12-16)
c′*	0.8	0.6	0.7	0.7	1.2	4.7**
V	–	–	52 ± 3.0 (46-56)	–	58 ± 3.3 (55-61)	–
Body diam.	145	138 ± 13.4 (105-161)	242 ± 13 (217-268)	78 ± 8.4 (64-89)	166 ± 10.2 (145-185)	34 ± 3.3 (28-38)
EP	99	112 ± 10.7 (95-139)	103 ± 6.2 (89-112)	85 ± 5.8 (75-91)	82 ± 5.6 (69-91)	66 ± 4.5 (60-72)
NR	103	111 ± 8.2 (101-133)	147 ± 8.8 (129-162)	100 ± 4.5 (93-107)	118 ± 5.0 (109-127)	–
ES	139	154 ± 11.6 (139-182)	210 ± 9.6 (186-220)	134 ± 4.8 (119-141)	169 ± 4.9 (158-176)	132 ± 8.9 (116-148)
Testis reflexion	151	297 ± 87 (99-392)	–	191 ± 11.9 (170-208)	–	–
Tail length (T)	30	29 ± 2.2 (24-32)	37 ± 6.2 (28-46)	25 ± 2.8 (20-30)	47 ± 6.3 (40-63)	70 ± 3.7 (64-78)
Hyaline (H)	–	–	–	–	–	31 ± 1.2** (30-33)
Anal body diam. (ABD)	40	47 ± 5.2 (38-55)	56 ± 9.5 (42-79)	36 ± 1.9 (32-40)	41 ± 3.3 (32-46)	14 ± 0.9** (13-16)
Spicule length (SL)	69	71 ± 1.9 (65-73)	–	59 ± 2.2 (54-61)	–	–
Spicule width	12	12 ± 0.7 (10-12)	–	9.3 ± 0.9 (7.9-12)	–	–
Gubernaculum length (GL)	55	56 ± 3.4 (52-59)	–	45 ± 3.5 (42-52)	–	–
Gubernaculum width	7.9	8.2 ± 0.9 (5.9-10)	–	7.3 ± 0.5 (6.9-7.9)	–	–

temperature (*ca* 22°C), it takes 3-4 days for infective juveniles to reach the first generation adult stage and another 2 days to reach the second adult generation. It usually takes 1 week for the nematodes to complete two generations and they emerge from the insect cadaver as infective

Table 41. *(Continued).*

Character	First generation			Second generation		Infective juvenile
	Holotype (male)	Male	Female	Male	Female	
D% = EP/ES × 100	71	73 ± 4.0 (64-75)	49 ± 3.0 (43-57)	64 ± 4.0 (55-68)	48 ± 3.0 (43-54)	50 ± 4.0 (40-60)
E% = EP/T × 100	330	390 ± 77 (310-640)	290 ± 60 (210-400)	340 ± 46 (250-380)	180 ± 20 (140-210)	100 ± 10 (90-110)
SW% = SL/ABD* × 100	173	152	–	164	–	–
GS% = GL/SL* × 100	11	7.9	–	7.7	–	–
H% = H/T × 100	–	–	–	–	–	0.47**

EP = distance from anterior end to excretory pore; NR = distance from anterior end to nerve ring; ES = distance from anterior end to end of pharynx.

* Recalculated according to Liu and Berry (1986).

** Calculated from own culture.

juveniles. Laboratory evaluation showed this nematode was capable of invading *Galleria mellonella* larvae within 1 h and killing the host in 48 h at 18°C, thereby indicating that the species is adapted to low temperatures.

DISTRIBUTION

The species is known only from the mountains of Oregon, USA (Liu & Berry, 1996). It may occur in California and Washington, USA, and British Columbia, Canada.

References

HOMINICK, W.M., BRISCOE, B.R., DEL PINO, F.G., HENG, J., HUNT, D.J., KOZODOY, E., MRÁČEK, Z., NGUYEN, K.B., REID, A.P., SPIRIDONOV, S., STURHAN, D., WATURU, C. & YOSHIDA, M. (1997). Biosystematics of entomopathogenic nematodes: current status, protocols and definitions. *Journal of Helminthology* 71, 271-298.

LIU, J. & BERRY, R.E. (1996). *Steinernema oregonensis* n. sp. (Rhabditida: Steinernematidae) from Oregon, USA. *Fundamental and Applied Nematology* 19, 375-380.

SPIRIDONOV, S.E., KRASOMIL-OSTERFELD, K. & MOENS, M. (2004). *Steinernema jollieti* sp. n. (Rhabditida: Steinernematidae), a new entomopathogenic nematode from the American Midwest. *Russian Journal of Nematology* 12, 85-95.

Steinernema pakistanense[*,**] Shahina, Anis, Reid, Rowe & Maqbool, 2001
(Figs 144, 145)

MEASUREMENTS

See Table 42.

DESCRIPTION

Male (first generation)

Body curved posteriorly, J-shaped when heat-relaxed. Cuticle with faint transverse striae. Lateral field and phasmids inconspicuous. Cephalic extremity truncate to slightly rounded, not set off from rest of body. Lips united, bearing one internal circle of six labial papillae and one external circle of four submedial cephalic papillae, amphidial apertures located at base of internal lateral papillae. Stoma shallow, triangular at base. Pharynx extending close to oral aperture. Cheilorhabdions located posterior to lips, composed of thick, cuticularised ring. Posterior to this, presumably representing the prorhabdions, is another cuticularised ring; meso-, meta- and telorhabdions vestigial. Pharynx muscular and cylindrical, metacorpus slightly swollen; non-valvate isthmus distinctly narrowing in males of both generation, terminating in a muscular basal bulb with small, distinct valve. Pharynx set off from intestine, posteriorly inserted into anterior portion of intestine. Cardia well developed. Excretory pore opening usually anterior to nerve ring at level of metacorpus. Nerve ring distinct, situated anterior to basal bulb. Genital system monorchic, reflexed, consisting of germinal growth zone leading into seminal vesicle. *Vas deferens* with inconspicuous muscular walls. Spicules paired, arcuate, symmetrical, distinct light golden yellow in colour. Each spicule with two internal ribs of variable length; dorsal rib curved ventrally and extending to posterior end. Calomus with a small rounded protrusion. From this point a distinct velum extending almost to proximal end of lamina. Shaft very short or absent. Spicule head elongate, broad, not continuous with blade. Blade (lamina) tapering to form a cylindrical part and then tapering again to sharply pointed distal tip.

[*] Named after *Pakistan*, the type locality.

[**] Description after Shahina *et al.* (2001). Revised by David J. Hunt.

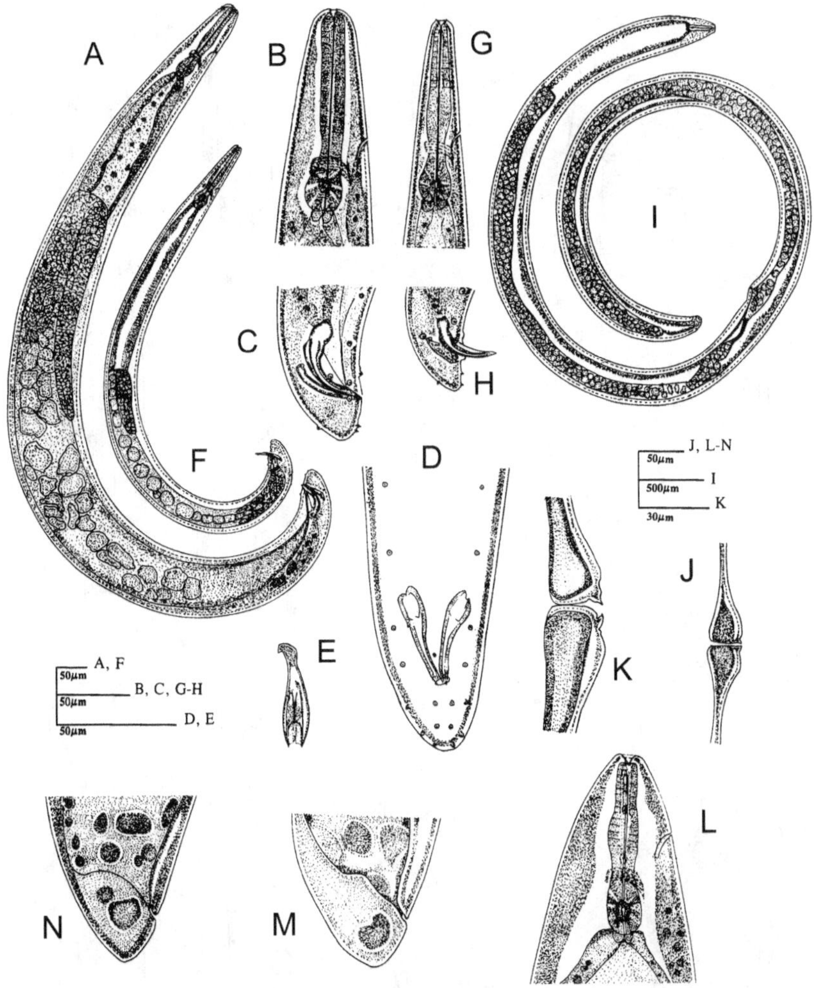

Fig. 144. Steinernema pakistanense. *A-E: Male, first generation. A: Entire body; B: Pharyngeal region; C: Tail region; D: Posterior region in ventral view; E: Gubernaculum, ventral view. F-H: Male, second generation. F: Entire body; G: Pharyngeal region; H: Tail. I-N: Female, first generation. I: Entire body; J, K: Variation in vulval region; L: Pharyngeal region; M, N: Variation in tail shape. After Shahina* et al. *(2001), courtesy of* International Journal of Nematology.

Gubernaculum boat-shaped in lateral view. In dorsal or ventral view anterior end of gubernaculum ventrally curved, enlarging gradually to form corpus; bifurcated to form two wings. Gubernaculum length/greatest

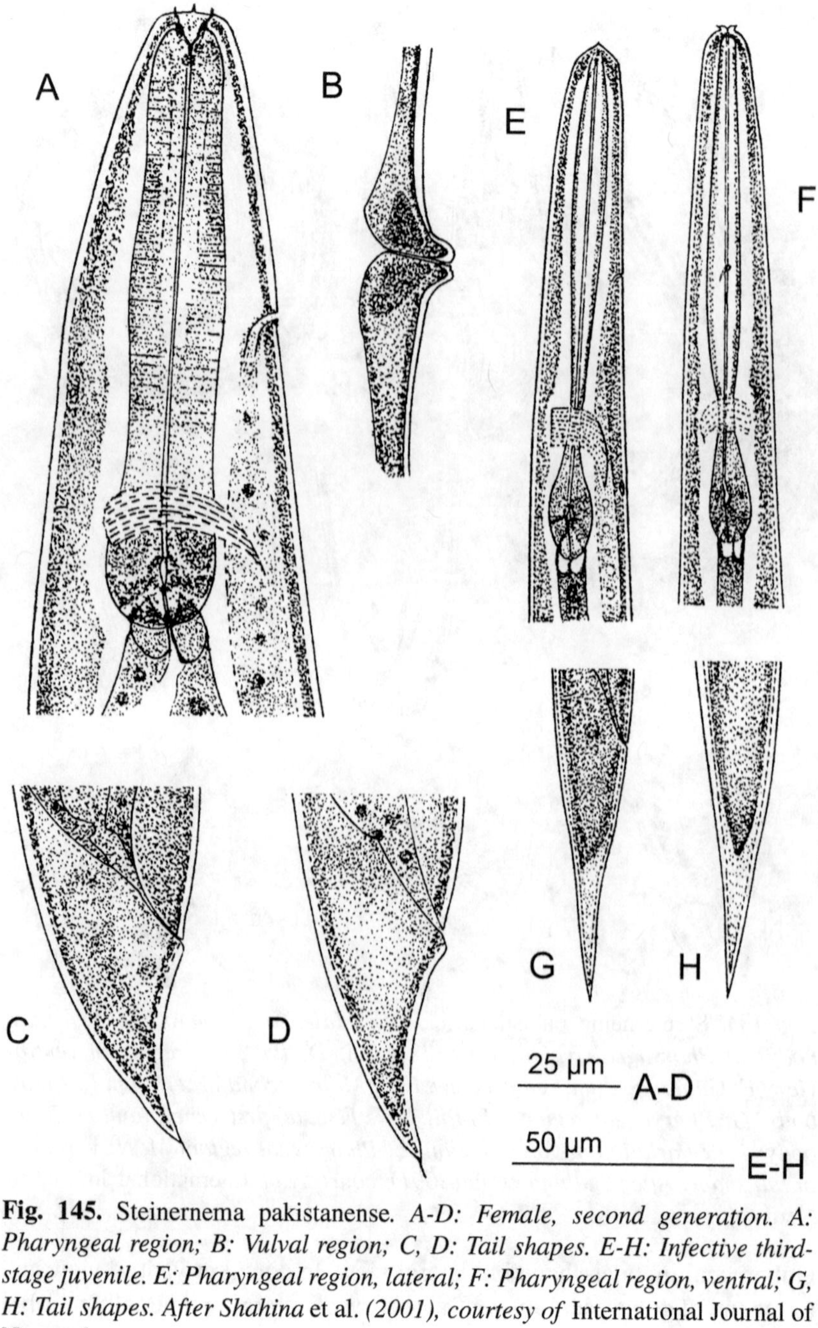

Fig. 145. Steinernema pakistanense. *A-D: Female, second generation. A: Pharyngeal region; B: Vulval region; C, D: Tail shapes. E-H: Infective third-stage juvenile. E: Pharyngeal region, lateral; F: Pharyngeal region, ventral; G, H: Tail shapes. After Shahina* et al. *(2001), courtesy of* International Journal of Nematology.

Table 42. *Morphometrics of type population of* Steinernema pakistanense. *Measurements are in μm and in the form: mean ± SD (range).*

Character	First generation			Second generation		Infective juvenile
	Holotype (male)	Male	Female	Male	Female	
n	–	20	20	20	20	20
L	1074	1357 ± 89 (1163-1505)	9901 ± 1591 (6461-12 092)	908 ± 64 (788-1018)	1829 ± 258 (1442-2339)	683 ± 21 (649-716)
a	11.1	13.4 ± 1.8 (11-16)	37 ± 4.8 (25-43)	16 ± 1 (14.1-18)	19 ± 1.3 (18-22)	24 ± 1.5 (21-27)
b	9.1	10.4 ± 0.6 (8.7-11)	57 ± 11.4 (37-62)	7.4 ± 0.6 (5.8-8)	12 ± 1.5 (10-15)	6 ± 0.3 (5-6)
c	40	55 ± 3.3 (52-62)	200 ± 21 (170-236)	37 ± 4 (31-43)	29 ± 4.1 (23-38)	11 ± 0.5 (10-12)
V	–	–	49 ± 1.7 (47-52)	–	52 ± 2.4 (47-55)	–
Body diam.	96	102 ± 10.2 (80-128)	265 ± 25 (240-310)	56 ± 5 (48-71)	92 ± 8.3 (77-101)	27 ± 1.2 (24-29)
EP	68	81 ± 4.8 (72-92)	81 ± 8.6 (72-94)	72 ± 6.3 (64-86)	72 ± 6 (64-80)	54 ± 2.2 (49-58)
NR	88	99 ± 6.3 (88-107)	112 ± 11.4 (96-126)	82 ± 3.1 (80-88)	104 ± 4 (97-110)	80 ± 2.1 (76-83)
ES	118	132 ± 5.8 (126-146)	179 ± 8.3 (165-194)	121 ± 5 (110-137)	146 ± 4 (140-152)	113 ± 4.2 (108-122)
Tail length (T)	26	25 ± 0.8 (24-27)	49 ± 9.8 (35-63)	24 ± 1.3 (21-30)	61 ± 2.2 (57-64)	58 ± 2.1 (53-62)
Anal body diam. (ABD)	35	36 ± 2.3 (32-40)	–	31 ± 1.5 (28-35)	–	–
Spicule length (SL)	64	68 ± 3.6 (62-73)	–	52 ± 3.5 (47-59)	–	–
Gubernaculum length (GL)	41	41 ± 3.2 (36-45)	–	29 ± 2 (25-32)	–	–
D% = EP/ES × 100	50	60 ± 3 (50-60)	40 ± 7 (30-40)	60 ± 5 (40-60)	50 ± 3 (40-50)	47 ± 2.7 (42-53)
E% = EP/T × 100	260	310 ± 40 (210-370)	160 ± 20 (130-200)	300 ± 30 (250-370)	110 ± 8 (100-130)	91 ± 5 (87-102)
SW% = SL/ABD × 100	183	189	–	167.7	–	–
GS% = GL/SL × 100	64.1	60.3	–	55.8	–	–
H% = H/T × 100	–	–	–	–	–	~50

EP = distance from anterior end to excretory pore; NR = distance from anterior end to nerve ring; ES = distance from anterior end to end of pharynx.

width = 3.6. Cuneus short, fine, needle-shaped, posteriorly pointed to form Y-shaped structure. Bursa absent. Complement of 23 genital papillae (11 pairs and one single) arranged as: six precloacal subventral pairs, single papilla located midventrally anterior to cloaca; one precloacal subdorsal pair; one adcloacal subventral pair; one postcloacal subdorsal pair and two postcloacal subventral pairs located near tail tip. Tail short, bluntly conoid, convex. Mucron present in 10% of males and indistinct in remaining 90%, being only visible with SEM.

Male (second generation)

Similar to first generation, differing by smaller size. Spicules and gubernaculum smaller than first generation. Tail short, dorsally convex, lacking a mucron.

Female (first generation)

Heat-relaxed specimens usually spiral- or C-shaped. Cuticle with faint transverse striae. Lateral field and phasmids inconspicuous. Cephalic extremity truncate to slightly rounded, not set off from rest of body. Lips united, bearing one internal circle of six labial papillae and one external circle of four submedial cephalic papillae, amphidial apertures located at base of internal lateral papillae. Stoma shallow, triangular at base. Pharynx extending close to oral aperture. Cheilorhabdions located beneath lips, composed of thick, cuticularised ring. Posterior to this is another cuticularised ring (presumably prorhabdions); meso-, meta- and telorhabdions vestigial. Pharynx muscular, cylindrical; metacorpus slightly swollen; non-valvate; isthmus slightly narrowing, terminating in muscular basal bulb with small, distinct, valve. Pharynx set off from intestine, posteriorly inserted into anterior portion of intestine. Cardia well developed. Excretory pore opening usually anterior to nerve ring at level of metacorpus. Nerve ring distinct, located anterior to basal bulb. Genital tract amphidelphic; opposed reflexed ovaries, in dorsal portion; ovaries leading to well developed glandular oviduct; spermatheca and uterus located ventrally between vagina and flexure. Vulva median in position, in form of a transverse slit, vulval lips usually protruding slightly from body surface with weakly developed, double flapped, epiptygmata. Vagina short, leading into paired genital tracts. Eggs commonly developing to juvenile stage, and juveniles hatching inside female body and boring their way out. Tail short, not constricted, dorsally convex with bluntly rounded terminus, mucron absent. Tail shorter than anal body diam.

Female (second generation)

Similar in general aspects to first generation except for smaller length and diam. Vulva generally protruding from body surface; vulval lips with transverse crease. Tail proportionately longer, conoid, tapering evenly to salient posterior anal lip; terminus with fine, mucronate, process.

Infective juvenile

Body cylindroid, tapering smoothly at both extremities (from base of pharynx to anterior end and from anus to terminus). Cuticle with fine transverse annulations. Newly moulted infective juvenile often ensheathed in second-stage cuticle. Heat-relaxed specimens die straight to slightly ventrally arcuate. Lateral field distinct at mid-body region, marked by eight distinct lines (*i.e.*, seven ridges). Phasmid inconspicuous. Cephalic extremity hemispherical, bearing two protuberant horn-like structures. Labial region with four distinct cephalic papillae, two subdorsal and two subventral, arranged medially on head. Cephalic region not set off. Oral aperture and anus closed. Excretory pore cuticularised, situated *ca* halfway between anterior end and base of pharynx, with long, cuticularised, excretory duct extending posteriorly, excretory glands usually distinct and displacing basal bulb, bacterial pouch and anterior end of intestine dorsally. Nerve ring distinct, located anterior to basal bulb. Pharynx long and narrow, distinctly narrower at level of nerve ring, terminating in a valvate basal bulb displaced dorsally. Cardia present. Sharply delineated bacterial pouch enclosed by intestinal wall, located posterior to cardia. Lumen of intestine narrow and closed. Rectum long and narrow, anus distinct. Tail conoid, tapering to a fine pointed terminus. Hyaline region comprising *ca* half total tail length, terminus convex, rounded, mucron absent.

TYPE HOST AND LOCALITY

Natural host unknown. Isolated by *Galleria mellonella* baiting from soil around the roots of vegetables in the Malir area, Karachi, Sindh, Pakistan.

TYPE MATERIAL

Holotype male and numerous paratypes (males, females, first and second generations and infective stage juveniles) deposited in the Nematode Collection at the National Nematological Research Centre, University of

Karachi, Karachi 75270, Pakistan. Other paratypes deposited in the nematode type collection at Rothamsted Research, Harpenden, Herts, UK, and in the CABI Europe-UK Nematode Collection, Egham, Surrey, UK.

RELATIONSHIPS

Steinernema pakistanense is one of a small number of nominal species to have two prominent, horn-like, structures on the cephalic region of the infective juveniles, the others being *S. abbasi*, *S. bicornutum*, *S. ceratophorum*, *S. riobrave* and *S. yirgalemense*. *Steinernema pakistanense* is principally characterised by the infective juvenile body length of 683 (649-716) μm compared to 635 μm or less for *S. yirgalemense*, *S. riobrave* and *S. abbasi* and 706 μm or more for *S. ceratophorum* and *S. bicornutum*; the tail length of 58 (53-62) μm is closer to that of *S. abbasi* (56 μm) but different from other species in the group, and lateral field with six or eight ridges (*i.e.*, seven to nine incisures). For the first generation male, the D% of 60 is similar to that of *S. abbasi* but different from the other species and SW% = 180 compared to 170 or less for the other species. The spicule shape of *S. pakistanense* also differs from that of the other species.

MOLECULAR CHARACTERISATION

The unique RFLP profile of the ITS region of rDNA confirmed that the species was distinct when compared to a database of 40 other steinernematid species. The sequence of the ITS region is deposited at GenBank under accession no. AY748449. This sequence is characterised by its length (flanked by the primers 18S and 26S), 998 bp, ITS1 = 291, ITS2 = 300 bp and its composition, A = 0.29058, C = 0.16734, G = 0.22545, T = 31663.

BIONOMICS

Shahina *et al.* (2001) confirmed that *S. pakistanense* had a life cycle similar to other steinernematids. After infection at 30-35°C, first generation adults appeared at 39 h and second generation adults at 98 h. Infective juveniles emerged from the cadavers only 6 days after initial infection. Shahina *et al.* (2004) reported on the symbiotic bacterium associated with *S. pakistanense* whilst Tabassum and Shahina (2004) discussed *in vitro* rearing of the nematode for mass production. Bajaj and

Walia (2005) reported isolating spores of *Pasteuria*, a bacterial parasite of nematodes, from this species. The infected nematodes were extracted from soil around the roots of *Zizyphus jujuba* from Hisar, Haryana, India, and represented the first report of this well known parasite of soil nematodes from an entomopathogenic nematode.

DISTRIBUTION

In addition to the type locality in Sindh, Pakistan, the species has also been isolated from grass and ornamental plants collected at Sonmiani beach and Mubarak village, Balochistan, Pakistan (Shahina *et al.*, 2001) and from soil around the roots of *Ziziphus jujuba* from Hisar, Haryana, India (Bajaj & Walia, 2005; Walia *et al.*, 2006).

References

BAJAJ, H.K. & WALIA, K.K. (2005). Studies on a *Pasteuria* isolate from an entomopathogenic nematode, *Steinernema pakistanense* (Nematoda: Steinernematidae). *Nematology* 7, 637-640.

SHAHINA, F., ANIS, M., REID, A.P., ROWE, J. & MAQBOOL, M.A. (2001). *Steinernema pakistanense* sp. n. (Rhabditida: Steinernematidae) from Pakistan. *International Journal of Nematology* 11, 124-133.

SHAHINA, F., MANZAR, H. & TABASSUM, K.A. (2004). Symbiotic bacteria *Xenorhabdus* and *Photorhabdus* associated with entomopathogenic nematodes in Pakistan. *Pakistan Journal of Nematology* 22, 117-128.

TABASSUM, K.A. & SHAHINA, F. (2004). *In vitro* mass rearing of different species of entomopathogenic nematodes in monoxenic solid culture. *Pakistan Journal of Nematology* 22, 167-175.

WALIA, K.K., WALIA, R.K. & BAJAJ, H.K. (2006). Occurrence, host range, pathogenicity and population build-up of *Steinernema pakistanense* Shahina *et al.*, 2001 (Nematoda: Rhabditida). *International Journal of Nematology* 16, 164-168.

Steinernema puertoricense[*,**] Román & Figueroa, 1994
(Figs 146, 147)

MEASUREMENTS

See Table 43.

DESCRIPTION

Male (first generation)

Body C-shaped, curved posteriorly when heat-relaxed. Cuticle smooth. Cephalic extremity truncate to slightly rounded. Labial papillae and lips barely discernible in lateral view, amphidial apertures not observed. Stoma short, lined by protorhabdions. Pharynx muscular with slightly swollen, non-valvate, metacorpus, short isthmus, slightly enlarged basal bulb. Cardia present. Nerve ring surrounding isthmus just anterior to basal bulb. Excretory pore usually situated anterior to nerve ring. Genital system monorchic, reflexed. Spicules paired, symmetrical, arcuate, yellow-orange in colour. Distal tips of spicules generally lacking ventral concavity or hook, which, when present, is very indistinct. Gubernaculum long, boat-shaped. Under SEM spicule well curved, slender. Spicule head somewhat longer than wide. Blade arcuate with spicule tip curved ventrally. Velum thin, sometimes invisible. In ventral view, gubernaculum tapering anteriorly. Cuneus somewhat Y-shaped, reaching end of corpus. Corpus not separated posteriorly. Complement of 23 genital papillae (seven pairs ventrolateral, three pairs postcloacal, one pair dorsolateral plus single midventral precloacal). Tail terminus rounded.

Male (second generation)

Resembling male of first generation but slightly smaller. Excretory pore located anterior to nerve ring.

Female (first generation)

Body spiral-shaped when heat-relaxed. Cuticle smooth. Cephalic extremity truncate to slightly rounded, continuous with body contour. Labial papillae and lips visible in lateral view, amphidial apertures not

[*] Named after the island of *Puerto Rico*, the type locality.

[**] Description after Román and Figueroa (1994). Revised by Khuong B. Nguyen.

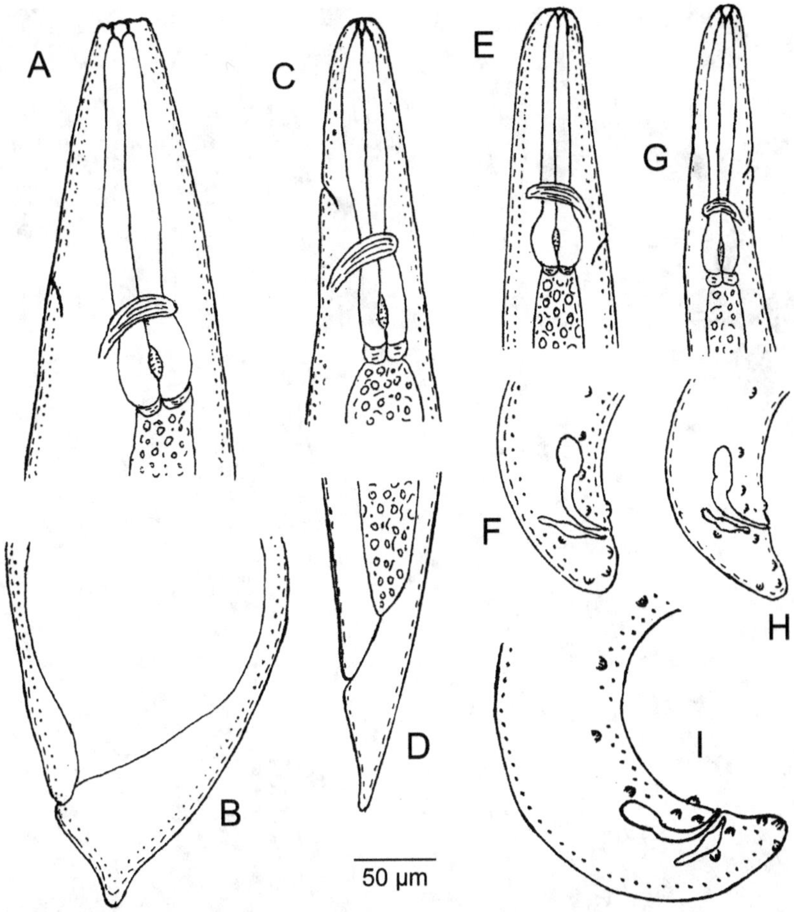

Fig. 146. Steinernema puertoricense. *First and second generation females and males. A, B: Anterior and posterior regions of first generation female; C, D: Anterior and posterior regions of second generation female; E, F: Anterior and posterior regions of first generation male; G, H: Anterior and posterior regions of second generation male; I: Posterior region of a first generation male showing genital papillae. Modified after Román and Figueroa (1994).*

observed under light microscope. Stoma short, lined by protorhabdions. Pharynx muscular with slightly swollen, non-valvate, metacorpus, short isthmus, slightly enlarged basal bulb with refractive valve plate ridges. Cardia present. Nerve ring surrounding isthmus just anterior to basal bulb. Excretory pore usually positioned anterior to nerve ring. Genital system amphidelphic, usually reflexed and filled with eggs from which

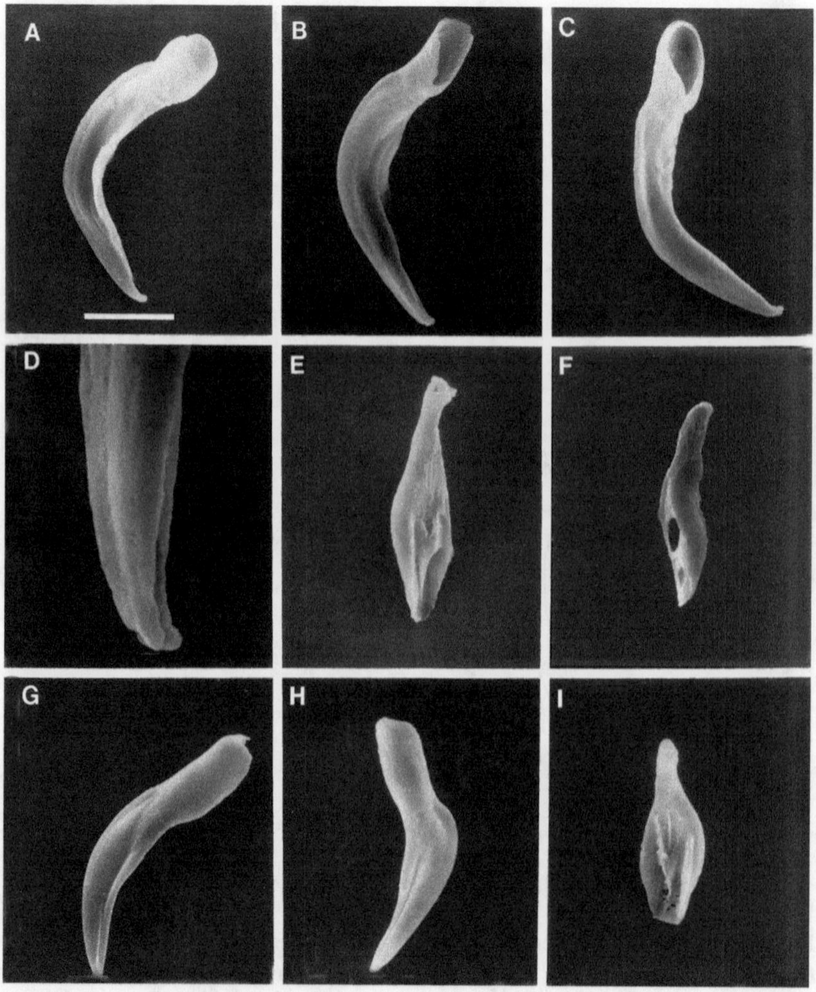

Fig. 147. Steinernema puertoricense. *SEM of spicules and gubernacula. A-C: Typical spicule; D: Spicule tip; E: Typical gubernaculum, subventral view; F: Gubernaculum, lateral view; G-I: Other forms of spicule and gubernaculum. (Scale bar: A-C, E, F, H, I = 20 μm; D = 4.3 μm; G = 17.6 μm.) After Nguyen and Smart (1997), courtesy of* Nematologica.

juveniles commonly hatch inside uterus. Vulva in form of a transverse slit protruding from body surface, double-flapped epiptygmata present. Vagina short, muscular. Tail usually shorter than anal body diam., conical or terminating in a slightly swollen, rounded projection.

Table 43. *Morphometrics of* Steinernema puertoricense. *Measurements are in μm and in the form: mean ± SD (range).*

Character	First generation			Second generation		Infective juvenile
	Holotype (female)	Male	Female	Male	Female	
n	–	10	10	20	10	10
L	7600	1461 ± 165 (1214-1667)	8627 ± 1591 (6090-11 090)	1203 ± 46 (1119-1262)	2256 ± 266 (1666-2690)	1171 ± 60 (1057-1238)
a	–	–	–	–	–	23 ± 1 (20-24)
b	–	–	–	–	–	8.2 ± 0.4 (7.4-8.6)
c	–	–	–	–	–	12.4 ± 0.8 (11.6-13.6)
V	50	–	50 ± 4 (40-53)	–	53 ± 3 (47-57)	–
Body diam.	375	101 ± 21 (67-143)	375 ± 39 (309-454)	64 ± 4 (57-67)	131 ± 17 (95-162)	51 ± 2 (47-54)
EP	142	123 ± 14 (95-143)	150 ± 35 (119-236)	122 ± 6 (112-133)	145 ± 16 (119-169)	95 ± 4 (90-102)
NR	200	122 ± 13 (98-138)	193 ± 32 (162-269)	131 ± 6 (124-143)	154 ± 11 (138-169)	117 ± 4 (111-121)
ES	238	159 ± 12 (143-178)	249 ± 26 (217-298)	166 ± 4 (157-171)	189 ± 9 (176-202)	143 ± 3 (138-147)
Tail length (T)	45	34 ± 5 (24-43)	59 ± 11 (40-74)	37 ± 3 (33-40)	83 ± 8 (74-98)	94 ± 6 (88-107)
Spicule length (SL)	–	78 ± 5 (71-88)	–	63 ± 8 (48-71)	–	–
Gubernaculum length (GL)	–	40 ± 3 (36-45)	–	34 ± 3 (28-36)	–	–
D% = EP/ES × 100	–	77	–	73		66 ± 3 (62-74)
E% = EP/T × 100	–	–	–	–	–	101 ± 7 (88-108)
SW% = SL/ABD × 100	–	–	–	–	–	–
GS% = GL/SL × 100	–	51	–	54	–	–

EP = distance from anterior end to excretory pore; NR = distance from anterior end to nerve ring; ES = distance from anterior end to end of pharynx.
After the original authors.

Pygmy females (first generation)

Body swollen, C-shaped when heat-relaxed. Resembling female but much smaller. Uncommon, observed only in one of three life cycle studies conducted.

Female (second generation)

Resembling female of first generation but *ca* 25% as long. Body usually C-shaped when heat-relaxed. Tail conical, *ca* twice as long as anal body diam.

Infective juvenile

Usually enclosed in second-stage cuticle. Body long (>1000 μm) and narrow. Oral aperture and anus generally closed. Pharynx and intestine usually indistinct. Lateral field marked with six incisures. (Note: This number was reported by the original authors, although it is probably not correct.) Tail pointed, twice as long as body diam.

TYPE HOST AND LOCALITY

Natural host unknown. Obtained from sand from a coconut plantation at Lofza, Puerto Rico.

TYPE MATERIAL

Holotype first generation female (P.R. Slide number 1), deposited in the USDA Nematode Collection, Beltsville, MD, USA. Allotype (first generation male, Slide number 2) and paratype first generation adults and infective juveniles in the same collection (first generation females, vial number 1; first generation males, vial number 2; infective juveniles, vial number 3).

NOTE

Steinernema puertoricensis, the original species name, was changed to *S. puertoricense*, by Hominick *et al.* (1997) in order to comply with International Code of Zoological Nomenclature (ICZN) rules. The original authority is unchanged.

RELATIONSHIPS

Infective juveniles are over 1000 μm in length and have a smaller pharynx and ratio a but greater tail length than related species. The male has a typical spicule and gubernaculum shape (Fig. 147). The female vulva bears epiptygmata.

Steinernema puertoricense can be distinguished from its most closely related species, *S. glaseri*, by the first generation male of *S. puertoricense* having the excretory pore located anterior or posterior to nerve ring, as compared with *S. glaseri* where the excretory pore is always posterior to the nerve ring. The second generation male has a longer tail at 37 (33-40) *vs* 28 (25-31) μm in *S. glaseri*. *Steinernema puertoricense* males usually lack the hook located at the ventral side of the spicule tip, a feature that is characteristic of *S. glaseri*, and the spicules are distinctly arcuate (Fig. 147). Infective juveniles have a shorter pharynx, 143 (138-147) *vs* 162 (158-168) μm in *S. glaseri*; greater tail length of 94 (88-107) *vs* 78 (62-87) μm in *S. glaseri*; and smaller ratio a of 23 (20-24) *vs* 29 (26-35) in *S. glaseri*. *Steinernema puertoricense* can be differentiated from other related species (*i.e.*, those with a long IJ) by ratio a, E% of the infective juvenile and SW% and GS% values of the male.

MOLECULAR CHARACTERISATION

D2D3 regions 870 bp (accession no. AF331903); 12S ribosomal RNA gene, mitochondrial, 460 bp (accession no. AY944022); cytochrome *c* oxidase subunit I (*cox1*) gene mitochondrial, 568 bp (accession no. AY943996).

BIONOMICS

Steinernema puertoricense entered larvae of *Galleria mellonella* and completed its life cycle in 5 days at 27-28°C. It took 2 days for the IJ to enter, kill the host and develop into first generation males and females. Second generation adults appeared 4 days after infection. Infective juveniles began emerging 5 days after infection.

DISTRIBUTION

This species is known only from Puerto Rico (Román & Figueroa, 1994; Fischer-Le Saux *et al.*, 1998).

References

FISCHER-LE SAUX, M., MAULEON, H., CONSTANT, P., BRUNEL, B. & BOEMARE, N. (1998). PCR ribotyping of *Xenorhabdus* and *Photorhabdus* isolates from the Caribbean region in relation to the taxonomy and geographic distribution of their nematode hosts. *Applied and Environmental Microbiology* 64, 360-366.

HOMINICK, W.M., BRISCOE, B.R., DEL PINO, F.G., HENG, J., HUNT, D.J., KOZODOY, E., MRÁČEK, Z., NGUYEN, K.B., REID, A.P., SPIRIDONOV, S., STURHAN, D., WATURU, C. & YOSHIDA, M. (1997). Biosystematics of entomopathogenic nematodes: current status, protocols and definitions. *Journal of Helminthology* 71, 271-298.

NGUYEN, K.B. & SMART JR, G.C. (1997). Scanning electron microscopic studies of spicules and gubernacula for *Steinernema* spp. (Nemata: Steinernematidae). *Nematologica* 43, 465-480.

ROMÁN, J. & FIGUEROA, W. (1994). *Steinernema puertoricensis* n. sp. (Rhabditida: Steinernematidae), a new entomopathogenic nematode from Puerto Rico. *Journal of Agriculture, University of Puerto Rico* 78, 167-175.

Steinernema rarum[*,**] (de Doucet, 1986) Mamiya, 1988

= *Neoaplectana rara* de Doucet, 1986

(Figs 148-153)

Measurements

See Table 44.

Description

Male (first generation)

Body C-shaped when heat-relaxed. Cuticle smooth, lateral fields and phasmid not observed. Cephalic extremity truncate; lips united, oral aperture subtriangular. Six labial papillae, four prominent cephalic papillae and amphidial apertures present. Stoma, cephalic region and pharyngeal region as for genus. Genital system monorchic, reflexed. Spicules paired, heavy, curved, with internal ribs. Capitulum round. Spicule tip usually blunt. Spicules and gubernacula of unusual shape for genus. Spicule head short, velum prominent, extending for more than two-third of blade. Posterior third of spicule straight, ventral side with depression close to spicule tip. Gubernaculum tapering anteriorly to a somewhat ventrally curved end. Corpus separated posteriorly. Cuneus rod-like. General shape of spicules of Argentina isolates similar to those of USA isolates. Complement of 10-11 pairs of genital papillae: 4-5 pairs precloacal, one pair lateral, two pairs adcloacal, three pairs postcloacal, plus a single midventral precloacal papilla. Poinar *et al.* (1988) noted that instead of the single ventral precloacal papilla seen in other species of *Steinernema*, double papillae were observed in *S. rarum*, but Nguyen *et al.* (2006) found that males of *S. rarum*, both from the United States and Argentina, have only one precloacal, midventral papilla. Two unique, slit-like, sometimes pore-like, apertures present on posterior lip of cloaca of both USA and Argentina strains. Tail conical, tapering to a fine mucronate tip.

[*] From the Latin *rarus* = scarce.

[**] Description after de Doucet (1986) and Nguyen *et al.* (2006). Revised by Khuong B. Nguyen.

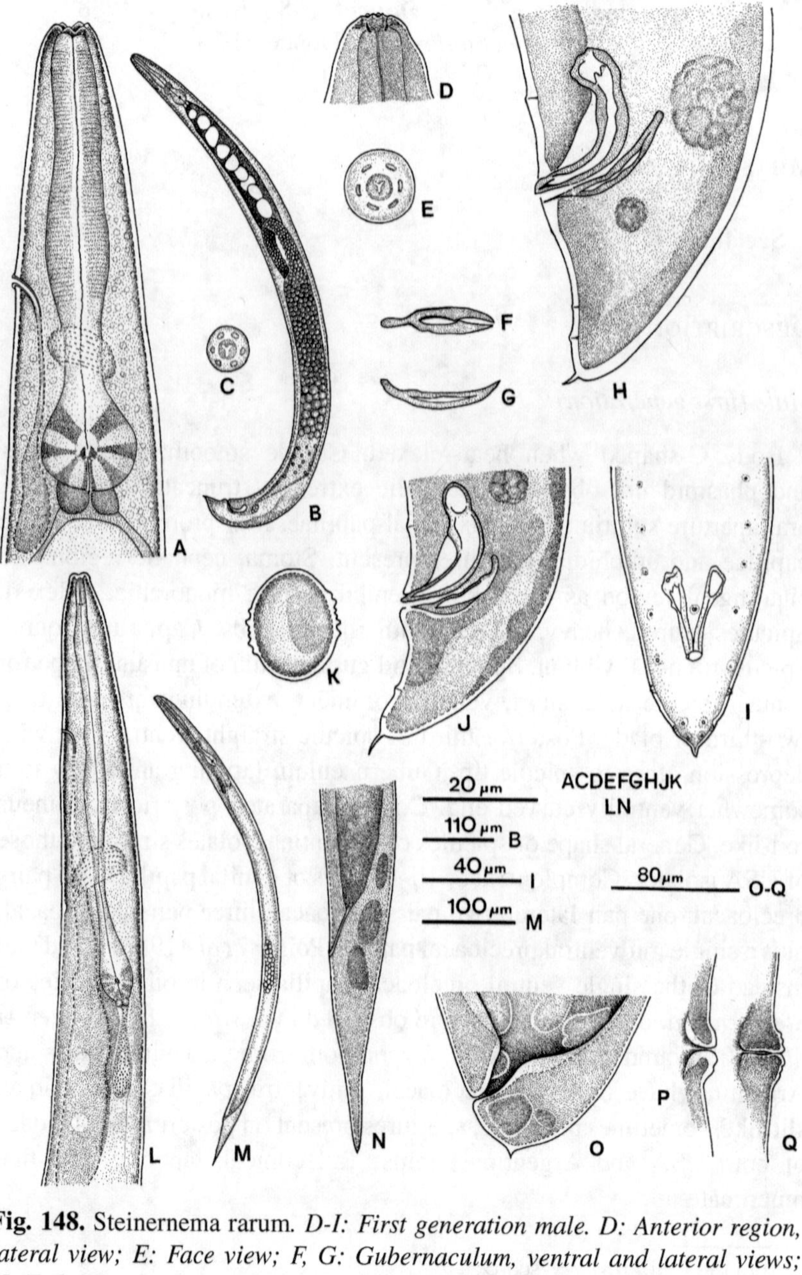

Fig. 148. Steinernema rarum. *D-I: First generation male. D: Anterior region, lateral view; E: Face view; F, G: Gubernaculum, ventral and lateral views; H: Tail, lateral view; I: Tail, ventral view. A-C, J: Second generation male. A: Pharyngeal region; B: Entire body; C: Face view; J: Tail, lateral view.*

Male (second generation)

Similar to first generation but smaller with thinner spicules and tail with a well developed mucron.

Female (first generation)

Cuticle smooth, lateral field and phasmids not observed. Cephalic papillae absent in a proportion of specimens (10% for USA isolate 17C). Cephalic papillae present in many specimens but small, sometimes difficult to detect. Perioral disc hexagonal or star-shaped. Labial papillae six, prominent. Hexagonal or star-shaped perioral disc making each lip resemble an eye. Amphidial apertures inconspicuous. Vulva with annules around its aperture. Rectum and anus distinct. Tail conoid to dome-shaped, sometimes with short spine on tip.

Female (second generation)

Similar to first generation but body smaller and tail longer.

Infective juvenile

Body slender, tapering to both ends, spiral shaped, usually enclosed in sheath formed from second-stage cuticle. Cuticle with fine striae. Cephalic extremity not annulated, labial papillae not observed, four prominent cephalic papillae, amphidial apertures pronounced. Oral aperture and anus closed. Lateral field very variable. Lateral field pattern beginning with two ridges. Near excretory pore level, there are eight ridges. Two submarginal ridges at mid-body not as raised as others, difficult to see; only six visible under light microscope. Posterior part of lateral field very variable. Sometimes (5% of specimens) number of ridges becomes ten. Near anus, number of ridges usually becoming six, then two at some annules anterior or posterior to phasmid. In posterior region of some individuals (*ca* 10% of specimens) pattern changing to two marginal bands and a central broad, smooth, band. Finally lateral field reducing to single smooth band. In *ca* 10% of population, lateral field with two wide ridges at beginning, turning into smooth band extending to end of lateral field. Pharynx long

Fig. 148. *(Continued). K-N: Infective juvenile. K: Cross-section through mid-body; L: Pharyngeal region; M: Entire body; N: Tail, lateral view. O, P: First generation female. O: Tail; P: Vulvar region. Q: Second generation female, vulva region. After de Doucet (1986), courtesy of* Revue de Nématologie.

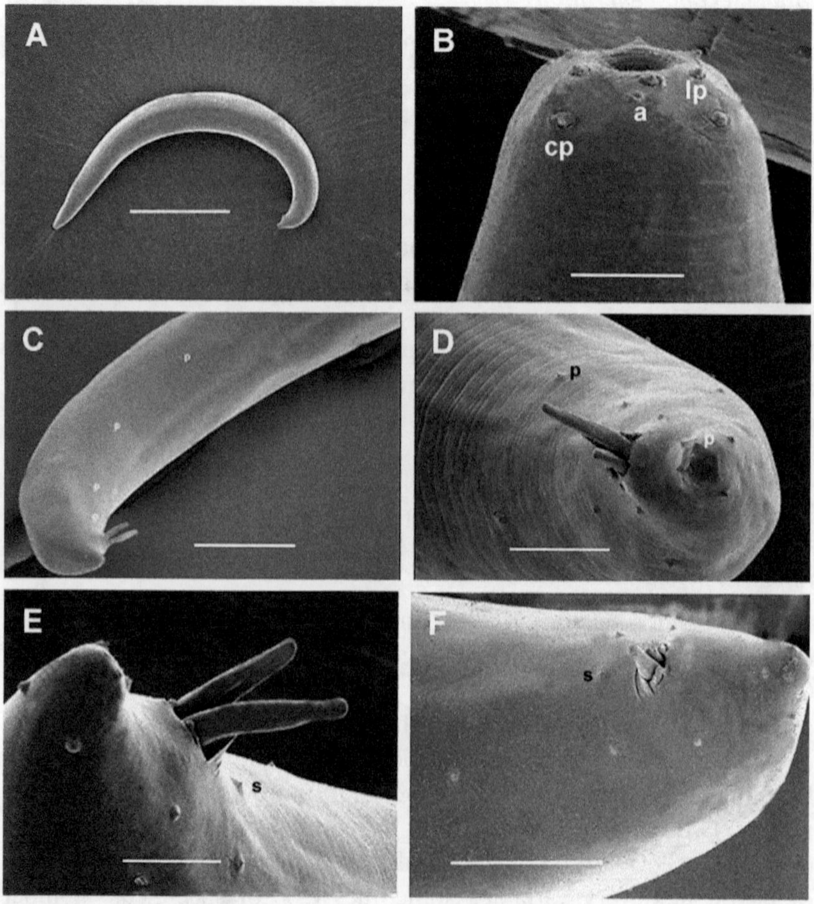

Fig. 149. Steinernema rarum. *SEM photographs of some structures of first generation males, USA isolate 17C. A: Entire, C-shaped, body; B: Anterior region showing oral aperture, six labial papillae (lp), two of four cephalic papillae (cp), and one of two amphidial apertures (a); C: Posterior region showing four preanal papillae and protruding spicules; D: Ventral view of posterior region of a male showing seven pairs of genital papillae and a small mucron; E: Subventral view of posterior region of a male showing lateral, subventral, caudal dorsal papillae, and a single precloacal papilla (s); F: Ventral view of posterior region of a male showing pairs of genital papillae and a single precloacal papilla (s). (Scale bars: A = 268 μm; B = 6 μm; C = 67 μm; D = 16.8 μm; E = 11.9 μm; F = 27.3 μm.) After Nguyen* et al. *(2006), courtesy of* Journal of Nematology.

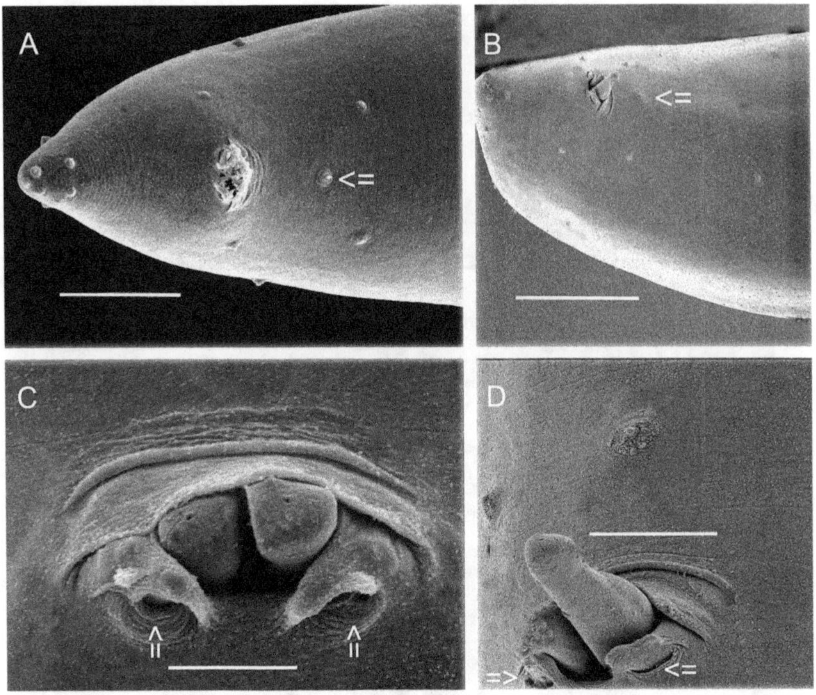

Fig. 150. Steinernema rarum. *Comparison of SEM photographs of posterior regions of male of Argentina (first column) and USA isolate 17C (second column). A, B: Both isolates showing only single preanal genital papilla (arrows); C, D: Cloaca showing spicule tips and two unique openings posterior to edge of cloaca. Note that the openings in Argentina (arrows) strains are wide but those of USA strains (arrows) are slit-like in shape. (Scale bars: A = 16.7 μm; B = 27.3 μm; C = 3 μm; D = 6 μm.) After Nguyen* et al. *(2006), courtesy of* Journal of Nematology.

and narrow, distinctly narrower at level of nerve ring, terminating in a dorsally displaced, valvate, basal bulb. Excretory pore located in anterior half of pharynx. Excretory duct long, usually distinct; ventral excretory glands displacing basal bulb, bacterial pouch and anterior end of intestine dorsally. Nerve ring distinct, located anterior to basal bulb. Hemizonid at level of basal bulb. Cardia present. Bacterial pouch present in intestine posterior to cardia. Lumen of intestine narrow, appearing closed. Rectum long, narrow. Anus distinct. Genital primordium well developed, in middle of posterior half of body. Tail conical, terminus pointed.

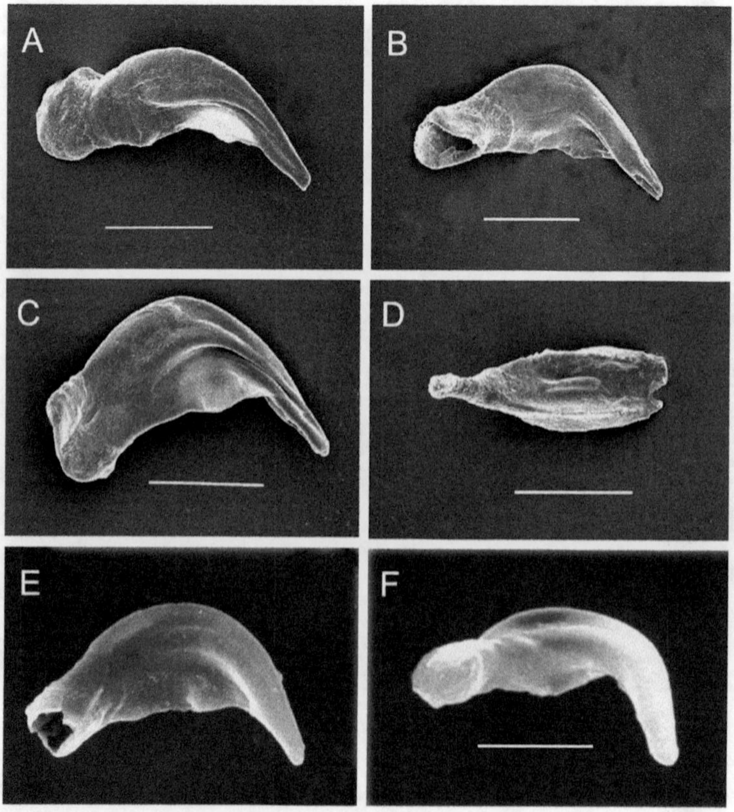

Fig. 151. Steinernema rarum. *SEM photographs of spicules and gubernacula of USA and Argentina isolates. A-C: USA strain 17C, spicules are dissimilar in shape, with short head, prominent velum; D: Gubernaculum, ventral view with rod-like cuneus, and posterior end separated; E, F: Argentina strain, spicules dissimilar in shape. (Scale bars: A-D = 15 μm; E, F = 12 μm.) After Nguyen* et al. *(2006), courtesy of* Journal of Nematology.

TYPE HOST AND LOCALITY

Body cavity of *Heliothis* sp. (Lepidoptera: Noctuidae), from Rio Cuarto, Province of Cordoba, Argentina.

TYPE MATERIAL

Holotype: female, on slide RAX/171-6, deposited in the Centro de Zoologia Aplicada, Universidad Nacional de Cordoba, Argentina. Paratypes: specimens distributed as follows: five females and five males

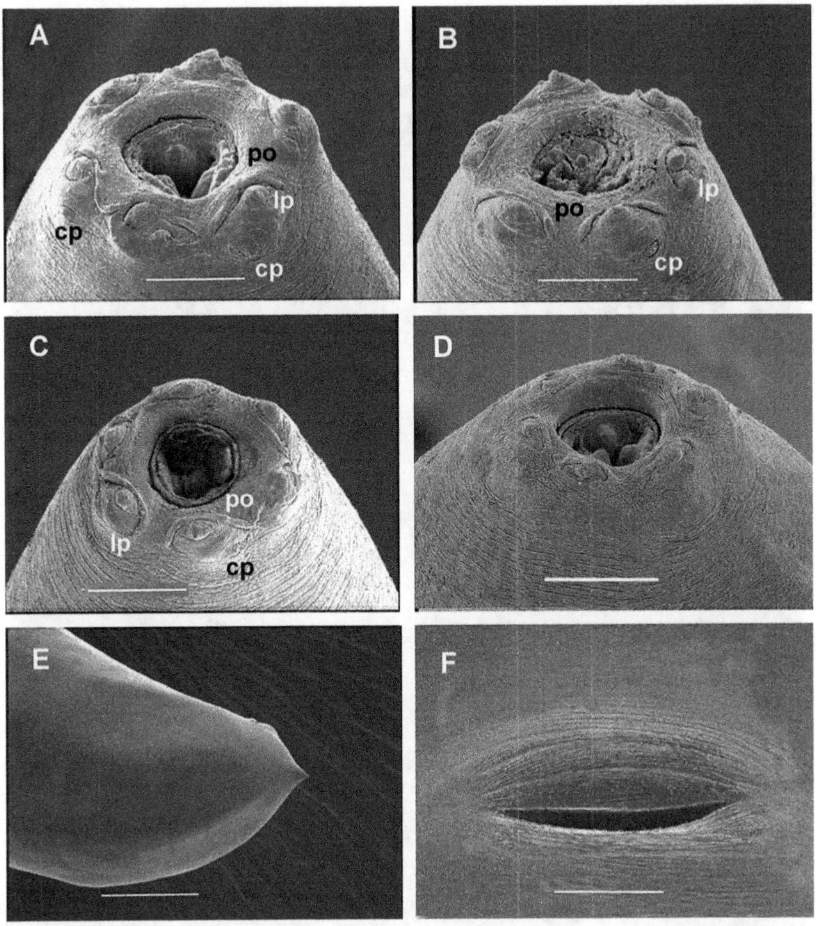

Fig. 152. Steinernema rarum. *SEM photographs of some morphological structures of first generation females. A-C: Variation in face views of four females showing oral aperture, hexagonal star-shaped perioral disc (po), six prominent eye-shaped lips with papillae (lp) and small cephalic papillae (cp); D: Head of female showing labial papillae but not cephalic papillae; E, F: First generation female tail and vulva. (Scale bars: A, B = 6.67 μm; C = 7.5 μm; D, F = 10 μm; E = 30 μm.) After Nguyen* et al. *(2006), courtesy of* Journal of Nematology.

of first generation in Station de Recherches sur les Nematodes, INRA, Antibes, France, and in the University of California-Davis Nematode Collection, Davis, CA, USA. Remaining specimens deposited in the Centro de Zoologia Aplicada, Argentina.

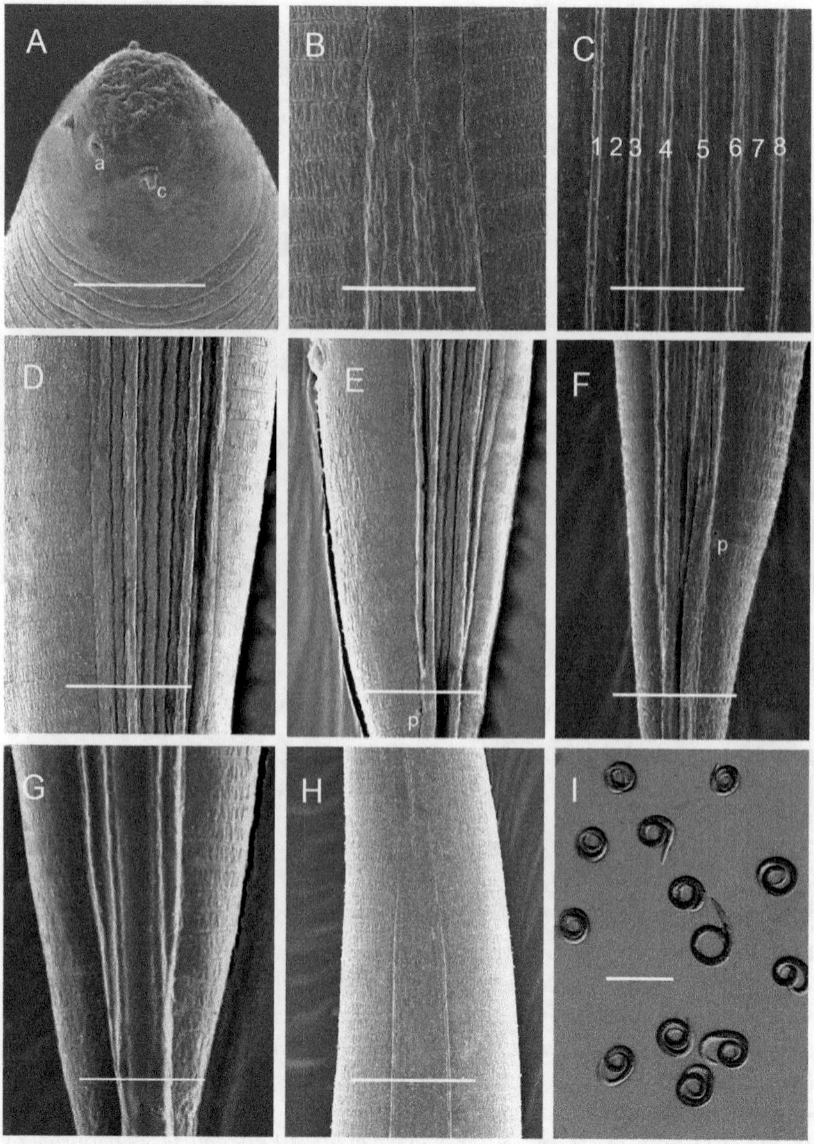

Fig. 153. Steinernema rarum. *SEM photographs of infective juveniles of USA isolate 17C. A: Head showing three of four prominent cephalic papillae (c), one of two* amphidial apertures *(a); B: Lateral field showing two ridges anteriorly followed by eight ridges; C: Lateral field at mid-body with eight ridges, two submarginal ridges (two, seven) difficult to see; D: Posterior part of body showing lateral field with ten ridges.*

NOTE

Steinernema rara, the original species name, was changed to *S. rarum* by Hominick *et al.* (1997) in order to comply with International Code of Zoological Nomenclature (ICZN) rules. The original authority is unchanged.

RELATIONSHIPS

Steinernema rarum is characterised by the short body length of the infective juveniles, size and shape of the spicules and rod-like cuneus of the gubernaculum. The male of this species has two pore-like apertures on the posterior cloacal lip. These two structures have not been observed in other *Steinernema* species. The female has a hexagonal or star-shaped perioral disc, the form of which makes each lip resemble an eye, another unique character for this species.

Steinernema rarum can be differentiated from other species of *Steinernema* by the above diagnostic characteristics.

MOLECULAR CHARACTERISATION

This nematode species is characterised by sequences of ITS and D2D3 regions. Length of the ITS regions = 935 bp, ITS1 = 251 bp, ITS2 = 312 bp and its composition (A = 0.274, C = 0.187, G = 0.235, T = 0.304) (accession no. AY253296). Length of D2D3 regions = 914 bp (accession no. AY253296). More information is available in Nguyen *et al.* (2006).

BIONOMICS

The life cycle of this nematode is similar to that of other *Steinernema*. As reported by Koppenhöfer and Kaya (1999), *S. rarum* infects a wide range of arthropod species under laboratory conditions but appears to

Fig. 153. *(Continued). E, F: Near anus, number of ridges lateral field reducing to six, then to four and two at some annules anterior (E) and posterior (F) to phasmid (p); G: In one form, lateral field changing into three bands, two marginal and a broad central band, becoming one near tail tip; H: Lateral field with two ridges and smooth band; I: Spiral IJ. (Scale bars: A = 3.7 μm; B = 3.8 μm; C-E = 6 μm; F = 6.0 μm; G = 4.2 μm; H = 10 μm; I = 250 μm.) After Nguyen* et al. *(2006), courtesy of* Journal of Nematology.

Table 44. *Morphometrics of* Steinernema rarum. *Measurements are in μm and in the form: mean ± SD (range).*

Character	First generation			Second generation		Infective juvenile*
	Holotype (female)	Male	Female	Male	Female	
n	–	20	20	20	20	20
L	7400	1600 ± 1100 (1300-1700)	7390 ± 1400 (5200-10 400)	800 ± 20 (700-1100)	4000 ± 280 (3100-4100)	510 ± 31 (446-578)
a	–	–	–	–	–	20.2 ± 1.7 (18.1-23.2)
b	–	–	–	–	–	5.0 ± 0.3 (4.4-5.7)
c	–	–	–	–	–	10.4 ± 0.7 (9.4-11.5)
V	56	–	57 ± 3.1 (52-66)	–	58 ± 1.9 (55-62)	–
Body diam.	181	123 ± 9.8 (100-142)	169 ± 23 (132-216)	53 ± 6.5 (45-63)	156 ± 11 (143-177)	13 ± 1 (11-15)
EP	83	67 ± 8.2 (55-77)	93 ± 11.2 (65-113)	60 ± 6.4 (47-70)	88 ± 5.7 (75-95)	37 ± 1.6 (33-40)
NR	160	103 ± 7.9 (91-120)	163 ± 12 (132-183)	86 ± 4.9 (75-92)	150 ± 5.8 (138-157)	67 ± 4 (62-76)
ES	223	135 ± 9.1 (125-150)	219 ± 14 (205-250)	114 ± 4 (106-120)	190 ± 7.6 (177-202)	99 ± 8 (88-117)
Tail length (T)	41	43 ± 4.7 (35-42)	44 ± 4.6 (37-50)	39 ± 2.5 (32-42)	46 ± 3.3 (38-51)	48 ± 3 (42-52)
Anal body diam. (ABD)	93	50 ± 4.2 (40-57)	84 ± 13 (65-110)	33 ± 2.6 (30-38)	62 ± 6.1 (53-73)	–
Spicule length (SL)	–	47 ± 3.3 (42-52)	–	36 ± 2.5 (32-40)	–	–
Spicule width	–	7 ± 0.8 (5.5-9)	–	6 ± 0.4 (5-6.5)	–	–
Gubernaculum length (GL)	–	34 ± 7.1 (23-38)	–	27 ± 1.4 (25-30)	–	–
Gubernaculum width	–	–	–	–	–	–

be best adapted to lepidopteran larvae as hosts. The infective juvenile employs an intermediate foraging strategy, efficiently infecting both mobile hosts on the soil surface and sedentary hosts within the soil. The acceptable range of soil moisture suggests an adaptation to fluctuating soil moisture in the upper soil layers. The temperature optimum of 25°C corresponded well to the warm temperate climate of its area of origin in Argentina. The optimal storage temperature was around 15°C. This study suggested that *S. rarum* is adapted to lepidopteran hosts that have

Table 44. *(Continued).*

Character	First generation			Second generation		Infective juvenile
	Holotype (female)	Male	Female	Male	Female	
D% = EP/ES × 100	–	50	–	53	–	40.7 ± 3.3 (36-43)
E% = EP/T × 100	–	–	–	–	–	81 ± 5.5 (67-91)
SW% = SL/ABD × 100	–	94	–	109	–	–
GS% = GL/SL × 100	–	72	–	75	–	–
H% = Hyaline/T × 100	–	–	–	–	–	56 ± 5 (45-63)

EP = distance from anterior end to excretory pore; NR = distance from anterior end to nerve ring; ES = distance from anterior end to end of pharynx.
After the original author.
* After Nguyen *et al.* (2006).

larvae spending part of their life on the soil surface and/or that pupate in the soil

DISTRIBUTION

This species is known from Argentina (de Doucet, 1986; Doucet & de Doucet, 1997) and from the USA (Nguyen *et al.*, 2006).

References

DOUCET, M.E. & DE DOUCET, M.M.A. (1997). Nematodes and agriculture in continental Argentina. An overview. *Fundamental and Applied Nematology* 20, 521-539.

DE DOUCET, M.M.A. (1986). A new species of *Neoaplectana* Steiner, 1929 (Nematoda: Steinernematidae) from Cordoba, Argentina. *Revue de Nématologie* 9, 317-323.

HOMINICK, W.M., BRISCOE, B.R., DEL PINO, F.G., HENG, J., HUNT, D.J., KOZODOY, E., MRÁČEK, Z., NGUYEN, K.B., REID, A.P., SPIRIDONOV, S., STURHAN, D., WATURU, C. & YOSHIDA, M. (1997). Biosystematics of entomopathogenic nematodes: current status, protocols and definitions. *Journal of Helminthology* 71, 271-298.

KOPPENHÖFER, A.M. & KAYA, H.K. (1999). Ecological characterization of *Steinernema rarum*. *Journal of Invertebrate Pathology* 73, 120-128.

MAMIYA, Y. (1988). *Steinernema kushidai* n. sp. (Nematoda: Steinernematidae) associated with scarabaeid beetle larvae from Shizuoca, Japan. *Applied Entomology and Zoology* 23, 313-320.

NGUYEN, K.B., SHAPIRO-ILAN, D.I., FUXA, J.R., WOOD, B.W., BERTOLOTTI, M.A. & ADAMS, B.J. (2006). Taxonomic and biological characterization of *Steinernema rarum* found in the Southeastern USA. *Journal of Nematology* 38, 28-40.

POINAR JR, G.O., MRÁČEK, Z. & DOUCET, M.M.A. (1988). A re-examination of *Neoaplectana rara* Doucet, 1986 (Steinernematidae: Rhabditida). *Revue de Nématologie* 11, 447-449.

*Steinernema riobrave**,** Cabanillas, Poinar & Raulston, 1994
(Figs 154-159)

MEASUREMENTS

See Table 45.

DESCRIPTION

Male (first generation)

Cuticle smooth, head rounded, continuous with body. Six distinct lips each with terminal papilla. Four cephalic papillae. Amphidial apertures distinct, located posterior to lateral cephalic papillae. Stoma partially collapsed. Pharyngeal collar absent, pharynx extending close to oral aperture. Cheilorhabdions represented by thickening sclerotised structure just posterior to lips. Posterior to this, another sclerotised ring representing prorhabdions. Meso-, meta- and telorhabdions vestigial, occurring in collapsed area of stoma. Pharynx muscular with cylindrical procorpus, a slightly swollen non-valvate metacorpus, narrow isthmus, and basal bulb with valve. Nerve ring usually surrounding isthmus. Excretory pore opening usually anterior to nerve ring, but location variable. Lateral fields and phasmids inconspicuous. Genital system monorchic, reflexed. Spicules paired, sickle-shaped, with distinct, dark golden yellow colour. Spicule head elongate, tapering anteriorly. Shaft present but short. Lamina moderately curved with two internal ribs tapering gradually to bluntly pointed tip. Velum present. Gubernaculum 70% as long as spicules, boat-shaped in lateral view, ventrally curved with proximal knob or hook. In ventral view, gubernaculum neck almost same width as posterior corpus. Cuneus Y-shaped or needle-shaped, long, pointed posteriorly. Complement of 25 genital papillae (12 pairs plus a single midventral precloacal) observed under scanning electron microscope. In addition to 11 pairs normally present in *Steinernema* (six pairs precloacal, one pair lateral, one pair adcloacal, three pair postcloacal) an additional pair occurring at edge of cloacal aperture.

* Named after the *Rio Bravo* (the Mexican name for the Rio Grande), Weslaco, Texas, USA, the locality where the nematode was collected.

** Description after Cabanillas *et al.* (1994) and Nguyen and Adams (2003). Revised by Khuong B. Nguyen.

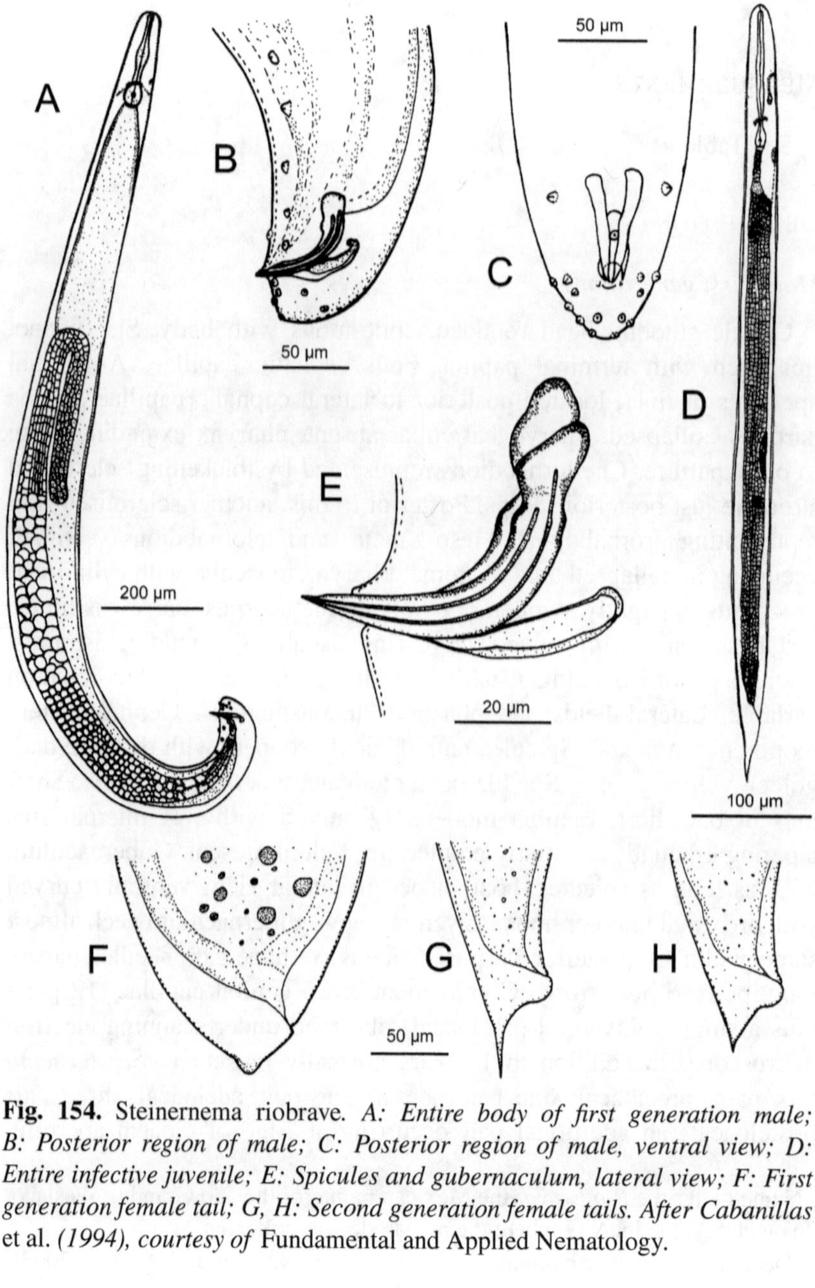

Fig. 154. Steinernema riobrave. *A: Entire body of first generation male; B: Posterior region of male; C: Posterior region of male, ventral view; D: Entire infective juvenile; E: Spicules and gubernaculum, lateral view; F: First generation female tail; G, H: Second generation female tails. After Cabanillas* et al. *(1994), courtesy of* Fundamental and Applied Nematology.

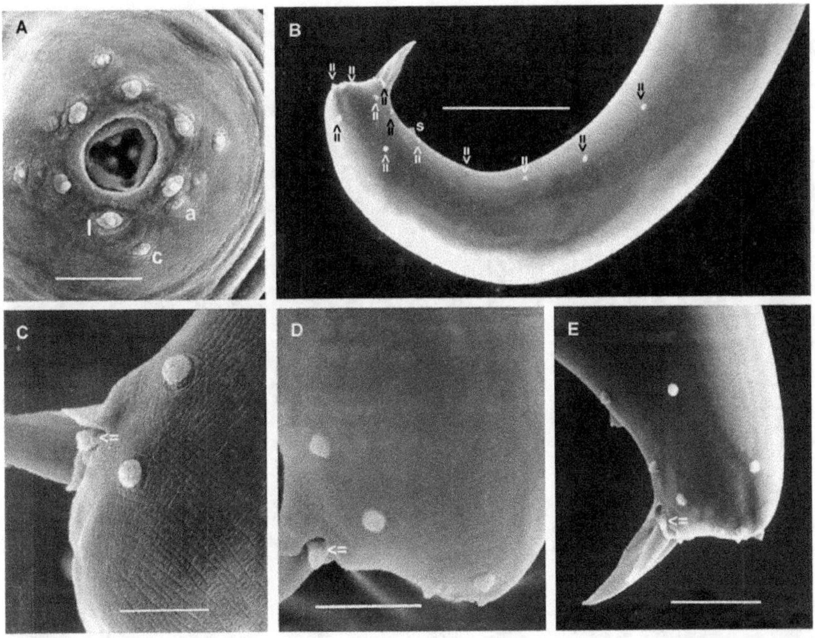

Fig. 155. Steinernema riobrave. *SEM photographs of first generation female and male. A: Labial region of female showing mouth, labial papillae (l), amphidial aperture (a), and cephalic papillae (c); B: Posterior region showing 12 pairs (arrows) and single (s) genital papillae; C-E: Posterior region of male showing presence of unique pair of genital papillae at edge of cloaca (arrow). (Scale bars: A = 7.5 μm; B = 60 μm; C = 8.6 μm; D = 10 μm; E = 17.6 μm.) After Nguyen and Adams (2003), courtesy of* Zootaxa.

Ventral portion of tail usually straight. Anterior cloacal flap present. Mucron on tail tip absent.

Male (second generation)

Similar to, but smaller than, first generation.

Female (first generation)

Anterior region similar to male, but larger and longer. Pygmy form occurring in some instances. Face view with six labial papillae, four cephalic papillae. Amphidial apertures present. Oral aperture round, subtriangular internally. Genital system amphidelphic, reflexed. Vulva in form of a transverse slit, usually protruding slightly from body surface. Vagina short, leading into paired uteri. Eggs initially deposited, but later

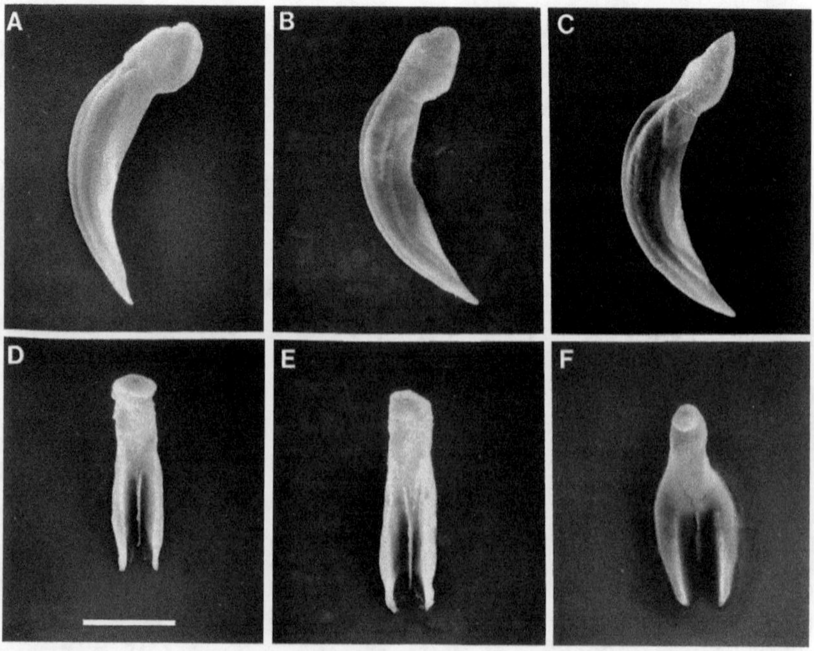

Fig. 156. Steinernema riobrave. *SEM photographs of spicules and gubernacula. A-C: Spicule; D, E: Gubernacula, ventral view; F: Variation in gubernaculum form. (Scale bar = 20 μm.) After Nguyen and Smart (1997), courtesy of* Nematologica.

juveniles hatching inside females and boring their way out. Tail with rounded projection terminally.

Female (second generation)

Similar to, but smaller than, first generation. Tail straight, pointed, with prominent postanal swelling.

Infective juvenile

Body narrower than corresponding parasitic juvenile. Labial region of exsheathed infective juveniles with two horn-like structures (but not observed on ensheathed juveniles). Oral aperture and anus closed, pharynx degenerate. Lateral field beginning anteriorly with one line, usually at first or second annule. At annule eight or nine, two other lines appearing to form two ridges. Near excretory pore, lateral field changing gradually from two to seven ridges. A short distance posteriorly, central ridge dividing into two making a total of eight ridges, the maximum

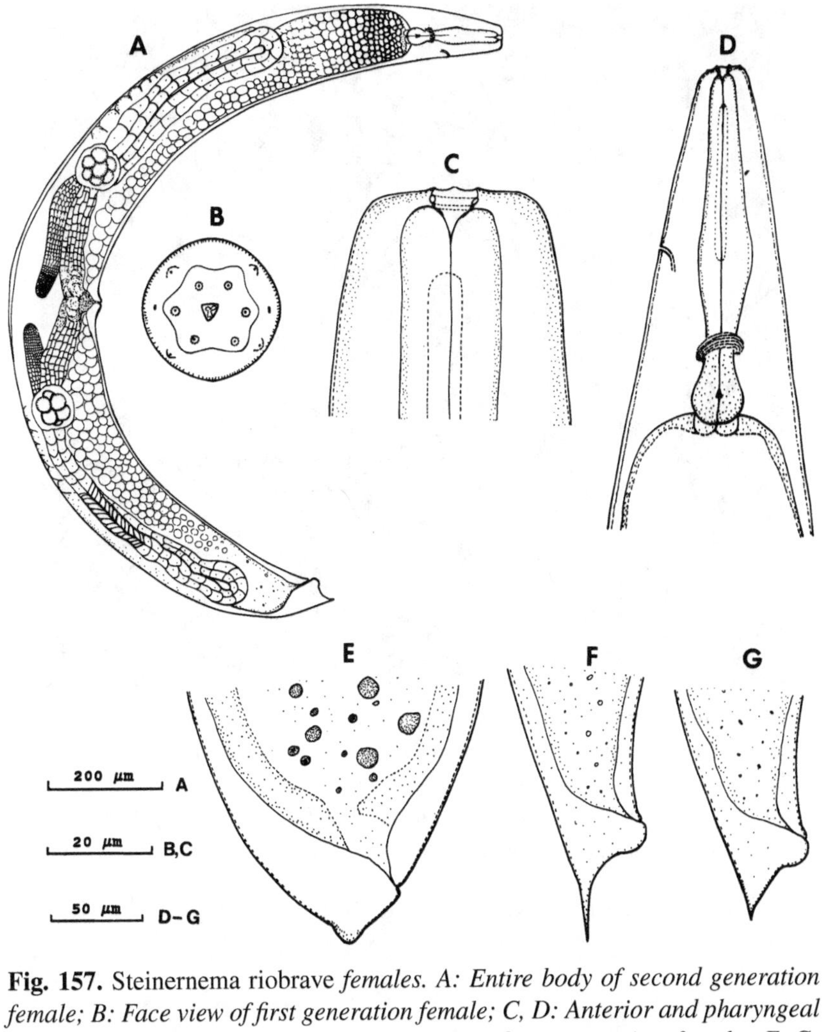

Fig. 157. Steinernema riobrave *females. A: Entire body of second generation female; B: Face view of first generation female; C, D: Anterior and pharyngeal region of first generation female; E: Tail of first generation female; F, G: Variation in tail of second generation female. After Cabanillas, Poinar and Raulston (1994), courtesy of* Fundamental and Applied Nematology.

number in lateral field. Near anus, two dorsal marginal ridges reducing in width then disappearing, making a total of six ridges. Finally, near phasmid, two submarginal ridges on both sides fusing, ventral external ridge and two central ridges disappearing to form two large ridges in lateral field. With above arrangement, formula of lateral field is: 2, 7, 8, 6, 2. Phasmid present, located ventral to lateral field. Tail pointed,

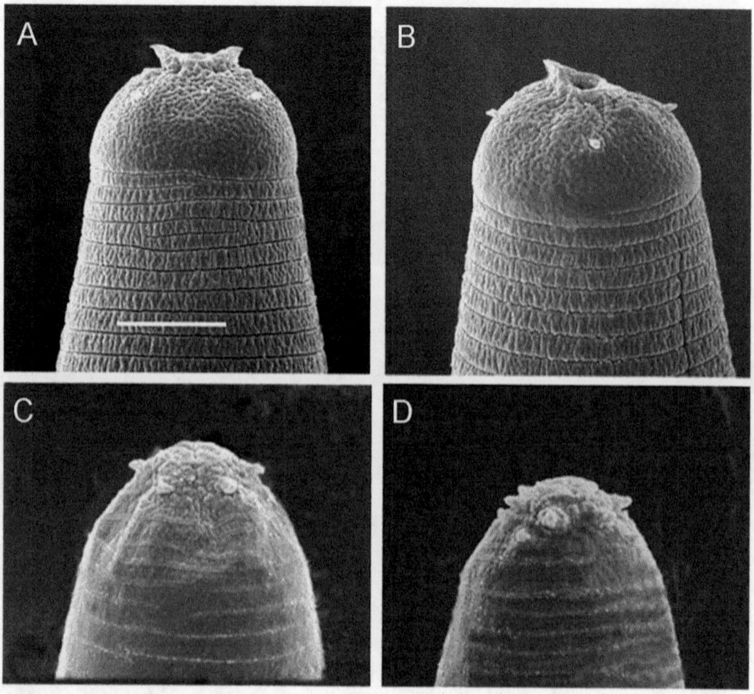

Fig. 158. Steinernema riobrave. *SEM of anterior region of infective juvenile. A, B: Anterior end of exsheathed infective juvenile showing two horn-like structure on labial region; C, D: Anterior end of ensheathed infective juvenile showing labial region lacking horn-like structures. (Scale bar = 4.3 μm.) After Nguyen and Adams (2003), courtesy of* Zootaxa.

usually curved ventrally when relaxed forming an angle of 110° with body.

REMARK

The horn-like structures on the IJ cephalic region and the number of ridges in the lateral field were not reported in the original description.

TYPE HOST AND LOCALITY

Natural host unknown as *S. riobrave* was recovered from corn earworms, *Helicoverpa zea* (Boddie) (Lepidoptera: Noctuidae) that had been placed in soil as trap insects in a corn field at the United States Department of Agriculture South Farm in the lower Rio Grande Valley

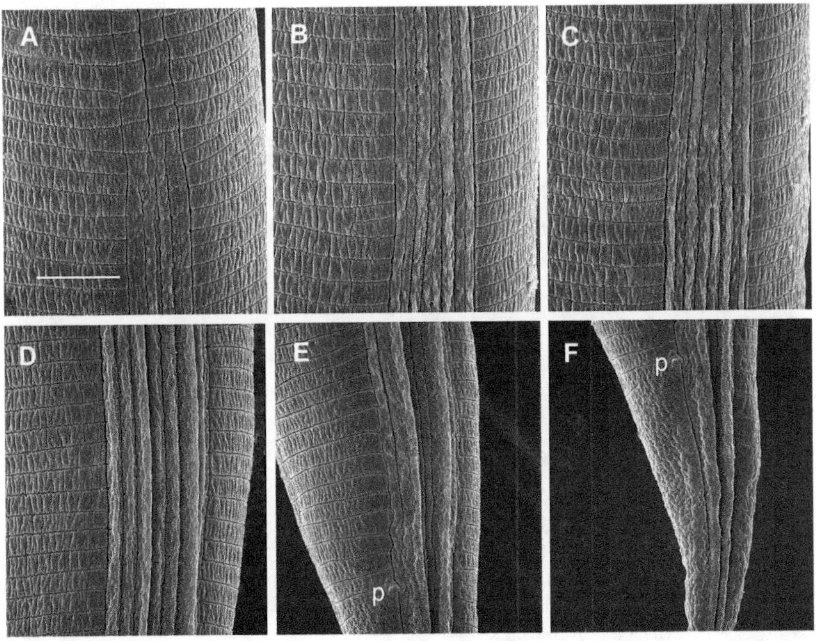

Fig. 159. Steinernema riobrave. *SEM of lateral field. A: Anterior region of infective juvenile showing change in lateral field from two to seven ridges; B, C: Anterior region of infective juvenile showing change in lateral field from seven to eight ridges; D: Lateral field changing from eight to six ridges; E, F: Posterior region of lateral field of infective juvenile showing phasmid (p) and change in ridge number from six to two. (Scale bar = 4.3 μm.) After Nguyen and Adams (2003), courtesy of* Zootaxa.

near Weslaco, Texas, USA. The coordinates of the location where the soil samples were taken were recorded by GPS (Global Positioning System) receivers as latitude 26°08.155′N, longitude 97°57.366′W, and altitude 21.7 m a.s.l. (Cabanillas *et al.*, 1994).

TYPE MATERIAL

Holotype (first generation male), allotype (first generation female) deposited in the University of California-Davis Nematode Collection, Davis, CA, USA. Paratype deposited in the USDA Nematode Collection, Beltsville, MD, USA.

Table 45. *Morphometrics of* Steinernema riobrave. *Measurements are in μm and in the form: mean ± SD (range).*

Character	First generation			Second generation		Infective juvenile
	Holotype (male)	Male	Female	Male	Female	
n	–	10	10	10	10	20
L	–	1700 ± 9 (1500-19 000)	6500 ± 1600 (3700-8300)	900 ± 100 (800-1100)	1700 ± 200 (1400-2000)	622 ± 39.5 (361-701)
a	–	–	–	–	–	22 ± 1.1 (19.9-23.5)
b	–	–	–	–	–	5.4 ± 0.3 (4.9-6.0)
c	–	–	–	–	–	11.6 ± 0.7 (10.1-12.4)
c′	–	–	–	–	–	3.3
V	–	–	52 ± 3.3 (49-56)	–	59 ± 3.2 (55-63)	–
Body diam.	–	133 ± 13.5 (116-160)	275 ± 54 (200-390)	72 ± 11.8 (57-88)	152 ± 18.9 (125-191)	27.6 ± 1.7 (25.6-30.0)
EP	–	103 ± 4.8 (94-111)	96 ± 12.7 (80-118)	88 ± 10.1 (76-94)	94 ± 3.3 (90-98)	56.2 ± 3.2 (51.2-63.7)
NR	–	103 ± 7.6 (106-134)	147 ± 13.3 (131-168)	113 ± 11.7 (94-126)	124 ± 5.6 (113-131)	77.2 ± 1.4 (83.7-88-7)
ES	–	144 ± 7.9 (128-154)	193 ± 13.9 (171-211)	146 ± 11.5 (119-156)	168 ± 9.5 (156-189)	113.5 ± 2.1 (108.7-116.2)
Testis reflexion	–	226 ± 26.5 (185-257)	–	154 ± 36.8 (100-182)	–	–
Tail length (T)	–	31 ± 1.7 (29-35)	45 ± 3.5 (41-50)	26 ± 2.7 (21-31)	52 ± 4.3 (44-59)	53.5 ± 3.5 (46.2-58.7)
Anal body diam. (ABD)	–	59 ± 4.7 (50-640)	93 ± 17.6 (63-115)	47 ± 8.1 (35-63)	54 ± 6.7 (45-63)	16 ± 0.4 (15-16.5)
Spicule length (SL)	–	67 ± 4.2 (63-75)	–	54 ± 4.0 (49-63)	–	–
Spicule width	–	12.4 ± 0.7 (11.2 13.7)	–	10 ± 1.3 (8.7-12.5)	–	–
Gubernaculum length (GL)	–	51 ± 2.6 (48-56)	–	38 ± 2.2 (35-43)	–	–
Gubernaculum width	–	8.1 ± 0.6 (7.1-8.7)	–	6.3 ± 0.2 (6.2-6.8)	–	–

NOTE

This species was named *S. riobravis* by the original authors but was changed to *S. riobrave* by Hominick *et al.* (1997) in order to comply with International Code of Zoological Nomenclature (ICZN) rules. The original authority is unchanged.

Table 45. *(Continued).*

Character	First generation			Second generation		Infective juvenile
	Holotype (male)	Male	Female	Male	Female	
D% = EP/ES × 100	–	71 ± 8 (60-80)	–	61 ± 7 (53-67)	56 ± 2 (53-59)	49 ± 2 (45-55)
E% = EP/T × 100	–	–	–	–	–	105 ± 5 (93-111)
SW% = SL/ABD × 100	–	114	–	115	–	–
GS% = GL/SL × 100	–	76	–	70	–	–

EP = distance from anterior end to excretory pore; NR = distance from anterior end to nerve ring; ES = distance from anterior end to end of pharynx.
After the original authors.

RELATIONSHIPS

This nematode is characterised by the presence of two horn-like structures on the anterior end of the infective juvenile. It is also characterised by the presence of 12 pairs of genital papillae with one pair located near the edge of cloaca, by the sickle-shaped spicules and the gubernaculum having almost the same width in its ventral view from anterior to posterior end.

Steinernema riobrave can be differentiated from other species in the *bicornutum*-group (*S. abbasi*, *S. bicornutum*, *S. ceratophorum*, *S. pakistanense* and *S. yirgalemense*) by the body length of its infective juvenile which is 622 μm (compared to >635 or <555 for other species) and the E% value of 105 (compared to $\leqslant$91 in other species). In the male, the presence of 12 pairs of genital papillae is similar to *S. yirgalemense* but different from all other species in the *bicornutum*-group. *Steinernema riobrave* can be distinguished from all other species of *Steinernema* (*i.e.*, those not in the *bicornutum*-group) by the presence of horn-like structures on the anterior end of the exsheathed infective juvenile.

MOLECULAR CHARACTERISATION

Steinernema riobrave is characterised molecularly by sequences of ITS and D2D3 regions (Nguyen & Adams, 2003): *i*) length ITS regions = 1004 base pairs (bp), ITS1 = 281 bp, ITS2 = 316 bp and its composition, A = 0.279, C = 0.148, G = 0.228, T = 0.345

(accession no. DQ835613); *ii*) length D2/D3 region = 869 bp and its composition, A = 0.250, C = 0.176, G = 0.303, T = 0.271 (accession no. AF331893).

More details are available in Nguyen and Adams (2003).

Bionomics

This nematode has been found parasitising pre-pupae and pupae of *Helicoverpa zea* and *Spodoptera frugiperda* (Raulston *et al.*, 1992). The soil type from which this nematode was isolated was a Hidalgo sandy clay loam (47.9% sand, 35.6% clay, 16.5% silt, 1.1% organic matter). Daytime soil temperatures at 5 cm deep during the corn growing season were *ca* 35°C.

The life cycle of *S. riobrave* is comparable to that of other species of *Steinernema*. The infective juveniles enter the haemocoel of the insect host, deliver their associated bacteria and usually complete two generations before emerging from the insect cadaver as infective juveniles. Infective juveniles transferred to insect blood drops (*Galleria mellonella* and *H. zea*) reached the adult stage in 48 h and produced eggs in 72 h at 24°C. In Petri dish assays, *S. riobrave* reached the pre-adult and/or adult stages in 48 h in fall armyworm larvae (9-day-old) exposed to infective juveniles at 29.5°C (Cabanillas *et al.*, 1994). This indicates that the development of this nematode is markedly influenced by temperature. Experiments at the type locality showed that this nematode moved up and down throughout the soil profile and survived well in the dry and hot (35°C) environment as found in Texas, USA.

Bacterial associate

The mutualistic bacterium associated with *S. riobrave* was isolated on nutrient agar and produced a brownish-cream colour. The primary form of the bacterium was characterised by its adsorption of bromothymol blue from NTBA (nutrient bromothymol blue-triphenyl tetrazolium chloride agar), and adsorption of neutral red from MacConkey agar (red colonies). These characteristics do not differ from the description of *Xenorhabdus nematophila*, but further studies are needed (Cabanillas *et al.*, 1994).

Biocontrol capability

The efficacy of *S. riobrave* to control corn earworm (*Helicoverpa zea*) pre-pupae has been evaluated in glasshouse and field experiments. It was shown that the efficacy of *S. riobrave* was influenced by nematode concentration and placement method under glasshouse conditions (Cabanillas & Raulston, 1995). These studies indicated that *S. riobrave* is effective for suppressing corn earworm populations under field conditions of high soil temperature with irrigation and appropriate timing (Cabanillas & Raulston, 1995, 1996). Recently, this nematode has been commercialised to control various insects, especially citrus root weevil, in Florida, USA.

DISTRIBUTION

Steinernema riobrave has been recorded only from the United States (Cabanillas *et al.*, 1994).

References

CABANILLAS, H.E. & RAULSTON, J.R. (1995). Impact of *Steinernema riobravis* (Rhabditida: Steinernematidae) on the control of *Helicoverpa zea* (Lepidoptera: Noctuidae) in corn. *Journal of Economic Entomology* 88, 58-64.

CABANILLAS, H.E. & RAULSTON, J.R. (1996). Evaluation of *Steinernema riobravis, S. carpocapsae*, and irrigation timing for the control of corn earworm, *Helicoverpa zea. Journal of Nematology* 28, 75-82.

CABANILLAS, H.E., POINAR JR, G.O. & RAULSTON, J.R. (1994). *Steinernema riobravis* n. sp. (Rhabditida: Steinernematidae) from Texas. *Fundamental and Applied Nematology* 17, 123-131.

HOMINICK, W.M., BRISCOE, B.R., DEL PINO, F.G., HENG, J., HUNT, D.J., KOZODOY, E., MRÁČEK, Z., NGUYEN, K.B., REID, A.P., SPIRIDONOV, S., STURHAN, D., WATURU, C. & YOSHIDA, M. (1997). Biosystematics of entomopathogenic nematodes: current status, protocols and definitions. *Journal of Helminthology* 71, 271-298.

NGUYEN, K.B. & ADAMS, B.J. (2003). SEM and systematic studies of *Steinernema abbasi* Elawad *et al.*, 1997 and *S. riobrave* Cabanillas *et al.*, 1994 (Rhabditida: Steinernematidae). *Zootaxa* 179, 1-10.

RAULSTON, J.R., PAIR, S.D., LOERA, J. & CABANILLAS, H.E. (1992). Prepupal and pupal parasitism of *Helicoverpa zea* and *Spodoptera frugiperda* (Lepidoptera: Noctuidae) by *Steinernema* sp. in corn fields in the Lower Rio Grande valley. *Journal of Economic Entomology* 85, 1666-1670.

***Steinernema ritteri*[*,**] de Doucet & Doucet, 1990**
(Figs 160, 161)

MEASUREMENTS

See Table 46.

DESCRIPTION

Male (first generation)

Cuticle, labial region and pharyngeal region similar to those of female. Genital system monorchic, reflexed; consisting of germinal growth zone leading into seminal vesicle. *Vas deferens* with inconspicuous muscular walls. Spicules paired, arcuate, symmetrical, with ribs and thin velum. Capitulum rounded, lamina bearing ridges on edges and surface, ridges thicker in first generation than in second. Distal tips of spicule sharply pointed with ventral concavity giving them a hooked appearance. Gubernaculum proximally rounded. Complement of 23 genital papillae, consisting of nine ventrolateral pairs, one lateral pair, one caudal subdorsal pair and a single midventral precloacal papilla. Of nine ventrolateral pairs, six ventrolateral pairs located precloacal, one pair adcloacal, and two pairs postcloacal. Tail terminus bluntly conical, mucron absent.

Male (second generation)

Spicule lamina with thinner ridges than in first generation. Gubernaculum with a small straight or recurved projection. Tail terminus with fine mucron.

Female (first generation)

Cuticle smooth, lateral lines and phasmids indistinct. Labial region truncate to slightly rounded, lips united, oral aperture subtriangular. Labial region bearing internal circle of six labial papillae and external circle of four, submedial cephalic papillae; amphidial apertures located at base of internal lateral papillae. Stoma reduced; cheilostom and lumen of pharynx triangular in face view. Prorhabdions barely visible, cheilorhabdions represented by thick ring of sclerotised material, meso-,

---

[*] Named after Dr Maurice *Ritter*.

[**] Description after de Doucet and Doucet (1990). Revised by Zdeněk Mráček.

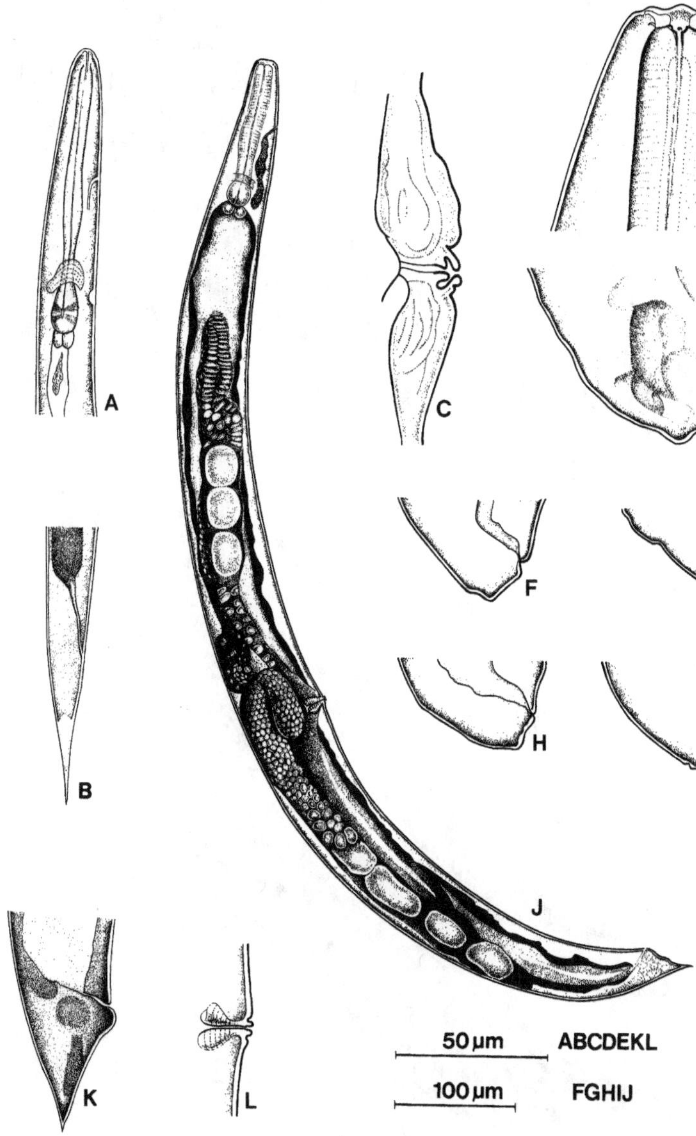

Fig. 160. Steinernema ritteri. *A, B: Infective juvenile. A: Head and pharyngeal region; B: Tail. C-I: First generation female. C: Vulvar region; D: Head region; E: Tail; F-I: Variation of tail shape. J-L: Second generation female. J: Female entire body; K: Tail; L: Vulval region. After de Doucet and Doucet (1990), courtesy of* Nematologica.

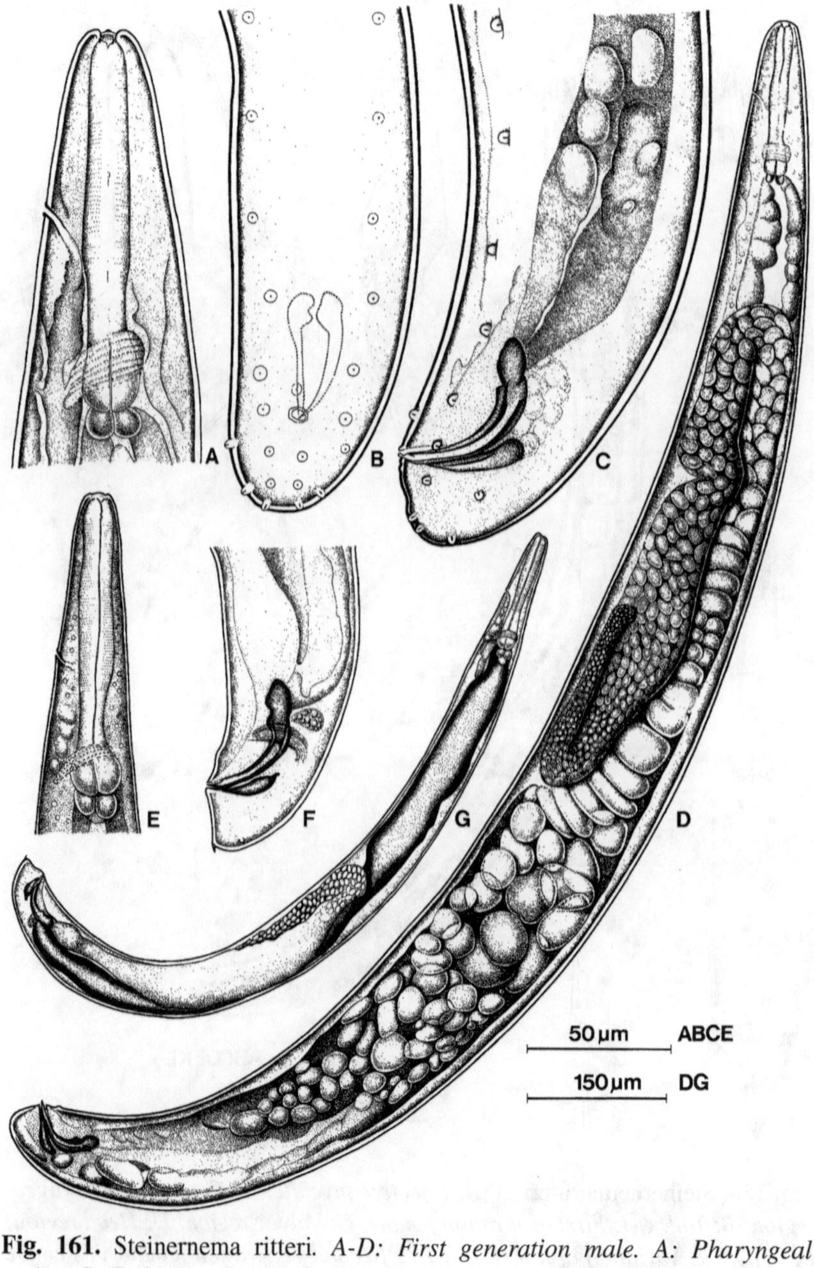

Fig. 161. Steinernema ritteri. *A-D: First generation male. A: Pharyngeal region; B: Tail, ventral view; C: Tail, lateral view; D: Entire body. E-G: Second generation male. E: Pharyngeal region; F: Tail; G: Entire body. After de Doucet and Doucet (1990), courtesy of* Nematologica.

Table 46. *Morphometrics of* Steinernema ritteri. *Measurements are in μm and in the form: mean ± SD (range).*

Character	First generation			Second generation		Infective juvenile
	Holotype (female)	Male	Female	Male	Female	
n	–	30	23	26	31	20
L	7800	1480 ± 180 (120-1900)	8600 ± 1400 (5200-11 500)	660 ± 40 (600-700)	1400 ± 180 (920-1480)	510 ± 30 (470-590)
a	34	11	34	13	–	24 ± 2.2 (20-31)
b	38	11	44	6.6	–	5.5 ± 0.4 (4.9-6.3)
c	325	55	307	33	–	11 ± 0.9 (9.2-13)
c′	0.4	0.6	0.4	0.6	–	3.5
V	53	–	51 ± 3.2 (44-56)	–	56 ± 1.5 (54-60)	–
Body diam.	231	130 ± 16 (110-176)	252 ± 23 (210-300)	51 ± 4.1 (43-60)	88 ± 16 (65-118)	22 ± 1.3 (19-24)
EP	48	66 ± 6.9 (53-78)	75 ± 27 (40-135)	55 ± 3.5 (49-63)	54 ± 5.4 (36-62)	43 ± 1.8 (40-46)
NR	156	105 ± 8.4 (93-124)	141 ± 20 (108-187)	78 ± 3.8 (70-90)	92 ± 5.8 (71-99)	73 ± 3.9 (68-85)
ES	204	137 ± 9.2 (121-155)	196 ± 26 (130-232)	100 ± 4.3 (90-406)	116 ± 3.8 (102-122)	92 ± 3.1 (85-95)
Tail length (T)	24	27 ± 3.6 (21-32)	28 ± 5.8 (19-36)	20 ± 2.3 (15-25)	44 ± 5.0 (38-58)	49 ± 2.9 (44 54)
Anal body diam. (ABD)	65	44 ± 3.7 (37-52)	66 ± 13 (45-95)	34 ± 3.5 (28-40)	33 ± 3.7 (27-44)	14 ± 1.4 (12-17)
Spicule length (SL)	–	69 ± 4.8 (58-75)	–	47 ± 2.9 (41-53)	–	–
Spicule width	–	12 ± 1.2 (10-15)	–	8.0 ± 1.8 (6.0-10)	–	–
Gubernaculum length (GL)	–	44 ± 4.4 (33-50)	–	25 ± 3.7 (20-32)	–	–
D% = EP/ES × 100	20	50	38	55		46 ± 1.0 (44-50)
E% = EP/T × 100	200	244	267	275	–	88 ± 5 (79-97)
SW% = SL/ABD × 100	–	157	–	138	–	–
GS% = GL/SL × 100	–	64	–	53	–	–

EP = distance from anterior end to excretory pore; NR = distance from anterior end to nerve ring; ES = distance from anterior end to end of pharynx.

meta- and telorhabdions vestigial. Pharynx muscular, with cylindrical non-valvate metacorpus, narrowing at level of isthmus, terminating in basal bulb containing reduced valve. Pharynx set off from intestine, rarely inserted into anterior portion of intestine, cardia well developed. Nerve ring usually surrounding anterior portion of basal bulb. Excretory pore circular, located at level of metacorpus. Ovaries opposed, reflexed, in dorsal position; oviduct well developed, glandular; spermatheca and uterus in ventral position; vagina short, with muscular walls; vulval lips protruding, covered by distinct double epiptygmata. Tail with postanal swelling, terminus bluntly conical to dome shaped, sometimes with rounded projection.

Female (second generation)

Tail conical with a well developed mucron.

Infective juvenile

Ensheathed within cuticle of second-stage. Body slender, tapering regularly from base of pharynx to anterior end and from anus to terminus. Cuticle with faint transverse annulation. Lateral fields with six ridges (*i.e.*, seven lines). Labial region smooth, usually continuous. Oral aperture closed. Pharynx long, narrow, distinctly narrower at nerve ring level, terminating in valvate basal bulb. Excretory pore located at *ca* middle of pharynx. Nerve ring distinct, situated immediately anterior to basal bulb. Hemizonid located at level of basal bulb. Cardia present. Anterior portion of intestine with dorsally displaced pouch containing symbiotic bacteria. Lumen of intestine narrow, rectum long, narrow; anus distinct. Genital primordium evident. Tail conical with pointed terminus.

TYPE HOST AND LOCALITY

Natural host unknown. Isolated by baiting with *Galleria mellonella* larvae from soil from Estancia Sol de Mayo, Rio Cuarto, Cordoba, Argentina.

TYPE MATERIAL

Holotype female deposited in the Muséum national d'Histoire naturelle, Paris, France. Paratypes: four females and seven males of the first generation, ten females and ten males of the second generation and

14 free-living infective juveniles in the mentioned institution and in the University of California-Davis Nematode Collection, Davis, CA, USA. Eighteen females and 23 males of the first generation, 21 females and 16 males of the second generation and six free-living infective juveniles in the Centro de Zoología Aplicada, Universidad Nacional de Córdoba, Argentina

NOTE

Z. Mráček asked to see the paratypes from the University of California-Davis Nematode Collection, Davis, CA, USA, but they were not available (Rhode, pers. comm.). It seems likely that these paratypes were never sent by de Doucet to this institution. Stock (pers. comm.) also tried, unsuccessfully, to look for paratypes at Universidad Nacional de Córdoba, Argentina and to re-isolate the species.

RELATIONSHIPS

First generation females possess double flapped vulval epiptygmata. According to the length of the infective juveniles (470-590 μm), this species belong to the group of species characterised by the shortest IJ, *i.e.*, the *carpocapsae*-group. D% varies from 44-50. The hyaline region comprises *ca* 45% of tail length (assessed from the original figure). The spicules are slender and arcuate, the manubrium being ellipsoid with a length to width ratio of *ca* 1 : 1. The rostrum is prominent.

According to de Doucet and Doucet (1990) the females of both generations of *S. ritteri* differ from the rest of the species of the genus, except for *S. carpocapsae* and *S. scapterisci* in having well developed epiptygmata. In *S. carpocapsae* this structure is less developed whereas in *S. scapterisci* it is actually more developed than in *S. ritteri*. *Steinernema ritteri* differs from *S. carpocapsae* by the absence of a mucron on the tail tip of males and first generation females, and by characters of the infective juveniles: L = 510 (470-590) *vs* 558 (438-650) μm; anterior end to excretory pore distance = 43 (40-46) *vs* 38 (30-60) μm; pharynx length = 92 (85-95) *vs* 120 (103-190) μm; tail length = 49 (44-54) *vs* 53 (46-61) μm. *Steinernema ritteri* differs from *S. scapterisci* by the absence of a mucron on the tail tip in first generation adults and by mean values of most measurements in all stages. Except for a few measurements that have greater values: body diam. at anus level (males of both generations and first generation females), tail length (first generation males), body length and greatest diam. (first generation females), distance from anterior end to excretory pore (infective juveniles) the remainder of the morphometrics have lower values in *S. ritteri*. This species is most similar to *S. carpocapsae*, *S. rarum* and *S. kushidai* in general morphology of the infective stage, males and females. It differs from the first two species by the absence of a mucron in adults of the first generation and by greater values of ratios D% and E% of the infective juvenile (D% = 44-50 *vs* 23-28 and 38-44; E% = 79-97 *vs* 54-66 and 84-95, respectively). It differs from the latter by the rounded tail in first generation adults and by all morphometric characters having a lower value in all stages.

COMMENTS

The status of this nematode needs to be confirmed by DNA data. This species may be close to *S. carpocapsae* but differs mainly by having vulval epiptygmata in the females. Infective juveniles of *S. ritteri* possess a more posterior excretory pore (ratio D% = 46 *vs ca* 35 for *S. carpocapsae*) and ratio c' = 3.5 compared to *ca* 4.0 for *S. carpocapsae*.

MOLECULAR CHARACTERISATION

No data are available.

BIONOMICS

Entry into the host takes place through the mouth with the food substrate. After entering, the infective juveniles penetrate the wall of the digestive system and gain access to the body cavity where the associated bacteria are liberated into the haemocoel, the developmental process begins, and the insect dies. The colour of the dead insect remains unchanged. The first generation develops 4-5 days after inoculation. Young females lay many eggs. In older females ovoviviparity is common. The second generation starts *ca* 6 days after inoculation and results in the development of the infective juveniles.

DISTRIBUTION

Recorded only from Cordoba province, Argentina (de Doucet & Doucet, 1990).

Reference

DE DOUCET, M.M.A. & DOUCET, M.E. (1990). *Steinernema ritteri* n. sp. (Nematoda: Steinernematidae) with a key to the species of the genus. *Nematologica* 36, 257-265.

Steinernema robustispiculum[*,**] Phan, Subbotin, Waeyenberge & Moens, 2005

(Fig. 162)

MEASUREMENTS

See Table 47.

DESCRIPTION

Male (first generation)

Body C-shaped when heat-relaxed. Cuticle smooth under light microscope. Cephalic extremity rounded, or somewhat offset from body. Face view with six pointed labial papillae and four cephalic papillae. Amphidial apertures not observed. Stoma funnel-shaped, shallow. Pharynx muscular; procorpus cylindrical; metacorpus slightly swollen nonvalvate; isthmus distinct; basal bulb prominent, valvate. Nerve ring located just anterior to basal bulb. Pharyngo-intestinal valve prominent, usually protruding into intestinal lumen. Excretory pore situated near mid-pharynx. Excretory duct cuticularised. Genital system monorchic, reflexed. Spicules paired, yellow-brownish in colour, strongly curved, large; length/width ratio = 4.5 (3.8-5.6); spicule head as long as wide; blade arcuate, with straight tip and three well defined lobes; dorsal lobe enlarged anteriorly and distinctly curved, narrowed gradually posteriorly to spicule tip; lateral lobe prominent, usually narrowed anteriorly and terminated a short distance posterior to shaft, narrowed gradually posteriorly to spicule tip; ventral lobe enlarged anteriorly on ventral side to form prominent rostrum, narrowed posteriorly to spicule tip; velum large, not reaching spicule tip. Gubernaculum boat-shaped in lateral view, swollen in middle, with ventrally curved knob at proximal end; in ventral view, cuneus long, bifurcate, not reaching end of corpus; corpus separated posteriorly. Complement of 11 pairs of genital papillae plus single midventral precloacal papilla arranged as follows: five pairs subventral precloacal, one pair lateral at same level as single ventral precloacal papilla; two pairs subventral adcloacal; two pairs subventral caudal and one pair subdorsal caudal. Tail conoid, mucron present. Phasmids not observed.

[*] The specific epithet refers to the *robust spicules* of the male.

[**] After Phan *et al.* (2005). Revised by Khuong B. Nguyen.

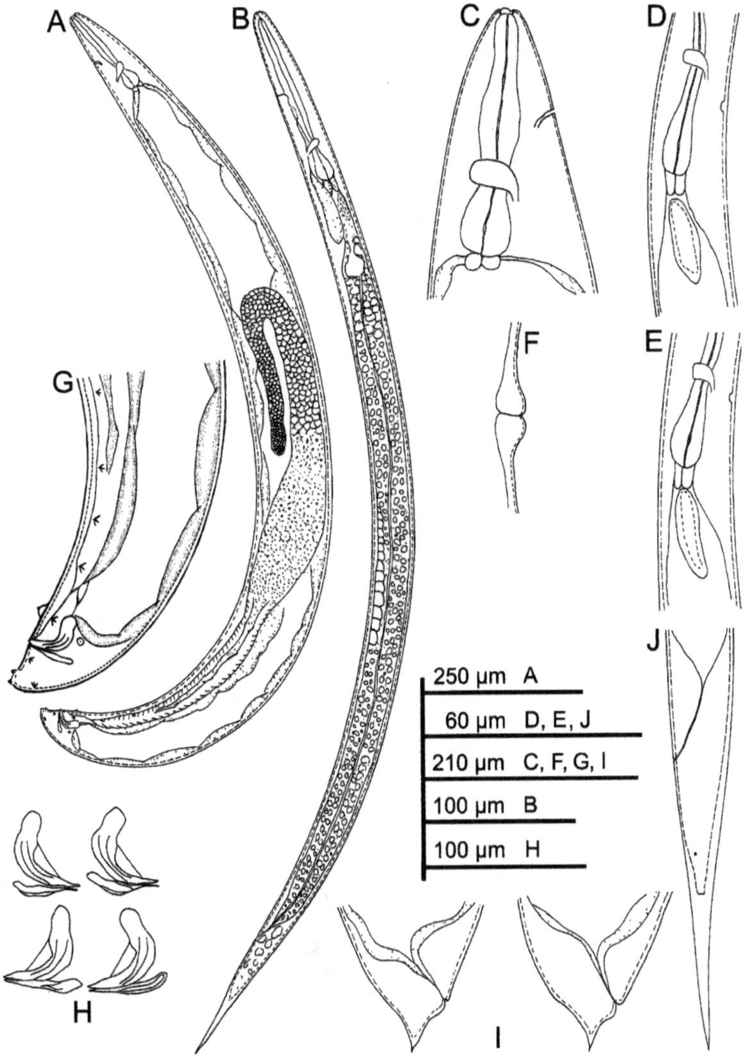

Fig. 162. Steinernema robustispiculum. *A, G, H: First generation male. A: Entire; G: Arrangement of genital papillae; H: Spicules and gubernaculum. B, D, E, J: Infective juvenile. B: Entire body; D, E: Bacterial vesicle; J: Tail, lateral view. C, F, I: First generation female. C: Pharyngeal region; F: Vulval region; I: Tail, lateral view. Reproduced from* Systematic Parasitology, *60 (2005), p. 26; A new entomopathogenic nematode,* Steinernema robustispiculum *n. sp. (Rhabditida: Steinernematidae), from Chumomray National Park in Vietnam, by Phan, L.K., Subbotin, S.A., Wayenberge, L. and Moens, M., Figure 1. With kind permission of Springer Science and Business Media.*

Table 47. *Morphometrics of* Steinernema robustispiculum. *Measurements are in μm and in the form: mean ± SD (range).*

Character	First generation			Infective juvenile
	Holotype (male)	Male	Female	
n	–	20	20	25
L	1560	1433 ± 106 (1320-1665)	3206 ± 249 (2745-3765)	712 ± 43 (642-778)
a	–	–	–	25 ± 3 (18-29)
b	–	–	–	6 ± 1 (3.6-6.3)
c	–	–	–	10 ± 1 (6-11)
c′	–	–	–	4.7
V	–	–	55 ± 2 (50-58)	–
Body diam.	150	127 ± 15 (105-150)	193 ± 18 (165-240)	28 ± 3 (26-35)
EP	92	96 ± 5 (89-104)	108 ± 8 (90-117)	56 ± 4 (50-68)
NR	119	120 ± 5 (110-129)	148 ± 9 (132-165)	84 ± 4 (80-1000)
ES	179	173 ± 7 (162-186)	226 ± 6 (216-239)	120 ± 7 (115-152)
Testis reflexion	–	264 ± 58 (165-360)	–	–
Tail length (T)	33	29 ± 5 (23-33)	54 ± 5 (47-63)	75 ± 5 (68-92)
Anal body diam. (ABD)	50	46 ± 4 (39-56)	73 ± 9 (59-87)	16 ± 1 (14-18)
Spicule length (SL)	65	58 ± 3 (51-65)	–	–
Spicule width	–	13 ± 1 (11-150)	–	–
Gubernaculum length (GL)	44	41 ± 3 (36-44)	–	–
Gubernaculum width	–	6 ± 1 (5-8)	–	–

Male (second generation)

Data not available.

Female (first generation)

Body C-shaped when heat-relaxed. Cuticle appearing smooth under light microscope. Cephalic extremity broadly rounded. Face view with six pointed labial papillae and four cephalic papillae. Amphidial aper-

Table 47. *(Continued).*

Character	Holotype (male)	First generation Male	First generation Female	Infective juvenile
D% = EP/ES × 100	51	56 ± 5 (50-63)	48 ± 3 (39-53)	46 ± 3 (43-59)
E% = EP/T × 100	–	–	–	75 ± 5 (67-87)
SW% = SL/ABD × 100	130	129 ± 12 (111-150)	–	–
GS% = GL/SL × 100	67	70 ± 5 (64-79)	–	–
H% = Hyaline/T × 100	–	–	–	54 ± 3 (49-62)

EP = distance from anterior end to excretory pore; NR = distance from anterior end to nerve ring; ES = distance from anterior end to end of pharynx.
After the original authors.

tures difficult to discern. Stoma funnel-shaped, shallow. Pharynx with cylindrical procorpus; metacorpus slightly swollen, non-valvate; isthmus indistinct; basal bulb pyriform; valve present. Excretory pore located at level of mid-pharynx. Excretory duct cuticularised and excretory gland swollen. Pharyngo-intestinal valve prominent, protruding into intestinal lumen. Didelphic, amphidelphic genital system, reflexed, filled with eggs. Vulva in form of a transverse slit protruding from body, epiptygmata not observed. Vagina short, oblique, with muscular walls. Postanal swelling present. Tail dome-shaped, shorter than anal body diam., with terminal peg.

Female (second generation)

Data not available.

Infective juvenile

Body moderately ventrally arcuate, often enclosed in cuticle of second-stage. Oral aperture and anus closed. In SEM, labial papillae not observed; pore-like amphidial apertures situated posterior to labial disc just anterior to cephalic papillae. Lateral fields at mid-body with eight ridges (*i.e.*, nine incisures); submarginal and central pairs of ridges less distinct; submarginal ridges sometimes not observed. Pharynx long, narrow; isthmus distinct, surrounded by nerve-ring; basal bulb elongate with valve. Pharyngo-intestinal valve prominent. Excretory pore situated at level of mid-pharynx. Hemizonid distinct, located anterior to basal bulb. Bacterial vesicle elliptical or elongate (26-28 × 7-10 μm). Tail long, constricted in hyaline region, especially dorsally. Hyaline region distinct, comprising 54% of tail length. Phasmids distinct, located in anterior half of tail.

TYPE HOST AND LOCALITY

Natural host unknown. The nematode was isolated from a soil sample collected in a forest in the Chumomray National Park, Kontum province, Vietnam.

TYPE MATERIAL

Holotype and six paratype males deposited at Ghent University, Institute of Zoology, K.L. Ledeganckstraat 35, Belgium, and at the University of California, Riverside, CA, USA.

RELATIONSHIPS

Steinernema robustispiculum is characterised by an IJ body length of 712 (642-778) μm, a distance from the anterior extremity to the excretory pore of 56 (50-68) μm, a tail length of 75 (68-92) μm and an E% of 75 (67-87). In addition, the lateral fields at mid-body have eight ridges (submarginal and central pair less distinct) and the male has robust spicules.

The morphometrics of the IJ of *S. robustispiculum* are close to those of *S. monticolum*, except for the position of the excretory pore (at level of middle of pharynx *vs* opposite anterior third of pharynx). Moreover, *S. robustispiculum* can be distinguished from *S. monticolum* by male characters, such as a shorter spicule length (58 *vs* 70 μm), broader spicule (SL/SPW = 4.5 *vs* 8.75) and the spicule head (manubrium) being elongate *vs* round. Both *S. intermedium* and *S. robustispiculum* have very large spicules, but the latter can be separated by its longer IJ tail length (75 *vs* 66 μm), lower E% (75 *vs* 96), shorter spicules (58 *vs* 91 μm), the shape of the spicules (arcuate *vs* distinctly curved proximally, distally almost straight), the number of genital papillae in the caudal region (four *vs* six pairs) and the presence of a mucron on the male tail. *Steinernema robustispiculum* has lateral fields resembling those of *S. sangi*, another species from Vietnam, but can be separated from this species by a higher E% (75 *vs* 62), higher D% (46 *vs* 40), broader spicules (ratio SL/SPW = 4.5 *vs* 5.25), shorter spicule length (58 *vs* 63 μm), the shape and size of the spicule head (elongate, forming 25% spicule length *vs* short, blunt, 20% spicule length) and the dorsal lobe of the spicule (not terminating at spicule tip *vs* terminating at spicule tip).

MOLECULAR CHARACTERISATION

Steinernema robustispiculum can be separated from related species by RFLP of ITS regions of rDNA using 17 restriction enzymes. The ITS sequence is deposited at GenBank (accession no. AY35542).

DISTRIBUTION

The nematode has been found only in Kontum, Vietnam (Phan *et al.*, 2005).

Reference

PHAN, L.K., SUBBOTIN, S.A., WAEYENBERGE, L. & MOENS, M. (2005). A new entomopathogenic nematode, *Steinernema robustispiculum* n. sp. (Rhabditida: Steinernematidae), from Chumomray National Park in Vietnam. *Systematic Parasitology* 60, 23-32.

Steinernema sangi[*,**] Phan, Nguyen & Moens, 2001
(Figs 163, 164)

MEASUREMENTS

See Table 48.

DESCRIPTION

Male (first generation)

Body ventrally curved, J- or C-shaped when heat-relaxed. Cuticle smooth under light microscope. Cephalic extremity rounded, offset from body with distinct labial papillae. SEM face view with six labial and four cephalic papillae. Amphidial apertures inconspicuous. Stoma usually funnel-shaped with prominent cheilorhabdions. Pharynx muscular with cylindrical procorpus; metacorpus distinct, swollen and non-valvate; isthmus distinct; basal bulb pyriform, valve not observed. Nerve ring surrounding isthmus just anterior to basal bulb. Cardia prominent and protruding into intestine lumen. Excretory pore located anterior to nerve ring at mid-pharynx. Excretory duct less cuticularised than excretory pore, excretory glands swollen. Genital system monorchic, reflexed. Spicules paired, curved, dark golden yellow in colour. Spicule head short, comprising *ca* 20% of spicule length. Velum large. Gubernaculum *ca* 65% of spicule length long, boat-shaped, swollen at middle, proximal end with knob, ventrally curved. Complement of 12 pairs of genital papillae plus single midventral precloacal papilla, seven pairs of which subventral precloacal, one pair lateral, three pairs caudal, subventral, and one pair caudal, subdorsal. Phasmids inconspicuous. Tail shorter than anal body diam.; small mucron present, not always observed in permanent mounts.

Male (second generation)

Data not available.

Female (first generation)

Body robust, spiral or C-shaped when heat-relaxed. Labial region rounded, not offset from the body. SEM face view with six labial and

---

[*] Named after Nguyen van *Sang* who collected the nematode.

[**] Description after Phan *et al.* (2001). Revised by Khuong B. Nguyen.

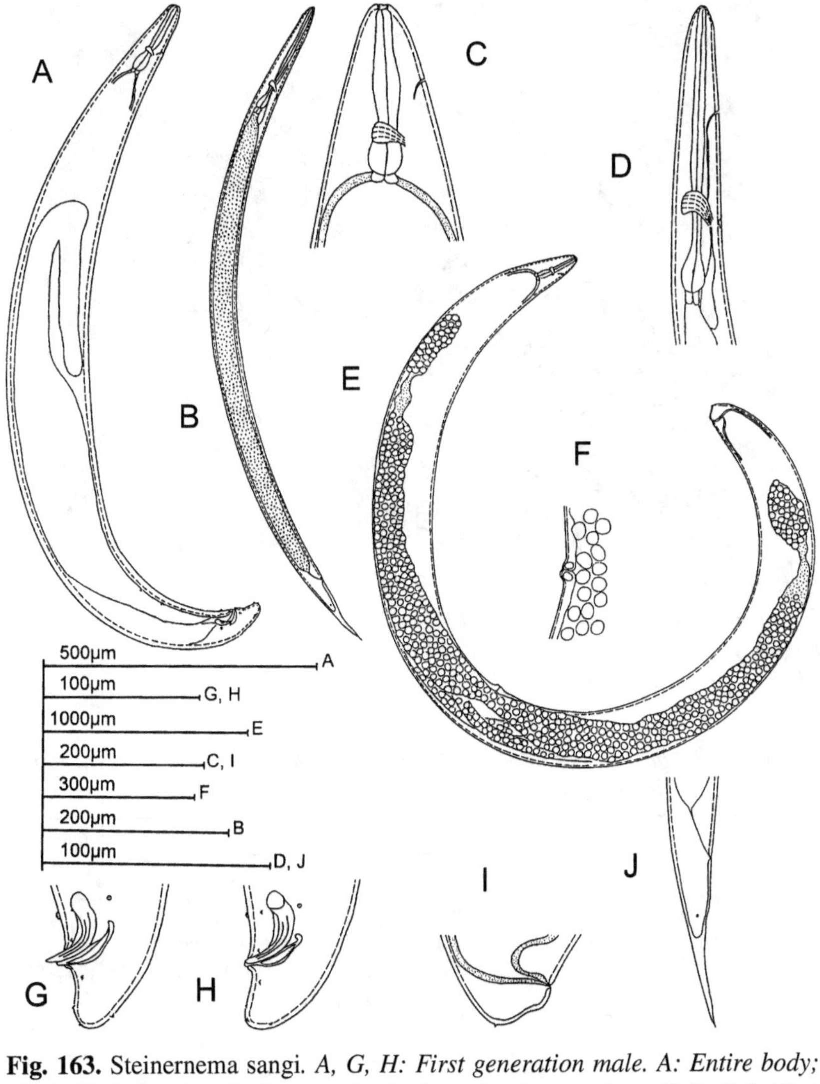

Fig. 163. Steinernema sangi. *A, G, H: First generation male. A: Entire body; G, H: Variation in tail shape and spicule and gubernaculum. C, E, F, I: First generation female. C: Pharyngeal region; E: Entire body; F: Vulval region; I: Tail, lateral view. B, D, J: Infective juvenile. B: Entire body; D: Pharyngeal region; J: Tail, lateral view. After Phan, Nguyen and Moens (2001), courtesy of* Russian Journal of Nematology.

four cephalic papillae. Pore-like amphidial apertures located posterior to lateral labial papillae. Stoma cup-shaped to funnel-shaped. Cuticle

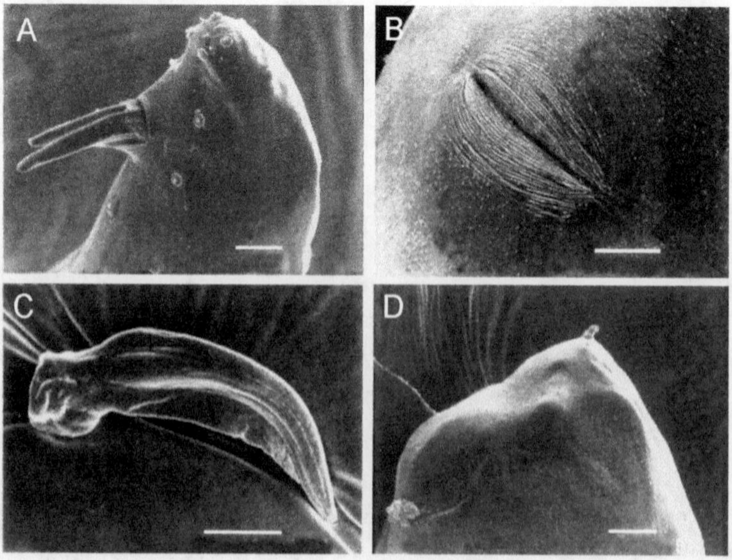

Fig. 164. Steinernema sangi. *SEM of first generation male and female. A: Posterior region of male; B: Vulval region; C: Spicule, lateral view; D: Female tail showing mucron on tail terminus. (Scale bars = 10 μm.) Modified from Phan, Nguyen and Moens (2001).*

Table 48. *Morphometrics of* Steinernema sangi. *Measurements are in μm and in the form: mean ± SD (range).*

Character	First generation			Infective juvenile
	Holotype (male)	Male	Female	
n	–	20	20	50
L	1590	1774 ± 221 (1440-2325)	6030 ± 680 (4830-7200)	753 ± 18 (704-784)
a	–	–	–	22 ± 1.3 (19-250)
b	–	–	–	5.9 ± 0.2 (5.6-6.30)
c	–	–	–	9.3 ± 0.3 (8.7-10.2)
c′	–	–	–	4.5
V	–	–	51 ± 2.2 (43-530)	–
Body diam.	166	159 ± 28 (120-225)	336 ± 23 (270-360)	35 ± 2.3 (30-400)
EP	90	82 ± 9 (67-99)	101 ± 10 (80-121)	51 ± 1.8 (46-540)

Table 48. *(Continued).*

Character	First generation			Infective juvenile
	Holotype (male)	Male	Female	
NR		126 ± 13 (109-166)	158 ± 8.2 (140-170)	91 ± 3 (78-970)
ES	168	166 ± 15 (150-221)	229 ± 7.8 (216-240)	127 ± 3.9 (120-138)
Testis reflexion	–	332 ± 50 (244-428)	–	–
Tail length (T)	39	32 ± 4 (27-42)	49 ± 7.3 (936-62)	81 ± 3 (76-89)
Anal body diam. (ABD)	45	43 ± 2.6 (40-50)	111 ± 13.6 (84-140)	18 ± 0.6 (17-19)
Spicule length (SL)	65	63 ± 4.4 (58-80)	–	–
Spicule width	–	12 ± 1.1 (10-14)	–	–
Gubernaculum length (GL)	38	40 ± 3.1 (34-46)	–	–
Gubernaculum width	–	7 ± 8 (5-9)	–	–
D% = EP/ES × 100	50	49 ± 6 (42-63)	44 ± 4.7 (35-51)	40 ± 1.7 (36-44)
E% = EP/T × 100	–	–	–	62 ± 3.1 (56-70)
SW% = SL/ABD × 100	140	150 ± 10 (120-160)	–	–
GS% = GL/SL × 100	60	60 ± 4 (50-70)	–	–
H% = Hyaline/T × 100	–	–	–	49 ± 2.6 (44-52)

EP = distance from anterior end to excretory pore; NR = distance from anterior end to nerve ring; ES = distance from anterior end to end of pharynx.
After the original authors.

smooth, lateral field not observed. Excretory pore located anterior to nerve ring near mid-pharynx. Pharynx muscular; procorpus cylindrical, metacorpus swollen non-valvate, isthmus distinct, basal bulb pyriform, valve not observed. Nerve ring surrounding isthmus or anterior to basal bulb. Cardia prominent, protruding into intestine lumen. Didelphic, amphidelphic genital system, reflexed. Vulva in form of a median transverse slit. Vulval lips protuberant. Epiptygmata absent. Vagina short, straight, with muscular walls. Tail shorter than anal body diam., mucron present. Phasmids inconspicuous.

Female (second generation)

Data not available.

Infective juvenile

Body ventrally curved when heat-relaxed, striated. Sheath (second-stage cuticle) present, sometimes lost. Cephalic extremity narrow. SEM face view with four prominent cephalic papillae. Pore-like amphidial apertures located anterior to cephalic papillae. Lateral field with eight ridges (*i.e.*, nine lines) at mid-body, submarginal and central pair less distinct. Oral aperture and anus closed. Pharynx narrowing to slender isthmus surrounded by nerve ring. Basal bulb distinct, pyriform, valve not observed. Excretory pore located anterior to mid-pharynx. Hemizonid distinct at isthmus level. Cardia prominent. Tail elongate, conoid and straight, mucron absent. Hyaline region comprising *ca* 39% of tail length. Phasmids distinct, located in anterior half of tail.

TYPE HOST AND LOCALITY

Natural host unknown as the species was isolated using *Galleria mellonella* as bait. Isolated from the forest of Xuanmy, Thuongxuan, Thanhhoa province, Vietnam (longitude 105°14′E, latitude 19°57′N, altitude 600 m a.s.l.).

TYPE MATERIAL

Holotype male (first generation) and paratypes deposited in Ghent University, Institute for Zoology, K.L. Ledeganckstraat 35, Belgium. Paratype slides with six males, six females of first generation and ten infective juveniles deposited in the Department of Nematology, Institute of Ecology and Biological Resources, National Centre for Science and Technology, Nghiado, Caugiay, Hanoi, Vietnam.

RELATIONSHIPS

Steinernema sangi is characterised by the presence of eight ridges in the lateral field of the IJ with the submarginal and central pairs less distinct. This characteristic makes *S. sangi* most similar to *S. kraussei* but it can be distinguished from the latter by the shorter body length of the infective juvenile (753 *vs* 951 μm), shorter distance from anterior end to excretory pore (51 *vs* 63 μm) and longer spicule length (63 *vs* 49 μm).

MOLECULAR CHARACTERISATION

Steinernema sangi is genetically characterised by sequence length of ITS regions, 721 bp, and its composition, A = 0.23162, C = 0.18724, G = 0.23440, T = 0.34674 (accession no. AY355441).

BIONOMICS

The symbiotic bacterium has not been identified, but emits a characteristic aroma.

DISTRIBUTION

Known only from the type locality, Xuanmy, Thuongxuan, Thanhhoa province, Vietnam (Phan *et al.*, 2001).

Reference

PHAN, K.L., NGUYEN, N.C. & MOENS, M. (2001). *Steinernema sangi* sp. n. (Rhabditida: Steinernematidae) from Vietnam. *Russian Journal of Nematology* 9, 1-7.

Steinernema sasonense[*,**] Phan, Spiridonov, Subbotin & Moens, 2006
(Fig. 165)

MEASUREMENTS

See Table 49.

DESCRIPTION

Male (first and second generation not distinguished in original description)

Body fusiform, obese, C-shaped in heat-relaxed specimens, tapering towards anterior end. Cuticle annulation 1.2-1.5 μm wide, poorly distinguishable throughout the body. Lateral fields absent. Anterior extremity rounded with flattened area just around oral aperture. Six labial and four prominent, 2-3 μm high, cephalic papillae. Cheilostom cuticle connected to thick, strongly cuticularised, 2.5 μm long ring of 6-6.3 μm diam., forming middle part of stoma. Additional thinner cuticular ring situated posterior to thicker one, embedded in pharynx tissue. Pharynx corpus 16 μm diam., cylindrical, without visible swelling around stoma, metacorpal swelling up to 18-19 μm diam. Basal bulb subspherical. Nerve ring situated on anterior slope of basal bulb in most specimens. Valves in form of platelets of thin folded cuticle. Cardia with vacuoles, protruding for 8 μm into intestine proventriculus. Deirids visible as thin channels situated posterior to level of pharyngo-intestinal junction. Excretory channel strongly cuticularised for 10 μm from pore. Intestinal lumen broad for 150-200 μm from bulb, then with narrow meandering lumen up to 130-160 μm long posterior to ventricle. Separate spermatocytes quickly disappearing in flexed testis region. *Vas deferens* walls vacuolated. Characteristic steinernematid set of genital papillae with a row of usually nine pairs of subventral papillae extending from tail terminus to mid-body, one large, unpaired, precloacal papilla (not so protruding in this species as other ones described here) in midventral position, one pair of papillae in lateral position at manubrium level when spicules retracted, and one pair of papillae in dorsolateral position in postcloacal area. Latter pair and third

[*] Named after *Sason*, the type locality.

[**] Description after Phan *et al.* (2006). Revised by Zdeněk Mráček.

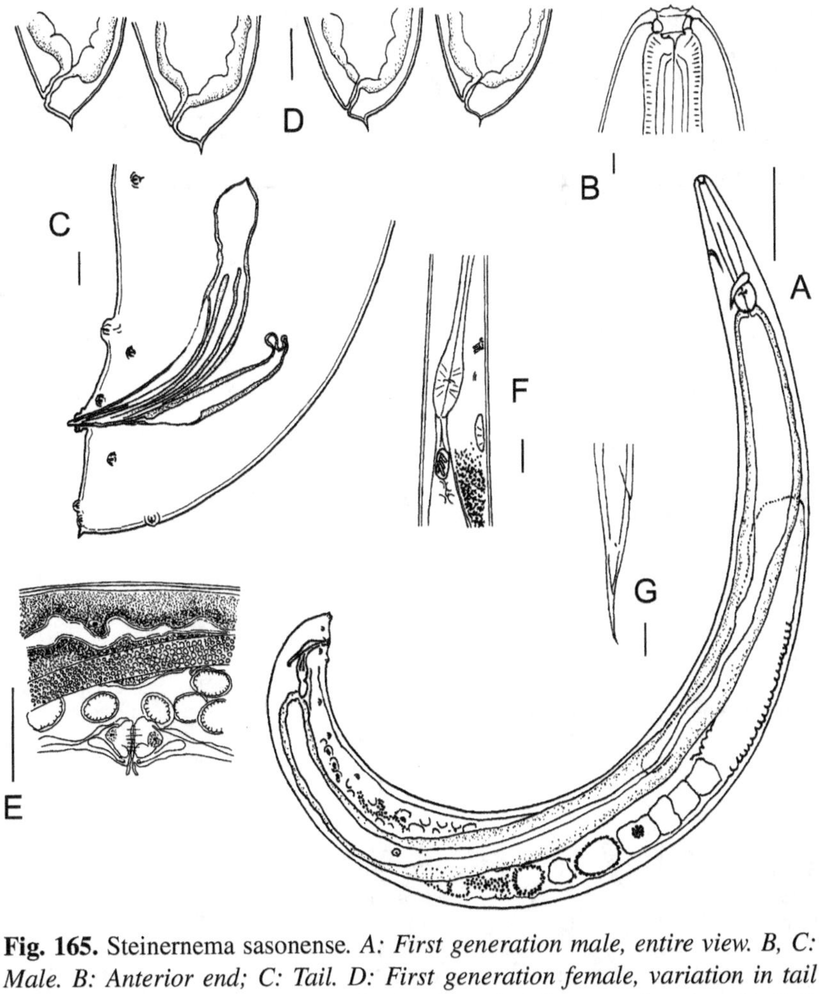

Fig. 165. Steinernema sasonense. *A: First generation male, entire view. B, C: Male. B: Anterior end; C: Tail. D: First generation female, variation in tail shape; E: Vulvar region. F, G: Infective juvenile, pharyngeal region. G: Tail. (Scale bars: A, E = 100 μm; B, C, F, G = 10 μm; D = 50 μm.) After Phan* et al. *(2006), courtesy of* Russian Journal of Nematology.

subventral pair from tail close to equatorial part of tail. Pair of postdeirids with pointed subcuticular ending situated dorsolaterally on level of anteriormost pair of precloacal papillae. Spicules with elongated, funnel-shaped manubrium, tapering towards distal end, with round extremity and more or less pronounced terminal duct. No bifurcation of outer line of velum and ventral rib of spicule observed near rostrum. All four main ribs of spicule discernible, but ventralmost rib tightly adpressed

Table 49. *Morphometrics of* Steinernema sasonense. *Measurements are in μm and in the form: mean ± SD (range).*

Character	First generation			Infective juvenile
	Holotype (male)	Male	Female	
n	–	15	15	25
L	1365	1344 ± 59 (1245-1440)	4417 ± 343 (3930-5070)	471 ± 15 (437-502)
a	11	12 ± 1 (11-13)	20 ± 2 (17-25)	22 ± 1 (21-24)
b	9	9.3 ± 0.4 (8.4-9.9)	20 ± 2 (17-22)	4.7 ± 0.2 (4.4-5.0)
c	83	77 ± 8 (64-94)	213 ± 47 (146-327)	10 ± 0.3 (9.7-11)
V	–	–	54 ± 1 (52-56)	–
Body diam.	120	113 ± 8 (105-120)	224 ± 18 (195-240)	21 ± 1 (20-22)
EP	65	58 ± 4 (53-68)	85 ± 14 (50-105)	30 ± 1 (27-34)
NR	102	89 ± 7 (80-102)	135 ± 11 (111-155)	68 ± 2 (65-72)
ES	149	145 ± 5 (134-152)	227 ± 19 (205-248)	100 ± 3 (94-104)
Testis reflexion	360	355 ± 52 (225-420)	–	–
Tail length (T)	17	18 ± 2 (15-21)	29 ± 5 (21-36)	46 ± 2 (43-49)
Anal body diam. (ABD)	30	29 ± 2 (26-32)	62 ± 6 (53-71)	11 ± 0.4 (10-11)
Spicule length (SL)	71	71 ± 3 (66-78)	–	–
Spicule width	12	12 ± 1 (11-14)	–	–
Gubernaculum length (GL)	54	53 ± 2 (50-56)	–	–
Gubernaculum width	6	6.3 ± 0.8 (4.5-7.5)	–	–

to inner one. Spicule body gradually narrowing from rostrum to distal end. Cloacal aperture covered by flap when spicules retracted. Two papilla-like structures on edges of cloacal aperture. Distal half of velum edge strongly cuticularised. Gubernaculum usually straight with slightly curved proximal end, prominent 10-12 μm long hole present. Tail mucron present, in form of short, rounded spike, 2 μm long.

Table 49. *(Continued).*

Character	First generation			Infective juvenile
	Holotype (male)	Male	Female	
D% = EP/ES × 100	43	40 ± 3 (37-45)	37 ± 6 (22-48)	30 ± 1 (28-33)
E% = EP/T × 100	–	–	–	66 ± 4 (59-74)
SW% = SL/ABD × 100	235	250 ± 20 (219-306)	–	–
GS% = GL/SL × 100	77	75 ± 4 (69-82)	–	–
H% = H/T × 100	–	–	–	42 ± 3 (35-50)

EP = distance from anterior end to excretory pore; NR = distance from anterior end to nerve ring; ES = distance from anterior end to end of pharynx.

Female (first and second generation not distinguished in original description)

Cuticle annulation poorly discernible. Cephalic extremity with six labial and four cephalic papillae. Cuticular ring forming main part of stoma walls 5-6 μm thick and 7-7.5 μm long. Stoma 12 μm diam. Excretory duct strongly cuticularised, entering body at angle of *ca* 45°. Vulva equatorial with petal-like, 12-14 μm long structures protruding from aperture. Vulval lips less than 10 μm high, equal in size. Tail terminus dome-shaped, bearing 12-14 μm long, spike-like mucron.

Infective juvenile

Body of heat-relaxed juveniles quite obese, moderately C-shaped, tapering towards anterior end. Cephalic extremity hemispherical with four protruding cephalic papillae. Lining of buccal cavity forming protruding plug with triradiate lumen. Cuticular annulation *ca* 1 μm wide, visible throughout body. Lateral field in anterior body half with four ridges in centre, two pronounced marginal ridges and poorly discernible submarginal lines. Such regular pattern not observed in posterior half of body with ventral gap between central and marginal ridges disappearing. Lateral field lines disappearing at level of phasmids. Lateral field 8-9 μm wide at mid-body. Pharynx corpus slender, slightly widening to metacorpus. Nerve ring located slightly posterior to corpus-isthmus junction. Hemizonid on isthmus level. Basal bulb with valves poorly visible. Cardia *ca* 10 μm long, cylindrical, slightly narrowing to posterior end. Bacterial vesicle round or nearly round, 5-8 μm long, loosely filled with 2-4 μm long, rod-like, bacterial cells. Rectum *ca* two anal body diam. long. Genital primordium 60 μm long. Phasmids located at 25% of tail length posterior to anus (phasmid-mucron distance *ca* 35-36 μm, with aperture near ventral edge of ventral marginal line. Tail with or without slightly pronounced ventral concavity, conical, with some irregularities in shape, terminus sharply pointed. Hyaline region comprising less than half of tail length.

TYPE HOST AND LOCALITY

Natural host unknown as species was obtained by the *Galleria mellonella* baiting method from the southern slope of the Chumomray mountain near Sason, Sathay in Kontum province, Vietnam (longitude 107°43′, latitude 14°27′, altitude 700 m a.s.l.). Soils were humic, fernalie acrisols of primary broad-leaved forest growing on the slopes of granite mountains.

TYPE MATERIAL

Holotype male and six paratype males deposited at Institute of Parasitology of Russian Academy of Sciences, Leninskii prospect 33, Moscow, 119071, Russia. Slides with paratype males, first generation females and infective juveniles deposited in Department of Nematology, Institute of Ecology and Biological Resources, National Center for Science and Technology, Hoang Quoc Viet 18, Nghiado, Caugiay, Hanoi, Vietnam.

RELATIONSHIPS

Males have two papilla-like structures on the margin of the cloacal aperture, the spicules have an elongate manubrium and a well developed rostrum, and there is a short, rounded, spike-like mucron present on the tail terminus. The first generation females have a pronounced conical tail mucron and long, petal-like epiptygmata. The body length of the infective juvenile is, on average, shorter than 500 μm, the lateral field has four ridges in the mid-body region and the hyaline region comprises less than 50% of the tail length.

This species can be distinguished from *S. scapterisci* by the IJ length (437-502 *vs* 517-609 μm), and from *S. cumgarense* and *S. eapokense* by the longer IJ (437-502 *vs* 384-432 and 370-434 μm, respectively) and lower H% of 42 (35-50) *vs* 53 (48-59) and 53 (47-57), respectively. The IJ pharynx is shorter than in *S. carpocapsae* (94-104 *vs* 103-190 μm) and D% = 28-33 *vs* 23-28 in *S. carpocapsae.*

COMMENTS

More precise description of taxonomically important characters, such as the number of ridges in the infective juvenile lateral field, presence or absence of a postanal swelling in the first generation female and occurrence or otherwise of papilla-like mucrons on the female tail tip, are missing or unsatisfactorily described.

MOLECULAR CHARACTERISATION

Steinernema sasonense is characterised by sequence lengths of ITS regions (ITS1 + 5.8S + ITS2) 742 bp, ITS1 = 278 bp, ITS2 = 307 bp and its composition, A = 0.22642, C = 0.16576, G = 0.21832, T = 0.38948. The sequence of ITS regions is deposited in GenBank (accession no. AY487919).

DISTRIBUTION

In addition to the type locality, this species was also isolated from primary, closed, broad-leaved forest mixed with some conifers. This habitat was situated on the slope of a sandstone and gneiss mountain in Thachnham (Mangla) near Konplong, Kontum province (longitude 108°25′, latitude 14°38′, altitude 1210 m a.s.l.). Soils were humic, fernalie acrisols (Phan *et al.*, 2006).

Reference

PHAN, K.L., SPIRIDONOV, S.E., SUBBOTIN, S.A. & MOENS, M. (2006). Four new species of *Steinernema* Travassos, 1928 with short infective juveniles from Vietnam. *Russian Journal of Nematology* 14, 11-29.

Steinernema scapterisci[*,**] Nguyen & Smart, 1990
(Figs 166-171)

MEASUREMENTS

See Table 50.

DESCRIPTION

Male (first generation)

Habitus spiral shape when heat-relaxed, posterior part of body curved ventrally. Cephalic extremity rounded, continuous with body. Lips not seen. Six prominent labial papillae, and four well developed cephalic papillae (much larger than labial papillae) present. Pharynx procorpus cylindrical, metacorpus slightly swollen, isthmus distinct, basal bulb swollen with small valve. Nerve ring located in isthmus region of pharynx, but position variable. Excretory pore located anterior to mid-pharynx. Excretory duct not forming elliptically-shaped structure as in females. Genital system monorchic, reflexed. Spicules averaging 83 μm long, dark brown in colour, paired, uniformly curved, with large, somewhat angular, head. Angle formed by shaft and blade of spicules = 110 (100-120)°. Shaft of spicules long when compared to other species of genus, apparently encased in sheath; blade tapering smoothly to tip with posterior portion thinner than that for other species of *Steinernema*. Small aperture visible on ventral side close to tip of blade. Each spicule with two internal ribs with variable termination point proximally. Gubernaculum boat-shaped, anterior part thin, long and ventrally curved. Spicules glide along gubernaculum in two grooves separated by needle-shaped cuneus. Cloacal aperture with anterior flap, located on elevation. Complement of 11 or 12 pairs of genital papillae plus single midventral papilla. An additional, unique, pair located at end of cloacal aperture. Tail with mucron. No phasmid observed.

---

[*] Named after the genus of the mole cricket host, *Scapteriscus*.

[**] Description after Nguyen and Smart (1990a, 1992b). Revised by Khuong B. Nguyen.

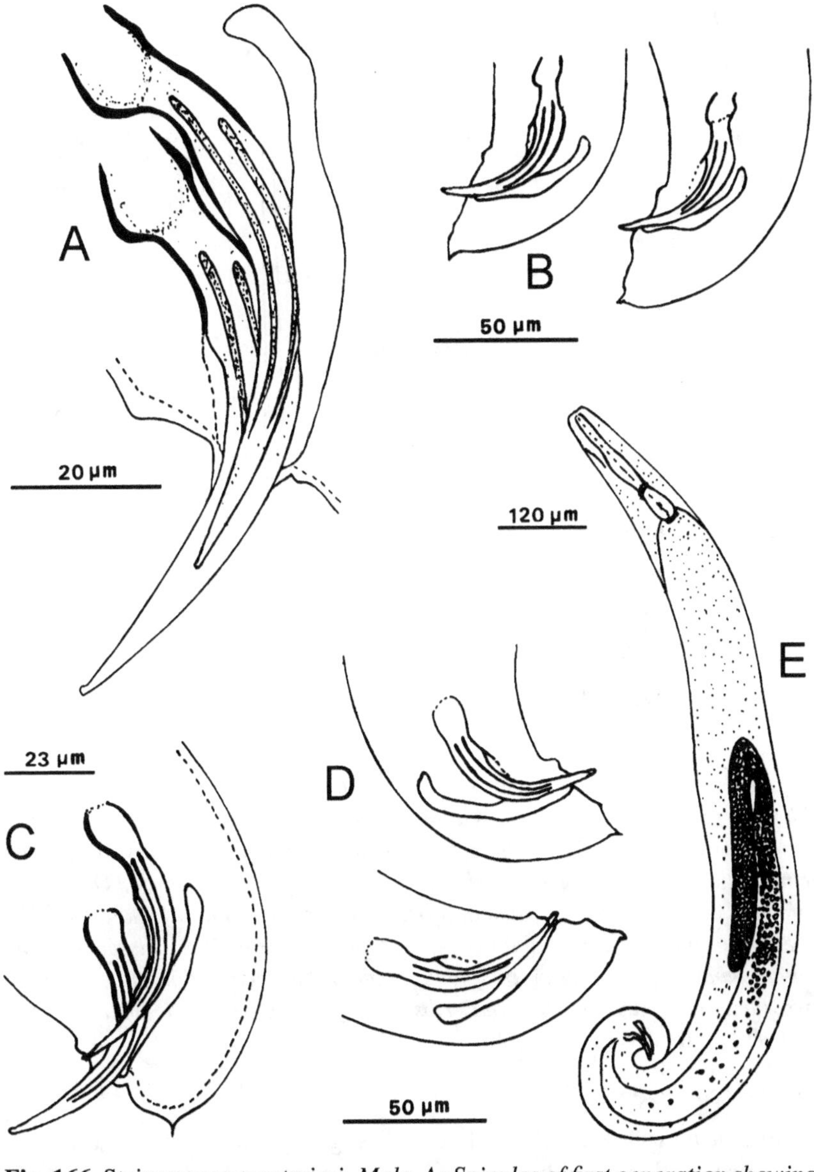

Fig. 166. Steinernema scapterisci. *Male. A: Spicules of first generation showing angular head, internal ribs and gubernaculum with anterior end bent upward; B: Variation in tail shape of first generation; C: Tail of second generation showing elongate spicule head; D: Variation of tail shape of second generation; E: Entire body of second generation male. After Nguyen and Smart (1990), courtesy of* Journal of Nematology.

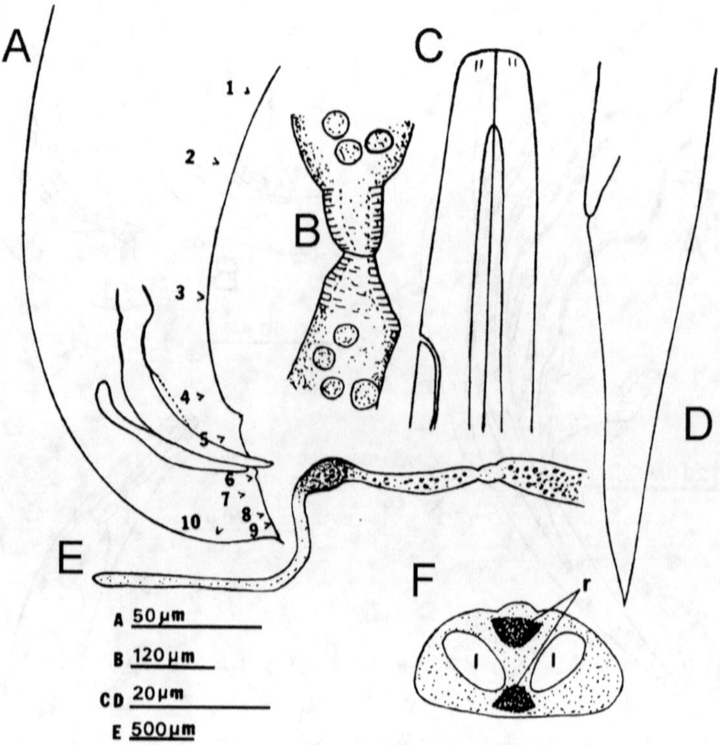

Fig. 167. Steinernema scapterisci. *A: First generation male showing ten pairs of papillae and single preanal midventral papilla; B: Gonad of first generation female at point where ovary reflexes (gonad, when dissected from female, lies straight, not reflexed); C: Head of infective juvenile; D: Tail of infective juvenile; E: Diagrammatic ovary of first generation female showing spermatheca (*in situ *the gonad is reflexed but no flexure after dissection); F: Diagrammatic cross-section of spicule blade of first generation male showing two lumina (l), and two internal ribs (r). After Nguyen and Smart (1990), courtesy of* Journal of Nematology.

Male (second generation)

Similar morphologically to that of first generation except that it is *ca* two-thirds as long and half as broad and the spicules have an elongate head.

Female (first generation)

Anteriorly similar to male but much larger. Lateral fields and phasmids not observed. Six labial papillae and four cephalic papillae present.

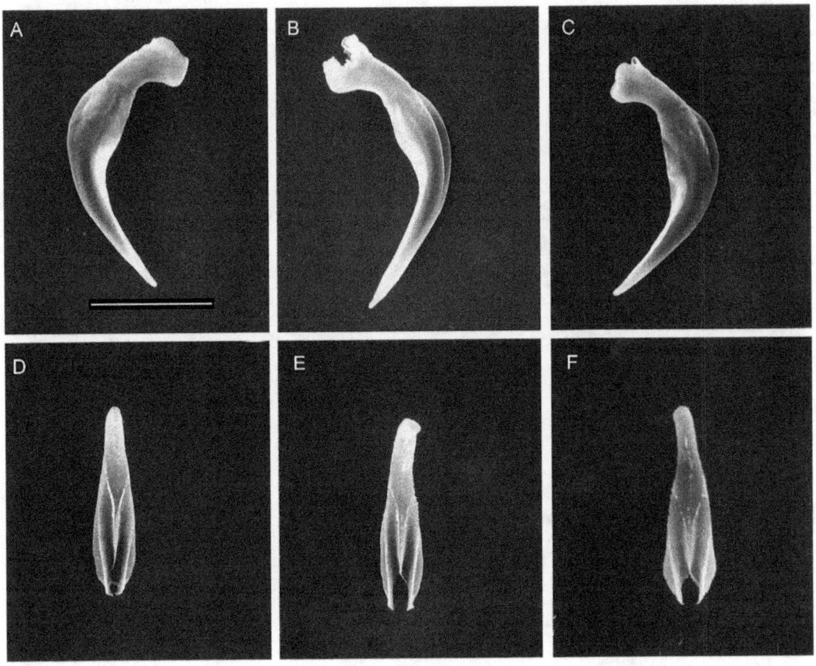

Fig. 168. Steinernema scapterisci. *SEM of spicules and gubernacula. A-C: Spicules of first generation males showing somewhat angular spicule head, prominent shaft and very small or absent velum; D-F: Gubernaculum, ventral view showing Y- or arrowhead-shaped cuneus. (Scale bar: A-F = 34 μm.)*

Stoma with prominent, unusually thickened, cheilorhabdions, appearing as a circular or hexagonal ring in face view. Pharynx typical of family. Pharyngo-intestinal valve large. Excretory pore located anterior to mid-metacorpus. Excretory duct unusually prominent forming an elliptically shaped structure. Genital system didelphic, amphidelphic, reflexed. Vulva in form of a transverse slit with prominent, double-flapped, epiptygmata. Tail length shorter than anal body diam., terminal mucron present.

Female (second generation)

Similar morphologically to that of first generation with following exceptions: *ca* half as long and two-thirds as broad, valve in basal bulb of pharynx more prominent, elliptically shaped structure formed by excretory duct less prominent, tail longer than anal body diam., tapering to a point bearing a mucron.

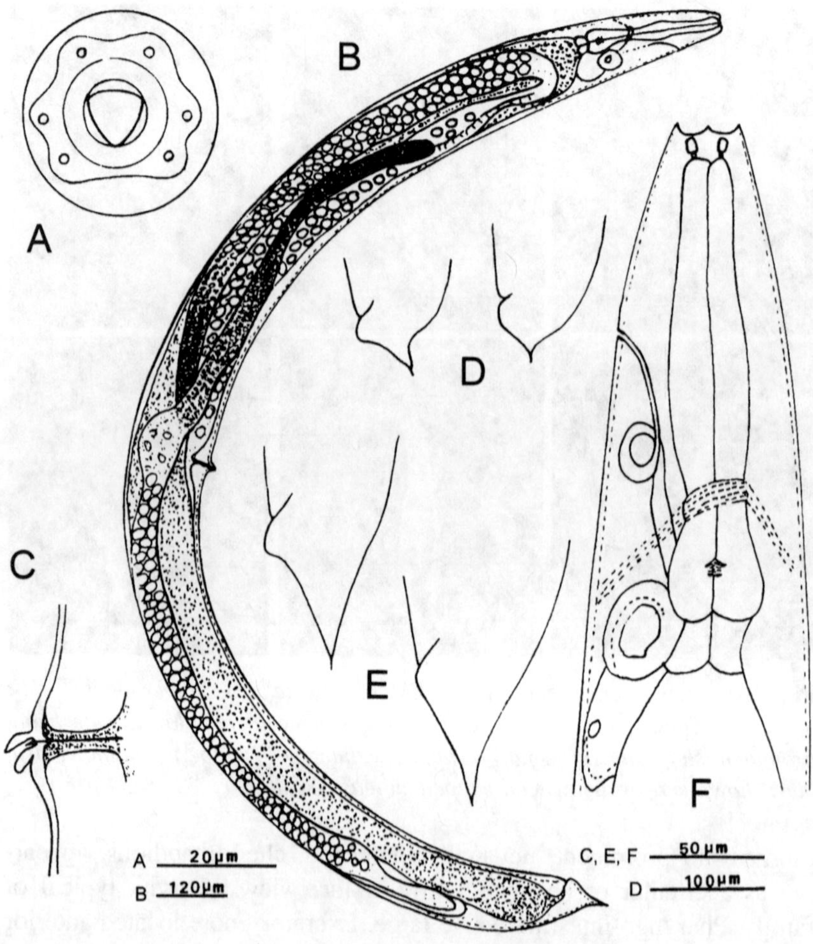

Fig. 169. Steinernema scapterisci *females. A: Diagrammatic face view of first generation female; B: Entire body of second generation female; C: Double-flapped epiptygmata on vulva of first generation female; D: Variation in tail of first generation females; E: Variation in tail of second generation females; F: Anterior region of first generation female showing large cheilorhabdions, pharynx, nerve ring, excretory pore and duct, elliptically shaped structure and gland cell associated with excretory system.*

Infective juvenile

Infective stage, when newly formed, always ensheathed in cuticle of second-stage juvenile. Sheath lost rather easily, even in storage, and thus may not always be present. Body slender, labial region not offset,

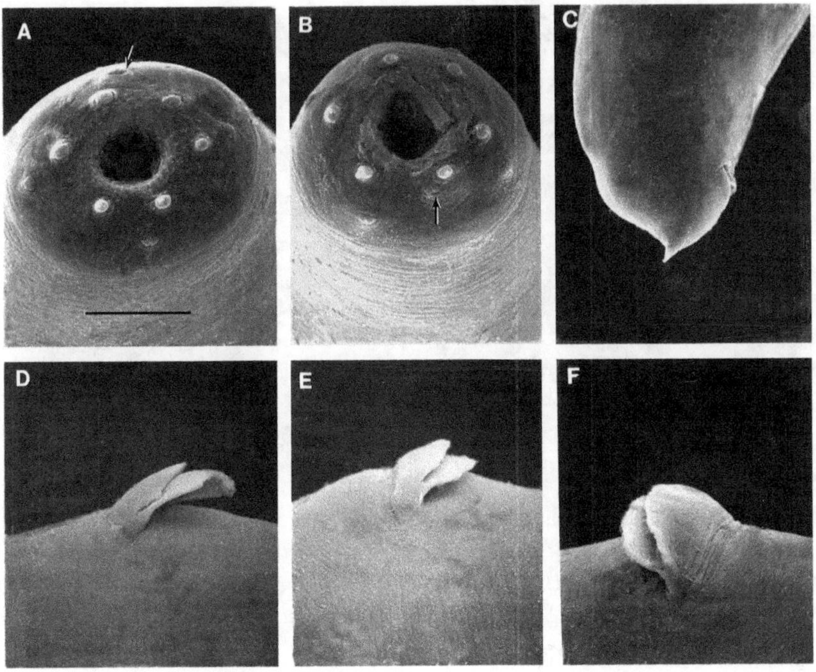

Fig. 170. Steinernema scapterisci, *first generation females. A, B: Variation of face view showing labial, cephalic papillae (arrow in A), amphidial aperture (arrow in B) and stoma; C: Tail with mucron; D-F: Variation in shape of epiptygmata. (Scale bar: A, B, D-F = 15 μm; C = 38 μm.) Modified after Nguyen and Smart (1992b).*

oral aperture not observed; six labial and four cephalic papillae present, prominent elevated oral disc visible on ensheathed nematode. Lateral field with six ridges (*i.e.*, seven lines or incisures). Pharynx degenerate, basal bulb elongate and valvate. Tail tapering gradually dorsally but abruptly ventrally.

TYPE HOST AND LOCALITY

Body cavity of the mole cricket, *Scapteriscus vicinus*, from Uruguay, South America.

TYPE MATERIAL

Holotype, allotype and paratype deposited in the USDA Nematode Collection, Beltsville, MD, USA.

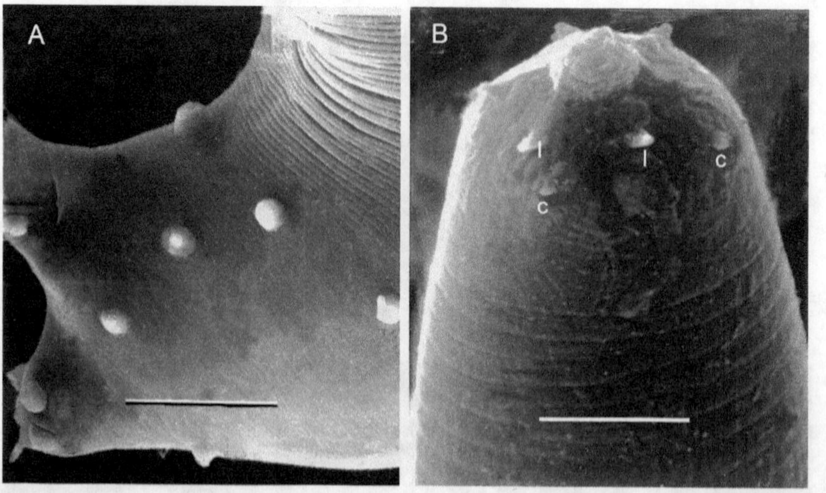

Fig. 171. Steinernema scapterisci. *A: Posterior region of male showing genital papillae and mucron; B: Anterior region of ensheathed infective juvenile showing labial papillae (l), cephalic papillae (c) and elevated oral disc. (Scale bars: A = 15 μm; B = 3.8 μm.) Modified after Nguyen and Smart (1992b).*

RELATIONSHIPS

Steinernema scapterisci is characterised by morphometric of the infective juvenile: body length = 572 μm; E% = 73 (60-80); D% = 31 (27-40). The male has a spicule 83 (72-92) μm long with angular head, prominent shaft and very pointed tip. In ventral view, the gubernaculum enlarges gradually posteriorly; the cuneus is needle-shaped, Y-shaped or arrowhead-shaped. A unique pair of genital papillae is situated near the end of the cloacal aperture. The female has large cheilorhabdions; the prominent excretory duct forms an elliptically-shaped structure and the epiptygmata are prominent.

Steinernema scapterisci is close to *S. carpocapsae*, but differs from this and other species in the *carpocapsae*-group by the above characters of the infective juvenile, male and female.

MOLECULAR CHARACTERISATION

Steinernema scapterisci is genetically characterised by the length of ITS regions, 959 bp, and its composition, A = 0.25235, C = 0.16580, G = 0.22836, T = 0.35349 (accession no. AF122020), the length of

Table 50. *Morphometrics of* Steinernema scapterisci. *Measurements are in μm and in the form: mean ± SD (range).*

Character	First generation			Second generation		Infective juvenile
	Holotype (male)	Male	Female	Male	Female	
n	–	10	10	10	10	20
L	1554	1728 ± 358 (1319-2271)	4162 ± 540 (3531-5156)	1147 ± 95 (1031-1342)	2209 ± 223 (1841-2530)	572 ± 27 (517-609)
a	–	–	–	–	–	24 (20-31)
b	–	–	–	–	–	4.5 (4.4.6)
c	–	–	–	–	–	10.7 (9.2-11.7)
V	–	–	53 ± 2 (50-54)	–	52 ± 2 (52-60)	–
Body diam.	131	156 ± 49 (97-231)	179 ± 13 (159-203)	73 ± 8 (62-84)	123 ± 14 (94-141)	24 ± 4 (18-30)
EP	77	71 ± 11 (63-98)	89 ± 5 (78-94)	68 ± 7 (50-75)	78 ± 6.8 (66-88)	39 ± 4 (36-48)
NR	130	136 ± 11 (120-152)	174 ± 13 (153-194)	121 ± 10 (103-131)	169 ± 12 (147-184)	97 ± 1.1 (83-106)
ES	189	187 ± 21 (164-216)	242 ± 17 (219-269)	168 ± 13 (138-181)	241 ± 15 (222-266)	127 ± 6 (113-134)
Testis reflexion	416	374 ± 52 (306-447)	–	205 ± 19 (176-234)	–	–
Tail length (T)	28	25 ± 3 (21-30)	46 ± 8 (34-59)	25 ± 3 (22-30)	58 ± 4 (48-64)	54 ± 3 (48-60)
Anal body diam. (ABD)	38	33 ± 5 (31-45)	58 ± 9 (41-72)	33 ± 4 (28-41)	47 ± 2.8 (43-52)	–
Spicule length (SL)	89	83 ± 5 (72-92)	–	78 ± 3 (75-83)	–	–
Spicule width	13.5	13 ± 4 (13-14)	–	12 ± 1 (11-14)	–	–
Gubernaculum length (GL)	65	65 ± 5 (59-75)	–	54 ± 3 (47-59)	–	–
Gubernaculum width	7.8	8 ± 0.5 (8-9)	–	6 ± 0.7 (5-8)	–	–
D% = EP/ES × 100	41	36 ± 2 (32-39)	–	40 ± 6 (29-52)	–	31 ± 3 (27-40)
E% = EP/T × 100	–	–	–	–	–	73 ± 6 (60-80)
SW% = SL/ABD × 100	234	252 (204-280)	–	236	–	–
GS% = GL/SL × 100	73	78 (69-84)	–	69	–	–

EP = distance from anterior end to excretory pore; NR = distance from anterior end to nerve ring; ES = distance from anterior end to end of pharynx.

After the original authors.

D2D3 regions, 887 bp and its composition, A = 0.26156, C = 0.19391, G = 0.29200, T = 0.25254 (accession no. AF331898).

BIONOMICS

This species has been reported to infect and kill insects of the order Orthoptera, including *Scapteriscus* (mole cricket), *Neocurtilla* (house cricket), *Acheta domesticus* (field cricket), *Gryllus* sp., and the grasshopper *Romalea guttata*. Insects in the orders Coleoptera, Lepidoptera, Hymenoptera are poor hosts or non-hosts (Nguyen & Smart, 1991a; Smart & Nguyen, 1995).

The nematode can survive in the absence of hosts, yet still be infective, for at least 10 weeks in moist soil (Nguyen & Smart, 1990b). When infective juveniles of *S. scapterisci* were released on the soil surface in the field and in the laboratory, they moved downward through the soil for at least 10 cm in 5 days, infecting and killing mole crickets (Nguyen & Smart, 1990c). When IJ were placed at the centre of a 16 cm soil column they moved in both directions, three times more nematodes moving downward than upward. In the field, nematode-infected mole crickets in the early stage of infection dispersed the nematode during flight (Smart *et al.*, 1990). As a consequence, the nematode can be spread by the infected host from inoculative field releases on golf courses and pastures (Parkman *et al.*, 1993, 1994).

The IJ enter the host through the mouth or spiracles (Nguyen & Smart, 1991a, b; 1992a). They develop in the host in the same way as other species of *Steinernema*. If the nematode is not overcrowded in a host (*i.e.*, nutrient supply is sufficient) the IJ develop to males and females of the first generation. Most eggs from these females hatch and the juveniles develop to become males and females of the second generation. Eggs produced from these females develop to IJ. This cycle takes 8-10 days. If the nematode number in a host is overcrowded (*i.e.*, nutrient supply is insufficient), they develop to males and females of the first generation and eggs produced by these females develop directly to IJ; this cycle takes 6-7 days. The nematode can develop at temperatures between 15 and 33°C.

Bacterial associates

Bacteria from *Steinernema scapterisci* cultured *in vivo*, *in vitro* and collected from fields, were isolated and identified (Anguillera & Smart,

1993a). Many bacteria were associated with this nematode (*Ochrobactrum anthropi*, *Paracoccus denitrificans*, *Xanthomonas maltophilia*, *Xenorhabdus bovienii*, *X. nematophila*, *Xenorhabdus* spp.). These bacteria and three others, *Pseudomonas fluorescens*, *P. aureofacians* and *Escherichia coli*, were used for *in vitro* culture of the nematode. The nematode developed normally and produced IJ with all studied bacteria (Anguillera & Smart, 1993b). This shows that the nematode can feed on a range of bacteria and can therefore survive very well under field conditions. The bacterium typically isolated from *S. scapterisci* was later named *Xenorhabdus innexi* by Lengyel *et al.* (2005).

Biocontrol capability

Steinernema scapterisci is specific to mole crickets and other orthopterans. The nematode was first released in Alachua County, Florida, where it became established and colonised surrounding areas. Three years after the nematode was released, a few mole crickets collected in a farm ten miles from the nearest released site were found to be infected with the nematode. By the 5th year, 25-65% of the mole crickets collected at this farm were infected (Smart *et al.*, 1990). The nematode was successfully introduced by inoculative applications in golf courses and pastures (Parkman *et al.*, 1993, 1994) in Florida. Since 1993, this nematode has been commercialised in Florida to control mole crickets in golf courses and pastures.

DISTRIBUTION

The nematode was originally found in Uruguay (Smart & Khuong, 1986; Nguyen & Smart, 1990a), but is also known from Argentina, Brazil and, more recently, the USA.

References

ANGUILLERA, M.M. & SMART JR, G.C. (1993a). Bacterial symbionts of *Steinernema scapterisci*. *Journal of Invertebrate Pathology* 62, 68-72.

ANGUILLERA, M.M. & SMART JR, G.C. (1993b). Development, reproduction, and pathogenicity of *Steinernema scapterisci* in monoxenic culture with different species of bacteria. *Journal of Invertebrate Pathology* 62, 289-294.

LENGYEL, K., LANG, E., FODOR, A., SZALLAS, E., SCHUMANN, P. & STACKEBRANDT, E. (2005). Description of four novel species of *Xenorhabdus*, family Enterobacteriaceae: *Xenorhabdus budapestensis* sp.

nov., *Xenorhabdus ehlersii*, *Xenorhabdus innexi* sp. nov., and *Xenorhabdus szentirmaii* sp. nov. *Systematic and Applied Microbiology* 28, 115-122.

NGUYEN, K.B. & SMART JR, G.C. (1990a). *Steinernema scapterisci* n. sp. (Steinernematidae: Nematoda). *Journal of Nematology* 22, 187-199.

NGUYEN, K.B. & SMART JR, G.C. (1990b). Preliminary studies on survival of *Steinernema scapterisci* in soil. *Soil and Crop Science Society of Florida, Proceedings* 49, 230-233.

NGUYEN, K.B. & SMART JR, G.C. (1990c). Vertical dispersal of *Steinernema scapterisci. Journal of Nematology* 22, 574-578.

NGUYEN, K.B. & SMART JR, G.C. (1991a). Pathogenicity of *Steinernema scapterisci* to selected invertebrates. *Journal of Nematology* 23, 7-11.

NGUYEN, K.B. & SMART JR, G.C. (1991b). Mode of entry and sites of development of *Steinernema scapterisci. Journal of Nematology* 23, 267-268.

NGUYEN, K.B. & SMART JR, G.C. (1992a). Life cycle of *Steinernema scapterisci* Nguyen & Smart, 1990. *Journal of Nematology* 24, 160-169.

NGUYEN, K.B. & SMART JR, G.C. (1992b). Addendum to the morphology of *Steinernema scapterisci. Journal of Nematology* 24, 478-481.

PARKMAN, J.P., FRANK, J.H., NGUYEN, K.B. & SMART JR, G.C. (1993). Dispersal of *Steinernema scapterisci* (Rhabditida: Steinernematidae) after inoculative applications for mole cricket (Orthoptera: Gryllotalpidae) control in pastures. *Biological Control* 3, 226-232.

PARKMAN, J.P., FRANK, J.H., NGUYEN, K.B. & SMART JR, G.C. (1994). Inoculative release of *Steinernema scapterisci* (Rhabditida: Steinernematidae) to suppress pest mole crickets (Orthoptera: Gryllotalpidae) of golf courses. *Environmental Entomology* 23, 1331-1337.

SMART JR, G.C. & KHUONG, N.B. (1986). Biological control of mole crickets with a neoaplectanid nematode. *Proceedings of the Soil and Crop Science Society of Florida* 45, 192. [Abstr.]

SMART JR, G.C. & NGUYEN, K.B. (1995). Biological control of Orthoptera pest insects. United States Patent, patent number 5,466,448, date of patent: November 14, 1995.

SMART JR, G.C., NGUYEN, K.B., PARKMAN, J.P. & FRANK, J.H. (1990). Biological control of mole crickets in the genus *Scapteriscus* with the nematode *Steinernema scapterisci* Nguyen & Smart, 1990. In: *Rencontres Caraibes en lutte biologique, Guadeloupe, 5-7 November 1990*. Ed. INRA, Paris, 1991.

Steinernema scarabaei[*],[**] Stock & Koppenhöfer, 2003

(Fig. 172)

MEASUREMENTS

See Table 51.

DESCRIPTION

Male (first generation)

Body curved posteriorly, J-shaped when heat-relaxed. Cuticle smooth under light microscopy, but with fine transverse striae visible under SEM. Lateral field and phasmids inconspicuous. Cephalic extremity truncate to slightly rounded, continuous with body contour. Six lips amalgamated but tips distinct and with one labial papilla each. Four cephalic papillae present. Amphidial apertures small, located posterior to lateral labial papillae. Stoma reduced, short and broad, with inconspicuous sclerotised walls. Cheilorhabdions and telorhabdions vestigial. Pharynx muscular; procorpus cylindrical; metacorpus slightly swollen and non-valvate; indistinct isthmus followed by pyriform basal bulb containing reduced valve. Pharynx set off from intestine. Nerve ring usually surrounding isthmus or anterior part of basal bulb. Excretory pore opening circular, located at posterior 33% of metacorpus, just anterior to nerve ring. Genital system monorchic, reflexed, consisting of germinal growth zone leading to seminal vesicle. *Vas deferens* with inconspicuous walls. Spicules paired, symmetrical, curved, of ochre-brown colour. Manubrium rhomboid. Lamina with rostrum (retinaculum) and two internal ribs. Velum present. Gubernaculum arcuate, *ca* 75% length of spicules. Complement of 23 genital papillae present (11 paired and one single) arranged as follows: six precloacal subventral pairs, one single ventral precloacal papilla (located between precloacal pairs 5 and 6), one precloacal subdorsal pair; one pair ventrosublateral adcloacal (or postcloacal in some specimens); two pairs subventral postcloacal, one postcloacal subdorsal pair. A minute pair of papilla-like structures may be present (40% of specimens examined) on anterior cloacal lip. Tail conoid and mucronate.

[*] The specific epithet derives from its natural host, the larvae of *scarabaeid* beetles.

[**] Description after Stock and Koppenhöfer (2003). Revised by David J. Hunt.

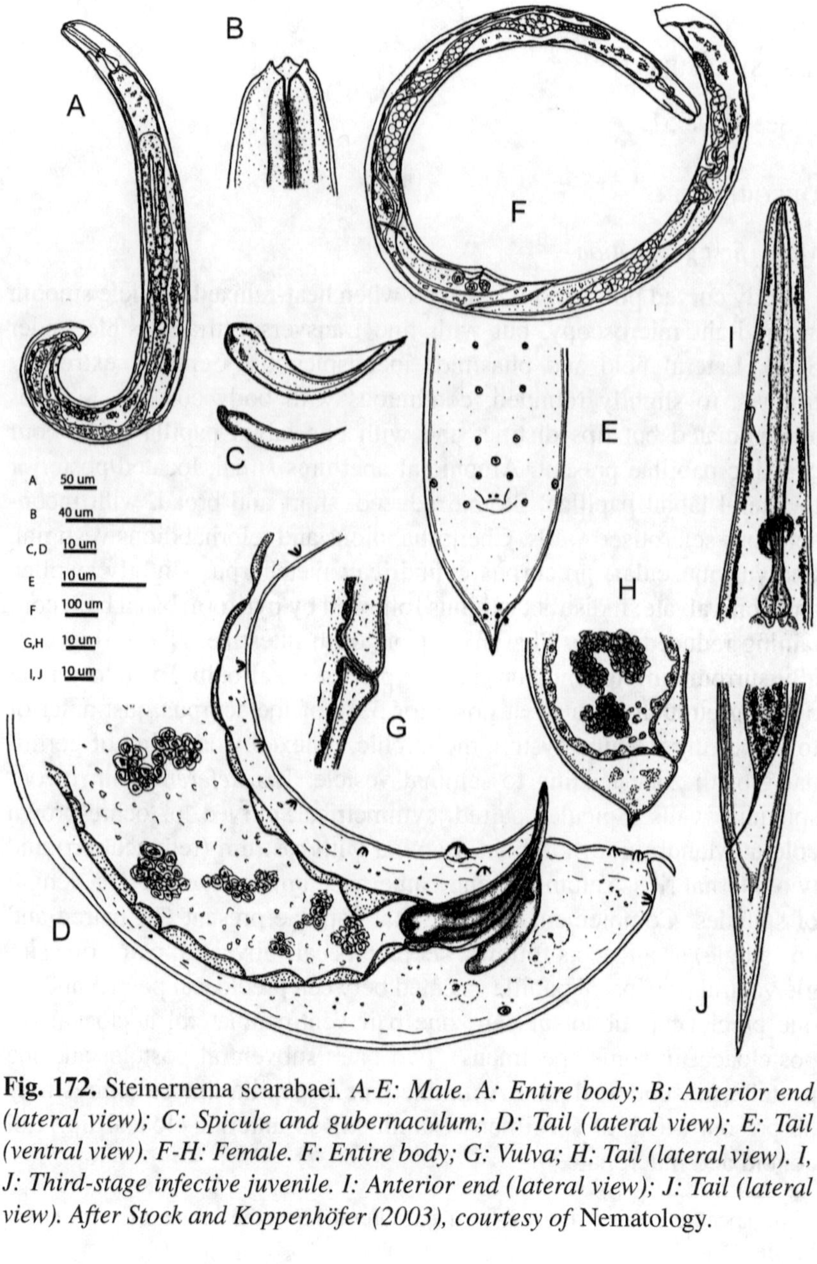

Fig. 172. Steinernema scarabaei. *A-E: Male. A: Entire body; B: Anterior end (lateral view); C: Spicule and gubernaculum; D: Tail (lateral view); E: Tail (ventral view). F-H: Female. F: Entire body; G: Vulva; H: Tail (lateral view). I, J: Third-stage infective juvenile. I: Anterior end (lateral view); J: Tail (lateral view). After Stock and Koppenhöfer (2003), courtesy of* Nematology.

Table 51. *Morphometrics of* Steinernema scarabaei. *Measurements are in μm and in the form: mean ± SD (range).*

Character	First generation			Second generation		Infective juvenile
	Holotype	Male	Female	Male	Female	
n	–	19	19	20	20	20
L	1826	1699 ± 236 (1039-2030)	3991 ± 531 (3355-5043)	1219 ± 115 (875-1470)	2318 ± 229 (1900-2600)	918 ± 20 (890-959)
a	24	22 ± 2 (18-26)	30 ± 2 (28-33)	18 ± 2 (17-24)	24 ± 1.3 (21-26)	30.5 ± 4 (26-36)
b	11	9.8 ± 0.5 (6.5-11)	17±1.5 (16-20)	8.5 ± 1 (7.5-10)	13 ± 1.6 (11.5-14.5)	6.6 ± 2 (5.4-7.6)
c	53	45.5 ± 1.5 (35-50)	104 ± 6 (98-126)	40 ± 1 (39-42)	45 ± 2 (42-47)	12 ± 0.5 (11-13)
c′	0.8	0.74	–	–	–	4
V	–	–	58 ± 3 (54-63)	–	58 ± 1.5 (56-60)	–
Body diam.	76	96 ± 17 (55-120)	132 ± 17.5 (102-181)	67 ± 12 (50-80)	95 ± 3 (89-102)	31 ± 4 (25-37)
EP	88	114 ± 14 (88-129)	126 ± 13 (109-148)	93 ± 13 (78-115)	85 ± 2 (80-91)	77 ± 3 (72-81.5)
NR	132	127 ± 13 (109-146)	177 ± 20 (141-205)	109 ± 8 (99-120)	124 ± 4 (115-130)	120 ± 3 (116-127)
ES	166	173 ± 9 (161-184)	234 ± 18 (212-266)	140 ± 18 (120-166)	170 ± 5 (165-178)	156 ± 7.5 (145-171)
Tail length (T)	34	34 ± 4 (29-40)	51 ± 7 (39-60)	29 ± 6 (22-37)	50 ± 4 (45-56)	76 ± 3 (71-80)
Hyaline (H)	–	–	–	–	–	55 ± 1.5 (50-60)
Anal body diam. (ABD)	43	46 ± 6 (38.5-61)	68 ± 9 (52-76)	39 ± 9 (24-52)	44 ± 3 (39-48)	19 ± 2 (16-21)
Spicule length (SL)	71	75 ± 5 (67-83)	–	48 ± 6 (40-58)	–	–
Gubernaculum length (GL)	44	44 ± 4 (36-50)	–	37 ± 8 (30-45)	–	–
D% = EP/ES × 100	53	66 ± 8 (53-77)	53 ± 1 (51-56)	66 ± 1.2 (65-69)	50 ± 2 (48-51)	60 ± 5 (50-75)
E% = EP/T × 100	257	336 ± 51 (257-445)	260 ± 25 (240-300)	293 ± 50 (237-362)	170 ± 19 (163-177)	100 ± 10 (90-110)
SW% = SL/ABD × 100	165	170 ± 20 (150-200)	–	150 ± 30 (150-210)	–	–
GS% = GL/SL × 100	62	60 ± 9 (50-65)	–	69 ± 1.4 (53-88)	–	–
H% = H/T × 100	–	–	–	–	–	68 ± 2 (65-72)

EP = distance from anterior end to excretory pore; NR = distance from anterior end to nerve ring; ES = distance from anterior end to end of pharynx.

Male (second generation)

Smaller than first generation, tail rounded, lacking a mucron; lacking minute pair of papillae on anterior cloacal lip.

Female (first generation)

Cuticle, labial region, stoma and pharyngeal region as in male. Ovaries opposed, reflexed in dorsal position; oviduct well developed; glandular spermatheca and uterus in ventral position. Vagina short, with muscular walls. Vulva located near middle of body. Vulval lips slightly protruding. Tail conoid with a ventrally bifurcate mucron (present in 70% of specimens examined). Bifurcate mucron more conspicuous in mature females. No postanal swelling.

Female (second generation)

Smaller than first generation and with postanal swelling and blunt, conoid tail lacking a mucron.

Infective juvenile

Body of heat-relaxed specimens slightly C-shaped to almost straight, slender, slightly tapering towards anterior and posterior ends. Cuticle with transverse striae; lateral field distinct with eight longitudinal ridges (= nine lines or incisures) in mid-body region. Cephalic extremity continuous with body, slightly truncate. Labial and cephalic papillae distinct. Amphidial apertures not visible. Labial region smooth; stoma closed. Pharynx long, narrow, with slightly expanded procorpus, narrower isthmus and subpyriform basal bulb with valve. Nerve ring located at level of isthmus. Excretory pore located *ca* middle of pharynx. Cardia present. Anterior portion of intestine with dorsally displaced pouch containing symbiotic bacterium. Bacterial pouch 'peanut-shaped' (in lateral view), distinct in live specimens. Intestine filled with numerous fat globules, lumen of intestine narrow. Rectum long and straight; anus distinct. Genital primordium evident. Tail conoid with pointed terminus. Hyaline region comprising *ca* 66% of tail length.

TYPE HOST AND LOCALITY

Third stage larvae of *Anomala* (= *Exomala*) *orientalis* Waterhouse (Coleoptera: Scarabaeidae) isolated from turf grass (tall fescue, *Festuca arundinacea* Schreb) fields near Freehold, in central New Jersey, USA.

OTHER HOST

Also isolated from *Popillia japonica* Newman (Coleoptera: Scarabaeidae) from the same locality as the type host.

TYPE MATERIAL

Holotype male, paratype first generation males, first generation females and infective juveniles deposited in the Nematology Collection at the University of California, Davis, CA, USA; additional paratype first generation males, first generation females and infective juveniles deposited in the USDA Nematode Collection, Beltsville, MD, USA.

RELATIONSHIPS

Steinernema scarabaei is characterised by the infective juvenile characters of body length of 918 (890-959) μm, the presence of eight longitudinal ridges (*i.e.*, nine lines or incisures) in the lateral field, excretory pore position, tail length, D% and E%. Males are characterised by the morphology and size of the spicules and gubernaculum, the disposition of the genital papillae and the form of the tail. First generation females are characterised by the presence of a ventrally bifurcate mucron on the tail terminus.

Steinernema scarabaei belongs to the *glaseri*-group and can be distinguished from the majority of species in this group by the IJ body length of 918 (890-959) μm, which is shorter than that of *S. arenarium*, *S. glaseri*, *S. longicaudum*, *S. puertoricense* and *S. cubanum*. The IJ most closely resembles that of *S. karii* and *S. loci*, many values, such as body length, excretory pore distance form the anterior end and tail length being similar. However, the male of *S. scarabaei* can be separated from *S. karii* by spicule morphology and the arrangement of the pre- and adcloacal genital papillae. In addition, the spicules of *S. scarabaei* are slightly shorter at 75 (67-83) *vs* 83 (73-91) μm. *Steinernema scarabaei* males differ from *S. loci* in the shape of the manubrium (rhomboid and *ca* as long as wide *vs* longer than wide) and disposition of the genital papillae.

MOLECULAR CHARACTERISATION

Ribosomal sequence of LSU rDNA deposited at GenBank under accession no. AY172023.

BIONOMICS

Steinernema scarabaei is highly pathogenic to scarab larvae (Stock & Koppenhöfer, 2003). Stock and Koppenhöfer (2003) reported that nematode development was slower in *Galleria mellonella* than in *A. orientalis*, the natural host, suggesting that the species was better adapted to the scarabaeid beetle. Koppenhöfer and Fuzy (2003a) discussed the ecological characterisation of *S. scarabaei* and re-iterated that the species was best suited to scarabaeid larvae. Infective juveniles were active over a wide range of temperatures, optimum infectivity occurring within the range 17.5-25°C. The infective juveniles exhibited a wide ranging foraging strategy with excellent infection of sedentary hosts such as scarab larvae. As a result, a number of studies involving the use of this nematode against a variety of scarab pests has been published (see Cappert & Koppenhöfer, 2003; Koppenhöfer & Fuzy, 2003b, c, 2004; Koppenhöfer *et al.*, 2004).

DISTRIBUTION

Reported only from New Jersey, USA, the type locality (Stock & Koppenhöfer, 2003).

References

CAPPERT, D.L. & KOPPENHÖFER, A.M. (2003). *Steinernema scarabaei*, an entomopathogenic nematode for control of the European chafer. *Biological Control* 28, 379-386.

KOPPENHÖFER, A.M. & FUZY, E.M. (2003a). Ecological characterisation of *Steinernema scarabaei*, a scarab-adapted entomopathogenic nematode from New Jersey. *Journal of Invertebrate Pathology* 83, 139-148.

KOPPENHÖFER, A.M. & FUZY, E.M. (2003b). *Steinernema scarabaei* for the control of white grubs. *Biological Control* 28, 47-59.

KOPPENHÖFER, A.M. & FUZY, E.M. (2003c). Biological and chemical control of the Asiatic garden beetle, *Maladera castanea* (Coleoptera: Scarabaeidae). *Journal of Economic Entomology* 96, 1076-1082.

KOPPENHÖFER, A.M. & FUZY, E.M. (2004). Effect of white grub developmental stage on susceptibility to entomopathogenic nematodes. *Journal of Economic Entomology* 97, 1842-1849.

KOPPENHÖFER, A.M., FUZY, E.M., CROCKER, R.L., GELERNTER, W.D. & POLAVARAPU, S. (2004). Pathogenicity of *Heterorhabditis bacteriophora*,

Steinernema glaseri, and *S. scarabaei* (Rhabditida: Heterorhabditidae, Steinernematidae) against 12 white grub species (Coleoptera: Scarabaeidae). *Biocontrol Science and Technology* 14, 87-92.

STOCK, S.P. & KOPPENHÖFER, A.M. (2003). *Steinernema scarabaei* n. sp. (Rhabditida: Steinernematidae), a natural pathogen of scarab beetle larvae (Coleoptera: Scarabaeidae) from New Jersey, USA. *Nematology* 5, 191-204.

Steinernema siamkayai [*,**] **Stock, Somsook & Reid, 1998**
(Fig. 173)

MEASUREMENTS

See Table 52.

DESCRIPTION

Male (first generation)

Body curved posteriorly, J-shaped when heat-relaxed. Cuticle smooth under light microscopy, but with fine transverse striae visible under SEM. Lateral field and phasmid inconspicuous. Cephalic extremity truncate to slightly rounded, continuous with body contour. Six lips, united but with distinct tips, each bearing single labial papilla. Four cephalic papillae present. Amphidial apertures small, located posterior to lateral labial papillae. Stoma reduced, short and broad, with inconspicuous, sclerotised, walls. Cheilorhabdions and telorhabdions vestigial. Pharynx muscular; procorpus cylindrical; metacorpus slightly swollen, non-valvate; isthmus indistinct followed by pyriform basal bulb containing reduced valve. Pharynx set off from intestine. Nerve ring usually surrounding isthmus or anterior part of basal bulb. Excretory pore opening circular, located at posterior third of metacorpus, just anterior to nerve ring. Genital system monorchic, reflexed, consisting of germinal growth zone leading to seminal vesicle. *Vas deferens* with inconspicuous walls. Spicules paired, symmetrical, curved, orange-brown in colour. Manubrium rhomboid. Lamina with rostrum or retinaculum and two internal ribs. Velum present. Gubernaculum long and slender, *ca* 75% of spicule length long, arcuate, distally forked in ventral or dorsal view. Complement of 23 genital papillae (11 pairs and one single) arranged as follows: six precloacal subventral pairs, one precloacal lateral pair; one single precloacal papilla (located between precloacal pairs 5 and 6); one pair postcloacal subventral; one pair postcloacal subdorsal, two subventral pairs near tail tip, one of which is very close to mucron and almost terminal (the original description is in error as to the distribution of the postcloacal papillae. The information presented here is as confirmed by

[*] Named from *Siam*, the old designation of the country of origin, and Harry K. *Kaya*, a leading scientist in insect pathology and biocontrol.
[**] Description after Stock *et al.* (1998). Revised by David J. Hunt.

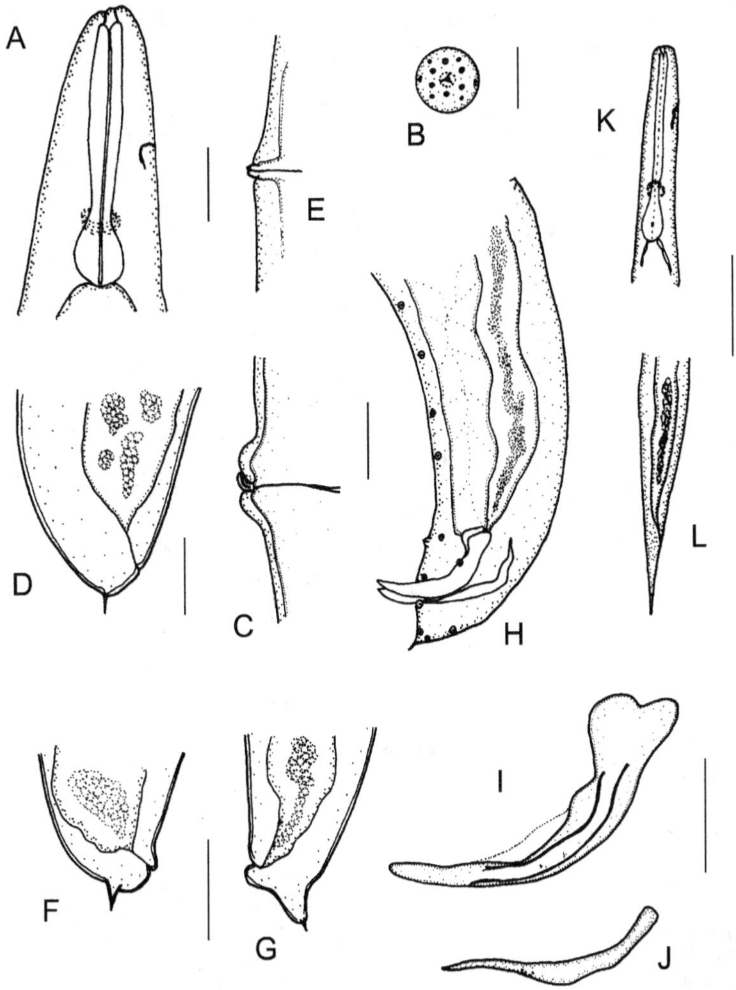

Fig. 173. Steinernema siamkayai. *A: Anterior end of first generation female; B:* En face *view of first generation female; C: Vulva of first generation female; D: Tail of first generation female; E: Vulva of second generation female; F, G: Tail of second generation female; H: Tail of first generation male, lateral view; I: Spicule, lateral view; J: Gubernaculum, lateral view; K: Anterior end of infective juvenile; L: Posterior end of infective juvenile. (Scale bars: A, C-J = 50 μm; B = 10 μm; K-L = 40 μm.) Reproduced from* Systematic Parasitology, *41 (1998), p. 107;* Steinernema siamkayai *n. sp. (Rhabditida: Steinernematidae), an entomopathogenic nematode from Thailand, by Stock, S.P., Somsook, V. and Reid, A.P., Figure 1. With kind permission of Springer Science and Business Media.*

Table 52. *Morphometrics of* Steinernema siamkayai*. *Measurements are in μm and in the form: mean (range).*

Character	First generation		Second generation		Infective juvenile
	Male	Female	Male	Female	
n	20	20	20	20	20
L	1135 (1035-1278)	3937 (3161-5172)	830 (716-952)	1836 (1410-2560)	446 (398-495)
a	–	–	–	–	21 (19-23)
b	–	–	–	–	4.7 (4.0-6.1)
c	–	–	–	–	11.3 (10.3-14.8)
V	–	51 (48-53)	–	56 (53-58)	–
Body diam.	140 (107-159)	198 (170-280)	57 (47-75)	105 (83-128)	21 (18-24)
EP	57 (47-67)	66 (54-84)	58 (54-68)	68 (65-71)	35 (29-38)
NR	91 (81-100)	133 (116-147)	85 (80-91)	123 (113-130)	72 (68-80)
ES	134 (123-141)	177 (152-19)	113 (108-121)	151 (141-169)	95 (80-107)
Testis reflexion	341 (296-355)	–	110 (90-120)	–	–
Tail length (T)	28 (22-32)	31 (22-38)	22 (19-23)	41 (28-52)	36 (31-41)
Anal body diam. (ABD)	45 (37-54)	51 (48-53)	29 (25-31)	38 (30-49)	11.5 (9-15.5)
Spicule length (SL)	78 (75-80)	–	62 (55-73)	–	–
Spicule width	8 (7.5-9)	–	5.5 (5-6.5)	–	–
Gubernaculum length (GL)	54 (47-65)	–	42 (37-55)	–	–
Gubernaculum width	5 (4-7)	–	3.5 (3-4.5)	–	–
D% = EP/ES × 100	42 (35-49)	–	51 (49-56)	–	37 (31-43)
E% = EP/T × 100	207 (166-257)	–	270 (260-290)	–	96 (85-112)
SW% = SL/ABD × 100	170 (140-220)	–	108 (97-117)	–	–
GS% = GL/SL × 100	70 (60-80)	–	69 (66-75)	–	–
H% = H/T × 100	–	–	–	–	–

EP = distance from anterior end to excretory pore; NR = distance from anterior end to nerve ring; ES = distance from anterior end to end of pharynx.

* No holotype data given in original description.

P. Stock, pers. comm., 2000). Tail conoid, mucronate, mucron 3 (2.5-3.5) μm long.

Male (second generation)

Smaller than first generation male. Tail mucron = 3 (2-4) μm long.

Female (first generation)

Body C-shaped when heat-relaxed. Cuticle, labial region, stoma and pharyngeal region as in male. Ovaries opposed, reflexed dorsally; oviduct well developed; glandular spermatheca and uterus in ventral position. Vagina short, with muscular walls. Vulva located near middle of body, lips protruding. Double-flapped epiptygmata present. Tail conoid, blunt, mucronate, mucron 7 (4.5-11) μm long. Postanal swelling absent.

Female (second generation)

Smaller than first generation female. Postanal swelling present. Tail mucron = 4.5 (3-6) μm long, noticeably shorter than in first generation.

Infective juvenile

Body slender, tapering regularly from base of pharynx to anterior end and from anus to terminus. Cuticle with transverse striae; lateral field distinct with six or eight longitudinal ridges in mid-body region (*i.e.*, seven or nine incisures). Labial region smooth; oral aperture closed. Pharynx long, narrow. Nerve ring located at isthmus level. Excretory pore located in anterior third of pharynx. Basal bulb valvate. Cardia present. Anterior portion of intestine with dorsally displaced pouch containing symbiotic bacterium. Lumen of intestine narrow; rectum long; anus distinct. Genital primordium evident. Tail conoid with pointed terminus.

TYPE HOST AND HABITAT

Type host unknown in nature as the isolate was collected using *Galleria mellonella* (L.), a bait insect. Isolated from sandy-clay loam soil under sweet tamarind, *Tamarindus indicus* L., from Lohmsak district (16.4°N, 101.2°E), Petchabun Province, northern Thailand.

TYPE MATERIAL

Holotype male first generation (UCDW 3645), allotype female first generation (UMNC 3646), five paratype males first generation

(UC13NC 3647), five paratype females first generation (UC13NC 3648), and five paratype infective juveniles (UMNC 3649) deposited in the University of California-Davis Nematode Collection, Davis, CA, USA. Five paratype males first generation, five paratype females first generation; and five paratype infective juveniles deposited in Division of Entomology and Zoology, Department of Agriculture, Bangkok, Thailand. Field collection number: T9.

RELATIONSHIPS

Steinernema siamkayai is characterised by the following combination of morphological features: infective juvenile with a body length of *ca* 445 (398-495) μm, short tail of 35.5 (31-41) μm, and lateral field with 6-8 longitudinal ridges (*i.e.*, 7-9 incisures); tail of both first and second generation adults with a cuticular mucron; males with long spicules, *ca* 77.5 and 73 μm in first and second generation, respectively; gubernaculum long and slender, comprising *ca* 75% of the spicule length; 23 genital papillae; first generation female with protruding vulval lips; second generation female lacking protruding vulval lips; first and second generation female with double flapped epiptygmata; first and second generation female tail conoid with a cuticular mucron *ca* 7 and 4.5 μm long, respectively; postanal swelling lacking in first generation female, but present in second generation.

MOLECULAR CHARACTERISATION

Ribosomal sequence of D2D3 (LSU rDNA) and ITS regions deposited at GenBank under accession no. AF331907 and AF331917, respectively.

BIONOMICS

The life-cycle of *S. siamkayai* is similar to other *Steinernema* spp. As reported in the original description, the infective juveniles enter the insect host through the mouth and spiracles (entrance through the anus was not observed) and then penetrate into the haemocoel. Once in the haemocoel, the infective juveniles release a symbiotic bacterium killing the insect within 36-48 h. Two amphimictic generations occur inside the host, and progeny of the second generation emerge from the insect cadaver as infective juveniles. Observations done in blood drops of *G. mellonella* showed that the infective juveniles are able to reach the adult stage (Stock *et al.*, 1998).

The influence of insect associated cues on the jumping behaviour of *S. siamkayai* infective juveniles was studied by Campbell and Kaya (2000). *Steinernema siamkayai* is an ambush forager and responded to small air movements by increased waving and jumping towards the source of movement. Volatile host cues did not increase jumping behaviour in *S. siamkayai*.

Tests to assess whether *S. siamkayai* and its bacterial symbiont could suppress root-knot nematode (*Meloidogyne incognita*) were done by Sasnarukkit *et al.* (2002). This study reported some evidence that the nematode and cell-free culture filtrates could be antagonistic to the infective juveniles of root-knot nematode, but only in the soil phase – once the nematodes had penetrated the root there was no further effect.

Umamaheswari *et al.* (2004) studied the host range of *S. siamkayai* isolated from Tamil Nadu, India. Host species tested included: *Spodoptera litura*, *Helicoverpa armigera*, *Plutella xylostella*, *Sylepta derogata*, *Leucinodes orbonalis*, *Chilo sacchariphagus indicus*, *Eupterote mollifera*, *Oryctes rhinoceros*, *Holotrichia serrata*, *Mantis religiosa* and *Periplaneta americana*. All species of test insect were killed by the nematode, although no nematode multiplication was observed in *Eupterote mollifera* or *Oryctes rhinoceros*.

Distribution

Known from the type locality in northern Thailand (Stock *et al.*, 1998) and from Tamil Nadu, India (see Umamaheswari *et al.*, 2004) and Kerala, India (Banu *et al.*, 2005).

References

BANU, J.G., NGUYEN, K.B. & RAJENDRAN, G. (2005). Occurrence and distribution of entomopathogenic nematodes in Kerala, India. *International Journal of Nematology* 15, 9-16.

CAMPBELL, J.F. & KAYA, H.K. (2000). Influence of insect associated cues on the jumping behaviour of entomopathogenic nematodes (*Steinernema* spp.). *Behaviour* 137, 591-609.

SASNARUKKIT, A., GAUGLER, R. & SONTIRAT, S. (2002). Effects of entomopathogenic nematode *Steinernema siamkayai* and its bacterial symbiont on *Meloidogyne incognita*. *International Journal of Nematology* 12, 169-174.

STOCK, S.P., SOMSOOK, V. & REID, A.P. (1998). *Steinernema siamkayai* n. sp. (Rhabditida: Steinernematidae), an entomopathogenic nematode from Thailand. *Systematic Parasitology* 41, 105-113.

UMAMAHESWARI, R., SIVAKUMAR, M. & SUBRAMANIAN, S. (2004). Host range of the native entomopathogenic nematodes from Tamil Nadu. *Insect Environment* 10, 151-152.

Steinernema sichuanense[*,**] Mráček, Nguyen, Tailliez, Boemare & Chen, 2006

(Figs 174-179)

MEASUREMENTS

See Table 53.

DESCRIPTION

Male (first generation)

Body curved ventrally posteriorly, tapering towards labial region, J-shaped in heat-relaxed specimens. Cuticle with very faint transverse striae, lateral fields absent. Cephalic extremity rounded, almost continuous with body. Six lips fused at base, each bearing a labial papilla. Four cephalic papillae present. Amphidial apertures distinct, located laterally between labial and cephalic circle of papillae. Excretory pore mostly located at metacorpus level, always anterior to nerve ring (ranging from 45 to 56% of length from anterior body end to base of pharynx). Stoma short, shallow, cheilorhabdions distinct, sclerotised, followed by sclerotised smaller, sometimes indistinct, prorhabdions. Pharynx muscular with procorpus extending into slightly swollen non-valvate metacorpus, a narrower isthmus region surrounded by a nerve ring, and a rounded basal bulb. Cardia rather long, intestine generally with wide lumen. Distinct, papilliform deirids situated between pharyngeal bulb to nerve ring level. Genital system monorchic, reflexed. Spicules paired symmetrical, finely curved, distally bluntly pointed; colourless. Spicule robust, manubrium rounded, with length/width ratio of 1 : 1; manubrium 18% of spicule length; calomus sometimes less distinct. Lamina with two ribs, slender and curved, almost straight posteriorly. Rostrum well developed and very prominent, at least as wide as manubrium width; velum bow-like, distinct, extending from base of rostrum to distal part of lamina. In lateral view, gubernaculum boat-shaped, anterior part mostly enlarged or bluntly rounded, sometimes slightly curved up, posterior part usually tapering abruptly. In ventral view, cuneus small, needle-shaped, or absent, corpus with two wings. Complement of 23 genital papillae comprising 11 pairs plus single ventral papilla located just anterior to cloacal aper-

[*] Named after the type locality, the province of *Sichuan*.

[**] Description after Mráček *et al.* (2006). Revised by Zdeněk Mráček.

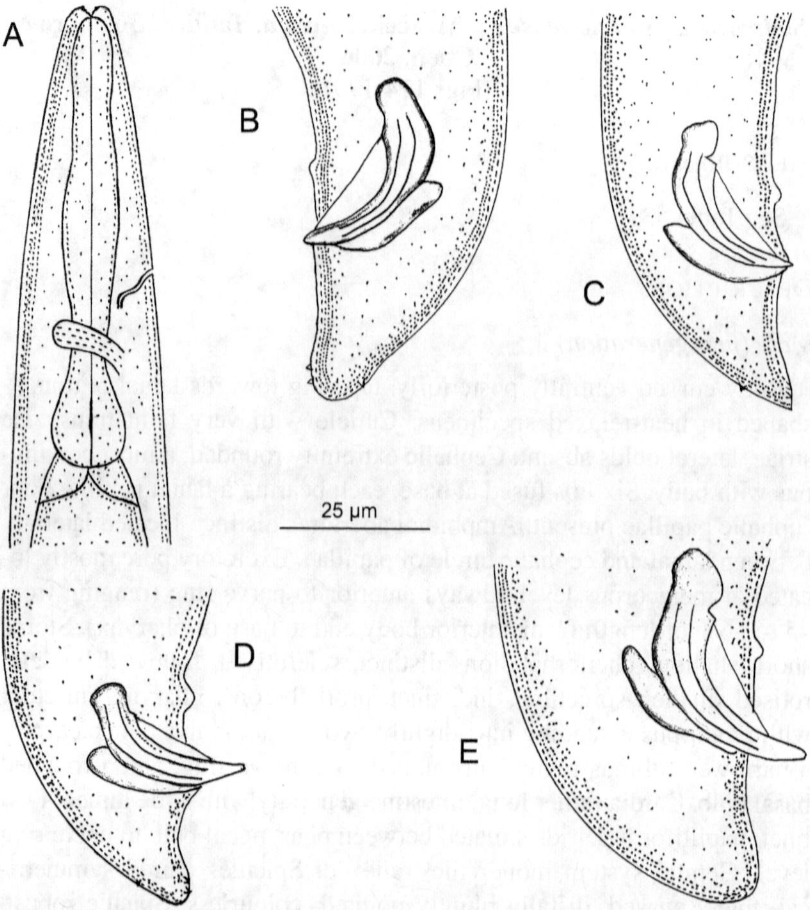

Fig. 174. Steinernema sichuanense. *A-E: Male. A: Anterior region; B-E: Variations of tail shape, spicule and gubernaculum. After Mráček* et al. *(2006), courtesy of* Journal of Invertebrate Pathology.

ture. Of these, six pairs located precloacal (five pairs subventral and one pair lateral), two pairs adcloacal, two pairs subterminal, and one pair subdorsal. Last three pairs consistently located near tail tip. Tail bluntly conoid, usually concave ventrally, mucron absent.

Male (second generation)

Similar to first generation, but slightly smaller in body length and other measurements. Greatest body diam. much smaller. Tail mucron mostly papillate rather than filamentous.

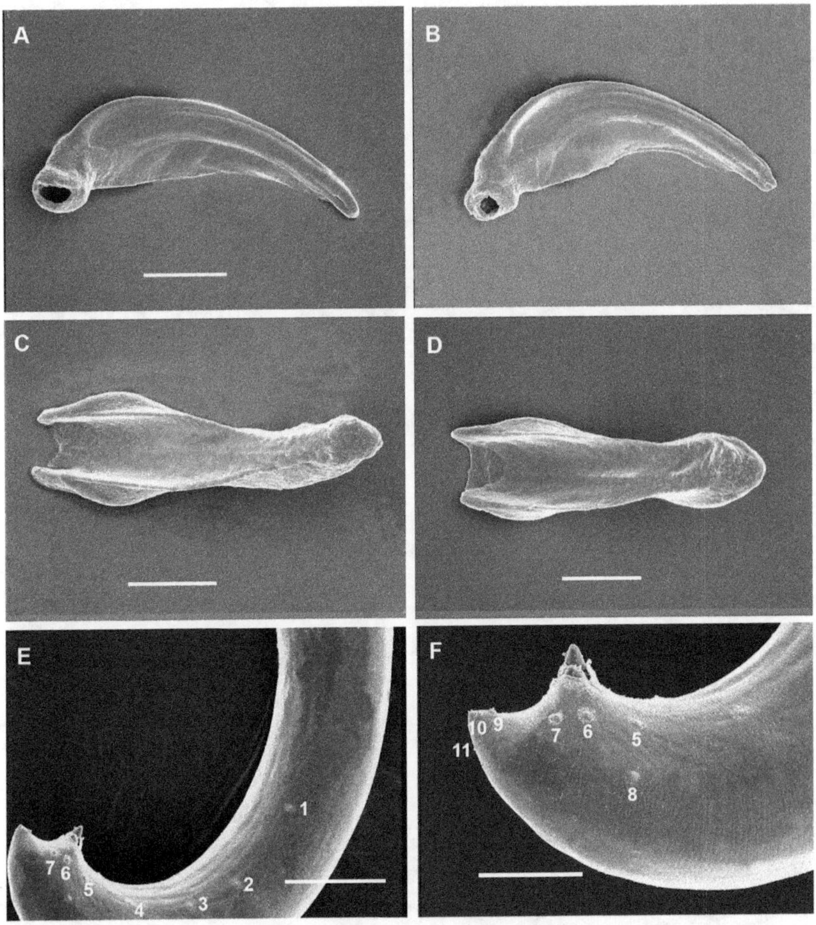

Fig. 175. Steinernema sichuanense. *A-F: Male, SEM micrographs. A, B: Spicules showing large velum; C, D: Gubernacula, ventral view showing very short cuneus and thick gubernaculum head; E, F: Posterior region showing 11 pairs and a single genital papilla. (Scale bars: A, B = 20 µm; C = 12 µm; D = 11 µm; E = 43 µm; F = 15 µm.) After Mráček* et al. *(2006), courtesy of* Journal of Invertebrate Pathology.

Female (first generation)

Body relatively short, rarely longer than 3 mm, usually C-shaped when heat-relaxed. Cuticle only faintly striated. Lateral field not observed. Cephalic extremity bluntly rounded, slightly tapering anteriorly. Morphology of head region and pharynx similar to male. Stoma *ca*

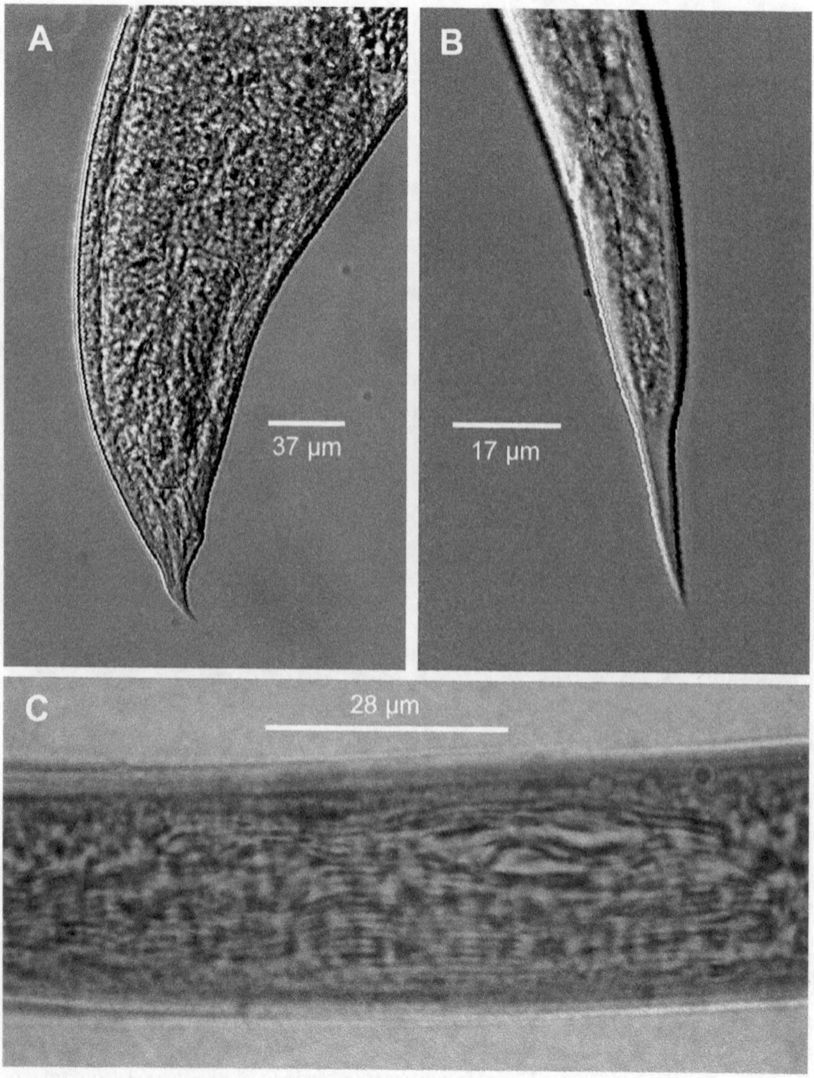

Fig. 176. Steinernema sichuanense. *A: Second generation female tail. B, C: Infective juvenile. B: Tail showing depression in dorsal contour; C: Bacterial vesicle. After Mráček* et al. *(2006), courtesy of* Journal of Invertebrate Pathology.

5 μm long × 10 μm diam. Excretory pore located anterior to nerve ring. Genital tract amphidelphic with reflexed ovaries. Vulva mostly symmetrical and in form of a transverse slit, slightly (3 μm) to moder-

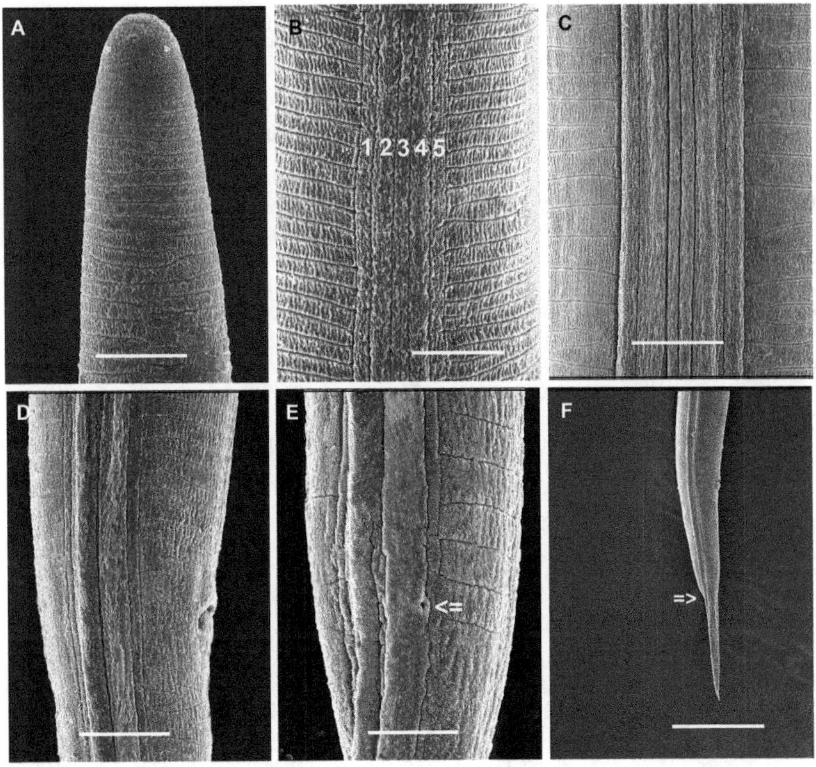

Fig. 179. Steinernema sichuanense. *A-F: Infective juvenile, SEM photographs of lateral field showing lateral field pattern. A: Anterior region showing two cephalic papillae and an amphidial aperture; B: Lateral field with five ridges with central ridge divided into two posteriorly; C: Lateral field with six ridges, the two submarginal ones being prominent; D: Lateral field pattern near anus showing lateral field with four ridges; E: Lateral field near phasmid (arrow). The lateral field changes from four to two ridges; F: Posterior region showing lateral field, anus and depressed area (arrow). (Scale bars: A = 6 μm; B-D = 5 μm; E = 3 μm; F = 23.1 μm.) After Mráček* et al. *(2006), courtesy of* Journal of Invertebrate Pathology.

ation indistinct with light microscope, but distinct with SEM, 1 μm thick in mid-body region. Cephalic extremity finely rounded, not truncate and usually slightly offset from body contour, not annulated. Stoma closed. Labial papillae not observed; amphidial apertures prominent; four distinct cephalic papillae, posterior to amphidial apertures. Lateral field beginning with one line from annules 17-20 from anterior end. A short

Table 53. *Morphometrics of* Steinernema sichuanense. *Measurements are in μm and in the form: mean ± SD (range).*

Character	Holotype (IJ)	First generation		Second generation		Infective juvenile
		Male	Female	Male	Female	
n	–	20	20	20	20	20
L	727	1852 ± 226 (1555-2202)	2278 ± 640 (2060-3796)	1129 ± 119 (808-1313)	1498 ± 194 (1232-1939)	710 ± 54 (606-787)
a	25	15 ± 1.5 (12-18)	17 ± 2.5 (12-21)	17 ± 1.1 (14-19)	16 ± 2.0 (14-20)	25 ± 1.7 (23-28)
b	5.7	9.9 ± 1.1 (8.1-12)	14 ± 2.3 (11-17)	7.4 ± 0.7 (5.6-8.4)	8.8 ± 1.1 (7.2-11)	5.4 ± 0.3 (4.9-6.0)
c	10	40 ± 4.6 (33-49)	50 ± 11 (35-68)	32 ± 3.2 (24-38)	33 ± 5.1 (23-42)	10 ± 0.5 (8.9-11)
c′	4.5	0.9 ± 0.04 (0.8-0.9)	1.1±0.1 (0.9-1.3)	0.9 ± 0.01 (0.9-1.0)	1.3 ± 0.2 (1.1-1.8)	4.3 ± 0.4 (3.8-5.0)
V	–	–	54 ± 3.1 (45-56)	–	55 ± 1.8 (53-58)	–
Body diam.	29	122 ± 16 (91-141)	157 ± 24 (124-199)	68 ± 5.0 (54-75)	93 ± 7.3 (80-107)	–
EP	64	95 ± 4.3 (88-105)	94 ± 8.6 (80-108)	82 ± 3.2 (75-88)	78 ± 4.8 (72-86)	64 ± 2.8 (57-68)
NR	97	–	–	–	–	99 ± 4.6 (91-108)
ES	127	188 ± 6.2 (172-199)	196 ± 16 (177-219)	153 ± 5.7 (137-161)	170 ± 10 (151-188)	131 ± 5.5 (121-142)
Tail length (T)	72	46 ± 2.7 (43-54)	54 ± 7.2 (46-72)	36 ± 2.5 (32-40)	46 ± 4.0 (37-54)	72 ± 3.8 (64-76)
Hyaline (H)	35	–	–	–	–	31 ± 2.2 (27-35)
Anal body diam. (ABD)	16	53 ± 2.4 (49-59)	48 ± 8.1 (38-68)	39 ± 1.9 (37-41)	37 ± 3.3 (27-45)	17 ± 1.7 (14-19)
Spicule length (SL) (chord)	–	68 ± 2.1 (65-72)	–	55 ± 3.3 (48-60)	–	–
Gubernaculum length (GL)	–	47 ± 3.3 (40-51)	–	39 ± 3.5 (32-46)	–	–
Gubernaculum width		8.2 ± 0.5 (7.2-8.7)	–	7.8 ± 0.7 (6.4-8.7)	–	–

distance posteriorly, two additional lines appearing forming two ridges. Close to base of pharynx, lateral field changing from two to five ridges, more posteriorly, central ridge dividing into two, making a total of six ridges, the maximum number in lateral field. Lateral field form is typical with two less distinct central and marginal ridges and two unusually prominent submarginal ridges. In posterior part of body, lateral field reducing gradually. At *ca* two anal body diam. anterior to anus, two

Table 53. *(Continued).*

Character	Holotype (IJ)	First generation		Second generation		Infective juvenile
		Male	Female	Male	Female	
D% = EP/ES × 100	–	51 ± 2.5 (45-56)	48 ± 4 (36-54)	54 ± 1.9 (51-58)	46 ± 4 (40-50)	49 ± 1.7 (46-53)
E% = EP/T* × 100	89	207	174	228	170	89
SW% = SL/ABD × 100	–	130 ± 10 (120-140)	–	140 ± 10 (130-170)	–	–
GS% = GL/SL × 100	–	80 ± 4 (60-80)	–	70 ± 8 (60+90)	–	–
H% = H/T × 100	49	–	–	–	–	44 ± 3.3 (38-51)

EP = distance from anterior end to excretory pore; NR = distance from anterior end to nerve ring; ES = distance from anterior end to end of pharynx.

* Recalculated according to Mráček *et al.* (2006).

middle ridges disappearing forming four ridges in lateral field. Near phasmid, two marginal ridges disappearing forming two ridges in lateral field. With above arrangement, formula of lateral field is: 2, 5, 6, 4, 2. Deirid not observed. Pharynx with slightly expanded procorpus, narrower isthmus and subpyriform basal bulb with very weakly developed valve. Hemizonid distinct, lens-like shaped, located at level of anterior margin of basal bulb. Excretory pore located in mid-pharynx region, cuticularised, *ca* one μm diam. Cardia present. Bacterial vesicle slender, unusually long, 81-115 × 23-28 μm. Gut filled with numerous fat globules, lumen not visible. Rectum straight, anus sickle-shaped. Phasmids pore-like, located posterior to anus at ventral edge of lateral fields at *ca* 35-40% of tail length from anus. Tail conoid with characteristic dorsal constriction present slightly anterior to proximal portion of hyaline region; terminus pointed; hyaline region comprising less than 50% of tail length.

TYPE HOST AND LOCALITY

Natural host unknown. The species was collected along the road from Yajing to Litang, province of Sichuan, approximately 20 km East of Yajing, People's Republic of China. The type locality was characterised by a dense vegetation that was formed by deciduous forest on the valley slopes (spruce, oaks) and deciduous trees and shrubs in the valley bottom (alders, poplars, willows, roses, *etc.*) were combined with pastures and poor fields. In the type locality, an insect fauna was rich in flies, butterflies and many small, flying, adult beetles belonging to the family Scarabaeidae, which could serve as EPN hosts in their larval stage.

Type material

Holotype: infective juvenile; paratype: infective juveniles, males and females of first and second generation deposited in the EPN collection at the Laboratory of Insect Pathology, Institute of Entomology, Czech Academy of Sciences, České Budějovice, The Czech Republic; paratype infective juveniles in the Nematode Collection at the Department of Entomology and Nematology, University of Florida, Gainesville, USA.

Relationships

Steinernema sichuanense belongs to a group of steinernematids including *S. affine*, *S. intermedium* and *S. beddingi* – *i.e.*, the *intermedium*-group. This group is morphologically characterised by features of the infective juveniles, male and females. For the IJ, the head is rounded, the lateral fields have six ridges with the two submarginal ridges prominent. For males, the spicules are robust with a well developed rostrum, a mucron is absent in the first generation but present, yet short, in the second generation. For females, those of the first generation have a conical tail with papilla-like terminal projections.

In the *intermedium*-group, the IJ of *S. sichuanense* differs from *S. affine* by its absence of the internal tail spine; it differs from *S. beddingi* by its shorter tail (72 *vs* 77 μm), its smaller D% (49 *vs* 57) and by its lateral field having six ridges compared to four ridges in *S. beddingi*. For the male, the mucron is absent in both generations of males of *S. affine*, *S. intermedium* and *S. beddingi*, whereas it is present in the second generation of *S. sichuanense*. Based on SEM photographs, the morphology and morphometrics of the spicules and gubernaculum of the four species in the *intermedium*-group are quite different. For females, the postanal swelling is absent in the first generation females of *S. sichuanense*, whereas it is present in *S. affine*, *S. intermedium* and *S. beddingi*.

MOLECULAR CHARACTERISATION

The ITS sequence, including primers TW81 and AB28 is characterised by its sequence length 892 base pairs (bp), ITS1 = 307 bp, ITS2 = 350 bp and its composition. The sequence length of ITS1 is different from that of all other species in the *intermedium*-group, and ITS2 length is identical with that of *S. beddingi* but differs from that of other species in the group by only one base pair. The percentage of A in the composition is the highest and G is the lowest among four closely related species. For more interspecific relationships, the alignments and pairwise distances show that this species has 21 diagnostic characters and differs from its closest taxon, *S. affine*, at 68 aligned positions and from the most divergent species *S. kraussei* at 387 aligned positions. The data show that this nematode is different from all other species studies.

For D2D3 regions, the sequence of *S. sichuanense* is 867 bp, but only the D3 domain is reported for *S. beddingi*, so data from D3 fragment were presented in Mráček *et al.* (2006). Those data show that the species has ten diagnostic characters and differs from its closest taxon, *S. affine*, at 20 aligned positions and from the most divergent species, *S. siamkayai*, at 259 aligned positions. Pairwise distances also show that *S. sichuanense* is different from closely related nematode species by 20-30 bp. This indicates that this species is valid when comparing these distances with those of others in the *carpocapsae*-group. The distances between these species are from 1-6 bp. There are some other described species that have lower pairwise distances; for example, pairwise distances between *S. longicaudum* and *S. cubanum*, and between *S. glaseri* and *S. cubanum*, are only 4 bp.

Both morphological and molecular characteristics show that *S. sichuanense* belongs to the *intermedium*-group.

BIONOMICS

The type isolate of *S. sichuanense* reproduced slowly either at laboratory temperature of 20-22°C or in an incubator at 15°C. *Galleria mellonella* larvae died in 2-3 days. The first generation adults appeared no earlier than 1 week after IJ invasion. The second generation was found in cadavers after 3 weeks, although adult nematodes usually died in cadavers of *G. mellonella*. After 1 month, only 5-10% of cadavers yielded infective juveniles.

Slightly more successful reproduction was found in *Spodoptera littoralis* larvae when seven out of 20 died in 3 days and 13 out of 20 in 1 week. However, only 10-20% cadavers yielded a new generation of IJ (Mráček *et al.*, 2006).

DISTRIBUTION

This species is only known from the type locality in Sichuan, People's Republic of China (Mráček *et al.*, 2006).

Reference

MRÁČEK, Z., NGUYEN, K.B., TAILLIEZ, P., BOEMARE, N. & CHEN, S. (2006). *Steinernema sichuanense* n. sp. (Rhabditida, Steinernematidae) a new species of entomopathogenic nematode from the province of Sichuan, east Tibetan Mountains, China. *Journal of Invertebrate Pathology* 93, 157-169.

*Steinernema silvaticum**,** Sturhan, Spiridonov & Mráček, 2005
(Figs 180-184)

MEASUREMENTS

See Table 54.

DESCRIPTION

Male (first generation)

Obvious cases of malformations observed in all three giant males studied with shortened spicules and gubernaculum with thinner walls, absence of some genital papillae, absence of normal spermatogenesis in testis. Deirid located well posterior to pharynx-intestine junction. Mucron absent, or present as tiny, vestigial spike.

Male (second generation)

Body quite slender, J-shaped in heat-relaxed specimens, rarely anterior end curved dorsad, tapering towards anterior end. Cuticle annulation virtually indistinguishable throughout body, apart from dorsal surface of posterior end, where an irregular pattern of annulation is prominent. Lateral protrusion of cuticle visible at mid-body level, 2-3 μm wide. Labial region rounded. Six 1 μm high labial papillae present as pointed tubercles arranged in circle around oral aperture. Circle of four, more protuberant, cephalic papillae, 1.5-2 μm long, located posterior to labial papillae. Cheilostom cuticle connected to a 2 μm thick $\times$ 2.5 μm long strongly cuticularised ring, forming middle part of stoma walls. Stoma base formed by thick cuticular lining of pharynx, appearing two-layered in most specimens. Short, funnel-shaped, section leading from stoma to pharynx lumen. Pharynx lumen completely collapsed up to 8-9 μm from stoma base, where its triradial structure starts. Pharynx corpus of almost uniform diam. throughout its length, with orifices of pharyngeal glands visible in posterior part. Isthmus narrowing, poorly developed. Basal bulb subpyriform; valves present as thin folded platelets. Cardia protruding 4-7 μm into intestine proventriculus, often asymmetrical. Nerve ring mostly situated on anterior slope of basal bulb. Deirids situated posterior to pharyngo-intestinal junction in most males. Excretory

* From Latin *silva* = forest (the species was isolated from woodlands).

** Description after Sturhan *et al.* (2005). Revised by Zdeněk Mráček.

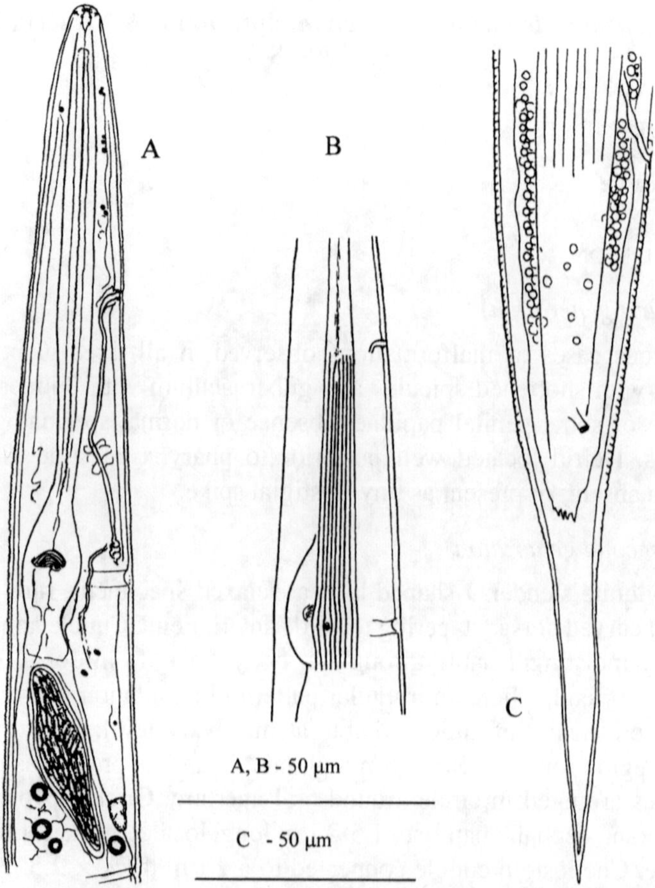

Fig. 180. Steinernema silvaticum. *A-C: Infective juvenile. A: Anterior region; B: Lateral field in posterior pharynx region with position of deirid; C: Tail. After Sturhan* et al. *(2005), courtesy of* Nematology.

pore located on small cuticular elevation, often accompanied by minute cuticular folds. Lumen of terminal part of excretory duct collapsed in most males, deeper part collapsed in all specimens studied. Excretory gland body situated posterior to pharyngo-intestinal junction, with large nucleus in posterior extremity. Anterior part of intestine with broadened lumen and pronounced microvillar layer. Genital system monorchic, reflexed. Testis flexure 250-300 μm posterior to basal bulb. Spermatocytes with round nucleus 3-4 μm diam. and transparent cytoplasm filling testis near flexure in mature males; nuclei disappearing completely

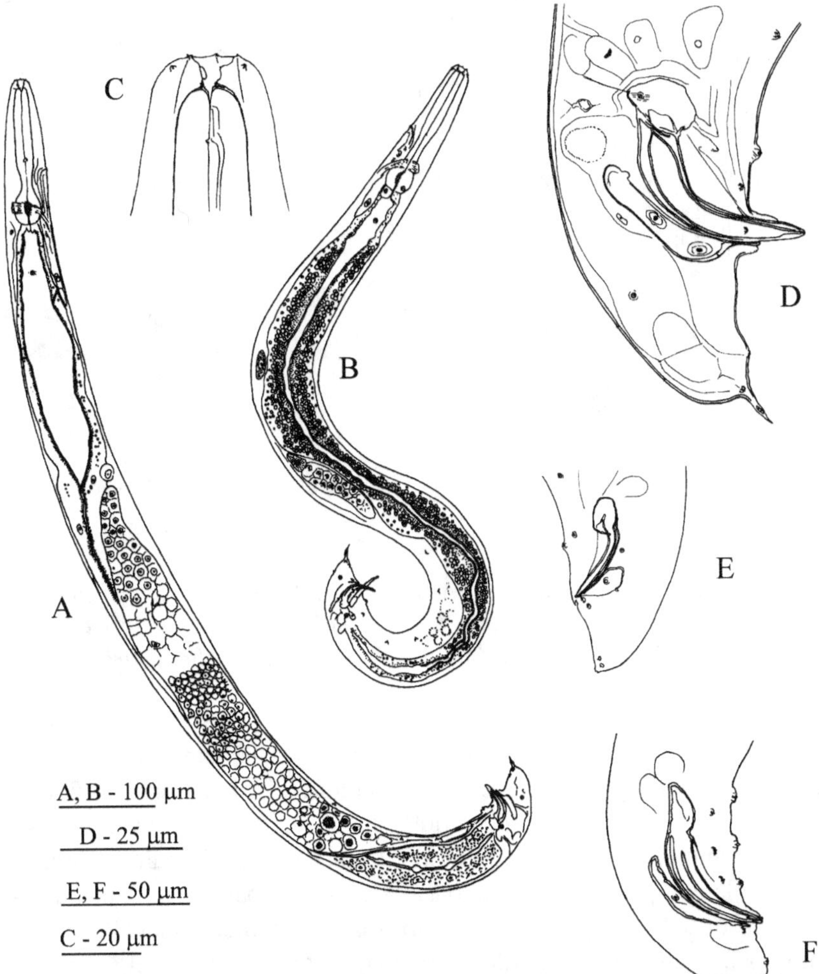

Fig. 181. Steinernema silvaticum. *A-F: Male. A, B: Entire view; C: Cephalic end; D-F: Posterior end showing variation of spicule, gubernaculum and tail terminus. After Sturhan* et al. *(2005), courtesy of* Nematology.

100-150 μm posteriad; spermatocytes 7-9 μm diam. with globular cytoplasm and round or rod-like, 2 μm long nucleus in middle part of testis and inside thin-walled *vas deferens*. Rectum-*vas deferens* junction *ca* 25-30 μm from cloaca, located near spicule manubrium in retracted state. Posterior part of intestinal lumen collapsed. Rectal glandular cells present around spicule muscles. Characteristic steinernematid comple-

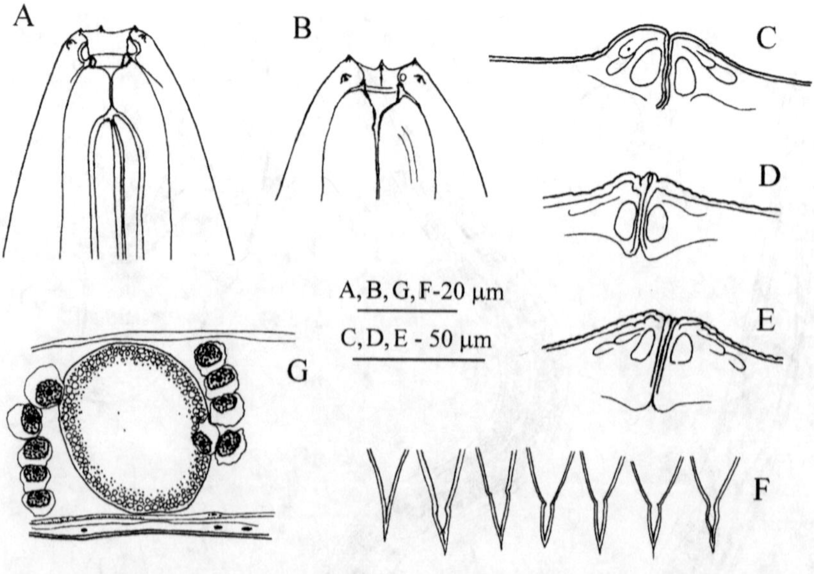

Fig. 182. Steinernema silvaticum. *A-G: Female. A, B: Cephalic end showing variation in stoma wall structure; C-E: Variation in vulva lip structure; F: Variation in tail terminus; G: Chains of spermatozoa. After Sturhan* et al. *(2005), courtesy of* Nematology.

ment of genital papillae with a row of usually nine pairs of subventral papillae extending from tail terminus to mid-body, large, unpaired, precloacal papilla in midventral position, one pair of papillae in lateral position at manubrium level when spicules retracted, and one pair of papillae in dorsolateral position in postcloacal area. Latter pair usually in equatorial part of tail, being near tail terminus in few studied males. Posteriormost pair of subventral papillae close to mucron base, with neighbouring pair situated 7-8 μm anteriorly. Next pair situated precisely at cloacal aperture level. Posteriormost pair of precloacal subventral papillae situated at or posterior to level of unpaired midventral papilla; other precloacal papillae separated by more or less irregular gaps of 30-45 μm. In a few cases ten pairs of subventral papillae were observed with seven pairs in precloacal position. Pair of postdeirids situated dorsolaterally at level of anteriormost pair of precloacal papillae. Spicules slightly yellowish in colour, moderately to strongly curved, tapering towards distal end, with rounded tip and more or less pronounced subterminal duct. Manubrium mostly rounded with average length/width ratio of 1 : 1, but often variable in shape. Velum indistinct, with straight, strongly cuticularised edge

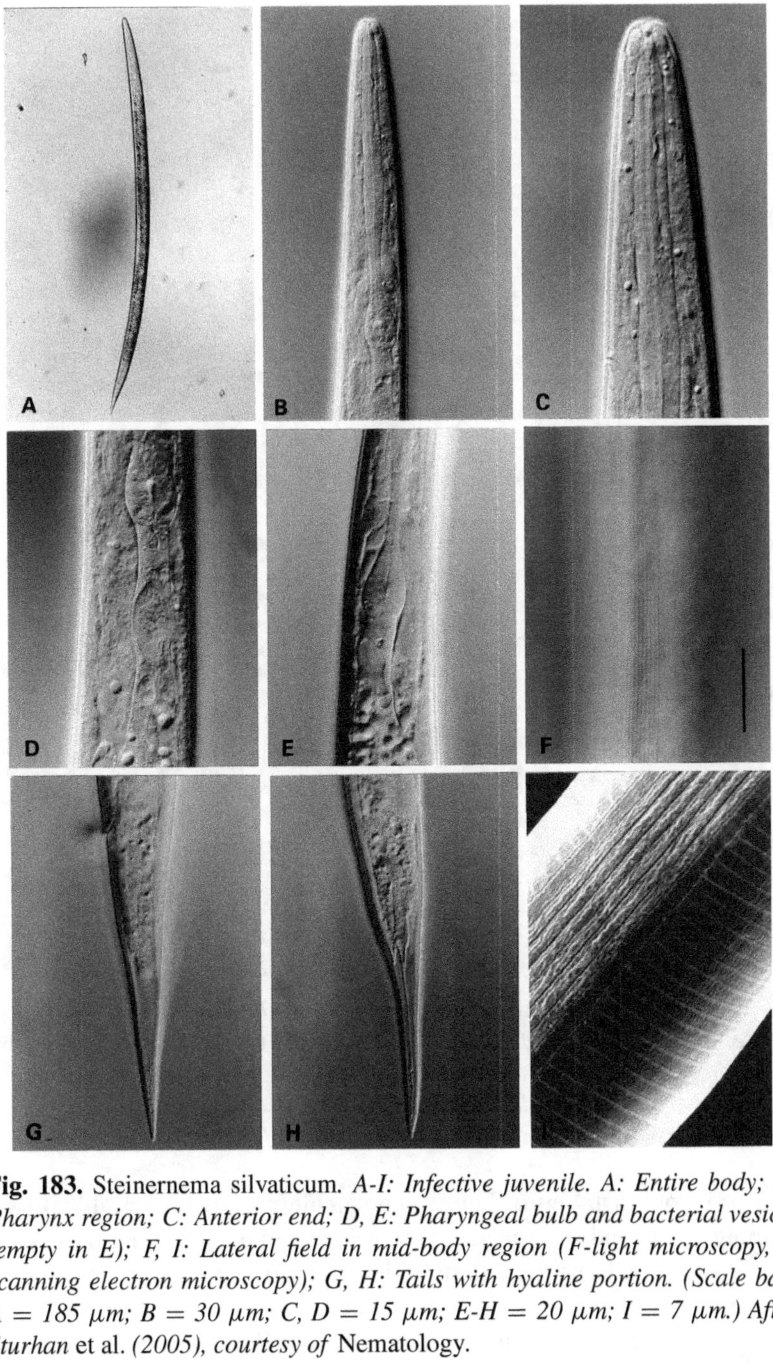

Fig. 183. Steinernema silvaticum. *A-I: Infective juvenile. A: Entire body; B: Pharynx region; C: Anterior end; D, E: Pharyngeal bulb and bacterial vesicle (empty in E); F, I: Lateral field in mid-body region (F-light microscopy, I-scanning electron microscopy); G, H: Tails with hyaline portion. (Scale bar: A = 185 μm; B = 30 μm; C, D = 15 μm; E-H = 20 μm; I = 7 μm.) After Sturhan* et al. *(2005), courtesy of* Nematology.

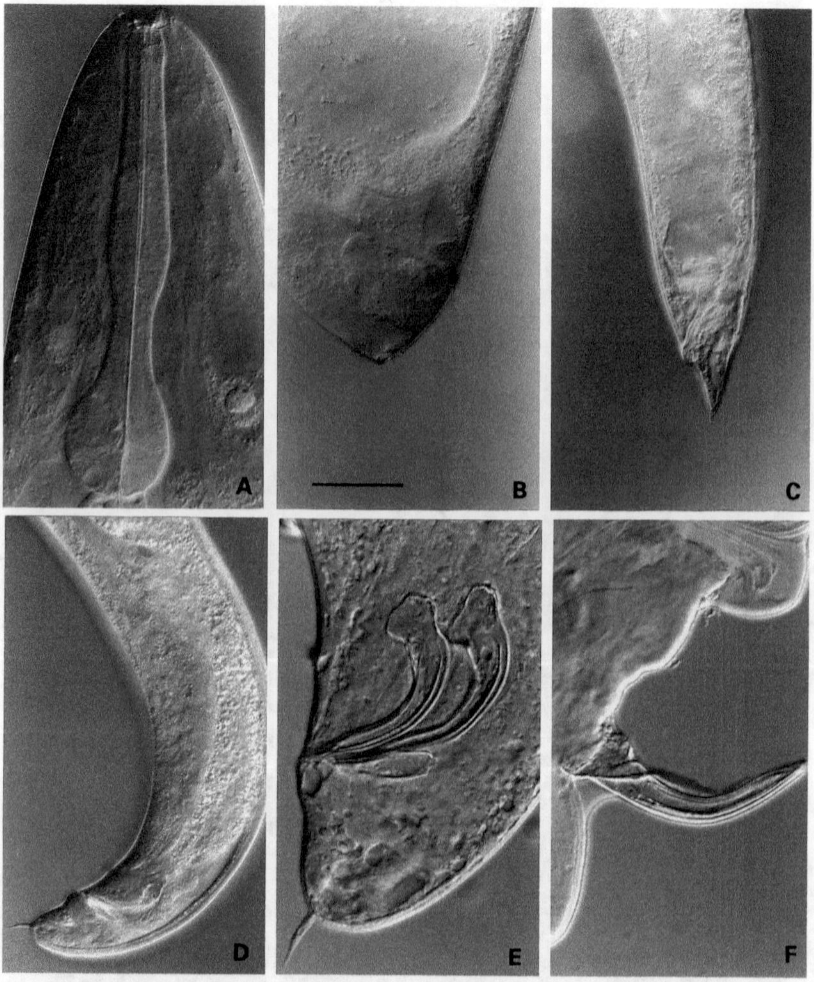

Fig. 184. Steinernema silvaticum. *A-C: Female. A: Pharynx of first generation female; B: Posterior end of first generation female; C: Posterior end of second generation female. D-F: Male. D: Posterior end; E: Tail region with spicules and gubernaculum; F: Spicule. (Scale bar: A-D = 50 μm; B = 30 μm; E, F = 20 μm.) After Sturhan* et al. *(2005), courtesy of* Nematology.

(5-6 μm from distal end). Four main spicule ribs (two central and two forming dorsal and ventral sides) traceable from manubrium up to distal end. Boat-shaped gubernaculum with narrowed and flattened distal end and wide median inner space; usually one or two nuclei discernible in inner space of gubernaculum. Tail with mucron. Hypoderm tissue usually

Table 54. *Morphometrics of* Steinernema silvaticum. *Measurements are in μm and in the form: mean ± SD (range).*

Character	Holotype (IJ)	First generation		Second generation		Infective juvenile
		Male	Female	Male	Female	
n	–	no data	no data	26	24	21
L	940	–	–	1090 ± 65 (975-270)	2150 ± 419 (1520-3290)	860 ± 93 (670-975)
a	30	–	–	17 ± 1.3 (14-20)	19 ± 5.1 (15-41)	29 ± 2.8 (23-33)
b	7.5	–	–	7.7 ± 0.7 (5.9-9.2)	15 ± 1.8 (10-18)	7.1 ± 0.4 (6.3-7.7)
c	12	–	–	34 ± 7.1 (24-55)	49 ± 10 (34-80)	11 ± 0.8 (9.9-13)
c′	4.0	–	–	1.0 ± 0.2 (0.8-1.4)	1.3 ± 0.2 (1.0-1.8)	4.0 ± 0.4 (3.1-4.9)
V	–	–	–	–	52 ± 3.1 (44-57)	–
Body diam.	31	–	–	65 ± 5.9 (52-78)	116 ± 28 (50-175)	30 ± 2.6 (26-35)
EP	65	–	–	79 ± 5.8 (71-92)	69 ± 12 (55-109)	62 ± 6.7 (51-73)
NR	98	–	–	113 ± 8.2 (90-126)	113 ± 14 (96-150)	96 ± 9.3 (75-109)
ES	125	–	–	142 ± 12 (116-168)	146 ± 17 (121-188)	121 ± 12 (100-141)
Tail length (T)	75	–	–	34 ± 6.3 (20-47)	45 ± 4.3 (38-54)	75 ± 7.5 (63-86)
Hyaline (H)	30	–	–	–	–	35 ± 4.3 (26-43)
Anal body diam. (ABD)	19	–	–	33 ± 3.0 (25-38)	35 ± 5.9 (26-53)	19 ± 1.7 (15-24)
Spicule length (SL)	–	–	–	51 ± 5.3 (42-64)	–	–
Gubernaculum length (GL)	–	–	–	37 ± 2.5 (30-43)	–	–
D% = EP/ES × 100	52	–	–	60 ± 5 (45-63)	47 ± 8 (33-79)	50 ± 3 (46-56)
E% = EP/T* × 100	87	–	–	232	153	83
SW% = SL/ABD × 100	–	–	–	155*	–	–
GS% = GL/SL × 100	–	–	–	73*	–	–
H% = H/T × 100	40	–	–	–	–	46 ± 4 (37-53)

EP = distance from anterior end to excretory pore; NR = distance from anterior end to nerve ring; ES = distance from anterior end to end of pharynx.

* Recalculated according to Sturhan *et al.* (2005).

entering mucron and forming an axial thread either of microglobular appearance or, at least in half of all examined male specimens, of rounded, refractile inclusions.

Female (first generation)

A single giant female examined showing obvious signs of failed genital system formation: no eggs and signs of differentiation of uterus walls, which can be expected in females of such size. Vulva slightly protruding. Tail broadly rounded, with only slight to moderate terminal projection. Postanal swelling reduced. Minute terminal protuberances not observed.

Female (second generation)

Body C-shaped to semicircular. Cuticle annulation discernible only near vulval lips. Lateral field 3-4 μm wide, with two prominent marginal and two or three poorly visible central lines. Deirids located at pharyngeal bulb level in young females. Cephalic capsule demarcated by slight swelling in young females. Labial papillae 1.5 μm high. Stoma as in males, but with comparatively less cuticularised base; thick cuticular ring forming stoma walls more or less pronounced. Excretory duct in vicinity of pore sometimes not collapsed, 2-3 μm diam., but deeper part of excretory duct always collapsed as in males. Vulva equatorial, with or without epiptygmata. Vulval lips more or less protuberant, equal in size. Vagina anteriorly directed. Genital tract amphidelphic with reflexed ovaries (in one female both branches posteriorly directed). Short chains of 3-6 spermatozoa, 6-7 μm diam., in distal part of uterus lumen. Anus mostly with postanal swelling. Tail conical, pointed, with hypodermal projection extending to tail terminus and forming an axial thread-like, or more expanded, terminal inclusion.

Infective juvenile

Body of heat-relaxed juveniles straight or only slightly curved ventrad, tapering towards anterior and posterior ends. Cuticular annulation (*ca* 1.5 μm wide at mid-body) visible throughout body, except labial region and tail terminus. Labial region flatly rounded. Cephalic capsule 7-8 μm long, occasionally discernible; cephalic region continuous with adjoining body or only slightly demarcated. Frontal surface of anterior end sometimes slightly skewed in dorsoventral aspect. Shallow (1 μm deep) axial depression leading to collapsed stoma. Labial papillae indistinct, four cephalic papillae and amphidial apertures occasionally visible. Lateral field with nine equally spaced and developed lines (= eight ridges or bands); all nine lines developed at level of pharyngeal bulb or more anterior; extending beyond level of anus, with generally only marginal lines running deeper into caudal region. Deirids located in middle of ridge three or four of lateral field from dorsal side at basal bulb level; hemizonid a 4 μm cuticular depression usually just anterior to deirids. Postdeirids located close to dorsal edge of lateral fields at *ca* 30% of total body length anterior to tail terminus. Pharynx corpus slender, of almost uniform diam. throughout its length. Nerve ring always around isthmus. Excretory pore located at mid-pharynx level; excretory duct traceable from pore up to strongly cuticularised, 3-4 μm long, tube lying inside excretory gland, mostly located just posterior to pharyngo-intestine junction; large, 6-7 μm diam., nucleus of excretory cell located posterior to this structure. Basal bulb with valves of thin folded platelets. Cardia *ca* 10 μm long, with rounded posterior end. Bacterial vesicle mostly with well developed 1-1.5 μm thick walls. In 1- to 3-month-old juveniles, bacteria completely filling lumen of vesicle. Genital primordium 100-160 μm long. Intestine without visible lumen. Rectum *ca* 1.5 anal body diam. long. Anus 7-8 μm wide, sickle-shaped. Phasmids located at 35-40% of tail length posterior to anus, with aperture near posterior end of ventral marginal lateral line; phasmidial channel more or less pronounced. Tail conical, mostly straight, but often bent to dorsal side, generally with irregular outline; terminus pointed to finely rounded. Hyaline region mostly occupying slightly less than half of tail length. Tail hypodermis ending with several spike-like projections, occasionally one or two hypodermal threads extending up to tail terminus.

TYPE HOST AND LOCALITY

The natural host is unknown. Infective juveniles were isolated from a soil sample collected 11 November 1998 and were inoculated to *Galleria mellonella* larvae and subsequently reared on this host. The sample came from Kleinmachnow near Berlin, south of Teltow canal, UTM grid UU 70, Germany, at low elevation in an *Acer-Quercus-Robinia* forest with *Hedera helix* as subvegetation, loamy-sand soil covered by fallen leaves.

Cultivation of this steinernematid in *Galleria* indicates that Lepidoptera are probably unsuitable hosts for this nematode (Sturhan *et al.*, 2005).

TYPE MATERIAL

Holotype infective juvenile and paratypes (12 infective juveniles, 17 males and 27 females) deposited in the German Nematode Collection, Biologische Bundesanstalt, Muenster, Germany (slides DNST 131/18/1-11). Additional paratypes in the nematode collections of the Laboratory of Insect Pathology, Institute of Entomology, Czech Academy of Sciences, České Budějovice (ten infective juveniles, ten males, ten females); Institute of Parasitology, Russian Academy of Sciences, Moscow (eight infective juveniles, 21 males, nine females), and CABI Europe-UK Nematode Collection, Egham, Surrey, UK (one infective juvenile, three males, three females).

RELATIONSHIPS

Steinernema silvaticum is characterised by a combination of the following characters of the infective-stage juveniles: body mostly straight and of medium length (mean = 700-900 μm, range = 580-975 μm), lateral field with nine equally spaced and developed lines (= eight ridges), cephalic region rather broad, flatly rounded and continuous with adjoining body, excretory pore located at level of middle of pharynx (mean D% = 45-50, range = 40-56) and hyaline region comprising *ca* half of total tail length (mean H% = 46-54, range = 37-60). Males have a mucronate tail, yellowish spicules of 48-51 μm mean length (range = 38-64 μm), with mostly broad manubria, strongly curved lamina and a boat-shaped gubernaculum of 35-37 μm mean length (range = 28-43 μm). Females have a short conoid tail with a pointed, non-mucronate tip.

Molecular data place *S. silvaticum* in the *feltiae*-group. *Steinernema kraussei* and *S. feltiae* differ in the appearance of the lateral field in the infective juveniles with the submarginal lines being only poorly developed and the total number of lines being only eight (seven ridges) in *S. kraussei.* In both species infective juveniles are in general more slender, often curved ventrad, their cephalic region is narrower and the hyaline region mostly comprises distinctly less than half the tail length. In *S. oregonense* the submarginal lines in the lateral field of the infective juveniles are also indistinct or often even absent; the gubernaculum is longer (52-59 μm in first generation males and 42-52 μm in second generation males *vs* 28-43 μm in *S. silvaticum* males). In *S. jollieti* the lateral field of the infective juveniles has only seven lines (six ridges).

Steinernema weiseri has a similar appearance of the lateral field in the infective juveniles, but the labial region is angular and flattened, the hyaline region much shorter (mostly *ca* 33 *vs ca* 50% of tail length in *S. silvaticum*) and the body shorter (average length = 640-740 *vs* 700-900 μm). The shape of the spicules of *S. silvaticum* is different from those of *S. feltiae*, *S. oregonense*, *S. jollieti* and *S. weiseri*, all of which possess a noticeably oblong manubria with a length to width ratio of 1.5-2, whilst *S. silvaticum* usually has a finely rounded manubrium with length/width ratio of 1 : 1. A similar spicule shape is characteristic for *S. kraussei* but the lamina is much more curved in *S. silvaticum*.

Molecular characterisation

The ITS1 + 5.SS + ITS2 sequences of German and British isolates of *S. silvaticum* are nearly identical, both consisting of 727 bp with difference in only two nucleotides in the ITS1 region. The phylogenetic maximum parsimony analysis of alignments of *S. silvaticum* and other steinernematids with known rDNA sequences consistently placed this species together with *S. feltiae*, *S. jollieti*, *S. kraussei*, *S. oregonense* and *S. weiseri*. In all obtained trees, *S. kraussei* and *S. silvaticum* were sister taxa (with strong bootstrap support), both species forming a single clade with *S. oregonense* (moderate bootstrap support). No significant change in bootstrap support was observed when gaps were treated as 'fifth base'. Maximum likelihood analysis demonstrated similar levels of bootstrap support for the same nodes. Despite stable clustering of *S. silvaticum* with *S. kraussei*, significant nucleotide differences were revealed between these two, rather close, species. Pairwise nucleotide difference between these two species was 27 bp or 3.7% of total ITS-region nucleotide count (gaps treated as missing data).

Bionomics

Steinernema silvaticum shows a remarkable habitat specificity, it being restricted almost exclusively to woodland. Of the 153 records in Germany 143 (= 93%) were from woodland (42 coniferous, 48 deciduous, 53 mixed), two from hedgerows, four from grassland (most of them close to forests), one from an abandoned field (close to a pine forest), two from coastal dunes and one from heath land (Sturhan, 1999 and data unpubl.). All nine prior records of this species from the UK and

The Netherlands were from woodland (Hominick *et al.*, 1995). A recent record from The Netherlands is from an *Alnus-Betula* forest.

The records for the Czech Republic are from a deciduous forest and from around an apple tree (Mráček & Bečvář, 2000; Sturhan & Mráček, 2000), and the record for Sweden is from a *Pinus-Picea* forest. Two each of the Belgian records are from road side verges and from 'cultivated land' (Miduturi *et al.*, 1997).

The occurrence of *S. silvaticum* is clearly restricted to light sandy soils. The soil type at the sampling sites in Germany ranged from almost pure sand and loamy sand to sandy loam, in part with high humus content. *Steinernema silvaticum* is thus more common in northern Germany, where light sandy soils are prevailing, than in the central and southern regions of the country (see also distribution map in Sturhan, 1999). The records from Belgium (Miduturi *et al.*, 1997) and records from The Netherlands and Sweden are also from sandy soils.

DISTRIBUTION

Steinernema silvaticum is known to occur in six European countries (Belgium, Czech Republic, Germany, Sweden, The Netherlands, UK). A survey on occurrence and distribution of entomopathogenic nematodes employing direct extraction of nematodes from soil samples revealed that *S. silvaticum* is one of the most common *Steinernema* species in Germany (18.3% of all *Steinernema* records). So far, the species is known from a total of 153 sites and is distributed all over the country (Sturhan, 1997, 1999, data unpubl.). Based on *Galleria* baiting, a total of nine records have been reported in the UK and four in The Netherlands (Hominick *et al.*, 1995). In The Netherlands, *S. silvaticum* was found at Renesse, Zeeland. In Belgium, several isolates, classified by biochemical methods as *Steinernema* spec. B3, were recovered by the *Galleria* baiting technique from three woodland sites in East Flanders, two roadside verge sites and two sites in cultivated land in Antwerpen and Luxembourg provinces (Miduturi *et al.*, 1996a, b, 1997). In the Czech Republic, two records are known from the Southern Bohemian region (Mráček & Bečvář, 2000; Sturhan & Mráček, 2000). By direct extraction from soil, *S. silvaticum* has been isolated from a soil sample collected in southern Sweden near Växjö.

At 25% of the sites in Germany, *S. silvaticum* co-occurred with other *Steinernema* species, most often with *S. intermedium* and *S. kraussei*

(Sturhan & Mráček, 2000; Sturhan, data unpubl.). From soil samples at four of these locations two *Steinernema* species besides *S. silvaticum* were isolated by direct extraction and, at one site, even three species were found. At the sampling site in Sweden, *S. silvaticum* co-existed with *S. kraussei*.

References

HOMINICK, W.M., REID, A.P. & BRISCOE, B.R. (1995). Prevalence and habitat specificity of steinernematid and heterorhabditid nematodes isolated during soil surveys of the UK and the Netherlands. *Journal of Helminthology* 69, 72-32.

MIDUTURI, J.S., MATATA, G.J.M., WAEYENBERGE, L. & MOENS, M. (1996a). Naturally occurring entomopathogenic nematodes in the province of East-Flanders, Belgium. *Nematologia Mediterranea* 24, 287-293.

MIDUTURI, J.S., MOENS, M., HOMINICK, W.M., BRISCOE, B.P. & REID, A.P. (1996b). Naturally occurring entomopathogenic nematodes in the province of West-Flanders, Belgium. *Journal of Helminthology* 70, 319-327.

MIDUTURI, J.S., WAEYENBERGE, L. & MOENS, M. (1997). Natural distribution of entomopathogenic nematodes (Heterorhabditidae and Steinernematidae) in Belgian soils. *Russian Journal of Nematology* 5, 55-65.

MRÁČEK, Z. & BEČVÁŘ, S. (2000). Insect aggregations and entomopathogenic nematode occurrence. *Nematology* 2, 297-301.

STURHAN, D. (1997). Untersuchungen zum Artenspektrum entomopathogener Nematoden in verschiedenen Biotopen. Schriften des Bundesministeriums für Ernährung. *Landwirtschaft und Forsten, Reihe A: Angewandte Wissenschaft* 465, 372. [Abstr.]

STURHAN, D. (1999). Prevalence and habitat specificity of entomopathogenic nematodes in Germany. In: Gwynn, R.L., Smith, P.H., Griffin, C., Ehlers, R.-U., Boemare, N. & Masson, J.P. (Eds). *COST 819, Entomopathogenic nematodes: Application and persistence of entomopathogenic nematodes.* Luxembourg, European Commission, pp. 123-132.

STURHAN, D. & MRÁČEK, Z. (2000). Comparison of the *Galleria* baiting technique and a direct extraction method for recovering *Steinernema* (Nematoda: Rhabditida) infective-stage juveniles from soil. *Folia Parasitologica* 47, 315-318.

STURHAN, D., SPIRIDONOV, S.E. & MRÁČEK, Z. (2005). *Steinernema silvaticum* sp. n. (Rhabditida: Steinernematidae), a new entomopathogenic nematode from Europe. *Nematology* 7, 227-241.

Steinernema tami[*,**] Luc, Nguyen, Reid & Spiridonov, 2000
(Figs 185-187)

MEASUREMENTS

See Table 55.

DESCRIPTION

Male (first generation)

Body of heat-relaxed specimens C- or J-shaped. Cuticle with faint transverse annulation poorly visible under light microscope. Cephalic extremity rounded with six low lips, six labial papillae and four prominent cephalic papillae. Oral aperture more or less circular. Stoma *ca* 5-6 μm deep with two sclerotised rings. Anterior ring lightly sclerotised, almost rounded, posterior ring more sclerotised, smaller, elliptical in shape and located on frontal border of pharyngeal tissue. Pharynx with procorpus, slightly swollen metacorpus, indistinct isthmus and pyriform basal bulb. Nerve ring surrounding isthmus. Cardia prominent. Excretory pore located slightly anterior to mid-pharynx, excretory duct *ca* 20 μm long, cuticularised. Excretory glands swollen, 12-18 μm wide sometimes displacing basal bulb to dorsal side. Pseudocoelom in pharyngeal region filled with granular material. Genital system monorchic, reflexed. Distance from anterior end to testis flexure ranging from 280-350 μm. Different spermatogenesis stages visible in testis. In mid-testis region primarily small spermatocytes disappearing and replaced by *ca* 30 large (40-50 μm) cells, variable in appearance, and often with wart-like structure on cell membrane. Closer to *vas deferens* these structures appearing nearly separated from membrane, producing groups of refractive granules around large cells. *Vas deferens* tightly packed with both large cells and granules. Spicules curved, yellow-brownish in colour. Oblique muscles well developed in precloacal zone. Spicule head (manubrium) length *ca* 1-1.5 times as long as wide, rarely twice. Spicule shaft (calomus) prominent. Blade (lamina) moderately curved bearing ribs reaching spicule tip. Velum thin. Spicule distal end bluntly pointed. Gubernaculum boat-shaped, proximal end slightly curved, in ventral view, corpus

[*] Patronym in honour of Truong Quang *Tam*.

[**] Description after Luc *et al.* (2000). Revised by Khuong B. Nguyen.

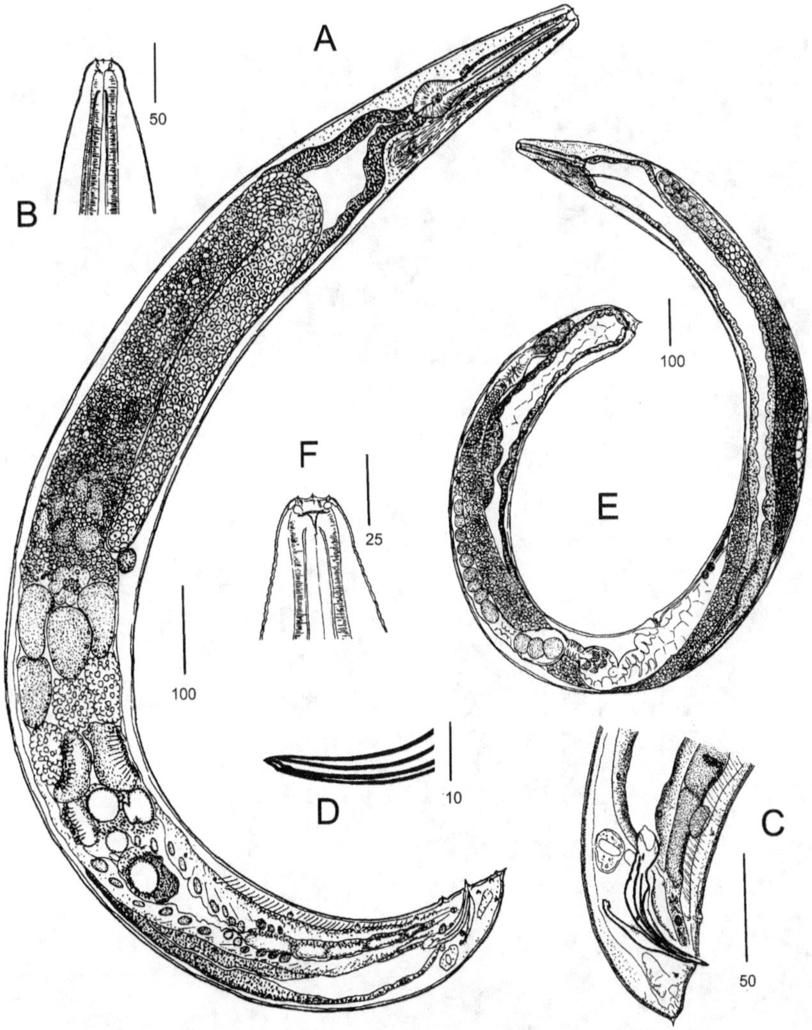

Fig. 185. Steinernema tami *first generation adults. A: Male, entire body; B: Male cephalic end; C: Male tail; D: Spicule, distal end; E: Female, entire body; F: Female, cephalic end. (Scale bars in μm.) After Luc* et al. *(2000), courtesy of* Russian Journal of Nematology.

tapering gradually anteriorly, cuneus Y-shaped. Complement of 23 genital papillae present comprising 11 pairs plus a single precloacal midventral papilla. Papillae arranged as six pairs subventral, precloacal, one pair sublateral precloacal, at level of mid-lamina, one pair subventral adcloa-

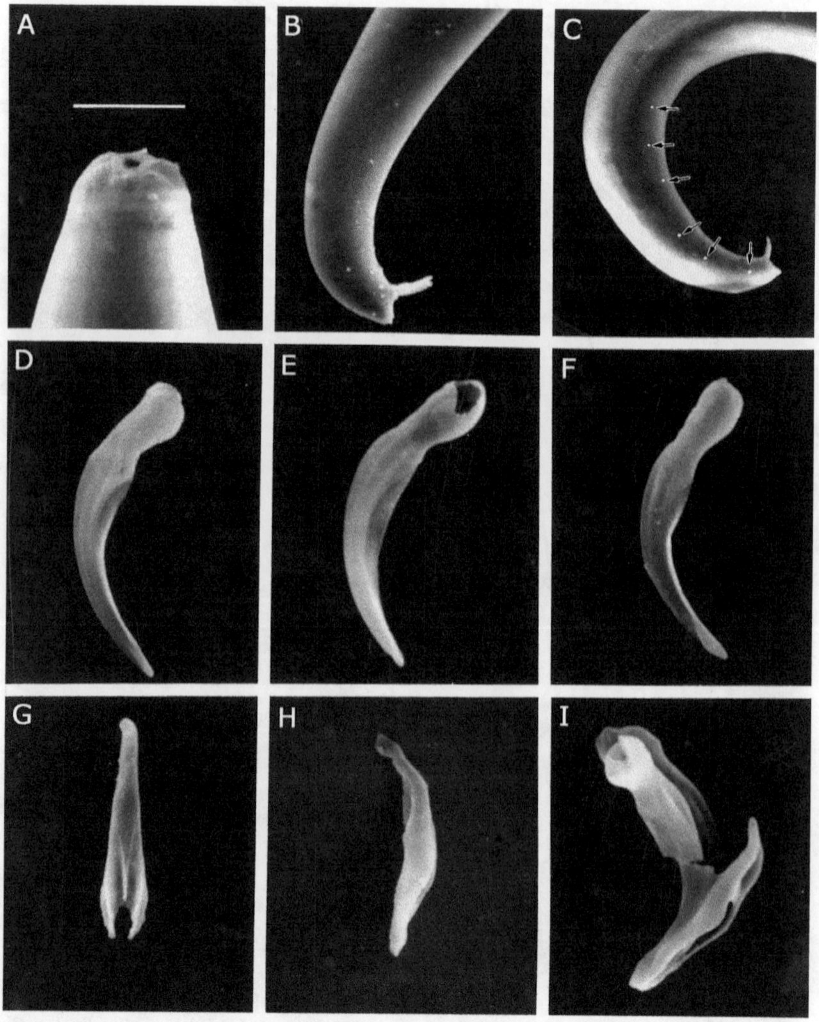

Fig. 186. Steinernema tami. *SEM micrographs of first generation male. A: Anterior region showing labial and cephalic papillae; B, C: Posterior region showing precloacal genital papillae (arrows) and spicules; D-F: Spicules showing head, shaft and blade; G: Gubernaculum, ventral view with Y-shaped cuneus, ventrally curved anterior end and forked posterior end; H: Gubernaculum lateral view; I: Spicules (left one broken) and gubernaculum. (Scale bar: A = 17.6 μm; B = 61 μm; C = 120 μm; D-I = 25 μm.) Modified after Luc* et al. *(2000), courtesy of* Russian Journal of Nematology.

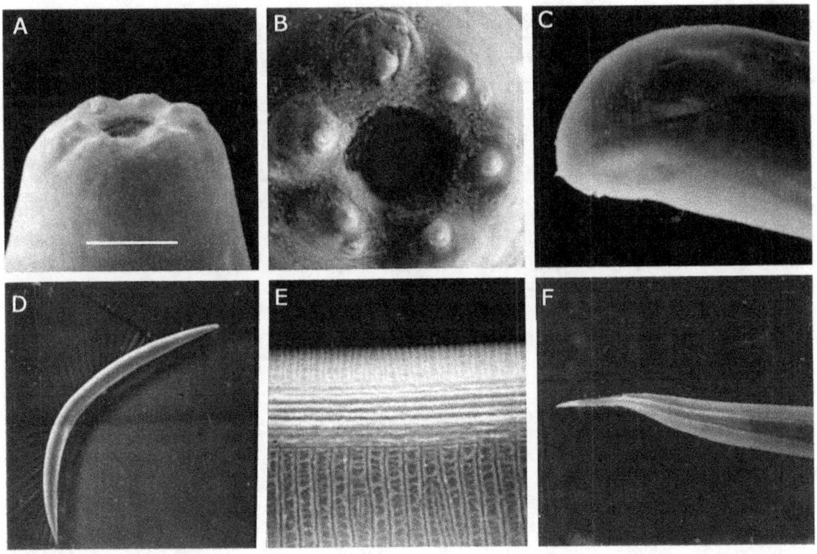

Fig. 187. Steinernema tami. *SEM micrographs of first generation female and infective juveniles. A: Anterior region of female showing mouth, lips, labial and cephalic papillae; B: Face view of female showing six labial papillae; C: Rounded tail of female with mucron; D: Entire infective juvenile; E: Lateral field showing eight ridges with four raised central ridges and four lower marginal ridges; F: Infective juvenile showing two prominent submarginal ridges and marginal ridges gradually disappearing posteriorly. (Scale bar: A = 15 μm; B = 7.5 μm; C = 50 μm; D = 120 μm; E = 4.3 μm; F = 17.6 μm.) Modified after Luc* et al. *(2000), courtesy of* Russian Journal of Nematology.

cal, two pairs subventral postcloacal, and one pair subdorsal, postcloacal. Tail dorsally convex, conoid, terminus bluntly rounded. Mucron 3-4 μm long, sometimes not observed. Phasmids inconspicuous.

Male (second generation)

Heat-relaxed specimens C-shaped. Anterior region similar to that of first generation, but smaller, lips relatively more prominent. Excretory duct cuticularised, discernible only to within 3-4 μm from excretory pore. Testis reflexion usually located posterior to mid-body. Spermatogenesis similar to that of first generation with up to 18 large cells bearing superficial granule-like structures in *vas deferens*. Spicules and gubernaculum of same colour and shape as in first generation. Distribution of genital papillae resembling that in first generation, but very often post-

Table 55. *Morphometrics of* Steinernema tami. *Measurements are in μm and in the form: mean ± SD (range).*

Character	First generation			Second generation		Infective juvenile
	Holotype (male)	Male	Female	Male	Female	
n	–	20	20	20	20	20
L	1337	1600 ± 20 (1200-1900)	5100 ± 1000 (2800-6300)	900 ± 50 (800-1000)	1600 ± 10 (1400-1900)	530 ± 4 (400-600)
a	–	–	–	–	–	23 ± 2 (19-28)
b	–	–	–	–	–	5 ± 0.3 (3.7-5.1)
c	–	–	–	–	–	11 ± 0.5 (9-11)
c′	–	–	–	–	–	4.2
V	–	–	52 ± 2 (47-55)	–	54 ± 3 (48-58)	–
Body diam.	–	129 ± 17 (89-161)	196 ± 23 (145-240)	59 ± 3 (55-65)	93 ± 10 (70-107)	23 ± 2 (19-29)
EP	–	68 ± 12 (43-92)	81 ± 17 (50-115)	64 ± 4 (58-72)	70± (64-77)	36 ± 2 (34-41)
ES	–	153 ± 8 (137-166)	221 ± 16 (187-245)	142 ± 6 (131-155)	166 ± 10 (150-184)	117 ± 4 (110-123)
Tail length (T)	–	23 ± 5 (10-32)	32 ± 10 (19-48)	18 ± 2 (15-22)	28 ± 6 (21-40)	50 ± 4 (42-57)
Anal body diam. (ABD)	–	37 ± 6 (25-40)	62 ± 13 (45-85)	29 ± 2 (25-32)	31 ± 5 (25-42)	12 ± 1 (11-13)
Spicule length (SL)	82	77 ± 4 (71-84)	–	67 ± 4 (62-76)	–	–
Gubernaculum length (GL)	52	48 ± 5 (38-55)	–	37 ± 3 (31-45)	–	–
D% = EP/ES × 100	–	44 ± 10 (30-60)	–	45 ± 0 (4-5)	–	31 ± 0 (28-34)
E% = EP/T × 100	–	–	–	–	–	73 ± 0 (67-86)
SW% = SL/ABD × 100	197	200 ± 40 (140-300)	–	210 ± 20 (170-250)	–	–
GS% = GL/SL × 100	63	62	–	55	–	–
H% = Hyaline/T × 100	–	–	–	–	–	40 ± 0.1 (30-50)

EP = distance from anterior end to excretory pore; NR = distance from anterior end to nerve ring; ES = distance from anterior end to end of pharynx.
After the original authors.

cloacal subventral papillae displaced on tail terminus with minute 2-3 μm long mucron protruding between papillae of posteriormost pair.

Female (first generation)

Body usually C-shaped on heat relaxation. Cephalic extremity truncate to rounded. Face view with rounded oral aperture, six labial papillae and four cephalic papillae immediately posterior. Cervical part of body with coarse annulation. Lateral field inconspicuous. Stoma *ca* 10-15 μm long × 8-10 μm diam., with two cuticular rings in walls: anterior ring thick, posterior one smaller, more refractive, located in pharyngeal tissue. Pharynx more or less cylindrical with isthmus. Cardia prominent, protruding into intestinal lumen. Excretory pore located anterior to mid-pharynx. Excretory duct heavily cuticularised up to 30 μm from pore. Excretory gland large, elongated. Anterior ovary flexure situated *ca* 100-200 μm posterior to basal bulb. Between 10-20 amoeboid spermatozoa, 20-25 × 30-45 μm in size, located in lumen between ovary and uterus. Vagina short, sclerotised. Vulva lips usually not protuberant. Developing oocytes arranged in multiple rows. Tail short, convex-conoid, with peg-like terminal mucron.

Female (second generation)

General morphology of anterior end similar to that of first generation. Excretory duct visibly cuticularised for 5-8 μm posterior to pore. Cardia relatively more developed than in first generation, especially when genital system less developed. Spermatozoa in oviduct having same morphology as in first generation female. Tail relatively longer than that in first generation, conoid, terminus with 2 μm long, spike-like, mucron. Posterior anal swelling prominent.

Infective juvenile

Heat-relaxed specimens assume ventrally arcuate habitus. Body smoothly tapering towards both ends. Cephalic extremity rounded. Cuticle marked with prominent transverse annulation *ca* 1 μm wide at mid-body. Lateral fields at mid-body with eight ridges (= nine incisures). At mid-body, central ridges more elevated than marginal ones. On tail, most ridges gradually disappearing and submarginal ones becoming prominent. Pharynx narrow with a slender isthmus. Basal bulb subpyriform, usually displaced towards dorsal side by development of excretory gland. Hemizonid distinct, located in isthmus level. Excretory pore cuticularised, 2 μm diam. Excretory duct less cuticularised than pore. Bacterial pouch rarely seen. *Xenorhabdus* cells, when found, with indistinct walls, tightly compressed in bacterial chamber. Tail conoid, tapering to a finely pointed terminus. No refractive inclusion in tip. Hyaline region pronounced, comprising *ca* 39% of tail length. Phasmids distinct, located in anterior half of tail.

TYPE HOST AND LOCALITY

Natural hosts unknown as the nematode was recovered by *Galleria* baiting from soil samples collected on 6-12 April 1997 from primary forest in Cat Tien National Park along the Dong Nai river, Vietnam.

TYPE MATERIAL

Holotype male (first generation) and numerous paratypes deposited in the Collection of Invertebrates of the Institute of Parasitology, Russian Academy of Sciences, Moscow, Russia. Slides with one male and one female of first generation each deposited in the Hanoi Institute of Ecology and Biological Resources, Vietnam; German Nematode Collection, Institute of Nematology and Vertebrate Studies, Münster, Germany; CABI Europe-UK Nematode Collection, Egham, Surrey, UK. According to Luc *et al.* (2000) living cultures are maintained in Hanoi and Moscow.

RELATIONSHIPS

Morphological and molecular data indicate that *S. tami* belongs to the group of species related to *S. carpocapsae*, namely the *carpocapsae*-group (body length < 600 μm). *Steinernema tami* can be distinguished from other species of the group by the average body length of infective juveniles: *S. tami* = 530 (400-550) μm; *S. kushidai* (589 μm); *S. scapterisci* (572 μm); *S. carpocapsae* (558 μm); the IJ is longer than that of *S. rarum*, *S. ritteri* and *S. siamkayai* (511, 510, 446 μm, respectively). *Steinernema tami* tail length, EP and E% are very different to *S. siamkayai*: tail length = 50 (42-57) *vs* 35.5 (31-41) μm; EP = 36 (34-41) *vs* 57 (48-67) μm; and E% = 73 (67-86) *vs* 96 (85-112). D% distinguishes *S. tami* D% = 31 (28-34) from *S. ritteri* D% = 46 (44-50). The infective juvenile of *S. tami* is quite different from that of *S. kushidai* but close to *S. carpocapsae* and *S. scapterisci*. *Steinernema tami* can be separated from *S. carpocapsae* and *S. scapterisci* by the shape of the spicule and gubernaculum (see Hominick *et al.*, 1997, Fig. 4).

Steinernema tami can be distinguished from other species in the group by male characteristics. With the exception of *S. scapterisci*, spicule length (SL) = 83 μm, and *S. siamkayai*, SL = 75.5 μm, the spicule length of *S. tami*, at 72 μm, is longer than that of all other species in the *carpocapsae*-group (*S. ritteri* = 69 μm; *S. carpocapsae* = 66 μm; *S. kushidai* = 63 μm; *S. rarum* = 47 μm). The ratio SW% can also be used to distinguish *S. tami* from other species. The SW% of *S. tami*, at 200, is smaller than that of *S. scapterisci* (= 250) but larger than that of all other species of the group (*S. carpocapsae* = 170; *S. siamkayai* = 170; *S. ritteri* = 160; *S. kushidai* = 150; and *S. rarum* = 94) (Nguyen & Smart, 1996).

MOLECULAR CHARACTERISATION

The RFLP patterns of ITS regions yielded from each of three species and three new isolates (*S. tami*, *S. carpocapsae*, *S. siamkayai*, and isolates Malaysia, SSL1 and SSL2) with 17 different restriction enzymes showed *S. tami* to be different from other species but molecularly closely related to them. The species is also characterised by the ITS region sequence length 739 bp (ITS1 + 5.8S + ITS2), ITS1 = 269 bp, ITS2 = 304 bp and its composition, A = 0.22598, C = 0.16779, G = 0.22192, T = 0.38431 (accession no. AY171280).

BIONOMICS

Optimal temperatures for the infection of *Galleria mellonella* with infective juveniles of *S. tami* were between 22-27°C. Between 22-24°C the first generation males and females were observed inside *Galleria* cadavers 3-4 days after inoculation. First generation adults disappeared on day 6 or 7, and numerous feeding juvenile stages of the second generation were found in *Galleria* bodies. Early males and females of the second generation were present in cadavers on day 8 or 9. Mass migration of the infective juveniles occurred on days 13-16 depending on humidity. Prolonged storage at 3-5°C increases juvenile mortality drastically.

DISTRIBUTION

Steinernema tami has been found in Cat Tien National Park, Dong Nai, Vietnam (Luc *et al.*, 2000). It may also occur in India (see Vidya & Nagesh, 2006).

References

HOMINICK, W.M., BRISCOE, B.R., DEL PINO, F.G., HENG, J., HUNT, D.J., KOZODOY, E., MRÁČEK, Z., NGUYEN, K.B., REID, A.P., SPIRIDONOV, S., STURHAN, D., WATURU, C. & YOSHIDA, M. (1997). Biosystematics of entomopathogenic nematodes: current status, protocols and definitions. *Journal of Helminthology* 71, 271-298.

LUC, P.V., NGUYEN, K.B., REID, A.P. & SPIRIDONOV, S.E. (2000). *Steinernema tami* sp. n. (Rhabditida: Steinernematidae) from Cat Tien Forest, Vietnam. *Russian Journal of Nematology* 8, 33-43.

NGUYEN, K.B. & SMART JR, G.C. (1996). Identification of entomopathogenic nematodes in the Steinernematidae and Heterorhabditidae (Nemata: Rhabditida). *Journal of Nematology* 28, 286-300.

VIDYA, H.S. & NAGESH, M. (2006). RAPD analysis of Indian isolates of *Xenorhabdus* spp. associated with entomopathogenic nematodes, *Steinernema* spp. *International Journal of Nematology* 16, 178-185.

Steinernema thanhi[*,**] Phan, Nguyen & Moens, 2001
(Figs 188, 189)

MEASUREMENTS

See Table 56.

DESCRIPTION

Male (first generation)

Body curved ventrally or J-shaped when heat-relaxed. Cuticle smooth. Cephalic extremity rounded and offset from body. SEM face view with six prominent labial papillae and four cephalic papillae. Amphidial apertures pore-like, situated posterior to labial papillae. Stoma funnel- or cup-shaped. Pharynx with cylindrical procorpus; metacorpus slightly swollen, non-valvate. Isthmus indistinct; basal bulb pyriform, valve not observed. Nerve ring usually surrounding isthmus or situated anterior to basal bulb. Excretory pore located between nerve ring and middle of pharynx. Excretory duct cuticularised and excretory gland swollen. Genital system monorchic, reflexed. Spicule paired, yellow-brownish in colour and ventrally curved; spicule head (manubrium) longer than wide. Velum thin. Gubernaculum boat-shaped, swollen at middle and proximal end with knob, ventrally curved. Complement of 27 genital papillae comprising a single ventral precloacal papilla and 13 pairs arranged as follows: six pairs subventral, precloacal and two pairs lateral at same level of first and second subventral precloacal papillae; one pair postcloacal subventral close to cloacal aperture and one pair lateral at same level; two pairs caudal subventral close to tail terminus and one pair caudal, subdorsal. Tail shorter than anal body diam., lacking mucron. Phasmids inconspicuous.

Female (first generation)

Body robust, spiral- or C-shaped when heat-relaxed. Cuticle smooth under light microscope. Cephalic extremity rounded, offset from body. SEM face view with six prominent labial and four cephalic papillae. Stoma funnel- or cup-shaped. Stoma shallow. Pharynx with cylindrical procorpus; metacorpus slightly swollen. Isthmus indistinct. Basal bulb

[*] Named after Hai *Thanh*, the type locality.

[**] After Phan *et al.* (2001). Revised by Khuong B. Nguyen.

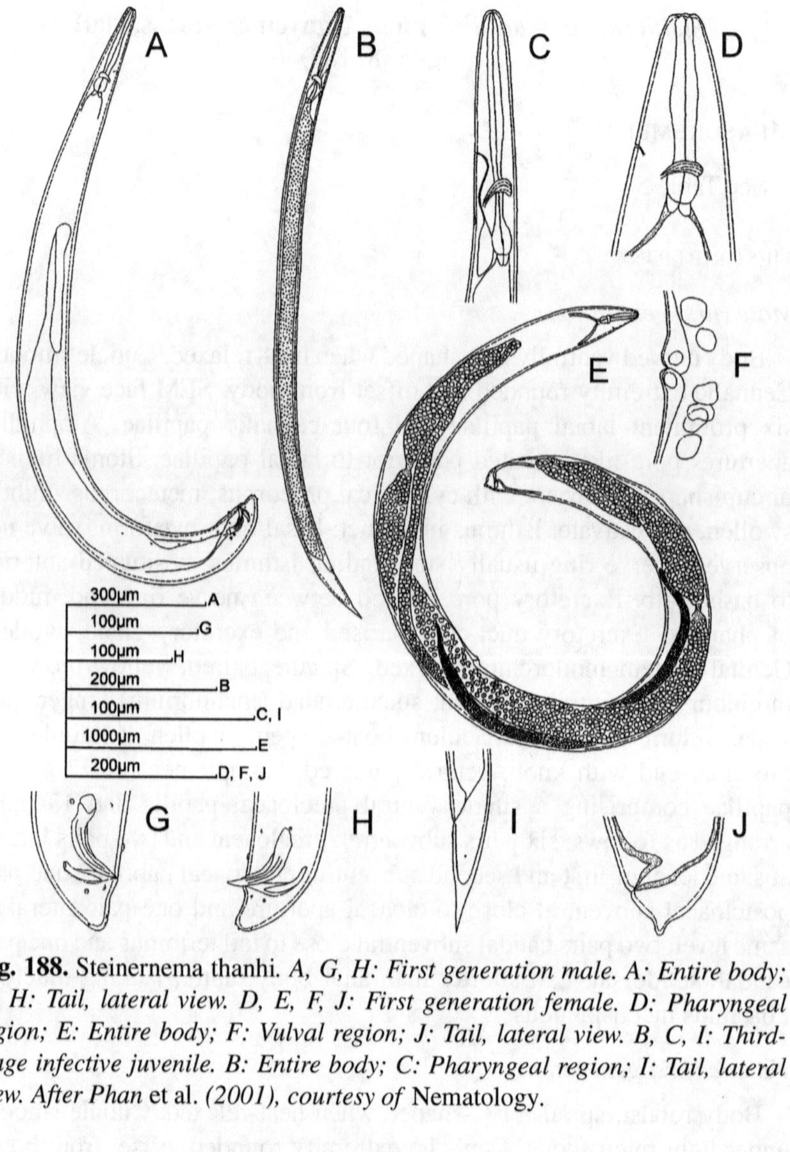

Fig. 188. Steinernema thanhi. *A, G, H: First generation male. A: Entire body; G, H: Tail, lateral view. D, E, F, J: First generation female. D: Pharyngeal region; E: Entire body; F: Vulval region; J: Tail, lateral view. B, C, I: Third-stage infective juvenile. B: Entire body; C: Pharyngeal region; I: Tail, lateral view. After Phan* et al. *(2001), courtesy of* Nematology.

pyriform, valve not observed. Nerve ring surrounding isthmus or situated anterior to basal bulb. Excretory duct cuticularised and excretory gland swollen. Excretory pore located between nerve ring and mid-pharynx. Cardia prominent. Genital system didelphic, amphidelphic, reflexed; tightly packed with eggs. Vulva in form of a transverse slit located on

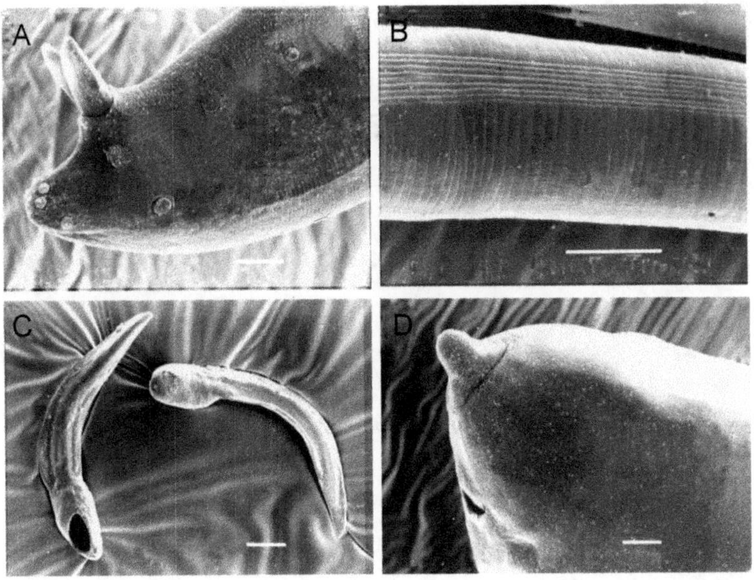

Fig. 189. Steinernema thanhi. *SEM photographs. A: Posterior region of first generation male; B: Lateral field of infective juvenile showing number of ridges changing from seven to eight; C: Spicules of first generation male; D: First generation female tail showing terminal peg. (Scale bars = 10 μm.) Modified after Phan* et al. *(2001).*

a protuberance, epiptygmata absent. Vagina short, oblique. Tail usually shorter than anal body diam, terminal peg present.

Infective juvenile

Often enclosed in cuticle of second-stage. Body usually straight or slightly curved when heat-relaxed. Cephalic extremity smoothly rounded, slightly offset from body. In SEM face view labial papillae not observed but four prominent cephalic papillae and amphidial apertures visible, latter pore-like, located post-labial, but anterior to cephalic papillae. Oral aperture and anus closed. At mid-body, lateral field with eight ridges (= nine incisures). Pharynx long and narrow, isthmus distinct with nerve ring surrounding it; basal bulb elongate. Valve not observed. Cardia prominent. Excretory pore located between nerve ring and mid-pharynx. Tail conoid without mucron. Hyaline region comprising *ca* 40% of tail length. Phasmids distinct, located near mid-tail.

Table 56. *Morphometrics of* Steinernema thanhi. *Measurements are in μm and in the form: mean ± SD (range).*

Character	First generation			Infective juvenile
	Holotype (male)	Male	Female	
n	–	20	20	50
L	1605	1591±93 (1432-1752)	718 ± 1110 (5370-9180)	851 ± 57 (720-960)
a	–	–	–	27 ± 2 (23-33)
b	–	–	–	7 ± 1 (6-8)
c	–	–	–	14 ± 1 (12-16)
c′	–	–	–	3
V	–	–	53 ± 5 (48-73)	–
Body diam.	113	97 ± 15 (76-124)	287 ± 40 (210-360)	31 ± 3 (27-39)
EP	156	144 ± 9 (129-163)	145 ± 14 (134-184)	75 ± 4 (68-84)
NR	–	167 ± 11 (134-182)	174 ± 14 (152-204)	96 ± 6 (83-112)
ES	197	198 ± 10 (182-214)	221 ± 13 (198-248)	129 ± 6 (115-146)
Testis reflexion	–	263 ± 57 (178-396)	–	–
Tail length (T)	36	29 ± 3 (24-35)	48 ± 8 (33-64)	63 ± 5 (52-72)
Anal body diam. (ABD)	47	41 ± 5 (34-48)	85 ± 10 (60-104)	21 ± 2 (18-24)
Spicule length (SL)	77	72 ± 3 (67-78)	–	–
Spicule width	–	13 ± 2 (10-18)	–	–
Gubernaculum length (GL)	57	49 ± 4 (40-56)	–	–
Gubernaculum width	–	10 ± 2 (8-13)	–	–

TYPE HOST AND LOCALITY

Type host unknown as the species was obtained by baiting sandy soil of a garden near the beach in Haithanh, Tinhgia, Thanhhoa, Vietnam, with *Galleria mellonella* larvae.

Table 56. *(Continued).*

Character	First generation			Infective juvenile
	Holotype (male)	Male	Female	
D% = EP/ES × 100	80	73 ± 4 (64-82)	66 ± 6 (58-85)	58 ± 4 (52-67)
E% = EP/T × 100	–	–	–	119 ± 8 (101-138)
SW% = SL/ABD × 100	160	180 ± 20 (150-210)	–	–
GS% = GL/SL × 100	70	70 ± 10 (50-80)	–	–
H% = Hyaline/T × 100	–	–	–	40 ± 4 (33-46)

EP = distance from anterior end to excretory pore; NR = distance from anterior end to nerve ring; ES = distance from anterior end to end of pharynx.
After the original authors.

TYPE MATERIAL

Holotype deposited at Ghent University, Institute of Zoology, K.L. Ledeganckstraat 35, 9000, Belgium.

RELATIONSHIPS

Steinernema thanhi is characterised by the body length of the infective juvenile (851 μm), the slightly offset labial region and lateral fields with eight ridges. In the male, the spicule head is longer than wide.

This species is placed in the *feltiae*-group and is closely related to *S. feltiae*, *S. loci* and *S. neocurtillae*. It can be distinguished from *S. feltiae* by its more posterior excretory pore (75 *vs* 63 μm), shorter tail length (63 *vs* 81 μm) and higher E% (119 *vs* 78); from *S. loci* by shorter body length of the infective juvenile (851 *vs* 985 μm), shorter tail (63 *vs* 75 μm) and the higher number of genital papillae (13 *vs* 11 pairs); and from *S. neocurtillae* by a more posterior excretory pore (80 *vs* 18 μm), shorter tail (63 *vs* 80 μm) and a higher E% (119 *vs* 23).

MOLECULAR CHARACTERISATION

Steinernema thanhi is molecularly characterised by the sequence length of ITS regions, 765 bp (ITS1+5.8S+ITS2), and its composition, A = 0.25098, C = 0.17778, G = 0.23922, T = 0.33203 (accession no. AY355444).

DISTRIBUTION

Steinernema thanhi is only known from the type locality in Ha Tinh, Vietnam (Phan *et al.*, 2001).

Reference

PHAN, K.L., NGUYEN, N.C. & MOENS, M. (2001). *Steinernema loci* sp. n. and *Steinernema thanhi* sp. n. (Rhabditida: Steinernematidae) from Vietnam. *Nematology* 3, 503-514.

Steinernema websteri[*,**] Cutler & Stock, 2003
(Figs 190, 191)

MEASUREMENTS

See Table 57.

DESCRIPTION

Male (first generation)

Body slender, ventrally curved posteriorly, J-shaped when heat-relaxed. Cuticle smooth under light microscopy, but fine transverse striae visible with SEM. Lateral field and phasmids inconspicuous. Cephalic extremity truncate to slightly rounded, continuous with body. Six lips amalgamated but tips distinct, each bearing single labial papilla. Four cephalic papillae present. Amphidial apertures small, located posterior to lateral labial papillae. Stoma reduced (cheilo-, gymno- and stegostom vestigial), short and broad, with inconspicuous sclerotised walls. Pharynx muscular; procorpus cylindrical; metacorpus slightly swollen, non-valvate; indistinct isthmus followed by pyriform basal bulb containing reduced valve. Pharynx set off from intestine. Nerve ring usually surrounding isthmus or anterior part of basal bulb. Excretory pore opening circular, located anterior to nerve ring at anterior third of metacorpus. Genital system monorchic, reflexed, consisting of germinal growth zone leading to seminal vesicle. *Vas deferens* with inconspicuous walls. Spicules paired, symmetrical, curved, of ochre-brown colour. Manubrium continuous dorsally with lamina and rounded. Calomus (shaft) inconspicuous. Lamina with rostrum or retinaculum and two internal ribs. Velum present, extending to terminus. Gubernaculum arcuate, *ca* two-thirds length of spicules. Complement of 23 genital papillae (11 pairs and one single) arranged as follows: six precloacal subventral pairs, single midventral precloacal papilla (located between precloacal pairs 5 and 6); one pair subventral adcloacal (or precloacal in some specimens); one pair subventral postcloacal, one pair subdorsal postcloacal, two pairs terminal postcloacal. Tail conoid, mucronate.

---

[*] Patronym in honour of John M. *Webster* (Simon Fraser University, Burnaby, BC, Canada), a leading scientist in nematology.

[**] Description after Stock and Cutler (2003). Revised by David J. Hunt.

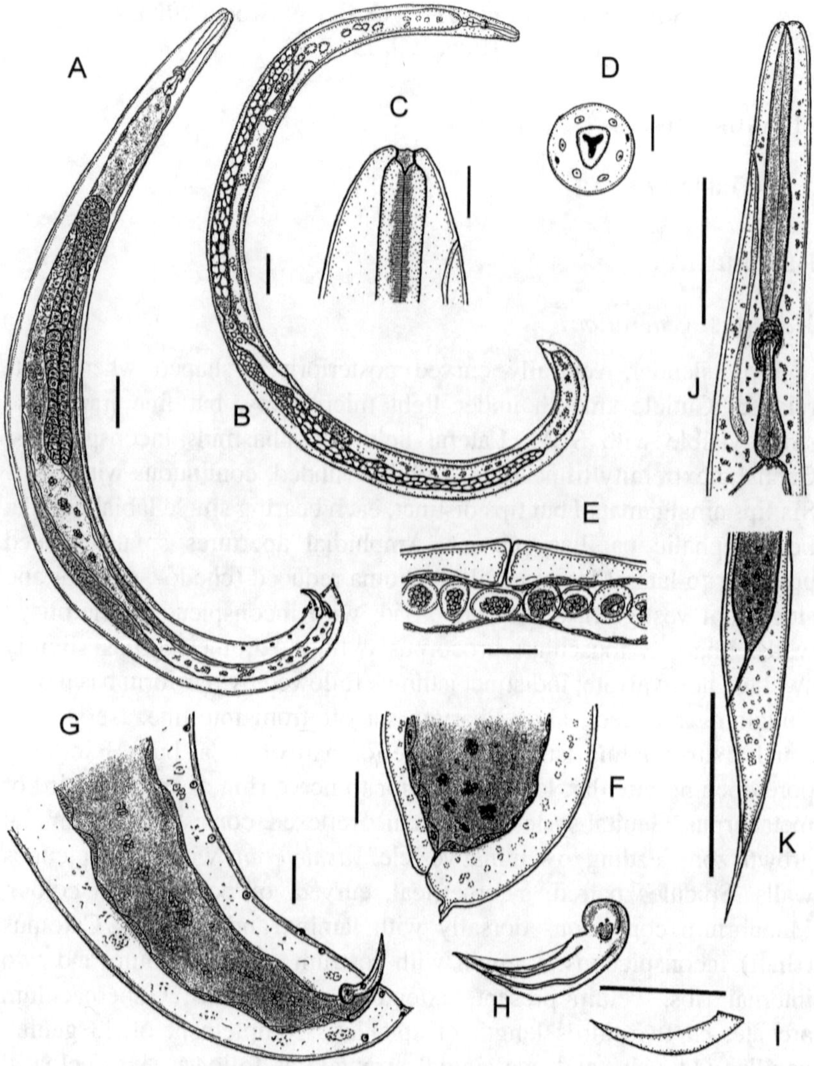

Fig. 190. Steinernema websteri. *A: First generation male, entire body; B: First generation female, entire body; C: Female anterior end, lateral view; D: Female* en face*; E: Vulva, first generation female; F: Tail, first generation female; G: Tail, first generation male, lateral view; H: Spicule; I: Gubernaculum, lateral view; J: Infective juvenile, anterior region; K: Infective juvenile, tail region. (Scale bars: A, B, J = 100 μm; C, E = 50 μm; D = 25 μm; F = 35 μm; G = 30 μm; H, I, K = 50 μm.) After Cutler and Stock (2003), courtesy of* Nematologia Mediterranea.

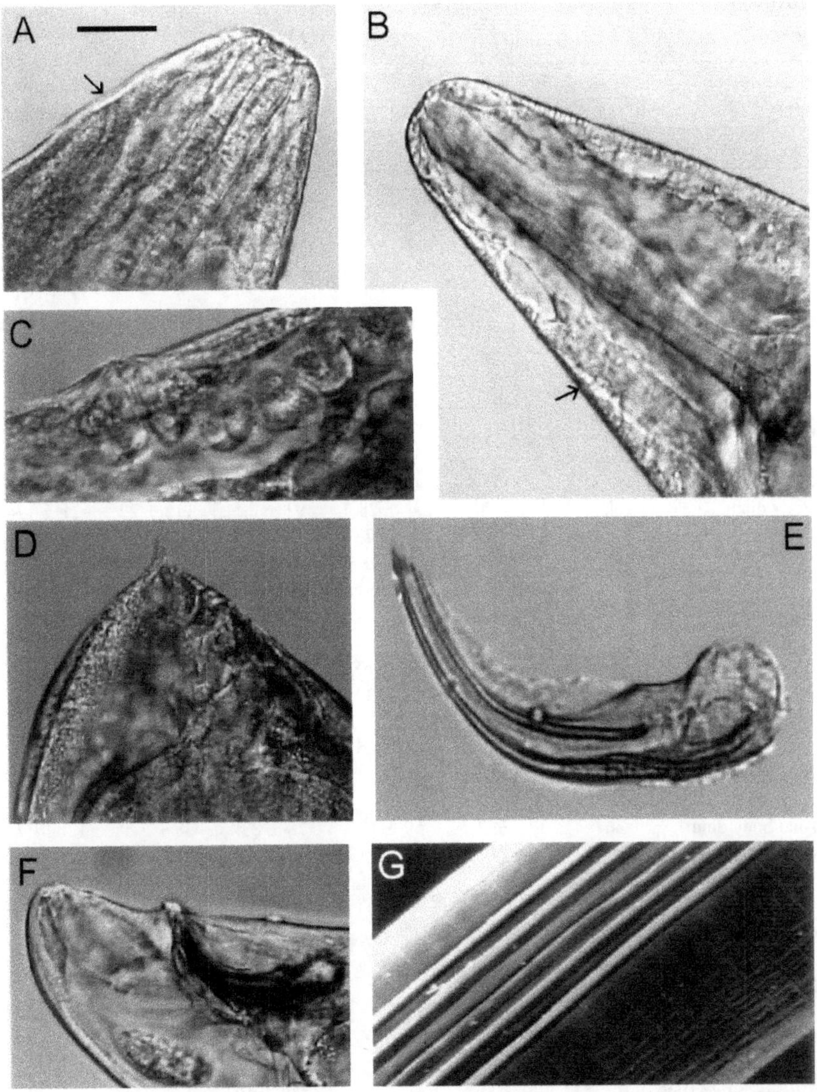

Fig. 191. Steinernema websteri. *Light micrographs. A: First generation female anterior end showing excretory pore (arrow); B: First generation male showing excretory pore (arrow); C: First generation female, vulva; D: First generation female tail; E: Spicule; F: First generation male tail; G: Infective juvenile lateral field. (Scale bar: A-C, E = 45 μm; D = 55 μm; F = 30 μm; G = 10 μm.) After Cutler and Stock (2003), courtesy of* Nematologia Mediterranea.

Table 57. *Morphometrics of type population of* Steinernema websteri. *Measurements are in μm and in the form: mean ± SD (range).*

Character	First generation			Second generation		Infective juvenile
	Holotype (male)	Male	Female	Male	Female	
n	–	19	19	20	20	20
L	1542	1712 ± 92 (1523-1865)	5542 ± 1559 (3497-10 450)	1122 ± 109 (900-1220)	2523 ± 116 (1895-3119)	584 ± 13 (553-631)
a	10.6	–	–	–	–	28 ± 2.7 (24-35)
b	9.3	–	–	–	–	5.1 ± 0.2 (4.8-5.6)
c	61.7	–	–	–	–	12.6 ± 1.3 (11-15.5)
c′	0.7	–	–	–	–	–
V	–	–	53 ± 2.8 (49-59)	–	52 ± 1.4 (51-54)	–
Body diam.	145	147 ± 17 (119-175)	177 ± 28.4 (116-223)	91 ± 8.5 (53-75)	115 ± 22 (99-132)	21 ± 17 (17-25)
EP	55	62 ± 6 (54-73)	64 ± 20.8 (57-68)	49 ± 1.5 (47-53)	60 ± 7 (55-68)	36 ± 3.1 (29-40)
NR	125	119 ± 10 (100-139)	144 ± 14 (119-174)	118 ± 2.3 (86-124)	118 ± 11 (91-133)	88 ± 3.6 (83-95)
ES	166	163 ± 12 (135-180)	181 ± 58 (169-237)	131 ± 9.5 (100-165)	178 ± 8 (154-200)	115 ± 4.4 (107-122)
Tail length (T)	25	29 ± 2 (25-33)	32 ± 11.7 (3.6-45.5)	22 ± 1.5 (18-25)	33 ± 4 (25-39)	47 ± 4.5 (37-56)
Hyaline (H)	–	–	–	–	–	11 ± 0.5 (10-14)
Anal body diam. (ABD)	36	37.5 ± 2 (34-41)	69 ± 13 (51-87)	28 ± 2.4 (25-35)	38 ± 3.5 (31-42)	12 ± 1.2 (10-14)
Spicule length (SL)	67	68 ± 2 (64-72)	–	58 ± 4.2 (54-64)	–	–
Gubernaculum length (GL)	56	49 ± 3.2 (42-56)	–	40 ± 3.1 (37-42)	–	–
D% = EP/ES × 100	33	40 ± 10 (30-50)	–	49 ± 3.5 (47-53)	–	31 ± 2.7 (24-34)
E% = EP/T × 100	217	210 ± 20 (180-250)				77 ± 11 (62-102)
SW% = SL/ABD × 100	190	180 ± 10 (160-210)	–	160 ± 10 (150-170)	–	–
GS% = GL/SL × 100	80	70 ± 10 (60-80)	–	70 ± 10 (60-80)	–	–
H% = H/T × 100	–	–	–	–	–	33 ± 0.6 (30-34)

EP = distance from anterior end to excretory pore; NR = distance from anterior end to nerve ring; ES = distance from anterior end to end of pharynx.

Male (second generation)

Smaller than first generation; tail with or without mucron.

Female (first generation)

Body C-shaped when heat-relaxed. Cuticle, labial region, stoma and pharyngeal region as in male. Excretory pore located at *ca* mid-region of procorpus. Ovaries opposed, reflexed in dorsal position; oviduct well developed; glandular spermatheca and uterus located ventrally. Vagina short, with muscular walls. Vulva located near middle of body, with non-protruding lips. Tail blunt, conoid, with a mucron. Postcloacal swelling absent.

Female (second generation)

Smaller than first generation. Vulval lips slightly protruding. Tail conoid, with or without postanal swelling.

Infective juvenile

Body of heat-relaxed specimens almost straight, slender, gradually tapering posteriorly. Labial region smooth, slightly truncate, continuous with body contour. Labial and cephalic papillae distinct. Amphidial apertures visible. Cuticle with transverse striae. Lateral field distinct, bearing six longitudinal ridges (*i.e.*, seven lines or incisures) in mid-body region. Stoma closed. Pharynx long, narrow, with slightly expanded procorpus, narrower isthmus and pyriform basal bulb with valve. Nerve ring located at isthmus level. Excretory pore located *ca* mid-corpus. Cardia present. Anterior portion of intestine with dorsally displaced pouch containing symbiotic bacteria. Intestine filled with numerous fat globules, lumen of intestine narrow. Rectum long, straight; anus distinct. Genital primordium evident. Tail conoid with pointed terminus. Hyaline region comprising *ca* 33% of tail length.

TYPE HOST AND LOCALITY

Type host unknown as isolate was recovered by baiting soil with *Galleria mellonella* larvae in Beijing, People's Republic of China (precise locality unknown).

TYPE MATERIAL

Holotype male first generation, first generation males, first generation females and infective juveniles deposited in the University of California-Davis Nematode Collection, Davis, CA, USA.

REMARKS

This species was originally identified as *S. carpocapsae* and was referred to as the BJ isolate. However, the latter species is an ambush-type forager whereas laboratory observations showed that the present isolate exhibited a cruiser-type foraging behaviour. As a result of this difference, the material was characterised more thoroughly, leading to its recognition as a distinct species.

RELATIONSHIPS

Steinernema websteri is characterised by the infective juvenile maximum body diam. of 21 (17-25) μm, excretory pore position of 36 (29-40 μm), D% value = 31 (24-34), E% value = 77 (62-102), and the presence of six longitudinal ridges (= seven lines) in the IJ lateral field. First generation males are distinguished by the position of the excretory pore (average = 62 μm), cloacal body diam. (average = 37.5 μm), tail length (average = 29 μm), D% (average = 40) and morphology of the spicules and gubernaculum, as well as by the number and arrangement of the genital papillae. First generation females are characterised by having a mucronate tail and non-protruding to slightly protuberant vulval lips.

The IJ and first generation males of *S. websteri* most resemble *S. carpocapsae* in many morphological/morphometric traits. However, these two species can be differentiated by the tail length of the IJ, which is shorter in the former (average = 47 *vs* 53 μm), and the E% value (average = 77 *vs* 60). Males of *S. websteri* can be separated from *S. carpocapsae* by the morphology of the spicules which have a more rounded manubrium and an inconspicuous calomus in the former. The arrangement of the genital papillae in these two species is also slightly different: *S. websteri* has one subdorsal pair in a postcloacal position, whereas *S. carpocapsae* has two postcloacal pairs located subdorsally.

Steinernema websteri also resembles *S. abbasi* in the total body length of the IJ (although *S. abbasi* has horn-like projections on the IJ cephalic region whereas *S. websteri* does not). However, IJ of the former can be distinguished by having a narrower body diam. (average = 21 *vs* 29 μm), and a more anteriorly located excretory pore (average = 36 *vs* 48 μm). These morphometric differences are also reflected in the values of D% and E%. Males of *S. websteri* can be separated from those of *S. abbasi* by the absence of a tail mucron, the overall morphology of the spicules, the morphology and size of the gubernaculum (average = 49 *vs* 45 μm), and disposition of the genital papillae. Females of *S. websteri* can be distinguished from *S. abbasi* by the morphology of the vulval lips (non- to slightly protruding *vs* protruding).

The IJ of *S. websteri* can be separated from those of *S. riobrave* by their body length and diam., which are shorter and narrower, respectively, in the former. The excretory pore of *S. websteri* is more anteriorly located. First generation males of *S. websteri* can be separated from those of *S. riobrave* by having a narrower cloacal body diam. (average = 37.5 *vs* 50 μm) and by the position of the excretory pore, which is more anteriorly located in the former (average = 62 *vs* 103 μm). The morphology and size of the spicules and gubernaculum, and the arrangement of the genital papillae also differ.

MOLECULAR CHARACTERISATION

Steinernema websteri was characterised by analysis of the large subunit (28S) of ribosomal DNA (LSU rDNA) sequences and by RFLP profiling. Phylogenetic analysis of LSU sequence data placed *S. websteri* close to a clade characterised by the presence of horn-like cephalic papillae – the *bicornutum*-group. However, *S. websteri* can be differentiated from these taxa by the absence of this feature and by several morphometric differences. The RFLP profile for *S. websteri* is similar to that for *S. carpocapsae*. Differences are mainly observed in the digest pattern of *Alu*I, *Hin*fI and *Sau*3AI.

Adjusted distance matrix showed that the BJ isolate and *S. carpocapsae* differ in 156 characters (base pairs).

BIONOMICS

Attempts to cross-hybridise *S. websteri* with *S. riobrave* and *S. abbasi* yielded no progeny. Control crosses using individuals of the same species produced viable offspring. When crossing *S. websteri* with *S. carpocapsae*, progeny production was observed in about 30% of the replicates. However, when pre-adults of the resulting hybrids were crossed with each other no offspring were produced (Cutler & Stock, 2003).

DISTRIBUTION

Known from the type locality, Beijing, People's Republic of China (Cutler & Stock, 2003). Stock (pers. comm., 2007) records this species from Costa Rica, Colombia, Peru, USA and the West Indies.

Reference

CUTLER, G.C. & STOCK, S.P. (2003). *Steinernema websteri* sp. n. (Rhabditida: Steinernematidae), a new entomopathogenic nematode from China. *Nematologia Mediterranea* 31, 215-224.

Steinernema weiseri[*,**] Mráček, Sturhan & Reid, 2003
(Figs 192-194)

MEASUREMENTS

See Table 58.

DESCRIPTION

Male (first generation)

Body J-shaped in heat-relaxed specimens. Cuticle with very faint transverse striae, lateral fields indistinct. Cephalic extremity flattened, almost continuous with body contour, with six lips fused at base, each lip bearing a terminal papilla. Outer circle of four cephalic papillae present. Amphidial apertures distinct, located laterally between labial and cephalic circle of papillae. Stoma short, shallow, comprising prominent, sclerotised cheilorhabdions anteriorly and then smaller, sclerotised prorhabdions. Pharynx muscular with procorpus extending into slightly swollen non-valvate metacorpus, narrower isthmus region surrounded by nerve ring, and rounded basal bulb. Excretory pore variable in position, usually located anterior to nerve ring at metacorpal swelling, rarely slightly posterior to nerve ring. Cardia rather long, intestine generally with a wide lumen. Genital system monorchic, reflexed. Spicules paired, symmetrical, slightly curved, distally bluntly pointed, light-brown in colour. Spicule manubrium distinctly elongated, with length/width ratio of 1.7-2.6 : 1; manubrium almost one-third of spicule length long; calomus indistinct, lamina slightly arched, ventral rostrum poorly developed; velum narrow, extending from calomus or rostrum almost to tip of lamina. Gubernaculum boat-shaped in lateral view, distally much attenuated, proximal part varying from straight and bluntly pointed to hook-like. Complement of 23 genital papillae comprising 11 pairs plus a single midventral papilla located just anterior to cloacal aperture. Of paired papillae, six pairs located precloacal and subventral and five pairs postcloacal, of which two pairs located posterior to gubernaculum and three consistently near tail tip. Tail blunt conoid, usually concave at ventral side, mucron mostly absent, 1.5-2 μm long if developed (present in 35% of specimens).

[*] Named after Prof. Jaroslav *Weiser*, a famous insect pathologist.

[**] Description after Mráček *et al.* (2003). Revised by Zdeněk Mráček.

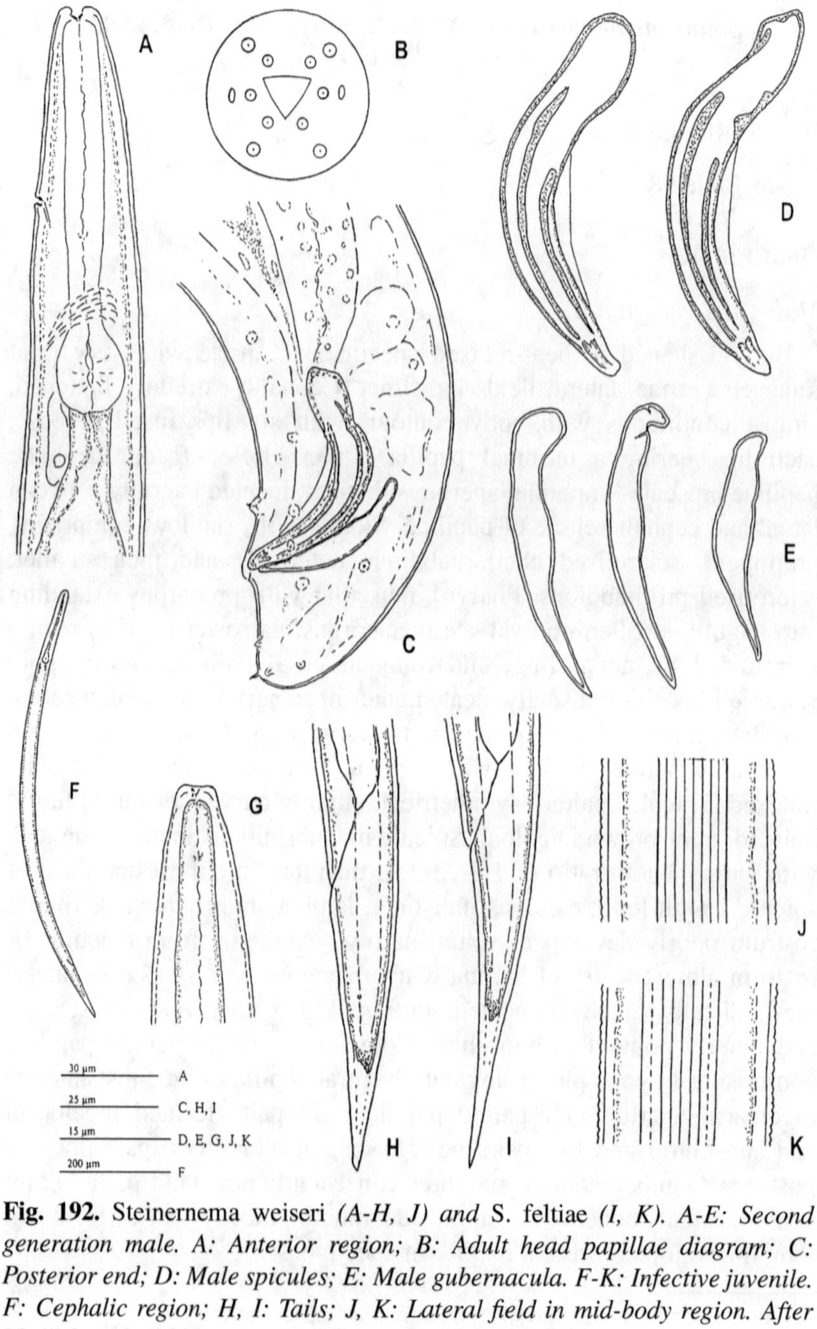

Fig. 192. Steinernema weiseri *(A-H, J) and* S. feltiae *(I, K). A-E: Second generation male. A: Anterior region; B: Adult head papillae diagram; C: Posterior end; D: Male spicules; E: Male gubernacula. F-K: Infective juvenile. F: Cephalic region; H, I: Tails; J, K: Lateral field in mid-body region. After Mráček* et al. *(2003), courtesy of* Systematic Parasitology.

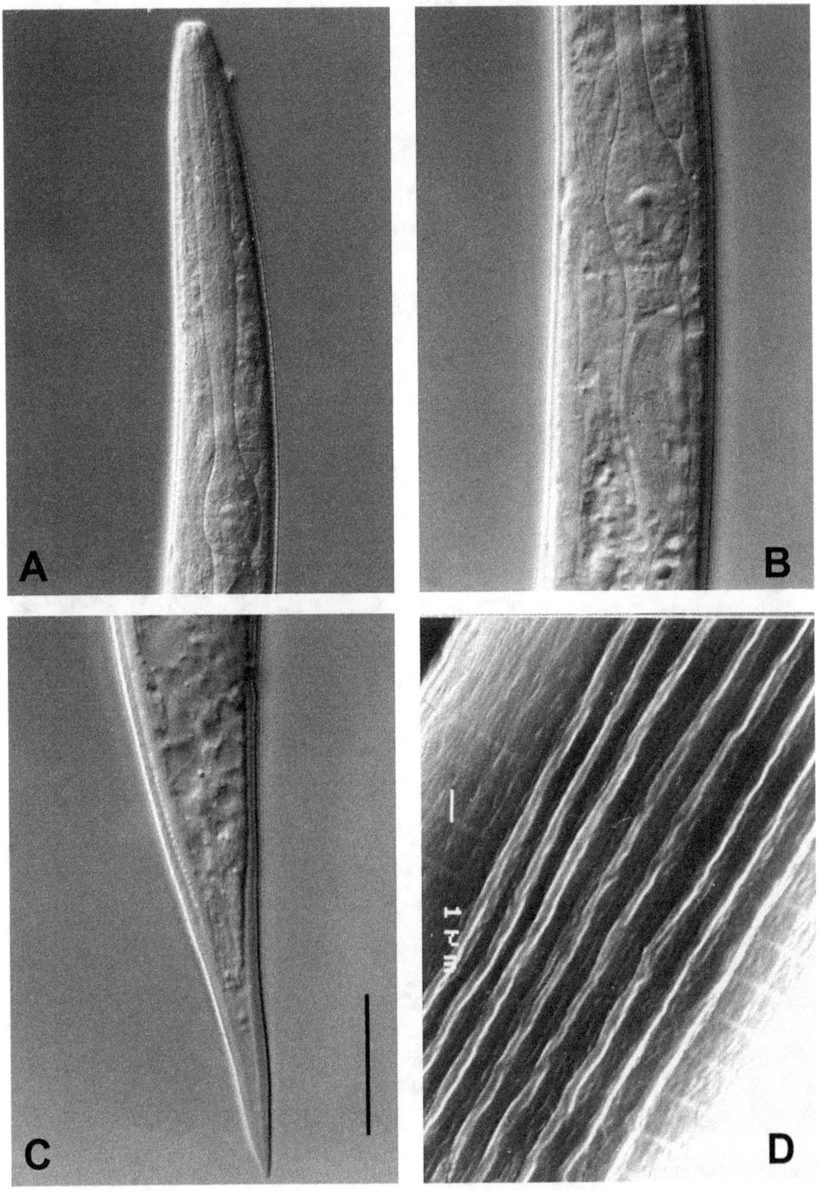

Fig. 193. Steinernema weiseri. *Infective juvenile. A: Pharyngeal region; B: Pharyngeal bulb and bacterial vesicle; C: Tail with hyaline portion; D: SEM of lateral field in mid-body region. (Scale bars: A = 30 μm; B, C = 20 μm; D = 1 μm.) After Mráček* et al. *(2003), courtesy of* Systematic Parasitology.

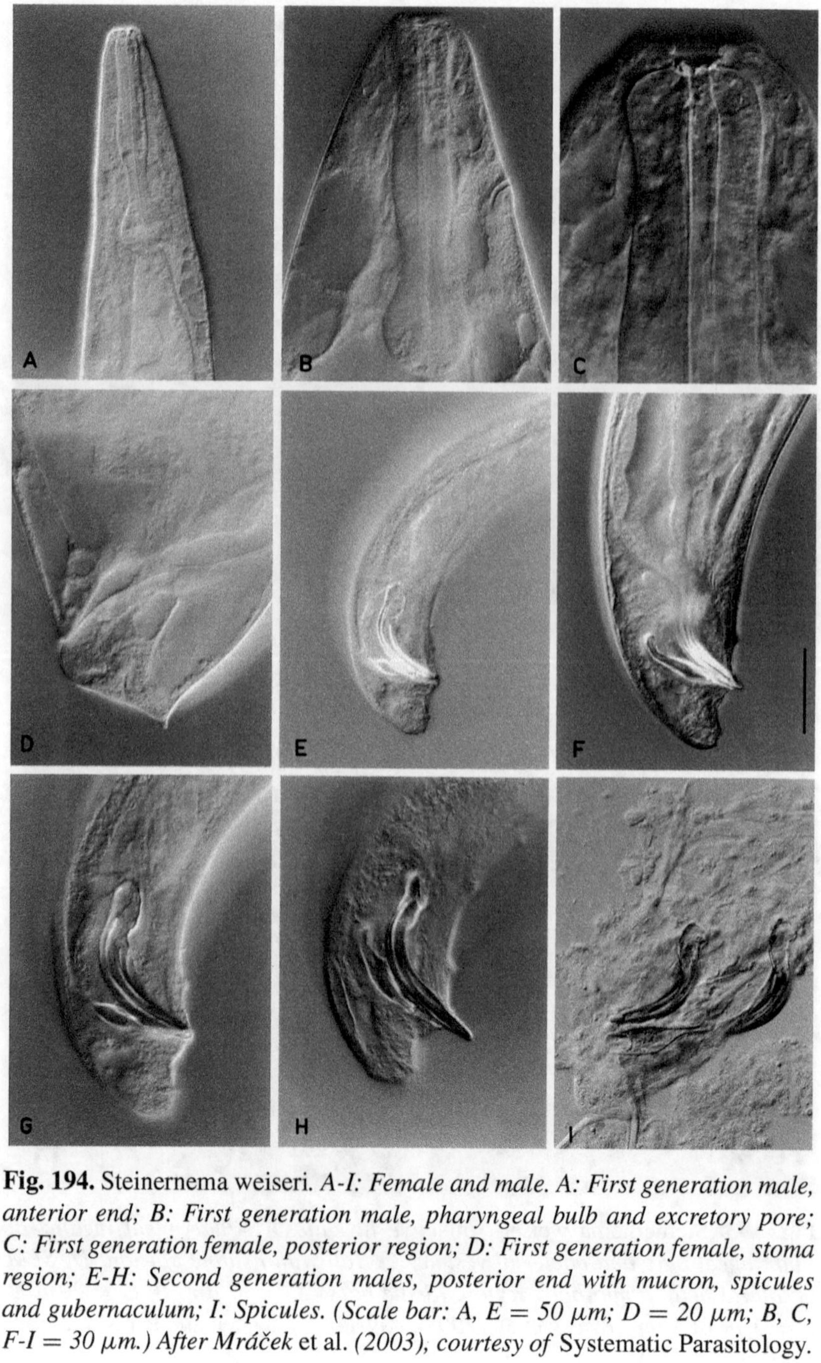

Fig. 194. Steinernema weiseri. *A-I: Female and male. A: First generation male, anterior end; B: First generation male, pharyngeal bulb and excretory pore; C: First generation female, posterior region; D: First generation female, stoma region; E-H: Second generation males, posterior end with mucron, spicules and gubernaculum; I: Spicules. (Scale bar: A, E = 50 μm; D = 20 μm; B, C, F-I = 30 μm.) After Mráček* et al. *(2003), courtesy of* Systematic Parasitology.

Male (second generation)

Similar to first generation, but slightly smaller in body length and other morphometric values, in particular, greatest body diam. much less. Excretory pore more posteriorly located than in first generation. Tail mucron present, but short, rarely longer than 3 μm, exceptionally lacking (absent in 5% of specimens). Deirids situated between level of pharyngeal bulb and *ca* one corresponding body diam. posterior to pharynx, papilliform; postdeirids *ca* five body diam. anterior to tail tip, indistinct.

Female (first generation)

Body long and obese, usually C-shaped in heat-relaxed specimens. Cuticle only faintly striated. Cephalic extremity bluntly rounded, slightly tapering anteriorly. Structure of head region and pharynx similar to male. Excretory pore located anterior to nerve ring, often at metacorpal swelling. Genital tract amphidelphic with reflexed ovaries. Vulva in form of a transverse slit, slightly protruding from body contour, symmetrical. Vagina short, leading into paired uteri. Females oviparous, but in aging specimens juveniles hatching in female body (= *endotokia matricida*). Rectum narrow, anal aperture distinct. Tail of mature females short and obtuse, without postanal swelling; in young females tail longer with spine-like tip.

Female (second generation)

Similar to first generation in body shape and shape of morphological structures, but smaller in dimension. Pygmy forms observed. Excretory pore located more posterior than in first generation. Tail tip pointed, but not mucronate, tail longer than anal body diam.

Infective juvenile

Body of heat-relaxed specimens slightly C-shaped to almost straight, slender, slightly tapering towards anterior and posterior ends. Specimens from *Galleria mellonella* cultures often enclosed in second-stage cuticle, but cuticle never retained in specimens isolated from soil. Cuticle indistinctly annulated, 1 μm thick in mid-body region. Cephalic extremity mostly distinctly off-set from body contour by a depression, 6-8 μm diam., truncate. Cephalic papillae and amphidial apertures indistinct. Lateral fields at mid-body region with nine lines (*i.e.*, eight ridges), all lines distinct and approximately spaced at even distances from one other.

Table 58. *Morphometrics of* Steinernema weiseri. *Measurements are in μm and in the form: mean ± SD (range).*

Character	Holotype (IJ)	First generation		Second generation		Infective juvenile
		Male	Female	Male	Female	
n	–	20	20	20	20	25
L	820	1180 ± 113 (990-1395)	4610 ± 650 (3780-5940)	1006 ± 104 (870-1295)	2020 ± 320 (1560-2890)	740 ± 68 (586-828)
a	34	11 ± 1.0 (9.0-12)	21 ± 3.3 (17-29)	15 ± 1.3 (12-16)	16 ± 1.6 (12-19)	(29 ± 2.0) 25-33
b	7.5	8.0 ± 0.8 (7.0-10)	25 ± 2.4 (22-31)	8.0 ± 0.7 (7.0-9.0)	14 ± 2.1 (12-21)	6.6 ± 0.5 (5.7-7.2)
c	13	48 ± 6.8 (36-64)	111 ± 19 (87-156)	41 ± 6.9 (30-53)	46 ± 5.6 (40-64)	12 ± 0.9 (10-14)
c′	4.1	0.7 ± 0.1 (0.6-0.9)	0.7 ± 0.1 (0.5-0.8)	0.7 ± 0.1 (0.6-0.9)	1.3 ± 0.1 (1.0-1.5)	3.7 ± 0.3 (3.2-4.1)
V	–	–	53 ± 1.7 (50-58)	–	52 ± 1.1 (50-54)	–
Body diam.	24	112 ± 16.2 (84-138)	223 ± 18 (202-263)	69 ± 8.7 (59-92)	126 ± 18 (99-167)	25 ± 1.4 (24-29)
EP	56	70 ± 7.4 (57-84)	80 ± 3.1 (75-86)	75 ± 8.4 (64-92)	73 ± 5.9 (64-88)	57 ± 5.3 (43-65)
NR	84	99 ± 5.0 (94-115)	125 ± 13 (108-154)	89 ± 8.3 (76-102)	96 ± 7.0 (84-108)	84 ± 5.0 (72-92)
ES	110	141 ± 5.0 (134-154)	184 ± 19 (162-226)	126 ± 7.8 (110-142)	141 ± 8.0 (127-156)	113 ± 6.2 (95-119)
Tail length (T)	63	25 ± 3.6 (19-32)	42 ± 4.5 (38-59)	25 ± 3.6 (18-30)	44 ± 5.6 (35-56)	60 ± 5.2 (49-68)
Hyaline (H)	21	–	–	–	–	22 ± 1.8 (18-24)
Anal body diam. (ABD)	15	38 ± 3.7 (29-43)	63 ± 6.7 (51-80)	34 ± 2.6 (29-38)	34 ± 4.6 (29-45)	17 ± 1.3 (14-19)
Spicule length (SL)	–	68 ± 2.9 (62-72)	–	57 ± 3.8 (51-62)	–	–
Gubernaculum length (GL)	–	53 ± 3.2 (46-57)	–	42 ± 2.5 (37-46)	–	–
Gubernaculum width	–	8.3 ± 0.8 (6.4-9.5)	–	7.2 ± 0.8 (6.4-8.0)	–	–

Deirids located in centre of lateral fields at level of pharyngeal bulb or slightly posterior, with process not extending to cuticle surface. Stoma closed. Pharynx with slightly expanded procorpus, narrower isthmus and subpyriform basal bulb with very weakly developed valve apparatus. Basal bulb usually slightly displaced towards dorsal side by excretory gland. Hemizonid distinct, 4 μm long, located at level of basal bulb or immediately posterior. Excretory pore situated in mid-pharynx region,

Table 58. *(Continued).*

Character	Holotype (IJ)	First generation		Second generation		Infective juvenile
		Male	Female	Male	Female	
D% = EP/ES × 100	56	49 ± 5.0 (39-60)	–	59 ± 5.0 (53-73)	–	51 ± 3.0 (44-55)
E% = EP/T* × 100	89	280	300	190	165	95
SW% = SL/ABD × 100	–	180 ± 20 (150-240)	–	170 ± 10 (150-190)	–	–
GS% = GL/SL × 100	–	80 ± 0 (70-85)	–	75 ± 10 (70-85)	–	–
H% = H/T × 100	33	–	–	–	–	36 ± 1.3 (34-39)

EP = distance from anterior end to excretory pore; NR = distance from anterior end to nerve ring; ES = distance from anterior end to end of pharynx.

* Recalculated according to Mráček *et al.* (2003).

cuticularised, *ca* 1 μm diam., excretory duct extending posteriorly to cardia region. Intestine separated from pharynx by 5-9 μm long cardia, in turn followed by a slender bacterial vesicle up to 2.5 body diam. long and displaced dorsad. Gut filled with numerous fat globules, lumen not visible. Rectum straight, slightly more than one anal body diam. long. Anus 4 μm wide, sickle-shaped. Tail conoid, straight, with terminus finely rounded or pointed, tip devoid of refractive inclusion. Hyaline region comprising *ca* 33% of tail length. Phasmids weakly developed, located at *ca* 40% of tail length posterior to anus at ventral edge of lateral fields.

TYPE HOST AND LOCALITY

The natural host is unknown, the type population being recovered by baiting using *G. mellonella* larvae from a soil sample collected in March 1999 from a roadside at the village Kaliště near České Budějovice, Czech Republic. The sample was taken beneath the canopy of an apple tree, where a hibernating aggregation of sciarid fly larvae (Diptera: Sciaridae) was found.

TYPE MATERIAL

Holotype infective juvenile, paratype infective juveniles, males and females of first and second generation deposited in the German Nematode Collection, Biologische Bundesanstalt, Münster, Germany. Additional paratypes (infective juveniles, males and females) at the Laboratory of Insect Pathology, Institute of Entomology, Czech Academy of Sciences, České Budějovice, Czech Republic; paratype infective juveniles in University of California-Davis Nematode Collection, Davis, CA, USA; in the USDA Nematode Collection, Beltsville, MD, USA; and in the CABI Europe-UK Nematode Collection, Egham, Surrey, UK.

Remarks

Males from an isolate originating from an experimental field of the Biologische Bundesanstalt at Berlin-Dahlem and reared on *G. mellonella* closely agree in morphological characters with paratype males. Infective juveniles isolated from soil samples collected at many sites in Germany and the Slovak Republic also agree in morphological characteristics with the type specimens, but there are certain differences in morphometrics, *e.g.*, the average body length is less and the hyaline region of the tail is shorter.

Infective juveniles of *Steinernema* sp. D1 from Britain, which had been recorded under this designation by Reid and Hominick (1992), are also morphologically similar to *S. weiseri*, and the British *Steinernema* sp. D1 is considered conspecific with *S. weiseri*.

Relationships

Steinernema weiseri is mainly characterised by the following morphological characteristics: males with light brown, slightly curved spicules with long manubrium and presence of a short tail mucron in the second generation (rarely present also in first generation males). Infective juveniles with body and tail of medium length, hyaline region of tail short (*ca* 33% of tail length), excretory pore located about mid-pharynx region, labial region mostly distinctly offset, angular and flattened, lateral fields with nine equally-developed lines (equivalent to eight ridges) spaced at even distances.

Steinernema weiseri most closely resembles *S. feltiae* in morphometrics and other morphological characters. However, in the second generation males of *S. feltiae* the mucron is much longer (>12 μm *vs* 2.5 μm) and the manubrium is shorter (length/width ratio 1.6 *vs* 2.0 in *S. weiseri*). In *S. feltiae* infective juveniles the body, tail and hyaline region are generally longer, the labial region rather rounded and mostly offset from body contour by only a very weak narrowing; the submarginal lines of the nine lines in the lateral fields are only weakly developed, *i.e.*, the marginal two ridges at both sides of the lateral field are indistinct, the inner four ridges mostly prominent. Moreover, the hemizonid and deirids are situated anterior to the pharyngeal bulb in *S. feltiae* but at bulb level and posterior to the bulb, respectively, in *S. weiseri*. *Steinernema kraussei* infective juveniles have only eight lines in the lateral field and *S. kushidai* infective juveniles only seven. The shape of the cephalic region of both species is also different from that of *S. weiseri*. The precise number of lateral lines in the juveniles of *S. oregonense* is unknown, but in first generation males the excretory pore is situated more posteriorly than in *S. weiseri* (D% = 70 *vs* 49).

MOLECULAR CHARACTERISATION

The RFLP profiles yielded by digestion of the ITS region with 17 restriction enzymes for *S. weiseri* were compared to a database containing over 50 different *Steinernema* isolates (including 22 described species). The closest match was with the UK isolate 541, previously referred to as the D1 RFLP type which displayed identical RFLP profiles for all 17 enzymes (Mráček *et al.*, 2003). The next nearest match was with *S. feltiae* which matched *S. weiseri* in 13 of the 17 enzymes. Of these, seven (*Bst*OI, *Eco*RI, *Hind*III, *Kpn*I, *Pvu*II, *Sal*I and *Xba*I) have no recognition sites within these two species. The enzymes which do have recognition sites within the ITS region of these species and which give identical RFLP profiles are *Alu*I, *Hae*III, *Hpa*II, *Pvu*II and *Sau*3AI and *Sau96*I. Sequence data for *S. feltiae* and the UK isolate D1 of *S. weiseri* prove that the cut sites for these six enzymes are located in the same positions within the ITS regions. Restriction enzymes yielding polymorphisms were *Dde*I, *Hha*I, *Hin*fI and *Rsa*I. The sequence of ITS region of rDNA is deposited at GenBank (accession no. AY171268).

BIONOMICS

Steinernema weiseri was easily reared on *G. mellonella* larvae at temperatures of 15-20°C. Death of wax moth larvae occurred within 3 days. After 1 week from host inoculation the first, giant, generation of nematodes was found in the host cadavers. Adults of the second generation were found on days 10-12. Migration of infective juveniles to a water trap occurred after 2 weeks. Infective juveniles stored at 4-8°C in a refrigerator survived for several months (Mráček *et al.*, 2003).

DISTRIBUTION

Steinernema weiseri is known from the Czech Republic, the Slovak Republic and Germany. In the Czech Republic the nematode was isolated in the south-western part of the country from the following habitats, apple tree roadsides and deciduous forest edges, where flies of the families Sciaridae and Bibionidae formed aggregations and where various species of Noctuidae and Geometridae were observed (Mráček *et al.*, 1999; Mráček & Bečvář, 2000). In Slovakia, *S. weiseri* was reported from forests and river banks from six sites from the western to the eastern regions of the country (Sturhan & Lišková, 1999); recently, it was also isolated by the authors from agricultural soil in eastern Slovakia and from a meadow in the northern part of the country. The records reported from Germany (Sturhan, 1995, 1997, 1999) increased the total of records to 29. Most records were from the River Main region, Berlin and Brandenburg and one record was from a coastal region (Hiddensee island, Baltic Sea).

Despite its wide distribution, *S. weiseri* is comparatively rare in Germany; the records amounting to less than 4% of all identified *Steinernema* populations. The species was isolated mainly from grassland and soil under various fruit trees; it was less often found in forest, river bank and agricultural soil. The soil type ranged from sand to loam; there is no record from clay or peaty soil so far. At 13 of the 29 sites in Germany, *S. weiseri* co-occurred with other *Steinernema* species and at one site with *Heterorhabditis megidis*.

The clear conspecificity of British *Steinernema* sp. D1 with *S. weiseri* extends the range of known distribution to north-western Europe. In the UK, the species appears to be rare and during soil surveys it was found only once in a verge (Hominick *et al.*, 1995). Subsequently, two further isolates were reported from a field and a verge in Scotland (Gwynn & Richardson, 1996).

References

GWYNN, R.L. & RICHARDSON, P.N. (1996). Incidence of entomopathogenic nematodes in soil samples collected from Scotland, England and Wales. *Fundamental and Applied Nematology* 19, 427-431.

HOMINICK, W.M., REID, A.P. & BRISCOE, B.R. (1995). Prevalence and habitat specificity of steinernematid and heterorhabditid nematodes isolated during soil surveys of the UK and the Netherlands. *Journal of Helminthology* 69, 72-32.

MRÁČEK, Z. & BEČVÁŘ, S. (2000). Insect aggregations and entomopathogenic nematode occurrence. *Nematology* 2, 297-301.

MRÁČEK, Z., BEČVÁŘ, S. & KINDLMANN, P. (1999). Survey of entomopathogenic nematodes from the families Steinernematidae and Heterorhabditidae (Nematoda: Rhabditida) in the Czech Republic. *Folia Parasitologica* 46, 145-148.

MRÁČEK, Z., STURHAN, D. & REID, A. (2003). *Steinernema weiseri* sp. n. (Rhabditida: Steinernematidae), a new entomopathogenic nematode from Europe. *Systematic Parasitology* 56, 37-47.

REID, A.P. & HOMINICK, W.H. (1992). Restriction fragment length polymorphisms within the ribosomal repeat unit of British entomopathogenic nematodes (Rhabditida: Steinernematidae). *Parasitology* 105, 317-323.

STURHAN, D. (1995). Untersuchungen über sympatrisches Vorkommen entomopathogener Nematoden. *Nachrichtenblatt des Deutschen Pflanzenschutzdienstes* 47, 54.

STURHAN, D. (1997). Untersuchungen zum Artenspektrum entomopathogener Nematoden in verschiedenen Biotopen. *Schriftenreihe des Bundesministeriums für Ernährung Landwirtschaft und Forsten, Reihe A: Angewandte Wissenschaft* 465, 372.

STURHAN, D. (1999). Prevalence and habitat specificity of entomopathogenic nematodes in Germany. In: *Proceedings of a workshop held at Todi, Perugia, Italy, 16-29 May 1995.* Luxembourg, European Commission, COST 819, pp. 123-132.

STURHAN, D. & LIŠKOVÁ, M. (1999). Occurrence and distribution of entomopathogenic nematodes in the Slovak Republic. *Nematology* 1, 273-277.

Steinernema yirgalemense[*,**] Nguyen, Tesfamariam, Gozel, Gaugler & Adams, 2005

(Figs 195-199)

MEASUREMENTS

See Table 59.

DESCRIPTION

Male (first generation)

Body curved ventrally posteriorly, C-shaped when heat-relaxed. Cephalic extremity rounded, usually slightly swollen; perioral disc not observed, six labial papillae, four larger cephalic papillae and two amphidial apertures present. Stoma shallow, cheilorhabdions visible as small, sclerotised, structures at anterior end, sometimes indistinct. Excretory pore situated anterior to nerve ring, slightly posterior to mid-pharynx. Pharynx with cylindrical procorpus, metacorpus absent or slightly swollen, isthmus present, nerve ring located around isthmus, just anterior to distinct basal bulb. Pharyngo-intestinal valve present, usually weak. Genital system monorchic, reflexed. Distance from base of pharynx to anterior end of testis variable. Spicules paired, brown in colour. Cephalic extremity (manubrium) of spicules elongate, length/width ratio *ca* 1.5 (1.4-1.7); shaft (calomus) present; blade (lamina) thick, variable in shape tapering gradually posteriorly to a point; velum prominent. Each spicule with two internal ribs. Gubernaculum boat-shaped in lateral view, gubernaculum neck short. Cuneus present but not prominent. Complement of 12 pairs of genital papillae plus single midventral precloacal papilla. Eleven pairs arranged in normal position for *Steinernema* and twelfth pair located close to edge of cloaca, as reported for *S. scapterisci* (Nguyen & Smart, 1992) and *S. riobrave* (Nguyen & Adams, 2003). There are six precloacal subventral pairs (1-6 in Fig. 196B), one adcloacal pair (7), one lateral pair (11), two subterminal pairs (8, 9) and one subdorsal pair (10). Tail conoid; mucron absent.

[*] Named after *Yirgalem*, Ethiopia, the type locality.

[**] Description after Nguyen *et al.* (2004). Revised by Khuong B. Nguyen.

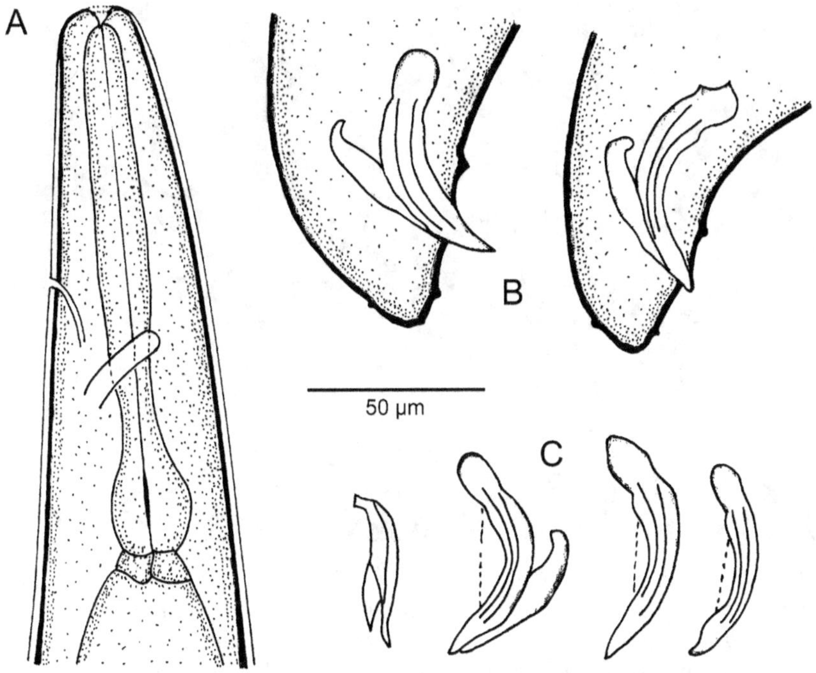

Fig. 195. Steinernema yirgalemense, *male. A: Anterior region; B: Two different shapes of tail. B, C: Variation of spicules and gubernaculum. After Nguyen* et al. *(2004), courtesy of* Nematology.

Male (second generation)

Second generation male similar to first generation but body size smaller, spicule and gubernaculum shorter and thinner, excretory pore much more anterior, and isthmus narrower. Mucron on tail terminus rarely seen.

Female (first generation)

Body cuticle smooth or with faint annules with SEM. Lateral fields and phasmids not observed. Cephalic extremity rounded, continuous with body; six labial papillae, four cephalic papillae present. Lips indistinct. Amphidial apertures present. Stoma shallow, subtriangular anteriorly; triradiate internally. Cheilorhabdions strongly sclerotised but small. Smaller sclerotised structure (presumably the prorhabdions) situated posterior to cheilorhabdions. Pharynx with cylindrical, muscular, procorpus; metacorpus slightly swollen; isthmus distinct; valvate basal bulb. Excretory pore located near mid-pharynx. Nerve ring sur-

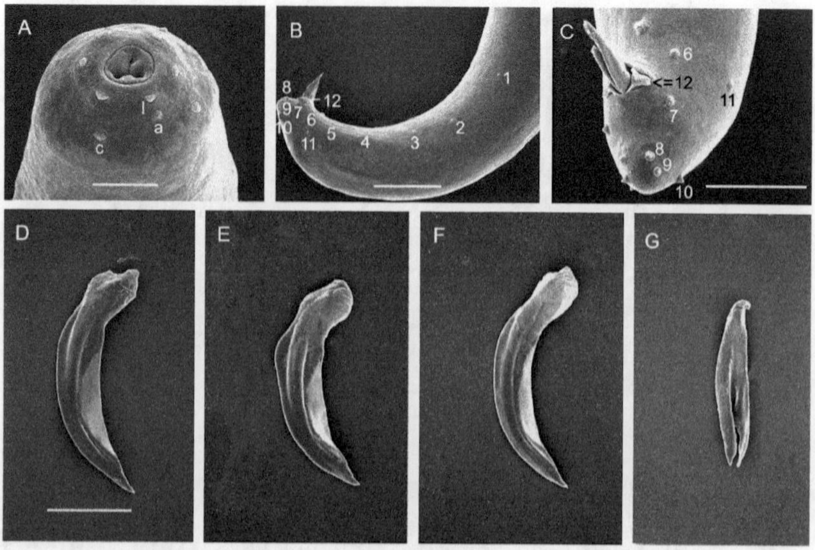

Fig. 196. Steinernema yirgalemense. *SEM of male. A: Face view showing mouth, labial papillae (l), amphidial aperture (a) and cephalic papillae (c); B: Posterior region showing 12 pairs of genital papillae; C: Tail at higher magnification showing positions of pairs 6-12; D-F: Variation in shape of spicules; G: Gubernaculum, subventral view showing curved anterior end and needle-shaped cuneus. (Scale bar: A = 6 μm; B = 60 μm; C = 43 μm; D-G = 23 μm.) After Nguyen* et al. *(2004), courtesy of* Nematology.

rounding isthmus just anterior to basal bulb. Pharyngo-intestinal valve present. Genital system amphidelphic, reflexed, often containing many eggs. Vulva in form of a transverse slit; usually protuberant; low, thick, epiptygmata present in adult females. Vagina sclerotised, short. Body diam. anterior and posterior to vulva equal. Tail terminus rounded, ventral postanal swelling present but small, tail shorter than anal body diam.

Female (second generation)

Similar to first generation female but smaller. Tail tapering gradually to terminus, longer than anal body diam.; terminal mucron present; ventral postanal swelling present.

Infective juvenile

Body elongate. Sheath (second-stage cuticle) present immediately after harvesting, but many IJ losing sheath in storage. Ensheathed juvenile with six labial and four cephalic papillae. Labial region smooth, contin-

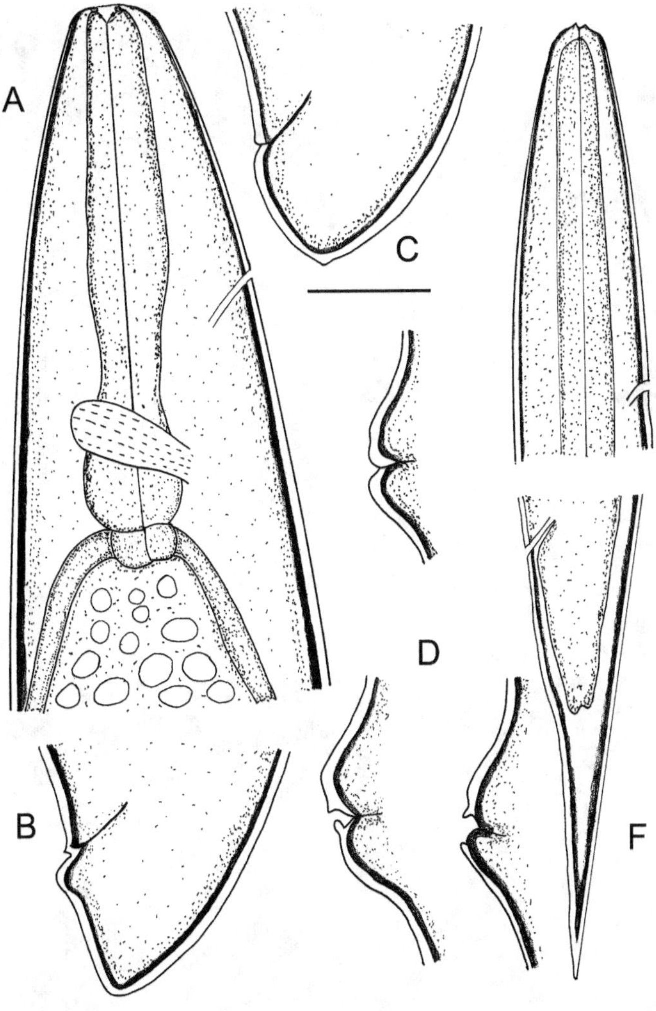

Fig. 197. Steinernema yirgalemense, *female and infective juvenile. A: Anterior region of female; B, C: Variation in shape of female tail; D: Variation in shape of vulva and epiptygmata; E, F: Head and tail of infective juvenile. (Scale bar: A-D = 50 μm; E, F = 20 μm.) After Nguyen* et al. *(2004), courtesy of* Nematology.

uous. Two horn-like structures present on cephalic region of exsheathed juvenile. Amphidial apertures prominent, located immediately posterior to horns. Cuticle marked with prominent transverse striations. Lateral field beginning anteriorly from annules 3 or 4 with one line. Two addi-

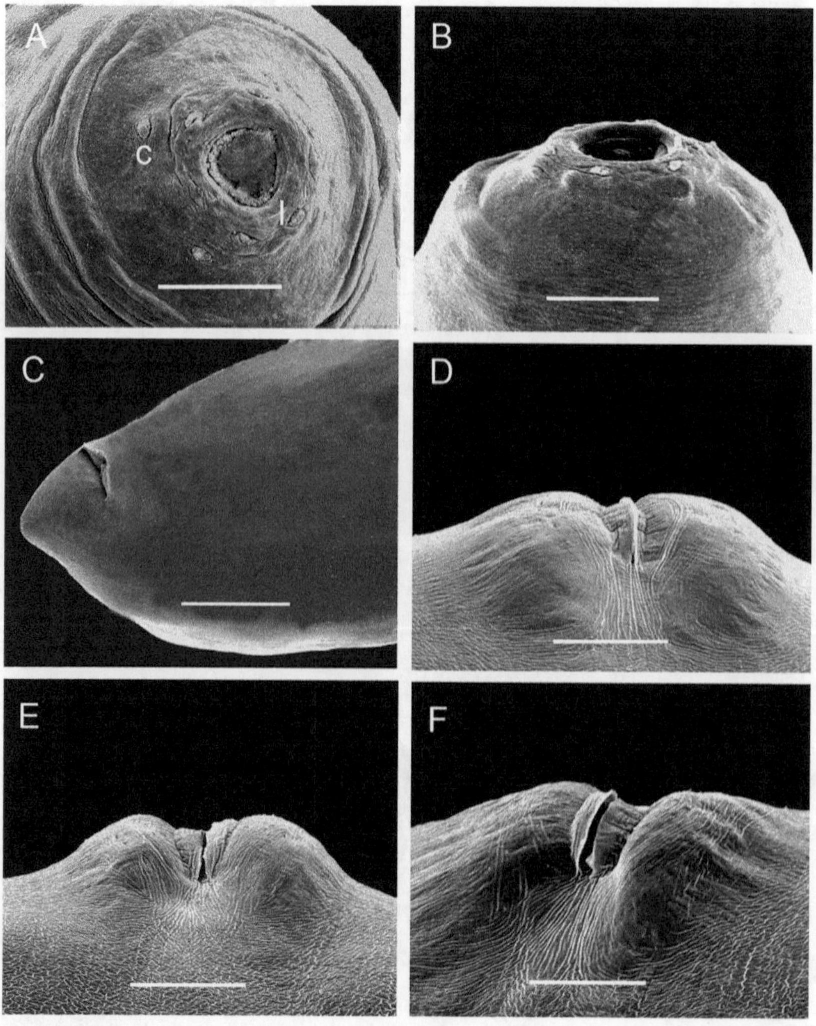

Fig. 198. Steinernema yirgalemense, *SEM of female. A, B: Anterior end showing labial (l) and cephalic papillae (c); C: Tail; D-F: Variation of double-flapped epiptygmata of young to mature females. (Scale bars: A, B = 10 μm; C = 37 μm; D-F = 15 μm.) After Nguyen* et al. *(2004), courtesy of* Nematology.

tional lines appearing at annules 20-24 forming two ridges. Near excretory pore level, number of ridges in lateral fields increasing from two to six. Two submarginal ridges appearing more posteriorly, making total of eight ridges, *i.e.*, maximum number in lateral field. Portion with eight

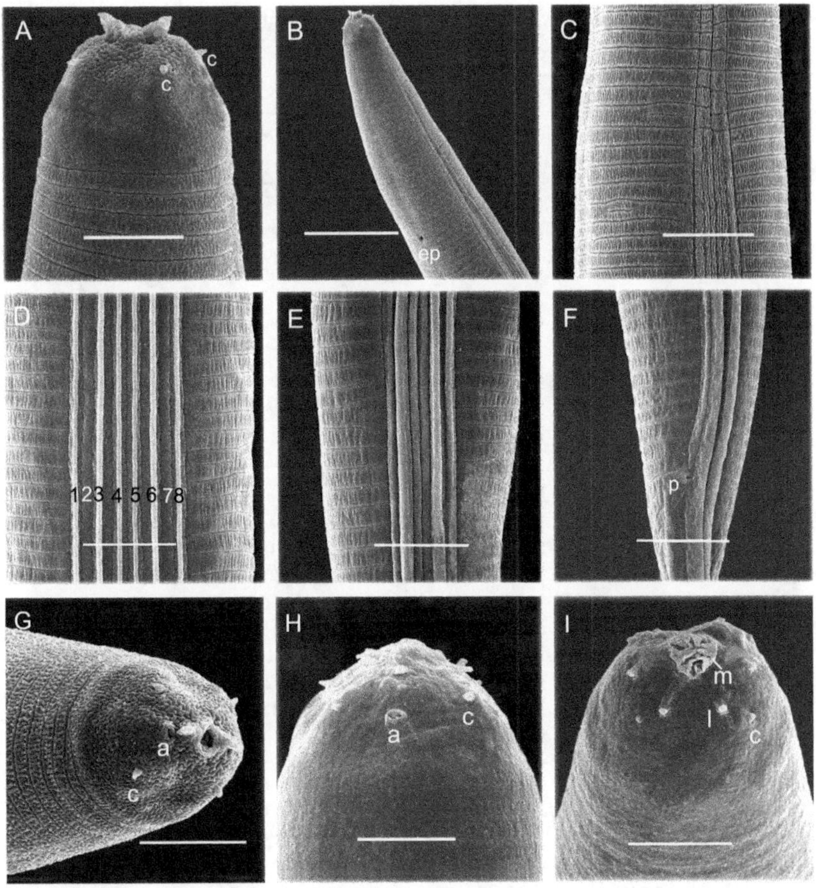

Fig. 199. Steinernema yirgalemense. *SEM of infective juvenile. A, B: Anterior region showing two horn-like structures, cephalic papillae (c) in labial region and anterior part of lateral field and excretory pore (ep); C: Lateral field showing change from two to six ridges; D: Lateral field with eight ridges (note: ridges two and seven are not as prominent as others and will disappear posteriorly); E: Lateral field, ridges two and seven reduced and disappearing gradually to form six ridges; F: Lateral field in posterior region showing two ridges and phasmid (p); G: Anterior end of exsheathed infective juvenile showing amphidial aperture (a) and cephalic papillae; H: Anterior end of ensheathed infective juvenile showing elevated labial region with amphidial aperture (a), cephalic (c) and labial papillae; I: Anterior region of pre-infective juvenile showing rib-like structure on top of mouth (m), cephalic (c) and labial (l) papillae. (Scale bars: A = 3.7 μm; B = 15 μm; C, D = 6.7 μm; E, F = 6 μm; G = 4.3 μm; H, I = 2.7 μm.) After Nguyen* et al. *(2004), courtesy of* Nematology.

Table 59. *Morphometrics of* Steinernema yirgalemense. *Measurements are in μm and in the from: mean ± SD (range).*

Character	First generation				Second generation		Infective juvenile
	Holotype (male)	Male	Female	Allotype (female)	Male	Female	
n	–	20	20	1	20	20	25
L	1572	1566 ± 39 (1331-1777)	6144 ± 78 (4940-7029)	6686	999 ± 32 (789-1144)	1859 ± 43 (1367-2199)	635 ± 36 (548-693)
a	–	–	–	–	–	–	21 ± 1.3 (20-25)
b	–	–	–	–	–	–	5.2 ± 0.3 (4.8-5.9)
c	–	–	–	–	–	–	10.3 ± 0.6 (9.2-11.2)
c′	–	–	–	–	–	–	3.3
V	–	–	51 ± 7 (49-52)	52	–	56 ± 7.5 (52-59)	–
Body diam.	118	112 ± 11 (97-138)	205 ± 14 (180-247)	277	59 ± 7.7 (45-67)	103 ± 10 (89-115)	29 ± 2.2 (24-33)
EP	89	86 ± 9 (74-107)	100 ± 10 (83-117)	98	80 ± 9 (66-94)	75 ± 8.7 (64-85)	51 ± 3.4 (45-59)
NR	111	108 ± 10 (98-136)	150 ± 12 (129-209)	145	104 ± 10 (91-115)	122 ± 12 (103-139)	88 ± 3.6 (82-93)
ES	151	148 ± 12 (132-165)	210 ± 14 (182-278)	198	132 ± 11 (115-147)	165 ± 13 (145-182)	121 ± 3.7 (115-128)
Testis reflexion	295	278 ± 17 (205-331)	–	–	–	–	–
Tail length	23	20 ± 4 (17-27)	46 ± 7 (35-71)	33	22 ± 5 (18-27)	38 ± 6 (32-47)	62 ± 2.7 (57-67)
Anal body diam. (ABD)	40	38 ± 6 (32-45)	75 ± 9 (59-90)	71	32 ± 6 (27-34)	42 ± 6.5 (32-53)	19 ± 4.3 (17-21)
Spicule length (SL)	68	64 ± 8 (51-77)	–	–	54 ± 7 (47-64)	–	–
Spicule width	14	14 ± 3.6 (12-15)	–	–	11 ± 3 (9-12)	–	–
Gubernaculum length (GL)	48	48 ± 6.9 (42-54)	–	–	33 ± 6 (27-38)	–	–
Gubernaculum width	11	11 ± 3 (9-12)	–	–	–	–	

ridges is longest part of lateral field. In posterior third of body, two submarginal ridges disappearing gradually, with only six ridges remaining in lateral field. Near phasmid, six ridges in lateral field becoming two ridges. Formula of lateral field = 2, 6, 8, 6, 2. Pharynx with slender corpus, basal bulb more or less elongate with visible valve. Tail atten-

Table 59. *(Continued).*

Character	First generation				Second generation		Infective juvenile
	Holotype (male)	Male	Female	Allotype (female)	Male	Female	
D% = EP/ES × 100	59	58 ± 7.6 (50-66)	–	–	60 ± 8 (53-66)	–	42 ± 2.7 (38-48)
E% = EP/T × 100	–	–	–	–	–	–	83 ± 7 (67-98)
SW% = SL/ABD × 100	170	171 ± 13 (121-213)	–	–	1.71 ± 1.3 (1.35-2.00)	–	–
GS% = GL/SL × 100	71	74 ± 8 (65-85)	–	–	0.62 ± 0.08 (0.53-0.74)	–	–
H% = Hyaline/T × 100	–	–	–	–	–	–	54 (47-61)

EP = distance from anterior end to excretory pore; NR = distance from anterior end to nerve ring; ES = distance from anterior end to end of pharynx.

uate, tapering gradually. Hyaline region comprising *ca* 54 (47-61)% of tail length.

Type host and locality

Type host unknown. The isolate was collected in an area with banana and sugarcane in Yirgalem, Ethiopia, altitude 1660 m, latitude 6°46.221′N, longitude 38°22.649′E.

Type material

Holotype (male, first generation) isolated from haemocoel of *Galleria mellonella* and deposited in the USDA Nematode Collection, Beltsville, MD, USA. Allotype (female, first generation): same data as holotype, deposited in the same collection. Paratypes: same data as holotype. Many first generation males and females and several infective juveniles in TAF deposited in the same collection; two first generation males, two first generation females and ten infective juveniles deposited in the Florida Collection of Nematodes, Florida Department of Agriculture and Consumer Services, Gainesville, FL, USA; one male and one female of the first generation, and ten infective juveniles deposited in the University of California-Davis Nematode Collection, Davis, CA, USA.

RELATIONSHIPS

Steinernema yirgalemense is characterised by the presence of horn-like structures on the cephalic region of the IJ and, therefore, is placed in the *bicornutum*-group. This species can be recognised by the IJ body length averaging 635 (578-693) μm, EP = 51 (45-59) μm; tail length = 62 (57-67) μm and E% = 83 (67-90). The lateral field pattern is variable, the formula for the arrangement of ridges from head to tail being 2, 6, 8, 6, 2 (Fig. 199). The species can be further recognised by the spicule having a prominent velum, tapering to a pointed terminus (Fig. 196); gubernaculum with a short neck; the male SW and GS ratios; and, especially, by the presence of 12 pairs of genital papillae (compared to 11 in other species). The twelfth pair is positioned at the edge of the cloacal aperture (Fig. 196).

Steinernema yirgalemense differs from all other species (except for the *bicornutum*-group) by the presence of horn-like structures in the IJ cephalic region. In the *bicornutum*-group, the body length of the IJ (635 μm) is similar to that of *S. riobrave* (622 μm) but different from that of *S. abbasi*, *S. bicornutum*, *S. ceratophorum* and *S. pakistanense*. Tail length, c′ ratio, D% and E% of the IJ of *S. yirgalemense* are also different and can be used to distinguish between nematode species in this group. The formula of the lateral field pattern (2, 6, 8, 6, 2) is unique, and can be differentiated from the other two most similar species, *S. abbasi* (2, 6, 8, 7, 6, 2) and *S. riobrave* (2, 7, 8, 6, 2). Males can be differentiated from other steinernematids by the shape (Figs 195, 196) and length of spicules and gubernaculum (Nguyen & Smart, 1997) D% = 58, and SW = 1.8. *Steinernema yirgalemense* has 12 pairs of genital papillae with pair 12 located at the edge of the cloacal aperture. Females can be distinguished from other species by their small epiptygmata.

MOLECULAR CHARACTERISATION

The species is characterised genetically by sequence lengths of ITS regions (960 bp) (accession no. AY748450), the length of the ITS1 (270 bp) and ITS2 (284 bp), by composition of its sequence (A = 0.274, C = 0.152, G = 0.237, T = 0.337) and by numerous unique, derived nucleotide character states.

More DNA information is available in Nguyen *et al.* (2004).

DISTRIBUTION

This nematode species has been found in the type locality of Yirgalem, Ethiopia (Nguyen *et al.*, 2004) and has also been identified from Kenya (Nguyen, unpubl.).

References

NGUYEN, K.B. & ADAMS, B.J. (2003). SEM and systematic studies of *Steinernema abbasi* Elawad *et al.*, 1997 and *S. riobrave* Cabanillas *et al.*, 1994 (Rhabditida: Steinernematidae). *Zootaxa* 179, 1-10.

NGUYEN, K.B. & SMART JR, G.C. (1992). Addendum to the morphology of *Steinernema scapterisci*. *Journal of Nematology* 24, 478-481.

NGUYEN, K.B. & SMART JR, G.C. (1997). Scanning electron microscopic studies of spicules and gubernacula for *Steinernema* spp. (Nemata: Steinernematidae). *Nematologica* 43, 465-480.

NGUYEN, K.B., TESFAMARIAM, M., GOZEL, U., GAUGLER, R. & ADAMS, B.J. (2004). *Steinernema yirgalemense* n. sp. (Rhabditida: Steinernematidae) from Ethiopia. *Nematology* 6, 839-856.

The genus *Neosteinernema* Nguyen & Smart, 1994

Neosteinernema longicurvicauda[*,**] Nguyen & Smart, 1994
(Figs 200-206)

MEASUREMENTS

See Table 60.

DESCRIPTION

Male

Body ventrally curved or C-shaped when heat-relaxed. Lateral field with single, sometimes indistinct, incisure. Stoma shallow, cheilorhabdions prominent; other parts of stoma appearing funnel-shaped. Under light microscopy, walls of stoma, especially dorsal side, appearing thickened anteriorly and raised to form a subtriangular structure in lateral view resembling a dorsal and ventral tooth. Pharynx steinernematid in form but somewhat thinner and more elongate than in *Steinernema*. Pharyngo-intestinal valve pronounced. Excretory pore located at, or anterior to, mid-pharynx; excretory duct sclerotised anteriorly. Genital system monorchic, reflexed. Spicules paired, foot-like in shape, prominent hump present on dorsal side of spicules; spicule head enlarged anteriorly; spicule tip pointed, slightly curved ventrally; two prominent ribs present in each spicule. Gubernaculum unusually long when compared to spicule (gubernaculum/spicule ratio 97%), 32% of males with gubernaculum equal to or longer than spicule. Proximal end of gubernaculum usually difficult to see *in situ*. Cuticle of cloacal area forming *tubus* around spicules. Complement of genital papillae comprising 13 or 14 pairs plus a single, midventral, precloacal papilla arranged as eight pairs precloacal, mostly subventral, more or less variable in position, one pair in cloacal area, two or three pairs caudal, subventral and two pairs caudal, subdorsal. Phasmids prominent. Tail conoid, tail tip digitate with some variations, especially in young males.

No second generation males were found.

[*] From Latin *longus* = long, *curvus* = curved and *cauda* = tail.

[**] Description after Nguyen and Smart (1994). Revised by Khuong B. Nguyen.

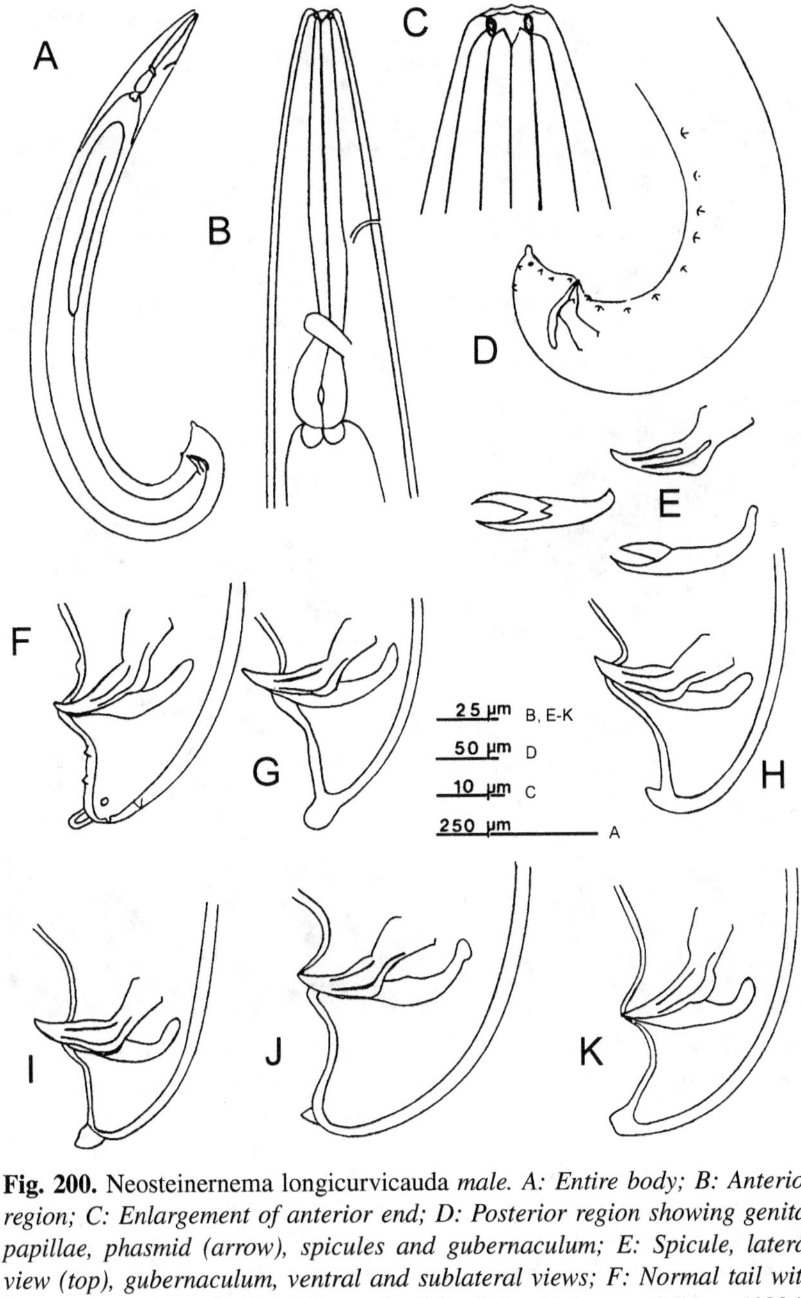

Fig. 200. Neosteinernema longicurvicauda *male. A: Entire body; B: Anterior region; C: Enlargement of anterior end; D: Posterior region showing genital papillae, phasmid (arrow), spicules and gubernaculum; E: Spicule, lateral view (top), gubernaculum, ventral and sublateral views; F: Normal tail with phasmid; F-K: Variation in shape of tail tip. After Nguyen and Smart (1994), courtesy of* Journal of Nematology.

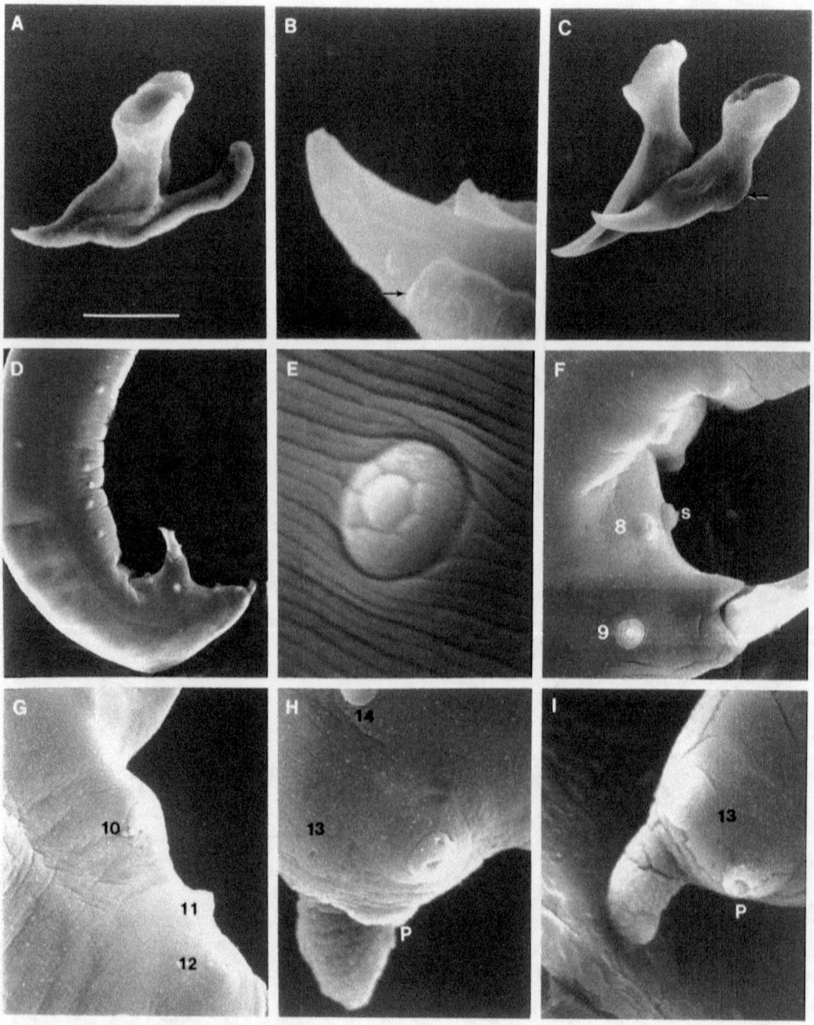

Fig. 201. Neosteinernema longicurvicauda. *SEM photographs. A: Spicule and gubernaculum; B: Enlargement of distal end of spicule and gubernaculum, tip of left spicule extending beyond gubernaculum (arrow), tip of right spicule at level of end of gubernaculum; C: Left and right spicules, note dorsal hump (arrow); D: Posterior region showing genital papillae (note that the second papilla, anteriorly to posteriorly, is subdorsal), cloacal tube (= tubus), protruding spicules and digitate tail terminus; E: Enlargement of most anterior genital papilla; F: Enlargement of part of male tail showing single ventral genital papilla (s), and one each of pairs 8 and 9; G: Genital papillae, one each of pairs 10, 11 and 12.*

Female

Anterior region similar to male but much larger. Body spiral, usually found in surrounding medium next to termite cadavers. Cuticle annulated. Lateral fields not observed. Cephalic extremity truncate or rounded. Labial papillae six, prominent, many additional papilla-like structures often present together with labial papillae. Stoma as in male but structures larger and more prominent; raised ring prominent in stoma appearing tooth-like in lateral view. Pharynx typical of steinernematids. Genital system didelphic, amphidelphic, reflexed. Epiptygmata absent. Tail curved, longer than anal body diam.; $c' = 1.40$ (1.10-1.70). Phasmids prominent, located on protuberance in posterior half of tail. Eggs not laid, juveniles hatching inside body and attaining infective stage before exiting female.

No second generation females were found.

Infective juvenile

Second-stage juveniles moulting to infective stage juveniles before emerging from female. Body thin, elongate, well striated. Sheath (second-stage cuticle) present but sometimes lost. Cephalic extremity smooth, mostly slightly enlarged. Oral aperture closed, triradiate; lips six. Labial papillae not observed; four prominent cephalic papillae situated well posterior to anterior extremity. Amphidial apertures large, slit like, post-labial in position but anterior to cephalic papillae. Lateral field at mid-body wider than one-third of body diam., usually only two border ridges seen clearly; eight smooth bands and nine incisures prominent in anterior and posterior regions. Pharynx degenerate, basal bulb elongate with valve, pharyngo-intestinal valve present but sometimes obscure. Excretory pore prominent; excretory duct long, sinuous, clearly visible on living specimens. Sclerotised structure associated with excretory duct just anterior to excretory cell, appearing as two black dots in lateral view. Tail about as long as pharynx, attenuate, ten times (8-11) as long as anal body diam., always curved ventrally at posterior end. Phasmids promi-

Fig. 201. *(Continued). H: Genital papillae, pairs 13 and 14, phasmid and digitate tail terminus; I: Genital papillae, pair 13, phasmid (p) and digitate tail terminus. (Scale bar: A = 23 μm; B = 2.5 μm; C = 20 μm; D = 43 μm; E = 2.4 μm; F = 12 μm; G = 10 μm; H = 6 μm; I = 5 μm.) After Nguyen and Smart (1994), courtesy of* Journal of Nematology.

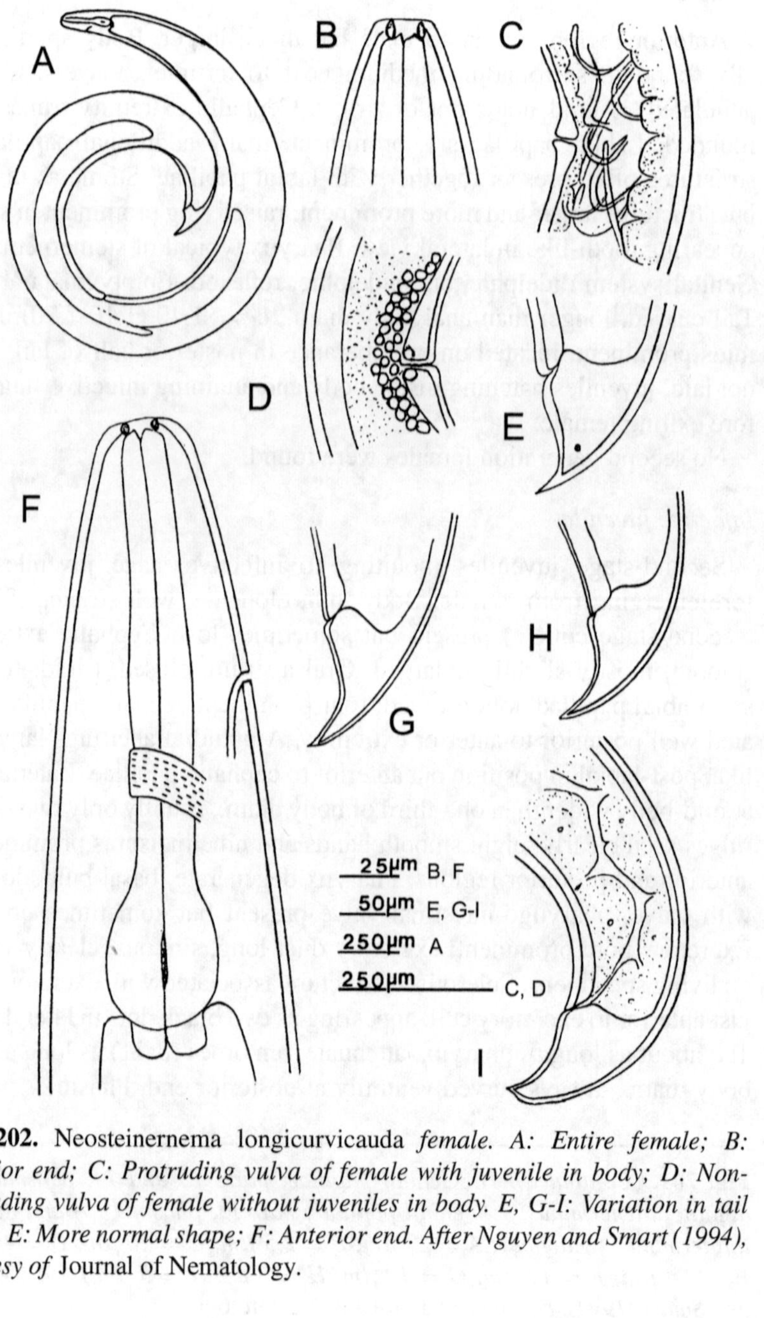

Fig. 202. Neosteinernema longicurvicauda *female. A: Entire female; B: Anterior end; C: Protruding vulva of female with juvenile in body; D: Non-protruding vulva of female without juveniles in body. E, G-I: Variation in tail shape. E: More normal shape; F: Anterior end. After Nguyen and Smart (1994), courtesy of* Journal of Nematology.

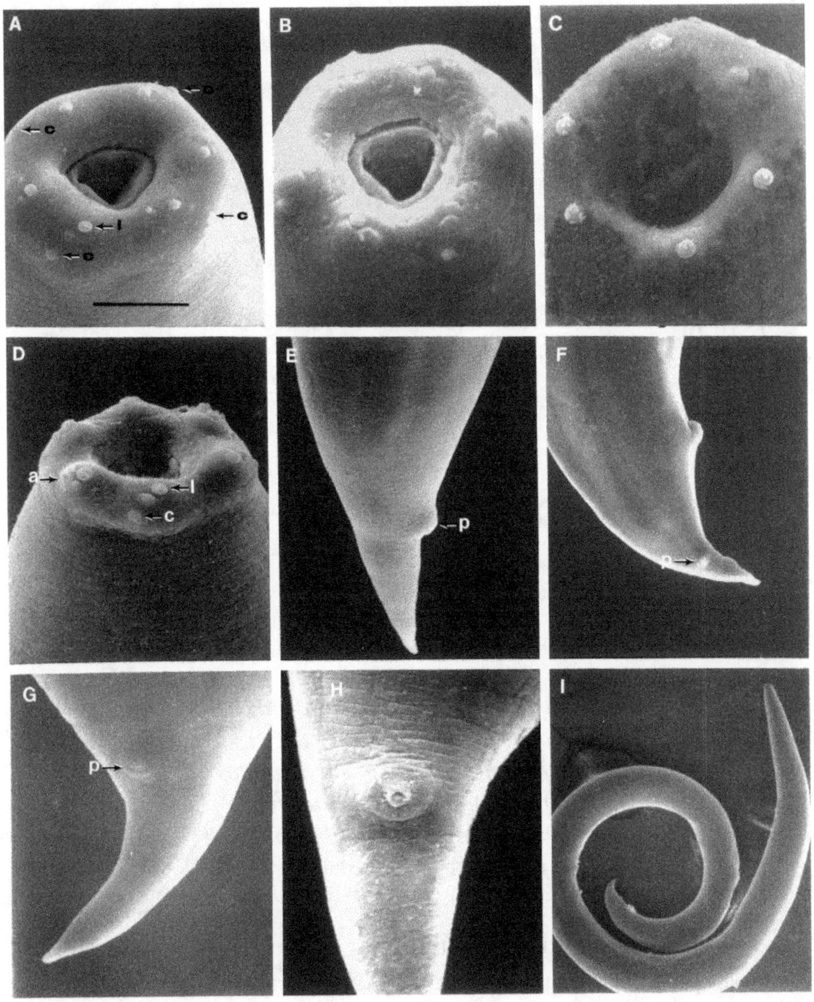

Fig. 203. Neosteinernema longicurvicauda. *SEM photographs of female. A-C: Face view, note several papilla-like structures between labial (l) and cephalic papillae (c) and raised triradiate opening of stoma in A and B; D: Anterior region of young female showing offset swollen lip region, amphidial aperture (a), labial papillae (l) and cephalic papillae (c). Note papilla-like structures near labial papillae; E-G: Variation in tail showing prominent phasmids (p) on protuberance; H: Enlargement of phasmid showing raised, round area with opening at centre; I: Entire body. (Scale bar: A = 12 μm; B, C = 7.5 μm; D = 10 μm; E = 13.6 μm; F = 30 μm; G = 17.6 μm; H = 7.5 μm; I = 475 μm.) After Nguyen and Smart (1994), courtesy of* Journal of Nematology.

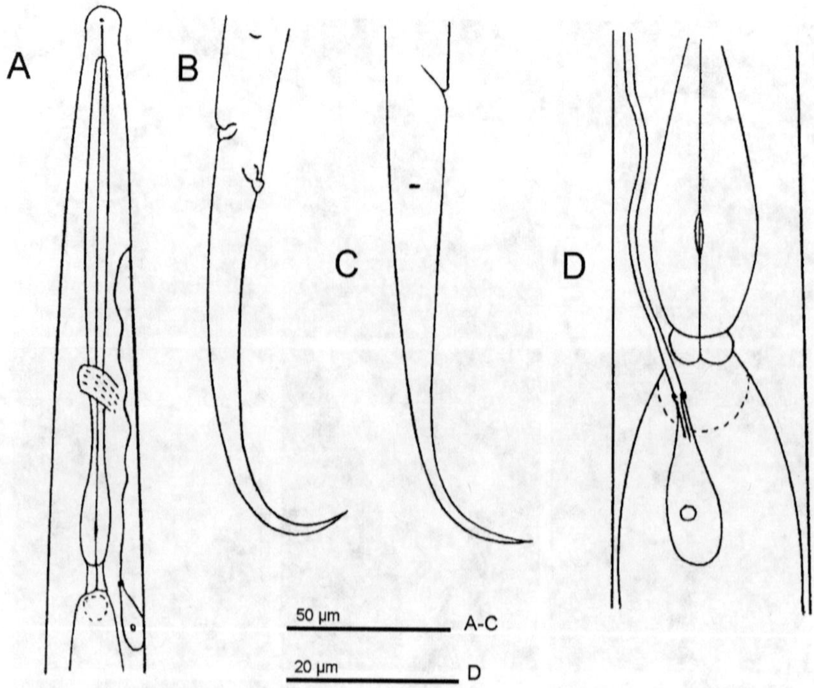

Fig. 204. Neosteinernema longicurvicauda. *Third-stage infective juvenile. A: Anterior region showing long, sinuous excretory duct terminating in excretory cell; B: Tail, ventral view showing two prominent phasmids; C: Posterior region with long, curved tail; D: Enlargement of a portion of pharyngeal region showing sclerotised structure where excretory duct and cell join. After Nguyen and Smart (1994), courtesy of* Journal of Nematology.

nent, slit-like. Left and right phasmids not at same level, usually situated 25-50 μm posterior to anus, occupying three or four lateral bands.

TYPE HOST AND LOCALITY

The nematode was collected in the head of the termite, *Reticulitermes flavipes* (Koller), near Lake Okeechobee, FL, USA.

TYPE MATERIAL

Holotype male, allotype and paratype in glycerin deposited in the USDA Nematode Collection, Beltsville, MD, USA.

RELATIONSHIPS

Males of *N. longicurvicauda* are characterised by the foot-shaped spicule (which bears a dorsal hump), the prominent phasmids, digitate tail terminus, genital papillae comprising a single midventral papilla plus 13-14 pairs of which eight pairs are precloacal, and by the gubernaculum being approximately the same length as the spicule: G/S = 97 (84-108)%. Females can be recognised by the prominent phasmids, a curved tail which is longer than the anal body diam., a spiral body, and by the retention of all eggs from which juveniles hatch and develop to the IJ *in situ*, the latter emerging from the then dead female. Females are usually found outside the termite cadaver. The IJ is characterised by the thin, elongate body (a = 39), the attenuated and ventrally curved tail which is ten times (8-11) as long as the anal body diam. and about the same length as the pharynx, and by the prominent phasmids.

MOLECULAR CHARACTERISATION

No molecular data have been reported for *N. longicurvicauda*.

BIONOMICS

The bionomics and host relationships were reported briefly by Nguyen and Smart (1994). The nematode developed only in the head of the termite. The adults emerged when the tissue of the termite began to decay. The number of adults in each termite was reported as from 1-4 (26% with one nematode, 40% with two, 20% with three and 14% with four). Twenty out of 35 termites studied harboured both males and females, and 15 harboured either males or females. Where nematodes of only one sex were present, no reproduction occurred. When both males and females were present, the females, after emerging from the termite, moved a short distance away, assumed a spiral shape and became immobile. The female retained all the eggs from which juveniles hatch, then undergo two moults to become infective juveniles. When most of the juveniles had reached the infective stage they moved vigorously, breaking through and emerging from the now dead female in the region of the tail or stoma. No first- or second-stage juveniles were observed in, or surrounding, the termite cadavers. During juvenile development, large number of bacterial cells were observed in the female bodies. No second generation was observed.

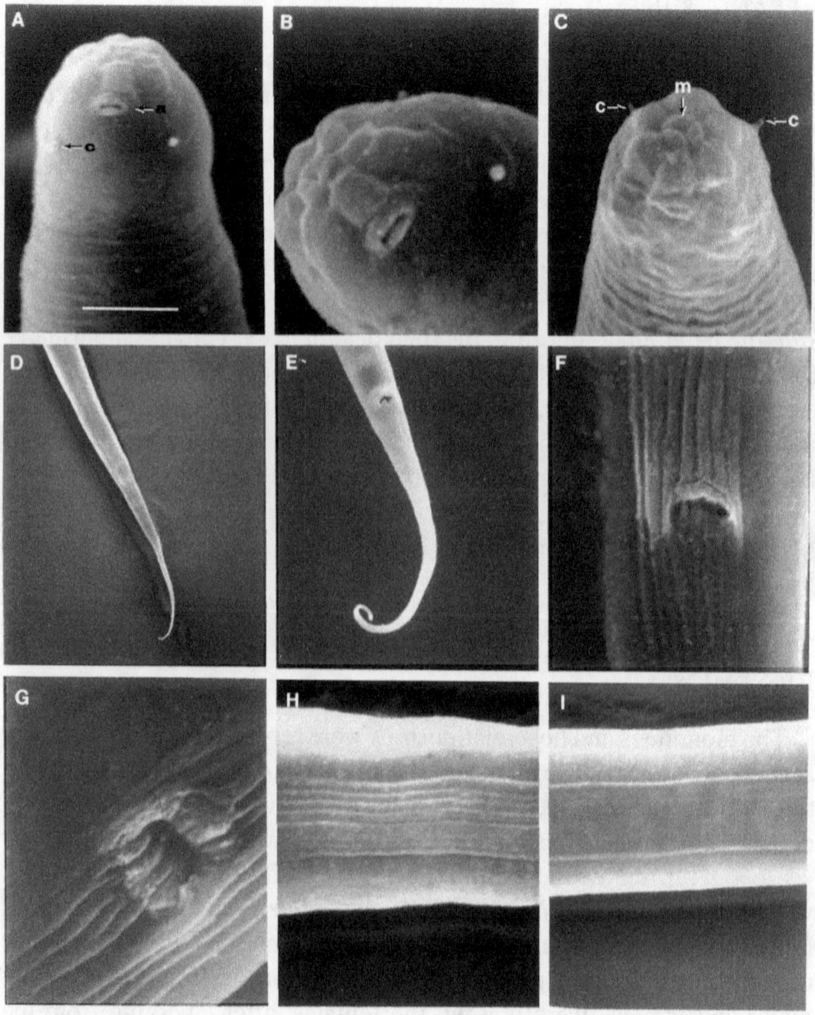

Fig. 205. Neosteinernema longicurvicauda. *SEM photographs of infective juveniles. A: Anterior region showing smooth, swollen head region, amphidial aperture (a), and two of four cephalic papillae (c); B: Enlargement of anterior end showing six lips, oral aperture, prominent amphidial aperture and two of four cephalic papillae; C: Anterior region with sheath showing amphidial aperture, triradiate oral aperture (m), and cephalic papillae (c); D: Posterior region showing filiform, ventrally curved tail; E: Tail showing large phasmid and hook-shaped terminus; F: Lateral field with eight lateral ridges (nine incisures) and large phasmid; G: Lateral field and large phasmid.*

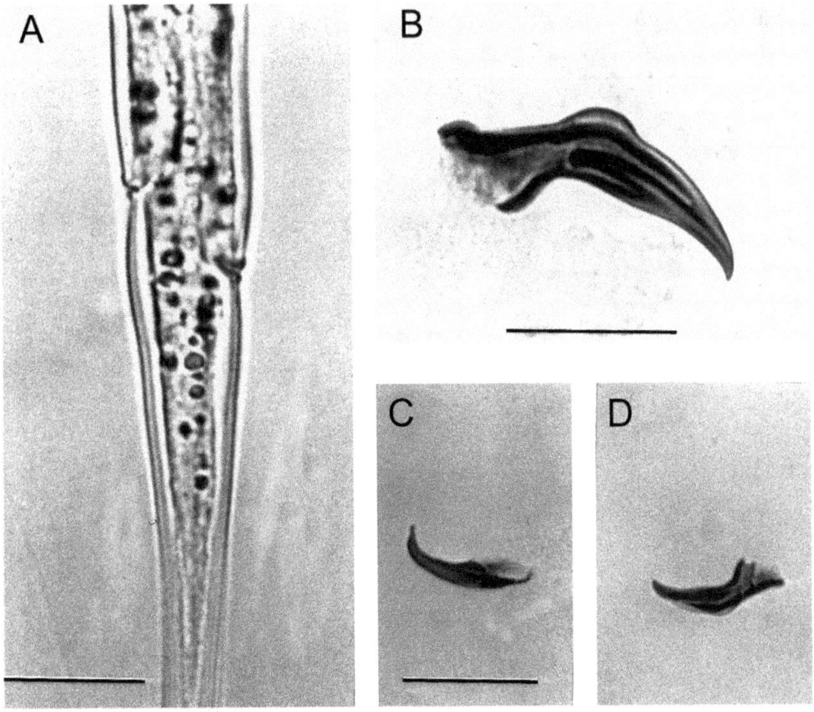

Fig. 206. Neosteinernema longicurvicauda. *Light micrographs. A: Tail of infective juvenile (ventral view) showing two prominent phasmids at different levels; B: Spicule, lateral view, showing foot-shaped form with dorsal hump; C: Gubernaculum (subventral view); D: Spicule, lateral view; note that spicule is shorter than gubernaculum in C (C and D were from the same male). (Scale bars: A = 17 μm; B = 30 μm; C, D = 31 μm.) After Nguyen and Smart (1994), courtesy of* Journal of Nematology.

Biocontrol capability

A patent was granted to Miles Inc., Vero Beach, FL, USA, for the use of the nematode to control termites. So far, no biological control details have been released.

Fig. 205. *(Continued). H: Lateral field with eight ridges; I: Lateral field at mid-body with two border ridges and vague internal ridges. (Scale bar: A = 3 μm; B = 1.8 μm; C = 2.5 μm; D = 43 μm; E = 17 μm; F = 4.3 μm; G = 1.8 μm; H = 5 μm; I = 10 μm.) After Nguyen and Smart (1994), courtesy of* Journal of Nematology.

Table 60. *Morphometrics of* Neosteinernema longicurvicauda. *Measurements are in μm and in the form: mean ± SD (range).*

Character	Holotype (male)	Male	Female	Infective juvenile
n	–	25	25	50
L	1079	1236 ± 228 (854-1713)	3444 ± 784 (2250-4766)	926 ± 60 (789-1048)
a	–	–	–	39 ± 3 (30-46)
b	–	–	–	5.6 ± 0.5 (5-7)
c	–	–	–	5.5 ± 0.3 (4.7-6.5)
c′	–	–	1.4 ± 0.1 (1.1-1.1.7)	10 ± 1 (8-11)
V	–	–	55 ± 3 (49-60)	–
Body diam.	80	97 ± 18 (67-140)	177 ± 37 (122-250)	24 ± 2 (20-31)
EP	82	82 ± 12 (59-103)	112 ± 16 (84-147)	68 ± 4 (61-76)
NR	128	126 ± 15 (103-156)	177 ± 21 (150-225)	107 ± 7 (92-125)
ES	172	187 ± 22 (147-227)	262 ± 28 (219-334)	164 ± 10 (144-188)
Testis reflexion	238	277 ± 55 (178-394)	–	–
Tail length (T)	48	48 ± 6 (36-63)	105 ± 12 (88-134)	167 ± 11 (141-190)
Anal body diam. (ABD)	47	60 ± 10 (41-78)	76 ± 10 (63-103)	17 ± 1 (14-20)
Spicule length (SL)	55	61 ± 4 (52-67)	–	–
Spicule width	12	13 ± 1 (11-15)	–	–
Gubernaculum length (GL)	50	59 ± 4 (52-66)	–	–
Gubernaculum width	9	9 ± 1 (8-13)	–	–
D% = EP/ES × 100	48	44 ± 6 (30-54)	–	41 ± 2 (38-46)
E% = EP/T × 100	–	–	–	41 ± 3 (37-48)
SW% = SL/ABD × 100	117	103 ± 18 (80-150)	–	–
GS% = GL/SL × 100	91	97 ± 6 (84-108)	–	–

EP = distance from anterior end to excretory pore; NR = distance from anterior end to nerve ring; ES = distance from anterior end to end of pharynx.

DISTRIBUTION

Neosteinernema longicurvicauda has been reported only from Florida, USA (Nguyen & Smart, 1994).

Reference

NGUYEN, K.B. & SMART JR, G.C. (1994). *Neosteinernema longicurvicauda* n. gen. n. sp. (Rhabditida: Steinernematidae), a parasite of the termite *Reticulitermes flavipes* (Koller). *Journal of Nematology* 26, 162-174.

Nematology Monographs & Perspectives, 2007, Vol. 5, 611-692

Chapter 5

Heterorhabditidae: species descriptions

Khuong B. NGUYEN [1] and David J. HUNT [2]

[1] *Entomology and Nematology Department, Institute of Food and Agricultural Sciences, University of Florida, Gainesville, FL 32611-0620, USA*

[2] *CABI Europe-UK, Bakeham Lane, Egham, Surrey TW20 9TY, UK*

Descriptions and illustrations of all species regarded as valid herein are presented in alphabetical order. Morphological and morphometric data are based largely on the original description (and hence differ to some extent depending on the detail originally provided) but have been expanded, corrected and/or updated where necessary. Illustrations have been taken from the original description wherever possible. In some cases, illustrations have had to be re-drawn. The author responsible for producing each of the species data sheets presented below is indicated in the footnotes. Other data provided include the etymology of the specific epithet, type locality, type host (when known), deposition of types, an up-to-date geographical distribution for each species, accession codes for DNA sequences, a brief account of the bionomics, *etc.*, of each species, plus relevant major references. Descriptions of *species inquirendae* (see Chapter 2 for a list) are not provided.

In the genus *Heterorhabditis*, there are two broad clades: the *indica*-group (*H. amazonensis*, *H. baujardi*, *H. floridensis*, *H. indica*, *H. mexicana* and *H. taysearae*) and the *megidis*-group (*H. bacteriophora*, *H. downesi*, *H. marelatus*, *H. megidis* and *H. zealandica*).

The genus *Heterorhabditis* Poinar, 1976

Heterorhabditis amazonensis[*,**] Andaló, Nguyen & Moino, 2007
(Figs 207-212)

MEASUREMENTS

See Table 61.

DESCRIPTION

Male

Body ventrally curved when heat-relaxed. Cephalic extremity truncate, sometimes slightly swollen. Face view with six labial papillae, cephalic papillae not observed. Amphidial apertures prominent. Stoma with sclerotised rod- or barrel-shaped cheilorhabdions; other rhabdions not distinguishable, forming a funnel-shaped structure posterior to cheilorhabdions. Pharynx with cylindrical corpus, metacorpus sometimes slightly enlarged. Nerve ring surrounding isthmus just anterior to basal bulb. Basal bulb with reduced valve. Cardia present, protruding into intestine. Excretory pore usually located posterior to basal bulb. Genital system monorchic, reflexed; distance from end of basal bulb to testis flexure variable. *Vas deferens* well developed. Spicules paired, separate, rostrum absent, ventral side flattened. Gubernaculum strongly curved ventrally in 20% of the population, and slightly curved ventrally in 80%. Bursa peloderan. Bursal papillae numbering 7-8 pairs. Moving from anterior to posterior, pair 1 located well anterior with tips extending beyond bursal rim; pairs 2 and 3 forming group anterior to cloacal aperture, also extending beyond bursal rim. Pairs 4, 5 and 6 forming a group posterior to cloacal aperture with pair 4 usually not curved outward (laterally). Number of genital papillae from pairs 1-6 (from anterior end) is unchanged and typical for *Heterorhabditis* spp. (Nguyen *et al.*, 2004). Pairs 7 and 8 forming a terminal group at end of bursa. Number and distribution of papillae in terminal group variable. Observation of papillae in terminal group of 24 bursas gave the following results: 50% (12/24) with two pairs of papillae (Fig. 209A); 25% (6/24) with two papillae on

[*] Named after the state of *Amazonas*, Brazil.
[**] Description after Andaló *et al.* (2007). Revised by Khuong B. Nguyen.

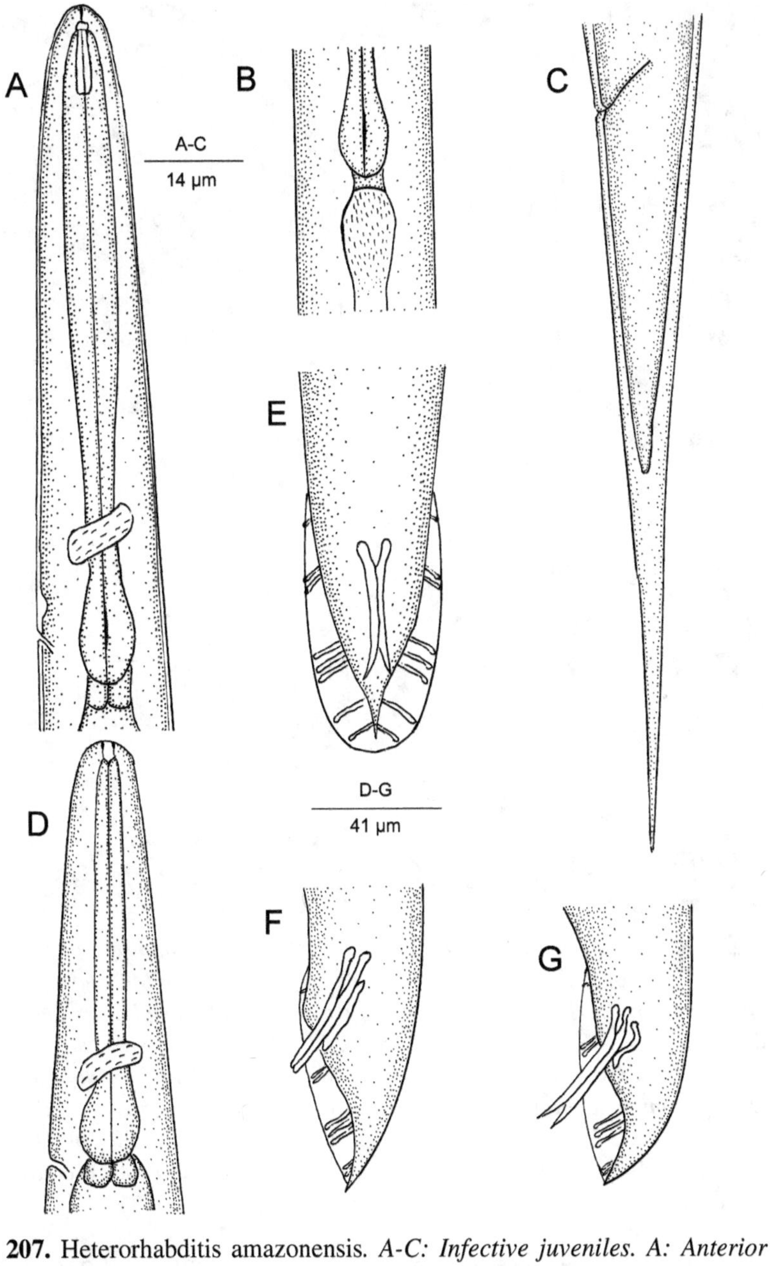

Fig. 207. Heterorhabditis amazonensis. *A-C: Infective juveniles. A: Anterior region; B: Basal bulb region; C: Posterior region. D-G: Males. D: Anterior region; E: Posterior region, ventral view; F, G: Variation of tail and gubernacula. After Andaló* et al. *(2007), courtesy of* Nematology.

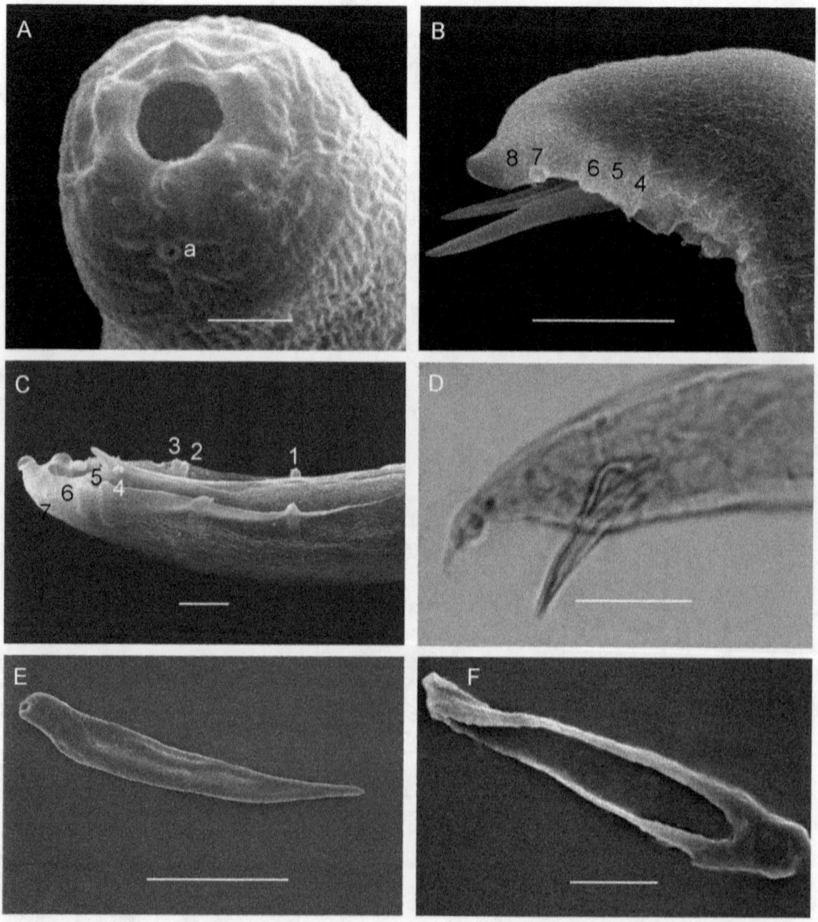

Fig. 208. Heterorhabditis amazonensis. *A-C, E, F: SEM photographs of male. A: Head region showing mouth opening, six labial papillae and amphidial aperture (a); B: Posterior region showing protruding spicules and bursa with papillae 4-8; C: Posterior region showing bursa with papillae 1-7, note that papilla 4 does not curve laterally; E: Spicule; F: Gubernaculum. D: Light microscope photograph of posterior region showing a well curved gubernaculum. (Scale bars: A = 2.4 μm; B = 16.8 μm; C = 10 μm; D = 25.8 μm; E = 16.8 μm; F = 4.3 μm.) After Andaló* et al. *(2007), courtesy of* Nematology.

one side and single papilla on other side (Fig. 209B) and 25% (6/24) with one pair of papillae (Fig. 209C). Tail conoid, slightly curved ventrally.

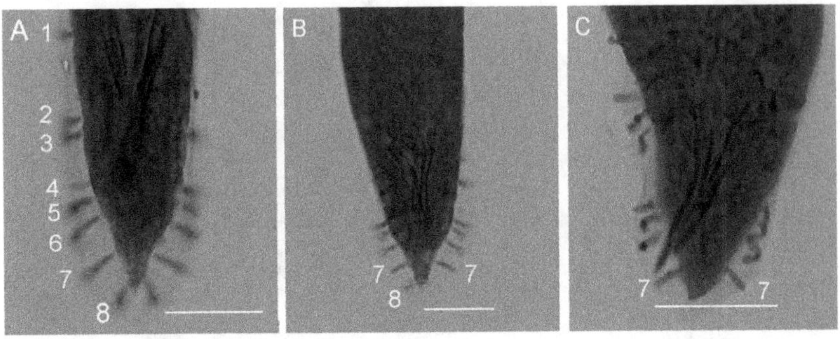

Fig. 209. Heterorhabditis amazonensis. *Male tails with bursal papillae. A: Bursa with two pairs of bursal papillae in terminal group; B: Bursa with two pairs of bursal papillae on one side and one papilla on other side in terminal group; C: Bursa with one pair of bursal papillae in terminal group. (Scale bars: A = 28.7 μm; B = 27 μm; C = 31.3 μm.) After Andaló* et al. *(2007), courtesy of* Nematology.

Hermaphroditic female

Body C-shaped when heat-relaxed, robust, always with many eggs in young worms and many eggs and juveniles in mature worms. Cuticle smooth under light microscope but finely annulated with SEM. Lateral field and phasmid not observed. Cephalic extremity tapering anteriorly. Labial region with six prominent lips, each bearing a terminal labial papilla. Cephalic papillae not observed with SEM. In face view, oral aperture hexagonal in shape. Amphidial apertures pore-like. Stoma with refractile cheilorhabdions. Posterior part of stoma funnel-shaped, enclosed by anterior part of pharynx. Pharynx with cylindrical corpus. Isthmus present. Nerve ring situated at middle of isthmus. Basal bulb prominent with inconspicuous valve, but lumen of pharynx in basal bulb well sclerotised. Cardia present. Genital system didelphic, amphidelphic. Vulva median, in form of a transverse slit, located on a slightly protuberant area. In superficial ventral view, vulval region smooth, elliptical with unique surrounding pattern. Pattern appearing as a rough area around vulva and radiating from aperture with many wart-like structures close to aperture, pattern present on all hermaphroditic females and different from that of closely related species. Vagina short. Tail longer than anal body diam., conoid, with pointed terminus. Tail sometimes widening near end then narrowing to a pointed tip. Postanal

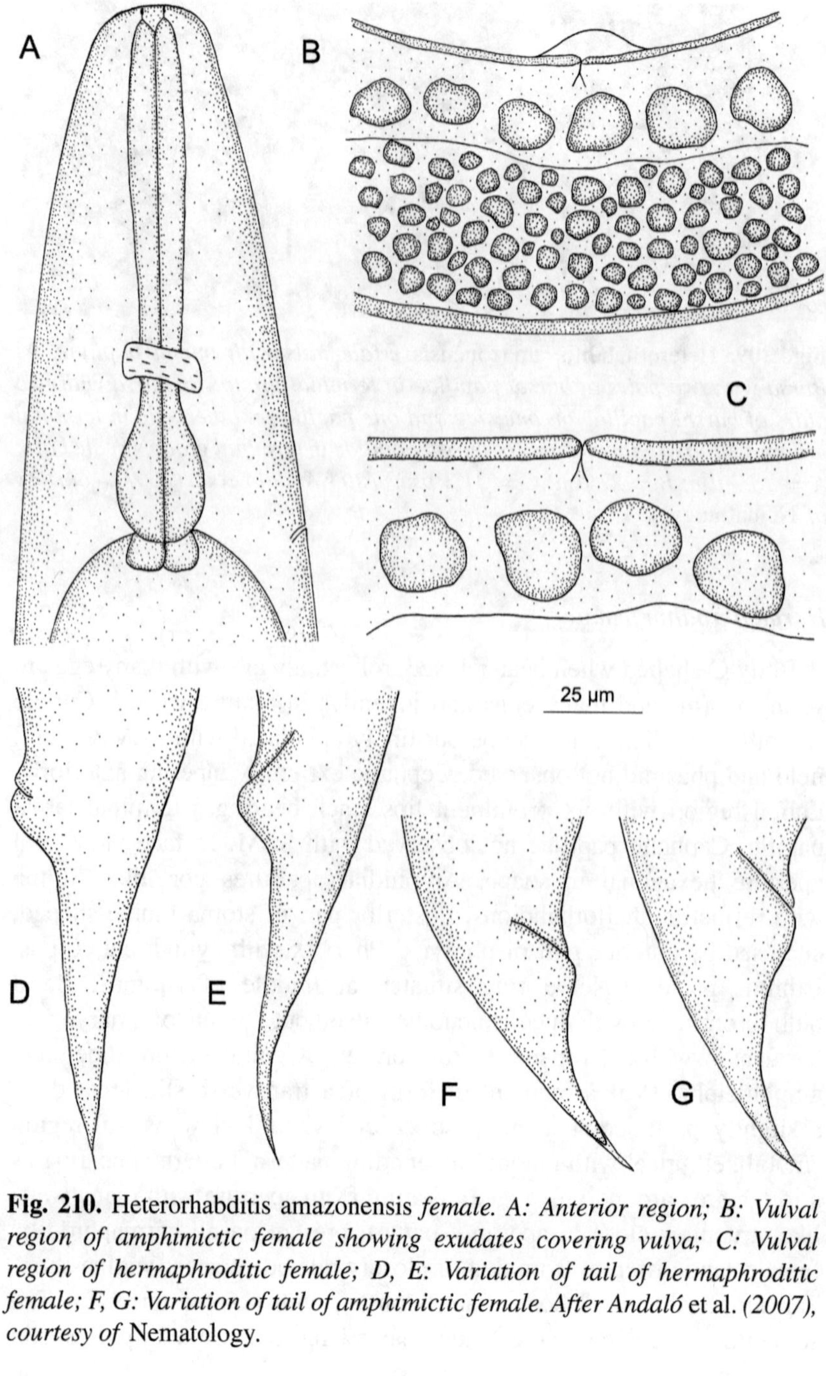

Fig. 210. Heterorhabditis amazonensis *female. A: Anterior region; B: Vulval region of amphimictic female showing exudates covering vulva; C: Vulval region of hermaphroditic female; D, E: Variation of tail of hermaphroditic female; F, G: Variation of tail of amphimictic female. After Andaló* et al. *(2007), courtesy of* Nematology.

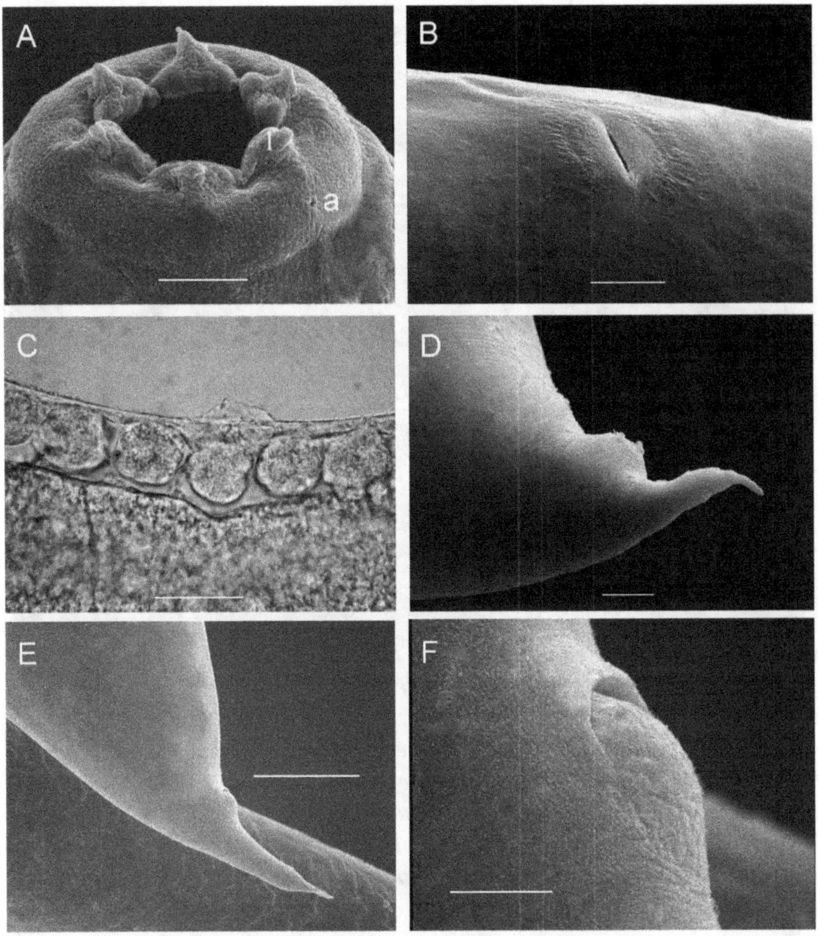

Fig. 211. Heterorhabditis amazonensis. *A, B, D: SEM photographs of hermaphroditic females. A: Anterior region showing six prominent lips (l), six labial papillae, and amphidial aperture (a); B: Vulva showing elliptical pattern; D: Tail showing postanal swelling with two papilla-like structures. C: Light microscope photograph of vulva of amphimictic female covered with exudates or copulation plug after mating; E: Another form of tail of hermaphroditic female; F: Anus of hermaphroditic female. (Scale bars: A = 8.6 μm; B = 20 μm; C = 23 μm; D = 10 μm; E = 93 μm; F = 4.9 μm.) After Andaló* et al. *(2007), courtesy of* Nematology.

swelling present, sometimes (20% of specimens) with two papilla-like structures. Phasmid inconspicuous.

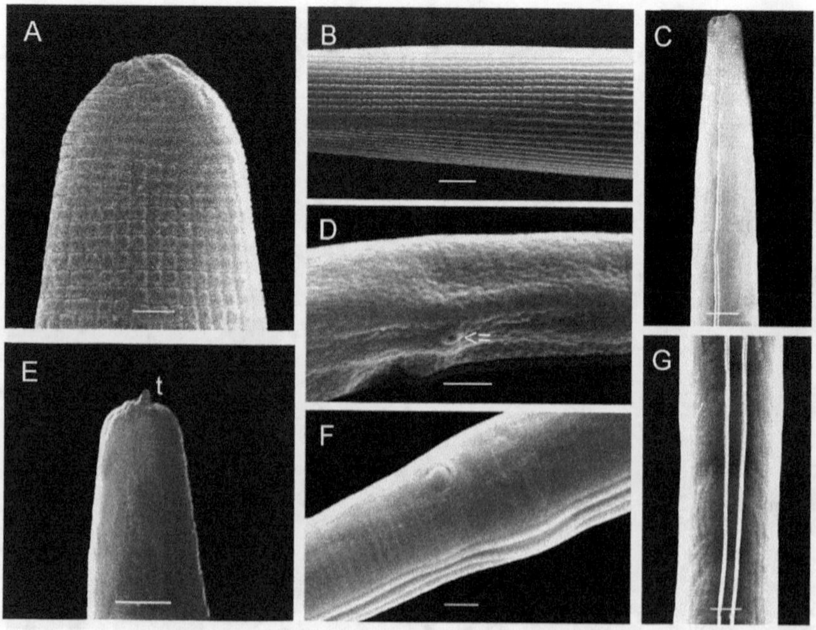

Fig. 212. Heterorhabditis amazonensis. *SEM photographs of infective juveniles. A: Anterior region of ensheathed infective juvenile (IJ) showing head with seven annules and body with tessellate pattern; B: Portion of body showing tessellate pattern and longitudinal ridges; C: Anterior region of exsheathed IJ showing lateral field in anterior region with one ridge, then changing to two ridges; D: Phasmid near mid-tail of infective juvenile (arrow); E: Anterior region of exsheathed IJ showing dorsal tooth (t); F: Lateral field and anus in posterior region; G: Lateral field at mid-body with two ridges. (Scale bars: A, B, C, E, G = 4 μm; D, F = 2 μm.) After Andaló* et al. *(2007), courtesy of* Nematology.

Amphimictic female

Amphimictic female similar to hermaphrodite but smaller. Vulva rarely protuberant, usually (80% of specimens) covered with exudates or copulation plug after mating. Postanal swelling much smaller.

Infective juvenile

Body elongate. Sheath (= second-stage cuticle) present immediately after harvesting, but many infective juveniles (IJ) losing sheath in storage. Labial region bearing seven annules but no longitudinal incisures. Anterior part of body (*i.e.*, *ca* 3-5 body diam. from anterior end) with tessellate pattern. Posterior part with longitudinal ridges. Exsheathed IJ

Table 61. *Morphometrics of* Heterorhabditis amazonensis. *Measurements are in μm and in the form: mean ± SD (range).*

Character	Holotype	Male	Hermaphrodite	Female	Infective juvenile
n	–	20	20	20	20
L	843	752 ± 43 (692-826)	4734 ± 664 (3517-5587)	1580 ± 180 (1279-2070)	589 ± 12 (567-612)
a	–	–	–	–	26 ± 1.3 (24-29)
b	–	–	–	–	4.9 ± 0.3 (4.4-5.5)
c	–	–	–	–	5.5 ± 0.2 (5.1-6.1)
V	–	–	44 ± 1.5 (42-47)	48 ± 1 (46-50)	–
Greatest body diam.	45	41 ± 2.3 (36-43)	282 ± 25 (220-316)	100 ± 13 (70-122)	23 ± 1.2 (20-24)
EP	109	109 ± 6 (96-116)	211 ± 14 (184-238)	114 ± 7.8 (103-126)	107 ± 6.1 (89-115)
NR	85	79 ± 5 (71-88)	149 ± 12 (128-171)	85 ± 9 (68-100)	85 ± 4.9 (76-93)
ES	109	105 ± 5 (97-114)	206 ± 15 (180-225)	130 ± 7.1 (119-142)	121 ± 6.6 (107-132)
Testis reflexion	75	59 ± 10 (42-85)	–	–	–
Tail length with sheath (T)	–	–	–	–	107 ± 4.7 (98-115)
Tail length without sheath	30.6	33 ± 2.7 (29-41)	128 ± 14 (104-154)	74 ± 12 (58-91)	69 ± 4.4 (59-74)
Anal body diam. (ABD)	23	27 ± 2.6 (23-33)	70 ± 6.5 (59-83)	30 ± 3.7 (25-38)	14 ± 1.4 (13-17)
Spicule length (SP)	40.6	41 ± 2.9 (35-45)	–	–	–
Gubernaculum length (GL)	20	21 ± 1.5 (19-23)	–	–	–
D% = EP/ES × 100	100	103 ± 3.7 (95-109)	–	–	88 ± 2.7 (83-92)
E% = EP/T × 100	–	–	–	–	100 ± 6.0 (89-109)
SW% = SP/ABD × 100	177	152 ± 20 (120-187)	–	–	–
GS% = GL/SP × 100	49	51 ± 3.2 (44-56)	–	–	–
Hyaline/Tail × 100	–	–	–	–	64.7 ± 5.3 (52-74)

EP = distance from anterior end to excretory pore; NR = distance from anterior end to nerve ring; ES = distance from anterior end to end of pharynx.

body annulated, lacking longitudinal ridges. Labial region with dorsal tooth, labial and cephalic papillae absent. Amphidial apertures prominent. Excretory pore located posterior to nerve ring but just anterior to base of pharynx. Hemizonid, when observed, located two or three annules anterior to excretory pore. Pharynx typical for genus. Lateral field beginning anteriorly with single ridge, but a short distance posteriorly (2-3 body diam.), a second ridge appearing, giving two ridges in all. Two-ridge lateral field pattern unchanged for remainder of body. Phasmids prominent, located in lateral field near mid-tail, pore-like. Tail elongate conoid with pointed terminus. Tail length without sheath equivalent to *ca* 65% of tail length plus sheath.

TYPE HOST AND LOCALITY

Natural host not known. Nematodes were collected by baiting with *Galleria mellonella* larvae from a forest soil in the northern part of the state of Amazonas, near the city of Benjamin Constant, Brazil.

TYPE MATERIAL

Holotype (male) and paratypes (hermaphrodites, females, male and infective juveniles in 3% formalin) isolated from haemocoel of *G. mellonella* deposited in the USDA Nematode Collection, Beltsville, MD, USA. Slide of one male and one female of the second generation, and several infective juveniles deposited in the University of California-Davis Nematode Collection, Davis, CA, USA. Several hermaphrodites, females, male, and all stages in 3% formalin deposited in the Entomology and Nematology Department, University of Florida, Gainesville, FL, USA.

RELATIONSHIPS

Heterorhabditis amazonensis is characterised by a combination of morphological and morphometric characters of the male, female and infective juvenile. For males, the number of papillae in the terminal group of bursae is variable either with two pairs of papillae (50%, Fig. 209A), with two papillae on one side and single papilla on the other side (25%, Fig. 209B), or with one pair of papillae (25%, Fig. 209C). Twenty percent of the population has a well-curved gubernaculum. The percentage of the gubernaculum to spicule length (GS%) is 51, similar to *H. baujardi* but lower than *H. mexicana* (50 *vs* 56), and the length of the spicule relative to anal body diam. (SW%) is 152, lower than that of *H. mexicana* (167) and *H. baujardi* (182). The D% value of the three nematodes is also different. Females possess a typical vulva pattern (Fig. 211B). This pattern is different from other closely related species as in Figure 5 of Nguyen *et al.* (2007). The infective juvenile has a body length of 589 (567-612) μm, EP = 107 (89-115) μm, ES = 121 (107-132) μm, tail length = 107 (98-115) μm, and a = 26 (24-29).

Heterorhabditis amazonensis is morphologically similar to *H. floridensis*, *H. baujardi*, *H. mexicana* and *H. indica*, but can be distinguished from these species mainly by male and infective juvenile characters. *Heterorhabditis amazonensis* can be distinguished from *H. bacteriophora*, *H. indica*, *H. megidis*, *H. zealandica* and *H. marelatus* by the body length, pharynx length of the IJ, D%, GS% and SW% of males and can be further characterised by molecular characteristics of the ITS regions of ribosomal DNA (see below).

MOLECULAR CHARACTERISATION

DNA analysis

The ITS rDNA regions, flanked by primers 18S and 26S, of *H. amazonensis* are characterised by the sequence lengths (1010 bp, ITS1 = 395 bp, ITS2 = 211 bp), and nucleotide usage composition, A = 0.25941, C = 0.20198, G = 0.25545, T = 0.28317 (accession no. DQ665222). Compared to sequences of nine other species, in the genus the sequence length of *H. amazonensis* is longer than that of six, the same length as that of *H. mexicana* and shorter than that of *H. bacteriophora* (1021 bp) and *H. floridensis* (1012 bp). The ITS1 sequence length of *H. amazonensis* (395 bp) is longer than that of all other species by 1-25 bp; it is only 1 bp longer than that of *H. mexicana* (ITS1 = 394 bp). The sequence length of the ITS2 region of this species, at 211 bp, is the same as that of *H. baujardi* and *H. marelatus* but shorter than that of all other species (Andaló *et al.*, 2007). The reconstructed nucleotide character transformations show that *H. amazonensis* differs from its closest taxon, *H. baujardi*, at 15 aligned positions, three of which are unambiguous, polarised, autapomorphies. It is most divergent from *H. zealandica* (184 aligned positions).

Phylogenetic relationships

The maximum parsimony (MP) tree includes two monophyletic groups that are well supported (⩾90%) by bootstrap analysis. *Heterorhabditis amazonensis*, *H. baujardi*, *H. floridensis*, *H. indica* and *H. mexicana* comprise a monophyletic group (100% support) – the *indica*-group – in which *H. indica* is a sister taxon to the group formed by the other four species, whilst *H. amazonensis* is a sister taxon to a group formed by *H. floridensis*, *H. baujardi* and *H. mexicana*. All other species, *viz.*, *H. bacteriophora*, *H. downesi*, *H. marelatus*, *H. megidis* and *H. zealandica*, constitute another monophyletic assemblage – the *megidis*-group. Maximum likelihood analysis of ITS regions generated a tree topology identical to that of the MP tree, but branch support was weak for some nodes, especially in the *megidis*-group (Andaló *et al.*, 2007).

DISTRIBUTION

The species has been found in several localities in Brazil, namely Lavras, Piaui and Amazonas (Andaló *et al.*, 2007).

References

ANDALÓ, V., NGUYEN, K.B. & MOINO, A. (2007). *Heterorhabditis amazonensis* n. sp. (Rhabditida: Heterorhabditidae) from Amazonas, Brazil. *Nematology* 8 (2006), 853-867.

NGUYEN, K.B., SHAPIRO-ILAN, D.I., STUART, R.J., MCCOY, C.W., JAMES, R.R. & ADAMS, B.J. (2004). *Heterorhabditis mexicana* n. sp. (Rhabditida: Heterorhabditidae) from Tamaulipas, Mexico, and morphological studies of the bursa of *Heterorhabditis* spp. *Nematology* 6, 231-244.

NGUYEN, K.B., GOZEL, U., KOPPENHÖFER, H.S. & ADAMS, B.J. (2006). *Heterorhabditis floridensis* n. sp. (Rhabditida: Heterorhabditidae) from Florida. *Zootaxa* 1177, 1-19.

***Heterorhabditis bacteriophora**,** Poinar, 1976**

= *Chromonema heliothidis* Khan, Brooks & Hirschmann, 1976
= *H. heliothidis* (Khan, Brooks & Hirschmann, 1976) Poinar, Thomas & Hess, 1977
= *H. argentinensis* Stock, 1993
(Figs 213-215)

MEASUREMENTS

See Table 62.

DESCRIPTION

Male

Cephalic extremity truncate or slightly rounded. Labial region bearing six distinct protruding lips surrounding oral aperture, each with a terminal labial papilla. Cephalic papillae not observed. Amphidial apertures located posterior to labial papillae. Under light microscope, cheilorhabdions visible as lightly refractive areas lining anterior (non-collapsed) portion of stoma; metarhabdions immediately posterior to cheilorhabdions attached to fused pro- and mesorhabdions, each metarhabdial segment bearing a small tooth; telorhabdions absent. Basal bulb often surrounded by anterior portion of intestine. Nerve ring surrounding isthmus just anterior to basal bulb. Excretory pore usually located posterior to the basal bulb. Genital system monorchic, reflexed, leading into a seminal vesicle containing sperm cells; *vas deferens* well developed. Spicules paired and separate; shape of capitulum variable, from pointed to flat. Gubernaculum with proximal portion curving ventrally between spicules. Bursa peloderan, open, normally supported by nine pairs of papillae: a small anterior pair, two pairs adjacent to spicules and six pairs distal to cloacal aperture. Latter six pairs arranged in two sets of three; rarely, terminal group consisting of only two or even four papillae. More frequently, one or two of the three may be branched. Papillae pairs 4, 7, 8 and 9 not reaching bursal rim.

* The species epithet means 'bacteria bearing' and was based on the fact that the nematode is always associated with a bacterium.

** Description after Poinar (1976). Revised by Khuong B. Nguyen.

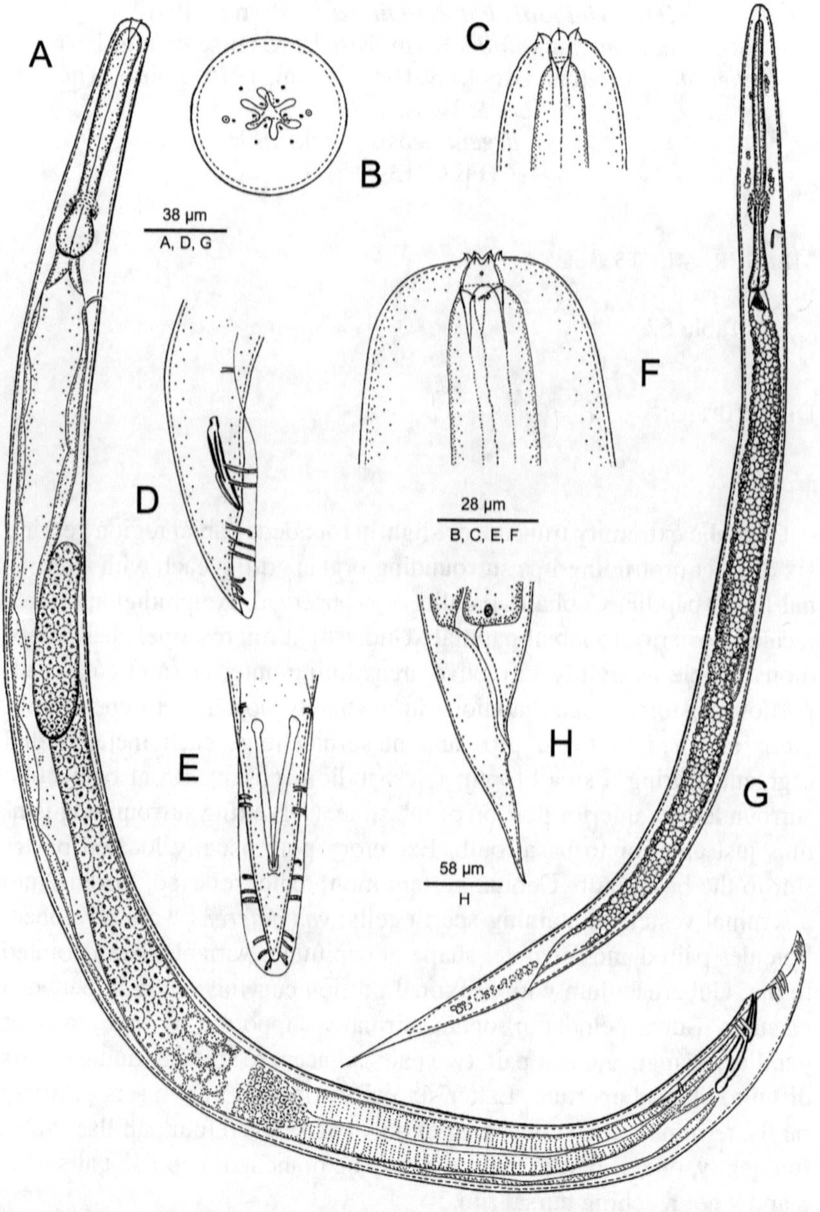

Fig. 213. Heterorhabditis bacteriophora. *A, D, E: Male. A: Entire worm; D: Tail region, lateral view; E: Tail region, ventral view. B, C, H: Amphimictic female. B: Face view; C: Head region; H: Tail region. F: Head of hermaphroditic female; G: Infective juvenile. After Poinar (1975), courtesy of* Nematologica.

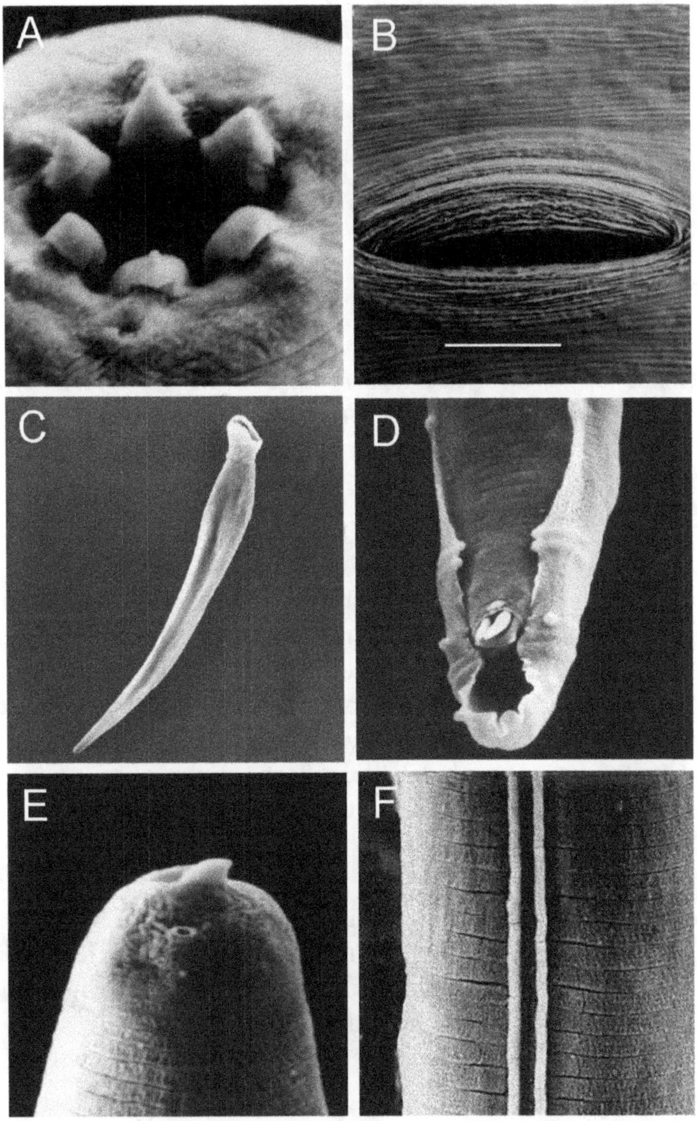

Fig. 214. Heterorhabditis bacteriophora. *SEM photographs. A: Face view of hermaphroditic female showing lip, labial papillae and amphidial aperture; B: Vulva of hermaphroditic female showing elliptical ring around vulva; C: Spicule, lateral view; D: Male tail with bursa, ventral view; E: Infective juvenile head showing dorsal tooth, and amphidial aperture; F: Lateral field with two ridges. (Scale bar: A = 5 μm; B = 8.6 μm; C, D = 15 μm; E = 3.8 μm; F = 7.5 μm.)*

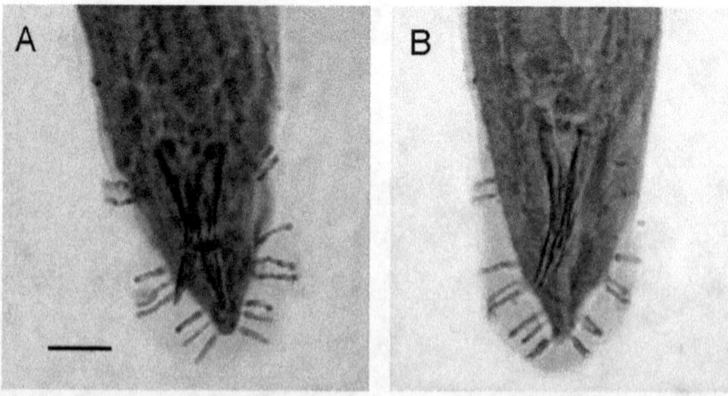

Fig. 215. Heterorhabditis bacteriophora. *A, B: Light micrographs of bursa showing terminal group of bursal papillae with three equal ribs. (Scale bar = 20 μm.)*

Hermaphroditic female

Anterior region similar to that of male but much larger. In face view, oral aperture hexagonal in shape, lips very prominent, pointed anteriorly. Labial papillae prominent. Cephalic papillae not observed. Basal bulb often surrounded by anterior portion of intestine. Excretory pore usually located posterior to basal bulb. Intestine with few large cells. Lateral fields not seen, phasmids inconspicuous. Vulva located near mid-body. Genital system amphidelphic, reflexed. Vulva open, protuberant, functional for oviposition. Tail pointed.

Amphimictic female

Similar to hermaphrodite but differing in head and vulva region: head subconical, labial region often faintly set off from remainder of head; vulva non-functional for oviposition, often covered with hardened deposit. Anal region slightly protruding.

Infective juvenile

Anterior part of body of ensheathed juvenile with tessellate pattern. Posterior part with longitudinal ridges; second-stage cuticle lost soon after juvenile exits host cadaver. Exsheathed juvenile with transverse striae; lateral field with two longitudinal ridges; anterior end with dorsal tooth; labial and cephalic papillae absent; amphidial apertures prominent. Excretory pore located posterior to nerve ring but anterior to base of pharynx. Oral aperture and anus closed. Pharynx typical for

Table 62. *Morphometrics of* Heterorhabditis bacteriophora. *Measurements are in μm and in the form: mean (range).*

Character	Male	Hermaphrodite	Female	Infective juvenile
n	15	15	15	15
L	820 (780-960)	4030 (3630-4390)	3500 (3180-3850)	588 (512-671)
a	–	–	–	25 (17-30)
b	–	–	–	4.5 (4.0-5.1)
c	–	–	–	6.2 (5.5-7.0)
V	–	44 (41-47)	47 (42-53)	–
Greatest body diam.	43 (38-46)	165 (160-180)	190 (160-220)	23 (18-31)
EP	121 (114-130)	209 (189-217)	192 (174-214)	103 (87-110)
NR	72 (65-81)	126 (121-130)	103 (93-118)	85 (72-93)
ES	103 (99-105)	197 (189-205)	168 (155-183)	125 (100-139)
Testis reflexion	79 (59-87)	–	–	–
Tail length with sheath (T)	–	–	–	98 (83-112)
Tail length without sheath	28	90 (81-93)	82 (71-93)	–
Anal body diam. (ABD)	23 (22-25)	46 (40-53)	28 (22-31)	–
Spicule length (SP)	40 (36-44)	–	–	–
Gubernaculum length (GL)	20 (18-25)	–	–	–
D% = EP/ES × 100	117	–	–	84 (76-92)
E% = EP/T × 100	–	–	–	112 (103-130)
SW% = SP/ABD × 100	174	–	–	–
GS% = GL/SP × 100	50	–	–	–

EP = distance from anterior end to excretory pore; NR = distance from anterior end to nerve ring; ES = distance from anterior end to end of pharynx.
After Poinar (1975, 1990).

genus. Intestine collapsed; tail pointed; cells of a rod-shaped bacterium visible in ventricular portion of intestine.

TYPE HOST AND LOCALITY

Found in the body cavity of *Heliothis punctigera* Hall (Noctuidae: Lepidoptera) collected from Brecon, South Australia.

TYPE MATERIAL

Holotype (hermaphroditic female), allotype (male) and paratype (amphimictic female) deposited in the University of California-Davis Nematode Collection, Davis, CA, USA.

RELATIONSHIPS

Heterorhabditis bacteriophora is characterised by the body length (588 μm) and E% (112) of the infective juvenile, spicule length and GS% of the male and vulval pattern of the female. It can be distinguished from other species of *Heterorhabditis* by a combination of the above diagnostic characters and is especially distinctive using molecular methodologies.

MOLECULAR CHARACTERISATION

The ITS rDNA regions, flanked by primers 18S and 26S, of *H. bacteriophora* are characterised by the sequence length, 1021 base pairs (bp), ITS1 = 389 bp, ITS2 = 228 bp and nucleotide usage composition, A = 0.26347, C = 0.19980, G = 0.25661, T = 0.28012 (accession no. AY321477; Nguyen *et al.*, 2004).

BIONOMICS

Infective juveniles of this species are capable of invading and killing larvae of *Galleria mellonella* in 48 h. Dissections during this period revealed the nematodes in the midgut of the host from where they probably enter the body cavity. Entry through the spiracles is also a possibility. Within 3-4 days after infection, the large hermaphroditic females produce offspring. Two days later, these develop into males and females of the bisexual generation. After mating, the juveniles hatch inside the females and develop to the second-, or even infective, stage before emerging from the mother nematode. The infective stage juveniles carry cells of a specific bacterium, *Photorhabdus luminescens* subsp. *luminescens*, contained in a bacterial pouch in the anterior part of the intestine. These cells are liberated when the juveniles are placed in hanging drops of insect blood and can easily be cultured on standard media. Examination of freshly killed insects shows that the bacterium is present during nematode development. The bacteria turn the freshly killed insect larvae a reddish colour. The host is undoubtedly killed by the bacteria associated with the nematodes.

Biocontrol capability

This species has been commercialised by many companies with the aim of controlling various types of insect pest.

DISTRIBUTION

Heterorhabditis bacteriophora has been found in many European countries (Azores, France, Germany, Hungary, Italy, Moldavia, Spain, Switzerland), North America (USA), South America (Argentina, Brazil, Colombia), Caribbean (Dominican Republic, Guadeloupe, Puerto Rico, Trinidad), Asia (China, Israel, Korea), Africa (Ethiopia, Kenya, South Africa) and from Australia and New Zealand (Poinar, 1990; Hominick, 2002). Several isolates have been identified using DNA sequences from the USA, Ethiopia, Kenya and South Africa (Nguyen, unpubl.).

References

HOMINICK, W.M. (2002). Biogeography. In: Gaugler, R. (Ed.). *Entomopathogenic nematology*. Wallingford, UK, CABI Publishing, pp. 115-143.

KHAN, A., BROOKS, W.M. & HIRSCHMANN, H. (1976). *Chromonema heliothidis* n. gen., n. sp. (Steinernematidae, Nematoda), a parasite of *Heliothis zeae* (Noctuidae, Lepidoptera), and other insects. *Journal of Nematology* 8, 159-168.

NGUYEN, K.B., SHAPIRO-ILAN, D.I., STUART, R.J., MCCOY, C.W., JAMES, R.J. & ADAMS, B.J. (2004). *Heterorhabditis mexicana* n. sp. (Rhabditida: Heterorhabditidae) from Tamaulipas, Mexico, and morphological studies of the bursa of *Heterorhabditis* spp. *Nematology* 6, 231-244.

POINAR JR, G.O. (1976). Description and biology of a new insect parasitic rhabditoid, *Heterorhabditis bacteriophora* n. gen. n. sp. (Rhabditida; Heterorhabditidae n. fam.). *Nematologica* 21, 463-470.

POINAR JR, G.O. (1990). Taxonomy and biology of Steinernematidae and Heterorhabditidae. In: Gaugler, R. & Kaya, H.K. (Eds). *Entomopathogenic nematodes in biological control*. Boca Raton, FL, USA, CRC Press, pp. 23-60.

POINAR JR, G.O., THOMAS, G.M. & HESS, R. (1977). Characteristics of the specific bacterium associated with *Heterorhabditis bacteriophora* (Heterorhabditidae: Rhabditida). *Nematologica* 23, 97.

STOCK, S.P. (1993). A new species of the genus *Heterorhabditis* Poinar, 1976 (Nematoda: Heterorhabditidae) parasitizing *Graphognathus* sp. larvae (Coleoptera: Curculionidae) from Argentina. *Research and Reviews in Parasitology* 53, 103-107.

Heterorhabditis baujardi *,** Phan, Subbotin, Nguyen & Moens, 2003
(Fig. 216)

MEASUREMENTS

See Table 63.

DESCRIPTION

Male

Body slender, curved ventrally or C-shaped when heat-relaxed. Cuticle appearing smooth under light microscope. Cephalic extremity truncate or slightly rounded. Amphidial apertures inconspicuous. Stoma funnel- or cup-shaped, shallow. Pharynx muscular; procorpus and metacorpus cylindrical; isthmus indistinct; basal bulb pyriform, valve distinct. Nerve ring surrounding isthmus or located just anterior to basal bulb. Cardia prominent and protruding into intestinal lumen. Genital system monorchic, reflexed. Spicules paired, slightly ventrally curved. Gubernaculum *ca* 50% of spicule in length, proximal end lacking knob, ventrally curved. Bursa peloderan with nine normal pairs of papillae arranged from anterior to posterior as follows: three pairs located anterior to cloacal aperture (two pairs very close together and anteriormost pair well anterior) and six caudal pairs with anteriormost approaching level of cloacal aperture. Tail conoid, lacking mucron. Phasmids inconspicuous.

Hermaphroditic female

Body spiral or C-shaped when heat-relaxed. Cuticle smooth. Cephalic extremity truncate or slightly rounded. Labial region with six distinct lips surrounding oral aperture, each bearing a labial papilla. Amphidial apertures inconspicuous. Stoma funnel- or cup-shaped, shallow. Pharynx with cylindrical procorpus and metacorpus; indistinct isthmus; pyriform basal bulb with distinct valve. Cardia prominent, protruding into intestinal lumen. Genital system didelphic, amphidelphic, reflexed. Vulva slightly protuberant, in form of transverse slit located near middle of body. Vagina short, straight, muscular. Tail longer than anal body diam. Phasmids inconspicuous.

* Named in memoriam of the late Dr Pierre *Baujard*.

** Description after Phan *et al.* (2003). Revised by David J. Hunt.

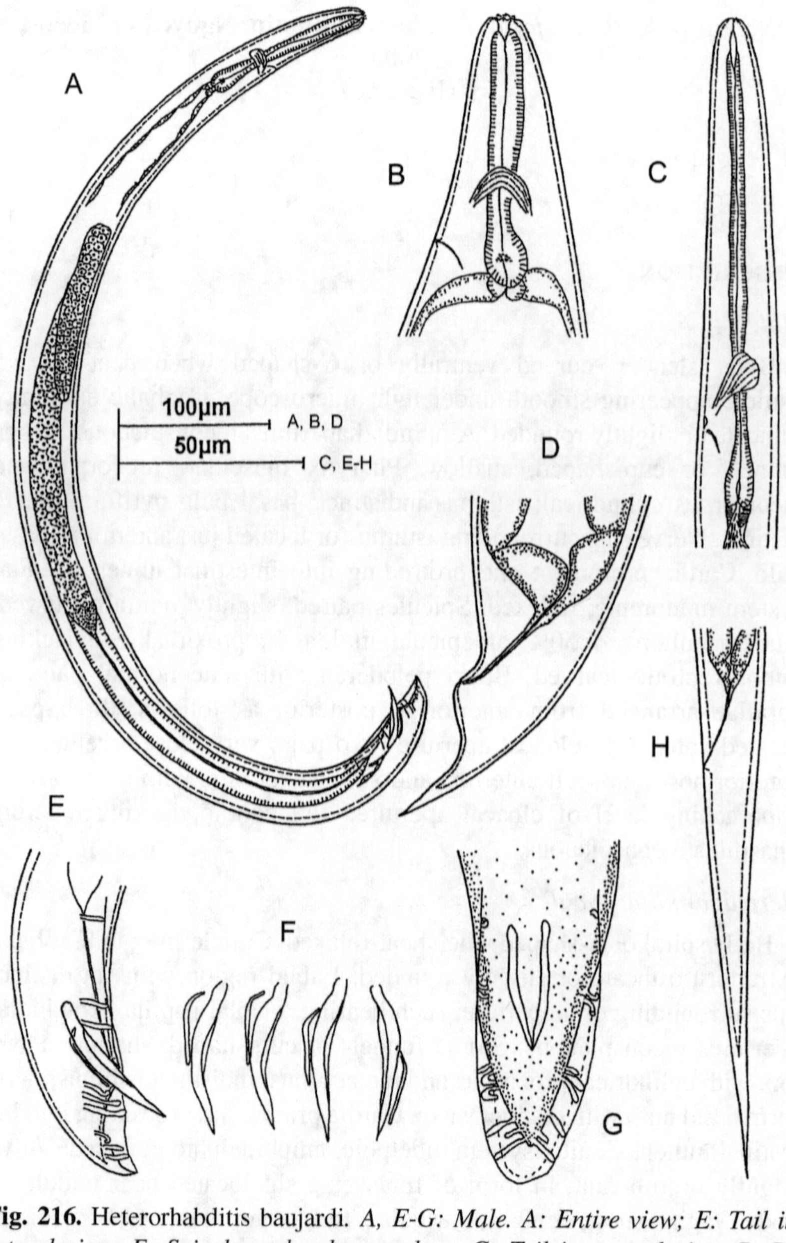

Fig. 216. Heterorhabditis baujardi. *A, E-G: Male. A: Entire view; E: Tail in lateral view; F: Spicule and gubernaculum; G: Tail in ventral view. B, D: Hermaphroditic female. B: Pharyngeal region; D: Tail in lateral view. C, H: Infective juvenile. C: Pharyngeal region; H: Tail in lateral view. After Phan* et al. *(2003) courtesy of* Nematology.

Table 63. *Morphometrics of type population of* Heterorhabditis baujardi. *Measurements are in μm and in the form: mean ± SD (range).*

Character	Holotype (male)	Male	Hermaphrodite	Female	Infective juvenile
n	–	14	15	15	25
L	840	889 ± 45 (818-970)	3571 ± 290 (3135-4170)	1647 ± 205 (1335-2130)	551 ± 27 (497-595)
a	18	18 ± 1 (16-22)	17 ± 1 (15-19)	14 ± 1 (12-16)	28 ± 1 (26-30)
b	7.8	7.7 ± 0.7 (6.4-8.8)	18 ± 1 (16-21)	11 ± 1 (10-12)	4.8 ± 0.2 (4.5-5.1)
c	27	27.2 ± 2 (24-33)	42 ± 4 (36-50)	21 ± 3 (19-32)	6 ± 0.3 (6-6.7)
V	–	–	45 ± 1 (43-48)	48 ± 1 (46-51)	–
Greatest body diam.	46	49 ± 2 (45-53)	208 ± 20 (18-240)	116 ± 17 (90-150)	20 ± 2 (18-22)
EP	85	81 ± 7 (71-93)	173 ± 12 (156-192)	121 ± 14 (104-149)	97 ± 3 (91-103)
NR	64	65 ± 7 (54-77)	136 ± 8 (119-147)	97 ± 13 (75-122)	81 ± 3 (75-86)
ES	107	116 ± 10 (105-132)	196 ± 7 (186-206)	149 ± 16 (131-185)	115 ± 3 (107-120)
Testis reflexion	94	91 ± 13 (63-106)	–	–	–
Tail length (IJ with sheath)	31	33 ± 3 (28-38)	86 ± 11 (66-114)	78 ± 7 (68-89)	90 ± 4 (83-97)
Anal body diam. (ABD)	21	22 ± 1 (20-24)	55 ± 5 (47-63)	32 ± 4 (27-41)	13 ± 0.7 (11-14)
Spicule length (SP)	42	40 ± 3 (33-45)	–	–	–
Gubernaculum length (GL)	18	20 ± 1.5 (18-22)	–	–	–
D% = EP/ES × 100	–	–	–	–	84 ± 3 (78-88)
E% = EP/T × 100	–	–	–	–	108 ± 4 (98-114)
SW% = SP/ABD × 100	203	182 ± 18 (138-208)	–	–	–
GS% = GL/SP × 100	44	50 ± 5 (44-61)	–	–	–

EP = distance from anterior end to excretory pore; NR = distance from anterior end to nerve ring; ES = distance from anterior end to end of pharynx.
After the original authors.

Amphimictic female

Similar to hermaphroditic female but much smaller and tail blunt.

Infective juvenile

Body long, slender, often enclosed in longitudinally ridged cuticle of second-stage juvenile. Cephalic region rounded, bearing large dorsal tooth. Oral aperture and anus closed. Pharynx long, narrow; isthmus distinct, surrounded by nerve ring, basal bulb elongate, with valve. Cardia prominent. Hemizonid distinct, located at isthmus level. Excretory pore located just posterior to hemizonid. Anterior intestinal region forming pouch filled with rod-like symbiotic bacteria. Bacteria also occurring in lumen of intestine. Phasmids inconspicuous.

TYPE HOST AND LOCALITY

Type host unknown as the type isolate was collected by baiting soil collected in Sanhon, Sathay, Kontum forest, Vietnam (longitude 107°43′E, latitude 14°28′N, altitude 850 m a.s.l.).

TYPE MATERIAL

Holotype male and paratype males deposited at Ghent University, Institute for Zoology, K.L. Ledeganckstraat 35, 9000, Belgium. Slides with paratype males, hermaphroditic females, amphimictic females and infective juveniles deposited at the Department of Nematology, Institute of Ecology and Biological Resources, National Centre for Science and Technology, Hoang Quoc Viet 18, Nghiado, Caugiay, Hanoi, Vietnam, and University of California Riverside, Riverside, CA, USA.

RELATIONSHIPS

Morphometrically and morphologically, *H. baujardi* resembles *H. indica*. It can be distinguished from this species by the shape of the gubernaculum (proximal end lacking knob, ventrally curved *vs* straight, narrow, not reflexed at tip) and nine pairs of normally developed genital papillae *vs* either seven pairs of normal papillae, or seven plus two pairs of atrophied, highly modified papillae (see Poinar *et al.*, 1992; Anis *et al.*, 2000).

The gubernaculum of *H. baujardi* resembles that of *H. bacteriophora*. However, *H. baujardi* has a slightly shorter infective juvenile body length of 563 (487-622) *vs* 588 (512-670) μm, slightly longer spicules of 40 (31-47) *vs* 36 (31-41) μm, slightly longer gubernaculum of 20 (14-

26) *vs* 18 (14-21) μm, and a higher male SW ratio of 1.94 (1.38-2.61) *vs* 1.28 (1.20-1.34).

Heterorhabditis baujardi differs from *H. brevicaudis* by a wider anal body diam. (13 *vs* 9 μm), a lower E ratio (1.1 *vs* 1.6) in the infective juvenile; and in the male by the shorter spicules (40 *vs* 47 μm), higher GS ratio (0.51 *vs* 0.4), and lower SW ratio (1.94 *vs* 2.3).

Canonical discriminant analysis (CDA)

According to Phan *et al.* (2003) CDA using 12 variables for infective juveniles and ten variables for males failed to discriminate their three *H. baujardi* isolates from *H. indica*, although it was easily distinguished from *H. downesi*, *H. marelatus*, *H. megidis* and *H. bacteriophora*. *Heterorhabditis baujardi* was only marginally different from *H. bacteriophora* by infective juvenile characters but was clearly distinguished by male characters.

Molecular characterisation

The polymerase chain reaction amplified a single DNA product of *ca* 850 bp from each sample of the three isolates studied. The ITS1 region of *Heterorhabditis* species ranges from 378 (*H. downesi*) to 406 bp (*H. baujardi*). The ITS1 region (accession no. AF548768) exhibited the following base composition: A = 0.25725, C = 0.19672, G = 0.25094, T = 0.29508.

According to Phan *et al.* (2003), the length of the alignment with nine *Heterorhabditis* sequences generated by Clustal X was 427 bp. The ITS1 sequence of *H. baujardi* differs from that of other species by 51 substitutions in *H. hawaiiensis* (which is now regarded as a junior synonym of *H. indica*) to 99 substitutions in *H. zealandica*. The pairwise divergence between taxa ranged from 0.5-26%. MP analysis of this alignment revealed 427 characters, of which 93 were parsimony informative. In an unrooted MP tree *H. baujardi* clustered with *H. indica* + *H. hawaiiensis* with high bootstrap value and was supported by 29 autapomorphies (unique substitutions).

The length of the *Heterorhabditis* sequence alignment with outgroup taxa was 491 bp. MP analysis of this alignment indicated that of the 492 characters, 141 were parsimony informative. The phylogenetic trees obtained from MP (tree length = 557) and ML (DNA model = GTR + G, − ln likelihood = 2687.38) had a similar topology. *Heterorhabditis baujardi* clustered with *H. indica* + *H. hawaiiensis* (clade of sub-tropical and tropical species) with high to moderate bootstrap support. The position of *H. marelatus* was not well resolved. The KHTest did not reveal significant difference (P = 0.4148) between the topology when *H. zealandica* clustered with the *H. megidis* + *H. downesi* clade, which required a two-steps-longer tree than the maximum parsimonious one (see Phan *et al.*, 2003).

Bionomics

Crossbreeding tests were restricted to one isolate of *H. baujardi* (H.TN40) and the morphologically and morphometrically closely related *H. indica* (LN2). Fertile progeny were not produced when the species were crossed with each other, but were obtained when each species was self-crossed.

The bacterial associate of this species is unknown.

Distribution

In addition to the type locality, this species was also isolated in Bong, Cucphuong, Ninhbinh (105°36′E, 20°20′N, 500 m a.s.l.), and Babe, Backan (105°34′E, 22°26′N, 700 m a.s.l.), Vietnam (Phan *et al.*, 2003).

References

PHAN, K.L., SUBBOTIN, S.A., NGUYEN, N.C. & MOENS, M. (2003). *Heterorhabditis baujardi* sp. n. (Rhabditida: Heterorhabditidae) from Vietnam and morphometric data for *H. indica* populations. *Nematology* 5, 367-382.

POINAR JR, G.O., KARUNAKAR, G.K. & DAVID, H. (1992). *Heterorhabditis indicus* n. sp. (Rhabditida, Nematoda) from India: separation of *Heterorhabditis* spp. by infective juveniles. *Fundamental and Applied Nematology* 15, 467-472.

Heterorhabditis downesi[*,**] Stock, Griffin & Burnell, 2002

= *Heterorhabditis* sp. Irish type
= *Heterorhabditis downesi* of Hass, Downes & Griffin, 2001 (*nomen nudum*)
(Fig. 217)

MEASUREMENTS

See Table 64.

DESCRIPTION

Male

Cephalic extremity truncate or slightly rounded. Six distinct, protruding, pointed lips surrounding oral aperture. Each lip bearing single labial papilla. Amphidial apertures inconspicuous, pore-like. Stoma short, broad. Cheilostom short with refractile rhabdions; gymnostom and stegostom reduced. Pharynx typical for genus. Corpus cylindrical; metacorpus not differentiated. Isthmus short. Basal bulb pyriform with reduced valve. Nerve ring located near mid-point of isthmus. Excretory pore located at level of basal bulb. Genital system monorchic, reflexed. Spicules paired, symmetrical, straight or slightly arcuate, with pointed tips. Manubrium (head) square, well set off from lamina. Calomus distinct. Lamina with single internal rib. Velum absent. Gubernaculum thin, *ca* half length of spicules. Bursa open, peloderan. Nine pairs of bursal rays (papillae) arranged as follows: 1 2 () 3 3. Pairs 4 and 7 curved dorsally, pair 8 not reaching rim of bursa.

Hermaphroditic female

Ovotestis present. Vulva located near middle of body. Vulval lips protruding, anterior lip larger than posterior. Postanal swelling well developed. Tail terminus blunt, mucronate.

Amphimictic female

Amphididelphic, reflexed. Vulva situated near middle of body. Vulval lips slightly protruding. Tail conoid, narrowing towards terminus, ending in refractile, spike-like structure. Postanal swelling slightly developed.

---

[*] Named in honour of Prof. Martin J. *Downes*.

[**] Description after Stock *et al.* (2002). Revised by David J. Hunt.

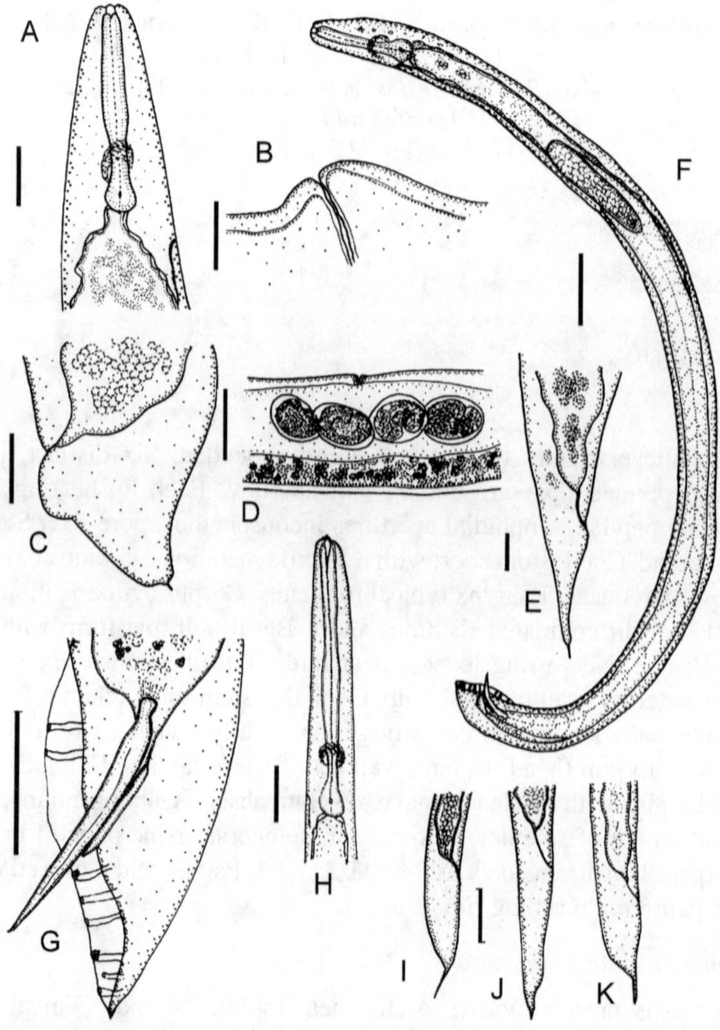

Fig. 217. Heterorhabditis downesi. *A-C: Hermaphrodite. A: Anterior end; B: Vulva; C: Tail, lateral view. D, E: Female. D: Vulva; E: Tail, lateral view. F, G: Male. F: Entire worm; G: Tail, lateral view. H-K: Infective juvenile. H: Anterior end; I-K: Tail. (Scale bars: A, D-F = 50 μm; B, C, G, H = 25 μm; I-K = 18 μm.) Reproduced from* Systematic Parasitology, *51 (2002), p. 98; Morphological characterisation of three isolates of* Heterorhabditis *Poinar, 1976 from the 'Irish group' (Nematoda: Rhabditida: Heterorhabditidae) and additional evidence supporting their recognition as a distinct species,* H. downesi *n. sp., Stock, S.P., Griffin, C.T. and Burnell, A.M., Figure 1. With kind permission of Springer Science and Business Media.*

Table 64. *Morphometrics of type population (K122) of* Heterorhabditis downesi. *Measurements are in μm and in the form: mean ± SD (range).*

Character	Holotype (male)	Male	Hermaphrodite	Female	Infective juvenile
n	–	19	19	20	20
L	877	800 ± 76 (669-876)	4362 ± 266 (3050-5051)	1747 ± 387 (1231-2728)	637 ± 32 (588-692)
a	–	–	–	–	35 ± 4 (29-42)
b	–	–	–	–	4.7 ± 0.3 (4.4-5.3)
c	–	–	–	–	9.5 ± 0.5 (8.5-10.5)
V	–	–	52 ± 0.5 (50-55)	51 ± 4 (47-60)	–
Greatest body diam.	33	36 ± 3 (33-40)	254 ± 84 (183-291)	102 ± 11 (74-131)	18 ± 2 (15-22)
EP	90	89 ± 2 (86-91)	235 ± 19 (200-254)	112 ± 7 (99-126)	115 ± 8 (96-128)
NR	62.5	70 ± 7 (62-78)	200 ± 19 (175-230)	131 ± 11 (117-151)	101 ± 3 (96-105)
ES	100	101 ± 3 (97-106)	240 ± 11 (230-244)	131 ± 13 (111-155)	135 ± 4 (126-141)
Testis reflexion	131	150 ± 22 (108-181)	–	–	–
Tail length without sheath	32	32 ± 2 (29-34)	65 ± 3 (60-70)	89 ± 18 (70-122)	68 ± 4 (62-74)
Anal body diam. (ABD)	23	24 ± 2 (21-28)	62 ± 2 (57-65)	33 ± 3.5 (25-38)	12 ± 1 (9-14)
Spicule length (SP)	42	43 ± 2 (41-47)	–	–	–
Gubernaculum length (GL)	19	18 ± 1 (17-19.5)	–	–	–
D% = EP/ES × 100	90	–	–	–	85 ± 5 (76-98)
E% = EP/T × 100	281	–	–	–	170 ± 10 (160-180)
SW% = SP/ABD × 100	180	180 ± 20 (170-220)	–	–	–
GS% = GL/SP × 100	45	43 ± 4 (36-47)	–	–	–

EP = distance from anterior end to excretory pore; NR = distance from anterior end to nerve ring; ES = distance from anterior end to end of pharynx.
After the original authors.

Infective juvenile

Body slender, gradually tapering posteriorly, ensheathed in cuticle of second-stage juvenile (J2). Cuticular sheath bearing longitudinal ridges over most of body length, tessellate pattern present in anterior region of body. Cephalic region bearing prominent cuticular dorsal tooth. Amphidial apertures small, pore-like. Lateral field marked by two ridges. Nerve ring located around middle of isthmus. Excretory pore located posterior to basal bulb. Tail short, conoid, tapering to small, spike-like, tip. Morphology of tail tip variable, with most characteristic shape (80% of examined specimens from three studied isolates) shown in Figure 217I, J.

TYPE HOST AND LOCALITY

Natural host unknown as the species was recovered by the soil baiting technique from grassland at Curracloe, County Wexford, Ireland. The type isolate, K-122, was recovered by Griffin *et al.* (1994).

TYPE MATERIAL

Holotype, paratype males, females and infective juveniles at the University of California-Davis Nematode Collection, Davis, CA, USA. Additional paratype material deposited at CABI Europe-UK Nematode Collection, Egham, UK.

RELATIONSHIPS

Heterorhabditis downesi can be distinguished from other described species by having a blunt, mucronate tail in the hermaphrodite, a more anteriorly located excretory pore in the male (both of which are highly distinctive characters) and by the morphology of the infective juvenile tail, which is pointed and tapers to a short, spike-like, terminus. The spicules of *H. downesi* resemble those of *H. marelatus* and *H. megidis* in having a square-shaped manubrium set off from the lamina by a thin calomus, but can be separated from these two species by the higher SW% value of 180 (170-220) *vs* 175 (173-178) and 140 (135-145), respectively.

MOLECULAR CHARACTERISATION

Joyce *et al.* (1994) showed that the so-called 'Irish group' of *Heterorhabditis* differed in RFLP profile from other described species of the genus. The 'Irish group' was subsequently described as *H. downesi* by Stock *et al.* (2002). The sequence of ITS regions was deposited at GenBank under accession no. AY321482.

NOTE

Hass *et al.* (2001) first published the combination '*Heterorhabditis downesi*' on 27 December 2001. They were presumably under the impression that the formal description of the species (and hence the binomen) would have been formally published before their paper, although this turned out not to be the case. As Hass *et al.* (2001) did not formally describe or provide any other means of diagnosing the species, their citation of the name should be regarded as a *nomen nudum*. The species name was validly proposed by Stock *et al.* (2002) in February 2002.

BIONOMICS

Hass *et al.* (2001) investigated the correlation between survival in water and persistence of infectivity in soil for a number of *Heterorhabditis* isolates, including *H. downesi*. They found that *H. downesi* tended to be less persistent than *H. megidis*. In a subsequent paper, Hass *et al.* (2002) reported upon the role of lipid reserves in persistence.

Rolston *et al.* (2005) investigated the distribution of *H. downesi* and other entomopathogenic nematodes in Irish sand dunes using *Galleria mellonella* larvae as bait insect. They found that *H. downesi* was most prevalent in the front 20 m of the dune system, particularly along sandy paths made by human traffic, and that two colour forms of the bait insects occurred, one purple and one yellow. *Heterorhabditis downesi* occurred in 1.9 and 3.8% of samples in 2001 and 2002, respectively.

Lola Luz *et al.* (2005) investigated the control potential of *H. downesi* against *Otiorhynchus sulcatus*, the black vine weevil.

DISTRIBUTION

In addition to the type locality in Ireland, the species has also been recorded from other localities in Ireland and from Britain, Denmark and Hungary (see Hominick *et al.*, 1995; Griffin *et al.*, 1999).

References

GRIFFIN, C.T., JOYCE, S.A., DIX, I., BURNELL, A.M. & DOWNES, M.J. (1994). Characterisation of the entomopathogenic nematode *Heterorhabditis* (Nematoda: Heterorhabditidae) from Ireland and Britain by molecular and cross-breeding techniques, and the occurrence of the genus in these islands. *Fundamental and Applied Nematology* 17, 245-253.

GRIFFIN, C.T., DIX, I., JOYCE, S.A., BURNELL, A.M. & DOWNES, M.J. (1999). Isolation and characterisation of *Heterorhabditis* spp. (Nematoda: Heterorhabditidae) from Hungary, Estonia and Denmark. *Nematology* 1, 321-332.

HASS, B., DOWNES, M.J. & GRIFFIN, C.T. (2001). Correlation between survival in water and persistence of infectivity in soil of *Heterorhabditis* spp. isolates. *Nematology* 3, 573-579.

HASS, B., DOWNES, M.J. & GRIFFIN, C.T. (2002). Persistence of four *Heterorhabditis* spp. isolates in soil: role of lipid reserves. *Journal of Nematology* 34, 151-158.

HOMINICK, W.M., REID, A.P. & BRISCOE, B.R. (1995). Prevalence and habitat specificity of steinernematid and heterorhabditid nematodes isolated during soil surveys of the UK and the Netherlands. *Journal of Helminthology* 69, 27-32.

JOYCE, S.A., GRIFFIN, C.T. & BURNELL, A.M. (1994). Characterisation of *Heterorhabditis* isolates by PCR amplification of segments of mtDNA and rDNA genes. *Journal of Nematology* 26, 260-270.

LOLA LUZ, T., DOWNES, M. & DUNNE, R. (2005). Control of black vine weevil larvae *Otiorhynchus sulcatus* (Fabricius) (Coleoptera: Curculionidae) in grow bags outdoors with nematodes. *Agricultural and Forest Entomology* 7, 121-126.

ROLSTON, A.N., GRIFFIN, C.T. & DOWNES, M.J. (2005). Distribution of entomopathogenic nematodes in an Irish sand dune system. *Nematology* 7, 259-266.

STOCK, S.P., GRIFFIN, C.T. & BURNELL, A.M. (2002). Morphological characterisation of three isolates of *Heterorhabditis* Poinar, 1976 from the 'Irish group' (Nematoda: Rhabditida: Heterorhabditidae) and additional evidence supporting their recognition as a distinct species, *H. downesi* n. sp. *Systematic Parasitology* 51, 95-106.

Heterorhabditis floridensis[*,**] Nguyen, Gozel, Koppenhöfer & Adams, 2006
(Figs 218-223)

MEASUREMENTS

See Table 65.

DESCRIPTION

Hermaphroditic female

Body C-shaped, robust, always with many eggs in young worms and many eggs and juveniles in mature worms. Cuticle smooth under light microscope but finely annulated with SEM. Labial region with six prominent lips apparently protruding into stoma, each lip with a terminal labial papilla. Cephalic papillae not observed with SEM. In face view, oral aperture hexagonal in shape. Amphidial apertures present, pore-like. Stoma with refractile cheilorhabdions. Posterior part of stoma funnel-shaped, enclosed by anterior part of pharynx. Pharynx with cylindrical corpus. Isthmus very short or indistinguishable. Nerve ring located near anterior margin of basal bulb. Basal bulb prominent with inconspicuous valve, but lumen of pharynx in basal bulb well sclerotised. In young hermaphrodite, anterior end of intestine with thick wall. Intestinal cells prominent, especially between basal bulb and flexure of anterior ovary, and from rectum to flexure of posterior ovary. Intestinal lumen well defined in these two regions. Cardia present. Genital system didelphic, amphidelphic. Vulva in form of a transverse slit, located on a slightly protuberant area, anterior to mid-body (V = 44-49%). In ventral view, vulva elliptical encircled by arch-shaped annules anteriorly and posteriorly. Vagina short. Tail longer than anal body diam., conoid with pointed terminus. Sometimes, near terminus, tail widening then narrowing to a pointed tip. Postanal swelling present. Phasmid inconspicuous.

Amphimictic female

Similar to hermaphroditic female but smaller. Vulva rarely protruding, usually (67%) covered with exudates or copulation plug after mating. Postanal swelling much smaller.

---

[*] Named after *Florida*, USA, where the species was isolated.

[**] Description after Nguyen *et al.* (2006). Revised by Khuong B. Nguyen.

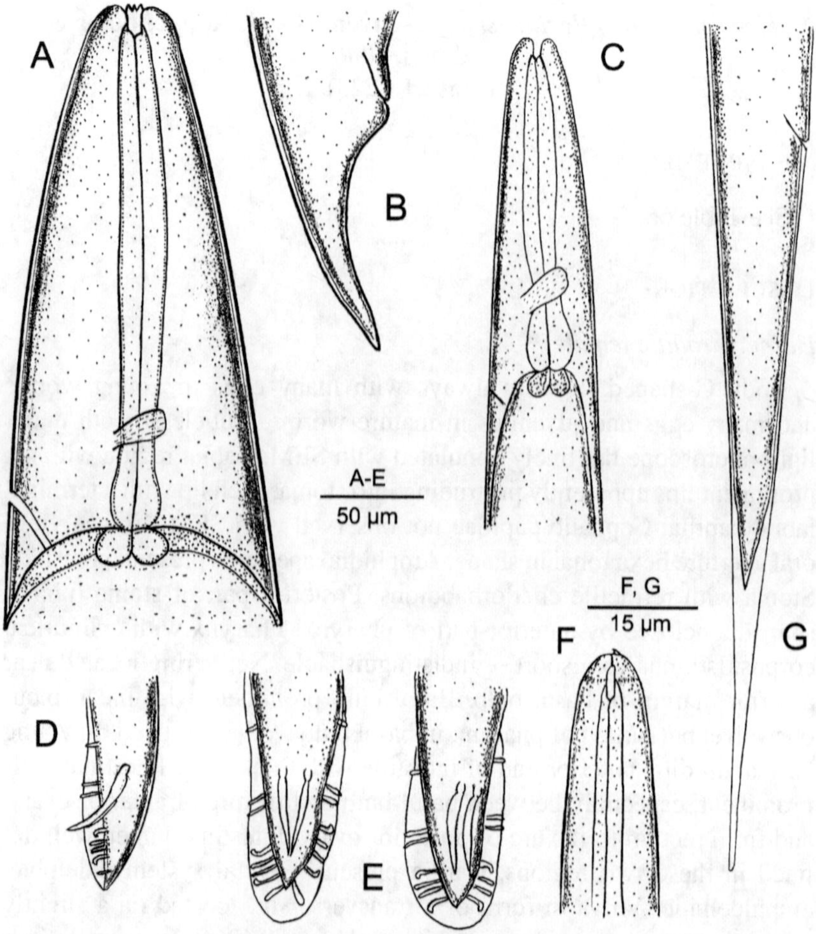

Fig. 218. Heterorhabditis floridensis. *A, B: Anterior and posterior regions of hermaphrodite; C-E: Head and tail of male. F, G: Infective juvenile. F: Anterior region showing dorsal tooth; G: Tail of ensheathed infective juvenile showing terminus of second-stage cuticle and third-stage juvenile. After Nguyen* et al. *(2006) courtesy of* Zootaxa.

Male

Body curved ventrally when heat-relaxed. Cephalic extremity truncate, sometimes slightly swollen. Labial papillae six, cephalic papillae not observed. Amphidial apertures prominent. Stoma with sclerotised rod- or barrel-shaped cheilorhabdions; other rhabdions not distinguishable forming funnel-shaped structure posterior to cheilorhabdions. Phar-

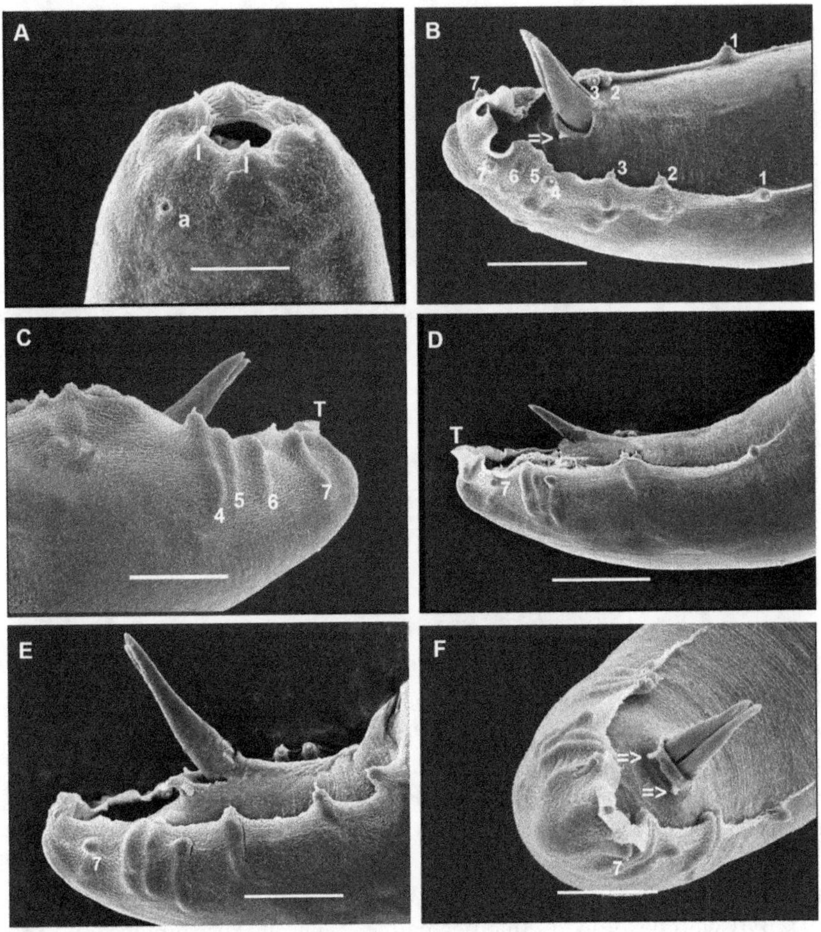

Fig. 219. Heterorhabditis floridensis. *SEM photographs of male. A: Head of male showing six labial papillae (l), and one of two amphidial apertures (a); B-F: Posterior region in different views showing bursal papillae (1, 2, 3, 4, 5, 6, 7, 8) with variation in shape and position, small papillae on posterior edge of cloacal aperture (arrows), tail tip (T), and spicules. (Scale bars: A = 3.75 μm; B = 12 μm; C = 8.75 μm; D = 15 μm; E = 12 μm; F = 10 μm.) After Nguyen* et al. *(2006) courtesy of* Zootaxa.

ynx with cylindrical corpus, metacorpus sometimes slightly enlarged. Nerve ring surrounding isthmus just anterior to basal bulb. Basal bulb with reduced valve. Cardia present, protruding into intestine. Excretory pore usually located posterior to basal bulb. Genital system monorchic,

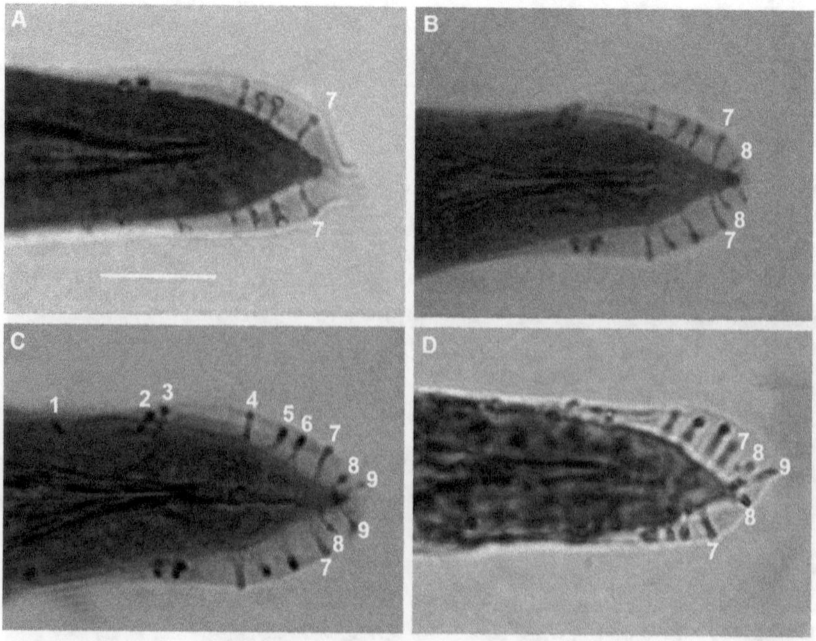

Fig. 220. Heterorhabditis floridensis. *Male tails with bursal papillae. A: Bursa with one pair of bursal papillae in terminal group; B: Bursa with two pairs of bursal papillae in terminal group; C: Bursa with three pairs bursal papillae in terminal group; D: Bursa with three papillae on one side and two papillae on other side in terminal group. (Scale bar = 20 μm.) After Nguyen* et al. *(2006), courtesy of* Zootaxa.

reflexed; distance from end of basal bulb to testis flexure variable. *Vas deferens* well developed. Spicules paired, separate, rostrum absent, ventral side usually flattened. Gubernaculum slightly curved ventrally. Bursa peloderan. Number of genital papillae from pairs 1 to 6 (from anterior end) are unchanged and typical for genus (Nguyen *et al.*, 2004). Distribution of papillae variable. From anterior to posterior, pair 1 is well anterior to cloaca, its tips reaching beyond bursal rim; pairs 2 and 3 usually (90%) occurring in a group, or sometimes (10%) separated. Two papillae located immediately anterior to cloaca also reaching beyond bursal rim. Pairs 4, 5 and 6 usually (40%) forming a group separately, or papillae 4 and 5 forming a group, with pair 6 standing separately (20%). These papillae situated just posterior to cloacal aperture, with pair 4 curved outward (laterally). Frequently (40%), papillae 4, 5, 6 and 7 form a group. The number and distribution of papillae in the terminal group are also

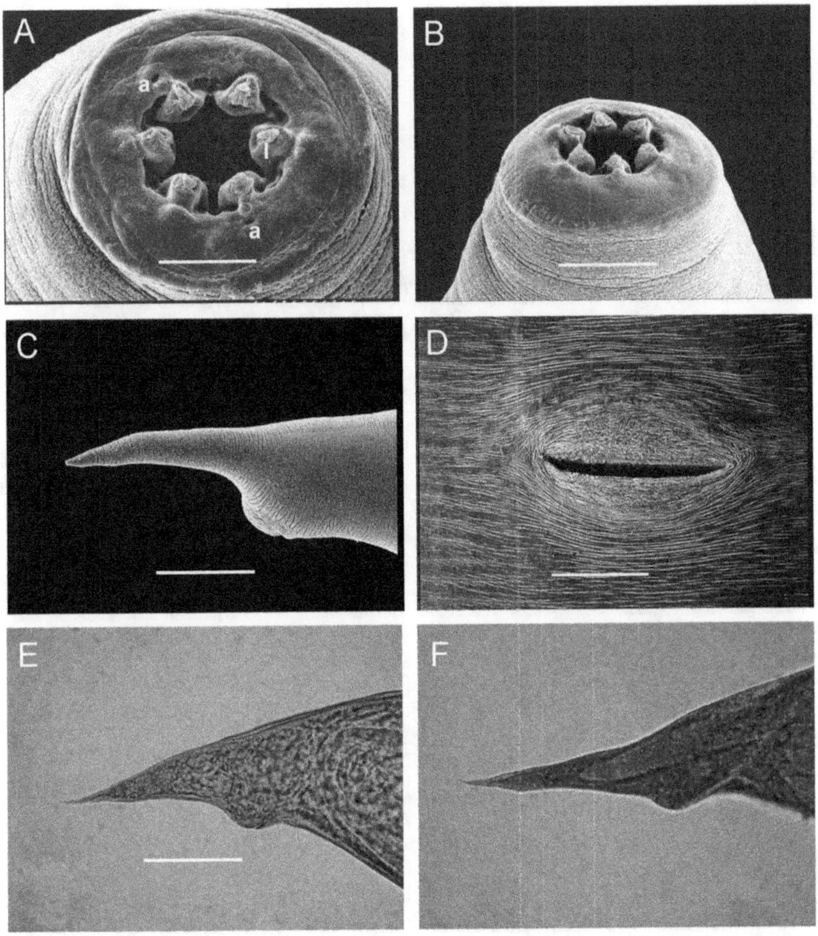

Fig. 221. Heterorhabditis floridensis. *A-C: SEM photographs of hermaphroditic and amphimictic females. A: SEM of face view of hermaphroditic female showing six lips protruding into stoma with labial papillae (l) terminally and two pore-like amphidial apertures (a); B: Face view of amphimictic female showing six prominent lips, six labial papillae on top, amphidial apertures not seen; C: SEM of hermaphroditic female showing post-anal swelling. D: Vulva pattern; E, F: Light micrographs of two different tail shapes of hermaphroditic female. (Scale bars: A = 6 μm; B = 10 μm; C = 20 μm; D = 10 μm; E, F (in E) = 60 μm.) After Nguyen* et al. *(2006), courtesy of* Zootaxa.

highly variable. Observation of papillae in terminal group of ten bursa yield following results: 4/10 (40%) with two pairs of papillae; 3/10

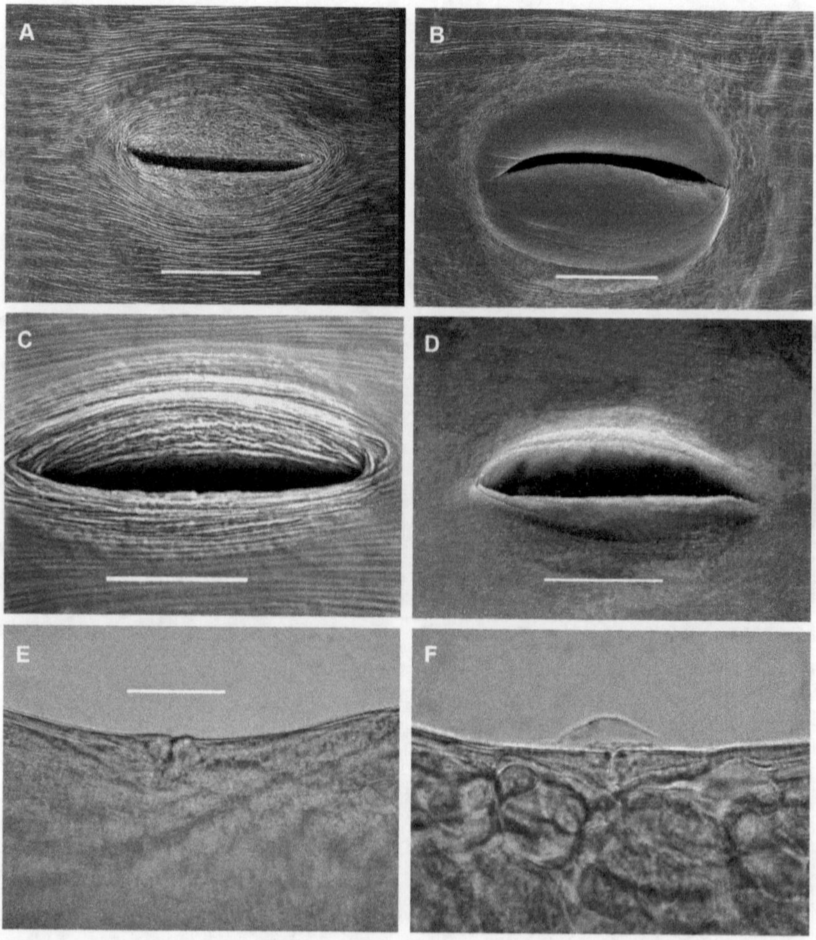

Fig. 222. *Comparison of vulva patterns of* Heterorhabditis floridensis *and three closely related species. A:* Heterorhabditis floridensis*; B:* H. mexicana*; C:* H. bacteriophora*; D:* H. indica*; E, F: Lateral views of vulva of hermaphroditic and amphimictic females, note that vulva of hermaphroditic female is slightly protuberant and that of amphimictic female is covered with exudates or copulation plug after mating. (Scale bars: A = 10 μm; B-D = 8.6 μm; E, F (in E) = 60 μm.) After Nguyen* et al. *(2006), courtesy of* Zootaxa.

(30%) with three papillae on one side and two papillae on other; 2/10 (20%) with one pair of papillae, and 1/10 (10%) with three pairs of papillae. In general, first papilla in terminal group prominent; last two papillae, if present, weaker, shorter, or appearing as small sclerotised

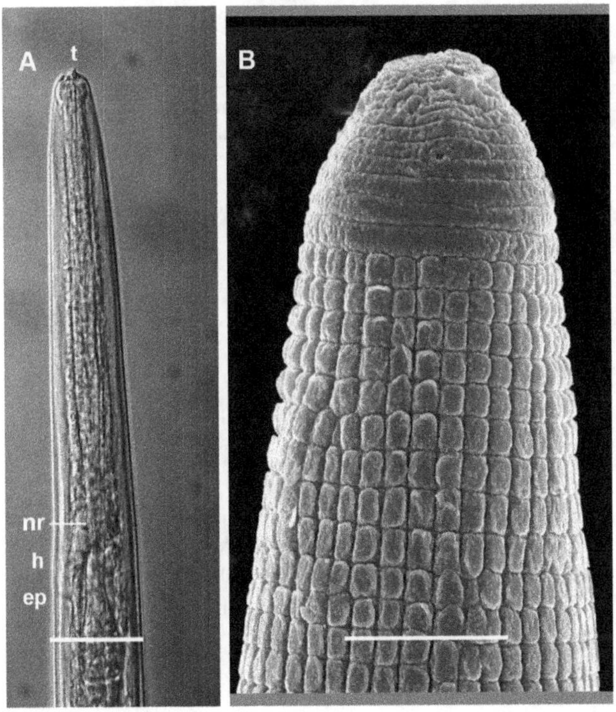

Fig. 223. Heterorhabditis floridensis *infective juvenile. A: Anterior region showing dorsal tooth (t), nerve ring (nr), hemizonid (h), excretory pore (ep) and excretory duct; B: SEM of anterior region showing labial region with six annules, and one of two pore-like amphidial apertures and anterior part of body with tessellate pattern. (Scale bars: A = 22 μm; B = 3.75 μm.) After Nguyen* et al. *(2006), courtesy of* Zootaxa.

dots. Under SEM observation, only pair 7 was seen in terminal group; in one male eight pairs were observed. In addition to bursal papillae, a pair of smaller papillae observed on posterior margin of cloacal aperture. Tail conoid, slightly curved ventrally.

Infective juvenile

Body elongate. Sheath (second-stage cuticle) present immediately after harvesting, but many infective juveniles losing sheath in storage. Labial region with six annules, lacking longitudinal incisures. Anterior part of body (*ca* 3-5 body diam. from anterior end) with tessellate pattern. Posterior region with longitudinal ridges. Exsheathed IJ body annulated, without longitudinal ridges. Labial region with dorsal tooth;

Table 65. *Morphometrics of* Heterorhabditis floridensis. *Measurements are in μm and in the form: mean ± SD (range).*

Character	Holotype	Male	Hermaphrodite	Female	Infective juvenile
n	–	20	20	20	25
L	837	862 ± 44 (785-924)	4885 ± 69 (3731-5865)	2302 ± 48 (2054-2548)	562 ± 24 (554-609)
a	–	–	–	–	27.6 ± 5.2 (25-32)
b	–	–	–	–	4.3 ± 2.1 (3.9-4.9)
c	–	–	–	–	5.6 ± 2.4 (5.3-6.6)
V	–	–	46 ± 7 (44-49)	48.1 ± 6.9 (44-50)	–
Greatest body diam.	42	47.6 ± 2.2 (43-50)	271 ± 16 (217-331)	139 ± 12 (120-156)	21.2 ± 4.6 (19-23)
EP	119	117 ± 6 (104-128)	246 ± 15 (211-301)	136 ± 12 (110-168)	109 ± 10.4 (101-122)
NR	79.7	80 ± 5 (73-90)	209 ± 14 (169-271)	107 ± 10.3 (86-122)	86 ± 9.2 (68-107)
ES	105.8	105 ± 4 (97-111)	308 ± 17 (271-391)	149 ± 12.2 (126-178)	135 ± 11.6 (123-142)
Testis reflexion	78	93 ± 11 (78-116)	–	–	–
Tail length with sheath (T)	–	–	–	–	103 ± 10.1 (91-113)
Tail length without sheath	34.8	34 ± 5.8 (29-40)	109 ± 10 (84-126)	77 ± 8.8 (69-87)	63 ± 7.9 (48-68)
Anal body diam. (ABD)	24.6	26.3 ± 3 (20-31)	57 ± 7 (42-78)	35.4 ± 5.9 (32-42)	14 ± 3.7 (12-16)
Spicule length (SP)	40.6	42 ± 3.5 (36-46)	–	–	–
Spicule width	4.3	4.8 ± 0.7 (4.3-5.8)	–	–	–
Gubernaculum length (GL)	21.7	23 ± 3.7 (17-30)	–	–	–
D% = EP/ES × 100	112	112 ± 4 (105-119)	–	–	81 ± 8.9 (71-90)
E% = EP/T × 100	–	–	–	–	105 ± 10.2 (95-134)
SW% = SP/ABD × 100	165	157 ± 25 (133-209)	–	–	–
GS% = GL/SP × 100	53.4	53.8 ± 6 (47-65)	–	–	–

EP = distance from anterior end to excretory pore; NR = distance from anterior end to nerve ring; ES = distance from anterior end to end of pharynx.
After the original authors.

labial and cephalic papillae absent. Amphidial aperture prominent, pore-like. Lateral field at mid-body region marked by two ridges. Pharynx typical for genus. Excretory pore located posterior to nerve ring but just anterior to base of pharynx. Excretory duct pronounced, well cuticularised except in anterior part, 3-5 μm from excretory pore. Hemizonid, when observed, located three or four annules anterior to excretory pore. Phasmid not observed under light microscope. Tail elongate conoid with pointed terminus. Tail length without sheath *ca* 61% of tail with sheath.

TYPE HOST AND LOCALITY

Natural host unknown as the nematode was obtained by baiting soil collected from a wooded area next to the highway (SR 70), GPS coordinates, N 27.20855 and W 81.75035, in Desoto County, FL, USA.

TYPE MATERIAL

Holotype (male), and paratype (hermaphrodites, females, males, and infective juveniles in 3% formalin) isolated from haemocoel of *G. mellonella* deposited in the USDA Nematode Collection, Beltsville, MD, USA.

Slides of one male and one female of the second generation, and several infective juveniles deposited in the University of California-Davis Nematode Collection, Davis, CA, USA.

Several slides of hermaphrodites, males, females and all stages in 4% formalin are at the Department of Entomology and Nematology, University of Florida, Gainesville, FL, USA.

RELATIONSHIPS

Heterorhabditis floridensis is characterised by unique male, female and infective juvenile characters. For males, the number of papillae in the terminal bursal group varies, either having two pairs of papillae (40%), three papillae on one side and two papillae on the other side (30%), one pair of papillae (20%), or with three pairs of papillae (10%). SW and GS values are 157 and 54, respectively. Females possess a typical vulva pattern (Fig. 221D). For infective juveniles, EP = 109 (101-122) μm, ES = 135 (123-142) μm, tail length = 103 (91-113) μm, and a = 27.6 (25-32) are characteristic. This species can be further characterised by molecular characteristics of the ITS regions of ribosomal DNA.

Heterorhabditis floridensis can be distinguished from other species of *Heterorhabditis* by morphological characters of the males, females and infective juveniles. *Heterorhabditis floridensis* is most similar to *H. mexicana*, *H. baujardi* and *H. indica*, but morphometrics of *H. floridensis* infective juveniles differ (L = 562 *vs* 578, 551, 528 μm, respectively; EP = 109 *vs* 102, 97, 98 μm, respectively; ES = 135 *vs* 122, 115, 117 μm, respectively; tail length = 103 *vs* 99, 90, 101 μm, respectively). For males of *H. floridensis*, the number of papillae in the terminal group is variable (see description), and measurements of some characteristics differ from the taxa listed above (EP = 117 *vs* 124, 81, 123 μm, respectively). For females, the vulval pattern (Fig. 222A) is quite different from that of the morphologically closely related species, *H. mexicana* (Fig. 222B), *H. bacteriophora* (Fig. 222C) or *H. indica* (Fig. 222D). Unfortunately, the vulval pattern of *H. baujardi* could not be obtained by the authors for comparison. *Heterorhabditis floridensis* can be distinguished from *H. megidis*, *H. zealandica* and *H. marelatus* by the presence of three pairs of bursal papillae in the terminal group (see Fig. 6, Nguyen *et al.*, 2004), and by the greater body length (>600 μm) of the infective juvenile of these three species.

Molecular characterisation

The ITS rDNA regions, flanked by primers 18S and 26S, of *Heterorhabditis floridensis* are characterised by length, 1012 base pairs (bp), ITS1 = 393 bp, ITS2 = 214 bp, and nucleotide usage composition, A = 0.25988, C = 0.20059, G = 0.25494, T = 0.28458 (GenBank accession no. DQ372922). Compared to sequences of eight other species in the genus *Heterorhabditis*, the sequence length of *H. floridensis* is longer than that of seven but shorter than that of *H. bacteriophora* (1021 bp). This species has evolved three autapomorphic nucleotide character states, and differs from its sister taxon, *H. mexicana*, at 16 aligned positions of the complete ITS rDNA array (Nguyen *et al.*, 2004, 2006).

More DNA information about this species is available in Nguyen *et al.* (2006).

Distribution

Known only from the type locality, Desoto County, FL, USA (Nguyen *et al.*, 2006).

References

NGUYEN, K.B., SHAPIRO-ILAN, D.I., STUART, R.J., MCCOY, C.W., JAMES, R.R. & ADAMS, B.J. (2004). *Heterorhabditis mexicana* n. sp. (Rhabditida: Heterorhabditidae) from Tamaulipas, Mexico, and morphological studies of the bursa of *Heterorhabditis* spp. *Nematology* 6, 231-244.

NGUYEN, K.B., GOZEL, U., KOPPENHÖFER, H.S. & ADAMS, B.J. (2006). *Heterorhabditis floridensis* n. sp. (Rhabditida: Heterorhabditidae) from Florida. *Zootaxa* 1177, 1-19.

Heterorhabditis indica[*,**] Poinar, Karunakar & David, 1992
= *H. hawaiiensis* Gardner, Stock & Kaya, 1994
(Figs 224-226)

MEASUREMENTS

See Table 66.

DESCRIPTION

Male

Anterior region similar to that of females, but smaller. Nerve ring near basal bulb. Genital system monorchic, reflexed. Spicules paired and separate, with pointed tips; capitulum usually set off from shaft; blade with single medial rib. Gubernaculum flat, narrow, approximately 50% of spicule length long, not reflexed at tip. Bursa open, peloderan, with a double membrane, one running external and other internal to bursal papillae; bursa with only seven pairs of normal papillae, last two (8 and 9) generally atrophied, highly modified (round or very slender) or absent. From anterior to posterior, bursal pair 1 normally located anterior to spicule heads, are straight and reach bursal rim, pairs 2 and 3 forming a group located near middle of spicules. They are straight and reach bursal rim. Fourth, fifth and sixth pairs forming another group located near cloacal aperture. Fourth pair turn outward (laterally) and do not reach bursal rim, fifth and sixth pair turn ventrally (inward) and usually reach inner bursal rim. Seventh pair often branched or swollen at their base, variable in form (straight, bent outward or inward) and may or may not reach bursal rim. Eighth and ninth pairs often absent but if present are modified in being short, narrow or swollen. They may also be branched and are sometimes asymmetrical, with one of pair 8 present and one of pair 9 absent on one side and both present on the opposite side (always reduced in length or width).

Hermaphroditic female

Cephalic extremity truncate, to slightly rounded; six distinct, protruding, lips curved outward, each bearing a prominent labial papilla, cephalic papillae not observed. Amphidial apertures pore-like, located

[*] Named after *India*, the country where the species was isolated.
[**] Description after Poinar *et al.* (1992). Revised by Khuong B. Nguyen.

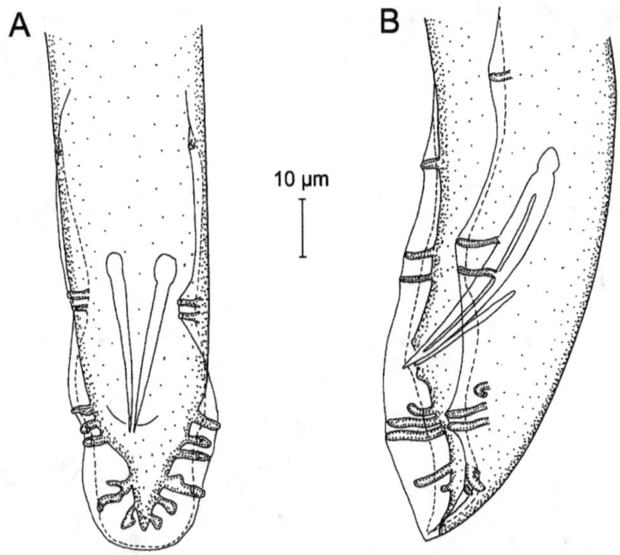

Fig. 224. Heterorhabditis indica. *A: Male tail, ventral and lateral views. After Poinar* et al. *(1992), courtesy of* Fundamental and Applied Nematology.

immediately posterior to lateral lips. Cheilorhabdions present as refractile ring just posterior to lips. Remaining meso-, meta- and telorhabdions fused and enclosed by anterior portion of pharynx. Pharynx cylindrical, without distinct metacorpus; isthmus present; basal bulb with fine striation in valve area. Nerve ring distinct, located in middle of isthmus. Genital system amphidelphic reflexed with sperm occurring in proximal portion of ovotestes. Tail longer than anal body diam. Anal swelling prominent. Anal aperture located in anterior half of anal swelling. Rectum strongly cuticularised.

Amphimictic female

Genital system with sperm in proximal portion of oviduct. Vulva narrow, surrounded by copulation plug after mating. Anal region conspicuously swollen. Anus located approximately at anterior third of anal swelling. Rectum heavily cuticularised with conspicuous valve separating it from intestine.

Infective juvenile

Anterior part of body of ensheathed infective juveniles (*ca* 3-5 body diam. from anterior end) with tessellate pattern. Infective juvenile

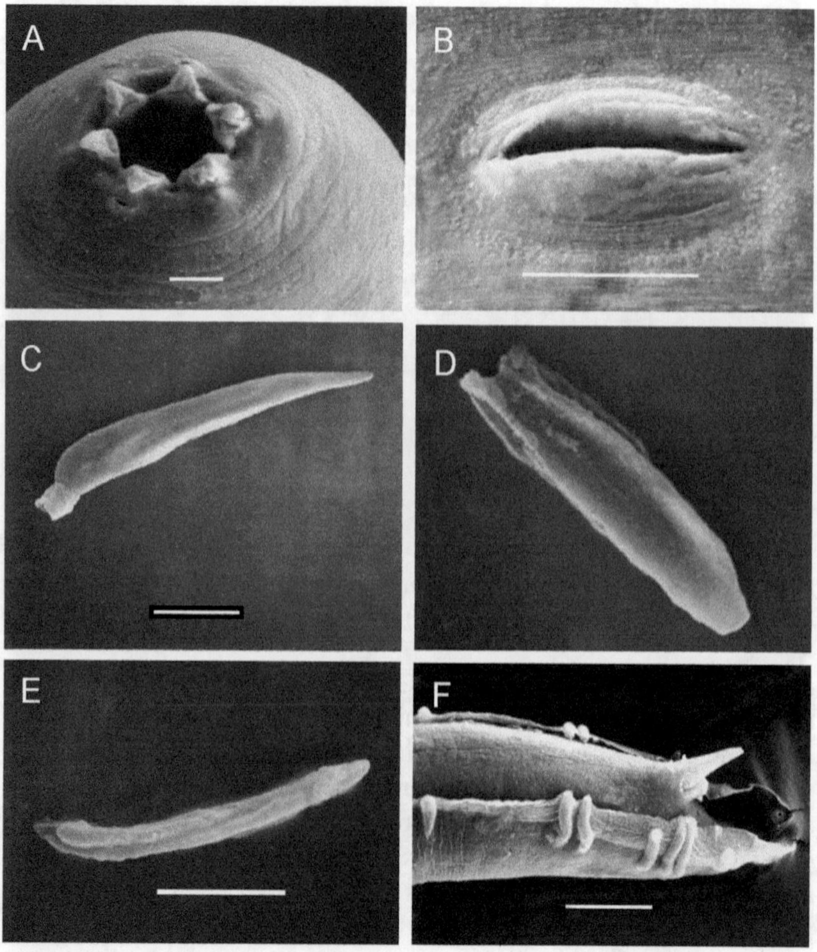

Fig. 225. Heterorhabditis indica. *SEM photographs. A: Hermaphroditic female showing lips curved outward; B: Protuberant vulva; C: Spicule showing well-curved dorsal side; D: Ventral view of gubernaculum; E: Gubernaculum, lateral view; F: Bursa showing prominent bursal ribs. (Scale bars: A = 10 μm; B, D, E = 8.6 μm; C, F = 15.2 μm.)*

(= third-stage juvenile ensheathed within second-stage cuticle) body short. Labial region bearing large dorsal tooth. Stoma walls open, not collapsed at base. Outer second-stage cuticle strongly ribbed longitudinally and closely adpressed to third-stage cuticle. Hemizonid quite distinct, located just anterior to excretory pore. Hemizonion smaller, less distinct, located in anterior intestinal region. Ventricular

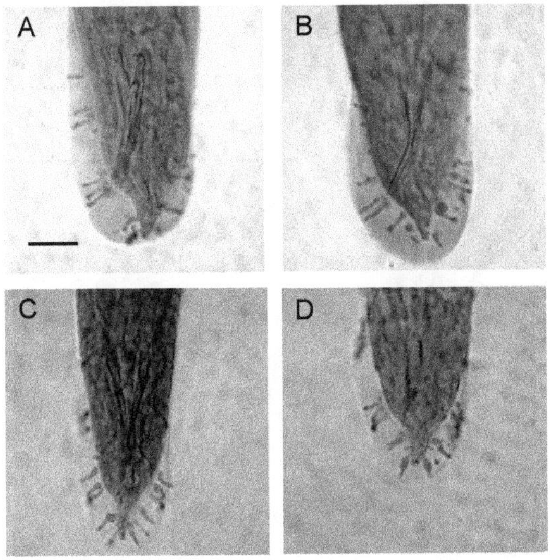

Fig. 226. Heterorhabditis indica *bursa showing irregular arrangement of bursal ribs. (Scale bar = 20 μm.)*

portion of intestine devoid of intestinal cells, forming an intestinal pouch filled with symbiotic bacteria. Bacteria also occurring in lumen of intestine, mainly anteriorly.

TYPE HOST AND LOCALITY

Natural host unknown as species was isolated from *Scirpophaga excerptalis* (Pyralidae: Lepidoptera) used as a trap insect in soil from Coimbatore, Tamil Nadu, India.

TYPE MATERIAL

Holotype (male) and allotype (hermaphroditic female) deposited in the University of California-Davis Nematode Collection, Davis, CA, USA. Paratypes (one male, one female) deposited in the Laboratoire des Vers, Muséum national d'Histoire naturelle, Paris, France.

RELATIONSHIPS

Heterorhabditis indica can be separated from the other species in the genus by the following characters: this species possesses short infective juveniles which separates it from all other described species

Table 66. *Morphometrics of* Heterorhabditis indica. *Measurements are in μm and in the form: mean ± SD (range).*

Character	Holotype (male)	Male	Hermaphrodite	Female	Infective juvenile
n	–	12	12	12	25
L	668	721 ± 64 (573-788)	2700 ± 1000 (2300-3100)	1600 ± 12 (1200-1800)	528 ± 26 (479-573)
a	–	–	–	–	26 ± 4 (25-27)
b	–	–	–	–	4.5 ± 0.34 (4.3-4.8)
c	–	–	–	–	5.3 ± 0.5 (4.5-5.6)
V	–	–	47 ± 3 (45-50)	48 ± 9 (40-53)	–
Greatest body diam.	38	42 ± 7 (35-46)	132 ± 9 (107-145)	95 ± 15 (76-113)	20 ± 6 (19-22)
EP	120	123 ± 7 (109-138)	173 ± 8 (163-187)	127 ± 4 (118-138)	98 ± 7 (88-107)
NR	80	75 ± 4 (72-85)	115 ± 5 (104-123)	92 ± 4 (88-96)	82 ± 4 (72-85)
ES	99	101 ± 4 (93-109)	172 ± 6 (163-179)	131 ± 4 (120-139)	117 ± 3 (109-123)
Testis reflexion	83	91 ± 26 (35-144)	–	–	–
Tail length with sheath (T)	–	–	–	–	101 ± 6 (93-109)
Tail length without sheath	29	28 ± 2 (24-32)	92 ± 11 (72-110)	76 ± 9 (66-88)	–
Anal body diam. (ABD)	–	23 ± 8 (19-24)	44 ± 9 (38-51)	26 ± 11 (22-32)	–
Spicule length (SP)	–	43 ± 3 (35-48)	–	–	–
Gubernaculum length (GL)	–	21 ± 3 (18-23)	–	–	–
D% = EP/ES × 100	–	121	–	–	84 ± 5 (79-90)
E% = EP/T × 100	–	–	–	–	94 ± 7 (83-103)
SW% = SP/ABD × 100	–	187	–	–	–
GS% = GL/SP × 100	–	49	–	–	–

EP = distance from anterior end to excretory pore; NR = distance from anterior end to nerve ring; ES = distance from anterior end to end of pharynx.
After the original authors.

except for *H. taysearae*, although there is some overlap between small individuals of *H. bacteriophora* and large individuals of *H. indica.*

However, the values of ratio E and ratio D separate *H. indica* from all other described *Heterorhabditis* species. This species can also be differentiated from other species by the arrangement of the terminal group of bursa ribs, which are variable, three pairs of rib in the terminal group is very rare compared to three prominent ribs in the terminal group in *H. bacteriophora*, *H. marelatus*, *H. megidis* and *H. zealandica*. The prominent lips in the female, which curve outward, and the appearance of the vulva are diagnostic of *H. indica*, and are different from that of other species.

MOLECULAR CHARACTERISATION

The ITS rDNA regions, flanked by primers 18S and 26S, of *H. indica* are characterised by the sequence length, 988 base pairs (bp), ITS1 = 370 bp, ITS2 = 215 bp, and nucleotide usage composition, A = 0.26316, C = 0.20547, G = 0.25506, T = 0.27632 (accession no. AY321483). This species has evolved ten autapomorphic nucleotide character states and differs from its sister taxon, *H. mexicana*, at 113 aligned positions of the complete ITS rDNA array (Nguyen *et al.*, 2004).

BIONOMICS

The bionomics of *H. indica* are similar to that of *H. bacteriophora* except the bacterium associated with this species is *Photorhabdus luminescens* subsp. *akhurstii*.

This species has been commercialised by some companies as a control agent for various kinds of insect pest.

DISTRIBUTION

Heterorhabditis indica was originally found in Coimbatore, India (Poinar *et al.*, 1992), its junior synonym *H. hawaiiensis* being isolated from Kauai, Hawaii, USA. Subsequently, the species has been reported from the Caribbean region (Cuba, Dominican Republic, Guadeloupe, Jamaica, Martinique, Puerto Rico, Trinidad, Virgin Islands), Venezuela, several other Asian countries (Indonesia, Israel, Japan, Malaysia, Pakistan, Palestine, Sri Lanka) and Africa (Egypt, Kenya) (see Hominick, 2002). Using DNA sequences, many isolates of *H. indica* have been identified (Nguyen, unpubl.) from Saudi Arabia and the USA (Florida and Georgia).

References

GARDNER, S.L., STOCK, S.P. & KAYA, H.K. (1994). A new species of *Heterorhabditis* from the Hawaiian islands. *Journal of Parasitology* 80, 100-106.

HOMINICK, W.M. (2002). Biogeography. In: Gaugler, R. (Ed.). *Entomopathogenic nematology*. Wallingford, UK, CABI Publishing, pp. 115-143.

NGUYEN, K.B., SHAPIRO-ILAN, D.I., STUART, R.J., MCCOY, C.W., JAMES, R.J. & ADAMS, B.J. (2004). *Heterorhabditis mexicana* n. sp. (Rhabditida: Heterorhabditidae) from Tamaulipas, Mexico, and morphological studies of the bursa of *Heterorhabditis* spp. *Nematology* 6, 231-244.

POINAR JR, G.O., KARUNAKAR, G.K. & DAVID, H. (1992). *Heterorhabditis indicus* n. sp. (Rhabditida: Nematoda) from India: separation of *Heterorhabditis* spp. by infective juveniles. *Fundamental and Applied Nematology* 15, 467-472.

Heterorhabditis marelatus[*,**] Liu & Berry, 1996
= *H. hepialius* Stock, Strong & Gardner, 1996
(Figs 227, 228)

MEASUREMENTS

See Table 67.

DESCRIPTION

Male

Cephalic extremity truncate or rounded; six lips surrounding oral aperture, each lip bearing a labial papilla. Cephalic papillae not observed. Cheilorhabdions visible as lightly refractile ring just posterior to lips. Meta-, pro-, meso- and metarhabdions fused and enclosed by anterior portion of pharynx. Pharynx cylindroid, without metacorpus; isthmus present, basal bulb pronounced, with valve. Nerve ring distinct, located mid-isthmus level. Genital system monorchic, reflexed. Spicules paired, separate, straight to slightly curved with pointed tips; capitulum usually set off from shaft; shaft with dorsal expansion; lamina with single rib. Gubernaculum <50% of spicule length. Bursa open, peloderan usually supported by nine pairs of genital papillae with enlarged tips, three pairs located anterior to cloacal aperture, six pairs posterior.

Hermaphroditic female

Anterior region similar to that of male but larger. Excretory pore positioned posterior to basal bulb. Genital system amphidelphic, reflexed. Hermaphroditic females (first generation) with ovotestes. Vulva with protruding lips, located near mid-body. Basal bulb of pharynx one-half to one-third body diam. in size. Tail pointed with postanal swelling.

Amphimictic female

Anterior region similar to that of hermaphroditic female but smaller. Vulva without protruding lips. Tail pointed, usually lacking a postanal swelling.

[*] From Latin: *mare* = sea, *latus* = side, and referring to the type locality of *Seaside*, OR, USA.

[**] Description after Liu and Berry (1996). Revised by Khuong B. Nguyen.

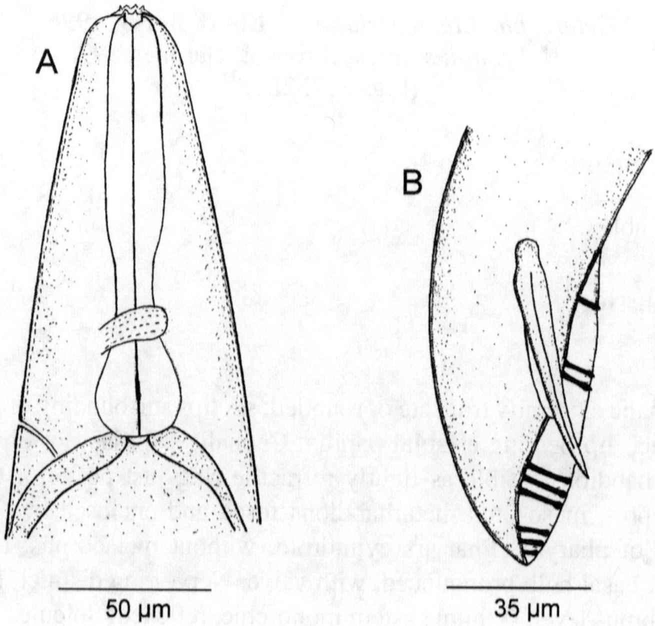

Fig. 227. Heterorhabditis marelatus. *A: Anterior region of amphimictic female (original); B: posterior region of male (redrawn and modified from Liu and Berry, 1996).*

Infective juvenile

Infective juvenile retaining shed cuticle of second-stage. Cuticle in posterior body with fine longitudinal striations. Oral aperture and anus closed. Cephalic region with large dorsal tooth. Hemizonid distinct, usually located just posterior to excretory pore. Nerve ring located at mid-isthmus level.

TYPE HOST AND LOCALITY

Natural host unknown as the species was obtained by baiting with *Galleria mellonella* larvae in a sandy soil sample from Seaside Beach, Seaside, OR, USA.

TYPE MATERIAL

Holotype (male) and allotype (hermaphroditic female) deposited in the USDA Nematode Collection, Beltsville, MD, USA. Paratype

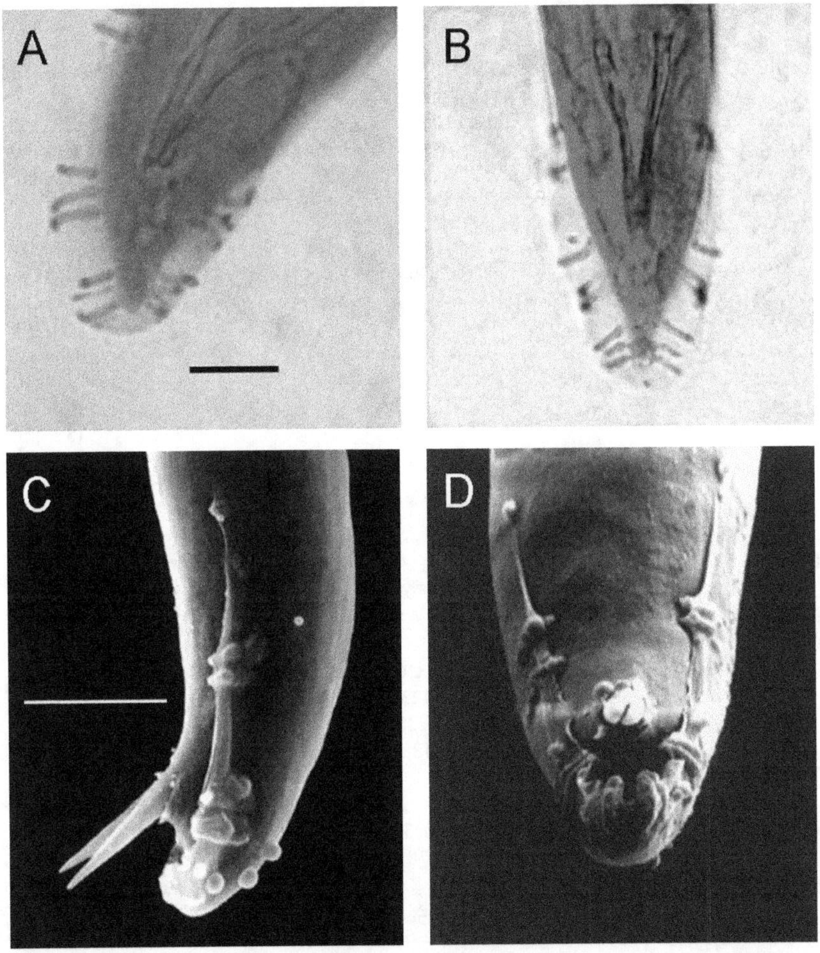

Fig. 228. Heterorhabditis marelatus. *A, B: Light micrographs of bursa showing three equal pairs of ribs in terminal group of bursal papillae; C, D: SEM photographs of bursa, lateral and ventral views. (Scale bars: A, B = 20 μm; C, D = 20 μm.)*

deposited in the University of California-Davis Nematode Collection, Davis, CA, USA.

RELATIONSHIPS

Heterorhabditis marelatus can be separated from the other species in the genus by the body length of infective juveniles (654 μm), the

Table 67. *Morphometrics of* Heterorhabditis marelatus. *Measurements are in μm and in the form: mean ± SD (range).*

Character	Holotype (male)	Male	Hermaphrodite	Female	Infective juvenile
n	–	20	20	20	20
L	869	942 ± 77 (805-1046)	3500 ± 40 (3000-4500)	2300 ± 30 (1600-2600)	654 ± 26 (588-700)
a	–	–	–	–	24 ± 2.5 (21-29)
b	–	–	–	–	4.9 ± 0.2 (4.7-5.4)
c	–	–	–	–	61 ± 0.3 (5.5-6.6)
V	–	–	48 ± 3 (45-50)	48 ± 2 (45-50)	
Greatest body diam.	56	51 ± 3.6 (48-56)	184 ± 21 (161-233)	153 ± 19 (113-177)	28 ± 2.7 (24-32)
EP	139	130 ± 30 (110-168)	246 ± 20 (212-287)	157 ± 13 (139-178)	102 ± 7.4 (81-113)
NR	71	73.7 ± 9 (61-95)	155 ± 11 (133-182)	97 ± 9 (79-119)	99 ± 7.8 (83-113)
ES	115	115 ± 6 (99-123)	218 ± 16 (190-244)	144 ± 5 (129-164)	133 ± 4.8 (121-139)
Testis reflexion	69	89 ± 22 (51-138)	–	–	–
Tail length with sheath (T)	–	–	–	–	107 ± 4.5 (99-117)
Tail length without sheath	30	31 ± 3 (24-38)	86 ± 7 (75-101)	67 ± 8 (55-81)	–
Anal body diam. (ABD)	28	23 ± 2.4 (20-28)	45 ± 5 (42-57)	31 ± 7 (29-48)	–
Spicule length (SP)	42	45 ± 3 (41-49.5)	–	–	–
Gubernaculum length (GL)	20	19 ± 1.2 (18-22)	–	–	–
D% = EP/ES × 100	–	113	–	–	77 ± 10 (60-86)
E% = EP/T × 100	–	–	–	–	96 ± 8 (89-110)
SW% = SP/ABD × 100	150	196	–	–	–
GS% = GL/SP × 100	47	41 ± 4 (36-50)	–	–	–

EP = distance from anterior end to excretory pore; NR = distance from anterior end to nerve ring; ES = distance from anterior end to end of pharynx.
After the original authors.

pharynx length (133 μm), tail length (107 μm) and E% (96). The males of *H. marelatus* have nine pairs of genital papillae, usually with enlarged tips. The spicule length of 45 μm is longer than that of *H. bacteriophora* (40 μm) and *H. indica* (43 μm), but shorter than that of *H. megidis* (49 μm) and *H. zealandica* (51 μm.) In addition, the presence of a dorsal expansion on the spicule shaft and the ratios GS% = 41 and D% = 113 also distinguish this nematode from other species.

MOLECULAR CHARACTERISATION

Heterorhabditis marelatus is characterised by the sequence length of ITS regions (accession no. AY321479), 995 bp, ITS1 = 379 bp, ITS2 = 211 bp, and the composition, A = 0.23417, C = 0.21608, G = 0.26432, T = 0.28543. This species has evolved four autapomorphic nucleotide character states, and differs from its sister taxon *H. megidis* at 58 aligned positions of the complete ITS rDNA array (Nguyen *et al.*, 2004, 2006). Additionally, RAPD fragments generated from 5-decamer oligonucleotides showed a clear distinction between *H. marelatus* and *H. bacteriophora*, *H. megidis* and *H. indica*, with a minimal RAPD polymorphism frequency of 0.85 to other *Heterorhabditis* species examined.

DISTRIBUTION

Reported from the type locality, Seaside, OR, USA (Liu & Berry, 1996) and also (as *H. hepialius*) from California (Stock *et al.*, 1996, 1999).

References

LIU, J. & BERRY, R.E. (1996). *Heterorhabditis marelatus* n. sp. (Rhabditida: Heterorhabditidae) from Oregon. *Journal of Invertebrate Pathology* 67, 48-54.

NGUYEN, K.B., SHAPIRO-ILAN, D.I., STUART, R.J., MCCOY, C.W., JAMES, R.R. & ADAMS, B.J. (2004). *Heterorhabditis mexicana* n. sp. (Rhabditida: Heterorhabditidae) from Tamaulipas, Mexico, and morphological studies of the bursa of *Heterorhabditis* spp. *Nematology* 6, 231-244.

NGUYEN, K.B., GOZEL, U., KOPPENHÖFER, H.S. & ADAMS, B.J. (2006). *Heterorhabditis floridensis* n. sp. (Rhabditida: Heterorhabditidae) from Florida. *Zootaxa* 1177, 1-19.

STOCK, S.P., STRONG, D. & GARDNER, S.L. (1996). Identification of *Heterorhabditis* (Nematoda: Heterorhabditidae) from California with a new species isolated from larvae of the ghost moth *Hepialus californicus* (Lepidoptera: Hepialidae) from the Bodega Bay Natural Reserve. *Fundamental and Applied Nematology* 19, 585-592.

STOCK, S.P., PRYOR, B.M. & KAYA, H.K. (1999). Distribution of entomopathogenic nematodes (Steinernematidae, Heterorhabditidae) in natural forests in California. *Biodiversity and Conservation* 8, 339-345.

Heterorhabditis megidis[*,**] Poinar, Jackson & Klein, 1987
(Figs 229-231)

MEASUREMENTS

See Table 68.

DESCRIPTION

Male

Cephalic extremity truncate or slightly rounded. Labial region with six elevated lips, each bearing a terminal labial papilla. Cephalic papillae indistinct. Amphidial apertures small, pore-like. Anterior portion of pharynx enveloping all of stoma except for cheilorhabdions. Cheilorhabdions seen as lightly refractile areas lining anterior (non-collapsed) portion of stoma; metarhabdions located immediately posterior to cheilorhabdions, attached to fused pro- and mesorhabdions, each metarhabdial segment bearing a small tooth; telorhabdions absent. Basal bulb often surrounded by anterior portion of intestine. Nerve ring surrounding isthmus just anterior to basal bulb. Nerve ring usually located on anterior portion of basal bulb. Four prominent pseudocoelomocytes: one pair near tip of testis and another pair near reflexed portion. Genital system monorchic, reflexed. Spicules paired and separate, slightly curved, spicule head variable, usually well set off from shaft and longer than wide; shaft lacking a rostrum; blade with a single rib usually dividing at distal portion; spicule tips pointed. Gubernaculum flat, shorter than half spicule length, both distal and proximal portions folded dorsally. Bursa open, usually pseudopeloderan with a fine tip extending beyond bursal membrane; bursa with nine pairs of papillae, three pairs anterior to cloacal aperture and six pairs posterior. Numbering from anterior to posterior, pair 1 is isolated from 2 and 3 (occasionally 2 and 3 are fused); 4, 5 and 6 normally form a group, as do 7, 8 and 9. Fifth and eighth pairs usually bent outward (laterally), whereas remainder are straight or bent inward (ventrally).

Hermaphroditic female

Anterior region similar to male but much larger. Labial region well developed. In face view, oral aperture appearing hexagonal star-shaped.

[*] From Greek *megistos* = largest.

[**] Description after Poinar *et al.* (1987). Revised by Khuong B. Nguyen.

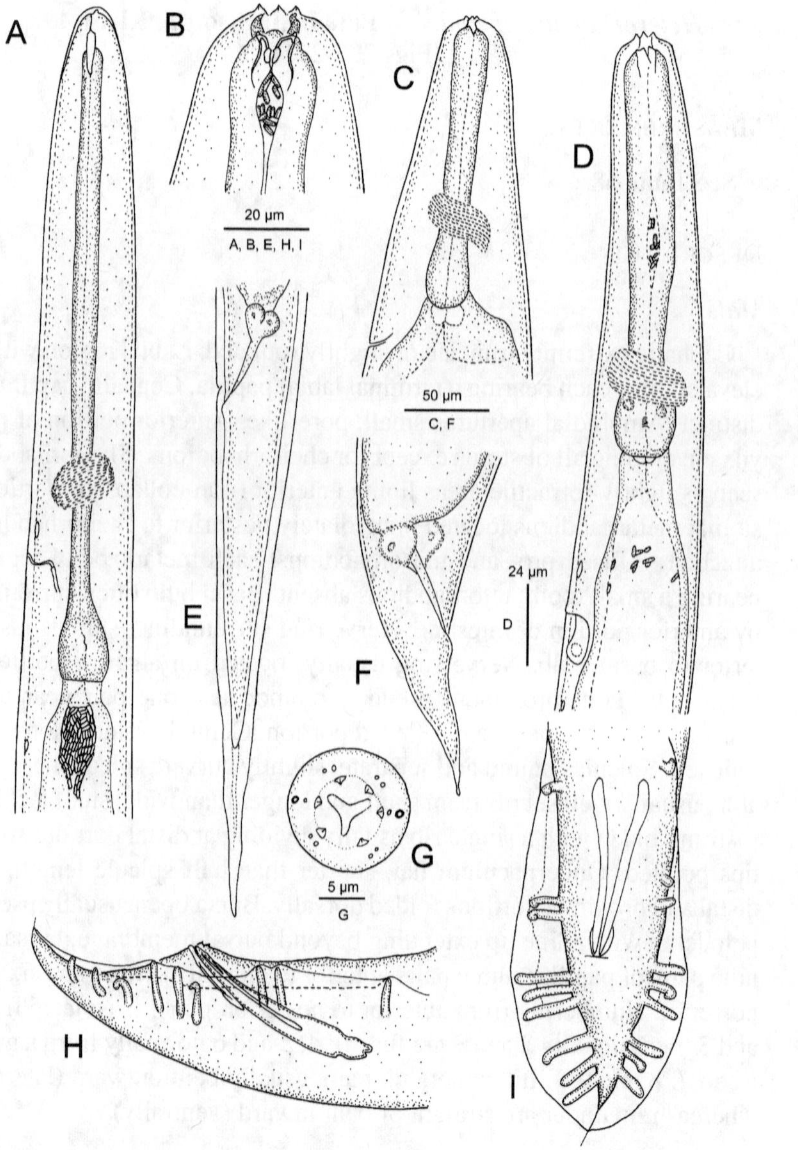

Fig. 229. Heterorhabditis megidis. *A: Anterior region of infective juvenile, lateral view; B: Anterior region of amphimictic female; C: Pharyngeal region of amphimictic female; D: Anterior region of male; E: Infective juvenile tail; F: Tail of amphimictic female; G: Face view of male; H: Male tail, lateral view; I: Male tail, ventral view. After Poinar* et al. *(1987), courtesy of* Proceedings of the Helminthological Society of Washington.

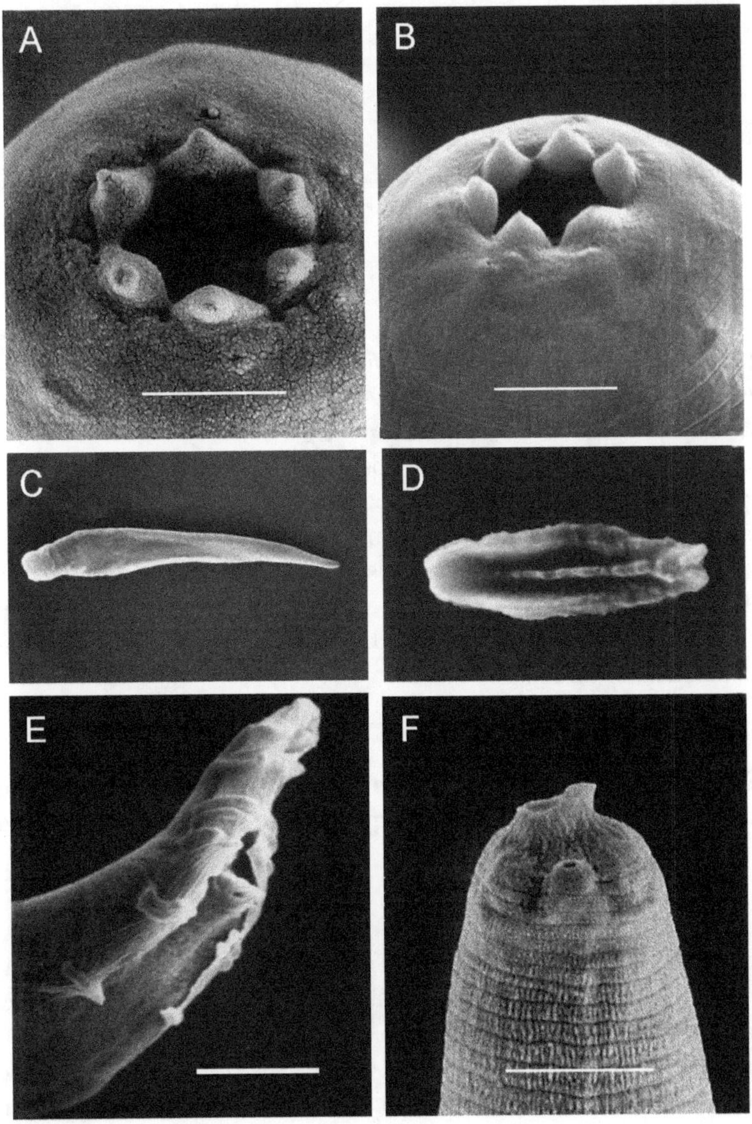

Fig. 230. Heterorhabditis megidis. *SEM photographs. A: Face view of hermaphroditic female; B: Face view of amphimictic female; C: Spicule showing almost square spicule head; D: Gubernaculum with needle-shaped cuneus; E: Bursa showing bursal ribs; F: Infective juvenile head showing one of two amphidial apertures and dorsal tooth with membranous ring around mouth. (Scale bars: A = 6.1 μm; B = 6 μm; C = 18.5 μm; D = 9 μm; E = 17.6 μm; F = 4.3 μm.)*

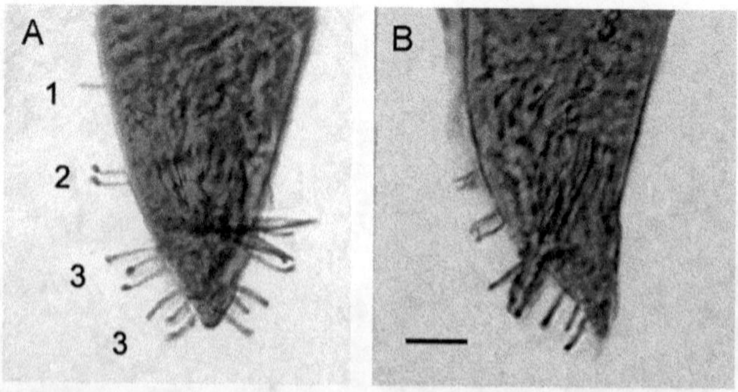

Fig. 231. Heterorhabditis megidis. *Light micrographs of bursa showing three equal pairs of ribs in terminal group of bursal papillae. (Scale bar = 20 μm.)*

Excretory pore usually positioned posterior to basal bulb. Genital system amphidelphic, reflexed. Reflexed portion often extending past vulval aperture. Sperm occurring in proximal portion of ovotestes. Vulva with slightly protruding lips. Rectum distinct, usually expanded and filled with bacteria when worm alive, three rectal glands surrounding junction of intestine and rectum. Tail pointed, normally wavy in outline, usually with postanal swelling.

Amphimictic female

Similar to hermaphrodite, but smaller. Vulva reduced, not protruding, usually covered by copulation plug. Sperm concentrated in proximal portion of oviduct. Tail attenuated.

Infective juvenile

Infective stage ensheathed inside second-stage cuticle. Second-stage cuticle strongly ribbed longitudinally, but also showing weak cross striations and closely adpressed to infective juvenile. If an infective juvenile with a tight sheath is placed directly in 70% alcohol, it contracts and pulls away from second-stage cuticle. Minute dorsal tooth always present on labial region.

TYPE HOST AND LOCALITY

The type host is the Japanese beetle, *Popillia japonica* Newm. (Scarabaeidae: Coleoptera) from Mohican Hills Golf Course, Jeromesville, OH, USA.

Table 68. *Morphometrics of* Heterorhabditis megidis. *Measurements are in μm and in the form: mean (range).*

Character	Male	Hermaphrodite	Female	Infective juvenile
n	15	10	15	25
L	1000 (800-1100)	3600 (2400-4900)	2100 (1500-2500)	768 (736-800)
a	–	–	–	26 (23-28)
b	–	–	–	5 (4.6-5.90)
c	–	–	–	6.5 (6.1-6.9)
V	–	48 (45-50)	49 (47-51)	–
Greatest body diam.	47 (44-50)	209 (120-333)	123 (95-140)	29 (27-32)
EP	156 (139-176)	209 (193-270)	178 (158-206)	131 (123-142)
NR	104 (96-112)	162 (139-178)	111 (105-120)	109 (104-115)
ES	128 (122-134)	229 (106-269)	160 (155-168)	155 (147-160)
Testis reflexion	138 (117-230)	–	–	–
Tail length with sheath (T)	–	–	–	119 (112-128)
Tail length without sheath	39 (35-43)	105 (95-124)	86 (70-101)	–
Anal body diam. (ABD)	26 (22-31)	63 (38-86)	31 (25-38)	19 (17-21)
Spicule length (SP)	49 (46-54)	–	–	–
Gubernaculum length (GL)	21 (17-24)	–	–	–
D% = EP/ES × 100	122	–	–	–
E% = EP/T × 100	–	–	–	85 (81-91)
SW% = SP/ABD × 100	188	–	–	110 (103-120)
GS% = GL/SP × 100	43	–	–	–

EP = distance from anterior end to excretory pore; NR = distance from anterior end to nerve ring; ES = distance from anterior end to end of pharynx.
After the original authors.

Type material

Holotype (male) and allotype (hermaphroditic female) deposited in the University of California-Davis Nematode Collection, Davis, CA, USA.

Relationships

The main difference between the infective stages of *H. megidis* and the other species is body length. The average length of *H. megidis* infective juveniles, as well as the average distance from the head to excretory pore and average length of the pharynx, separates this species from other described species of *Heterorhabditis*. Additionally, the dorsal tooth of the infective juvenile is always associated with a distinct membranous ring around the oral aperture (Fig. 230F).

Heterorhabditis megidis can be distinguished from other species of the genus by the long body, pharynx and tail lengths of infective juveniles and distance from anterior end to excretory pore. In the males, the ratio of the length of the gubernaculum to the length of the spicule, the ratio of the spicule length to the anal body diam. and shape of spicule head distinguish *H. megidis* from other species.

Molecular characterisation

The ITS rDNA regions, flanked by primers 18S and 26S, of *H. megidis* are characterised by the sequence length, 1005 base pairs (bp), ITS1 = 384 bp, ITS2 = 220 bp, and nucleotide usage composition, A = 0.23284, C = 0.21990, G = 0.27164, T = 0.27562 (accession no. AY321480). This species has evolved four autapomorphic nucleotide character states, and differs from its sister taxon *H. marelatus* at 58 aligned positions of the complete ITS rDNA array.

More molecular information is available in Nguyen *et al.* (2004, 2006).

Bionomics

The bionomics of *H. megidis* are similar to *H. bacteriophora* except that the bacterium associated with this species is *Photorhabdus temperata*. These cells are very characteristic in shape and can be found in the haemolymph of insects attacked by the nematodes. They produce a red pigment and impart a red colour to the infected host. They are also luminescent and freshly killed insects or agar plates with 1- to 2-day-old cultures are luminous in the dark.

This nematode species has been commercialised by several companies to control various kinds of insect pest.

DISTRIBUTION

Heterorhabditis megidis is known from the USA (Poinar *et al.*, 1987) and has also been found in Canada, Israel, Japan and European countries such as the Czech Republic, Greece and Switzerland. The 'north-west European type', generally accepted as being conspecific with *H. megidis*, has been recorded from Belgium, Denmark, England, Estonia, Germany, The Netherlands, Norway, Poland, Russia, Switzerland and the USA (see Hominick, 2002).

References

HOMINICK, W.M. (2002). Biogeography. In: Gaugler, R. (Ed.). *Entomopathogenic nematology*. Wallingford, UK, CABI Publishing, pp. 115-143.

NGUYEN, K.B., SHAPIRO-ILAN, D.I., STUART, R.J., MCCOY, C.W., JAMES, R.J. & ADAMS, B.J. (2004). *Heterorhabditis mexicana* n. sp. (Rhabditida: Heterorhabditidae) from Tamaulipas, Mexico, and morphological studies of the bursa of *Heterorhabditis* spp. *Nematology* 6, 231-244.

NGUYEN, K.B., GOZEL, U., KOPPENHÖFER, H.S. & ADAMS, B.J. (2006). *Heterorhabditis floridensis* n. sp. (Rhabditida: Heterorhabditidae) from Florida. *Zootaxa* 1177, 1-19.

POINAR JR, G.O., JACKSON, T. & KLEIN, M. (1987). *Heterorhabditis megidis* sp. n. (Heterorhabditidae: Rhabditida) parasitic in the Japanese beetle, *Popillia japonica* (Scarabaeidae: Coleoptera), in Ohio. *Proceedings of the Helminthological Society of Washington* 54, 53-59.

Heterorhabditis mexicana[*,**] Nguyen, Shapiro-Ilan, Stuart, McCoy, James & Adams, 2004
(Figs 232-236)

Measurements

See Table 69.

Description

Male

Body curved ventrally when heat-relaxed. Cephalic extremity truncate, sometimes slightly swollen. Labial papillae six, cephalic papillae not observed. Amphidial apertures prominent. Stoma with sclerotised, rod- or barrel-shaped cheilorhabdions; other rhabdions not distinguishable forming a funnel-shaped structure posterior to cheilorhabdions. Pharynx with cylindrical corpus, metacorpus sometimes slightly enlarged. Nerve ring surrounding isthmus just anterior to basal bulb. Basal bulb with reduced valve. Cardia present, protruding into intestine. Excretory pore usually posterior to basal bulb. Genital system monorchic, reflexed; distance from end of basal bulb to testis flexure variable. *Vas deferens* well developed. Spicules paired, separate, rostrum absent, ventral side usually flattened. Gubernaculum slightly curved ventrally. Bursa peloderan. Seventy percent (28/40) of observed males with eight pairs of bursal papillae. Three pairs located anterior to cloaca and five pairs posterior to cloaca. From anterior to posterior, pair 1 well anterior, its tips reaching beyond bursal rim; pairs 2 and 3 in a group, immediately anterior to cloaca, also reaching beyond bursal rim. Pairs 4, 5 and 6 forming a group, situated just posterior to cloacal aperture, pair 4 usually curved outward (laterally). Number of pairs from 1-6 are constant. Pairs 7 and 8 forming terminal group of papillae at end of bursa. Pair 7 always curved outward (laterally). *Circa* 30% of males lack two pairs in this last group of papillae. Instead, they have either three papillae on one side and one or two papillae on other side, or only one pair. In addition to bursal papillae, a pair of smaller papillae observed on posterior edge of cloaca. Tail conoid, slightly curved ventrally.

[*] Named after the country *Mexico*.

[**] Description after Nguyen *et al*. (2004). Revised by Khuong B. Nguyen.

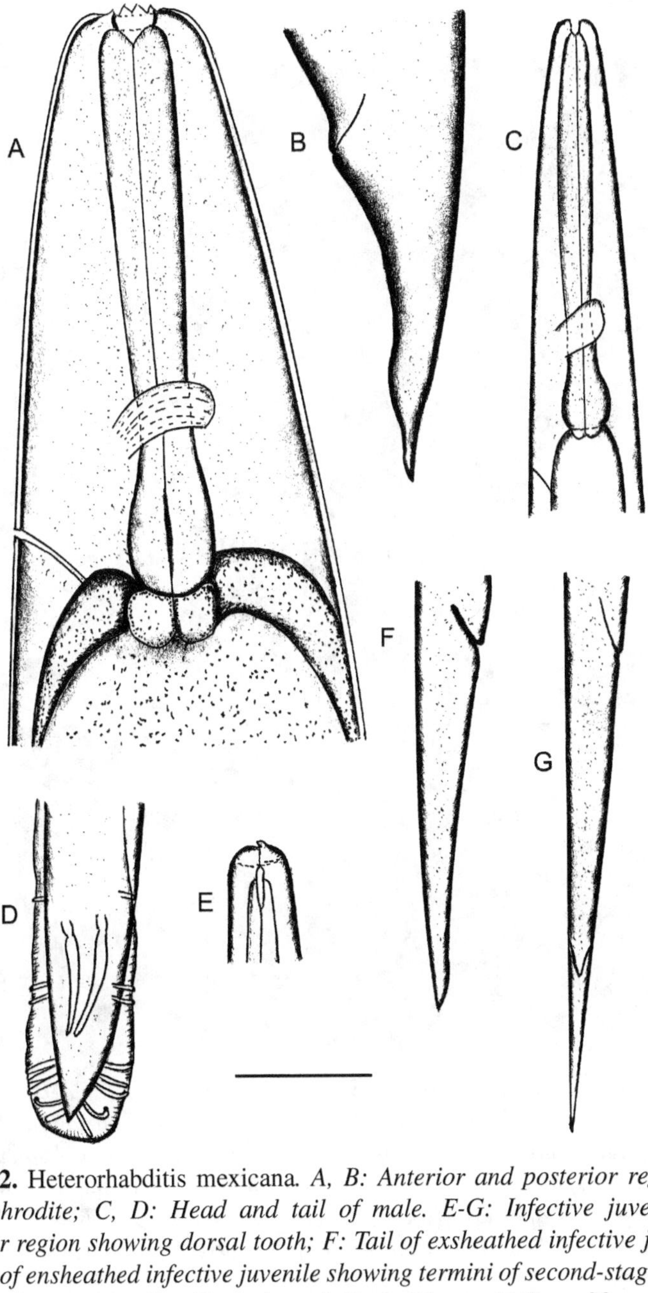

Fig. 232. Heterorhabditis mexicana. *A, B: Anterior and posterior regions of hermaphrodite; C, D: Head and tail of male. E-G: Infective juvenile. E: Anterior region showing dorsal tooth; F: Tail of exsheathed infective juvenile; G: Tail of ensheathed infective juvenile showing termini of second-stage cuticle and third-stage juvenile. (Scale bar: A-D = 50 μm; E-G = 29 μm.) After Nguyen* et al. *(2004), courtesy of* Nematology.

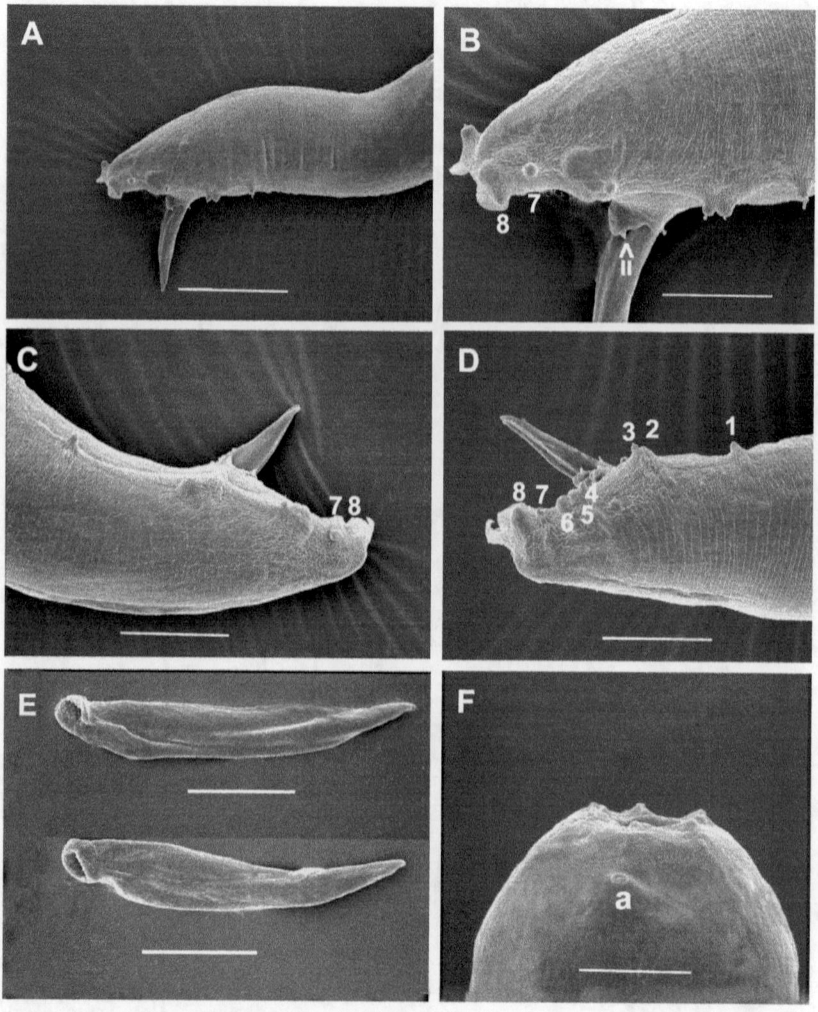

Fig. 233. Heterorhabditis mexicana. *SEM photographs of males. A-D: Posterior region showing bursal papillae (1, 2, 3, 4, 5, 6, 7, 8) with variation in shape, small papilla on posterior edge of cloacal aperture (arrow), and spicules; E: Two forms of spicule; F: Head region, showing six labial papillae, and one of two amphidial apertures (a). (Scale bars: A = 27.3 μm; B = 12 μm; C, D = 15 μm; E = 12 μm; F = 3.8 μm.) After Nguyen* et al. *(2004), courtesy of* Nematology.

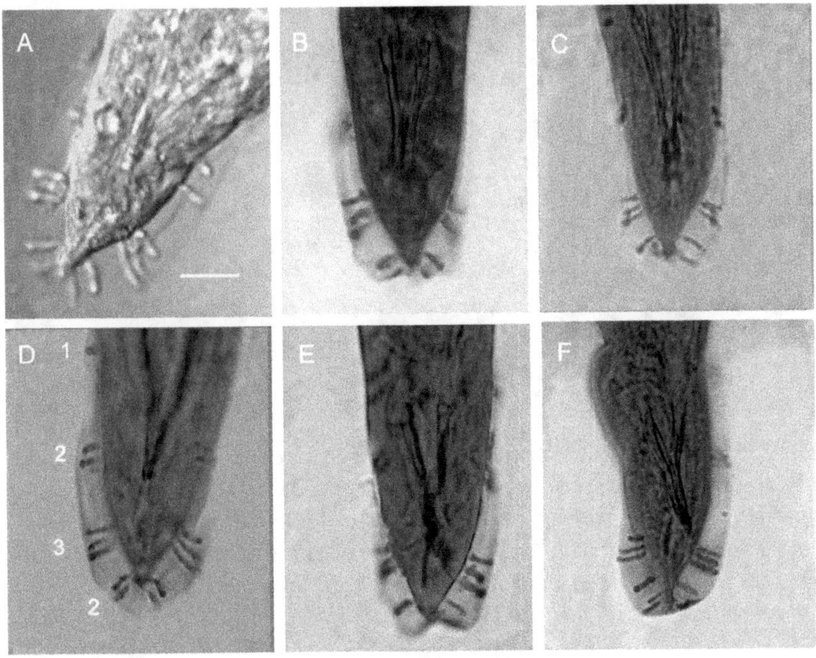

Fig. 234. Heterorhabditis mexicana. *Male tails with bursal papillae. A-D: Bursa with two pairs of bursal papillae in terminal group; E, F: Bursa showing variation in number of papillae in terminal group. (Scale bar = 20 μm.) After Nguyen* et al. *(2004), courtesy of* Nematology.

Hermaphroditic female

Body robust, C-shaped when heat-relaxed, always with many eggs in young worms and many eggs and juveniles in mature worms. Cuticle appearing smooth under light microscope but finely annulated with SEM. Labial region with six prominent lips, each with a terminal labial papilla. Cephalic papillae not observed with SEM. In face view, oral aperture hexagonal in shape as in amphimictic female. Amphidial apertures small. Stoma with refractile cheilorhabdions appearing as a circle posterior to lips. Posterior part of stoma funnel-shaped, enclosed by anterior part of pharynx. Pharynx with cylindrical corpus. Isthmus present. Nerve ring located at middle of isthmus. Basal bulb prominent with inconspicuous valve, lumen of pharynx in basal bulb well sclerotised. Cardia present. Genital system didelphic, amphidelphic. Vulva in form of a transverse slit, located on a protuberant area anterior to mid-body. In ventral view, vulval region appearing smooth, elliptical (this structure

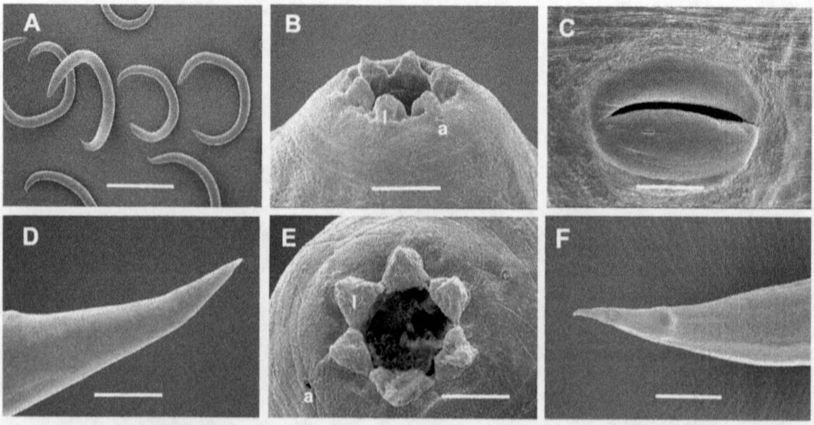

Fig. 235. Heterorhabditis mexicana. *A-D: SEM photographs of hermaphroditic female. A: C-shaped body of hermaphrodite; B: Anterior region showing six prominent lips (l) bearing terminal labial papillae, and one of two small amphidial apertures (a); C: Vulva showing elliptical pattern; D: Tail. E, F: SEM photographs of amphimictic female. E: Anterior region showing six prominent lips, six labial papillae and amphidial aperture (a); F: Tail. (Scale bars: A = 1000 μm; B = 6 μm; C = 8.6 μm; D = 23 μm; E = 3 μm; F = 30 μm.) After Nguyen* et al. *(2004), courtesy of* Nematology.

was observed on at least 20 specimens under SEM). Vagina short. Tail longer than anal body diam., conoid with pointed terminus. Sometimes, near terminus, tail widening then narrowing to a pointed tip. Postanal swelling present. Phasmid inconspicuous.

Amphimictic female

Similar to hermaphroditic female but smaller. Vulva not as wide open as in hermaphrodite.

Infective juvenile

Ensheathed infective juvenile (third-stage inside second-stage cuticle) with body length close to that of *H. bacteriophora* and *H. indica*. Labial region with seven annules, longitudinal incisures absent. Anterior part of body (*ca* 3-5 body diam. from anterior end) with tessellate pattern, posterior part with longitudinal ridges. Tail long, pointed. Exsheathed infective juvenile body annulated, without longitudinal ridges. Labial region bearing dorsal tooth; labial and cephalic papillae absent. Amphidial apertures prominent. Excretory pore located posterior to nerve ring but anterior to base of pharynx. Pharynx typical for

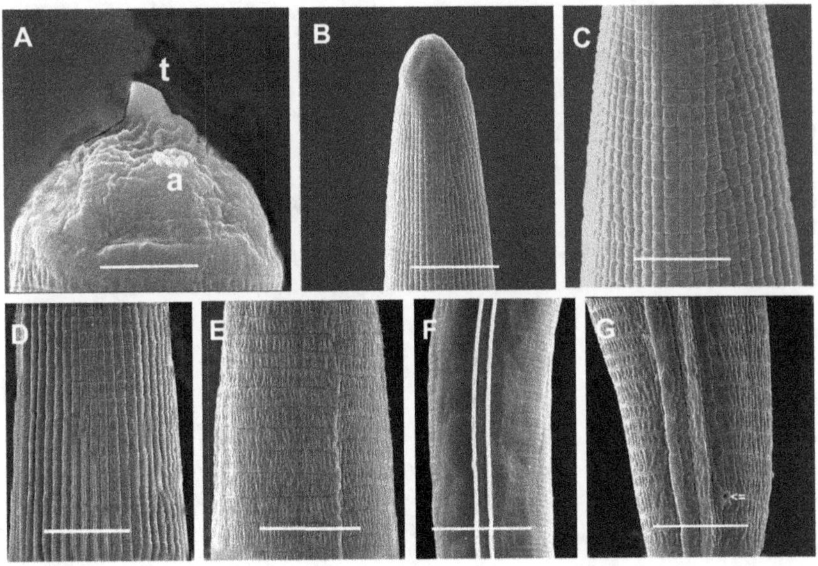

Fig. 236. Heterorhabditis mexicana. *SEM photographs of infective juveniles. A: Anterior region of exsheathed infective juvenile showing dorsal tooth (t) and amphidial aperture (a); B: Anterior region of ensheathed nematode showing head with seven annules and body with tessellate pattern; C: Enlargement of B showing more details of tessellate pattern; D: Portion of body showing parallel ridges and tessellates; E: Lateral field of exsheathed nematode beginning with one line; F: Lateral field at mid-body with two ridges; G: Lateral field in posterior region showing two ridges and a phasmid (arrow), just dorsal to lateral field. (Scale bars: A = 2 μm; B = 8.6 μm; C, E = 4.3 μm; D = 6 μm; F = 10 μm; G = 5 μm.) After Nguyen* et al. *(2004), courtesy of* Nematology.

genus. Lateral field beginning anteriorly with one ridge. A short distance posteriorly (2-3 body diam.), another ridge appearing, making two ridges in all. This pattern remaining unchanged for remainder of body. Phasmid located near mid-tail, pore-like, prominent, positioned just dorsal to lateral field (ventral to lateral field in *Steinernema*; see Nguyen & Smart, 1993 and Nguyen & Adams, 2003). Tail elongate conoid with bluntly pointed terminus. Tail length without sheath *ca* 67% of tail with sheath.

TYPE HOST AND LOCALITY

Sorghum field, approximately 4 miles south of Nuevo Progresso, in the northern part of the state of Tamaulipas, Mexico.

Table 69. *Morphometrics of* Heterorhabditis mexicana. *Measurements are in μm and in the form: mean ± SD (range).*

Character	Holotype (male)	Male	Hermaphrodite	Female	Infective juvenile
n	–	20	20	20	25
L	717	686 ± 38 (614-801)	3516 ± 543 (2440-4606)	1691 ± 253 (1144-2108)	578 ± 23 (530-620)
a	–	–	–	–	25.8 (23.6-28.4)
b	–	–	–	–	4.6 (4.2-5.1)
c	–	–	–	–	5.9 (5.5-6.3)
V	–	–	43 ± 1.6 (40-46)	47 ± 1.7 (44-51)	–
Greatest body diam.	33	42 ± 3 (38-47)	179 ± 31 (135-267)	99 ± 14 (65-123)	23 ± 0.8 (20-24)
EP	123	124 ± 10 (108-145)	171 ± 22 (103-201)	130 ± 12 (114-148)	102 ± 5.2 (83-109)
NR	77	71 ± 6 (61-83)	148 ± 17 (114-171)	94 ± 8 (76-103)	81 ± 4.2 (74-88)
ES	105	96 ± 5 (89-108)	192 ± 14 (168-221)	139 ± 8 (121-150)	122 ± 27 (104-142)
Testis reflexion	91	96 ± 20 (65-130)	–	–	–
Tail length with sheath (T)	–	–	–	–	99 ± 4.2 (91-106)
Tail length without sheath	26	27 ± 4 (21-36)	123 ± 22 (94-170)	87 ± 8 (76-106)	66 ± 3 (59-73)
Anal body diam. (ABD)	23	24 ± 1.3 (23-27)	45 ± 5 (30-58)	27 ± 5 (21-36)	15 ± 1.2 (12-17)
Spicule length (SP)	42	41 ± 3.8 (30-47)	–	–	–
Gubernaculum length (GL)	20	23 ± 3 (18-32)	–	–	–
D% = EP/ES × 100	–	129 ± 9 (114-149)	–	–	81 ± 3 (72-86)
E% = EP/T × 100	–	–	–	–	104 ± 5.2 (87-111)
SW% = SP/ABD × 100	183	167 ± 2 (130-196)	–	–	–
GS% = GL/SP × 100	48	56 ± 7 (43-70)	–	–	–

EP = distance from anterior end to excretory pore; NR = distance from anterior end to nerve ring; ES = distance from anterior end to end of pharynx.
After the original authors.

Type material

Holotype (male), allotype (female) and paratype (hermaphrodites, females, male, and infective juveniles in 3% formalin) isolated from haemocoel of *Galleria mellonella* deposited in the USDA Nematode Collection, Beltsville, MD, USA.

Relationships

Heterorhabditis mexicana can be distinguished from other species of *Heterorhabditis* by morphological characters of males and females. For males, 70% have a bursa with eight pairs of bursal papillae and they also have high GS and low SW values. Females have a typical vulva pattern. This species can be further distinguished from other species by molecular characteristics of the ITS regions of ribosomal DNA.

Morphologically, this species is most similar to *H. bacteriophora*, *H. brevicaudis* and *H. indica*, and can be distinguished from these species mainly by male and female characters. For males of *H. mexicana*, 70% have eight pairs of bursal papillae compared to nine or a variable number in other species (Fig. 234). GS values are higher than that of *H. bacteriophora*, *H. brevicaudis* and *H. indica*, and the SW value is lower than that of all other species. For females, the vulval pattern of *H. mexicana* (Fig. 235C) is quite different from that of the other morphologically similar species *H. bacteriophora* and *H. indica* (Nguyen & Smart, 1998). This pattern resembles that of *H. zealandica* as indicated by Adams and Nguyen (2002). The vulval pattern of *H. brevicaudis* was not available for comparison. *Heterorhabditis mexicana* can be distinguished from *H. megidis*, *H. zealandica* and *H. marelatus* by the number of bursal papillae in the terminal group.

Molecular characterisation

Heterorhabditis mexicana is characterised by sequence data of the ITS regions (accession no. AY321478), sequence length = 1010 bp, ITS1 = 394 bp, ITS2 = 213 bp, and composition, A = 0.25545, C = 0.20198, G = 0.26040, T = 0.28218 (Nguyen *et al.*, 2006). This species has evolved 13 autapomorphic nucleotide character states, and differs from its sister taxon, *H. indica*, at 113 aligned positions of the complete ITS rDNA array.

More molecular formation is available in Nguyen *et al.* (2004).

DISTRIBUTION

Heterorhabditis mexicana is only known from the type locality of Tamaulipas in Mexico (Nguyen *et al.*, 2004).

References

ADAMS, B.J. & NGUYEN, K.B. (2002). Taxonomy and systematics. In: Gaugler, R. (Ed.). *Entomopathogenic nematology*. Wallingford, UK, CABI Publishing, pp. 1-33.

NGUYEN, K.B. & ADAMS, B.J. (2003). SEM and systematic studies of *Steinernema abbasi* Elawad *et al.*, 1997 and *S. riobrave* Cabanillas *et al.*, 1994 (Rhabditida: Steinernematidae). *Zootaxa* 179, 1-10.

NGUYEN, K.B. & SMART JR, G.C. (1993). Location of the phasmids on infective juveniles of *Steinernema glaseri*. *Journal of Nematology* 25, 625-627.

NGUYEN, K.B. & SMART JR, G.C. (1998). Morphology of the life stages of three *Heterorhabditis* spp. from the infective juvenile to the hermaphrodite. *Soil and Crop Science Society of Florida Proceedings* 57, 101-107.

NGUYEN, K.B., SHAPIRO-ILAN, D.I., STUART, R.J., MCCOY, C.W., JAMES, R.J. & ADAMS, B.J. (2004). *Heterorhabditis mexicana* n. sp. (Rhabditida: Heterorhabditidae) from Tamaulipas, Mexico, and morphological studies of the bursa of *Heterorhabditis* spp. *Nematology* 6, 231-244.

NGUYEN, K.B., GOZEL, U., KOPPENHÖFER, H.S. & ADAMS, B.J. (2006). *Heterorhabditis floridensis* n. sp. (Rhabditida: Heterorhabditidae) from Florida. *Zootaxa* 1177, 1-19.

Heterorhabditis taysearae[*,**] Shamseldean, El-Sooud, Abd-Elgawad & Saleh, 1996
(Fig. 237)

MEASUREMENTS

See Table 70.

DESCRIPTION

Male

Body ventrally curved in posterior region when heat-relaxed. Cephalic extremity truncate, or rarely slightly rounded. Six distinct, protuberant lips surrounding oral aperture, each with terminal papilla. Amphidial apertures oval. Cheilorhabdions present as a refractile ring just posterior to lips. Pro- and mesorhabdions reduced, bearing large tooth near posterior end of protorhabdions. Pharynx with fairly indistinct metacorpus, short isthmus and prominent basal bulb containing fine striations in valve area. Nerve ring distinct, located around anterior part of isthmus. Genital system monorchic, reflexed, spicules paired, separate, curved, with pointed tips. Gubernaculum almost half spicule length long, relatively broad and curved. Bursa peloderan with nine pairs of genital papillae. From anterior to posterior, pair 1 well anterior, pairs 2 and 3 in a group immediately anterior to cloaca; pairs 4, 5 and 6 forming a group, just posterior to cloaca. Pairs 7 and 8 not reaching rim of bursal membrane. Remainder of genital papillae straight, extending to bursal rim.

Female (hermaphroditic and amphimictic stages not distinguished in original description)

Anterior region similar to that of male but much larger. Hermaphroditic and amphimictic females with paired, amphidelphic ovaries, reflexed portion always extending past vulval aperture. Hermaphroditic females with sperm occurring in proximal portion of ovotestes; amphimictic females with sperm in proximal portion of oviduct. Vulval region protuberant, copulation plug absent. Anal swelling more pronounced in hermaphroditic than amphimictic females. Rectum slightly sclerotised.

[*] Named after A.H. El-Kifl *Taysea*.

[**] Description after Shamseldean *et al.* (1996). Revised by Khuong B. Nguyen.

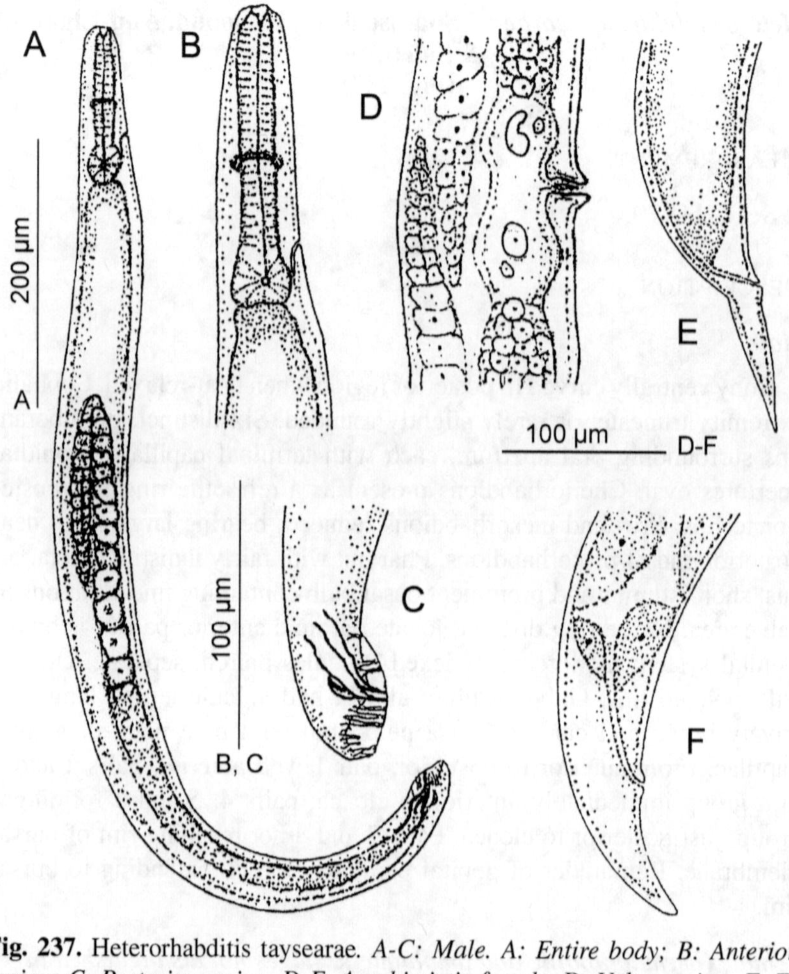

Fig. 237. Heterorhabditis taysearae. *A-C: Male. A: Entire body; B: Anterior region; C: Posterior region. D-F: Amphimictic female. D: Vulva region; E, F: Variation in tail shape. Modified after Shamseldean* et al. *(1996).*

Infective juvenile

Infective juvenile (ensheathed within second-stage cuticle) short. Second-stage cuticle closely adpressed to third-stage cuticle. Cuticle with fine longitudinal striations. Cephalic region with dorsal tooth. Stoma walls open, not collapsed at base. Oral aperture and anus closed. Opening of excretory pore located anterior to level of basal bulb. Nerve ring located in anterior part of isthmus. Hemizonid distinct. Ventricular portion of intestine with intestinal pouch filled with symbiotic bacteria.

Table 70. *Morphometrics of* Heterorhabditis taysearae. *Measurements are in μm and in the form: mean $\pm$ SD (range).*

Character	Holotype (female)	Male	Hermaphrodite	Female	Infective juvenile
n	–	20	20	20	30
L	950	703 ± 23 (648-736)	2510 ± 49 (2200-2800)	1200 ± 16 (830-1400)	418 ± 38 (332-499)
a	–	–	–	–	21 ± 2.2 (18-27)
b	–	–	–	–	3.8 ± 0.2 (3.4-4.2)
c	–	–	–	–	7.7 ± 0.7 (6.5-8.7)
V	54	–	49 ± 5.3 (40-64)	57 ± 8.3 (44-73)	–
Greatest body diam.	63	43.5 ± 2.3 (38-48)	134 ± 21 (116-170)	69 ± 15 (42-96)	20 ± 1.9 (17-23)
EP	130	95 ± 12.3 (78-120)	163 ± 13.2 (137-182)	137 ± 14 (120-166)	90 ± 9.1 (74-113)
NR	81	65 ± 11.6 (54-88)	92 ± 9.2 (83-120)	82 ± 6.8 (76-109)	64 ± 6.8 (58-87)
ES	147	112 ± 12.5 (85-123)	176 ± 14.5 (161-200)	151 ± 13 (129-179)	110 ± 8.4 (96-130)
Testis reflexion	–	122 ± 14.4 (100-146)	–	–	–
Tail length with sheath (T)	–	–	–	–	55 ± 6.6 (44-70)
Tail length without sheath	68	25 ± 5.4 (20-29)	85 ± 13.4 (72-100)	71 ± 7.4 (62-80)	–
Anal body diam. (ABD)	21	25 ± 2.9 (21-30)	50 ± 6.9 (41-67)	23 ± 2.5 (19-28)	–
Spicule length (SP)	–	39 ± 4.8 (30-42)	–	–	–
Gubernaculum length (GL)	–	(18 ± 3.2) (14-21)	–	–	–
D% = EP/ES × 100	–	88	–	–	82 ± 6 (71-96)
E% = EP/T × 100	–	–	–	–	180 ± 27 (110-230)
SW% = SP/ABD × 100	–	156	–	–	–
GS% = GL/SP × 100	–	46	–	–	–

EP = distance from anterior end to excretory pore; NR = distance from anterior end to nerve ring; ES = distance from anterior end to end of pharynx.
After the original authors.

Bacteria also found in intestinal lumen, mainly in anterior portion. Tail short.

TYPE HOST AND LOCALITY

The nematode was collected from soil in Al-Husane village in West Nubaria, Behera Governorate, Egypt.

TYPE MATERIAL

Holotype amphimictic female isolated from the haemocoel of *G. mellonella* last instar larvae (Lepidoptera: Galleridae) deposited in the Nematode Collection at the Faculty of Agriculture, University of Cairo, Egypt.

RELATIONSHIPS

This nematode is characterised by the short body length of the infective juvenile at 418 (332-499) μm, short tail length of 55 (44-70) μm, and high value of E% (180). The excretory pore of the male was reported to be anterior to the pharynx base.

The short body of the infective juvenile can be used to separate this species from all other species of the genus *Heterorhabditis*.

MOLECULAR CHARACTERISATION

RFLP patterns of *H. taysearae* were compared with those of *H. megidis*, *H. marelatus*, *H. indica* and *H. bacteriophora* and indicated *H. taysearae* to be a valid species (Sansour *et al.*, 2002).

DISTRIBUTION

The nematode is only known from Egypt (Shamseldean, El-Sooud, Abd-Elgawad & Saleh, 1996).

References

SANSOUR, M., SANDOUKA, B., ABU SA'DA, A., EHLERS, R.-U. & IRAKI, N. (2002). The Middle East: A region of high diversity of entomopathogenic nematodes. Available from: http://www.cost850.ch/publications/20020404_proceedings.html. (Accessed on 16 March 2006.)

SHAMSELDEAN, M.M., ABOU EL-SOOUD, A.B., ABD-ELGAWAD, M.M. & SALEH, M.M. (1996). Identification of a new heterorhabditid species from Egypt, *Heterorhabditis taysearae* n. sp. (Rhabditida: Heterorhabditidae). *Egyptian Journal of Biological Pest Control* 6, 129-138.

Heterorhabditis zealandica[*,**] Poinar, 1990

= *H. heliothidis* apud Wouts, 1979
(Figs 238, 239)

MEASUREMENTS

See Table 71.

DESCRIPTION

Male

Cuticle smooth, lateral field absent. Labial region and stoma as in pre-adult female. Pharynx more than half body diam. wide, narrowing distinctly at level of nerve ring, basal bulb pyriform with distinct valve. Cardia large. Excretory pore located posterior to pharynx base. Genital system monorchic, reflexed, filling body diam. posteriorly and *ca* half body diam. anteriorly. Spicules slender, paired. Gubernaculum small, single. Bursa enveloping tail, supported by nine pairs of papillae, three pairs precloacal, six pairs postcloacal. Tail tapering, tip with clear, digitate, blunt terminus.

Hermaphroditic female

Cuticle smooth, lateral lines not observed. In small females labial region with pronounced labial papillae. In large females lip with wide shoulders surrounding stoma. Labial papillae large, flap-like. Cheilostom as long as broad with thick, non-sclerotised, walls. Protostom/telostom initially as broad as cheilostom, tapering posteriorly, in larger females becoming wider and shorter, funnel shaped. Pharynx short in relation to total length, narrowing at level of nerve ring. Basal bulb pyriform, large when compared to other stages but small in relation to body size, valve distinct. Cardia present. Excretory pore located at level of basal bulb. Genital tracts paired, reflexed. Uterus and spermatheca located ventrally between vulva and reflex, ovaries located dorsally beyond reflex. Vulva lips protruding. Rectum and anus distinct. Body slightly widening near anus. Tail tapering rapidly over anterior half, especially in large females, more slowly over posterior half. Small females usually

[*] Named after New *Zealand*, the country where it was first found.

[**] Description after Wouts (1979). Revised by Khuong B. Nguyen.

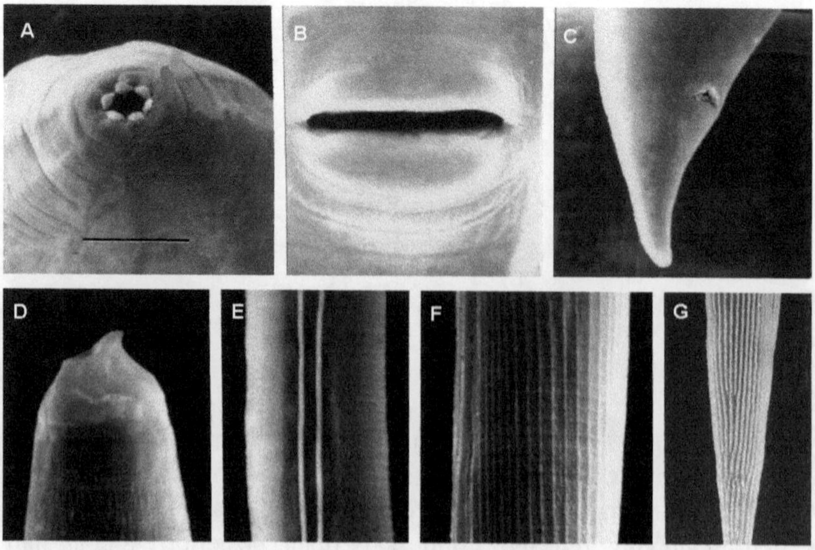

Fig. 238. Heterorhabditis zealandica. *SEM photographs. A: Face view of hermaphroditic female showing six prominent lips; B: Vulva pattern; C: Tail of hermaphroditic female; D: Anterior region of infective juvenile showing dorsal tooth; E: Lateral field of infective juvenile showing two ridges; F, G: Second-stage cuticle showing longitudinal ridges. (Scale bar: A = 25 μm; B = 17.8 μm; C= 30 μm; D = 5 μm; E = 12.1 μm; F, G = 10 μm.)*

with slight swelling in posterior half of tails, which tends to disappear in larger animals when tail swells, tail tip usually rounded.

Amphimictic female

Identical to small first generation females. Differ from large first generation females as follows: typical protruding labial papillae never develop into elaborate flaps; wide shouldered labial region and wide open, funnel shaped, protostom/telostom of large first generation female absent, usually no eggs are laid; no sperm observed; vulva occasionally excreting mucus, something not observed in first generation females.

Infective juvenile

Body slender, tapering regularly from base of pharynx to anterior end and from anal area to tail terminus. Cuticle with large number of longitudinal lines starting short distance posterior to head and terminating on tail. Transverse annulation present between labial region and level of excretory pore. Labial region smooth, not set off, bearing

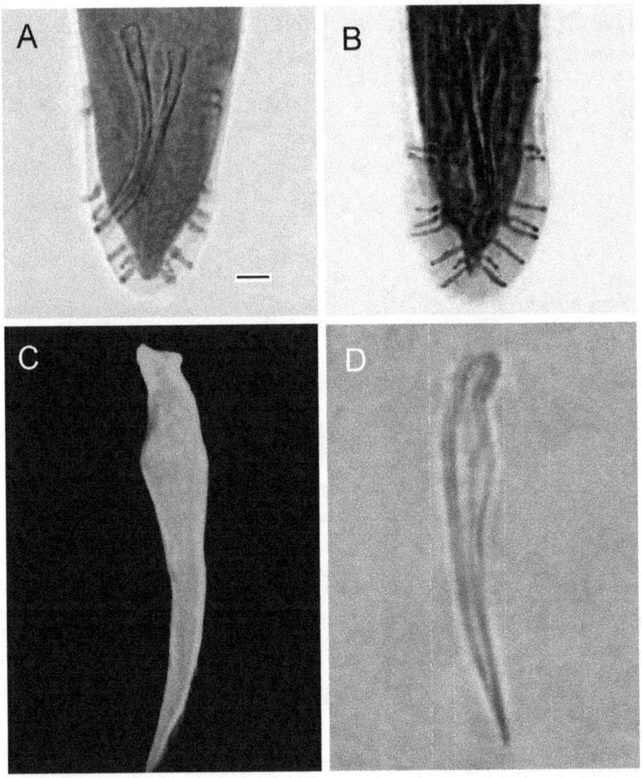

Fig. 239. Heterorhabditis zealandica *bursa and spicules. A, B: Bursa of strains NZH3 (A) and Florida (B) showing terminal group of bursal ribs with a short central rib; C, D: Spicules showing prominent rostrum. (Scale bar: A, B = 20 μm; C = 3.3 μm; D = 4.2 μm.)*

six lips, six labial papillae, and two amphidial apertures. Cheilostom narrow, rounded in face view, asymmetrically running between thick lips of cephalic region. Protostom/telostom comprising a wider triangular tube of sclerotised, indistinct, plates anteriorly and with continuous cylindrical lining posteriorly. Pharynx with triradiate lumen, long, narrow, one-third to one-fourth body diam. wide, narrowest at level of nerve ring, terminating in valvate bulb. Nerve ring distinct. Cardia present. Excretory pore and hemizonid obscure. Anterior lumen of intestine wide, containing bacteria. Intestine wall dark with globules. Genital primordium obscure. Sphincter muscle between rectum and intestine distinct. Rectum thick walled, anus distinct. Tail conical with very narrow pointed terminus. Phasmids and deirids not observed.

Table 71. *Morphometrics of* Heterorhabditis zealandica. *Measurements are in μm and in the form: mean (range).*

Character	Male	Infective juvenile
n	15	25
L	914 (848-1044)	685 (570-740)
a	–	25
b		4.9
c	–	6.7
Greatest body diam.	41 (36-45)	27 (22-30)
EP	139 (130-150)	112 (94-123)
NR	–	100 (90-107)
ES	118 (110-128)	140 (135-147)
Tail length with sheath (T)	–	102 (87-119)
Tail length without sheath	35 (30-41)	–
Anal body diam. (ABD)	20.7 (18-24)	–
Spicule length (SP)	51 (48-55)	–
Gubernaculum length (GL)	22.2 (19-25)	–
D% = EP/ES × 100	118	80
E% = EP/T × 100	–	109
SW% = SP/ABD × 100	246	–
GS% = GL/SP × 100	44	–

EP = distance from anterior end to excretory pore; NR = distance from anterior end to nerve ring; ES = distance from anterior end to end of pharynx.
After Wouts (1979).

TYPE HOST AND LOCALITY

The species was collected from an adult black beetle *Heteronychus arator* (F.) from pastures near Auckland, New Zealand.

TYPE MATERIAL

According to Poinar (1990), the holotype (male) and allotype (infective juvenile) are deposited in the National Museum in Wellington, New Zealand.

Remarks

The species was originally described by Wouts (1979) as a population of *H. heliothidis* (now regarded as a junior synonym of *H. bacteriophora*). Poinar (1990) recognised it as a separate species and formally proposed the specific epithet.

Relationships

Heterorhabditis zealandica is characterised by body length (685 μm), E% (108) and c ratio (6.6) of the infective juvenile. Other good diagnostic characters include the fact that the spicule has a prominent rostrum (Fig. 239C, D) and the terminal group of bursal ribs has three pairs of papillae with the central rib shorter than the others. For the female, the vulva shape is typical (Fig. 238B).

Heterorhabditis zealandica can be distinguished from IJ of the other species of the genus by the long body, distance from the anterior end to the excretory pore, and by the pharynx and tail length.

Molecular characterisation

The ITS rDNA regions, flanked by primers 18S and 26S, of *H. zealandica* are characterised by length, 1003 base pairs (bp), ITS1 = 387 bp, ITS2 = 212 bp, and nucleotide usage composition, A = 0.22233, C = 0.21834, G = 0.27318, T = 0.28614. The sequence of the ITS regions is deposited in GenBank under accession no. AY321481.

Bionomics

The bionomics of *H. zealandica* are similar to other *Heterorhabditis* spp. except that the associated bacterium is *Photorhabdus temperata.*

Distribution

According to Hominick (2002) the species has, in addition to New Zealand, been found in Lithuania, Russia and Australia. Using DNA sequences, Nguyen (unpubl.) has also identified it from the USA (Florida and South Carolina) and South Africa.

References

HOMINICK, W.M. (2002). Biogeography. In: Gaugler, R. (Ed.) *Entomopathogenic nematology.* Wallingford, UK, CABI Publishing, pp. 115-143.

POINAR JR, G.O. (1990). Taxonomy and biology of Steinernematidae and Heterorhabditidae. In: Gaugler, R. & Kaya, H.K. (Eds). *Entomopathogenic nematodes in biological control.* Boca Raton, FL, USA, CRC Press, pp. 23-60.

WOUTS, W.M. (1979). The biology and life cycle of a New Zealand population of *Heterorhabditis heliothidis* (Heterorhabditidae). *Nematologica* 25, 191-202.

Nematology Monographs & Perspectives, 2007, Vol. 5, 693-733

Chapter 6

Phylogeny and evolution

Byron J. ADAMS, Scott M. PEAT and Adler R. DILLMAN

Microbiology & Molecular Biology Department, and Evolutionary Ecology Laboratory, Brigham Young University, Provo, UT 84602-5253, USA

Introduction

Nematodes originated during the Precambrian or Cambrian explosion over 500 million years ago (Wray *et al.*, 1996; Ayala & Rzhetsky, 1998; Rodriguez-Trelles *et al.*, 2002). The phylogenetic position of the Nematoda relative to other metazoans has historically been one of contention. Originally circumscribed within the Vermes Linnaeus, 1758 and later the Aschelminthes Grobben, 1910 (Claus & Grobben, 1910), the Nematoda are now believed to belong to a clade of moulting animals, the Ecdysozoa (Aguinaldo *et al.*, 1997), and share a most recent common ancestor with arthropods, kinorhynchs, nematomorphs, onychophorans, priapulids and tardigrades.

ORIGINS OF ENTOMOPATHOGENIC NEMATODES (EPN)

Entomopathogenic nematodes of the Steinernematidae and Heterorhabditidae are not monophyletic, but likely began independently to explore biotic relationships with arthropods and Gram-negative enteric bacteria (Enterobacteriaceae) by the mid-Palaeozoic (approximately 350 million years ago) (Poinar, 1993). Their origins were probably not synchronous and the ages of their respective lineages appear to be significantly different. Evidence for the disparate origins and relative age of these Families is illustrated in Figure 240. Assuming even a somewhat sloppy molecular clock, the long branch lengths of *Steinernema*, both within the genus and relative to its most recent common ancestor, imply that it has been evolving independently of other nematode lineages for a longer period of time than *Heterorhabditis*. This could explain, in part,

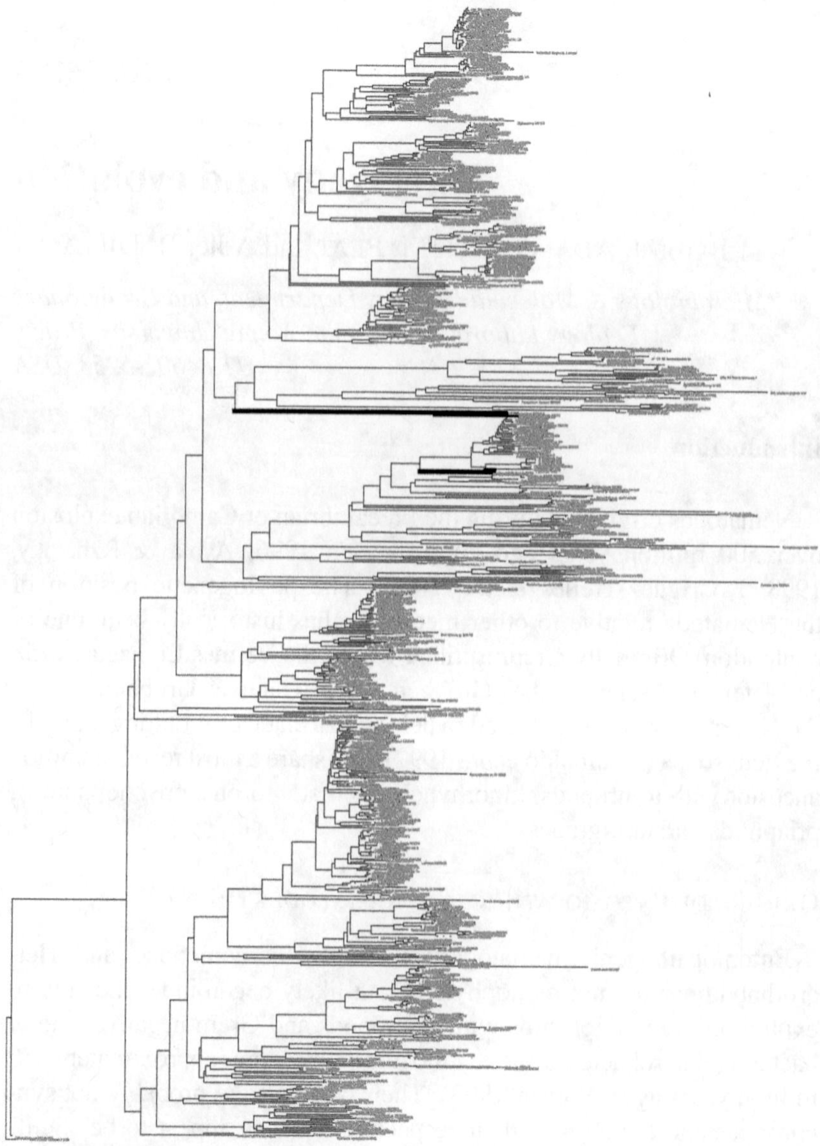

Fig. 240. *Phylogenetic tree of 434 representative taxa from the Rhabditida, Plectida and Monhysterida. The tree was generated from 18S rDNA sequences by the neighbour joining BioNJ algorithm using log determinant transformed distances in order to account for rate and nucleotide usage heterogeneity among lineages. EPN branches are in bold.*

the differences in species richness between the two genera (*i.e.*, *Heterorhabditis* has had less time to generate species). However, if EPN in general have been coevolving with their insect hosts for 350 million years, why do we not see numbers of entomopathogenic nematodes equal to the species richness of their insect hosts? Even if the number of species of insect hosts were whittled down to only those that have a soil-borne larval stage, the host-parasite species deficit would still be tremendous.

ORIGINS OF EPN SPECIES

Little is known of the modes of speciation that have produced the extant biodiversity of EPN. Because most EPN sister species appear to be (or have been) geographically isolated (Hominick *et al.*, 1996; Hominick, 2002), vicariant allopatric speciation, in most cases followed by dispersal, is the most likely mode of speciation to explain their distribution (Lynch, 1989). There are several hypothetical reasons as to why we might predict that there should be more species of EPN than have been described to date. First of all, there appears to be a correlation between species size and the probability of extinction (Jackson, 1974; Jablonski, 1987). Species with large range sizes, such as bison and woolly mammoths, require more resources and space and, therefore, have a lower probability of speciating and a higher probability of extinction. By contrast, animals such as EPN that have a small body size may be able to partition more niche space per unit area (Hutchinson & MacArthur, 1959; Morse *et al.*, 1985). Therefore, because nematodes have small spatial requirements, they should have higher speciation rates and greater niche filling or niche partitioning opportunities. Just such partitioning is evident in studies of habitat preference (Sturhan, 1999) and could be maintained phylogenetically (Spiridonov *et al.*, 2004). However, nematodes could have high rates of dispersal, as evidenced by the inter-

Fig. 240. *(Continued). The upper bold lines depict the lineages of* Steinernema carpocapsae *and* S. glaseri *relative to their most recent common ancestor. The lower bold lines illustrate the branching point of* Heterorhabditis marelatus *and* H. bacteriophora *relative to their most recent common ancestor. Note the long branch length of the steinernematid lineage relative to the heterorhabditid lineage, and the long branch lengths of the steinernematid species relative to the heterorhabditid species. The multiple sequence alignment for this clade was retrieved from NemaTOL (http://nematol.unh.edu).*

and transcontinental geographic ranges of many species (Adams *et al.*, 2006); their effective range requirements could be vast, effectively decreasing the probability of speciation. In the case of EPN, their small body size, coupled with an association with mobile insect hosts, could facilitate dispersal and gene flow across expansive geographic ranges, resulting in retarded speciation rates or extinction probabilities comparable to those of megafauna (but see Blouin *et al.*, 1999).

Additional explanations for the patterns of EPN biodiversity are legion but the rate of molecular evolution can probably be ruled out, as it has been shown that for multiple genes nematodes evolve at anomalously high rates relative to other metazoans (*e.g.*, Wray *et al.*, 1996). Rather, mechanisms of host specificity and virulence may exert greater influence on the origin of species (or lack thereof) in these genera than other factors (Figs 241, 242). For example, if the nematode depends on the survival or extended mobility of its insect host to vector it to other potential hosts (offspring, mates, *etc.*), then there will be strong selection pressure to reduce virulence because if the host is killed too quickly the nematode will have extinguished the resources or transmission mechanisms required to continue its life cycle (net positive differential reproduction). Reliance on the host as a vector should favour nematode-insect host specificity because the nematode must closely track the evolutionary trajectory of its host. According to this scenario, as the insect speciates, the nematode species must also adapt to changes in its host that involve the nematode's ability to maintain a symbiotic relationship. This would result in tight host-tracking and a longer-term relationship between an individual nematode and its insect host that leads to decreased opportunities for gene flow among other members of its population that have a broader host range. Such a scenario promotes evolutionary fidelity between the nematode and insect host such that if the insect has a high rate of speciation, the nematode must also speciate rapidly as it tracks its host over time. Thus, entomogenous nematodes with low virulence are predicted to have simultaneously narrower host ranges, lower cophylogenetic fidelity, lower rates of gene flow, and comparable rates of speciation relative to those of their insect host. Under this scenario, we would expect strong selection for the evolution of avirulence. Production of toxins or reproduction would increase only to the point that it hindered host mobility or longevity (Fig. 241).

Relative to other families of nematodes, or clades for that matter (Figs 240, 243), the Steinernematidae and Heterorhabditidae contain rel-

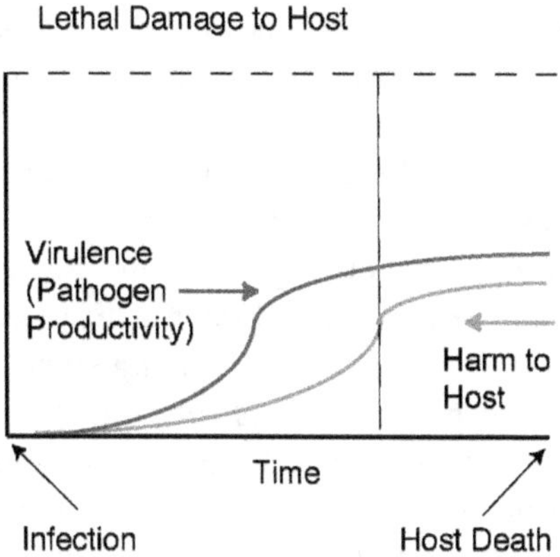

Fig. 241. *Selection between hosts, where inclusive fitness is a function of host mobility/survival. When insect host mobility or longevity is required for successful completion of the nematode life cycle, EPN/B virulence, as measured by either toxin production or reproduction, increases at a maximal rate to establish infection and reproduction, but tapers off as it begins to influence host survival. Such a scenario favours the evolution of increasing avirulence, narrow host ranges, increased nematode-insect host fidelity (cophylogeny), decreased gene flow among nematode populations, and increased rates of speciation.*

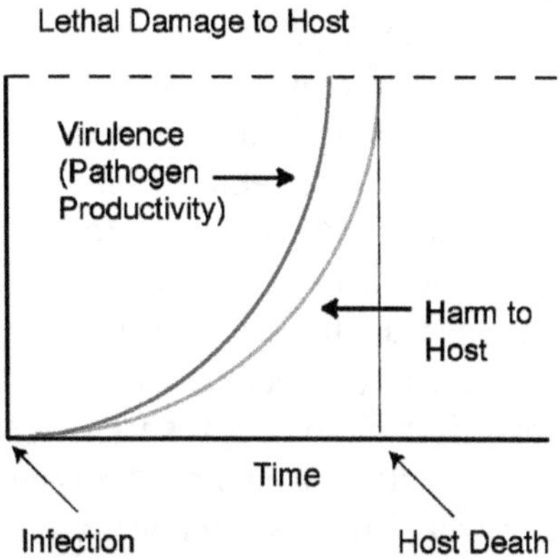

Fig. 242. *Selection within hosts, where inclusive fitness is a function of the rate of reproduction in the host. Competition for resources among different pathogens in an insect host can lead to selection pressures that favour increased virulence. Accordingly, virulence, as measured by toxin production or reproduction, increases at a maximum rate culminating with the rapid death of the host insect. Such a scenario favours the evolution of increasing virulence, broad host ranges, decreased host fidelity (cophylogeny), increased gene flow, and decreased rates of speciation.*

atively few species. This may be due, in part, to the discrepancy between the number of EPN species that currently exist and the number of EPN species that have been described (Liu *et al.*, 2000). Taxonomic efforts aside, it may be that the highly virulent nature of these nematode-bacterium complexes works to retard speciation in these nematodes. For example, the rapid rate at which the bacterium kills its insect host may have evolved in response to the highly competitive environment inside the host cadaver where bacterial colonies are under strong selection pressure to be as virulent as possible. Considering that the insect represents a rich carbon resource for soil microbes and the tremendous diversity of soil microbes positioned to utilise it, individual bacterial colonies that can get into the insect, utilise cadaver resources and replicate the fastest have the greatest chance of being vectored into the next insect by the emergent IJ. Under these conditions, natural selection for high rates of within-host competition results in virulence which continually increases to an optimum that is tempered only by the rate at which the nematode can evolve tolerance to the bacterial toxins (Fig. 242). The rapid death of the host obviates its ability to vector the nematodes (and their endosymbiotic bacteria) back to insect host nests, breeding sites, mates or offspring. The result is that as the infective juveniles emerge from their host's cadaver, they may be equally, if not more, likely to encounter a potential insect host that is a different species to the one from which they just emerged. Thus, the increase in virulence that is generated by competition within the host reinforces the requirement for a broad host range, and *vice versa*. Similarly, high virulence and broad host ranges decrease the requirement for nematodes to establish host-specific relationships. Subsequently, as hosts speciate, there is no selective pressure on the nematodes to do the same. Thus, nematodes that are highly virulent and have broad host ranges will tend to be less speciose, so long as dispersal is high and barriers to gene flow remain relatively inconsequential. Nematodes that have narrow host ranges and lower virulence should in general be more species rich. The prediction that an increase in host range is concomitant with increased gene flow is likely countered by the rate of mortality of infected hosts (low mobility) and the insularity of the soil environment (Blouin *et al.*, 1999). The effect (evolution of increasing virulence) is crudely reflected across larger phylogenetic space in Figure 243. That virulence could be maintained, and even increased over evolutionary time, counters the widely held view that virulence is largely only manifest in the initial stages of symbiosis, and that the pathogenic-

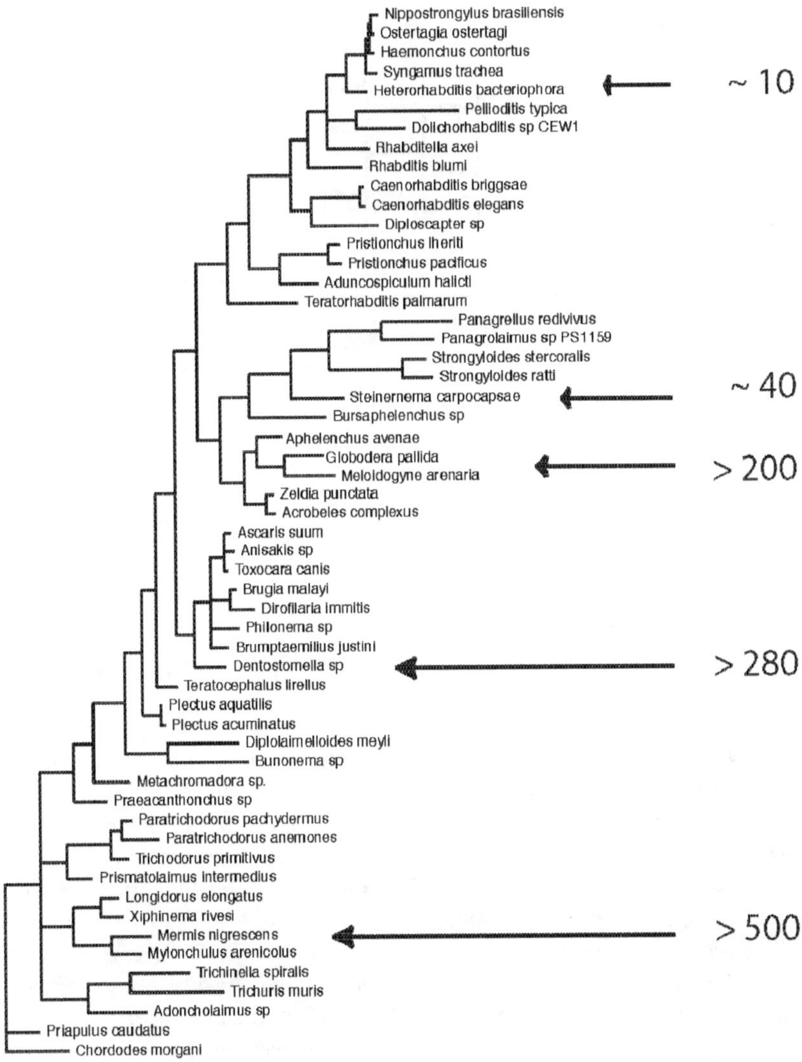

Fig. 243. *Species richness of select entomogenous nematode clades.* Heterorhabditis *and* Steinernema *have relatively few species, broad geographic ranges, little host fidelity and high virulence. Although it is predicted that an increase in host range is concomitant with increased gene flow, this is likely countered by the rate of mortality of infected hosts (low mobility) and the insularity of the soil environment (Blouin* et al., *1999). Note also the influence of sampling effort (some groups have been investigated more intensively than others) and species richness increases over evolutionary time (earlier branching taxa are also more speciose). Tree modified from Blaxter* et al. *(1998).*

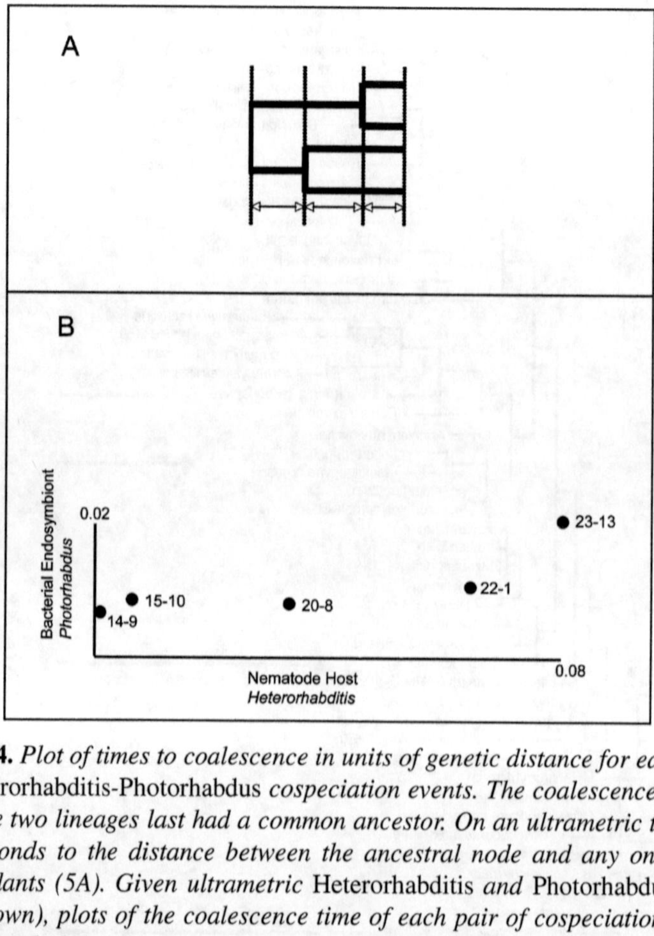

Fig. 244. *Plot of times to coalescence in units of genetic distance for each pair of* Heterorhabditis-Photorhabdus *cospeciation events. The coalescence time is the time two lineages last had a common ancestor. On an ultrametric tree this corresponds to the distance between the ancestral node and any one of its descendants (5A). Given ultrametric* Heterorhabditis *and* Photorhabdus *trees (not shown), plots of the coalescence time of each pair of cospeciation nodes (5B) reveal if the times to coalescence are correlated (they are: $r = 0.82$; $P < 0.05$) and also whether the timing of the codivergence events are pre-emptive (plotted line crosses a positive y-axis;* Photorhabdus *diverges earlier), synchronous (line crosses at the origin; they diverge at the same time) or delayed (line crosses a negative y-axis;* Heterorhabditis *diverges earlier (Page & Hafner, 1996). The number pairs of the plots refer to the paired nodes of the* Heterorhabditis-Photorhabdus *trees, respectively. The positive slope in the figure suggests pre-emptive evolution of the bacterium followed by a tracking response of the nematode host.*

ity of the parasite decreases over time. Work has shown that the evolution of virulence can be highly plastic and driven by adaptive virulence-associated responses by both host and parasite (Perlman & Jaenike,

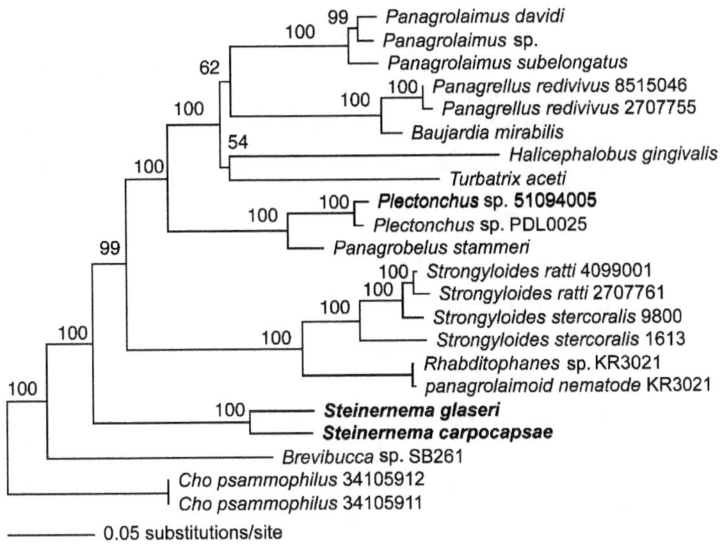

Fig. 245. *Phylogenetic position of* Steinernema *relative to 20 other most closely related nematodes for which comparable sequences are publicly available. Numbers on branches indicate node support (bootstrap, 1000 BioNJ reps) mapped on a tree produced with SSU sequences using the BioNJ neighbour-joining algorithm (Gascuel, 1997). Note sister taxon relationship with sampled Strongyloidoidea and Panagrolaimidae; see also Figure 243 and Nadler* et al. *(2006a).*

2003), that often these responses can involve increases in virulence (Herre, 1993) and are correlated with a lack of concerted host-parasite coevolution (Perlman *et al.*, 2003). In *Heterorhabditis* a piece of tenuous evidence supports the notion that the nematode is tracking changes that first occur in its bacterial symbiont. When times to coevolving coalescent events (time since two lineages shared a common ancestor)between species of *Heterorhabditis* and their bacterial endosymbionts *Photorhabdus* are plotted in terms of genetic distance, the correlation is significant (r $= 0.82$; $P < 0.05$) and the slope crosses the y axis positively, suggesting that the nematode is exhibiting a delayed response to evolutionary diversification of its symbiont even though the gene used to reconstruct the phylogenetic history of *Heterorhabditis* (ITS) is evolving at a faster rate than the gene used to reconstruct *Photorhabdus* phylogeny (16S) (Fig. 245). Such analyses provide insight into the degree and mechanisms by which nematode and endosymbiont can influence each other's evolutionary trajectory. The small sample sizes and somewhat crude analyses caution that it is still preliminary to suggest that much, if anything, at this point can be inferred with confidence about correlations between the evolution of virulence, coevolution, host-range, gene flow and biodiversity, but EPN are well positioned to serve as model organisms for comparative and experimental research programmes in this area.

Phylogeny

Numerous molecular approaches have been applied to the evolutionary relationships of EPN. Initial efforts included PCR RFLP analyses of the internal transcribed spacer (ITS) regions of the ribosomal DNA tandem repeat unit (Reid, 1994; Reid *et al.*, 1997). Other approaches included combinations of random amplified polymorphic DNA (RAPD) and morphological data (Liu & Berry, 1996). These papers still contain valid data, although their interpretations have been questioned due to analytical drawbacks, primarily insufficient numbers of phylogenetically informative characters, tenuous homology statements and tree building methods that are suboptimal for reconstructing evolutionary history (Stock *et al.*, 2001). While the use of morphological characters in phylogenetic analyses remains robust, the performance of RAPD markers has fallen out of favour, primarily because of problems associated with replication and sensitivity of the approach.

The ribosomal small subunit (SSU, or 18S gene) was the first to be sequenced in an effort to resolve phylogenetic relationships within and between *Steinernema* and *Heterorhabditis* (Liu *et al.*, 1997). While variation among species in both genera could be identified, the particular 5′ segment of the gene that was sequenced and analysed is evolving too slowly to produce the quantity of phylogenetically informative characters (substitutions) needed appropriately to infer relationships among species in their respective genera. Subsequently, the ribosomal large subunit (LSU), ITS and mitochondrial NADH dehydrogenase subunit 4 (ND4), and cytochrome oxidase I gene (Cox1) and small mitochondrial ribosomal RNA gene (16S) have been used to reconstruct the phylogenetic relationships among EPN (Adams *et al.*, 1998; Liu *et al.*, 1999; Szalanski *et al.*, 2000; Stock *et al.*, 2001; Nguyen *et al.*, 2001, 2004; Saeb *et al.*, unpubl.). As the SSU is too conserved to be informative of relationships among most EPN species, the ITS, 16S, COI and ND4 appear to be evolving more rapidly than is optimal for phylogenetic resolution at this taxonomic level (Adams *et al.*, 1998; Liu *et al.*, 1999;

Szalanski *et al.*, 2000; Nguyen *et al.*, 2001) where intraspecific and even intraindividual variability, although rare (see Adams *et al.*, 1998; Spiridonov *et al.*, 2004), exists. Nadler *et al.* (2006a) showed that for some *Steinernema* the 16S and COI mtDNA genes have incongruent evolutionary histories. Their fairly rapid rate of evolution means that the ITS, 16S, COI and ND4 genes perform well at revealing relationships among closely related species, and even populations (Adams *et al.*, 1998; Spiridonov *et al.*, 2004) but they are unable to resolve more basal nodes with confidence (Adams, 2001; Nguyen *et al.*, 2001; Spiridonov *et al.*, 2004). Over its entire length, the LSU is highly conserved, but within it are stretches of expansion regions that have relatively higher substitution rates than the gene overall. The LSU (Duncan *et al.*, 1999; Kaplan *et al.*, 2000) D2D3 segment is one such expansion region that has been shown to evolve more slowly than the above mentioned genes but more rapidly than the SSU, making it (in most cases) more appropriate for resolving more deeply branching nodes in phylogenetic trees, while at the same time being sensitive enough to reveal differences among most species (De Ley *et al.*, 1999; Duncan *et al.*, 1999; Kaplan *et al.*, 2000; Nadler *et al.*, 2000; Tenente *et al.*, 2004; He *et al.*, 2005). This property also makes it, along with some of the more variable regions of the SSU, an appealing tool for species delimitation (Adams, 2001; Nadler, 2002; Sites & Marshall, 2004; Gozel *et al.*, 2006b) and for use as a molecular barcode (Powers, 2004; Blaxter *et al.*, 2005).

Whilst phylogenetic analyses of these genes appears to be straightforward, the ND4 and COI mitochondrial genes are protein coding genes that undergo selection to maintain an open reading frame, whereas the ribosomal genes are non-protein coding, and are thought to be under less selective constraint to maintain their length (Hillis & Dixon, 1991) (but note that there is some apparent selection pressure to maintain certain sequence and size fidelity for the proper maturation of the complete subunit (Van Nues *et al.*, 1994)). Therefore, while insertion-deletion events (indels) are infrequent in the protein-coding genes, they are often the most common form of substitution in the ITS regions of EPN (Adams *et al.*, 1998; Nguyen *et al.*, 2001). Although robust models of nucleotide substitution have been developed for protein coding genes that incorporate different substitution parameters as a function of codon position, similar models for rDNA (*i.e.*, that can account for indels and differential rates of substitution in stems *vs* loops of the RNA secondary structure) await full development. This is a problem for parsimony and model-based analyses. At the moment, the stem-loop issue can be addressed by determining the empirical transition/transversion ratios in stem and loop partitioned datasets and then analysing them under mixed model assumptions (Ronquist & Huelsenbeck, 2003, but see also He *et al.*, 2005). Indels can be accounted for by gap coding (Adams *et al.*, 1998; Muller, 2006) and unreliably aligned regions can be identified and removed based on posterior probability approaches (Loytynoja & Milinkovitch, 2003; Nadler *et al.*, 2006b).

Optimal analyses of EPN phylogenetic relationships will incorporate every type of character for which homologous states can be identified across as many taxa as possible (Kluge, 1998, 2004; Giribet *et al.*, 2001). The analyses of Liu and Berry (1996), Stock *et al.* (2001), Spiridonov *et al.* (2004) and Nadler *et al.* (2006b) remain the most deliberate attempts to integrate morphological and molecular characters in a cohesive analysis. The multi locus approach taken by Nadler *et al.* (2006b) clearly illustrates the improved explanatory power that accompanies analyses of multiple loci and data types. As with all phylogenetic analyses, ongoing taxonomic work renders every previous solution obsolete as soon as it is published. Below are analyses that sample the taxa and comparable DNA sequence data currently available in public databases.

PHYLOGENY OF *STEINERNEMA* SPECIES

Spiridonov and Belostotskaya (1983) suggested that morphological features of *Steinernema* imply a most recent common ancestry with strongyloidids and alloionematids. Poinar (1993) subsequently inferred from morphological and bionomic characters that the Steinernematidae might have evolved from a proto-*Rhabditonema*-like ancestor. Subsequent analyses of SSU rDNA depict the Steinernematidae sharing common ancestry, if not sister taxon to the Strongyloidoidea and Panagrolaimidae (Blaxter *et al.*, 1998, 2000) (Figs 243, 245).

Phylogenetic relationships among species of *Steinernema* have been carried out using morphological characters and RAPDs (Liu & Berry, 1996), PCR-RFLP of the ITS region (Reid, 1994; Reid *et al.*, 1997), and DNA sequences of the SSU (Liu *et al.*, 1997), ITS, COII and 16S (Szalanski *et al.*, 2000), ITS (Nguyen *et al.*, 2001; Spiridonov *et al.*, 2004) and LSU (Stock *et al.*, 2001) genes. In the present analysis, we conservatively reconstruct phylogenetic relationships among nominal taxa using LSU sequences available in public databases (Fig. 246). Using the pro-

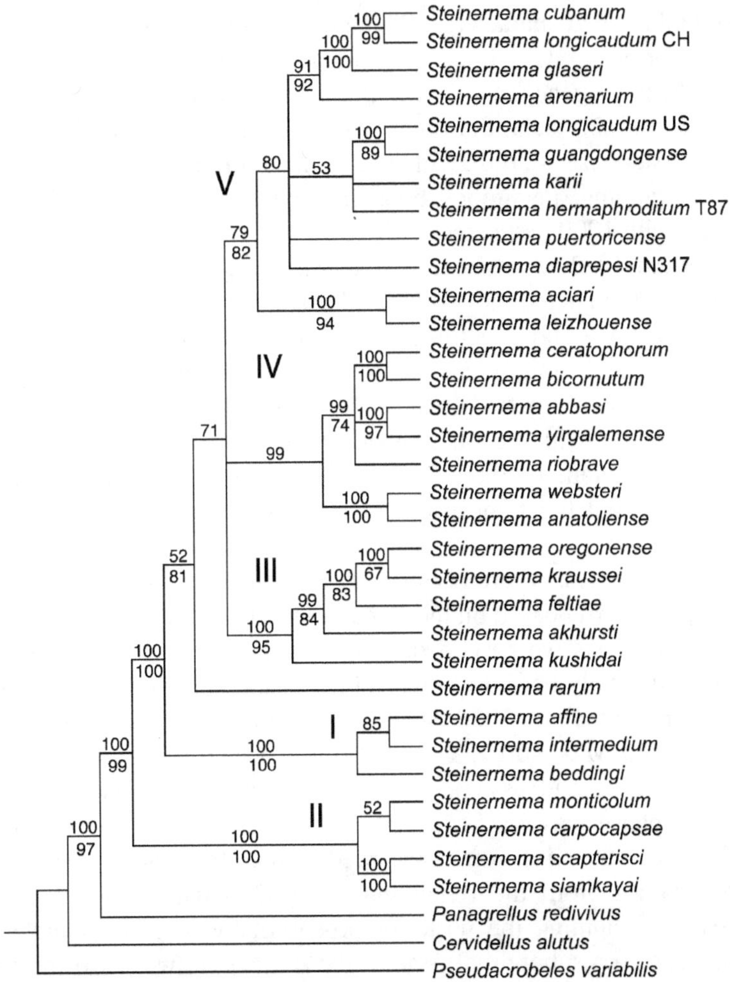

Fig. 246. *Phylogenetic relationships among species of* Steinernema *based on LSU rDNA sequence data, summarising Bayesian, likelihood and parsimony analyses. Values above each branch represent Bayesian posterior probabilities, parsimony bootstrap indices (1000 replicates), where concordant with the Bayesian analyses, appear below. Clade designations (I-V) follow Spiridonov* et al. *(2004).*

file alignment mode of Clustal X (Thompson *et al.*, 1997), individual sequences were aligned to the most complete alignment of Stock *et al.* (2001) available in TreeBase (Sanderson *et al.*, 1994). For the parsimony analyses all characters were considered unordered and unweighted. Of

1159 total aligned characters, 538 are constant and 438 are phylogenetically informative. One hundred and eighty-three characters varied but are likely phylogenetically uninformative autapomorphies. Gaps were treated as missing data. Starting trees for the heuristic search were obtained by stepwise addition with a random addition sequence, with TBR branch-swapping. Parsimony bootstrapping was carried out the same way, but with 1000 replicates.

Maximum likelihood searches were carried out based on the GTR + G + I model of nucleotide evolution (Posada & Crandall, 1998) with six substitution types and substitution rate matrix and nucleotide frequencies estimated from the empirical data (transformation series). The model assumed that approximately 15% of the sites (homologous nucleotides) were invariable, with the variable sites approximating a gamma distribution (shape parameter 0.6213, four rate categories). Heuristic searches were carried out the same way as described above for the parsimony analyses.

Bayesian analyses were performed using MrBayes 3.1 (Huelsenbeck & Ronquist, 2001) based on the same models used for the ML searches. For each analysis we carried out runs for 7 000 000 generations with four incrementally heated chains and then sampled at intervals of 1000 generations to include 7000 data points. Stationarity was estimated by plotting (with Microsoft Excel 2004) the log likelihood scores against generation time and assumed stationarity when the curves flattened out. This phase was reached between 5000 and 6000. The first three to six trees (burn-in) were discarded and 50% majority rule trees were obtained from the remaining 6994-6995 data points with the purpose of obtaining the posterior probability values. To avoid local entrapment on a suboptimal peak in the tree space, we performed two independent analyses and compared these for convergence to similar log likelihood mean values (Huelsenbeck & Bollback, 2001; Leache & Reeder, 2002). We also compared the posterior probabilities for individual clades obtained from the separate analyses for congruence to ensure convergence of the two analyses. All phylogenetic analyses were performed on a RackSaver 64 node dual Opteron processor computing cluster. The strength of support for nodes in the Bayesian tree is estimated by posterior probabilities. For Bayesian analyses, posterior probabilities greater than 0.95 are typically considered robust (Leache & Reeder, 2002), 0.70 for maximum parsimony and maximum likelihood bootstrap analyses (Hillis & Bull, 1993).

The phylogeny presented in Figure 246 presents few deviations from previous hypotheses (Nguyen *et al.*, 2001; Stock *et al.*, 2001; Spiridonov *et al.*, 2004) but includes several taxa previously unrepresented in comprehensive phylogenies of the genus, namely: *S. guangdongense*, *S. aciari*, *S. leizhouense*, *S. yirgalemense*, *S. websteri*, *S. akhursti* and *S. beddingi*. *Steinernema pakistanense*, *S. weiseri*, *S. tami* and *S. neocurtillae* are sampled in the analyses of Spiridonov *et al.* (2004) but are not included in the present analysis for lack of LSU sequence data for these taxa. *Steinernema pakistanense* most likely nests within clade IV, *S. weiseri* within clade III and *S. tami* within clade II. The position of *S. neocurtillae* within the Steinernematidae can be inferred to diverge somewhere between clades I and II, but its placement is one of the most enigmatic (in terms of support) of all the taxa in the genus (Nguyen *et al.*, 2001; Spiridonov *et al.*, 2004). Another problematic taxon is *S. rarum* (see discussion below). The addition of taxa to a dataset is typically seen as more important than adding more characters, particularly where there may be problems of long-branch attraction (LBA) (Graybeal, 1998; Poe, 1998). However, the only potential LBA event in *Steinernema* involves the branch between the species in clade II and the rest of the genus, and none of the additional taxa appears to intersect this branch.

The overall structure of the *Steinernema* tree is not completely resolved without some ambiguity; clades II, IV and V all contain taxa with weakly supported relationships, and *S. rarum* appears not to belong to any predetermined clade at all. Spiridonov *et al.* (2004) placed this taxon as sister to the remaining members of clade V, Stock *et al.* (2001) depict this taxon as sister to clades III and IV, Nadler *et al.* (2006b) as a member of clades III and IV, and the present analyses suggest it is the sister taxon to clades III, IV and V.

One of the earliest morphological trends to emerge from mapping character traits on phylogenetic trees was the length of the infective juveniles where there appeared to be clades of 'long' and 'short' morphs (Reid, 1994; Reid *et al.*, 1997) but, as analyses have since become inclusive of increasing numbers of taxa, this character has been shown to be fairly plastic and homoplasious (Stock *et al.*, 2001). Although Stock *et al.* (2001) suggest that the majority of morphological characters commonly used for classification and phylogenetic relationships are either homoplastic or plesiomorphic, Spiridonov *et al.* (2004) suggested that the structure of bacterial vesicles, colour of male copulatory structures, position of the excretory pore and sperm morphology could be useful synapomorphies for future phylogenetic analyses. Detailed information on steinernematid morphological characters is provided elsewhere (see Chapter 3), but in the context of the present phylogeny it would be interesting to see if, for example, *S. beddingi* has yellow to colourless spicules and whether *S. guangdongense* and *S. hermaphroditum* lack a true bacterial vesicle as predicted by Spiridonov *et al.* (2004). Still, Nadler *et al.* (2006b) are sceptical that such observations will be informative, suggesting that these two characters are likely too variable to be coded properly for phylogenetic analyses. Another testable hypothesis is that *S. websteri* and *S. anatoliense* actually have horn-like cephalic structures (in contradiction to their original descriptions), an observation that requires examination of exsheathed infective juveniles and is easily overlooked (Nguyen & Adams, 2003).

PHYLOGENY OF *HETERORHABDITIS* SPECIES

Poinar (1993) suggested that, based on morphological, physiological, distributional and biological evidence, *Heterorhabditis* most likely shared a most recent common ancestor with a *Pellioditis*-like ancestor. This argument is further corroborated by SSU gene genealogy (Fig. 247).

Virtually all molecular systematic approaches to relationships among species and populations of *Heterorhabditis* have been done using the ITS rDNA gene (Adams *et al.*, 1998, 2006; Phan *et al.*, 2003; Nguyen *et al.*, 2004, 2006b), although the ND4 gene has also been used effectively (Blouin *et al.*, 1999; Liu *et al.*, 1999). In the present analysis, we conservatively reconstruct phylogenetic relationships among nominal taxa using ITS and ND4 sequences available in public databases. Accordingly, individual ITS sequences were aligned using the profile alignment mode of Clustal X to the most complete alignment of Adams *et al.* (1998). Alignment of the ND4 sequences was straightforward as there were no insertion or deletion events.

For the parsimony analyses of the ITS sequences, all characters were considered unordered and unweighted. Of 1167 total aligned characters, 533 were constant and 389 were phylogenetically informative. Some 243 characters varied, but were probably phylogenetically uninformative autapomorphies. Gaps were treated as missing data. Starting trees for the heuristic search were obtained by stepwise addition with a random addition sequence with TBR branch swapping. Parsimony bootstrapping was carried out the same way, but with 1000 replicates. Likelihood analyses

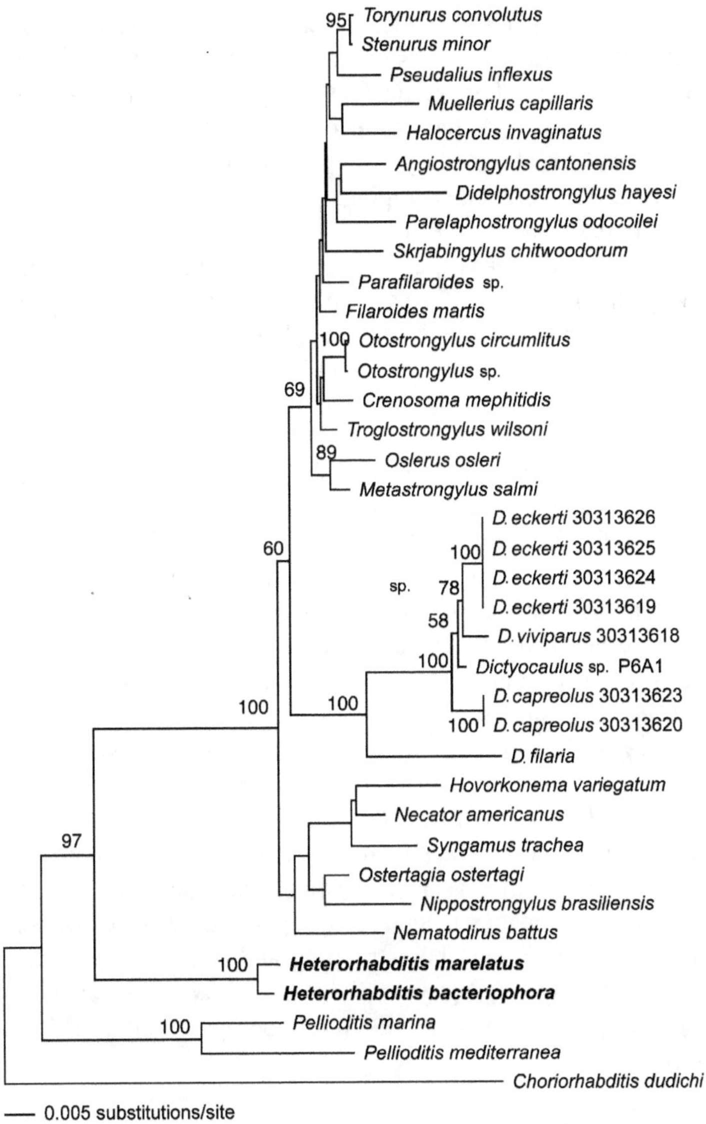

Fig. 247. *Phylogenetic position of* Heterorhabditis *relative to 35 other most closely related nematodes for which comparable sequence information is publicly available. Numbers on branches indicate node support (bootstrap, 1000 reps) mapped on a tree produced with SSU sequences using the BioNJ neighbour-joining algorithm (Gascuel, 1997). Note sister taxon relationship to the Filaroides and Strongyloidea; see also Figure 243 and Hoglund* et al. *(2003).*

were run using the GTR-G model selected by AIC (Posada & Crandall, 1998; Posada & Buckley, 2004) (six substitution types, empirical rate matrix and nucleotide frequencies with no invariable sites and a gamma distribution of rate shape parameter of 0.9 across four rate categories). Starting trees were obtained by stepwise addition with random addition sequence and TBR branch-swapping.

For the parsimony analyses of the ND4 sequences, of 495 total characters 173 were informative; 257 were constant, and 65 varied but were parsimony-uninformative. Tree search strategy was the same as that described above for the ITS analyses. Likelihood analyses of the ND4 data matrix assumed a GTR + I + G model of sequence evolution, as estimated by ModelTest (Posada & Crandall, 1998). Accordingly, six substitution types, their rate matrix and nucleotide usage frequencies were estimated empirically from the multiple sequence alignment. Approximately 4% of the inferred homologous sites were considered to be invariant, with the distribution of rates at the variable sites approximating a gamma distribution with a shape parameter of 0.36 and four rate categories (mean rate for each category). Bayesian analyses for the ITS and ND4 regions employed the same models of sequence evolution as were utilised in the ML analyses and proceeded as described above for the *Steinernema* LSU analyses.

There is general topological agreement between the two trees except for the placement of *H. zealandica* and *H. marelatus* to sister taxa *H. downesi* and *H. megidis* (Figs 248, 249). The ITS solution depicts *H. marelatus* as sister to *H. downesi* + *H. megidis*, whereas the ND4 solution places *H. zealandica* as sister taxon to this clade. The node associated with this discrepancy is tenuously supported by Bayesian posterior probability and parsimony bootstrapping and is consistent with previous analyses (Adams & Nguyen, 2002), which also document the discordance between the present ND4 tree topology and that of Liu *et al.* (1999).

Species delimitation

DELIMITATION OF NEMATODE SPECIES

The body of literature surrounding species and species concepts is amongst the largest in biology. Much progress has been made, both in terms of determining what species are (ontology) and how best to find them (epistemology) but, upon inspection, the majority of metazoan

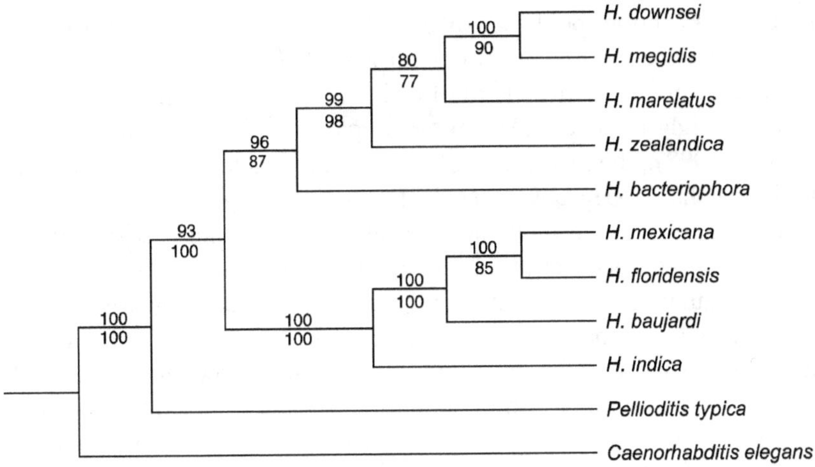

Fig. 248. *Phylogenetic relationships among species of* Heterorhabditis *based on ITS sequences, summarising Bayesian, maximum likelihood and parsimony analyses. Values above each branch represent Bayesian posterior probabilities. Parsimony bootstrap indices (1000 replicates) appear below (where trees are concordant).*

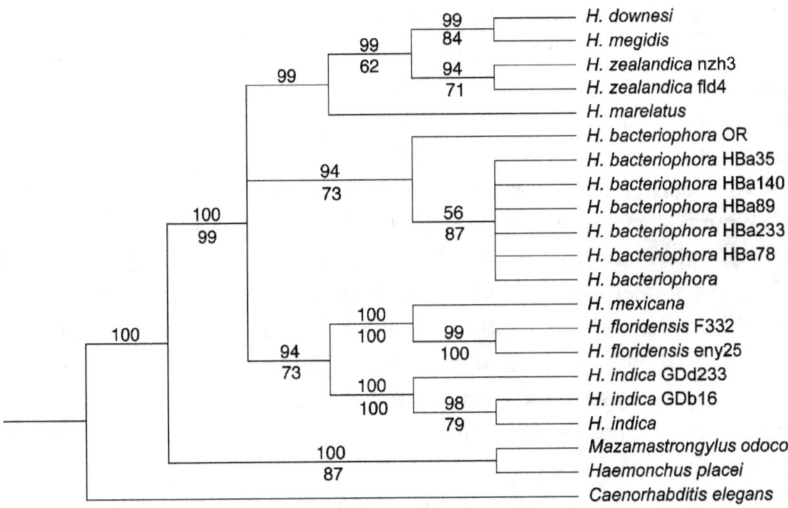

Fig. 249. *Phylogenetic relationships among species and some populations of* Heterorhabditis *based on ND4 mtDNA sequence data, summarising Bayesian, maximum likelihood and parsimony analyses. Values above each branch represent Bayesian posterior probabilities, parsimony bootstrap indices (100 replicates), where concordant, appear below.*

species descriptions published over the last 10 years do not readily reflect the progressive trend. This is not surprising. The language of the best ideas that this literature has to offer is dense and burdensome to read, if not outright impenetrable – for example, a sure cure for insomnia is Adams (1998). The result is that, despite the philosophical and analytical tools that have been developed to recover and represent evolutionary history accurately, few taxonomists are actually taking advantage of them. Few species descriptions are explicit as to what notion of species they are attempting to delimit, or the analytical tools they are using to do so. The confusion surrounding definitions of species is not unlike the definition of what is 'obscene' under United States law. In 1964, Justice Potter Steward summarised the court's position as indefinable, but that "I know it when I see it . . ." (Anon., 1964). As applied to taxonomy, this problem defers to the common quip that a species is "what a good taxonomist says it is" (Crum, 1985). Not state of the art, but *status quo* nonetheless. Thus it appears "[taxonomists] defend the *status quo* long past the time when the *quo* has lost its status" (with apologies to Peter; Peter & Hull, 1969).

Analytical methods for morphological and molecular data types vary somewhat, but there is no conceptual difference for their use in systematics and conflicts between morphological and molecular approaches have been overemphasised. The divisions that do exist are being bridged rapidly by research in developmental and molecular genetic processes and systematic theory. Both types of data can be used to identify, diagnose and delimit species. As with all scientific endeavours, systematic hypotheses (including the validity of species and their phylogenetic relationships) are supported or rejected based on the preponderance of evidence. The most robust hypotheses are those that withstand falsification by the greatest amount of corroborative evidence – whether molecular, morphological or behavioural. The only prerequisite for systematic data (characters) is that they have a heritable genetic basis, and demonstrate the level of fixation and variation appropriate to resolve the question at hand.

The identification and diagnosis of species is often confused with species delimitation, or the process of testing the hypothesis that the entity in question *is* a species. Identifying species involves the recognition of differences and similarities among described taxa. Diagnosis is merely the description of the taxon. Less intuitively, species delimitation involves making a determination and prediction about the history and fate of an evolutionary lineage. The theoretical difference between delimitation and diagnosis is that the former results in something that has meaning in a historical, evolutionary context. A diagnosis cannot make the same claim. In practice, species delimitation differs from diagnosis in that it requires character argumentation, or determining whether homologous characters are plesiomorphic (primitive) or apomorphic (derived). In contrast, diagnosis consists of simply determining whether homologous characters are 'similar' or 'different'. It is crucial that systematists make this distinction because the two activities are as disparate as saying, "this entity looks different" and "this entity is a species" – see Kiontke *et al.* (2002) and Sudhaus and Koch (2004) for clear examples of this exercise.

The most salient theoretical definition of species is the evolutionary species concept. According to this concept, a species is "an entity composed of organisms that maintains its identity from other such entities through time and over space, and which has its own independent evolutionary fate and historical tendencies" (Wiley & Mayden, 2000). The best way to find species is to identify independent evolutionary lineages. There are numerous ways to achieve this, depending on the biological properties of the nematodes, and the types of data that are accessible. The particular mechanics of these methods are described as those that emphasise the tokogeny-phylogeny interface (*i.e.*, incorporate inferences about gene flow) (Sites & Marshall, 2003). The gist of these approaches is that if a lineage (species) is evolving independently, it will accumulate fixed changes among its populations that do not appear in any other lineage. Conversely, shared or variable character states are indicative of non-exclusive reticulation and reflect genealogical patterns consistent with outcrossing populations. A cartoon illustrating these principles is presented in Figure 250.

Delimitation of EPN species

Although the first EPN species was described in 1923, over 80% of them have been described since 1990 (Adams *et al.*, 2006). The recent increase in species descriptions is driven primarily by the potential for these species to find applications in biological control, but they are quickly becoming the focus of basic biological questions of coevolution (ffrench-Constant *et al.*, 2003; Ciche *et al.*, 2006) and ecology (Lewis *et al.*, 2006), which also spurs the search for biodiversity.

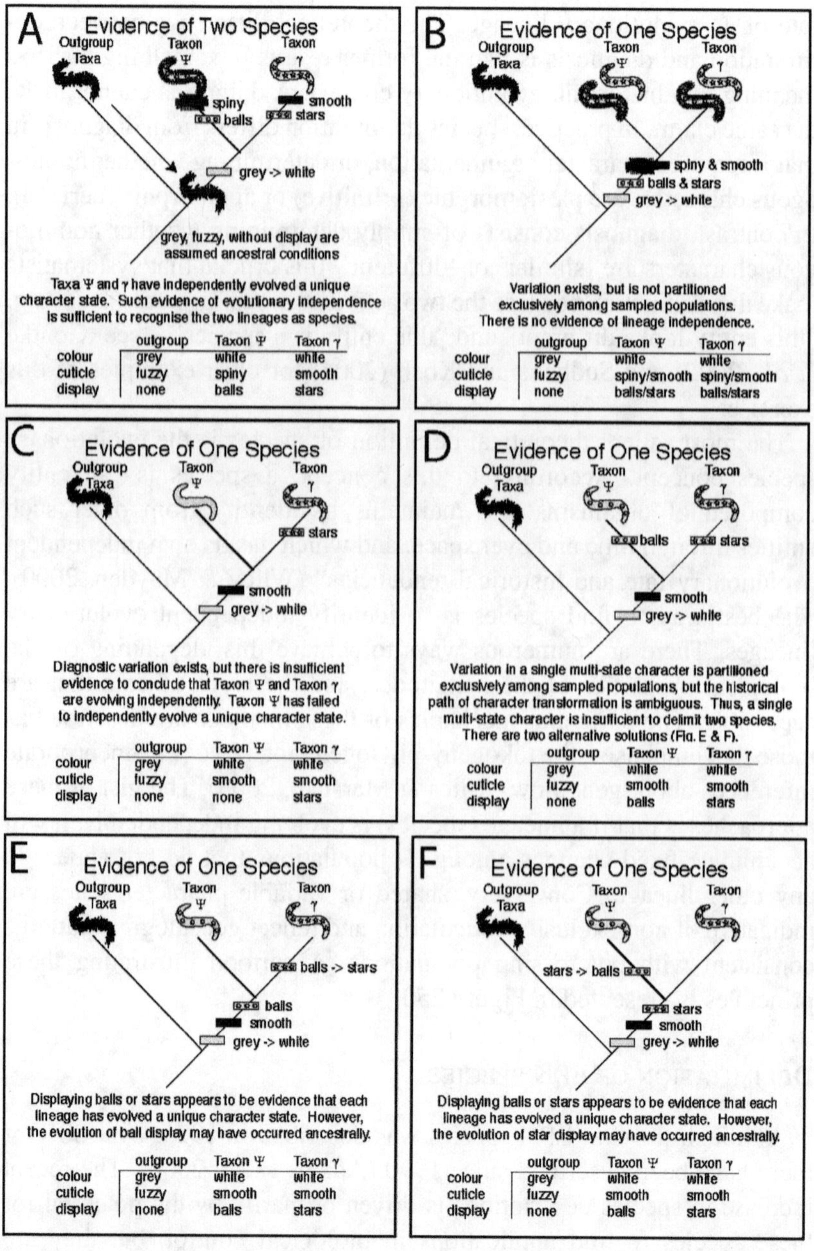

Fig. 250. *Examples of discovery operations required to delimit species. Character states are for illustrative purposes only and are not meant to depict those actually found in entomopathogenic nematodes.*

Thus far, virtually all species of entomopathogenic nematodes have been delimited in Linnaean (phenetic) fashion (Mayr, 1963) based on diagnosable differences in morphology, morphometrics, and even molecules. In some cases crossbreeding tests are done to examine potential reproductive isolation to clarify whether the criteria of biological species can be met (Mayr, 1964). However, despite the thoroughness of these descriptive investigations, delimiting species solely on the basis of overall similarity or potential to interbreed can misrepresent the actual number of unique evolutionary species that exist in nature and their inferred historical relationships. For example, delimitation in the same group based on reproductive compatibility can result in a situation where two sister populations can be reproductively isolated, but each is interfertile with a more distantly related population (Adams *et al.*, 1998). Delimitation based on overall similarity or reproductive compatibility betrays historical relationships and probably fails to accurately reflect the number of species that actually exist (Adams, 1998).

Adams (1998) argued that lineages that evolve unique character states show evidence of lineage independence. Identifying unique characters (autapomorphies) requires character argumentation (polarisation) *via* outgroup comparison with all the other known members of the group. The requirement that species possess autapomorphies protects against misrepresenting the actual number of species and ensures that all valid species are consistent with evolutionary history (Adams, 1998). Spiridonov *et al.* (2004) take exception to this point. As evidence, they point to instances where their molecule of choice (ITS) failed to yield autapomorphies for either of two sister taxa (in this case *S. karii* and *S. siamkayai*). This leads to their conclusion that "autapomorphies for some well established species may disappear as the number of species stud-

Fig. 250. *(Continued). A: Demonstration of lineage exclusivity (independent evolution of autapomorphic character states), evidence of two species; B: Sampled populations are polymorphic, no evidence of lineage independence; C: Diagnostic variation exists in the form of a unique combination of characters, but taxon Ψ does not show evidence of independent evolution (is a privative group). D-F: The reason a single multi-state character cannot be used to delimit two species. D: Character states for a fixed, multi-state character appear to be partitioned exclusively among taxa Ψ and γ, but there are two equally logical paths (E, F) that are operationally indistinguishable from the scenario presented in C.*

ied increase" and that therefore the utility of autapomorphies for species delimitation is not a "sound procedure" (see Spiridonov *et al.*, 2004, pp. 560-561). We disagree that sampling additional loci from additional populations and analysing them with an eye for autapomorphies implies that Adams' approach is flawed or will result in decreased systematic accuracy. Nor are we troubled by the fact that additional data might suggest that what we think are two 'well established' species in reality represent two or more populations of a single species. Where morphological data suggest otherwise, the addition of molecular characters usually supports the original species descriptions (Nadler *et al.*, 2000) or sheds light on cryptic speciation (Gozel *et al.*, 2006a). Also, if the two species are indeed well supported by other characters (autapomorphies?), then the absence of autapomorphies in the ITS data is irrelevant because there is no *a priori* reason to believe that the molecular characters are somehow superior to the morphological characters used in the original or subsequent species descriptions. As an alternative to phylogenetic methods for delimiting species, Spiridonov *et al.* (2004) suggest that sequence divergence is a better indication of lineage independence. Unfortunately, sequence divergence cannot reveal lineage independence and, therefore, is a suboptimal tool to find independent lineages (Farris, 1980, 1981, 1983). The use of sequence divergence alone to delimit species is arbitrary; no more informative of lineage divergence than say, body length divergence. It is a poor indicator of species boundaries (Ferguson, 2002). We caution those who wish to interpret them in this manner.

Molecular methods

The allure of molecular data for work in systematics is their clear genetic basis, a compelling advantage given the significant amount of environmental and host-induced morphological variation displayed by entomopathogenic nematodes (Nguyen & Smart, 1996; Hominick *et al.*, 1997). Potentially sensitive, objective and powerful, molecular characters can also be misleading when wielded carelessly. The use of common genetic markers for EPN research, together with a discussion of the relative merits of using DNA sequencing methods as opposed to RAPD, PCR-RFLP and AFLP markers, has been addressed previously (Adams & Nguyen, 2002; Adams *et al.*, 2006). Below we provide some basic protocols for acquiring DNA sequence data for use in the molecular systematics of EPN.

DNA EXTRACTION

DNA extraction from individual nematodes

For most applications, it is important that DNA template used for RFLP or DNA sequencing come from individual nematodes because variation within and between individual animals must be accounted for. This is especially important when working with variable markers that are informative at the species boundary. Individuals of any sex or stage may be used, but large females commonly yield the largest amounts of DNA.

DNAzol® DNA extraction (modified from Steve Nadler, pers. comm.)

Place a single nematode in a 0.2 ml PCR tube with no more than 5 μl of water. Add 100 μl of the digestion solution to the tube and incubate overnight at 56°C, vortexing the tube occasionally. Check tissues for digestion using a dissection or other low-powered light microscope. If undigested, add 1 μl proteinase K and incubate longer (eggs will not usually digest). 'Pump-mix' digested tissues. Centrifuge the sample for 5 min at 10 000 *g* in a microfuge.

Use a pipette to remove *ca* 90 μl of the supernatant and place it in a new 0.2 ml PCR tube (the important thing is to leave the last *ca* 10 μl containing the cellular debris at the bottom of the tube to be discarded). Heat kill the proteinase by incubating the tube at 95°C for 15 min in a thermal cycler with heated lid (hot bonnet).

Add the contents of the PCR tube from step 4 to 250 μl of DNAzol® in a new 1.7 ml Eppendorf tube. Add 4 μl of the Polyacryl carrier (stored in refrigerator at 4°C; mix the carrier well before using).

Mix the contents of the tube by inverting it *ca* five times. Add 250 μl of 100% ethanol. Mix by inversion ten times. Let sample sit at room temperature for 5 min.

Pellet DNA by centrifugation at 5000 *g* for 5 min at room temperature (spin with cap 'tabs' facing out).

Wash the DNA twice with 800 μl of 75% ethanol. To wash, invert the tube containing the sample and ethanol *ca* six times. Briefly centrifuge the tube to pellet the DNA/carrier pellet, if necessary (if it has broken loose). Carefully decant the ethanol (or remove by pipetting as required).

Remove remaining alcohol from the tube using a fine pipette tip/pipettor. Do NOT spin-dry the DNA/carrier. Once excess ethanol is removed, leave tubes open in the refrigerator (ensure that tubes are slightly covered or tilted on their side to prevent contamination) or on ice in a hood for 1-2 h to allow DNA pellet to dry. This should be done in the refrigerator or on ice under a hood. Once all ethanol has evaporated and the pellet has dried, resuspend the DNA pellet in 6-10 μl of TE. Try using 2 μl of the resuspension for a 25 μl PCR reaction, though more can be used if the DNA concentration is low.

- Preparation 100 mM Tris-HCl pH 7.6

Tris hydroxymethyl	15.8 g
Distilled water	to 1000 ml

Dissolve 15.8 g of Tris in 800 ml of distilled water. Adjust pH with NaOH. Add distilled water to bring the solution to 1000 ml. Sterilise by autoclaving at 121°C for 15 min (may turn yellow after autoclaving). Store at room temperature.

- Preparation 200 mM NaCl

Sodium chloride	11.7 g
Distilled water	to 1000 ml

Dissolve 11.7 g of NaCl in 800 ml of distilled water, and then add distilled water to bring the solution to 1000 ml. Sterilise by autoclaving at 121°C for 15 min. Store at room temperature.

- Preparation 0.5 M EDTA pH 8.0

EDTA	186.1 g
Distilled water	to 1000 ml

Add 186.1 g of EDTA in 500 ml of distilled water stir and heat at 30°C. Add *ca* 100 ml of 1 M NaOH to bring the pH to 8.0. Add distilled water to bring the solution to 1000 ml while adding 1 M NaOH as needed for pH. Sterilise by autoclaving at 121°C for 15 min. Store at room temperature.

- Preparation 1 M NaOH (for making EDTA)

Sodium hydroxide	40 g
Distilled water	to 1000 ml

Dissolve 40 g of sodium hydroxide in 800 ml of distilled water, and then add distilled water to bring the solution to 1000 ml. Store at room temperature.

- Preparation 10% Sarkosyl

Sodium N-lauroyl sarcosine	100 g
Distilled water	to 1000 ml

Dissolve 100 g of sodium N-lauroyl sarcosine in 800 ml of distilled water, then add distilled water to bring the solution to 1000 ml. Sterilise by autoclaving at 121°C for 15 min. Filter sterilise this solution with a 0.2 μm HT Tuffryn® Membrane.

- Preparation Proteinase K

Proteinase K	0.1 g
Distilled water	to 10 ml

Dissolve 0.1 g of proteinase K in 9 ml of distilled water, then add distilled water to bring the solution to 10 ml. Aliquot this into 0.2 PCR tubes and store at −20°C.

- Preparation Digestion Solution

100 mM Tris HCl pH 7.6	200 μl
200 mM NaCl	200 μl
10% Sarkosyl	200 μl
0.5 M EDTA pH 8.0	400 μl
Proteinase K (10 mg ml^{-1})	20 μl
Ultrapure water	980 μl

(Final solution is 10 mM Tris-HCl pH 7.6, 20 mM NaCl, 100 mM EDTA pH 8.0, 1% Sarkosyl, 0.1 mg ml^{-1} proteinase K.)

Worm Lysis Buffer DNA extraction (modified after Jones et al., *2006)*

This method is rapid and yields the largest volumes of template for PCR, but does not work equally well with all taxa (works poorly for some Tylenchomorpha). Place a single nematode in a 0.2 ml PCR tube with no more than 5 μl of water. Add 40 μl of the digestion solution to the tube and incubate at 60°C for 1 h or until the individual is completely digested. Vortex the tube occasionally. Manually rupturing the individual organism can aid in rapid and complete digestion. This can be done by using a pipette tip or pulled Pasteur pipette with a blunted end.

Check tissues for digestion using a dissection or light microscope. If undigested, add 1 μl proteinase K and incubate longer (eggs will not usually digest). 'Pump-mix' digested tissues.

Heat-kill the proteinase by incubating the tube at 95°C for 15 min in a thermal cycler with heated lid (hot bonnet). Use 2-5 μl of lysate as the DNA template in a 25 μl or 50 μl PCR reaction. Lysate should be stored at −20°C or −80°C.

- Preparation 100 mM Tris pH 8.2

Tris (hydroxymethyl) aminomethane	12.1 g
Distilled water	to 1000 ml

Dissolve 12.1 g of Tris in 800 ml of distilled water. Adjust pH with NaOH. Add distilled water to bring the solution to 1000 ml. Sterilise by autoclaving at 121°C for 15 min (may turn yellow after autoclaving). Store at room temperature.

- Preparation Worm Lysis Buffer 10 ml

Proteinase K	10 mg
100 mM Tris pH 8.2	1 ml
KCl	37.3 mg
$MgCl_2$	5.1 mg
Tween-20	45 μl
NP-40 (IGEPAL)	45 μl
Gelatin	5.0 mg
Ultrapure water	to 10 ml

(Final solution is 1 mg ml^{-1} proteinase K, 10 mM Tris pH 8.2, 50 mM KCl, 2.5 mM $MgCl_2$, 0.45% Tween-20, 0.45% NP-40, 0.05% gelatin.)

Dissolve 10 mg of proteinase K, 37.3 mg of KCl, 5.1 mg of $MgCl_2$, and 5.0 mg of gelatin in 8 ml of ultrapure water. Then add 1 ml of 100 mM Tris (pH 8.2), 45 μl of Tween-20, and 45 μl of NP-40. Add ultrapure water to bring the solution to 10 ml. Store at 4°C in the refrigerator.

DNeasy® DNA extraction

Another extraction method that is effective in isolating nematode DNA is the Qiagen DNeasy® tissue kit, which is our method of choice for ethanol preserved specimens. Use the protocol for purification of total DNA from animal tissues with the following modifications: In step 1, a single nematode is manually crushed using a pipette tip or pulled Pasteur pipette with a blunted end. Follow step 2 but incubate at 60°C for 2-3 h while vortexing every 15 min. Follow the rest of the steps as written in the DNeasy® tissue handbook. After the final step, dry the sample down using a vacuum speed dryer for 2 h or until dry and resuspend the DNA in 20 μl of ultrapure water or TE.

Sequence analysis

Polymerase chain reaction (PCR)

PCR is a powerful molecular technique used to amplify target genes, although the specifics vary depending on the primers and commercial DNA polymerase that is used. Here is a generic PCR protocol that works with many brands of *Taq* and several primer sets as well. All reagents should be stored at −20°C or −80°C. Thaw reagents on ice:

Buffer (commercially available *Taq* comes with its own specific buffer)
Thermus aquaticus (*Taq*) DNA Polymerase
DNA Primers (both forward and reverse)
Deoxynucleotide triphosphates (dNTPs)
Ultrapure water
DNA template (from extraction or other source)

Once the reagents have completely thawed, they are combined in a 1.5 ml Eppendorf tube in the following order and quantities, forming the reaction mixture also known as the master mix. This step avoids pipetting minute amounts into multiple individual tubes:

Reagent	1 Reaction	10 Reactions	x Reactions
ThermoPol buffer	2.5 μl	25 μl	$2.5 \times x$ μl
Ultrapure water	17.3 μl	173 μl	$17.3 \times x$ μl
Forward Primer	1.25 μl	12.5 μl	$1.25 \times x$ μl
Reverse Primer	1.25 μl	12.5 μl	$1.25 \times x$ μl
dNTPs	0.5 μl	5 μl	$0.5 \times x$ μl
Taq	0.2 μl	2 μl	$0.2 \times x$ μl
Total Master Mix Volume	23 μl	230 μl	$23 \times x$ μl

To allow for pipetting error, it is recommended that one extra reaction be added to the calculation for every ten reactions needed. The master mix should be mixed by flicking the tube or gently inverting.

Transfer 23 μl of reaction mixture into a PCR tube for each DNA template.

Add 2 μl of DNA template to each PCR tube for a total reaction volume of 25 μl. If more DNA template is used, subtract the extra volume from the water added so that each reaction still has a total volume of 25 μl.

Load tubes into thermal cycler. The exact cycling parameters used depend on the gene targeted, or in other words the primers being used. The following cycling parameters are generic and work with a variety of primer sets, although optimisation may be required in order to obtain the best results for your primer set.

Steps	*Step instructions*
1. Initial denaturation	Incubate at 95°C for 10:00 min
2. Denaturation	Incubate at 94°C for 1:00 min
3. Annealing	Incubate at 50°C for 1:00 min
4. Elongation	Incubate at 72°C for 2:00 min
5. Cycling step	Cycle to step 2 for 39 more times
6. Final elongation	Incubate at 72°C for 10:00 min
7. Cycle termination	Incubate at 4°C forever

Once the thermal cycler reaches step 7, the cycle can be stopped and the tubes with PCR product removed. PCR product should be stored at −20°C or −80°C and should always be kept on ice when in use.

- Preparation Primer Resuspension

DNA forward and reverse primer	lyophilised
Ultrapure water	variable

Primers are shipped lyophilised. To make a master stock, check the number of nanomoles on the tube. Multiply this number by 10 then add that number of μl of ultrapure water to the tube. For example, in a tube with 72.5 nm of primers, add 617 μl of ultrapure water. This creates a final stock of about 100 pmol μl^{-1} which should be stored at −20°C or −80°C. Mix by inversion or by gently vortexing. Working stocks of primer for PCR should be 20 pmol μl^{-1} by diluting at a ratio of 1 : 5. This is done by adding 20 μl of your final stock primer to 80 μl of ultrapure water in a 0.2 ml PCR tube.

- Preparation dNTP solution

100 mM dATP	62.5 μl
100 mM dCTP	62.5 μl
100 mM dGTP	62.5 μl
100 mM dTTP	62.5 μl
Ultrapure water	2.25 ml

Pipette 2.25 ml of ultrapure water into a 15 ml conical tube, then pipette 62.5 μl of each dNTP solution into the tube. Mix by inversion or by gently vortexing. Aliquot 500 μl of the mixture in five separate 0.5 ml tubes to minimise freeze thawing and to prevent contamination. This solution should be stored at −20°C or −80°C and should always be thawed on ice before use.

Agarose gel electrophoresis

PCR reactions can be verified and quantified using gel electrophoresis. Make a 1% agarose gel by adding 0.5 g agarose to 50 ml of 1X TBE. This is heated on a Bunsen burner or in a microwave until the solution is uniform with no visible agar grains. While still warm but no longer steaming add 4 μl of ethidium bromide, which allows the DNA to be visualised in the gel. Pour the still warm mixture into a gel mould. Allow the gel to cool, setting in combs that provide the amount of wells as needed.

Once solid, the gel can be transferred to a gel box and submerged in 1X TBE. PCR product, thawed on ice, can be added to the individual wells by pipetting 3 μl of DNA tracking dye onto a sheet of Parafilm® for each sample to be run. Then 5 μl of PCR product is added to the bead of tracking track, pump mixed, and then pipetted into a well. DNA ladder should be added in the same manner but using 1 μl of ladder (amount of ladder required will vary by manufacturer and concentration) and 1 μl of tracking dye. The purpose of the dye is to allow the otherwise clear PCR product to be visualised as it is pumped into each well. Before running the gel, add an additional 4 μl of ethidium bromide into the TBE at the cathode end of the gel box.

Be sure that the end of the gel with the wells is toward the anode and the bottom of the gel is positioned toward the cathode. The gel should be run using 90 volts for *ca* 30 min, allowing for the tracking dye to reach the bottom of the gel.

DNA bands are then visualised using UV light. Based on the DNA ladder used, both the fragment size and quantity of DNA in the PCR product can be quantified using the position of the band in the gel and the brightness of each band, respectively.

- Preparation 1X TBE Electrophoresis Buffer

Tris (hydroxymethyl) aminomethane	10.8 g
Boric acid	5.5 g
0.5 M EDTA	2 ml
Distilled water	to 1000 ml

Dissolve 10.8 g of Tris base and 5.5 g of boric acid in 800 ml of distilled water, then add 2 ml of 0.5 M EDTA. Add distilled water to bring the solution to 1000 ml. Store at room temperature.

- Preparation 1% Ethidium bromide

Ethidium bromide	0.1 g
Distilled water	100 ml

Dissolve 0.1 g of ethidium bromide in 100 ml of distilled water. Mix by shaking or inversion. Store this in a light-protected bottle in the refrigerator. This represents the stock solution. Make a working stain by diluting 0.4 ml of the stock solution with 400 ml of distilled water. This can be stored at room temperature. Exercise caution when making ethidium bromide as it is mutagenic.

- Preparation DNA Tracking Dye (with Glycerol or Sugar)

Bromophenol Blue	25 mg
Glycerol	3 ml
Sucrose	4 g
Distilled water	to 10 ml

Dissolve 25 mg bromophenol blue in 7 ml of distilled water. Add 3 ml of glycerol or 4 g of sucrose and mix by vortexing. Store at 4°C.

ExoSAP-IT® purification of PCR product

For sequencing and further analysis of the product resulting from PCR, excess primers and dNTPs need to be removed. There are several options to do this, but we find ExoSAP-IT® fast and relatively inexpensive. Transfer 5 μl of PCR product into a new 0.2 ml PCR tube and then add 2 μl of ExoSAP-IT® for a total volume of 7 μl. Place this tube in the thermal cycler and have it run the following programme:

Steps	*Step instructions*
1. Enzyme activation	Incubate at 37°C for 15 min
2. Enzyme denaturation	Incubate at 80°C for 15 min
3. Cycle termination	Incubate at 4°C forever

The resulting PCR product is free of excess nucleotides and primer and is now ready for further analysis (sequencing, SNP analyses, *etc.*). ExoSAP-IT® and cleaned PCR product should be stored at −20°C.

Big Dye cycle sequencing

This technique is used to determine the nucleotide sequence of genes amplified using polymerase chain reaction. It is very similar to PCR but utilises chain terminating, fluorescently labelled ddNTPs. Reagents should be stored at −20°C and thawed on ice:

5X Buffer (commercially available Big Dye comes with its own specific buffer) Big Dye

DNA Primers (only one direction per reaction; forward or reverse only)

Ultrapure water
Purified PCR product

Once the reagents have completely melted, they are combined in a 1.5 ml Eppendorf tube in the following order and quantities, forming the reaction mixture also known as the master mix. This step avoids pipetting minute amounts into multiple individual tubes:

Reagent	1 Reaction	10 Reactions	x Reactions
Ultrapure water	6.4 μl	64 μl	$6.4 \times x$ μl
5X Buffer	2.1 μl	21 μl	$2.1 \times x$ μl
Primer (forward or reverse)	1.0 μl	10 μl	$1.0 \times x$ μl
Big Dye	0.5 μl	5.0 μl	$0.5 \times x$ μl
Total Master Mix Volume	10 μl	100 μl	$10 \times x$ μl

To allow for pipetting error, it is recommended that one extra reaction worth of reagents be added for every ten reactions needed. The master mix should be mixed by flicking the tube or gently inverting.

Transfer 10 μl of reaction mixture into a PCR tube for each DNA template and primer direction. In order to obtain sequence for the forward and reverse directions for each sample, two reactions must be done, one with the forward primer and one with the reverse.

Add 2 μl of cleaned PCR product to each 0.2 ml PCR tube for a total reaction volume of 12 μl. Load tubes into thermal cycler. For Big Dye cycle sequencing reactions use the following sequencing parameters:

Steps	*Step instructions*
1. Initial denaturation	Incubate at 95°C for 10 min
2. Denaturation	Incubate at 94°C for 1 min
3. Annealing	Incubate at 50°C for 1 min
4. Elongation	Incubate at 72°C for 2 min
5. Cycling step	Cycle to step 2 for 39 more times
6. Final elongation	Incubate at 72°C for 10 min
7. Cycle Termination	Incubate at 4°C forever

Once the thermal cycler reaches step 7, the cycle can be stopped and the tubes with sequencing product removed. Sequencing product should be stored at −20°C or −80°C and should always be kept on ice when in use.

Cycle sequencing Sephadex™ clean-up

Sequencing reactions must be completely cleaned before submitted for electrophoresis. Much like PCR product, the extra primers and dNTPs must be removed from the reaction mix while preserving the sequenced target gene. This can be done by column purification using Sephadex™ G-50 Fine. The column is set up in a 96-well MultiScreen™ Millipore filter plate. This is done using a Sephadex™ loading plate and filling it with Sephadex™ G-50 Fine. The column is hydrated using 300 μl of distilled water and allowed to set up at room temperature for 20 min. Excess water is pulled from the column using centrifugation. The MultiScreen™ Millipore filter plate is attached to a 96-well collection plate by a MultiScreen™ centrifuge alignment frame and then centrifuged for 4 min at *ca* 500 *g*. The collection plate is emptied and new sterile collection plate is attached. Before adding the 12 μl sequencing reaction product to the column, 15 μl of distilled water is added to the sequencing product, adding volume. The now diluted 27 μl sequencing reaction mixture is added to one well of the 96-well MultiScreen™ Millipore filter plate where the Sephadex™ columns have been set up. This is then pulled through the column by centrifuging the plate at *ca* 500 *g* for 4 min. The collection plate is now ready for electrophoresis.

Acknowledgements

We wish to acknowledge Bishwo Adhikari, Jim Baldwin, Ernie Bernard, Dana Blackburn, Mark Blaxter, Dan Brooks, John Chaston, T.J. Bliss, Tim Davie, Paul De Ley, Virginia Ferris, Ugur Gozel, Martin Garcia, David Hull, Heather Koppenhöfer, Hugo Kons, John Lynch, Steve Nadler, Tom Powers, Rebecca Scholl, Sergei Subbotin, Patricia Stock, Don Strong, Walter Sudhaus and Sergei Spiridonov for helpful discussions on various topics in this paper (whether they agree with us or not) and/or their roles in developing and optimising the molecular protocols. Errors in the present work should be attributed to us, not them.

References

ANON. (1964). Jacobellis *v*. Ohio. Supreme Court of Ohio.

ADAMS, B.J. (1998). Species concepts and the evolutionary paradigm in modern nematology. *Journal of Nematology* 30, 1-21.

ADAMS, B.J. (2001). The species delimitation uncertainty principle. *Journal of Nematology* 33, 153-160.

ADAMS, B.J. & NGUYEN, K.B. (2002). Taxonomy and systematics. In: Gaugler, R. (Ed.). *Entomopathogenic nematology.* Wallingford, UK, CABI Publishing, pp. 1-33.

ADAMS, B.J., BURNELL, A.M. & POWERS, T.O. (1998). A phylogenetic analysis of *Heterorhabditis* (Nemata: Rhabditidae) based on internal transcribed spacer 1 DNA sequence data. *Journal of Nematology* 30, 22-39.

ADAMS, B.J., FODOR, A., KOPPENHÖFER, H.S., STACKEBRANDT, E., STOCK, S.P. & KLEIN, M.G. (2006). Biodiversity and systematics of nematode-bacterium entomopathogens. *Biological Control* 37, 32-49.

AGUINALDO, A.M.A., TURBEVILLE, J.M., LINFORD, L.S., RIVERA, M.C., GAREY, J.R., RAFF, R.A. & LAKE, J.A. (1997). Evidence for a clade of nematodes, arthropods and other moulting animals. *Nature* 387, 489-493.

AYALA, F.J. & RZHETSKY, A. (1998). Origin of the metazoan phyla: Molecular clocks confirm paleontological estimates. *Proceedings of the National Academy of Sciences of the United States of America* 95, 606-611.

BLAXTER, M., DORRIS, M. & DE LEY, P. (2000). Patterns and processes in the evolution of animal parasitic nematodes. *Nematology* 2, 43-55.

BLAXTER, M., MANN, J., CHAPMAN, T., THOMAS, F., WHITTON, C., FLOYD, R. & ABEBE, E. (2005). Defining operational taxonomic units using DNA barcode data. *Philosophical Transactions of the Royal Society B – Biological Sciences* 360, 1935-1943.

BLAXTER, M.L., DE LEY, P., GAREY, J.R., LIU, L.X., SCHELDEMAN, P., VIERSTRAETE, A., VANFLETEREN, J.R., MACKEY, L.Y., DORRIS, M., FRISSE, L.M., VIDA, J.T. & THOMAS, W.K. (1998). A molecular evolutionary framework for the phylum Nematoda. *Nature (London)* 392, 71-75.

BLOUIN, M.S., LIU, J. & BERRY, R.E. (1999). Life cycle variation and the genetic structure of nematode populations. *Heredity* 83, 253-259.

CICHE, T.A., DARBY, C., EHLERS, R.-U., FORST, S. & GOODRICH-BLAIR, H. (2006). Dangerous liaisons: the symbiosis of entomopathogenic nematodes and bacteria. *Biological Control* 38, 22-46.

CLAUS, C.F.W. & GROBBEN, K. (1910). *Lehrbuch der Zoologie.* Marburg/Lahn, Germany, Elwert, 1002 pp.

CRUM, H. (1985). Traditional make-do taxonomy. *The Bryologist* 88, 221-222.

DE LEY, P., FELIX, M.A., FRISSE, L.M., NADLER, S.A., STERNBERG, P.W. & THOMAS, W.K. (1999). Molecular and morphological characterisation of two reproductively isolated species with mirror-image anatomy (Nematoda: Cephalobidae). *Nematology* 1, 591-612.

DUNCAN, L.W., INSERRA, R.N., THOMAS, W.K., DUNN, D., MUSTIKA, I., FRISSE, L.M., MENDES, M.L., MORRIS, K. & KAPLAN, D.T. (1999). Molecular and morphological analysis of isolates of *Pratylenchus coffeae* and closely related species. *Nematropica* 29, 61-80.

FARRIS, J.S. (1980). Naturalness, information, invariance, and the consequences of phenetic criteria. *Systematic Zoology* 29, 360-381.

FARRIS, J.S. (1981). Distance data in phylogenetic analysis. In: Funk, F.A. & Brooks, D.R. (Eds). *Advances in cladistics. Proceedings of the First Meeting of the Willi Hennig Society.* New York, NY, USA, New York Botanical Garden, pp. 3-23.

FARRIS, J.S. (1983). The logical basis of phylogenetic analysis. In: Platnick, N.I. & Funk, V.A. (Eds). *Advances in cladistics, volume 2. Proceedings of the Second Meeting of the Willi Hennig Society.* New York, NY, USA, Columbia University Press, pp. 1-47.

FERGUSON, J.W.H. (2002). On the use of genetic divergence for identifying species. *Biological Journal of the Linnean Society* 75, 509-516.

FFRENCH-CONSTANT, R., WATERFIELD, N., DABORN, P., JOYCE, S., BENNETT, H., AU, C., DOWLING, A., BOUNDY, S., REYNOLDS, S. & CLARKE, D. (2003). *Photorhabdus*: towards a functional genomic analysis of a symbiont and pathogen. *FEMS Microbiology Reviews* 26, 433-456.

GASCUEL, O. (1997). BIONJ: An improved version of the NJ algorithm based on a simple model of sequence data. *Molecular Biology and Evolution* 14, 685-695.

GIRIBET, G., EDGECOMBE, G.D. & WHEELER, W.C. (2001). Arthropod phylogeny based on eight molecular loci and morphology. *Nature* 413, 157-161.

GOZEL, U., ADAMS, B.J., NGUYEN, K.B., INSERRA, R.N., GIBLIN-DAVIS, R.M. & DUNCAN, L.W. (2006a). A phylogeny of *Belonolaimus* populations in Florida inferred from DNA sequences. *Nematropica* 36, 155-171.

GOZEL, U., LAMBERTI, F., DUNCAN, L., AGOSTINELLI, A., ROSSO, L., NGUYEN, K. & ADAMS, B.J. (2006b). Molecular and morphological consilience in the characterisation and delimitation of five nematode species from Florida belonging to the *Xiphinema americanum*-group. *Nematology* 8, 521-532.

GRAYBEAL, A. (1998). Is it better to add taxa or characters to a difficult phylogenetic problem? *Systematic Biology* 47, 9-17.

HE, Y., SUBBOTIN, S.A., RUBTSOVA, T.V., LAMBERTI, F., BROWN, D.J.F. & MOENS, M. (2005). A molecular phylogenetic approach to Longidoridae (Nematoda: Dorylaimida). *Nematology* 7, 111-124.

HERRE, E.A. (1993). Population structure and the evolution of virulence in nematode parasites of fig wasps. *Science* 259, 1442-1445.

HILLIS, D.M. & BULL, J.J. (1993). An empirical test of bootstrapping as a method for assessing confidence in phylogenetic analysis. *Systematic Biology* 42, 182-192.

HILLIS, D.M. & DIXON, M.T. (1991). Ribosomal DNA: molecular evolution and phylogenetic inference. *The Quarterly Review of Biology* 66, 411-453.

HOGLUND, J., MORRISON, D.A., DIVINA, B.P., WILHELMSSON, E. & MATTSSON, J.G. (2003). Phylogeny of *Dictyocaulus* (lungworms) from eight species of ruminants based on analyses of ribosomal RNA data. *Parasitology* 127, 179-187.

HOMINICK, W.M. (2002). Biogeography. In: Gaugler, R. (Ed.). *Entomopathogenic nematology.* Wallingford, UK, CABI Publishing, pp. 115-143.

HOMINICK, W.M., REID, A.P., BOHAN, D.A. & BRISCOE, B.R. (1996). Entomopathogenic nematodes – biodiversity, geographical distribution and the convention on biological diversity. *Biocontrol Science and Technology* 6, 317-331.

HOMINICK, W.M., BRISCOE, B.R., DEL PINO, F.G., HENG, J.A., HUNT, D.J., KOZODOY, E., MRÁČEK, Z., NGUYEN, K.B., REID, A.P., SPIRIDONOV, S., STOCK, P., STURHAN, D., WATURU, C. & YOSHIDA, M. (1997). Biosystematics of entomopathogenic nematodes: current status, protocols and definitions. *Journal of Helminthology* 71, 271-298.

HUELSENBECK, J.P. & BOLLBACK, J.P. (2001). Empirical and hierarchical Bayesian estimation of ancestral states. *Systematic Biology* 50, 351-366.

HUELSENBECK, J.P. & RONQUIST, F. (2001). MRBAYES: Bayesian inference of phylogenetic trees. *Bioinformatics* 17, 754-755.

HUTCHINSON, G.E. & MACARTHUR, R.H. (1959). A theoretical ecological model of size distributions among species of animals. *The American Naturalist* 93, 117-125.

JABLONSKI, D. (1987). Heritability at the species level: Analysis of geographic ranges of Cretaceous mollusks. *Science* 238, 360-363.

JACKSON, J.B.C. (1974). Biogeographic consequences of eurytopy and stenotopy among marine bivalves and their evolutionary significance. *The American Naturalist* 108, 541-560.

JONES, K.L., TODD, T.C. & HERMAN, M.A. (2006). Development of taxon-specific markers for high-throughput screening of microbial-feeding nematodes. *Molecular Ecology Notes* 6, 712-714.

KAPLAN, D.T., THOMAS, W.K., FRISSE, L.M., SARAH, J.-L., STANTON, J.M., SPEIJER, P.R., MARIN, D.H. & OPPERMAN, C.H. (2000). Phylogenetic analysis of geographically diverse *Radopholus similis via* rDNA sequence reveals a monomorphic motif. *Journal of Nematology* 32, 134-142.

KIONTKE, K., HIRONAKA, M. & SUDHAUS, W. (2002). Description of *Caenorhabditis japonica* n. sp (Nematoda: Rhabditida) associated with the burrower bug *Parastrachia japonensis* (Heteroptera: Cydnidae) in Japan. *Nematology* 4, 933-941.

KLUGE, A.G. (1998). Total evidence or taxonomic congruence: Cladistics or consensus classification. *Cladistics – the International Journal of the Willi Hennig Society* 14, 151-158.

KLUGE, A.G. (2004). On total evidence: for the record. *Cladistics – the International Journal of the Willi Hennig Society* 20, 205-207.

LEACHE, A.D. & REEDER, T.W. (2002). Molecular systematics of the Eastern Fence Lizard (*Sceloporus undulatus*): a comparison of parsimony, likelihood, and Bayesian approaches. *Systematic Biology* 51, 44-68.

LEWIS, E.E., CAMPBELL, J., GRIFFIN, C., KAYA, H. & PETERS, A. (2006). Behavioral ecology of entomopathogenic nematodes. *Biological Control* 38, 66-79.

LIU, J. & BERRY, R.E. (1996). Phylogenetic analysis of the genus *Steinernema* by morphological characters and randomly amplified polymorphic DNA fragments. *Fundamental and Applied Nematology* 19, 463-469.

LIU, J., BERRY, R.E. & MOLDENKE, A.F. (1997). Phylogenic relationships of entomopathogenic nematodes (Heterorhabditidae and Steinernematidae) inferred from partial 18S rRNA gene sequences. *Journal of Invertebrate Pathology* 69, 246-252.

LIU, J., BERRY, R.E. & BLOUIN, M.S. (1999). Molecular differentiation and phylogeny of entomopathogenic nematodes (Rhabditida: Heterorhabditidae) based on ND4 gene sequences of mitochondrial DNA. *Journal of Parasitology* 85, 709-715.

LIU, J., POINAR JR, G.O. & BERRY, R.E. (2000). Control of insect pests with entomopathogenic nematodes: the impact of molecular biology and phylogenetic reconstruction. *Annual Review of Entomology* 45, 287-306.

LOYTYNOJA, A. & MILINKOVITCH, M.C. (2003). A hidden Markov model for progressive multiple alignment. *Bioinformatics* 19, 1505-1513.

LYNCH, J.D. (1989). The gauge of speciation: On the frequencies and modes of speciation. In: Otte, D. & Endler, J.A. (Eds). *Speciation and its consequences.* Sunderland, MA, USA, Sinauer Associates, pp. 527-553.

MAYR, E. (1963). *Animal species and evolution.* Cambridge, MA, USA, The Belknap Press of Harvard University Press, 797 pp.

MAYR, E. (1964). *Systematics and the origin of species, from the viewpoint of a zoologist.* New York, NY, USA, Dover Publications, 334 pp.

MORSE, D.R., LAWTON, J.H., DODSON, M.M. & WILLIAMSON, M.H. (1985). Fractal dimension of vegetation and the distribution of arthropod body lengths. *Nature* 314, 731-733.

MULLER, K. (2006). Incorporating information from length-mutational events into phylogenetic analysis. *Molecular Phylogenetics and Evolution* 38, 667-676.

NADLER, S.A. (2002). Species delimitation and nematode biodiversity: phylogenies rule. *Nematology* 4, 615-625.

NADLER, S.A., ADAMS, B.J., LYONS, E.T., DELONG, R.L. & MELIN, S.R. (2000). Molecular and morphometric evidence for separate species of *Uncinaria* (Nematoda: Ancylostomatidae) in California sea lions and northern fur seals: hypothesis testing supplants verification. *Journal of Parasitology* 86, 1099-1106.

NADLER, S.A., DE LEY, P., MUNDO-OCAMPO, M., SMYTHE, A.B., STOCK, S.P., BUMBARGER, D., ADAMS, B.J., DE LEY, I.T., HOLOVACHOV, O. & BALDWIN, J.G. (2006a). Phylogeny of Cephalobina (Nematoda): molecular evidence for recurrent evolution of probolae and incongruence with traditional classifications. *Molecular Phylogenetics and Evolution* 40, 696-711.

NADLER, S.A., BOLOTIN, E. & STOCK, S.P. (2006b). Phylogenetic relationships of *Steinernema* Travassos, 1927 (Nematoda: Cephalobine: Steinernematidea) based on nuclear, mitochondrial and morphological data. *Systematic Parasitology* 63, 161-181.

NGUYEN, K.B. & ADAMS, B.J. (2003). SEM and systematic studies of *Steinernema abbasi* Elawad *et al.*, 1997, and *S. riobrave* Cabanillas *et al.*, 1994 (Rhabditida: Steinernematidae). *Zootaxa* 179, 1-10.

NGUYEN, K.B. & SMART, G.C. (1996). Identification of entomopathogenic nematodes in the Steinernematidae and Heterorhabditidae (Nemata: Rhabditida). *Journal of Nematology* 28, 286-300.

NGUYEN, K.B., MARUNIAK, J. & ADAMS, B.J. (2001). Diagnostic and phylogenetic utility of the rDNA internal transcribed spacer sequences of *Steinernema*. *Journal of Nematology* 33, 73-82.

NGUYEN, K.B., SHAPIRO-ILAN, D.I., STUART, R.J., MCCOY, C.W., JAMES, R.R. & ADAMS, B.J. (2004). *Heterorhabditis mexicana* n. sp. (Rhabditida: Heterorhabditidae) from Tamaulipas, Mexico, and morphological studies of the bursa of *Heterorhabditis* spp. *Nematology* 6, 231-244.

NGUYEN, K.B., GOZEL, U., KOPPENHÖFER, H.S. & ADAMS, B.J. (2006). *Heterorhabditis floridensis* n. sp. (Rhabditida: Heterorhabditidae) from Florida. *Zootaxa* 1177, 1-19.

PAGE, R.D.M. & HAFNER, M.S. (1996). Molecular phylogenies and host-parasite cospeciation: gophers and lice as a model system. In: Harvey, P.H., Brown, L., Maynard Smith, J. & Nee, S. (Eds). *New uses for new phylogenies.* Oxford, UK, Oxford University Press, pp. 255-270.

PERLMAN, S.J. & JAENIKE, J. (2003). Evolution of multiple components of virulence in *Drosophila*-nematode associations. *Evolution* 57, 1543-1551.

PERLMAN, S.J., SPICER, G.S., SHOEMAKER, D.D. & JAENIKE, J. (2003). Associations between mycophagous *Drosophila* and their *Howardula* nematode parasites: a worldwide phylogenetic shuffle. *Molecular Ecology* 12, 237-249.

PETER, L.J. & HULL, R. (1969). *The Peter principle*. New York, USA, William Morrow & Company, Inc., 179 pp.

PHAN, K.L., SUBBOTIN, S.A., NGUYEN, N.C. & MOENS, M. (2003). *Heterorhabditis baujardi* sp. n. (Rhabditida: Heterorhabditidae) from Vietnam and morphometric data for *H. indica* populations. *Nematology* 5, 367-382.

POE, S. (1998). Sensitivity of phylogeny estimation to taxonomic sampling. *Systematic Biology* 47, 18-31.

POINAR JR, G.O. (1993). Origins and phylogenetic relationships of the entomophilic rhabditids, *Heterorhabditis* and *Steinernema*. *Fundamental and Applied Nematology* 16, 333-338.

POSADA, D. & BUCKLEY, T. (2004). Model selection and model averaging in phylogenetics: Advantages of Akaike information criterion and Bayesian approaches over likelihood ratio tests. *Systematic Biology* 53, 793-808.

POSADA, D. & CRANDALL, K.A. (1998). Modeltest: testing the model of DNA substitution. *Bioinformatics* 14, 817-818.

POWERS, T.O. (2004). Nematode molecular diagnostics: from bands to barcodes. *Annual Review of Phytopathology* 42, 367-383.

REID, A.P. (1994). Molecular taxonomy of *Steinernema*. In: Burnell, A.M., Ehlers, R.-U. & Mason, J.P. (Eds). *Genetics of entomopathogenic nematode-bacterium complexes. Proceedings of a symposium and workshops held at St Patrick's College, Maynooth, Co. Kildare, Ireland.* Luxembourg-Belgium, European Commission Directorate-General XII, Science, Research and Development Environment Research Programme, pp. 49-58.

REID, A.P., HOMINICK, W.M. & BRISCOE, B.R. (1997). Molecular taxonomy and phylogeny of entomopathogenic nematode species (Rhabditida: Steinernematidae) by RFLP analysis of the ITS region of the ribosomal DNA repeat unit. *Systematic Parasitology* 37, 187-193.

RODRIGUEZ-TRELLES, F., TARRIO, R. & AYALA, F.J. (2002). A methodological bias toward overestimation of molecular evolutionary time scales. *Proceedings of the National Academy of Sciences of the United States of America* 99, 8112-8115.

RONQUIST, F. & HUELSENBECK, J.P. (2003). MrBayes 3: Bayesian phylogenetic inference under mixed models. *Bioinformatics* 19, 1572-1574.

SANDERSON, M.J., DONOGHUE, M.J., PIEL, W. & ERIKSSON, T. (1994). TreeBASE: a prototype database of phylogenetic analyses and an interactive tool for browsing the phylogeny of life. *American Journal of Botany* 81, 183.

SITES, J.W. & MARSHALL, J.C. (2003). Delimiting species: a Renaissance issue in systematic biology. *Trends in Ecology and Evolution* 18, 462-470.

SITES, J.W. & MARSHALL, J.C. (2004). Operational criteria for delimiting species. *Annual Review of Ecology Evolution and Systematics* 35, 199-227.

SPIRIDONOV, S.E. & BELOSTOTSKAYA, F.N. (1983). [New data about morphology and biology of the nematodes of the genus *Neoaplectana* and their position in the system of rhabditids.] *Parasitologyia* 17, 119-125.

SPIRIDONOV, S.E., REID, A.P., PODRUCKA, K., SUBBOTIN, S.A. & MOENS, M. (2004). Phylogenetic relationships within the genus *Steinernema* (Nematoda: Rhabditida) as inferred from analyses of sequences of the ITS-5.8S-ITS2 region of rDNA and morphological features. *Nematology* 6, 547-566.

STOCK, S.P., CAMPBELL, J.F. & NADLER, S.A. (2001). Phylogeny of *Steinernema* Travassos, 1927 (Cephalobina: Steinernematidae) inferred from ribosomal DNA sequences and morphological characters. *Journal of Parasitology* 87, 877-889.

STURHAN, D. (1999). Prevalence and habitat specificity of mentioned nematodes in Germany. In: Gwynn, R.L., Smith, P.H., Griffin, C.T., Ehlers, R.-U., Boemare, N. & Mason, J.-P. (Eds). *Entomopathogenic nematodes: application and persistence of entomopathogenic nematodes.* Brussels-Luxembourg, European Commission, COST819, pp. 123-132.

SUDHAUS, W. & KOCH, C. (2004). The new nematode species *Poikilolaimus ernstmayri* sp. n. associated with termites, with a discussion on the phylogeny of *Poikilolaimus* (Rhabditida). *Russian Journal of Nematology* 12, 143-156.

SZALANSKI, A.L., TAYLOR, D.B. & MULLIN, P.G. (2000). Assessing nuclear and mitochondrial DNA sequence variation within *Steinernema* (Rhabditida: Steinernematidae). *Journal of Nematology* 32, 229-233.

TENENTE, G., DE LEY, P., DE LEY, I.T., KARSSEN, G. & VANFLETEREN, J.R. (2004). Sequence analysis of the D2/D3 region of the large subunit RDNA from different *Meloidogyne* isolates. *Nematropica* 34, 1-12.

THOMPSON, J.D., GIBSON, T.J., PLEWNIAK, F., JEANMOUGIN, F. & HIGGINS, D.G. (1997). The CLUSTAL_X windows interface: flexible strategies for multiple sequence alignment aided by quality analysis tools. *Nucleic Acids Research* 25, 4876-4882.

VAN NUES, R.W., RIENTJES, J.M.J., VAN DER SANDE, C.A.F.M., ZERP, S.F., SLUITER, C., VENEMA, J., PLANTA, J. & RAUE, H.A. (1994). Separate structural elements within internal transcribed spacer 1 of *Saccharomyces cerevisiae* precursor ribosomal RNA direct the formation of 17S and 26S rRNA. *Nucleic Acids Research* 22, 912-919.

WILEY, E.O. & MAYDEN, R.L. (2000). The evolutionary species concept. In: Wheeler, Q.D. & Meier, R. (Eds). *Species concepts and phylogenetic theory: a debate.* New York, Columbia University Press, pp. 70-89.

WRAY, G.A., LEVINTIN, J.S. & SHAPIRO, L.H. (1996). Molecular evidence for deep Precambrian divergences among metazoan phyla. *Science* 274, 568-573.

Nematology Monographs & Perspectives, 2007, Vol. 5, 735-808

Chapter 7

Bacterial symbionts of *Steinernema* and *Heterorhabditis*

Heather S. KOPPENHÖFER

Department of Entomology, Rutgers University, New Brunswick, NJ 08901-8524, USA

Introduction

Bacteria of the genera *Xenorhabdus* Thomas & Poinar, 1979 and *Photorhabdus* Boemare, Akhurst & Mourant, 1993 share a mutualistic relationship with nematodes of the families Steinernematidae and Heterorhabditidae, respectively. The bacteria are Gram-negative, nonfermentative rods, belonging to the family Enterobacteriaceae. *Photorhabdus* and *Xenorhabdus* form a unique group being phenotypically (Holt *et al.*, 1994) and genotypically (Brenner & Farmer, 2005) similar to no other genera within the family. However, they both produce the enterobacterial common antigen, which is universally present among the species of Enterobacteriaceae (Ramia *et al.*, 1982). Generally, the bacterial symbionts are pathogenic to insects, and one species, *Photorhabdus asymbiotica* Fischer-Le Saux, Viallard, Brunel, Normund & Boemare, 1999, has been identified as an opportunistic pathogen to humans (Farmer *et al.*, 1989; Peel *et al.*, 1999). The bacteria are transported by their associated nematodes into the haemocoel of a host insect and, once inside, the nematodes use their symbiotic bacteria to overcome the immune system of the host. The bacteria proliferate, release endo- and exotoxins, septicaemia ensues, and insect death generally occurs within 1-3 days (Poinar, 1990; Forst & Clarke, 2002). The bacteria support nematode growth and reproduction by providing nutrients, antimicrobial agents that inhibit the growth of other organisms, including bacteria, yeast and fungi (Poinar & Thomas, 1966; Akhurst & Boemare, 1988), and insect deterrent factors that protect the nematodes from scavenging insects (Zhou *et al.*, 2002). The nematodes produce up to three generations within the host and, once the host cadaver is depleted, a new cohort of infective juveniles (IJ) exits into the environment taking their respective symbionts with them.

Several of the symbiotic bacteria have been screened and assayed for their insecticidal properties, and their virulence factors have been well studied. Less research has been directed toward elucidating the taxonomy and systematics of these bacteria. Inter- and intrageneric relationships within *Photorhabdus* and *Xenorhabdus* have been analysed revealing considerable diversity among the symbionts. Some 14 new species of *Xenorhabdus* (Lengyel *et al.*, 2005a; Tailliez *et al.*, 2006) and three new subspecies of *Photorhabdus* (Akhurst *et al.*, 2004; Hazir *et al.*, 2004a) have been described since 2004, but many bacterial isolates from described nematode species have yet to be examined. The objectives of this chapter are to provide: *i*) an overview of the current knowledge on the taxonomy and phylogeny of *Xenorhabdus* and *Photorhabdus*; *ii*) bacterial and nematode associations; and *iii*) methods for the isolation and characterisation of bacterial symbionts.

Xenorhabdus and *Photorhabdus* life cycle

Nematodes of the families Heterorhabditidae and Steinernematidae are obligate insect pathogens in nature and are only able to persist in the environment outside a host as the specialised third-stage infective juvenile (IJ). Their bacterial symbionts are located within the intestinal tract of the IJ. In *Heterorhabditis*, *Photorhabdus* primarily colonises the anterior region of the intestine just posterior to the basal bulb and, to varying degrees, is also located throughout the remainder of the intestine (Endo & Nickle, 1991; Ciche & Ensign, 2003). *Steinernema* species have a specialised bilobed intestinal vesicle that is colonised by *Xenorhabdus* (Bird & Akhurst, 1983; Martens *et al.*, 2003). In *Steinernema carpocapsae*, the intestinal vesicle is formed by two perivesicular cells, and a subcellular structure, the IVS (intravesicular structure), is contained within the lumen unattached to the vesicle (Martens & Goodrich-Blair, 2005). The IVS and the colonising *Xenorhabdus nematophila* (Poinar & Thomas, 1965) Thomas & Poinar, 1979 co-localise and this IVS-bacterial interaction may be responsible for specific and efficient colonisation of the IJ (Martens & Goodrich-Blair, 2005). In both nematode-symbiont systems, the bacteria persist in a quiescent state while within the IJ. However, limited bacterial growth occurs in *S. carpocapsae* until the intestinal vesicle is colonised by *X. nematophila* (Martens *et al.*, 2003).

Once the IJ gain entry into the haemocoel of a suitable insect host, the nematode recovers from developmental arrest and liberates its bacterial symbiont. However, *S. carpocapsae* was found to recover when in contact with insect gut content prior to moving into the haemocoel (Sicard *et al.*, 2004a). While inside the insect gut, *X. nematophila* cells are released from the vesicle into the nematode intestine (Sicard *et al.*, 2004a) and, as the nematode enters the haemocoel, the bacteria are released by defecation (Poinar & Thomas, 1966; Wouts, 1984; Martens *et al.*, 2004; Sicard *et al.*, 2004a). *Photorhabdus* exit through the mouth of the nematode in a manner similar to regurgitation (Ciche & Ensign, 2003).

The bacteria and nematodes cooperate to overcome the host's immune response, allowing the bacteria to proliferate vegetatively (see Dowds & Peters, 2002). *Steinernema carpocapsae* is able to suppress the immune response by secreting proteins, which may facilitate the release of their symbionts (Götz *et al.*, 1981; Simões, 1998). It is unknown if similar proteins are produced by *Heterorhabditis* (Forst & Clarke, 2002). In *Manduca sexta*, *Photorhabdus luminescens* (Thomas & Poinar, 1979) Boemare, Akhurst & Mourant, 1993 cells secrete an antiphagocytic factor that allows the bacterial cells to impede their own phagocytosis (Silva *et al.*, 2002). In *Spodoptera exigua*, *X. nematophila* cells are able to hamper nodule formation by inhibiting the eicosanoid biosynthetic pathway (Park & Kim, 2000; Park *et al.*, 2003). Furthermore, *X. nematophila* inhibits transcription of insect genes encoding antimicrobial peptides in *S. exigua* and *M. sexta* (Ji & Kim, 2004; Park *et al.*, 2007). *Photorhabdus luminescens* and *X. nematophila* release lipopolysaccharide (LPS) or endotoxin, a component of the cell outer membrane (Dunphy & Thurston, 1990). *Xenorhabdus nematophila* LPS inhibits prophenoloxidase activity (Dunphy & Webster, 1991) and in both systems the lipid A moiety is thought to be cytotoxic to haemocytes (Dunphy & Webster, 1988, 1991). Both genera produce haemolysin activity (Brillard *et al.*, 2001, 2002). The bacteria proliferate within the insect host, whilst producing toxins and exoenzymes that result in septicaemia and bioconversion of the cadaver and provide nutrition for the developing nematodes (see Forst & Clarke, 2002). Early in infection, and prior to insect death, *X. nematophila* cells specifically proliferate extracellularly in the haemolymph and connective tissues surrounding the midgut in *Spodoptera littoralis* (Sicard *et al.*, 2004a). Similarly, in *M. sexta*, *Photorhabdus* proliferate in the haemocoel, destroying the immune system, and in the midgut where the bacteria release toxins and, late in infection, a metalloprotease that destroys the midgut epithelium and may facilitate bioconversion of the tissue (Bowen *et al.*, 1998; Silva *et al.*, 2002). Additionally, the *Photorhabdus mcf* (makes caterpillars floppy) gene encodes a toxin that also destroys haemocytes and the insect midgut (Daborn *et al.*, 2002).

Near the end of bacterial growth, the symbionts produce a variety of antimicrobial compounds that protect the cadaver from colonisation by other organisms. Such compounds include bacteriocins that are active against closely related bacteria and in *P. luminescens* are active against distantly related bacterial taxa (Thaler *et al.*, 1995; Sharma *et al.*, 2002). Other compounds produced by the bacteria are antibiotics that are active against other bacteria, fungi and yeasts (Akhurst, 1982; Boemare *et al.*, 1997b; Webster *et al.*, 2002).

The developing nematodes feed on the bacteria and bioconverted host tissue and reproduce in the cadaver for one to three generations until the food resources in the cadaver are exhausted. The nematodes develop a new generation of IJ that acquire symbiotic bacterial cells and emerge from the host cadaver in search of a new host.

Phenotypic variation

Both *Xenorhabdus* and *Photorhabdus* produce phenotypic variants. Form I, which is also known as phase I, is the primary form and is found naturally associated with the nematode, whilst the variant or secondary form (form II or phase II) spontaneously arises when the bacteria are in culture or occasionally in the cadaver of the insect during late stages of nematode reproduction (Akhurst, 1980). The two forms differ morphologically and physiologically.

Form I cells are motile by peritrichous flagella (usually absent in form II) and are larger than form II cells (Forst *et al.*, 1997). Form I cells are generally able to adsorb certain dyes, form crystalline inclusion bodies, produce antibiotics, lipase, protease, and in *Photorhabdus*, dependent upon the strain, bioluminescence (Akhurst, 1980, 1982; Couche *et al.*, 1987; Boemare & Akhurst, 1988; Forst *et al.*, 1997). These characteristics are greatly reduced or lost in the variant cells, and in *X. nematophila* the secondary form does not produce OpnB, a stationary phase outer membrane protein (Volgyi *et al.*, 2000). Other traits, which tend to be primary form-specific in many but not all strains, include the cell's ability to lyse red blood cells, to be motile and to produce fimbriae and pigment (Akhurst, 1980; Givaudan *et al.*, 1995; Forst *et al.*, 1997; Volgyi *et al.*, 1998). On nutrient agar *Xenorhabdus* form II colonies have little or no pigment and *Photorhabdus* form II have differential pigmentation that is strain/species-dependent (Akhurst, 1983a; Boemare & Akhurst, 1988; Boemare *et al.*, 1997a).

The secondary variants of both *X. nematophila* and *P. luminescens* have higher levels of respiratory enzyme activity, increased membrane potential and, after a period of starvation, are able to commence growth more quickly than their primary form counterparts, indicating that form II cells are able to take in nutrients more efficiently (Smigielski *et al.*, 1994). Pathogenicity is maintained in form II but differences in pathogenicity between phenotypic variants of *X. nematophila* have been reported in lepidopteran hosts. Against *M. sexta* pathogenicity of form II varies between strains, some of which are avirulent (Volgyi *et al.*, 1998, 2000), whereas form I and form II are equally pathogenic against *G. mellonella* (Akhurst, 1980). Form II of *Photorhabdus* does not support growth and reproduction of *Heterorhabditis* spp. (Gerritsen & Smits, 1993, 1997). Conversely, *X. nematophila* form II cells have the ability to support nematode reproduction *in vitro* (Ehlers *et al.*, 1990; Volgyi *et al.*, 2000) and *in vivo* (Sicard *et al.*, 2005). When *S. carpocapsae* was provided with an equal quantity of both cell variants, form II was almost always retained by the nematode (Sicard *et al.*, 2005). Form II cells allowed for equal rates of reproduction of the nematode and provided protection for the nematode, specifically against antagonistic bacteria sensitive to xenorhabdicin (Sicard *et al.*, 2005). The production of xenorhabdicin is maintained in the secondary variant. However; when confronted with antagonistic bacteria not sensitive to xenorhabdicin, form II cells were inferior to form I cells for providing protection for their nematode host (Sicard *et al.*, 2005). The production of antibiotics is a large part of metabolism and comes at great cost. Without this cost, bacteria can invest more into nutrient uptake and proliferation and therefore better adapt to conditions outside of the tripartite relationship (Smigielski *et al.*, 1994; Sicard *et al.*, 2005).

The mechanisms for phenotypic change are beginning to be understood. In *P. luminescens* HexA negatively regulates primary form characteristics (Joyce & Clarke, 2003) whilst in *X. nematophila* Lrp positively regulates primary form characteristics (Cowles *et al.*, 2007). In both bacteria the regulators affect both mutualism and pathogenesis. Occasionally, reversion from form II to form I occurs in *Xenorhabdus* but has not been documented in *Photorhabdus* (Givaudan *et al.*, 1995; Forst & Clarke, 2002). For detailed information about form-specific traits see Forst *et al.* (1997) and Forst and Clarke (2002).

Taxonomy and systematics of *Xenorhabdus* and *Photorhabdus*

The current definition of species for bacteria is arbitrary and artificial but is universally operational (Stackebrandt *et al.*, 2002). A species is described as a genomically coherent group of individual isolates that share a high degree of similarity in many independent features, and is diagnosable by a discriminative phenotype (Rosselló-Mora & Amann, 2001; Stackebrandt *et al.*, 2002; Adams *et al.*, 2006). DNA : DNA homology remains the main criterion for the delineation of bacterial species. Strains within a species are recommended to have a DNA : DNA relatedness value of 70% or higher and a ΔT_m of 5°C or less (Wayne *et al.*, 1987; Rosselló-Mora & Amann, 2001; Stackebrandt *et al.*, 2002). However, the values should not be absolute for the description of a new species (Rosselló-Mora & Amann, 2001) and other molecular methods are encouraged as long as there is a sufficient degree of congruence with DNA : DNA reassociation (Stackebrandt *et al.*, 2002). Based on the differences between 16S rRNA gene sequences and the low DNA : DNA relatedness values of previous studies (Boemare *et al.*, 1993; Nishimura *et al.*, 1994), *Xenorhabdus* species have been described without the incorporation of DNA : DNA reassociation (Lengyel *et al.*, 2005; Somvanshi *et al.*, 2006). Together with the comparison of 16S rRNA sequences, Tailliez *et al.* (2006) included molecular typing profiles combining randomly amplified polymorphic DNA (RAPD) and enterobacterial repetitive intergenic consensus (ERIC) sequences, a reliable alternative to DNA : DNA hybridisation. Profiles derived from ERIC sequences concatenated with REP (repetitive extragenic palindromic) and Box1A (BOX element) sequences have been used successfully to assess the diversity of *Xenorhabdus* and *Photorhabdus*

(Smith *et al.*, 2003). This method should also be considered as it has been shown with other bacteria to have a strong congruence with DNA : DNA reassociation (Rademaker *et al.*, 2000).

A highly discriminative method of great promise for delineation of species is multi-locus sequence typing (MLST). MLST uses partial sequences of internal fragments from multiple housekeeping genes subjected to stabilising selection, and the evolutionary distance between isolates is quantified based on the number of different loci (Maiden *et al.*, 1998; Adams *et al.*, 2006). The method is highly reproducible and databases containing sequences and software are publicly accessible for use in comparing the isolates of interest (see http://www.mlst.net and http://pubmlst.org). MLSTs have allowed for genomic affiliation of strains with much more confidence than by DNA : DNA reassociation (Lan & Reeves, 2001; Adams *et al.*, 2006).

Although DNA similarity or congruent molecular methodology persists as the main criterion for delineation of bacterial species, a polyphasic approach for species description is the preferred method. Species descriptions should include an almost complete 16S rDNA sequence ($>$1300 nucleotides and $<$0.5% ambiguity), the G + C mol% content of the type strain of the type genus (has been established for *Xenorhabdus* and *Photorhabdus*) and phenotype, including chemotaxonomic characters (Stackebrandt *et al.*, 2002). Phenotypic data can be quickly and easily obtained by using standardised systems. API® substrate panels (BioMérieux, Inc.; http://www.biomerieux.com) for carbohydrate metabolism and other biochemical tests have been used in all the descriptions of *Xenorhabdus* and *Photorhabdus* species. Some of these data are presented in Tables 72 and 73, respectively. In recent studies, Biolog GN™ (Biolog, http://www.biolog.com), another standardised method of substrate utilisation, has been employed for species descriptions and bacterial identification (Hazir *et al.*, 2004a; Lengyel *et al.*, 2005a; Gouge & Snyder, 2006; Somvanshi *et al.*, 2006). Phenotypes described by metabolism are useful but may be insufficient; therefore, the descriptions of other chemotaxonomic characters should be considered (Rosselló-Mora & Amann, 2001). Specific fatty acids can provide important taxonomic information (Dickstein *et al.*, 2001) and fatty acid methyl ester profiles have been helpful for symbiont characterisation (Janse & Smits, 1990; Suzuki *et al.*, 1990; Aguillera *et al.*, 1993). Taxonomic characterisation of bacteria can be done by using automated techniques (http://plantpath.ifas.ufl.edu/Fame/) and the data obtained from

Table 72. *Phenotypic characters for differentiating species of* Xenorhabdus.

	X. *beddingii*[1]	X. *bovienii*[1]	X. *japonica*[1]	X. *nematophila*[1]	X. *poinarii*[1]	X. *budapestensis*[2]	X. *cabanillasii*[3]	X. *doucetiae*[3]	X. *ehlersii*[2]	X. *griffiniae*[3]	X. *hominickii*[3]	X. *indica*[4]	X. *innexi*[2]	X. *koppenhoeferi*[3]	X. *kozodoii*[3]	X. *mauleonii*[3]	X. *miraniensis*[3]	X. *romanii*[3]	X. *stockiae*[3]	X. *szentirmaii*[2]
Upper limit for growth, °C (LB broth)[3]	39	32-33	34	34-35	40	37	39-40	40-42	38-40	39	33-35	39-41	40	33	38-41	40	38	37	39	37-39
Esculin hydrolysis	+	–	–	–	d[5]	v	–	v	[(+)]	+	+	v	+	–	[(–)]	+	+	–	+	+
Citrate (Simmons)	+	+[5]	–	+[5]	d	–	–	–	[(–)]	+	–	–	v	–	[(–)]	+	+	–	–	[(+)]
DNase	+	+	+	+	+ w[6]	+	+	v	[(–)]	+	+	+	+	+	–	+	–	–	+	+
Lecithinase	d	d	d	d	–[7]	v	v	+	[(+)]	+	+	v	+	+	+	+	+	–	+	–[6]
Lipase (Tween 80)	+ w	+	–	d	+ w	+	nr	nr	+	nr	nr	–	+	nr	nr	nr	nr	nr	nr	–
Acid production from:																				
Gluconate	+	d w[6]	–	–	d w[6]	–	–	–	–	–	–	–	+	–	+	+	w	w	+	–
myo-Inositol	–	d w	–	+ w[5]	–	+	v	v	–	–	–	+	v	–	–	+	–	w	+	–[8]
Maltose	+	+[8]	+	+[8]	+	–	v	v	[(+)]	–	[(+)]	w[6]	–	+	+	+	+	+	+	–[9]
Ribose	+	+[5]	–	–[5]	–	+[6]	–	v	[(+)]	+	[(–)]	w	+	–	[(+)]	+	+	–	+	[(–)]
Trehalose	+	+ w[5]	+	+[8]	+[8]	–	–	+	–	–	–	+[6]	–	–	[(+)]	+	+	+	–	–
Utilisation of:																				
Citrate	+	[+][5]	–	–[5]	+	–	+	+	[(+)]	+	–	–	v	–	+	+	+	+	–	+
Diaminobutane	–	[+][5]	–	–[5]	–	–	–	v	[(+)]	+	–	–	–	–	+	–	+	w	–	[(+)]

Table 72. *(Continued).*

	X. beddingii[1]	*X. bovienii*[1]	*X. japonica*[1]	*X. nematophila*[1]	*X. poinarii*[1]	*X. budapestensis*[2]	*X. cabanillasii*[3]	*X. doucetiae*[3]	*X. ehlersii*[2]	*X. griffiniae*[3]	*X. hominickii*[3]	*X. indica*[4]	*X. innexi*[2]	*X. koppenhoeferi*[3]	*X. kozodoii*[3]	*X. mauleonii*[3]	*X. miraniensis*[3]	*X. romanii*[3]	*X. stockiae*[3]	*X. szentirmaii*[2]
Utilisation of:																				
L-Histidine	+	+[5]	–	–	[+][5]	–	v	v	[(–)]	–	–	–	v	–	[(–)]	–	–	–	–	–
myo-Inositol	–	d w[6]	–	+ w[8]	–	+[6]	–	+	–	w	–	+	+	–	–	+	–	+	+	+
DL-Lactate	+	+[8]	+	+	+	+	–	+	[(+)]	+	[(+)]	+	+	–	+	+	+	+	+	+
Maltose	+	+	+	+	+	v	+	+	+	+	+	w	–	–	+	+	+	+	+	+
Proprionate	+	[–] w[6]	–[10]	[–]	+ w[5]	–	–	–	[(–)]	–	–	–	–	–	–	w	w	w	–	[(–)]
Ribose	+	[+][5]	–	–	–	–	–	v	[(–)]	–	[(–)]	–	v	–	[(–)]	+	+	–	+	[(+)]
L-Tyrosine	+	[+][5]	–	–	–	–	–	–	[(–)]	–	–	v	–	–	–	–	+	–	+	–

w, weak or delayed reaction; v, variable reaction; nr, not reported. For *X. beddingii*, *X. bovienii*, *X. japonica*, *X. nematophila* and *X. poinarii*: [+], 76-89% positive; d, 26-75% positive; [–], 11-25% positive. For all other *Xenorhabdus*: [(+)], 50-89% positive; [(–)], 11-49% positive.

[1] From Akhurst and Boemare (1988); Boemare and Akhurst (1988); Yamanaka *et al.* (1992); Nishimura *et al.* (1994); Fischer-Le Saux *et al.* (1999a); Bonifassi *et al.* (1999); Boemare and Akhurst (2000); Akhurst and Boemare (2001); Boemare (2002).

[2] From Lengyel *et al.* (2005); Tailliez *et al.* (2006).

[3] From Tailliez *et al.* (2006).

[4] From Somvanshi *et al.* (2006); Tailliez *et al.* (2006).

[5] Tailliez *et al.* (2006) reported 11-49% positive.

[6] Tailliez *et al.* (2006) reported ⩽10% positive.

[7] Lecithinase activity observed in *X. poinarii* by Lengyel *et al.* (2005).

[8] Tailliez *et al.* (2006) reported 50-89% positive.

[9] Tailliez *et al.* (2006) reported ⩾90% positive.

[10] Tailliez *et al.* (2006) reported a weak reaction.

Table 73. *Phenotypic characters differentiating species of* Photorhabdus.

Characteristic	*P. asymbiotica* subsp. *asymbiotica*[1,3]	*P. asymbiotica* subsp. *australis*[3]	*P. luminescens* subsp. *luminescens*[1]	*P. luminescens* subsp. *akhurstii*[1]	*Photorhabdus luminescens* subsp. *kayaii*[2]	*P. luminescens* subsp. *laumondii*[1]	*P. luminescens* subsp. *thracensis*[2]	*P. temperata*[1]
Maximum growth temperature, °C (nutrient broth)	37-38	40	38-39	38-39	35-36	35-36	34-35	33-35
Protoplasmic inclusions	produced poorly	rare	+	+	+	+	+	+
Dye adsorption	–	–	+	+	+	+	+	+
Bioluminescence	+	+	+	+	+ w[4]	+	+	+
Antimocrobial production	–[5]	–[5]	+	+	nr	+	nr	+
Indole production	–	–	+	+	–	+	–	[–]
Citrate (Simmons)	+ w	d	+	d	+	d	v	d
Urease (Christensen)	+	–	–	d	w	[+]	w	d
Phenylalanine deaminase	–	–	–	–	nr	d	nr	[+]
Tryptophan deaminase	–	–	–	–	nr	d	nr	[–] w
Esculin hydrolysis	+	+	+	+	+	+	+	[+]
Lecithinase (egg yolk agar)	–	–[6]	+	+	nr	+	nr	+
Lipase (Tween 80)	+	[–]	+	+	+	+	+	+
DNase	[+]	–	–	–	+	+	+	+
Acid production from:								
myo-Inositol	d w	d	+	[+]	NI	[+]	NI	d w
D-Mannitol	–	–	d w	+	–	–	–	[–]
Trehalose	+	–	+ w	[+] w	+	[+] w	–	[+]

Table 73. *(Continued).*

Characteristic	*P. asymbiotica* subsp. *asymbiotica*[1,3]	*P. asymbiotica* subsp. *australis*[3]	*P. luminescens* subsp. *luminescens*[1]	*P. luminescens* subsp. *akhurstii*[1]	*P. luminescens* subsp. *kayaii*[2]	*P. luminescens* subsp. *laumondii*[1]	*P. luminescens* subsp. *thracensis*[2]	*P. temperata*[1]
Utilisation of:								
L-Fucose	–	[–]	d	[+]	–	–	–	d
DL-Glycerate	d	–	d	–	–	–	–	+
L-Histidine	d	–	+	d	–	[+] w	+	[+]
myo-Inositol	d	[+]	+	+	v[7]	+	v[7]	[+]
DL-Lactate	–	–	–	d w	–	–	–	–
D-Mannitol	–	–	+	+	–	–	–	[–]
Annular hemolyis on:								
Sheep blood agar	+	d	+	+	NS	[–]	NS	+
Horse blood agar	+	–	+	d	nr	–	nr	+

[+], 76-89% positive; d, 26-75% positive; [–], 11-25% positive; w, weak or delayed reaction; nr, not reported; v, variable.

[1] From Akhurst *et al.* (1996); Fischer-Le Saux *et al.* (1999b); Boemare and Akhurst (2001).

[2] From Hazir *et al.* (2004).

[3] From Peel *et al.* (1999) and Akhurst *et al.* (2004).

[4] One strain is not bioluminescent.

[5] If antimicrobials produced, reactions are weak.

[6] Strain GCH001 produced a lecithinase weakly.

[7] *P. luminescens* subsp. *kayaii* produced weak or negative reactions with API 50CH and positive reactions when using a mineral salt medium. *Photorhabdus luminescens* subsp. *thracensis* produced positive reactions with Biolog GN and negative reactions with API 50CH and a mineral salt medium.

the strain(s) of interest are compared to databases containing information from other organisms.

For *Xenorhabdus* and *Photorhabdus* substantial differences in biochemical reactions of the same species have been reported in the literature (Holt *et al.*, 1994; Brenner & Farmer, 2005), which may lead to difficulty in comparisons of species. Such variation is most likely a result of the use of different bacterial strains and/or phenotypic variants (Akhurst & Boemare, 1988). Other possible factors for these differences may include weak and slow reactions and the media type used for biochemical characterisation (Holt *et al.*, 1994).

XENORHABDUS AND *PHOTORHABDUS* TAXONOMY

The first symbiont of entomopathogenic nematodes was isolated from the DD136 strain of *S. carpocapsae* and described as a new bacterial species, *Achromobacter nematophilus* Poinar & Thomas, 1965, by Poinar and Thomas (1965). The genus *Achromobacter* was later rejected (Hendrie *et al.*, 1974) and the described species were transferred to other genera. Since *A. nematophilus* did not fit into any of the accepted genera, the authors eventually erected a new genus, *Xenorhabdus*, to accommodate the bacteria *X. nematophilus* and *X. luminescens*, symbionts of *Steinernema* species and *Heterorhabditis* species, respectively (Thomas & Poinar, 1979). *Xenorhabdus luminescens* was unquestionably distinct from other *Xenorhabdus* strains by both phenotypic (Akhurst, 1983a; Akhurst & Boemare, 1988; Boemare & Akhurst, 1988) and molecular characters (Grimont *et al.*, 1984; Farmer *et al.*, 1989; Suzuki *et al.*, 1990). By using both the s1 nuclease and hydroxyapatite methods for determination of DNA : DNA similarity, Grimont *et al.* (1984) separated *X. luminescens* isolates into a DNA relatedness group distinct from the other *Xenorhabdus* isolates. These finding were supported by the DNA relatedness studies conducted by Boemare *et al.* (1993). The lack of DNA homology to other *Xenorhabdus* species supported the proposal of *Photorhabdus* as a new genus for the symbionts of *Heterorhabditis* (Boemare *et al.*, 1993). *Photorhabdus*, although comparatively species poor, was found to be more homogeneous than the species rich *Xenorhabdus* (Akhurst *et al.*, 1996). Currently, 20 species of *Xenorhabdus* and three species of *Photorhabdus* are recognised; however, the bacterial symbionts from the majority of recently isolated entomopathogenic nematode species have yet to be described.

Xenorhabdus and *Photorhabdus* are assigned to the family Enterobacteriaceae (Rahn, 1937) Ewing, Farmer & Brenner, 1980, within the gamma subdivision of the Proteobacteria. Members of this family are Gram-negative rods, motile by peritrichous flagella or nonmotile, facultatively anaerobic, negative for oxidase, asporogenous, nonacid fast, chemoorganic heterotrophs with respiratory and fermentative metabolisms (Brenner, 1999; Brenner & Farmer, 2005). *Xenorhabdus* and *Photorhabdus* are considered phenotypically atypical to the other members of this family (Holt *et al.*, 1994). *Xenorhabdus* isolates are negative for the production of catalase, similar only to *Shigella dysenteriae* O group 1. Most *Xenorhabdus* and *Photorhabdus* are negative for the reduction of nitrate to nitrite, a trait shared only by some strains of *Erwinia* and *Yersinia* (Brenner, 1999; Boemare, 2002a; Brenner & Farmer, 2005). The DNAs from species within most genera of Enterobacteriaceae are at least 20% homologous to each other and to *Escherichia coli*, the type species of the type genus of the Enterobacteriaceae; however, *Xenorhabdus* and *Photorhabdus* are among the few notable exceptions (Brenner & Farmer, 2005).

Principal differences between *Xenorhabdus* and *Photorhabdus* are that *Photorhabdus* isolates are catalase positive and most are bioluminescent while all *Xenorhabdus* isolates are negative for both traits (Forst *et al.*, 1997). In addition, *Photorhabdus* are distinguished from *Xenorhabdus* by the occurrence of signature sequences in the 16S small subunit rDNA. The sequence TTCG is at positions 208-211 (*E. coli* numbering) in *Xenorhabdus* 16S rDNA, whilst *Photorhabdus* have the longer TGAAAG at the same location (Szállás *et al.*, 1997).

The polymorphism of the genes coding for the ribosomal RNA small subunit allows for assessment of bacterial diversity and/or the rapid identification of a bacterium in question so as to avoid laborious phenotypic characterisation. Restriction fragment analysis of PCR amplified gene products is a method that has been used successfully for these purposes. *Xenorhabdus* and *Photorhabdus* can be quickly and accurately identified on the basis of restriction fragment length polymorphisms of the 16S rRNA gene sequence (Brunel *et al.*, 1997; Fischer-Le Saux *et al.*, 1998; Bonifassi *et al.*, 1999) and placed into genotypes that correspond to DNA relatedness groups (Fischer-Le Saux *et al.*, 1998). Phylogeny based on 16S rRNA sequences is also practical for the identification of *Xenorhabdus* and *Photorhabdus* strains (Liu *et al.*, 2001; Sergeant *et al.*, 2006).

The genus *Xenorhabdus* Thomas & Poinar, 1979

Xenorhabdus nematophila (= *X. nematophilus*) was divided into four subspecies based on variation in host nematode(s), pigmentation, maximum growth temperature, adsorption of dyes and other phenotypic characters (Akhurst, 1983a, 1986; Thomas & Poinar, 1983). After a comprehensive phenotypic study (Boemare & Akhurst, 1988), the subspecies were elevated to the species *X. nematophila*, *X. beddingii* (Akhurst, 1986) Akhurst & Boemare, 1993, *X. bovienii* (Akhurst, 1983) Akhurst & Boemare, 1993 and *X. poinarii* (Akhurst, 1983) Akhurst & Boemare, 1993. Genomic DNA was hybridised and bacterial isolates coalesced into different DNA relatedness groups that validated the described species (Boemare *et al.*, 1993). A fifth species, *X. japonica* Nishimura, Hagiwara, Suzuki & Yamanaka, 1995 (originally described as *X. japonicus*), was proposed by Nishimura *et al.* (1994) on the basis of both phenotypic characterisation and DNA : DNA homology. The bacterial nomenclature of *Xenorhabdus* was later revised by feminising the species name to correspond to the feminine ending *-rhabdus* (*i.e.*, *X. nematophila* and *X. japonica*) (Euzeby & Boemare, 2000). With the addition of the recently described species *X. budapestensis* Lengyel, Lang, Fodor, Szállás, Schumann & Stackebrandt, 2005; *X. cabanillasii* Tailliez, Pagès, Ginibre & Boemare, 2006; *X. doucetiae* Tailliez, Pagès, Ginibre & Boemare, 2006; *X. ehlersii* Lengyel, Lang, Fodor, Szállás, Schumann & Stackebrandt, 2005; *X. griffiniae* Tailliez, Pagès, Ginibre & Boemare, 2006; *X. hominickii* Tailliez, Pagès, Ginibre & Boemare, 2006; *X. indica* Somvanshi, Lang, Ganguly, Swiderski, Saxena & Stackebrandt, 2006; *X. innexi* Lengyel, Lang, Fodor, Szállás, Schumann & Stackebrandt, 2005; *X. koppenhoeferi* Tailliez, Pagès, Ginibre & Boemare, 2006; *X. kozodoii* Tailliez, Pagès, Ginibre & Boemare, 2006; *X. mauleonii* Tailliez, Pagès, Ginibre & Boemare, 2006; *X. miraniensis* Tailliez, Pagès, Ginibre & Boemare, 2006; *X. romanii* Tailliez, Pagès, Ginibre & Boemare, 2006; *X. stockiae* Tailliez, Pagès, Ginibre & Boemare, 2006 and *X. szentirmaii* Lengyel, Lang, Fodor, Szállás, Schumann & Stackebrandt, 2005) (Lengyel *et al.*, 2005a; Hazir *et al.*, 2006; Tailliez *et al.*, 2006), some 20 species have been proposed in the genus.

CHARACTERS OF *XENORHABDUS*

Xenorhabdus are asporogenous, rod-shaped Gram-negative cells measuring 0.3-2 μm by 2-10 μm and with occasional filaments that are 15-

20 μm long (Akhurst & Boemare, 2005). Spheroplasts that average 2.6 μm in diam. occur at the end of exponential growth (Akhurst & Boemare, 2005). Cells occur in two forms. They are facultatively anaerobic with both respiratory and fermentative types of metabolism, are negative for oxidase and catalase and generally do not reduce nitrate to nitrite (Akhurst & Boemare, 2005; Brenner & Farmer, 2005). Optimum growth temperature is 28°C or less, with some strains able to grow at temperatures as high as 42°C (Akhurst & Boemare, 2005; Tailliez *et al.*, 2006). Acid but not gas is produced from glucose. *N*-acetylglucosamine, glycerol and mannose are fermented and most species ferment fructose (Akhurst & Boemare, 2005; Lengyel *et al.*, 2005a). Form I cells are motile by peritrichous flagella (Givaudan *et al.*, 1995). Cells generally produce protoplasmic crystalline inclusions during the stationary phase of growth (Akhurst & Boemare, 2005). Most strains are lypolytic and positive for DNase and protease (Boemare & Akhurst, 1988). The G+C mol% of the DNA is 43-50 as measured by DNA equilibrium buoyant density. The bacteria associate with nematodes of the family Steinernematidae. The type species is *X. nematophila*. Some relevant phenotypic characters of species within *Xenorhabdus* are listed in Table 72.

SPECIES DESCRIPTIONS

*Xenorhabdus beddingii** (Akhurst, 1986) Akhurst & Boemare, 1993

(Effective publication: Akhurst & Boemare, 1988)
[Basonym: *Xenorhabdus nematophila* subsp. *beddingii* Akhurst, 1986]

The cells measure *ca* 0.5-1.6 × 2-13 μm and may occur as filaments measuring more than 20 μm long. The maximum growth temperature is 38°C in nutrient broth (Akhurst, 1986) and 39°C in tryptic soy broth (Lengyel *et al.*, 2005a). All isolates grow at 34°C. On nutrient agar, form I isolates produce a light brown pigment. Form I adsorbs dyes producing brown and blue colonies on MacConkey agar and NBTA, respectively. Both forms are motile, swimming in a semisolid medium and swarming on a solid medium (Givaudan *et al.*, 1995). The primary form is especially unstable, leading to a very stable secondary form (Boemare, 2002a). The cells do not produce indole from tryptophan and are weak

* Patronym in honour of R.A. *Bedding*, who made major contributions to the development of *Xenorhabdus/Steinernema* associations for the control of insect pests.

for the production of tryptophan deaminase (Boemare & Akhurst, 1988). Esculin is hydrolysed. Lecithinase and lipase are produced. L-histidine, ribose and L-tyrosine are assimilated, but inositol is not assimilated. Acid is produced from salicin and some strains produce acid from ribose. The cells are inhibited by cephaloridine and ampicillin. The G + C mol% of the DNA is 45.5-50 as measured by DNA equilibrium buoyant density, thermal denaturation and HPLC (Akhurst, 1983a, 1986; Nishimura *et al.*, 1994; Akhurst & Boemare, 2005). *Xenorhabdus beddingii* has been isolated from *S. longicaudum* from China and from two undescribed *Steinernema* species from Australia. The type strain is UQM 2871, which is the primary form of strain Q58 (= ATCC 49452, CCM 7079, DSM 4764). GenBank accession nos (16S rRNA): X82254, D78006.

***Xenorhabdus bovienii** (Akhurst, 1983a) Akhurst & Boemare, 1993**
(Effective publication: Akhurst & Boemare, 1988)
[Basonym: *Xenorhabdus nematophila* subsp. *bovienii* Akhurst, 1983a]

The mean cell diam. and length are 1.0 μm and 5.8 μm, respectively, with some cells measuring more that 20 μm long. Growth occurs until 32-33°C (Boemare, 2002a; Tailliez *et al.*, 2006) and some strains grow at temperatures as low as 5°C (Boemare & Akhurst, 2000; Boemare, 2002a). No growth occurs at 34°C. Most primary form colonies have yellow pigmentation when cultured on nutrient agar. Form I adsorbs neutral red, forming brown colonies on MacConkey agar (Akhurst & Boemare, 1988) and bromothymol blue, forming greenish to blue-green colonies on Tergitol 7 agar with TTC (Thomas & Poinar, 1983). Motility does occur but is rare in form II (Givaudan *et al.*, 1995). Weak reduction of nitrate to nitrite has been reported (Akhurst & Boemare, 2005; Lengyel *et al.*, 2005a). Indole pyruvic acid is produced from tryptophan. Esculin is not hydrolysed. Cells are strongly lypolytic. D-gluconate, DL-glycerate and maltotriose are assimilated. Most strains assimilate DL-lactate and maltose and produce acid from maltose (Akhurst & Boemare, 2005; Lengyel *et al.*, 2005a; Tailliez *et al.*, 2006). The G + C mol% of the DNA is 44.3-46.9 as measured by HPLC and DNA equilibrium buoyant density (Akhurst, 1983a; Nishimura *et al.*, 1994). *Xenorhabdus bovienii* is associated with the nematodes *S. kraussei*, *S. feltiae*, *S. affine*

* Patronym in honour of P. *Bovien*, who was the first to report the occurrence of bacteria in the intestinal vesicle of *S. feltiae*.

and *S. intermedium*, all of which are found in temperate climates. The bacterial species holotype, UQM 2210, is the primary form of strain T228 (=ATCC35271, CCM 7080, DSM4766). GenBank accession nos (16S rRNA): X82252, D78007, AY278673.

Xenorhabdus budapestensis[*] Lengyel, Lang, Fodor, Szállás, Schumann & Stackebrandt, 2005b

(Effective publication: Lengyel, Lang, Fodor, Szállás, Schumann & Stackebrandt, 2005a)

The cells measure 0.84-0.98 × 3.3-3.7 μm. The maximum temperatures for growth are 35°C in nutrient broth and 39°C in tryptic soy broth. The bacteria do not produce pigment when cultured on nutrient agar. The colonies measure 1.0-1.5 mm in diam., are circular, convex, undulate and granular. The primary form adsorbs dye producing blue colonies on NBTA. The cells do not hydrolyse esculin. Lipase is produced as determined by the hydrolysis of Tweens 40, 60 and 80; however, Tween 20 is not hydrolysed. Fructose and DL-lactate are assimilated (Lengyel *et al.*, 2005a; Tailliez *et al.*, 2006). Acid is weakly produced from inositol. *Xenorhabdus budapestensis* was isolated from *S. bicornutum*, which was found in Szabadka, Vojvodina, Serbia. The type strain is DSM 16342 (=CIP 108891, NCIMB14016). GenBank accession no. (16S rRNA): AJ810293.

Xenorhabdus cabanillasii[**] Tailliez, Pagès, Ginibre & Boemare, 2006

There are presently two known strains of this species. The upper threshold for growth in LB broth is between 39-40°C. Pigment production is variable. Cells produce DNase, do not hydrolyse esculin, and are negative for Simmons citrate. Lecithinase production is variable. Citrate is assimilated, but *cis*-aconitate, D-alanine, diaminobutane, D-gluconate, D-glucosamine, *myo*-inositol, DL-lactate, ribose and succinate are not. *Xenorhabdus cabanillasii* is associated with the nematode *S. riobrave* which was originally isolated in the Rio Grand Valley in Texas (USA).

[*] Named after *Budapest*, Hungary, a European centre of entomopathogenic research.

[**] Patronym in honour of H.E. *Cabanillas*, who described the host nematode.

The type strain is USTX62, CIP 109066 (=DSM 17905). GenBank accession no. (16S rRNA): AY521244.

Xenorhabdus doucetiae[*] Tailliez, Pagès, Ginibre & Boemare, 2006

Xenorhabdus doucetiae is more heat tolerant than most other species in the genus. The upper threshold for growth is between 40-42°C. Colonies are pigmented. Negative for Simmons citrate. Lecithinase is produced and the production of DNase is variable. Citrate, *cis*-aconitate, D-alanine, L-aspartate, D-gluconate and *myo*-inositol are assimilated. Acid is produced from trehalose. The two strains of *X. doucetiae* used for this description are associated with *S. diaprepesi* found in the Caribbean islands. The symbiont from the type host, isolated from a citrus grove in Polk County, FL, USA (Nguyen & Duncan, 2002), was not included in the description. The type strain is FRM16, CIP 109074 (=DSM 17909). GenBank accession no. (16S rRNA): DQ211709.

Xenorhabdus ehlersii[**] Lengyel, Lang, Fodor, Szállás, Schumann & Stackebrandt, 2005b

(Effective publication: Lengyel, Lang, Fodor, Szállás, Schumann & Stackebrandt, 2005a)

The cells measure 0.75 × 3.0-3.1 μm. The upper limit for growth in nutrient broth is 36°C, and for growth in tryptic soy broth is 41°C. On nutrient agar, the bacteria produce a yellow pigment; the colonies measure 3.0-3.5 mm in diam., are circular, convex, undulate and granular. Bromothymol blue is not adsorbed, resulting in red colonies when grown on NBTA. Esculin is hydrolysed. Cells are strongly lypolytic, hydrolysing Tweens 20, 40, 60 and 80. Gluconate, esculin, *cis*-aconitate, diaminobutane and citrate are assimilated by most strains (Bonifassi *et al.*, 1999; Lengyel *et al.*, 2005a; Tailliez *et al.*, 2006). Acid is produced from ribose and esculin. *Xenorhabdus ehlersii* is associated with the nematode *Steinernema serratum* (=*nomen nudum*,

[*] Patronym in honour of M.M.A. de *Doucet* who described the first *Steinernema* species from South America (*S. rarum*) which was isolated in Cordoba, Argentina.

[**] Patronym in honour of R.-U. *Ehlers*, a major contributor to the study of entomopathogenic nematodes and their bacterial symbionts.

see Chapter 2) from southern China. The type strain is DSM 16342 (=CIP 108893, NCIMB14016). GenBank accession no. (16S rRNA): AJ810294.

*Xenorhabdus griffiniae** Tailliez, Pagès, Ginibre & Boemare, 2006

In LB broth growth does not occur at temperatures above 39°C. Pigment is not produced. Cells are positive for the utilisation of Simmons citrate, esculin hydrolysis and the production of DNase and lecithinase. Cells assimilate *cis*-aconitate, D-alanine, L-aspartate, citrate, diaminobutane, D-gluconate, DL-lactate, but are unable to assimilate fructose. Acid is produced from ribose and weakly from 5-ketogluconate, but not from inositol, gluconate, maltose, or trehalose. *Xenorhabdus griffiniae* is associated with the entomopathogenic nematode *S. hermaphroditum* from Indonesia. The type strain is ID10 CIP 109073 (=DSM 17911). GenBank accession no. (16S rRNA): DQ211710.

*Xenorhabdus hominickii*** Tailliez, Pagès, Ginibre & Boemare, 2006

Three strains of this species are currently recognised. The maximum temperature for growth lies between 33-35°C. The colonies are yellow-pigmented. Indole is produced from tryptophan, a characteristic rarely seen in *Xenorhabdus* but which has also been observed in *X. innexi* (Lengyel *et al.*, 2005a) and, albeit weakly, in *X. indica* (Somvanshi *et al.*, 2006) Simmons citrate is not assimilated. Cells are positive for the hydrolysis of esculin and the production DNase and lecithinase. Acid is produced from 5-ketogluconate. D-malate and D-gluconate are used as substrates for growth, but not *cis*-aconitate, citrate, *myo*-inositol and D-glucosamine. *Xenorhabdus hominickii* is associated with the entomopathogenic nematodes *S. karii* from Kenya and *S. monticolum* from Korea. The type strain is KE01, CIP 109072 (=DSM 17903). GenBank accession no. (16S rRNA): DQ211719.

* Patronym in honour of C.T. *Griffin*, who contributed to the systematics of the nematode host and has made many important contributions to the ecology of entomopathogenic nematodes.

** Patronym in honour of W.M. *Hominick*, who has contributed greatly to the biodiversity and systematics of nematode-bacterium complexes.

Xenorhabdus indica[*] Somvanshi, Lang, Ganguly, Swiderski, Saxena & Stackebrandt, 2006

At the time of writing this chapter the species name has been effectively, but not yet validly, published. The cells measure 0.70-1.30 × 2.5-8.6 μm. The maximum growth temperature in nutrient broth is 36°C and in tryptic soy broth is 41°C. On nutrient agar, colonies have a brownish pigmented centre, appear shiny and opaque, measure 2.5-3.0 mm in diam., and are circular to irregular and convex. Form I colonies adsorb neutral red forming red colonies on MacConkey agar, and adsorb bromothymol blue forming blue colonies on NBTA Some strains weakly reduce nitrate to nitrite. Indole pyruvic acid is produced from tryptophan. Esculin, gelatine and starch are hydrolysed. Cells are lypolytic, hydrolysing Tweens 20 and 40 but not Tweens 60 or 80. Indole is produced weakly from tryptophan. Fructose, inositol, esculin and malate are assimilated. Maltose, gluconate and adipic acid are weakly assimilated, and the cells produce variable results for the assimilation of trehalose. Acid is produced from 5-ketogluconate. *Xenorhabdus indica* is associated with *S. thermophilum* (regarded as a junior synonym of *S. abbasi* in this book because of molecular congruence – see Chapter 2). The type strain is 28, DSM 17382 (=CIP 108830). GenBank accession no. (16S rRNA): AM040494.

Xenorhabdus innexi[**] Lengyel, Lang, Fodor, Szállás, Schumann & Stackebrandt, 2005b

(Effective publication: Lengyel, Lang, Fodor, Szállás, Schumann & Stackebrandt, 2005a)

The cells measure 0.87-1.1 × 3.3-3.4 μm. The maximum growth temperature is 36°C in nutrient broth and 41°C in tryptic soy broth. On nutrient agar, colonies appear cream-coloured, measure 2.0-2.5 mm in diam., are circular, convex, undulate and granular. The primary form adsorbs dye forming blue colonies on NBTA (Dr Elke Lang, pers. comm., 2006). Cells are lypolytic, hydrolysing Tweens 60 and 80 but not

[*] Named after *India*, the country of origin of the host nematode.

[**] The species is named for the company *Innexum* of Hanover, Germany. Innexum, meaning the interconnected one, guided DSMZ through the process of attaining ISO 90001:2000 certification.

Tweens 20 or 40. Indole is produced from tryptophan. Cells assimilate gluconate and malate but are unable to assimilate maltose (Bonifassi *et al.*, 1999; Lengyel *et al.*, 2005a). Acid is produced from ribose and gluconate. *Xenorhabdus innexi* is associated with the nematode *S. scapterisci*, which was isolated from Uruguay. The type strain is DSM 16336 (CIP 108894, NCIMB 14017). GenBank accession no. (16S rRNA): AJ810292.

Xenorhabdus japonica[*] Nishimura, Hagiwara, Suzuki & Yamanaka, 1995

(Effective publication: Nishimura, Hagiwara, Suzuki & Yamanaka, 1994)

Cell sizes are variable, measuring 0.5-1.0 × 2.0-10.0 μm. The maximum growth temperature reported is 36°C in nutrient broth and 39°C in tryptic soy broth (Lengyel *et al.*, 2005a). Form I produces a yellowish-brown pigment on nutrient agar and adsorbs bromothymol blue, forming blue colonies on NBTA. Cells are lypolytic as determined by the hydrolysis of Tweens 20, 40 and 60 and tributyrin, but Tween 80 is not hydrolysed. Lecithinase, DNase and protease are produced by form I but not by form II cells. Arginine dihydrolase activity is detected in form II (Akhurst & Boemare, 2005). Esculin is not hydrolysed. Assimilation of ribose is weak and fructose, malate and succinate are not assimilated (Bonifassi *et al.*, 1999; Lengyel *et al.*, 2005a). Acid is produced from maltose. The G + C mol% of the DNA is 45.9, measured by HPLC. *Xenorhabdus japonica* has been isolated only from *S. kushidai*. The type strain is SK-1, IAM 14265 (=DSM 16522). GenBank accession nos (16S rRNA): Z76739, D78008, DQ202310.

Xenorhabdus koppenhoeferi[**] Tailliez, Pagès, Ginibre & Boemare, 2006

The upper threshold for bacterial growth in LB broth is 33°C. The colonies produce a yellow pigment. The bacterium is positive for DNase

[*] Specific epithet formed from the Latin *japonicus* (=Japanese) and referring to its country of origin.

[**] Patronym in honour of A.M. *Koppenhöfer*, a major contributor to the ecology and application of entomopathogenic nematodes, and who discovered the host nematode and contributed to its systematics.

and lecithinase activities. Esculin is not hydrolysed, and Simmons citrate is not assimilated. Cells are unable to use carbon substrates commonly used by other *Xenorhabdus*. Acid is produced from maltose and 5-ketogluconate. *Xenorhabdus koppenhoeferi* is mutualistically associated with *S. scarabaei* isolated from *Popillia japonica* and *Anomala orientalis* larvae in turfgrass areas in central NJ (USA) (Stock & Koppenhöfer, 2003). The type strain is USNJ01, CIP 109199 (=DSM 18168). GenBank accession no. (16S rRNA): DQ205450.

Xenorhabdus kozodoii[*] Tailliez, Pagès, Ginibre & Boemare, 2006

Presently, there are three known strains of this species. The maximum temperature for growth in LB broth is between 38-41°C. Colonies are pigmented. DNase is not produced. The strains are positive for lecithinase activity and the hydrolysis of esculin. Citrate and diaminobutane are assimilated and acid is produced from maltose and gluconate. Acid is not produced from 5-ketogluconate. *Xenorhabdus kozodoii* is associated with the entomopathogenic nematodes *S. arenarium*, which was isolated in Russia, and *S. apuliae* from Italy. The type strain is SaV, CIP 109068 (=DSM 17907). GenBank accession no. (16S rRNA): DQ211716.

Xenorhabdus mauleonii[**] Tailliez, Pagès, Ginibre & Boemare, 2006

The upper limit for bacterial growth is 40°C. Colonies are yellow. The bacterium hydrolyses esculin, produces DNase and lecithinase and is positive for the utilisation of Simmons citrate. *cis*-aconitate, citrate, D-gluconate, D-glucosamine, *myo*-inositol, D-malate and ribose are assimilated. Acid is produced by gluconate, inositol, maltose, ribose trehalose, and produced weakly from 5-ketogluconate. *Xenorhabdus mauleonii* is associated with an undescribed species of *Steinernema* isolated in the Caribbean island of St Vincent. The type strain is VC01, CIP 109075 (=DSM 17908). GenBank accession no. (16S rRNA): DQ211715.

---

[*] Patronym in honour of E.M. *Kozodoi*, who described the nematode *S. anomali*, which later became a synonym of *S. arenarium*, the host of the bacterium.

[**] Patronym in honour of H. *Mauléon*, who has made important contributions to the distribution and biodiversity of entomopathogenic nematodes and their symbiotic bacteria.

Xenorhabdus miraniensis[*] Tailliez, Pagès, Ginibre & Boemare, 2006

The maximum temperature for growth in LB broth is 38°C. Colonies produce pigment. Cells are negative for DNase, and are positive for the hydrolysis of esculin and the production of lecithinase. Simmons citrate is assimilated. Cells are able to use *cis*-aconitate, citrate, diamonbutane, D-glucosamine, D-malate, ribose and tyrosine but unable to utilise *myo*-inositol as a carbon source. Acid is produced from maltose, ribose and trehalose, and is produced weakly from gluconate and 5-ketogluconate. *Xenorhabdus miraniensis* is symbiotically associated with a nematode of the family Steinernematidae found in the town of Mirani in Queensland, Australia (Akhurst & Boemare, 1988). The type strain is Q1, CIP 109069 (=DSM 17902). GenBank accession no. (16S rRNA): DQ211713.

Xenorhabdus nematophila[**] (Poinar & Thomas, 1965) Thomas & Poinar, 1979[***]

=*Achromobacter nematophilus* Poinar & Thomas, 1965

The cell size is variable with measurements ranging from 0.5-1.6 μm in diam. and 3.1-21.9 μm in length (Akhurst, 1983a; Yamanaka *et al.*, 1992). Two distinct intracellular inclusion bodies are produced (Goetsch *et al.*, 2006). The upper limits for growth in nutrient broth and tryptic soy broth are 35°C and 36°C, respectively (Boemare, 2002a; Lengyel *et al.*, 2005a). Neither form is pigmented. Form I adsorbs neutral red forming brown colonies on MacConkey agar, and adsorbs bromothymol blue forming blue to deep purple colonies on NBTA or T7-TTC agar (Thomas & Poinar, 1983; Akhurst & Boemare, 1988). Form II of some strains is motile in liquid media (Volgyi *et al.*, 1998). Lecithinase is produced in form I of most strains and also in some secondary variants (Akhurst & Boemare, 1988; Volgyi *et al.*, 2000). The production of indolpyruvic acid from tryptophan is variable and weak. Lipase is produced hydrolysing Tweens 20, 40 and 60, and some strains may hydrolyse Tween 85 and/or tributyrin. Esculin is not hydrolysed (Akhurst & Boemare, 1988;

[*] Named after the town of *Mirani*, Queensland, Australia, from where the nematode host was isolated.

[**] Specific epithet derived from *nema* and *philus* (=nematode loving) and referring to its close association with the nematode from which it was isolated.

[***] Approved lists of bacterial names 1980 (Skerman *et al.*, 1980).

Boemare & Akhurst, 1988). Inositol is assimilated but growth is weak. Some strains assimilate ribose but arbutin, L-arginine and salicin are not assimilated (Boemare & Akhurst, 1988). Most strains are sensitive to furazolidone (Akhurst & Boemare, 2005). The G + C mol% of the DNA is 43-48 as measured by DNA equilibrium buoyant density and HPLC (Thomas & Poinar, 1979; Akhurst, 1983a; Nishimura *et al.*, 1994; Akhurst & Boemare, 2005). *Xenorhabdus nematophila* has only been found associated with *S. carpocapsae*, a species of nematode with a global distribution. This bacterial species is appropriately named 'nematode-loving'. The type strain is AN6, ATCC 19061 (=CCM 7081, DSM 3370, LMG 1036). GenBank accession nos (16S rRNA): X82251, D78009, AY278674.

Xenorhabdus poinarii* (Akhurst, 1983a) Akhurst & Boemare, 1993
(Effective publication: Akhurst & Boemare, 1988)
[Basonym: *Xenorhabdus nematophila* subsp. *poinarii* Akhurst, 1983a]

Mean cell diam. and length are 1.2 μm and 4.5 μm, respectively, with some cells measuring up to 9.4 μm long. The cells are generally more heat tolerant than in other *Xenorhabdus*. The upper limits for growth in nutrient and tryptic soy broths are 36°C and 41°C, respectively. Form I colonies have a brown pigment that varies from light to reddish brown between strains (Boemare, 2002a; Akhurst & Boemare, 2005). Bromothymol blue is not adsorbed, but form I adsorbs neutral red forming red colonies on MacConkey agar. Form II variants may be motile (Givaudan *et al.*, 1995). Citrate and gluconate are assimilated and some strains assimilate glycerate and, although weakly, tyrosine (Fischer-Le Saux *et al.*, 1999a). Acid is produced from trehalose and some strains produce acid from gluconate. The G + C mol% of the DNA is 42.6-49 as measured by HPLC and DNA equilibrium buoyant density (Akhurst, 1983a; Nishimura *et al.*, 1994). *Xenorhabdus poinarii* is associated with *S. glaseri* and *S. cubanum* (Fischer-Le Saux, 1999a). The type strain is UQM 2216, which is strain G1 (=ATCC 35272, CIP 103468, DSM 4768). GenBank accession nos (16S rRNA): X82253, D78010.

* This species is named after G.O. *Poinar* Jr, a major contributor to the understanding of the nematode-bacterial association and involved in the isolation of the *S. glaseri* symbiont.

Xenorhabdus romanii[*] Tailliez, Pagès, Ginibre & Boemare, 2006

The upper threshold for growth in LB broth is 37°C. Colonies are yellow-pigmented. DNase and lecithinase are not produced and esculin is not hydrolysed. Simmons citrate, D-glucosamine and ribose are not assimilated, but *cis*-aconitate citrate D-gluconate and *myo*-inositol are assimilated. Acid is produced from maltose and trehalose but only weakly from inositol, gluconate and 5-ketogluconate. *Xenorhabdus romanii* is mutually associated with *S. puertoricense* which was described from Puerto Rico (Román & Figueroa, 1994). The type strain is Pr06-A, CIP 109070 (=DSM 17190). GenBank accession no. (16S rRNA): DQ211717.

Xenorhabdus stockiae[**] Tailliez, Pagès, Ginibre & Boemare, 2006

The maximum temperature for growth in LB broth is 39°C. Colonies are pink and the bacterium is negative for the utilisation of Simmons citrate and positive for the hydrolysis of esculin. Lecithinase and DNase are produced. Assimilates D-gluconate, D-glucosamine, *myo*-inositol, ribose and tyrosine. Acid is produced from gluconate, inositol, maltose and ribose but not from trehalose. *Xenorhabdus stockiae* is the bacterial symbiont of *S. siamkayai*, which was isolated in Thailand. The type strain is TH01, CIP 109067 (=DSM 17904). GenBank accession no. (16S rRNA): DQ202309.

Xenorhabdus szentirmaii[***] Lengyel, Lang, Fodor, Szállás, Schumann & Stackebrandt, 2005b

(Effective publication: Lengyel, Lang, Fodor, Szállás, Schumann & Stackebrandt, 2005a)

The cells measure 0.84-1.06 × 3.8-4.6 μm. The upper limit for growth is 35°C in nutrient broth and 39°C in tryptic soy broth. On nutrient

[*] Patronym in honour of J. *Román*, who described the host nematode.

[**] Patronym in honour of S.P. *Stock*, a leading figure in the phylogeny, taxonomy and systematics of entomopathogenic nematodes and who described this host nematode.

[***] Patronym in honour of the microbiologist, Atilla *Szentirmai*, of Budapest, Hungary, who has made many contributions to the field of microbiology, including work on *Photorhabdus* and *Xenorhabdus*.

agar, the colonies appear brownish, measure 2.5-3.0 mm in diam., are circular, convex, undulate and granular. The primary form adsorbs dye forming blue colonies on NBTA. Bromothymol blue is adsorbed, forming blue colonies when grown on NBTA. Esculin is hydrolysed. Cells are lypolytic, able to hydrolyse Tweens 20, 40 and 60 but not 80. Inositol, malate, gluconate, esculin, *cis*-aconitate and citrate are assimilated and fructose is not utilised as a sole carbon source (Lengyel *et al.*, 2005a; Tailliez *et al.*, 2006). Acid is produced from esculin. The G + C mol% of the DNA was not reported. *Xenorhabdus szentirmaii* is symbiotically associated with the nematode *S. rarum* isolated from Cordoba, Argentina. The type strain is DSM 16338 (=CIP 108895, NCIMB 14019). GenBank accession no. (16S rRNA): AJ810295.

The genus *Photorhabdus* Boemare, Akhurst & Mourant, 1993

The species *X. luminescens* was erected to accommodate the luminescent symbiotic bacteria isolated from the genus *Heterorhabditis* (Thomas & Poinar, 1979). Strains within this species were found to be phenotypically (Akhurst, 1983a) and genotypically (Grimont *et al.*, 1984) divergent from the symbionts of steinernematid nematodes. Grimont *et al.* (1984) demonstrated that *X. luminescens* isolates formed two DNA relatedness groups that were dissimilar to the DNA relatedness groups of *X. nematophila* and *X. bovienii*. Clinical isolates of *X. luminescens* revealed a fifth DNA hybridisation group, forming three DNA relatedness groups within *X. luminescens* (Farmer *et al.*, 1989). Today we know that these groups correspond to the species *P. luminescens*, *P. temperata* Fischer-Le Saux, Viallard, Brunel, Normand & Boemare, 1999 and *P. asymbiotica*. Chemotaxonomic characteristics and DNA relatedness data (Akhurst & Boemare, 1988; Boemare & Akhurst, 1988; Suzuki *et al.*, 1990) strongly supported the transfer of *X. luminescens* to a new genus and, accordingly, *Photorhabdus* was proposed by Boemare *et al.* (1993). Further study confirmed that *Photorhabdus* was clearly divided into two distinct groups: *i*) nematode symbionts; and *ii*) clinical specimens (Akhurst *et al.*, 1996). The symbiotic group appeared to be fairly heterogeneous, and phylogenetic analysis of 16S rRNA gene sequences supported by DNA relatedness provided evidence for heterogeneity within this group (Szállás *et al.*, 1997). The combined information derived from phenotypic characterisation, DNA relatedness and phylogenetic inference based on 16S rDNA, supported the proposal of three subspecies of *P. luminescens*, including *P. luminescens* subsp. *luminescens* (Thomas & Poinar, 1979) Boemare, Akhurst & Mourant, 1993, *P. luminescens* subsp. *akhurstii* Fischer-Le Saux, Viallard, Brunel, Normand & Boemare, 1999, and *P. luminescens* subsp. *laumondii* Fischer-Le Saux, Viallard, Brunel, Normand & Boemare, 1999, plus two additional species, *P. temperata* and the clinical strain, *P. asymbiotica* (Fischer-Le Saux *et al.*, 1999b). Peel *et al.* (1999) isolated *P. asymbiotica* from human specimens in Australia which, according to PCR-RFLP analysis of 16S rDNA, formed a homogeneous group that differed from the described American isolates. *Photorhabdus asymbiotica* subsp. *australis* Fischer-Le Saux, Viallard, Brunel, Normand & Boemare, 1999 was proposed to include these isolates (Akhurst *et al.*, 2004). Additionally, two new subspecies of *P. luminescens* have been described; *P. luminescens* subsp. *kayaii* Hazir, Stackebrandt, Lang, Schumann, Ehlers & Keskin, 2004 and *P. luminescens* subsp. *thracensis* Hazir, Stackebrandt, Lang, Schumann, Ehlers & Keskin, 2004 (Hazir *et al.*, 2004a). Currently, three species of *Photorhabdus* are recognised.

CHARACTERS OF *PHOTORHABDUS*

Photorhabdus are asporogenous, rod-shaped, Gram-negative cells measuring 0.5-2 × 1-10 μm with occasional filaments up to 30 μm in length (Boemare & Akhurst, 2005). The cell size is highly variable. Spheroplasts that average 2.6 μm in diam. occur at the end of exponential growth and during stationary growth (Boemare & Akhurst, 2005). Cells occur as two phenotypic variants. Optimum growth temperature is 28°C and some strains are able to grow at 40°C (Akhurst *et al.*, 2004). On nutrient agar, most strains produce pink, red, orange, yellow, or green pigment (Boemare & Akhurst, 2005). Most strains are bioluminescent, which is usually detectable by the dark-adapted human eye. They are facultatively anaerobic, chemo-organotrophs with both respiratory and fermentative types of metabolism, negative for oxidase, positive for catalase, and do not reduce nitrate to nitrite (Brenner & Farmer, 2005). A discriminating factor for *Photorhabdus* is haemolytic patterns on blood agar (see next paragraph). Most strains are haemolytic for sheep and/or horse blood agars. *Photorhabdus* produce protease as determined by the lysis of gelatin. Cells are lypolytic and able to hydrolyse Tween 20 with many strains able to hydrolyse Tweens 40, 60, 80 and 85. Acid, but not gas, is produced from glucose, *N*-acetylglucosamine, fructose, mannose, maltose and ribose, but weakly so from glycerol (Boemare & Akhurst, 2005; Akhurst *et al.*, 2004; Hazir *et al.*, 2004a). Fumarate, glucosamine, L-glutamate, L-malate, L-proline, succinate and L-tyrosine are assimilated. Cells are motile by peritrichous flagella. Most cells produce protoplasmic crystalline inclusions during the stationary phase of growth (Boemare & Akhurst, 2005). The G + C mol% of the DNA is 43-45. The type species is *P. luminescens*. For phenotypic characters of species within *Photorhabdus* see Table 73.

Farmer *et al.* (1989) described an unusual haemolytic reaction on sheep blood agar by *Photorhabdus* in which there was no haemolysis immediately surrounding the colony. Instead haemolysis occurs as a thin line (2 mm) 6-13 mm away from the colony edge (Farmer *et al.*, 1989; Peel *et al.*, 1999). It was referred to as "*Xenorhabdus luminescens* haemolysis" because it was thought to be unique to *Photorhabdus*, formerly *X. luminescens*. This phenotypic character is not observed in *Xenorhabdus* naturally and accordingly, Akhurst *et al.* (1996) proposed the new term "annular haemolysis" because of the formation of an annulus that encircles each colony. It should be noted that this type of haemolysis is not exclusive to *Photorhabdus* but also occurs in *Bacillus cereus* for which it is identified as a "discontinuous haemolytic pattern" (Beecher & Wong, 1994). Annular haemolysis, a descriptive character of most isolates of *P. asymbiotica* (Weissfeld *et al.*, 2005), does not occur in all strains of *Photorhabdus* and only rarely in *P. luminescens* subsp. *laumondii* (Fischer-Le Saux *et al.*, 1999b; Peel *et al.*, 1999; Akhurst *et al.*, 2004).

SPECIES DESCRIPTIONS

*Photorhabdus asymbiotica** Fischer-Le Saux, Viallard, Brunel, Normand & Boemare, 1999b

(Emended description: Akhurst, Boemare, Janssen, Peel, Alfredson & Beard, 2004)

From the descriptions of Fischer-Le Saux *et al.* (1999b) and Akhurst *et al.* (2004), cells are rod-shaped, motile by peritrichous flagella and measure 0.5-1.0 × 2.0-3.0 μm. Maximum temperature for growth in nutrient broth is 37-40°C. Cells are bioluminescent. Pigmentation is

* Named *asymbiotica* (= not symbiotic) because of the initial lack of a nematode association.

yellow, brown or absent. Dyes are not adsorbed. To date, only form II has been isolated and characterised. Tryptophan deaminase and indole are not produced. Some are weakly antibiotic. Strains are negative for lecithinase on egg yolk agar and are positive for Simmons citrate. On sheep blood agar at 25°C, annular haemolysis usually occurs (Akhurst *et al.*, 2004). Acid is produced from *N*-acetylglucosamine, fructose, glucose, glycerol, maltose, mannose and ribose. Cells are unable to use L-fucose, DL-lactate and mannitol as sole carbon sources. Proteinaceous inclusions are poorly produced. Until recently, isolates had only been collected from clinical specimens and a nematode host had not been found. However, Gerrard *et al.* (2006) have isolated *P. asymbiotica* from a heterorhabditid nematode that is closely related to *H. indica.* The nematode was baited in the locality where a patient had been infected, and the symbiotic bacterium isolated from this nematode was found to be the same strain as the clinical isolate (Gerrard *et al.*, 2006). Two subspecies have been described.

Photorhabdus asymbiotica subsp. *asymbiotica* Fischer-Le Saux, Viallard, Brunel, Normand & Boemare, 1999b

(Emended description: Akhurst, Boemare, Janssen, Peel, Alfredson & Beard, 2004)

Maximum limits for growth are 37-38°C in nutrient broth. Pigmentation is yellow or brown. Cells hydrolyse esculin, are positive for urease and negative for DNase. Positive for Christensen's citrate while only weakly positive for Simmons citrate. Cells are lypolytic, hydrolysing Tweens 60 and 80 and sometimes Tween 40. Annular haemolysis occurs on both sheep and horse blood agars. Acid is produced from trehalose and usually from esculin. Acid production from gluconate is variable. The type strain is 3265-86, ATCC 43950 (=CIP 106331, DSM 15149). GenBank accession no. (16S rRNA): Z76755.

Photorhabdus asymbiotica subsp. *australis** Akhurst, Boemare, Janssen, Peel, Alfredson & Beard, 2004

The maximum temperature for growth in nutrient broth is 40°C. Cells are weakly pigmented, producing yellow or no pigment. Most isolates do not hydrolyse esculin, are negative for urease and are positive for DNase.

* Named after *Australia*, the country from where the bacterium was isolated.

Most strains do not produce Tween 60 and Tween 80 esterases. Annular haemolysis is variable on sheep and horse blood agars. Acid is produced from gluconate by most isolates. Acid is not produced from trehalose and rarely from esculin. Most strains assimilate *myo*-inositol. The type strain is 9802892, ACM 5210 (=CIP 108025). GenBank accession no. (16S rRNA): AY280572.

Photorhabdus luminescens[*] (Thomas & Poinar, 1979) Boemare, Akhurst & Mourant, 1993

(Emended description: Fischer-Le Saux, Viallard, Brunel, Normand & Boemare, 1999b)

[Basonym: *Xenorhabdus luminescens* Thomas & Poinar, 1979[**]]

Unless otherwise noted, the following is from the description of Fischer-Le Saux *et al.* (1999b). Cells are large rods measuring 0.5-1.4 × 2-6 μm, occurring as two phenotypic variants. Named for its luminescence, cells are bioluminescent with luminescence more than 100-fold greater in form I. The upper limits for growth in nutrient broth range 34-39°C (Fischer-Le Saux *et al.*, 1999b; Hazir *et al.*, 2004a), strain dependent. In form I, pigment is produced and bromothymol blue is adsorbed from NBTA and neutral red from MacConkey agar (Akhurst & Boemare, 1988). Acid is weakly produced by most strains from *N*-acetylglucosamine, fructose, glucose, glycerol, maltose, mannose, ribose and trehalose. Some strains produce acid from mannitol. In form I, proteinaceous inclusions are produced in the protoplasm. The inclusions are produced poorly in form II. These bacteria are found naturally in the intestinal lumen of *H. bacteriophora* (Brecon and HP88 subgroups and isolates from Turkey) and of *H. indica* (Fischer-Le Saux *et al.*, 1999b; Hazir *et al.*, 2004a). Five subspecies have been described and are characterised below.

Photorhabdus luminescens subsp. *luminescens* (Thomas & Poinar, 1979) Boemare, Akhurst & Mourant, 1993

From the description of Fischer-Le Saux *et al.* (1999b), the maximum temperature for growth is 38-39°C in nutrient broth. The cells hydrolyse

[*] So named because of the luminescent ability of the bacterium.

[**] Approved lists of bacterial names 1980 (Skerman *et al.*, 1980).

esculin, weakly produce indole and are negative for the production of tryptophan deaminase, DNase and urease. The bacteria generate annular haemolysis on sheep and horse blood agars. DL-lactate is not assimilated, but L-histidine, *myo*-inositol and mannitol are utilised as sole carbon sources. This subspecies is symbiotically associated with the Brecon subgroup of *H. bacteriophora*. The type strain is Hb, ATCC 29999 (=CIP 106429, DSM 3368). GenBank accession nos (16S rRNA): X82248, D78005, AY278640, AY870658.

Photorhabdus luminescens subsp. *akhurstii** Fischer-Le Saux, Viallard, Brunel, Normand & Boemare, 1999b

The upper limits for growth in nutrient broth are 38-39°C. Esculin is hydrolysed but hydrolysis may be weak, indole and urease production are variable, negative for production of DNase and tryptophan deaminase. Annular haemolysis occurs on sheep blood agar and some strains produce annular haemolysis on horse blood agar. Mannitol and, in most strains, L-fucose are assimilated, assimilation of DL-lactate and L-histidine is variable and DL-glycerate is not utilised as a sole carbon source. The first strain, strain D1, was isolated by Dr R. Akhurst from Northern Territory, Australia (Akhurst, 1980). This subspecies is symbiotically associated with *H. indica*. The type strain is FRG04, CIP 105564 (=DSM 15138). GenBank accession no. (16S rRNA): AJ007359.

Photorhabdus luminescens subsp. *kayaii*** Hazir, Stackebrandt, Lang, Schumann, Ehlers & Keskin, 2004b

(Effective publication: Hazir, Stackebrandt, Lang, Schumann, Ehlers & Keskin, 2004a)

The maximum temperature for growth is 35-36°C in nutrient broth and 37-38°C in nutrient broth supplemented with sodium chloride (Hazir *et al.*, 2004a). Bioluminescence is weak or absent. On nutrient agar, colonies have some orange colouration, measure 1.5-1.7 mm in diam., are convex, granular and have an undulate margin. The adsorption of

* Subspecific patronym in honour of R. *Akhurst*, a major contributor to the study of the bacterial symbionts of entomopathogenic nematodes.

** Subspecific patronym in honour of H.K. *Kaya* for his scientific contributions to the biology of entomopathogenic nematodes and biological control agents.

bromothymol blue leads to the formation of blue-green colonies. Esculin is hydrolysed, indole is produced from tryptophan and urease is weakly produced. Degree and type of haemolysis on sheep blood agar is strain dependent. Acid is produced from trehalose but not from mannitol. *Myo*-inositol is assimilated but not L-fucose, L-histidine and D-mannitol. This subspecies is symbiotically associated with *H. bacteriophora* collected from grasslands or clover fields in Turkey. The type strain is 1121, DSM 15194 (=NCIMB 13951). GenBank accession no. (16S rRNA): AJ560630.

Photorhabdus luminescens subsp. *laumondii** Fischer-Le Saux, Viallard, Brunel, Normand & Boemare, 1999b

The upper temperature threshold for growth in nutrient broth is 35-36°C. The bacteria are positive for the hydrolysis of esculin and production of DNase and indole. Most strains produce urease, reactions may be weak. Production of tryptophan deaminase is variable. Haemolysis on sheep and horse blood agars is total. Annular haemolysis is rare. Most strains produce acid from trehalose. L-fucose, DL-glycerate, DL-lactate and D-mannitol are not assimilated. This subspecies is symbiotically associated with the HP88 subgroup of *H. bacteriophora*. The type strain is TT01, CIP 105565 (=DSM 15139). The genome of strain TT01 has been sequenced (Duchaud *et al.*, 2003). GenBank accession nos (16S rRNA): AJ007404, AY278646, BX571859, BX571860, BX571861, BX571863, BX571873, BX571874 (2 rRNA operons).

Photorhabdus luminescens subsp. *thracensis*** Hazir, Stackebrandt, Lang, Schumann, Ehlers & Keskin, 2004b

(Effective publication: Hazir, Stackebrandt, Lang, Schumann, Ehlers & Keskin, 2004a)

The maximum temperature for growth in nutrient broth is the lowest of all the subspecies (34-35°C). The maximum growth temperature in nutrient broth supplemented with sodium chloride is 37°C. Cells

* Subspecific patronym in honour of C. *Laumond*, a major contributor to the use of entomopathogenic nematode/bacterial complexes for insect control.

** Named for the location from which the type strain was isolated, *Thracia*, Thrace, Turkey.

are highly luminous. The colonies have an orange pigment that is apparent on nutrient agar, measure 0.3-1.5 mm in diam., and are convex and granular. The cells adsorb bromothymol blue forming olive green colonies on NBTA. The cells hydrolyse esculin, weakly produce urease and are negative for indole production. Degree and type of haemolysis on sheep blood agar is strain dependent. Acid is produced from trehalose. L-histidine is assimilated but not L-fucose, *myo*-inositol and D-mannitol. This subspecies is symbiotically associated with *H. bacteriophora* collected from fallow field, pine forest, or a sunflower field in Turkey. The type strain is 39-8, DSM 15199 (=NCIMB 13952). GenBank accession no. (16S rRNA): AJ560634.

Photorhabdus temperata[*] Fischer-Le Saux, Viallard, Brunel, Normand & Boemare, 1999b

Cells are large rods measuring 0.5-1.4 × 2-6 μm. Occurs as two phenotypes. The upper limits for growth are 33-35°C. Cells are highly bioluminescent. Form I produces pigment and adsorbs bromothymol blue on NBTA and neutral red on MacConkey agar. Cells are highly luminescent. Most strains are negative for indole production and are positive for tryptophan deaminase, hydrolysis of esculin and the production of DNase. Urease is variable. Annular haemolysis usually occurs on both sheep and horse blood agars. Acid is produced from N-acetylglucosamine, fructose, glucose, mannose and ribose and is produced weakly from glycerol and maltose. Cells assimilate DL-glycerate but not DL-lactate and mannitol is not assimilated by most strains. Proteinaceous inclusions are produced in form I and weakly in form II. This species is symbiotically associated with the NC group of *H. bacteriophora*, *H. megidis* and *H. zealandica*. The type strain is XlNach, CIP 105563. GenBank accession no. (16S rRNA): AJ007405.

Phylogenetic systematics

Evolutionary relationships of members of the family Enterobacteriaceae derived from small subunit RNA are shown in Figure 251 (Francino *et al.*, 2003). Phylogenetic analyses based on partial 16S rRNA gene

[*] The specific epithet refers to the moderate temperatures at which the bacterium grows.

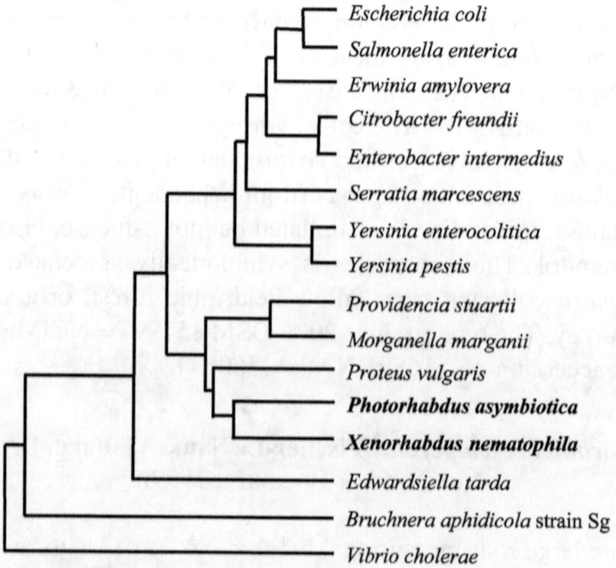

Fig. 251. *16S rRNA gene-based phylogeny for relationships of genera within the family Enterobacteriaceae. To view tree in its entirety, see Francino* et al. *(2003).*

sequences reveal that within Enterobacteriaceae, *Xenorhabdus* and *Photorhabdus* form a monophyly (Szállás *et al.*, 1997) with *Proteus* as their closest relative (Forst *et al.*, 1997; Szállás *et al.*, 1997; Francino *et al.*, 2003). This is in agreement with the phenotypic traits of *Xenorhabdus* and *Photorhabdus*. Both symbionts are atypical of the genera belonging to Enterobacteriaceae (Holt *et al.*, 1994), being similar only to each other. Branch lengths between *X. nematophila* and *P. asymbiotica* are long in comparison to those of other genera included in the phylogeny and this too is supported by chemotaxonomic characterisation.

Phylogenies investigating the inter- and intrageneric relationships of *Xenorhabdus* and *Photorhabdus* have been constructed using 16S rRNA gene sequences (Rainey *et al.*, 1995; Suzuki *et al.*, 1996; Liu *et al.*, 1997, 2001; Szállás *et al.*, 1997; Fischer-Le Saux *et al.*, 1999b; Marokhazi *et al.*, 2003; Akhurst *et al.*, 2004; Hazir *et al.*, 2004a; Lengyel *et al.*, 2005a; Sicard *et al.*, 2005; Sergeant *et al.*, 2006). Maximum likelihood trees were produced to incorporate different sequences into the phylogenies for *Xenorhabdus* and *Photorhabdus* and are presented in Figures 252 and 253, respectively. To construct the phylogenies, multiple alignments were downloaded from the RDP-II website (Cole

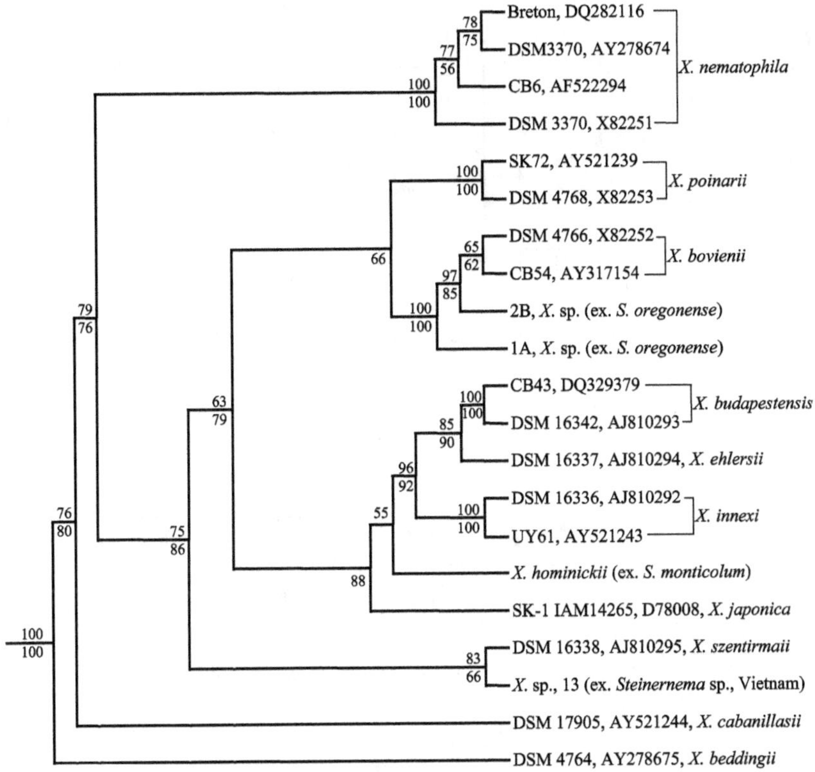

Fig. 252. *Phylogenetic relationships of various* Xenorhabdus *species based on maximum likelihood analysis of partial 16S rDNA sequences. Bootstrap support indices (1000 replications) for Neighbour Joining (above branch) and Maximum Parsimony (below branch) analyses are indicated where congruent with maximum likelihood tree. Species of* Photorhabdus *chosen for outgroup (Suzuki* et al., *1996; Sicard* et al., *2004). Tree rooted with* P. luminescens *ssp.* akhurstii *(AJ007359),* P. temperata *(AY278655) and* P. asymbiotica *ssp.* asymbiotica *(AY278672). TrN+I+G model of evolution (γ shape parameter = 0.5621; substitution rate matrix Ra = 1.0000, Rb = 2.8789, Rc = 1.0000, Rd = 1.0000, Re = 4.6016, Rf = 1.0000; base frequencies: freqA = 0.2459, freqC = 0.2210, freqG = 0.3271, freqT = 0.2060).*

et al., 2005) and optimised manually with BioEdit 7.05 (Hall, 1999). Parsimony, distance and maximum likelihood analyses were performed using PAUP* 4.0b10 (Swofford, 2002) and models of sequence evolution were evaluated using ModelTest 3.7 (Posada & Crandall, 1998). The TrN + I + G model was selected for the data sets by hierarchical

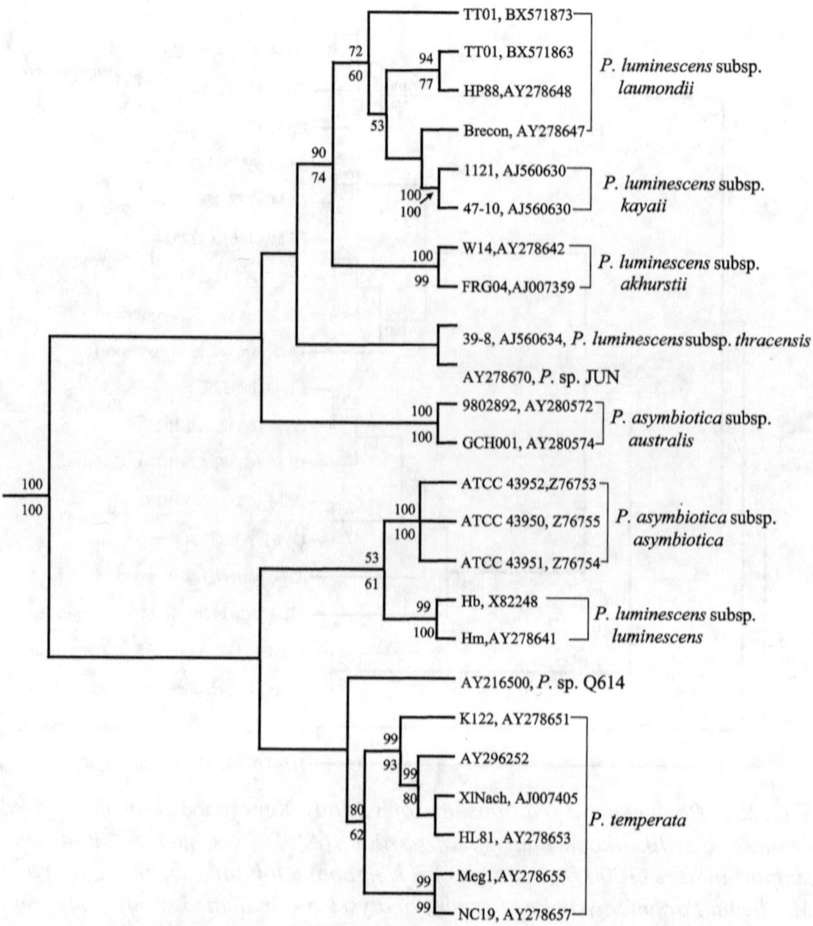

Fig. 253. *Phylogenetic relationships of* Photorhabdus *based on maximum likelihood analysis of partial 16S rDNA sequences. Bootstrap support indices (1000 replicates) for Neighbour Joining (above branch) and Maximum Parsimony (below branch) analyses are indicated where congruence resided with maximum likelihood tree. Tree rooted with* Proteus vulgaris *(X07652) and* Serratia proteomaculans *(AJ233435). TrN + I + G model of evolution (γ shape parameter = 0.6219; substitution rate matrix Ra = 1.0000, Rb = 2.9659, Rc = 1.0000, Rd = 1.0000, Re = 3.2786, Rf = 1.0000; base frequencies: freqA = 0. 2559, freqC = 0.2173, freqG = 0.3181, freqT = 0.2087).*

likelihood ratio tests and used for resolution of both the neighbour joining and maximum likelihood trees. The maximum parsimony and neighbour joining trees are not shown, but for both, the bootstrap support values are indicated on the maximum likelihood trees where the trees were concordant. All three phylogenetic tree reconstruction methods generated similar tree topologies for *Xenorhabdus* and including only the same taxa, the maximum likelihood tree (Fig. 252) is congruent with the tree of Lengyel *et al.* (2005). Not all of the branch nodes have high support values (*e.g.*, the clades with the symbionts of *S. oregonense* and *S. monticolum*), indicating that branching order is not necessarily resolved. The different reconstruction methods produced tree topologies that were less concordant for *Photorhabdus*. Taxonomic monophyly existed among most of the clades within the phylogeny, but for neighbour joining and maximum parsimony methods, *Photorhabdus* sp. Q614, *Photorhabdus* sp. JUN and *P. luminescens* subsp. *thracensis* were placed in different positions from the maximum likelihood tree. For the corresponding taxa the maximum likelihood tree (Fig. 253) is congruent with the tree of Marokhazi *et al.* (2003) and, excluding the position of *P. luminescens* subsp. *luminescens*, is similar to the trees of Akhurst *et al.* (2004) and Hazir *et al.* (2004a). In this scenario, *Photorhabdus* diverges into two clades that are inconsistent with species-level groupings (Akhurst *et al.*, 2004). *Photorhabdus asymbiotica* subsp. *asymbiotica* diverges from the other *Photorhabdus* as sister taxon to *P. luminescens* subsp. *luminescens*, while *P. asymbiotica* subsp. *australis* forms a monophyletic group with other *P. luminescens* subspecies. The branch support values are weak or nonexistent at nodes that join interspecific taxa. The phylogeny based on small subunit rRNA gene sequences supports the subspecies-level groupings of Fischer-Le Saux *et al.* (1999b) but is not suitable for distinguishing species within *Photorhabdus* (Akhurst *et al.*, 2004).

The use of 16S rRNA gene sequences for determining phylogenetic relationships has limitations. The sequences 1A and 2B from the symbiont of *S. oregonense* are representative of different operons within the same bacterium. The percent sequence divergence between 1A and 1B is 1.1% (Koppenhöfer & Adams, data unpubl.). These, together with two different sequences for strain DSM 3370, were included in the phylogenetic analyses of *Xenorhabdus* (Fig. 252). Similarly, two different operons from *P. luminescens* subsp. *laumondii* strain TT01 were included in the analyses of *Photorhabdus* (Fig. 253). In all three cases, the sequences clustered closely together but, particularly with the symbiont of *S. oregonense* did not necessarily appear to be representative of the same organism. This is a variable that should be considered for interpretation of evolutionary history between taxa. Like other bacterial genes, ribosomal genes can undergo lateral gene transfer across diverse taxonomic groups or be recombined, which could lead to artificial evolutionary data (Yap *et al.*, 1999). Consequently, resolving bacterial phylogeny based solely on 16S rRNA gene sequences may not be sufficient (Lerat *et al.*, 2003). Akhurst *et al.* (2004) were able to demonstrate that a phylogeny based on the *gyrB* gene provided a more robust tree for determining intrageneric relationships among *Photorhabdus*. This is due to the lower proportion of invariant nucleotide positions than in the 16S rRNA gene (Akhurst *et al.*, 2004). Figure 254 is a maximum likelihood tree based on the partial *gyrB* sequences of Akhurst *et al.* (2004). A multiple alignment was built using Clustal X (Thompson *et al.*, 1997) and optimised manually using BioEdit 7.05 (Hall, 1999). Parsimony, distance and maximum likelihood analyses were performed using PAUP* 4.0b10 (Swofford, 2002) and ModelTest 3.7 was used to select a model of sequence evolution (Posada & Crandall, 1998). The TrNef + I + G model was selected by hierarchical likelihood ratio tests and used to resolve the neighbour joining and maximum likelihood trees. All three phylogenetic reconstruction methods generated equivalent tree topologies for *Photorhabdus* and the maximum likelihood tree (Fig. 254) is similar to the tree of Akhurst *et al.* (2004). At deeper nodes, the taxa are divided by species, then as the branches near the terminus, the taxa are subdivided by subspecies. Unlike in the 16S rRNA gene-based phylogeny, *P. asymbiotica* subsp. *asymbiotica* and *P. asymbiotica* subsp. *australis* form a monophyletic group that diverges as sister taxon to *P. luminescens*. The subspecies of *P. luminescens* also form a monophyletic group. This is further supported by the *Photorhabdus* phylogeny based on concatenated sequences of *glnA* and *gyrB* constructed by Gerrard *et al.* (2006). Recently, Sergeant *et al.* (2006) characterised *Xenorhabdus* isolates by multi locus sequence typing in which a portion of four housekeeping genes were sequenced, concatenated and analysed. When compared to analysis by 16S rRNA sequences, the sequence typing was more informative and able to differentiate between strains that were indistinguishable by 16S rDNA sequence data alone (Sergeant *et al.*, 2006). Although 16S rRNA gene sequence data are informative and play an integral role for bacterial systematics, to resolve the intrageneric relationships for both *Xenorhabdus* and *Photorhabdus* more robust trees should be developed by comparing sequences from additional strains from carefully selected protein-coding genes.

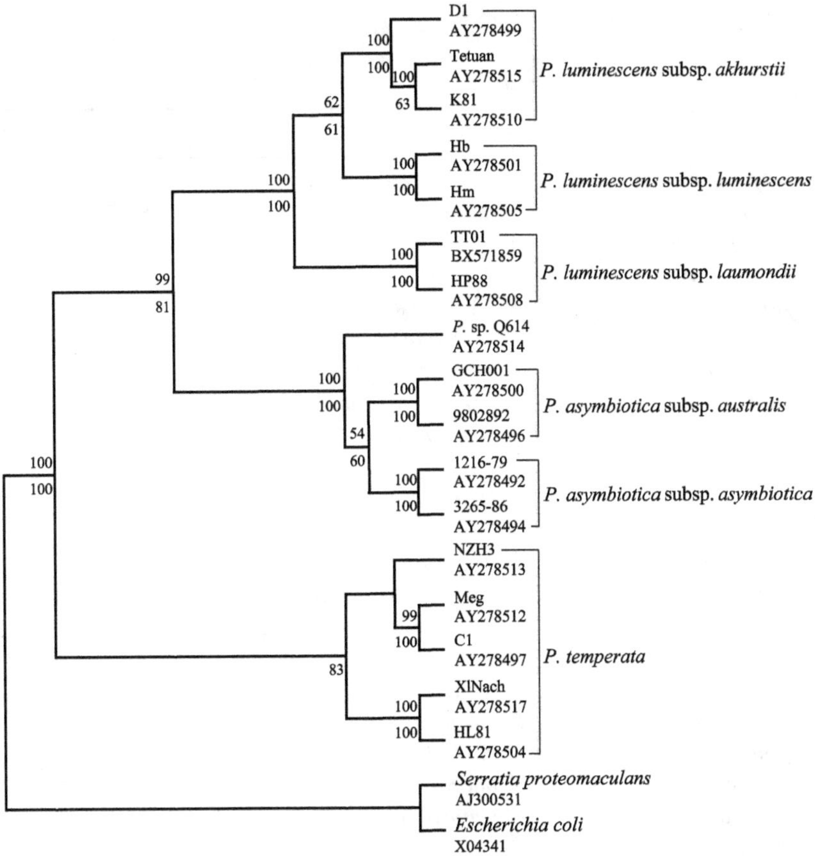

Fig. 254. *Phylogenetic relationships between species/strains of* Photorhabdus *based on Maximum likelihood analysis of partial* gyrB *gene sequences. Bootstrap support indices (1000 replications) for Neighbour Joining and Maximum Parsimony analyses are indicated above and below the branches, respectively where congruent with maximum likelihood tree. TrNef+I+G model of evolution by hLRT (γ shape parameter = 0.9677; substitution rate matrix Ra = 1.0000, Rb = 4.6683, Rc = 1.0000, Rd = 1.0000, Re = Rf = 1.0000; equal base frequencies).*

Nematode/bacteria association

The relationships between entomopathogenic nematodes and their symbiotic bacteria are summarised in Table 74. Each species of *Steinernema* is mutualistically associated with only one species of *Xenorhabdus*; however, some species of *Xenorhabdus* are symbionts of multiple

Table 74. *Symbiotic associations.*

Xenorhabdus/Steinernematidae	
X. beddingii	*S. longicaudum*
X. bovienii	*S. affine, S. feltiae, S. intermedium, S. kraussei*
X. budapestensis	*S. bicornutum*
X. cabanillasii	*S. riobrave*
X. doucetiae	*S. diaprepesi*
X. ehlersii	*S. serratum* (*nomen nudum*)
X. griffiniae	*S. hermaphroditum*
X. hominickii	*S. karii, S. monticolum*
X. innexi	*S. scapterisci*
X. indica	*S. thermophilum* (= junior syn. of *S. abbasi*)
X. japonica	*S. kushidai*
X. koppenhoeferi	*S. scarabaei*
X. kozodoii	*S. arenarium, S. apuliae*
X. mauleonii	*Steinernema* sp. isolated from St Vincent in the Caribbean Sea
X. miraniensis	Nematode of the family Steinernematidae isolated from Mirani, Queensland, Australia
X. nematophila	*S. carpocapsae*
X. poinarii	*S. glaseri, S. cubanum*
X. romanii	*S. puertoricense*
X. stockiae	*S. siamkayai*
X. szentirmaii	*S. rarum*
***Photorhabdus*/Heterorhabditidae**	
P. asymbiotica subsp. *asymbiotica*	unknown host
P. asymbiotica subsp. *australis*	*Heterorhabditis* sp. (similar to *H. indica*)
P. luminescens subsp. *luminescens*	*H. bacteriophora* Brecon
P. luminescens subsp. *akhurstii*	*H. indica*
P. luminescens subsp. *kayaii*	*H. bacteriophora* Turkey
P. luminescens subsp. *laumondii*	*H. bacteriophora* HP88
P. luminescens subsp. *thracensis*	*H. bacteriophora* Turkey
P. temperata	*H. zealandica, H. bacteriophora* NC1, *H. megidis*

steinernematid species. Symbionts with multiple host associations include: *Xenorhabdus beddingii*, the symbiont of *S. longicaudum* and an undescribed nematode species; *X. bovienii*, the symbiont of *S. affine*, *S. feltiae*, *S. intermedium* and *S. kraussei*; *X. poinarii*, the symbiont of *S. glaseri* and *S. cubanum*; *X. hominickii*, the symbiont of *S. karii* and *S. monticolum*; and *X. kozodoii*, the symbiont of *S. arenarium* and *S. apuliae* (Akhurst & Boemare, 1988; Fischer-Le Saux *et al.*, 1999a; Boemare, 2002a, b; Talliez *et al.*, 2006). In contrast to the steinernematid-*Xenorhabdus* complex, different strains of the nematode species *H. bacteriophora* are associated with two different species of *Photorhabdus* and subspecies thereof. The Brecon and HP88 groups are associated with the subspecies *P. luminescens* subspecies *luminescens* and *P. luminescens* subspecies *laumondii*, respectively, whereas the NC group is associated with *P. temperata* (Fischer-Le Saux *et al.*, 1999b). *Photorhabdus temperata* is also associated with *H. downesi*, *H. megidis* and *H. zealandica*. Other multiple symbiotic associations may exist but are not yet known for the symbiotic bacteria of the most recently described nematode species have yet to be characterised.

Some combinations of *Steinernema* spp. with non-native *Xenorhabdus* spp. support *in vitro* nematode growth and development (Akhurst, 1983b; Grewal *et al.*, 1997; Sicard *et al.*, 2004b). However, Sicard *et al.* (2004b) found that, within an insect host, non-native *Xenorhabdus* spp. other than *X. poinarii* were pathogenic to *S. carpocapsae*. This leads to the assumption that a host nematode ought to be resistant to virulence factors produced by its symbiotic bacteria but not necessarily to those produced by non-native symbionts. This selective resistance could be an advantage given to the nematode by their symbiotic bacteria when confronted with coinfection of an insect by another nematode species (Sicard *et al.*, 2004b). Other bacteria may produce conditional associations with a nematode, but the natural symbiont is the most efficient partner for pathogenicity, reproduction and development of the nematode (Bonifassi *et al.*, 1999; Sicard *et al.*, 2004b).

The specificity between the nematode and its bacterial symbiont is a result of the exclusion of bacterial competitors and a specific recruitment by the IJ of its symbiont. Both *Xenorhabdus* and *Photorhabdus* produce bacteriocins, which are proteinaceous compounds with antimicrobial activity against closely related bacterial strains or species. The bacteriocins are functionally important for the symbiotic relationship as they help the natural symbiont to out compete closely related bacteria (Boemare *et al.*, 1997b; Sicard *et al.*, 2005). Xenorhabdicin is a phage tail-like bacteriocin isolated from *X. nematophila* that has antibiotic activity against other *Xenorhabdus* spp., *P. luminescens* and species of *Proteus* (Boemare *et al.*, 1992; Thaler *et al.*, 1995). Lumicins are bacteriocins isolated from *P. luminescens* and are active against other *Photorhabdus* spp. as well as distantly related bacteria such as *E. coli* (Sharma *et al.*, 2002). The latter suggests that these bacteriocins may not be exclusively active against competing symbionts but may be involved in clearing of the insect gut microflora (Sharma *et al.*, 2002).

For the symbiotic relationship to persist, the bacteria must be able to colonise their host nematode. Fimbriae are bacterial cell surface appendages and have been proposed to be involved in host colonisation, specifically for association with the epithelial cells of the nematode gut (Binnington & Brooks, 1993; Moureaux *et al.*, 1995; Forst & Nealson, 1996); however, the role of the fimbriae has not yet been ascertained (Ciche *et al.*, 2006). *Xenorhabdus nematophila* produces mannose-resistant fimbriae, resembling the fimbriae of *Proteus mirabilis* (Moureaux *et al.*, 1995; Forst & Clarke, 2002). The genome of *P. luminescens* subsp. *laumondii* strain TT01 was sequenced and was found to have a large repertoire of fimbrial genes (Duchaud *et al.*, 2003). Among the fimbrial genes were two gene clusters encoding proteins similar to the mannose-resistant fimbriae and on another gene cluster, *ngrA*, which in *Photorhabdus* is important for bacteria-nematode interactions (Ciche *et al.*, 2001; Duchaud *et al.*, 2003). These genes too may play a role in nematode colonisation (Duchaud *et al.*, 2003).

Several genes involved in vesicle colonisation of the host nematode have been identified in *X. nematophila.* Among these is a novel class of genes, designated *nil* for nematode intestine localisation, which may have a specific role for allowing the bacterium to colonise the nematode host (Heungens *et al.*, 2002). A small transmembrane protein of unknown function is encoded by *nilA*; *nilB* encodes an outer membrane protein and *nilC* encodes an outer membrane associated lipoprotein (Heungens *et al.*, 2002; Cowles & Goodrich-Blair, 2004). *nilA* is in part required for vesicle colonisation whereas the gene products for *nilB* and *nilC* are essential (Heungens *et al.*, 2002; Cowles & Goodrich-Blair, 2004). Also, cells may be colonisation-defective due to mutations in the regulatory proteins RpoS, RpoE and Lrp (Vivas & Goodrich-Blair, 2001; Heungens *et al.*, 2002). In *E. coli*, the transcription factor, RpoS, controls regulons that can mediate stress resistance and survival. In *X.*

nematophila, RpoS is essential for nematode colonisation and helps to mediate resistance to peroxide stress. However, to date, the genes regulated by RpoS in *X. nematophila* are unknown (Vivas & Goodrich-Blair, 2001). Very few cells with defective extracytoplasmic stress sigma factor, σ^{E}, are able to colonise the vesicle, and this defect may be related to the reduced survival of *rpoE* mutants during the colonisation assay (Heungens *et al.*, 2002). Whilst RpoS and σ^{E} are not necessary for expression of the *nil* genes, they may regulate other factors involved in colonisation to promote survival within the intestinal vesicle (Vivas & Goodrich-Blair, 2001; Heungens *et al.*, 2002). The regulatory protein Lrp is required, synergistically with a small transcription factor NilR, to repress the expression of *nilA*, *B* and *C*, and cells defective in either the *nil* genes or *lrp* lose the ability to colonise the host (Cowles & Goodrich-Blair, 2004, 2006; Cowles *et al.*, 2007). The *nil* loci are not present in the *P. luminescens* subsp. *laumondii* genome (Ciche *et al.*, 2006); however, the *pbgPE* operon, which is required for pathogenicity, is also required by *P. luminescens* for colonisation of the IJ intestine (Bennett & Clarke, 2005). Our understanding of the nematode-bacterium symbiosis is growing but requires further study to explain the provisions given by the symbiont and the resilience of the association over many generations (Bonifassi *et al.*, 1999; Boemare, 2002b).

Even though the entomopathogenic nematodes share an exclusive mutualistic relationship with their respective symbiotic bacteria, other bacteria are occasionally isolated from the nematode or the host insect subsequent to infection. Some of these bacteria include members of the genera *Acinetobacter*, *Citrobacter*, *Ochrobactrum*, *Providencia*, *Pseudomonas*, *Serratia*, *Salmonella* and *Vibrio*, and are likely associated with the cuticle or found between the L2 and L3 layers of the nematode cuticle (Lysenko & Weiser, 1974; Aguillera & Smart, 1993; Jackson *et al.*, 1995; Bonifassi *et al.*, 1999; Babic *et al.*, 2000; Walsh & Webster, 2003; Gouge & Snyder, 2006). Some bacteria are able to persist alongside the symbiont despite the virulence factors produced by the symbiotic bacteria. *Acinetobacter*, a bacterium carried by the nematode, is able to endure the antimicrobial compounds produced by *X. bovienii* and proliferate inside the host insect cadaver (Walsh & Webster, 2003). *Ochrobactrum* spp. were found to associate naturally with *P. luminescens* subsp. *akhurstii*, the symbiont of *H. indica* (Babic *et al.*, 2000) and *Ochrobactrum anthropi* has been isolated with the symbiont of *H. floridensis* (Koppenhöfer & Jones, unpubl.). *Paenibacillus nematophilus* has a stable relationship with *Heterorhabditis* spp. The bacterial sporangia adhere to the cuticle of the IJ, are resistant to the *Photorhabdus* virulence factors and reproduce in sympatry in the insect host (Enright *et al.*, 2003; Enright & Griffin, 2004, 2005). It is possible for nematodes to transport other nonsymbiotic bacteria and these dixenic associations may be detrimental to the nematode (Enright & Griffin, 2005) or the symbiotic relationship (Bonifassi, 1999; Walsh & Webster, 2003).

Discussion

Unique to this complex is the involvement of three partners (bacterium, nematode and insect) and the presence of both mutualistic and parasitic symbiotic relationships. The bacterium is an ecologically obligate partner (Boemare *et al.*, 1997b) because it is unable to exist in the environment without the nematode host. In turn, it is required by the nematode to complete its life cycle efficiently (Forst & Nealson, 1996). Furthermore, the nematode and the bacterium are both implicated in the parasitic relationship with the insect. Also unique to this mutualism is that the bacteria are able to inhabit two different ecological niches: the nematode intestine (nutritionally poor) and the insect cadaver (nutritionally rich). Much progress has been made for the explanation of factors involved in the mutualistic symbiosis and fidelity of these relationships but more study is required.

Although the *Xenorhabdus-Steinernema* and *Photorhabdus-Heterorhabditis* symbioses are very similar in life cycle and pathogenicity, they differ in how the bacteria are released into the insect, colonisation of the host nematode and the specificity of the relationship. The bacteria differ from each other as the genome of *Photorhabdus* is about 1 Mb larger than that of *Xenorhabdus* (Forst & Goodner, 2006) and the bacteria produce different virulence and symbiotic factors. Therefore, it seems fair to assume that similarities between these two systems have most likely resulted from convergent evolution (Boemare, 2002a).

The genome of *P. luminescens* subsp. *laumondii* has been sequenced (Duchaud *et al.*, 2003) and sequencing projects for the genomes of *P. asymbiotica*, *X. bovienii* and *X. nematophila* are underway (http://www.sanger.ac.uk/Projects/P_asymbiotica and http://www1.xenorhabdus.org). These tools give insight into the symbiosis and pathogenicity of the bacteria and aid in the reconstruction of evolutionary and co-evolutionary history. In addition, new entomopathogenic nematodes are continuously being described, each presumably having a unique relationship with a bacterial partner, providing more opportunity for the use of taxonomic and phylogenic tools and the study of symbiosis and pathogenesis.

Methodology

Presented in this section are general methods and recipes for working with the bacterial symbionts. Techniques that are used in the initial isolation of the bacterium are important for the success of subsequent research. The need for aseptic techniques is emphasised throughout the protocols as contaminating bacteria are ever present. In the following section are procedures for the isolation, cultivation, phenotypic form determination and storage of the symbiotic bacteria, as well as for the extraction of genomic DNA and amplification of 16S rDNA for identification of the strain or for phylogenetic study. Although only a selection of methods is presented here, many other excellent techniques have been published.

SURFACE STERILISATION OF INFECTIVE JUVENILES (IJ)

Allow the IJ to settle out of suspension and, using a pipette, remove most of the water. Collect the IJ (25-100) and resuspend in 0.5 ml of sterile water in a 1.5 ml microcentrifuge tube.

Note: Before surface sterilising the IJ, ensure the nematode suspension is free of any extraneous matter; otherwise, the process may be ineffective leading to difficulty isolating the symbiont.

Surface sterilisation using methylbenzethonium chloride or benzethonium chloride (Hyamine® 1622)

Surface sterilise the IJ by adding 0.5 ml of 0.25% (w/v) methylbenzethonium chloride or benzethonium chloride (wash solution) to the suspension and rock gently on a platform rocker or other suitable mixer. Alternatively, the solution can be gently rocked by hand every 5-10 min. After 20-30 min, rinse the IJ by successive transfers to three sterile microcentrifuge tubes each containing 1.0 ml of sterile purified water using the following approach. Place the microcentrifuge tube with the wash solution so that it is standing upright and allow the nematodes to settle to the bottom. Using a sterile pipette, remove most of the wash solution being careful not to disturb the IJ at the bottom of the tube. Then transfer the IJ with a sterile pipette to one of the tubes with sterile purified water. Allow the IJ to settle to the bottom and transfer to the next tube following the same process. Repeat once more. It is not necessary to recover every IJ. Begin with enough to compensate for some loss at each step.

Note: Heterorhabditid IJ seem to be particularly sensitive to the wash solution and should be removed and the rinse process started within 30 min. Steinernematid IJ are less sensitive to the Hyamine and can remain in the wash solution for several hours.

Surface sterilisation using sodium hypochlorite

Kaya and Stock (1997) described a method for surface sterilisation using a solution of sodium hypochlorite. In brief, immerse the IJ in a solution of 0.1% (w/v) sodium hypochlorite for 1 h. Transfer the IJ to a fresh tube of 0.1% sodium hypochlorite for 3 h. Rinse the IJ as described previously for the Hyamine wash.

Note: Various methods have been published for the elimination of contaminating microorganisms. For alternatives to those described in this text, please see Lengyel *et al.* (2005a), Han and Ehlers (2001), Bonifassi *et al.* (1999) and Kaya and Stock (1997).

Isolation of symbiotic bacteria

Isolation from surface-sterilised IJ (Akhurst, 1980)

Surface sterilise the IJ and suspend in yeast salts broth or nutrient broth. Transfer the IJ to a sterile tissue grinder and homogenise the suspension. Alternatively, macerate the nematodes in the same microcentrifuge tube using a Kontes® pellet pestle. Spread the macerate onto an agar plate such as nutrient agar (NA) or nutrient agar supplemented with 0.004% (w/v) triphenyltetrazolium chloride and 0.025% (w/v) bromothymol blue (NBTA). Incubate at 27-28°C for 48 h.

Note: To reduce the possibility of contamination from bacteria which may remain on some nematodes through the sterilisation process, the symbiotic bacterium may be isolated from a single surface sterilised IJ as follows. Under a stereomicroscope, aseptically transfer an IJ to a drop of M9 buffer or nutrient broth on a sterile microscope slide. Using a sterile platinum wire (Lengyel *et al.*, 2005a) or a sterile scalpel (Vivas & Goodrich-Blair, 2001), cut the nematode and with a sterile inoculating loop, transfer the buffer/broth and spread onto an agar plate.

Isolation from haemolymph of infected Galleria mellonella *(Akhurst, 1980; Kaya & Stock, 1997)*

In a Petri dish with moist filter paper, inoculate last instar *G. mellonella* larvae with 100 IJ/larva. After 24-48 h, surface sterilise a *G. mellonella* larva by washing with 1.0% (w/v) sodium hypochlorite and rinsing three times with sterile water or by dipping the larva into 95% (v/v) ethanol, igniting and quickly plunging into sterile water. Using sterile scissors open the cadaver being careful not to break the midgut. With a sterile inoculating loop, transfer and spread a loopful of haemolymph onto an agar plate. Alternatively, with a sterile syringe, inject the surface sterilised larva with 100 μl of Ringer's solution, again being careful not to break the midgut. Withdraw some of the solution and haemolymph and spread on an agar plate. Incubate at 25-28°C for 48 h.

Isolation from insect haemolymph (Poinar & Thomas, 1966)

Surface sterilise a *G. mellonella* larva. Withdraw a drop of haemolymph and aseptically transfer to a sterile microscope slide cover. Place 5-10 surface-sterilised IJ in the haemolymph and invert the slide cover over the well of a depression slide. Incubate at 25°C for 24 h. The IJ exsheath and begin development, releasing their symbiotic bacteria. Touch a sterile inoculating loop to the haemolymph drop and spread on an agar plate. Incubate as described previously.

To separate the symbiotic bacteria from contaminates, the bacteria obtained from the isolation procedures should be cultured on a selective medium such as MacConkey agar or NBTA.

PHENOTYPIC VARIATION

Adsorption of bromothymol blue (BTB)

The adsorption of BTB can be examined on NBTA, the medium of Akhurst (1980) or on Tergitol 7 agar supplemented with 0.004% (w/v) triphenyltetrazolium chloride (T7-TTC) (Akhurst, 1983). Both forms of bacteria reduce TTC to formazan (Akhurst, 1980; Boemare *et al.*, 1997a) indicated by the production of a red colour. This reaction is hidden with form I cells due to the ability of this form to adsorb BTB. Form

Table 75. *Characteristic colouration of* Xenorhabdus *and* Photorhabdus *when cultured on media containing bromothymol blue + TTC or neutral red.*

Bacterial species	NBTA and/or T7-TTC agar	MacConkey Agar
X. beddingii	blue (Akhurst, 1986)	brown (Akhurst & Boemare, 1988)
X. bovienii	greenish to blue green (Thomas & Poinar, 1983)	brown (Akhurst & Boemare, 1988)
X. budapes-tensis	blue (Lengyel *et al.*, 2005)	
X. ehlersii	red (Lengyel *et al.*, 2005)	
X. indica	blue (Somvanshi *et al.*, 2006)	red (Somvanshi *et al.*, 2006)
X. innexi	blue (Dr Elke lang, pers. comm., 2006)	
X. japonica	blue (Nishimura, 1994)	
X. nematophila	blue to deep purple (Thomas & Poinar, 1983)	brown (Akhurst & Boemare, 1988)
X. poinarii	red (Akhurst, 1983; Thomas & Poinar, 1983)	red (Akhurst & Boemare, 1988)
X. szentirmaii	blue (Lengyel *et al.*, 2005)	
P. luminescens	green with orangish brown centres (Thomas & Poinar, 1979, 1983; Akhurst, 1983)	red (Akhurst & Boemare, 1988)
P. temperata	greenish darkening to dull olive green (Thomas & Poinar, 1979)	

I colonies take on a characteristic greenish to blue colour and as the BTB is adsorbed from the medium a decolourised zone surrounding the colonies becomes apparent. Form II colonies are unable to adsorb BTB and, therefore, are red in colour with no clear zone in the medium. Table 75 indicates colour produced by dye adsorption (or lack thereof) of various strains.

On either NBTA or T7-TTC agar, quadrant streak (Fig. 255) the bacteria to obtain isolated colonies. Incubate at 25-28°C for 3-5 days. Observe the colonies for the adsorption of BTB (Fig. 256). It should be noted that due to the ability of form I variants to reduce TTC, some reddish colour may be evident. Not all *Xenorhabdus* are able to adsorb BTB; consequently, if the colonies appear reddish in colour, this does not necessarily indicate that the bacteria are in the secondary form.

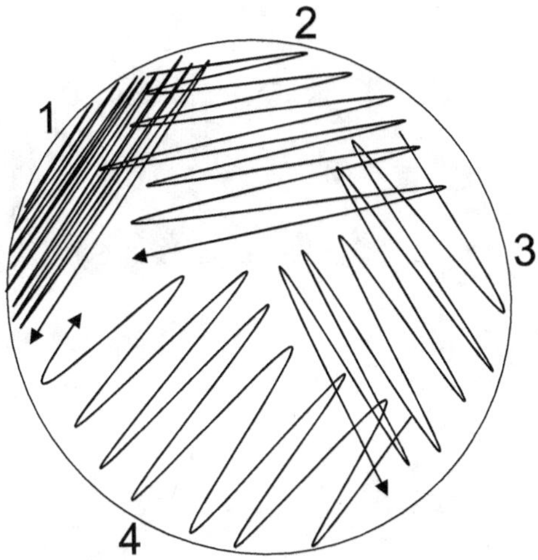

Fig. 255. *Bacteria can be isolated into single colonies by using a quadrant streak method. The inoculum is spread on a small part of the agar with a sterile loop (1). The bacteria are successively spread on the agar by taking a small part of the previous streak and carrying the bacteria onto a fresh part of the agar (2-4) each time using a sterilised loop. Eventually, by the end of the third or fourth streak, the bacteria should be present as single cells leading to the development of isolated colonies.*

Adsorption of neutral red

On MacConkey agar, quadrant streak the bacteria so as to obtain isolated colonies. Incubate at 25-28°C for 24 h. Form I colonies appear red or reddish brown due to the adsorption of neutral red. Form II colonies will appear off-white, grey or light yellow. The bile salts in this medium can hamper the growth of *Xenorhabdus* and *Photorhabdus*; therefore it is necessary inoculate the medium heavily (Boemare *et al.*, 1997a). Boemare *et al.* (1997a) further suggest the elimination of the bile salts from the medium composition. See Table 75 for the colour descriptions of some strains growing on MacConkey agar.

Antibiosis

Akhurst (1982) described the antibiotic activity of *Xenorhabdus* and *Photorhabdus* and found antibiotic activity to be restricted to form I. Form II bacteria show slight or no antibiotic activity. In a glass Petri dish with nutrient agar, spot inoculate the test organism using a 24 h broth culture of each variant (Boemare *et al.*, 1997a). Incubate at 25-

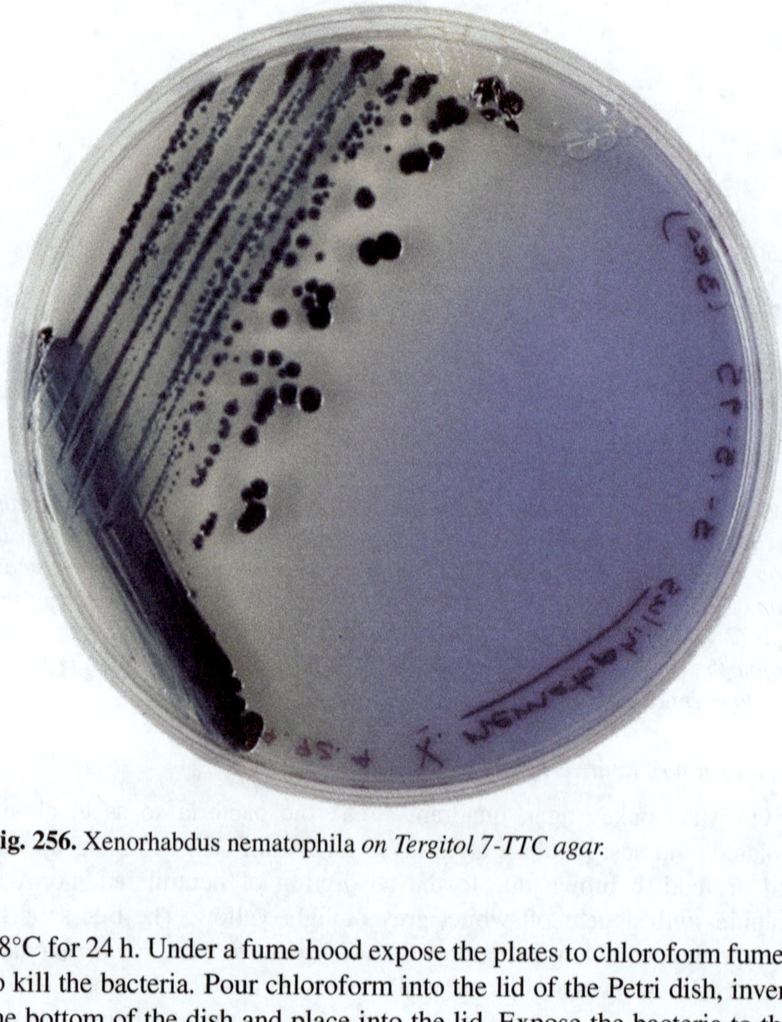

Fig. 256. Xenorhabdus nematophila *on Tergitol 7-TTC agar.*

28°C for 24 h. Under a fume hood expose the plates to chloroform fumes to kill the bacteria. Pour chloroform into the lid of the Petri dish, invert the bottom of the dish and place into the lid. Expose the bacteria to the fumes for 1 h. Transfer the Petri dish to a laminar flow hood and allow the chloroform to evaporate for 1 h. Mix 1 ml of an overnight culture of *Micrococcus luteus* or other suitable bacterium with 100 ml of sterile soft agar (nutrient broth 5.0 ml, Bacto-agar 7.0 g, 1000 ml water) that has been cooled to 45°C. Dispense the inoculum evenly onto the plates with the chloroform-killed bacteria. Incubate for 24 h and then observe for zones of inhibition. Form I variants produce a greater quantity of agar diffusible antibiotics and therefore will have larger zones of inhibition (Boemare *et al.*, 1997a).

Preservation of the bacteria

Short-term preservation

For short-term preservation, most bacteria can be maintained on agar plates or slants at 8-15°C and routinely subcultured on a monthly basis. Inoculate the medium (*e.g.*, nutrient or MacConkey agar) and incubate at 25°C only until ample bacterial growth appears. Then hold the plate or slant at 8-15°C. To select for form I clones, it is best to use agar plates so that well isolated colonies can be obtained. Select colonies that adsorb neutral red dye from MacConkey agar or, with the exception of *X. ehlersii* and *X. poinarii*, that adsorb bromothymol blue from NBTA and Tergitol-7 agar. To constrain phenotypic variation, subculture the cells on a weekly basis selecting for form I clones (Boemare & Akhurst, 2000).

Long term preservation

Collect a large loopful of cells from a 24 h YS, LB or nutrient agar culture. Deposit the cells into a 2.0 ml polypropylene cryogenic vial containing 1.5 ml of nutrient broth with 20% (v/v) glycerol as a cryoprotectant. Suspend the cells in the nutrient broth/glycerol as best as possible and store immediately at −70°C or below or in the gas phase of liquid nitrogen. To recover the cells, scrape a sample of the frozen material with a sterile loop and streak onto an agar plate.

Isolation of genomic DNA from the symbiotic bacteria

This protocol is fairly simple and yields large amounts of DNA. However; there are many other protocols as well as commercially available DNA extraction kits with which the chromosomal DNA can be easily purified.

Modified from Ausebel et al., *1994*

Using aseptic technique, inoculate 5 ml of LB broth with a single colony from an agar plate with a 48 h bacterial culture. Allow the bacteria to proliferate overnight at 25-28°C with shaking (150-200 rpm). With a sterile pipette, transfer 0.75-1.0 ml of the broth culture to a sterile 1.5 ml microcentrifuge tube. Using a table top microcentrifuge, centrifuge the bacteria at 5200 *g* for 3 min. Decant the broth. If the growth is weak, repeat by adding broth culture to the same tube, centrifuge and decant as before. Wash the cells by adding 1.0 ml of sterile TE buffer (pH 8.0). Resuspend the cells and centrifuge as before. Decant all but a very small amount of TE (20-30 μl), just enough to start allowing the cells to go into suspension, and briefly vortex. (This will make resuspending the cells a little easier in the next stage.)

Add 550 μl of TE to the cells and resuspend. Once cells are in suspension, add 30 μl of 10% (w/v) SDS (sodium dodecyl sulphate) and 3 μl of proteinase K (10 mg ml^{-1}). Incubate the suspension at 37°C until the solution is clear (cells have lysed) or up to 1 h.

After the cells have lysed, add 150 μl of 5M NaCl, mix well and add 80 μl of CTAB/NaCl (the CTAB solution is quite viscous and therefore the use of wide bore tips is helpful). Mix well and incubate at 65°C for 10 min.

Working in a fume hood with eye protection, a laboratory coat and protective gloves, add 650 μl of phenol/chloroform/isoamyl alcohol (25 : 24 : 1). Mix thoroughly but gently to homogenise the solution. Centrifuge at 11 600 *g* for 5 min. Carefully transfer the aqueous layer to a sterile microcentrifuge tube being sure not to disturb the white denatured protein interface. Add an equal volume of chloroform/isoamyl alcohol (24 : 1). Mix as before. Centrifuge at 11 600 *g* for 5 min and carefully transfer the aqueous layer to a sterile microcentrifuge tube while not disturbing the interface. Precipitate the nucleic acids by adding 0.6 volume of isopropyl alcohol (99%). Let stand briefly, and then gently rock back and forth. Collect the nucleic acids by centrifugation at 8000 *g* for 5 min. To remove residual salt from the nucleic acids, add 1.0 ml of 70% (v/v) ethanol and centrifuge at 8000 *g* for 2 min. Repeat the wash procedure.

Allow the pellet to dry briefly. When no moisture is apparent in the tube, add 100 μl of 10 mM Tris (pH 7.6-8.0) to the pellet and be sure to collect any of the precipitant adhering to the wall of the tube. Let stand at 4°C overnight (otherwise resuspending the nucleic acids may be difficult). Dissolve the nucleic acids and store at −20°C.

Optional step (for the digestion of RNA)

To the nucleic acid suspension, add 100 μl of 10 mM Tris (pH 7.6-8.0) to bring the suspension up to 200 μl. Gently mix. Add 2 μl of RNase A 10 mg ml^{-1} and incubate at 37°C for 30 min. Add an equal volume of chloroform/isoamyl alcohol (24 : 1). Mix thoroughly to homogenise the solution. Centrifuge at 11 600 *g* for 5 min and carefully transfer the upper aqueous layer to a sterile tube. Add 0.1 volume 3M sodium acetate (pH 6.0). Precipitate the DNA with 2 volumes of cold 100% ethanol and place on ice for 15 min. Collect the DNA by centrifugation at 11 800 *g* for 5-20 min. Wash the DNA pellet with 1.0 ml of 70% (v/v) ethanol and centrifuge at 11 800 *g* for 2 min. Repeat the wash procedure.

Allow the pellet to dry briefly. When no moisture is apparent in the tube, add 100 μl of 10 mM Tris (pH 7.6-8.0) to the pellet. Let stand overnight at 4°C. Resuspend the DNA and store at −20°C.

- TE Buffer, pH 8.0

Preparation 1.0 M Tris-HCl, pH 8.0

Trizma base	121.1 g
Distilled water	to 1000 ml

Dissolve 121.1 g of Tris in 700 ml of distilled water. Adjust pH to 8.0 with HCl. Add distilled water to bring the solution to 1000 ml. Sterilise by autoclaving at 121°C for 15 min. Store at room temperature.

Preparation 0.5M EDTA

Na_2EDTA	186.1 g
Sodium hydroxide	*ca* 20 pellets
Distilled water	to 1000 ml

Dissolve 186.1 g of Na_2EDTA in 700 ml of distilled water. Adjust the pH to 8.0 by adding the pellets (this allows the Na_2EDTA to go into solution). Once the Na_2EDTA is in solution, bring the final volume up to 1000 ml. Sterilise by autoclaving at 121°C for 15 min. Store at room temperature.

Preparation TE

1.0 M Tris, pH 8.0	10 ml
0.5 M EDTA, pH 8.0	2 ml
Sterile distilled water	to 1000 ml

To 800 ml sterile distilled water, add 10 ml of 1.0 M Tris and 2 ml 0.5 M EDTA. Adjust the pH to 8.0 if necessary. Bring the final volume up to 1000 ml. Can be autoclaved at 121°C for 15 min. Store at room temperature.

- CTAB/NaCl

Preparation

CTAB	10 g
Sodium chloride	4.1 g
Sterile distilled water	to 100 ml

Dissolve 4.1 g of sodium chloride in 70 ml of distilled water. Using a magnetic stir plate and stir bar, add 10 g CTAB. If necessary, heat the solution to 65°C to dissolve the CTAB and allow to cool to room temperature. Adjust solution volume to 100 ml with sterile distilled water.

- RNase solution

Preparation TEN

1.0 mM Tris-HCl, pH 8.0	1.0 ml
0.5 M EDTA	0.2 ml
Sodium chloride	0.087 g
Distilled water	to 100 ml

In 80 ml water, dissolve 0.087 g NaCl and add 1.0 ml 1.0 M Tris-HCl, pH 8 and 0.2 ml 0.5 M EDTA. Adjust pH to 7.5 with HCl and bring to volume to 100 ml with distilled water.

Preparation RNase solution

RNase A

TEN

Dissolve RNase A to 10mg ml^{-1} in TEN (if the RNase is not DNase-free, heat to 100°C for 15 min to inactivate DNases). Store in small aliquots at −20°C.

AMPLIFICATION OF 16S rDNA

Universal primers that have been used successfully for amplification of the ribosomal small subunit DNA are listed in Table 76.

Each 50 μl reaction contains the following: 0.2 mM of each deoxynucleotide triphosphate, 5 μl 10X PCR reaction buffer, 0.1 mM each primer, 2.5 U Taq polymerase and 1 μl template DNA (50 ng) (or sterile water for negative control). Prepare a master mix on ice and keep thin walled PCR tubes cold until placed on thermocycler.

Place tubes into a thermocycler that has been preheated to 94°C. Once all the tubes are loaded, start the PCR program.

Denature the DNA at 95°C for 3 min followed by 35 cycles of the thermal profile below, and then a final extension at 72°C for 5 min.

Denaturation	95°C for 1 min
Annealing	52-55°C for 1 min
Extension	72°C for 2 min

Table 76. *Universal primer sets for amplification of* Xenorhabdus *and* Photorhabdus *16S rDNA.*

Sense	5′-GGA GAG TTA GAT CTT GGC TC-3′	Brunel *et al.*, 1997
Antisense	5′-AAG GAG GTG ATC CAG CCG CA-3′	
Sense	5′-GAG AG T TTG ATC ATG GCT C-3′	Sicard *et al.*, 2004
Antisense	5′ AAG GAG GTG ATC CAG CCG CA-3′	
Sense	5′-GAA GAG TTT GAT CAT GGC TC-3′	Fischer-Le Saux
Antisense	5′ AAG GAG GTG ATC CAG CCG CA-3′	*et al.*, 1999
Sense	5′-GAG TTT GAT CCT GGC TCAG-3′	Rainey *et al.*, 1996
Antisense	5′-AGA AAG GAG GTG ATC CAG CC-3′	
Sense	5′-GAA GAG TTT GAT CAT GGC TC-3′	Babic *et al.*, 2000
Antisense	5′- AAG GAG GTG ATC CAG CCG CA-3′	

Alternatively, the methods of Brunel *et al.* (1997) and Rainey *et al.* (1996) for PCR amplification have been used with good success and the reader may prefer to use one of these methods.

For detection of the products (CARE! Wear protective gloves) prepare a 0.7-1.0% agarose gel containing 0.5 μg ml^{-1} ethidium bromide. Combine 5 μl of PCR product with 2 μl of gel loading buffer. Load the mixture into a well of the agarose gel and electrophorese at 75V for 30 min. Use a 1 kb DNA ladder to compare to the product size. The product should be about 1.5 kb. Using facial protection, protective clothing and gloves, visualise the bands over UV light. The PCR products can be purified using a commercially available kit and then either cloned for sequencing or sequenced directly.

Media and reagents for the isolation, propagation and testing of symbiotic bacteria

- 1% (w/v) 2,3,5-Triphenyltetrazolium Chloride (TTC) Stock Solution

Preparation

Triphenyltetrazolium chloride	1.0 g
Sterile distilled water	to 100 ml

Dissolve 1 g of TTC in 80 ml of sterile distilled water. Add sterile distilled water to bring the solution to 100 ml. Filter the solution through a sterile 0.22 μm membrane. Store the TTC solution in the dark at 4°C.

- M9 Buffer (Brenner, 1974)

Preparation

Potassium phosphate monobasic	3.0 g
Sodium phosphate dibasic	6.0 g
Sodium chloride	5.0 g
Magnesium sulphate heptahydrate	0.25 g
Sterile distilled water	to 1000 ml

Dissolve salts in 900 ml of sterile distilled water. Add sterile distilled water to bring the solution to 1 l. Autoclave at 121°C for 15 min.

- Ringer's Solution (Woodring & Kaya, 1988)

Preparation

Sodium chloride	9.0 g
Potassium chloride	0.4 g
Calcium chloride	0.4 g
Sodium carbonate	0.2 g
Sterile distilled water	to 1000 ml

Dissolve salts in 900 ml of sterile distilled water. Add sterile distilled water to bring the solution to 1 l. Autoclave at 121°C for 15 min.

Excluding YS broth and NBTA, the following media are commercially available already prepared or as dehydrated culture media. Separate ingredients need not be purchased.

- Luria Bertani (LB) Broth (Miller, 1972; Sambrook *et al.*, 1989)

Preparation

Bacto Tryptone	10.0 g
Bacto Yeast Extract	5.0 g
Sodium chloride	5.0 g
Purified water	1000 ml

Dissolve dehydrated ingredients in 900 ml of purified water and adjust pH to 7.0. Add water to 1 l. Dispense into tubes or flasks. Sterilise by autoclaving at 121°C for 15-20 min.

- MacConkey Agar (MacConkey, 1905; MacFaddin, 1985; Zimbro & Power, 2003)

Preparation

Peptone (Difco) or Gelysate™ peptone (BBL)	17.0 g
Proteose peptone (Difco) or Polypeptone peptone (BBL)	3.0 g

Lactose	15.0 g
Bile salts	1.5 g
Sodium chloride	5.0 g
Neutral red	0.03 g
Crystal violet	0.001 g
Agar	1.0 g
Purified water	1000 ml

Dissolve in 1 l of purified water. Boil briefly to dissolve the agar. Autoclave at 121°C for 15-20 min. Once agar has cooled to 50-55°C, dispense into Petri dishes.

- Nutrient Broth (NB) and Nutrient Agar (NA) (Zimbro & Power, 2003)

Preparation

Beef extract	3.0 g
Peptone	5.0 g
Purified water	1000 ml

Dissolve dehydrated ingredients in 1 l of purified water and adjust pH to 7.0-7.2. Dispense into tubes or flasks, or for preparation of NA, add 15 g of agar to the broth solution and melt by boiling briefly with agitation. Autoclave at 121°C for 15-20 min. For NA, once agar has cooled to 50-55°C, dispense into Petri dishes.

- *NBTA* (Akhurst, 1980)

Preparation

Nutrient agar	1000 ml
Bromothymol blue	0.025 g
1% TTC	4 ml

Mix the bromothymol blue into the melted NA. Sterilise by autoclaving at 121°C for 15-20 min. Allow the medium to cool to 50°C in a water bath and aseptically transfer 4 ml of 1% TTC to the agar base and gently mix. Dispense into Petri dishes.

- *Tergitol 7 (T7) Agar* (Chapman, 1951; MacFaddin, 1985)

Preparation

Peptone	
Casein peptone	2.5 g
Meat peptone (or 5.0 g Meat peptone only)	2.5 g
Yeast Extract	3.0 g

Lactose	10.0 g
Tergitol 7	
(Sodium heptadecyl sulphate)	0.1 g
Bromothymol blue	0.025 g
Agar	15.0 g
Purified water	1000 ml
1% TTC	4 ml

Dissolve all ingredients except agar in 800 ml of purified water. Adjust the pH to 7.0-7.2. Add the agar, then add purified water to 1 l. While stirring, heat the medium to boiling to melt the agar. Autoclave at 121°C for 15-20 min. Once the agar base has cooled to 50°C, aseptically transfer 4 ml of 1% TTC to the medium and mix gently. Dispense into Petri dishes.

If using the commercial dehydrated culture medium, suspend the dry mixture in 250 ml of purified water and dissolve completely by heating to boiling while stirring. Once the mixture is dissolved, add 700 ml of purified water. Quickly adjust the pH to 7.0-7.2 before the agar solidifies. Add purified water to 1 l. Sterilise and add TTC as described above. Dispense into Petri dishes.

- Trypticase/Tryptic Soy Broth (TSB) and Tryptic Soy Broth with Agar (TSBA) (Fredette *et al.*, 1961; MacFaddin, 1985; Zimbro & Power, 2003)

Preparation

Casein peptone	17.0 g
Soy peptone	3.0 g
Dextrose	2.5 g
Sodium chloride	5.0 g
Potassium phosphate dibasic	2.5 g
Purified water	1000 ml

Dissolve in 1 l of purified water and dispense into tubes or flasks. For preparation of TSBA, add 15 g of agar to the broth solution and melt by boiling briefly with agitation. Autoclave at 121°C for 15-20 min. Once the TSBA has cooled to 50-55°C, dispense into Petri dishes.

*Tryptic soy agar (TSA) has a different formulation from that of TSB and therefore is not prepared by the addition of agar to the broth. TSA contains 15.0 g peptone, 5.0 g soy peptone and no dextrose or dipotassium phosphate. For the growth of *Photorhabdus* and *Xenorhabdus*, TSBA is a suitable culture medium and it is not necessary to use TSA.

- Yeast Salts (YS) Broth (Dye, 1968)

Preparation

Ammonium phosphate monobasic	0.5 g
Potassium phosphate dibasic	0.5 g
Magnesium sulphate heptahydrate	0.2 g
Sodium chloride	5.0 g
Yeast extract	5.0 g
Purified water	1000 ml

Dissolve in 1 l of purified water and dispense into tubes or flasks. Autoclave the broth at 121°C for 15-20 min. To prepare YS agar, prior to autoclaving add 15 g of agar to the broth solution and melt by boiling briefly with agitation. Autoclave at 121°C for 15-20 min. After the YS agar has cooled to 50-55°C, dispense into Petri dishes.

Acknowledgements

I would like to thank Byron Adams, Alejandra Garcia-Maruniak, Heidi Goodrich-Blair, Albrecht Koppenhöfer and Khuong Nguyen for thoughtful discussion and all of their suggestions. I greatly appreciate their help.

References

ADAMS, B.J., FODOR, A., KOPPENHÖFER, H.S., STACKEBRANDT, E., STOCK, S.P. & KLEIN, M.G. (2006). Biodiversity and systematics of nematode-bacterium entomopathogens. *Biological Control* 37, 32-49.

AGUILLERA, M.M. & SMART JR, G.C. (1993). Development, reproduction, and pathogenicity of *Steinernema scapterisci* in monoxenic culture with different species of bacteria. *Journal of Invertebrate Pathology* 62, 289-294.

AGUILLERA, M.M., HODGE, N.C., STALL, R.E. & SMART JR, G.C. (1993). Bacterial symbionts of *Steinernema scapterisci. Journal of Invertebrate Pathology* 62, 68-72.

AKHURST, R.J. (1980). Morphological and functional dimorphism in *Xenorhabdus* spp., bacteria symbiotically associated with the insect pathogenic nematodes *Neoaplectana* and *Heterorhabditis. Journal of General Microbiology* 121, 303-309.

AKHURST, R.J. (1982). Antibiotic activity of *Xenorhabdus* spp., bacteria symbiotically associated with insect pathogenic nematodes of the families

Heterorhabditidae and Steinernematidae. *Journal of General Microbiology* 128, 3061-3065.

AKHURST, R.J. (1983a). Taxonomic study of *Xenorhabdus*, a genus of bacteria symbiotically associated with insect pathogenic nematodes. *International Journal of Systematic Bacteriology* 33, 38-45.

AKHURST, R.J. (1983b). *Neoaplectana* species: Specificity of association with bacteria of the genus *Xenorhabdus. Experimental Parasitology* 55, 258-263.

AKHURST, R.J. (1986). *Xenorhabdus nematophilus* subsp. *beddingii* (*Enterobacteriaceae*): A new subspecies of bacteria mutualistically associated with entomopathogenic nematodes. *International Journal of Systematic Bacteriology* 36, 454-457.

AKHURST, R.J. & BOEMARE, N.E. (1988). A numerical taxonomic study of the genus *Xenorhabdus* (*Enterobacteriaceae*) and proposed elevation of the subspecies of *X. nematophilus* to species. *Journal of General Microbiology* 134, 1835-1845.

AKHURST, R.J. & BOEMARE, N.E. (1993). In: Validation list no. 47. Validation of publication of new names and new combinations previously effectively published outside the IJSB. *International Journal of Systematic Bacteriology* 43, 864-865.

AKHURST, R.J. & BOEMARE, N.E. (2005). Genus 40. *Xenorhabdus*. In: Garrity, G.M., Brenner, D.J., Krieg, N.R. & Staley, J.T. (Eds). *Bergey's manual of systematic bacteriology, volume 2: The Proteobacteria, part B: The Gammaproteobacteria*, 2nd edition. New York, USA, Springer, pp. 831-838.

AKHURST, R.J., MOURANT, R.G., BAUD, L. & BOEMARE, N.E. (1996). Phenotypic and DNA relatedness between nematode symbionts and clinical strains of the genus *Photorhabdus* (*Enterobacteriaceae*). *International Journal of Systematic Bacteriology* 46, 1034-1041.

AKHURST, R.J., BOEMARE, N.E., JANSSEN, P.H., PEEL, M.M., ALFREDSON, D.A. & BEARD, C.E. (2004). Taxonomy of Australian clinical isolates of the genus *Photorhabdus* and proposal of *Photorhabdus asymbiotica* subsp. *asymbiotica* subsp. nov. and *P. asymbiotica* subsp. *australis* subsp. nov. *International Journal of Systematic and Evolutionary Microbiology* 54, 1301-1310.

AUSUBEL, F.M., BRENT, R., KINGSTON, R.E., MOORE, D.D., SEIDMAN, J.G., SMITH, J.A. & STRUHL, K. (EDS) (1994). *Current protocols in molecular biology, volume 1.* New York, NY, USA, John Wiley & Sons, pp. 2.4.1.-2.4.5.

BABIC, I., FISCHER-LE SAUX, M., GIRAUD, E. & BOEMARE, N. (2000). Occurrence of natural dixenic associations between the symbiont *Photorhabdus luminescens* and bacteria related to *Ochrobactrum* spp. in tropical entomopathogenic *Heterorhabditis* spp. (Nematoda, Rhabditida). *Microbiology* 146, 709-718.

BEECHER, D.J. & WONG, A.C.L. (1994). Identification of hemolysin bl-producing *Bacillus cereus* isolates by a discontinuous hemolytic pattern in blood agar. *Applied and Environmental Microbiology* 60, 1646-1651.

BENNETT, H.P.J. & CLARKE, D.J. (2005). The pbgPE operon in *Photorhabdus luminescens* is required for pathogenicity and symbiosis. *Journal of Bacteriology* 187, 77-84.

BINNINGTON, K.C. & BROOKS, L. (1993). Fimbrial attachment of *Xenorhabdus nematophilus* to the intestine of *Steinernema carpocapsae*. In: Bedding, R.A., Akhurst, R.J. & Kaya, H.K. (Eds). *Nematodes and the biological control of insect pests*. East Melbourne, Australia, CSIRO, pp. 147-155.

BIRD, A.F. & AKHURST, R.J. (1983). The nature of the intestinal vesicle in nematodes of the family Steinernematidae. *International Journal for Parasitology* 13, 599-606.

BOEMARE, N. (2002a). Biology, taxonomy and systematics of *Photorhabdus* and *Xenorhabdus*. In: Gaugler, R. (Ed.). *Entomopathogenic nematology*. Wallingford, UK, CABI Publishing, pp. 35-56.

BOEMARE, N. (2002b). Interactions between the partners of the entomopathogenic bacterium nematode complexes, *Steinernema-Xenorhabdus* and *Heterorhabditis-Photorhabdus*. *Nematology* 4, 601-603.

BOEMARE, N. & AKHURST, R.J. (2000). Genera *Photorhabdus* and *Xenorhabdus*. In: Dworkin, M., Falkow, S., Rosenberg, E., Schleifer, K.-H. & Stackebrandt, E. (Eds). *The Prokaryotes: An evolving electronic resource for the microbiological community*. Release 3.4. New York, USA, Springer-Verlag, http://link.springer-ny.com/link/service/books/10125/

BOEMARE, N., THALER, J.O. & LANOIS, A. (1997a). Simple bacteriological tests for phenotypic characterization of *Xenorhabdus* and *Photorhabdus* phase variants. *Symbiosis* 22, 167-175.

BOEMARE, N., GIVAUDAN, A., BREHELIN, M. & LAUMOND, C. (1997b). Symbiosis and pathogenicity of nematode-bacterium complexes. *Symbiosis* 22, 21-45.

BOEMARE, N.E. & AKHURST, R.J. (1988). Biochemical and physiological characterization of colony form variants in *Xenorhabdus* spp. (*Enterobacteriaceae*). *Journal of General Microbiology* 134, 751-761.

BOEMARE, N.E. & AKHURST, R.J. (2005). Genus 26. *Photorhabdus*. In: Garrity, G.M., Brenner, D.J., Krieg, N.R. & Staley, J.T. (Eds). *Bergey's manual of systematic bacteriology, volume 2: The Proteobacteria, part B: The Gammaproteobacteria*. New York, NY, USA, Springer, pp. 732-740.

BOEMARE, N.E., BOYER-GIGLIO, M.H., THALER, J.O., AKHURST, R.J. & BREHELIN, M. (1992). Lysogeny and bacteriocinogeny in *Xenorhabdus nematophilus* and other *Xenorhabdus* spp. *Applied and Environmental Microbiology* 58, 3032-3037.

BOEMARE, N.E., AKHURST, R.J. & MOURANT, R.G. (1993). DNA relatedness between *Xenorhabdus* spp. (*Enterobacteriaceae*), symbiotic bacteria of entomopathogenic nematodes, and a proposal to transfer *Xenorhabdus luminescens* to a new genus, *Photorhabdus* gen. nov. *International Journal of Systematic Bacteriology* 43, 249-255.

BONIFASSI, E., FISCHER-LE SAUX, M., BOEMARE, N., LANOIS, A., LAUMOND, C. & SMART, G. (1999). Gnotobiological study of infective juveniles and symbionts of *Steinernema scapterisci*: A model to clarify the concept of the natural occurrence of monoxenic associations in entomopathogenic nematodes. *Journal of Invertebrate Pathology* 74, 164-172.

BOWEN, D., ROCHELEAU, T.A., BLACKBURN, M., ANDREEV, O., GOLUBEVA, E., BHARTIA, R. & FFRENCH-CONSTANT, R.H. (1998). Insecticidal toxins from the bacterium *Photorhabdus luminescens*. *Science* 280, 2129-2132.

BRENNER, D.J. (1999). Additional genera of *Enterobacteriaceae*. In: Dworkin, M., Falkow, S., Rosenberg, E., Schleifer, K.-H. & Stackebrandt, E. (Eds). *The Prokaryotes: An evolving electronic resource for the microbiological community.* Release 3.0. New York, NY, USA, Springer-Verlag, http://link.springer-ny.com/link/service/books/10125/

BRENNER, D.J. & FARMER III, J.L. (2005). Family I. Enterobacteriaceae. In: Garrity, G.M., Brenner, D.J., Krieg, N.R. & Staley, J.T. (Eds). *Bergey's manual of systematic bacteriology, volume 2: The Proteobacteria, part B: The Gammaproteobacteria*, 2nd edition. New York, USA, Springer, pp. 587-607.

BRENNER, S. (1974). Genetics of *Caenorhabditis elegans*. *Genetics* 77, 71-94.

BRILLARD, J., RIBEIRO, C., BOEMARE, N., BREHELIN, M. & GIVAUDAN, A. (2001). Two distinct hemolytic activities in *Xenorhabdus nematophila* are active against immunocompetent insect cells. *Applied and Environmental Microbiology* 67, 2515-2525.

BRILLARD, J., DUCHAUD, E., BOEMARE, N., KUNST, F. & GIVAUDAN, A. (2002). The PhlA hemolysin from the entomopathogenic bacterium *Photorhabdus luminescens* belongs to the two-partner secretion family of hemolysins. *Journal of Bacteriology* 184, 3871-3878.

BRUNEL, B., GIVAUDAN, A., LANOIS, A., AKHURST, R.J. & BOEMARE, N. (1997). Fast and accurate identification of *Xenorhabdus* and *Photorhabdus* species by restriction analysis of PCR-amplified 16S rRNA genes. *Applied and Environmental Microbiology* 63, 574-580.

CHAPMAN, G.H. (1951). A culture medium for detecting and confirming *Escherichia coli* in 10 hours. *American Journal of Public Health* 41, 1381.

CICHE, T.A. & ENSIGN, J.C. (2003). For the insect pathogen *Photorhabdus luminescens*, which end of a nematode is out? *Applied and Environmental Microbiology* 69, 1890-1897.

CICHE, T.A., BINTRIM, S.B., HORSWILL, A.R. & ENSIGN, J.C. (2001). A phosphopantetheinyl transferase homolog is essential for *Photorhabdus luminescens* to support growth and reproduction of the entomopathogenic nematode *Heterorhabditis bacteriophora. Journal of Bacteriology* 183, 3117-3126.

CICHE, T.A., DARBY, C., EHLERS, R.-U., FORST, S. & GOODRICH-BLAIR, H. (2006). Dangerous liaisons: The symbiosis of entomopathogenic nematodes and bacteria. *Biological Control* 38, 22-46.

COLE, J.R., CHAI, B., FARRIS, R.J., WANG, Q., KULAM, S.A., MCGARRELL, D.M., GARRITY, G.M. & TIEDJE, J.M. (2005). The ribosomal database project (RDP-II): Sequences and tools for high-throughput rRNA analysis. *Nucleic Acids Research* 33, D294-D296.

COUCHE, G.A., LEHRBACK, P.R., FORAGE, R.G., COONEY, G.C., SMITH, D.R. & GREGSON, R.P. (1987). Occurrence of intracellular inclusions and plasmids in *Xenorhabdus* spp. *Journal of General Microbiology* 133, 967-973.

COWLES, C.E. & GOODRICH-BLAIR, H. (2004). Characterization of a lipoprotein, NilC, required by *Xenorhabdus nematophila* for mutualism with its nematode host. *Molecular Microbiology* 54, 464-477.

DABORN, P.J., WATERFIELD, N., SILVA, C.P., AU, C.P.Y., SHARMA, S. & FFRENCH-CONSTANT, R.H. (2002). A single *Photorhabdus* gene, makes caterpillars floppy (mcf), allows *Escherichia coli* to persist within and kill insects. *Proceedings of the National Academy of Sciences* 99, 10742-10747.

DICKSTEIN, E.R., JONES, J.B. & STEAD, D.E. (2001). Automated techniques. In: Schaad, N.W., Jones, J.B. & Chun, W. (Eds). *Laboratory guide for the identification of plant pathogenic bacteria*, 3rd edition. St Paul, MN, USA, APS Press, pp. 343-352.

DOWDS, B.C.A. & PETERS, A. (2002). Virulence mechanisms. In: Gaugler, R. (Ed.). *Entomopathogenic nematology*. Wallingford, UK, CABI Publishing, pp. 79-98.

DUCHAUD, E., RUSNIOK, C., FRANGEUL, L., BUCHRIESER, C., GIVAUDAN, A., TAOURIT, S., BOCS, S., BOURSAUX-EUDE, C., CHANDLER, M., CHARLES, J., DASSA, E., DEROSE, R., DERZELLE, S., FREYSSINET, G., GAUDRIAULT, S., MEDIGUE, C., LANOIS, A., POWELL, K., SIGUIER, P., VINCENT, R., WINGATE, V., ZOUINE, M., GLASER, P., BOEMARE, N., DANCHIN, A. & KUNST, F. (2003). The genome sequence of the entomopathogenic bacterium *Photorhabdus luminescens. Nature Biotechnology* 21, 1307-1313.

DUNPHY, G.B. & THURSTON, G.S. (1990). Insect immunity. In: Gaugler, R. & Kaya, H.K. (Eds). *Entomopathogenic nematodes in biological control*. Boca Raton, FL, USA, CRC Press, pp. 301-323.

DUNPHY, G.B. & WEBSTER, J.M. (1988). Virulence mechanisms of *Heterorhabditis heliothidis* and its bacterial associate, *Xenorhabdus luminescens*, in non-immune larvae of the greater wax moth, *Galleria mellonella*. *International Journal for Parasitology* 18, 729-737.

DUNPHY, G.B. & WEBSTER, J.M. (1991). Antihemocytic surface components of *Xenorhabdus nematophilus* var. Dutki and their modification by serum on nonimmune larvae of *Galleria mellonella*. *Journal of Invertebrate Pathology* 58, 40-51.

DYE, D.W. (1968). A taxonomic study of the genus *Erwinia*. I. The '*amylovera*' group. *New Zealand Journal of Science* 11, 590-607.

EHLERS, R.-U., STOESSEL, S. & WYSS, U. (1990). The influence of phase variants of *Xenorhabdus* spp. and *Escherichia coli* (Enterobacteriaceae) on the propagation of entomopathogenic nematodes of the genera *Steinernema* and *Heterorhabditis*. *Revue de Nématologie* 13, 417-424.

ENDO, B.Y. & NICKLE, W.R. (1991). Ultrastructure of the intestinal epithelium, lumen, and associated bacteria in *Heterorhabditis bacteriophora*. *Journal of the Helminthological Society of Washington* 58, 202-212.

ENRIGHT, M.R. & GRIFFIN, C.T. (2004). Specificity of association between *Paenibacillus* spp. and the entomopathogenic nematodes, *Heterorhabditis* spp. *Microbial Ecology* 48, 414-423.

ENRIGHT, M.R. & GRIFFIN, C.T. (2005). Effects of *Paenibacillus nematophilus* on the entomopathogenic nematode *Heterorhabditis megidis*. *Journal of Invertebrate Pathology* 88, 40-48.

ENRIGHT, M.R., MCINERNEY, J.O. & GRIFFIN, C.T. (2003). Characterization of endospore-forming bacteria associated with entomopathogenic nematodes, *Heterorhabditis* spp., and description of *Paenibacillus nematophilus* sp. nov. *International Journal of Systematic and Evolutionary Microbiology* 53, 435-441.

EUZEBY, J.P. & BOEMARE, N.E. (2000). The modern Latin word -rhabdus belongs to the feminine gender, inducing necessary corrections according to rules 65(2), 12c(1), and 13b of the bacteriological code (1990 revision). *International Journal of Systematic and Evolutionary Microbiology* 50, 1691-1692.

EWING, W.H., FARMER III, J.J. & BRENNER, D.J. (1980). Proposal of *Enterobacteriaceae* fam. nov., nom. rev. to replace *Enterobacteriaceae* Rahn 1937, nom. fam. cons. (Opin. 15, Jud. Comm. 1958), which lost standing in nomenclature on 1 January 1980. *International Journal of Systematic and Evolutionary Microbiology* 30, 674-675.

FARMER III, J.J., JORGENSEN, J.H., GRIMONT, P.A.D., AKHURST, R.J., POINAR JR, G.O., AGERON, E., PIERCE, G.V., SMITH, J.A., CARTER, G.P., WILSON, K.L. & HICKMAN-BRENNER, F.W. (1989). *Xenorhabdus luminescens* (DNA hybridization group 5) from human clinical specimens. *Journal of Clinical Microbiology* 27, 1594-1600.

FISCHER-LE SAUX, M., MAULEON, H., CONSTANT, P., BRUNEL, B. & BOEMARE, N. (1998). PCR-ribotyping of *Xenorhabdus* and *Photorhabdus* isolates from the Caribbean region in relation to the taxonomy and geographic distribution of their nematode hosts. *Applied and Environmental Microbiology* 64, 4246-4254.

FISCHER-LE SAUX, M., ARTEAGA-HERNANDEZ, E., MRÁČEK, Z. & BOEMARE, N.E. (1999a). The bacterial symbiont *Xenorhabdus poinarii* (*Enterobacteriaceae*) is harbored by two phylogenetic related host nematodes: the entomopathogenic species *Steinernema cubanum* and *Steinernema glaseri* (Nematoda: Steinernematidae). *FEMS Microbiology Ecology* 29, 149-157.

FISCHER-LE SAUX, M., VIALLARD, V., BRUNEL, B., NORMAND, P. & BOEMARE, N.E. (1999b). Polyphasic classification of the genus *Photorhabdus* and proposal of new taxa: *P. luminescens* subsp. *luminescens* subsp. nov., *P. luminescens* subsp. *akhurstii* subsp. nov., *P. luminescens* subsp. *laumondii* subsp. nov., *P. temperata* sp. nov., *P. temperata* subsp. *temperata* subsp. nov. and *P. asymbiotica* sp. nov. *International Journal of Systematic Bacteriology* 49, 1645-1656.

FORST, S. & CLARKE, D. (2002). Bacteria-nematode symbiosis. In: Gaugler, R. (Ed.). *Entomopathogenic nematology*. Wallingford, UK, CABI Publishing, pp. 57-77.

FORST, S. & GOODNER, B. (2006). Comparative bacterial genomics and its use in undergraduate education. *Biological Control* 38, 47-53.

FORST, S. & NEALSON, K. (1996). Molecular biology of the symbiotic-pathogenic bacteria *Xenorhabdus* spp. and *Photorhabdus* spp. *Microbiological Reviews* 60, 21-43.

FORST, S., DOWDS, B., BOEMARE, N. & STACKEBRANDT, E. (1997). *Xenorhabdus* and *Photorhabdus* spp.: Bugs that kill bugs. *Annual Review of Microbiology* 51, 47-72.

FRANCINO, M.P., SANTOS, S.R. & OCHMAN, H. (2003). Phylogenetic relationships of bacteria with special reference to endosymbionts and enteric species. In: Dworkin, M., Falkow, S., Rosenberg, E., Schleifer, K.-H. & Stackebrandt, E. (Eds). *The prokaryotes: An evolving electronic resource for the microbiological community*. Release 3.13. New York, NY, USA, Springer-Verlag, http://link.springer-ny.com/link/service/books/10125/

FREDETTE, V., AUGER, A. & FORGET, A. (1961). Anaerobic flora of chronic nasal sinusitis in adults. *Canadian Medical Association Journal* 84, 164-165.

GERRARD, J.G., JOYCE, S.A., CLARKE, D.J., FFRENCH-CONSTANT, R.H., NIMMO, G.R., LOOKE, D.F.M., FEIL, E.J., PEARCE, L. & WATERFIELD, N.R. (2006). Nematode symbiont for *Photorhabdus asymbiotica*. *Emerging Infectious Diseases* 12, 1562-1564.

GERRITSEN, L.J.M. & SMITS, P.H. (1993). Variation in pathogenicity of recombinations of *Heterorhabditis* and *Xenorhabdus luminescens* strains. *Fundamental and Applied Nematology* 16, 367-373.

GERRITSEN, L.J.M. & SMITS, P.H. (1997). The influence of *Photorhabdus luminescens* strains and form variants on the reproduction and bacterial retention of *Heterorhabditis megidis*. *Fundamental and Applied Nematology* 20, 317-322.

GIVAUDAN, A., BAGHDIGUIAN, S., LANOIS, A. & BOEMARE, N. (1995). Swarming and swimming changes concomitant with phase variation in *Xenorhabdus nematophilus*. *Applied and Environmental Microbiology* 61, 1408-1413.

GOETSCH, M., OWEN, H., GOLDMAN, B. & FORST, S. (2006). Analysis of the PixA inclusion body protein of *Xenorhabdus nematophila*. *Journal of Bacteriology* 188, 2706-2710.

GÖTZ, P., BOMAN, A. & BOMAN, H.G. (1981). Interactions between insect immunity and an insect-pathogenic nematode with symbiotic bacteria. *Proceedings of the Royal Society of London, Series B* 212, 333-350.

GOUGE, D.H. & SNYDER, J.L. (2006). Temporal association of entomopathogenic nematodes (Rhabditida: Steinernematidae and Heterorhabditidae) and bacteria. *Journal of Invertebrate Pathology* 91, 147-157.

GREWAL, P.S., MATSUURA, M. & CONVERSE, V. (1997). Mechanisms of specificity of association between the nematode *Steinernema scapterisci* and its symbiotic bacterium. *Parasitology* 114, 483-488.

GRIMONT, P.A.D., STEIGERWALT, A.G., BOEMARE, N., HICKMAN-BRENNER, F.W., DEVAL, C., GRIMONT, F. & BRENNER, D.J. (1984). Deoxyribonucleic acid relatedness and phenotypic study of the genus *Xenorhabdus*. *International Journal of Systematic Bacteriology* 34, 378-388.

HALL, T.A. (1999). Bioedit: A user friendly biological sequence alignment editor and analysis program for Windows 95/98/NT. *Nucleic Acids Symposium Serial* 41, 95-98.

HAN, R. & EHLERS, R.-U. (2001). Effect of *Photorhabdus luminescens* phase variants on the *in vivo* and *in vitro* development and reproduction of the entomopathogenic nematodes *Heterorhabditis bacteriophora* and *Steinernema carpocapsae*. *FEMS Microbiology Ecology* 35, 239-247.

HAZIR, S., STACKEBRANDT, E., LANG, E., SCHUMANN, P., EHLERS, R.-U. & KESKIN, N. (2004a). Two new subspecies of *Photorhabdus luminescens*, isolated from *Heterorhabditis bacteriophora* (Nematoda: Heterorhabditidae): *Photorhabdus luminescens* subsp. *kayaii* subsp. nov. and *Photorhabdus luminescens* subsp. *thracensis* subsp. nov. *Systematic and Applied Microbiology* 27, 36-42.

HAZIR, S., STACKEBRANDT, E., LANG, E., SCHUMANN, P., EHLERS, R.-U. & KESKIN, N. (2004b). In: Validation list no. 98. Validation of publication of new names and new combinations previously effectively published outside the IJSEM. *International Journal of Systematic and Evolutionary Microbiology* 54, 1005-1006.

HENDRIE, M.S., HOLDING, A.J. & SHEWAN, J.M. (1974). Emended descriptions of the genus *Alcaligenes* and of *Alcaligenes faecalis* and proposal that the generic name *Achromobacter* be rejected; status of the named species of *Alcaligenes* and *Achromobacter*. *International Journal of Systematic Bacteriology* 24, 534-550.

HEUNGENS, K., COWLES, C.E. & GOODRICH-BLAIR, H. (2002). Identification of *Xenorhabdus nematophila* genes required for mutualistic colonization of *Steinernema carpocapsae* nematodes. *Molecular Microbiology* 45, 1337-1353.

HOLT, J.G., KRIEG, N.R., SNEATH, P.H., STALEY, J.T. & WILLIAMS, S.T. (1994). *Bergey's manual of determinative bacteriology*, 9th edition. Baltimore, MD, USA, Williams & Wilkins, pp. 188-189.

JACKSON, T.J., WANG, H., NUGENT, M.J., GRIFFIN, C.T., BURNELL, A.M. & DOWDS, B.C.A. (1995). Isolation of insect pathogenic bacteria, *Providencia rettgeri*, from *Heterorhabditis* spp. *Journal of Applied Bacteriology* 78, 237-244.

JANSE, J.D. & SMITS, P.H. (1990). Whole cell fatty acid patterns of *Xenorhabdus* species. *Letters in Applied Microbiology* 10, 131-135.

JI, D. & KIM, Y. (2004). An entomopathogenic bacterium, *Xenorhabdus nematophila*, inhibits the expression of an antibacterial peptide, cecropin, of the beet armyworm, *Spodoptera exigua*. *Journal of Insect Physiology* 50, 489-496.

JOYCE, S.A. & CLARKE, D.J. (2003). A *hexA* homologue from *Photorhabdus* regulates pathogenicity, symbiosis and phenotypic variation. *Molecular Microbiology* 47, 1445-1457.

KAYA, H.K. & STOCK, S.P. (1997). Techniques in insect nematology. In: Lacey, L.A. (Ed.). *Manual of techniques in insect pathology*. London, UK, Academic Press, pp. 281-324.

KOZODOI, E.M. (1984). [New entomopathogenic nematode *Neoaplectana anomali* n. sp. (Rhabditida: Steinernematidae) and observations of its biology.] *Zoologichesky Zhurnal* 63, 1605-1609.

LAN, R. & REEVES, P.R. (2001). When does a clone deserve a name? A perspective on bacterial species based on population genetics. *Trends in Microbiology* 9, 419-424.

LENGYEL, K., LANG, E., FODOR, A., SZALLAS, E., SCHUMANN, P. & STACKEBRANDT, E. (2005a). Description of four novel species of *Xenorhabdus*, family *Enterobacteriaceae*: *Xenorhabdus budapestensis* sp. nov., *Xenorhabdus ehlersii* sp. nov., *Xenorhabdus innexi* sp. nov., and *Xenorhabdus szentirmaii* sp. nov. *Systematic and Applied Microbiology* 28, 115-122.

LENGYEL, K., LANG, E., FODOR, A., SZALLAS, E., SCHUMANN, P. & STACKBRANDT, E. (2005b). In: Validation list no. 104. Validation of publication of new names and new combinations previously effectively published outside the IJSEM. *International Journal of Systematic and Evolutionary Microbiology* 55, 1395-1397.

LERAT, E., DAUBIN, V. & MORAN, N.A. (2003). From gene trees to organismal phylogeny in prokaryotes: The case of the Gamma-Proteobacteria. *PLoS Biology* 1, 1-9.

LIU, J., BERRY, R., POINAR, G. & MOLDENKE, A. (1997). Phylogeny of *Photorhabdus* and *Xenorhabdus* species and strains as determined by comparison of partial 16S rRNA gene sequences. *International Journal of Systematic Bacteriology* 47, 948-951.

LIU, J., BERRY, R.E. & BLOUIN, M.S. (2001). Identification of symbiotic bacteria (*Photorhabdus* and *Xenorhabdus*) from the entomopathogenic nematodes *Heterorhabditis marelatus* and *Steinernema oregonense* based on 16S rDNA sequence. *Journal of Invertebrate Pathology* 77, 87-91.

LYSENKO, O. & WEISER, J. (1974). Bacteria associated with the nematode, *Neoaplectana carpocapsae* and the pathogenicity of this complex for *Galleria mellonella* larvae. *Journal of Invertebrate Pathology* 24, 332-336.

MACCONKEY, A.T. (1905). Lactose-fermenting bacteria in faeces. *The Journal of Hygiene* 5, 333-379.

MACFADDIN, J.F. (1985). *Media for isolation-cultivation-identification-maintenance of medical bacteria.* Baltimore, MD, USA, Williams & Wilkins, 928 pp.

MAIDEN, M., BYGRAVES, J., FEIL, E., MORELLI, G., RUSSELL, J., URWIN, R., ZHANG, Q., ZHOU, J., ZURTH, K., CAUGANT, D., FEAVERS, I., ACHTMAN, M. & SPRATT, B. (1998). Multilocus sequence typing: A portable approach to the identification of clones within populations of pathogenic microorganisms. *Proceedings of the National Academy of Sciences USA* 95, 3140-3145.

MAROKHAZI, J., WATERFIELD, N., LEGOFF, G., FEIL, E., STABLER, R., HINDS, J., FODOR, A. & FFRENCH-CONSTANT, R.H. (2003). Using a DNA microarray to investigate the distribution of insect virulence factors in strains of *Photorhabdus bacteria*. *Journal of Bacteriology* 185, 4648-4656.

MARTENS, E.C. & GOODRICH-BLAIR, H. (2005). The *Steinernema carpocapsae* intestinal vesicle contains a subcellular structure with which

Xenorhabdus nematophila associates during colonization initiation. *Cellular Microbiology* 7, 1723-1735.

MARTENS, E.C., HEUNGENS, K. & GOODRICH-BLAIR, H. (2003). Early colonization events in the mutualistic association between *Steinernema carpocapsae* nematodes and *Xenorhabdus nematophila* bacteria. *Journal of Bacteriology* 185, 3147-3154.

MARTENS, E.C., VIVAS, E.I., HEUNGENS, K., COWLES, C.E. & GOODRICH-BLAIR, H. (2004). Investigating mutualism between entomopathogenic bacteria and nematodes. In: Cook, R. & Hunt, D.J. (Eds). *Proceedings of the fourth international congress of nematology, Tenerife, Spain, 8-13 June 2002, Nematology monographs and perspectives, volume 2.* Leiden, The Netherlands, Brill, pp. 447-462.

MILLER, J.H. (1972). *Experiments in molecular genetics.* Cold Spring Harbor, NY, USA, Cold Spring Harbor Laboratory Press, 466 pp.

MOUREAUX, N., KARJALAINEN, T., GIVAUDAN, A., BOURLIOUX, P. & BOEMARE, N. (1995). Biochemical characterization and agglutinating properties of *Xenorhabdus nematophilus* f1 fimbriae. *Applied and Environmental Microbiology* 61, 2707-2712.

NGUYEN, K.B. & DUNCAN, L.W. (2002). *Steinernema diaprepesi* n. sp. (Rhabditida: Steinernematidae), a parasite of the citrus root weevil *Diaprepes abbreviatus* (L.) (Coleoptera: Curculionidae). *Journal of Nematology* 34, 159-170.

NISHIMURA, Y., HAGIWARA, A., SUZUKI, T. & YAMANAKA, S. (1994). *Xenorhabdus japonicus* sp. nov. associated with the nematode *Steinernema kushidai. World Journal of Microbiology and Biotechnology* 10, 207-210.

NISHIMURA, Y., HAGIWARA, A., SUZUKI, T. & YAMANAKA, S. (1995). In: Validation list no. 54. Validation of publication of new names and new combinations previously effectively published outside the IJSB. *International Journal of Systematic Bacteriology* 45, 619-620.

PARK, Y. & KIM, Y. (2000). Eicosanoids rescue *Spodoptera exigua* infected with *Xenorhabdus nematophilus*, the symbiotic bacteria to the entomopathogenic nematode *Steinernema carpocapsae*. *Journal of Insect Physiology* 46, 1469-1476.

PARK, Y., KIM, Y., PUTNAM, S.M. & STANLEY, D.W. (2003). The bacterium *Xenorhabdus nematophilus* depresses nodulation reactions to infection by inhibiting eicosanoid biosynthesis in tobacco hornworms, *Manduca sexta*. *Archives of Insect Biochemistry and Physiology* 52, 71-80.

PARK, Y., HERBERT, E.E., COWLES, C.E., COWLES, K.N., MENARD, M.L., ORCHARD, S.S. & GOODRICH-BLAIR, H. (2007). Clonal variation in *Xenorhabdus nematophila* virulence and suppression of *Manduca sexta* immunity. *Cellular Microbiology* 9, 645-656.

PEEL, M.M., ALFREDSON, D.A., GERRARD, J.G., DAVIS, J.M., ROBSON, J.M., MCDOUGALL, R.J., SCULLIE, B.L. & AKHURST, R.J. (1999). Isolation, identification, and molecular characterization of strains of *Photorhabdus luminescens* from infected humans in Australia. *Journal of Clinical Microbiology* 37, 3647-3653.

POINAR JR, G.O. (1990). Biology and taxonomy of Steinernematidae and Heterorhabditidae. In: Gaugler, R. & Kaya, H.K. (Eds). *Entomopathogenic nematodes in biological control.* Boca Raton, FL, USA, CRC Press, pp. 23-61.

POINAR JR, G.O. & THOMAS, G.M. (1965). A new bacterium, *Achromobacter nematophilus* sp. nov. (Achromobacteriaceae: Eubacteriales), associated with a nematode. *International Bulletin of Bacteriological Nomenclature and Taxonomy* 15, 249-252.

POINAR JR, G.O. & THOMAS, G.M. (1966). Significance of *Achromobacter nematophilus* Poinar and Thomas (*Achromobacteriaceae: Eubacteriales*) in the development of the nematode, DD-136 (*Neoaplectana* sp. Steinernematidae). *Parasitology* 56, 385-390.

POSADA, D. & CRANDALL, K.A. (1998). ModelTest: Testing the model of DNA substitution. *Bioinformatics* 14, 817-818.

RADEMAKER, J.L.W., HOSTE, B., LOUWS, F.J., KERSTERS, K., SWINGS, J., VAUTERIN, L., VAUTERIN, P. & DE BRUIJN, F.J. (2000). Comparison of AFLP and rep-PCR genomic fingerprinting with DNA-DNA homology studies: *Xanthomonas* as a model system. *International Journal of Systematic and Evolutionary Microbiology* 50, 665-677.

RAHN, O. (1937). New principles for the classification of bacteria. *Zentralblatt für Bakteriologie, Parasitenkunde, Infektionskrankheiten und Hygiene, Abteilung II* 96, 273-286.

RAINEY, F.A., EHLERS, R.-U. & STACKEBRANDT, E. (1995). Inability of the polyphasic approach to systematics to determine the relatedness of the genera *Xenorhabdus* and *Photorhabdus*. *International Journal of Systematic Bacteriology* 45, 379-381.

RAMIA, S., NETER, E. & BRENNER, D.J. (1982). Production of enterobacterial common antigen as an aid to classification of newly identified species of the families *Enterobacteriaceae* and *Vibrionaceae*. *International Journal of Systematic Bacteriology* 32, 395-398.

ROMÁN, J. & FIGUEROA, W. (1994). *Steinernema puertoricensis* n. sp. (Rhabditida: Steinernematidae), a new entomopathogenic nematode from Puerto Rico. *Journal of Agriculture of the University of Puerto Rico* 78, 167-175.

ROSSELLÓ-MORA, R. & AMANN, R. (2001). The species concept for prokaryotes. *FEMS Microbiology Reviews* 25, 39-67.

SAMBROOK, J., FRITSCH, E.F. & MANIATIS, T. (1989). *Molecular cloning: a laboratory manual*, 2nd edition. Cold Spring Harbor, NY, USA, Cold Spring Harbor Laboratory Press, 1659 pp.

SERGEANT, M., BAXTER, L., JARRETT, P., SHAW, E., OUSLEY, M., WINSTANLEY, C. & MORGAN, J.A.W. (2006). Identification, typing, and insecticidal activity of *Xenorhabdus* isolates from entomopathogenic nematodes in United Kingdom soil and characterization of the *xpt* toxin loci. *Applied and Environmental Microbiology* 72, 5895-5907.

SHARMA, S., WATERFIELD, N., BOWEN, D., ROCHELEAU, T., HOLLAND, L., JAMES, R. & FFRENCH-CONSTANT, R. (2002). The lumicins: novel bacteriocins from *Photorhabdus luminescens* with similarity to the uropathogenic-specific protein (USP) from uropathogenic *Escherichia coli*. *FEMS Microbiology Letters* 214, 241-249.

SICARD, M., BRUGIRARD-RICAUD, K., PAGES, S., LANOIS, A., BOEMARE, N.E., BREHELIN, M. & GIVAUDAN, A. (2004a). Stages of infection during the tripartite interaction between *Xenorhabdus nematophila*, its nematode vector, and insect hosts. *Applied and Environmental Microbiology* 70, 6473-6480.

SICARD, M., FERDY, J., PAGES, S., LE BRUN, N., GODELLE, B., BOEMARE, N. & MOULIA, C. (2004b). When mutualists are pathogens: An experimental study of the symbioses between *Steinernema* (entomopathogenic nematodes) and *Xenorhabdus* (bacteria). *Journal of Evolutionary Biology* 17, 985-993.

SICARD, M., TABART, J., BOEMARE, N.E., THALER, O. & MOULIA, C. (2005). Effect of phenotypic variation in *Xenorhabdus nematophila* on its mutualistic relationship with the entomopathogenic nematode *Steinernema carpocapsae*. *Parasitology* 131, 687-694.

SILVA, C.P., WATERFIELD, N.R., DABORN, P.J., DEAN, P., CHILVER, T., AU, C.P.Y., SHARMA, S., POTTER, U., REYNOLDS, S.E. & FFRENCH-CONSTANT, R.H. (2002). Bacterial infection of a model insect: *Photorhabdus luminescens* and *Manduca sexta*. *Cellular Microbiology* 4, 329-339.

SIMÕES, N. (1998). Pathogenicity of the complex *Steinernema carpocapsae-Xenorhabdus nematophilus*: Molecular aspects related with the virulence. In: Simões, N., Boemare, N. & Ehlers, R.-U. (Eds). *Pathogenicity of entomopathogenic nematode* versus *insect defence mechanisms: Impacts on selection of virulent strains*. Brussels, Belgium, European Commission, pp. 73-83.

SKERMAN, V.B.D., MCGOWEN, V. & SNEATH, P.H.A. (EDS). (1980). Approved lists of bacterial names. *International Journal of Systematic Bacteriology* 30, 225-420.

SMIGIELSKI, A.J., AKHURST, R.J. & BOEMARE, N.E. (1994). Phase variation in *Xenorhabdus nematophilus* and *Photorhabdus luminescens*:

Differences in respiratory activity and membrane energization. *Applied and Environmental Microbiology* 60, 120-125.

SMITH, H.L., JONES, J.B., LOUWS, F.J. & ADAMS, B.J. (2003). Genomic diversity of the bacterial symbionts of entomopathogenic nematodes. *Journal of Nematology* 35, 364. [Abstr.]

SOMVANSHI, V.S., LANG, E., GANGULY, S., SWIDERSKI, J., SAXENA, A.K. & STACKEBRANDT, E. (2006). A novel species of *Xenorhabdus*, family *Enterobacteriaceae*: *Xenorhabdus indica* sp. nov., symbiotically associated with entomopathogenic nematode *Steinernema thermophilum* Ganguly & Singh, 2000. *Systematic and Applied Microbiology* 29, 519-525.

STACKEBRANDT, E., FREDERIKSEN, W., GARRITY, G.M., GRIMONT, P.A.D., KÄMPFER, P., MAIDEN, M.C.J., NESME, X., ROSSELLÓ-MORA, R., SWINGS, J., TRÜPER, H.G., VAUTERIN, L., WARD, A.C. & WHITMAN, W.B. (2002). Report of the *ad hoc* committee for the re-evaluation of the species definition in bacteriology. *International Journal of Systematic and Evolutionary Microbiology* 52, 1043-1047.

STOCK, S.P. & KOPPENHÖFER, A.M. (2003). *Steinernema scarabaei* n. sp. (Rhabditida: Steinernematidae), a natural pathogen of scarab beetle larvae (Coleoptera: Scarabaeidae) from New Jersey, USA. *Nematology* 5, 191-204.

SUZUKI, T., YAMANAKA, S. & NISHIMURA, Y. (1990). Chemotaxonomic study of *Xenorhabdus* species-cellular fatty acids, ubiquinone and DNA-DNA hybridization. *Journal of General and Applied Microbiology* 36, 393-401.

SUZUKI, T., YABUSAKI, H. & NISHIMURA, Y. (1996). Phylogenetic relationships of entomopathogenic nematophilic bacteria: *Xenorhabdus* spp. and *Photorhabdus* sp. *Journal of Basic Microbiology* 36, 351-354.

SWOFFORD, D.L. (2002). *PAUP* Phylogenetic analysis using parsimony (*and other methods), version 4.0b 10.* Sunderland, MA, USA, Sinauer Associates.

SZÁLLÁS, E., KOCH, C., FODOR, A., BURGHARDT, J., BUSS, O., SZENTIRMAI, A., NEALSON, K.H. & STACKEBRANDT, E. (1997). Phylogenetic evidence for the taxonomic heterogeneity of *Photorhabdus luminescens*. *International Journal of Systematic Bacteriology* 47, 402-407.

TAILLIEZ, P., PAGÈS, S., GINIBRE, N. & BOEMARE, N. (2006). New insight into the diversity of the genus *Xenorhabdus*, including the description of ten novel species. *International Journal of Systematic and Evolutionary Microbiology* 56, 2805-2818.

THALER, J.O., BAGHDIGUIAN, S. & BOEMARE, N. (1995). Purification and characterization of xenorhabdicin, a phage tail-like bacteriocin, from the lysogenic strain f1 of *Xenorhabdus nematophilus*. *Applied and Environmental Microbiology* 61, 2049-2052.

THOMAS, G.M. & POINAR JR, G.O. (1979). *Xenorhabdus* gen-nov, a genus of entomopathogenic, nematophilic bacteria of the family Enterobacteriaceae. *International Journal of Systematic Bacteriology* 29, 352-360.

THOMAS, G.M. & POINAR JR, G.O. (1983). Amended description of the genus *Xenorhabdus* Thomas and Poinar. *International Journal of Systematic Bacteriology* 33, 878-879.

THOMPSON, J.D., GIBSON, T.J., PLEWNIAK, F., JEANMOUGIN, F. & HIGGINS, D.G. (1997). The clustalx windows interface: Flexible strategies for multiple sequence alignment aided by quality analysis tools. *Nucleic Acids Research* 24, 4876-4882.

VIVAS, E.I. & GOODRICH-BLAIR, H. (2001). *Xenorhabdus nematophilus* as a model for host-bacterium interactions: *rpoS* is necessary for mutualism with nematodes. *Journal of Bacteriology* 183, 4687-4693.

VOLGYI, A., FODOR, A., SZENTIRMAI, A. & FORST, S. (1998). Phase variation in *Xenorhabdus nematophilus*. *Applied and Environmental Microbiology* 64, 1188-1193.

VOLGYI, A., FODOR, A. & FORST, S. (2000). Inactivation of a novel gene produces a phenotypic variant cell and affects the symbiotic behavior of *Xenorhabdus nematophilus*. *Applied and Environmental Microbiology* 66, 1622-1628.

WALSH, K.T. & WEBSTER, J.M. (2003). Interaction of microbial populations in *Steinernema* (Steinernematidae, Nematoda) infected *Galleria mellonella* larvae. *Journal of Invertebrate Pathology* 83, 118-126.

WAYNE, L.G., BRENNER, D.J., COLWELL, R.R., GRIMONT, P.A.D., KANDLER, O., KRICHEVSKY, M.I., MOORE, L.H., MOORE, W.E.C., MURRAY, R.G.E., STACKEBRANDT, E., STARR, M.P. & TRÜPER, H.G. (1987). Report of the *ad hoc* committee on reconciliation of approaches to bacterial systematics. *International Journal of Systematic Bacteriology* 37, 463-464.

WEBSTER, J.M., CHEN, G., HU, K. & LI, J. (2002). Bacterial metabolites. In: Gaugler, R. (Ed.). *Entomopathogenic nematology*. Wallingford, UK, CABI Publishing, pp. 99-114.

WEISSFELD, A.S., HALLIDAY, R.J., SIMMONS, D.E., TREVINO, E.A., VANCE, P.H., O'HARA, C.M., SOWERS, E.G., KERN, R., KOY, R.D., HODDE, K., BING, M., LO, C., GERRARD, J., VOHRA, R. & HARPER, J. (2005). *Photorhabdus asymbiotica*, a pathogen emerging on two continents that proves that there is no substitute for a well-trained clinical microbiologist. *Journal of Clinical Microbiology* 43, 4152-4155.

WOODRING, J.L. & KAYA, H.K. (1988). *Steinernematid and heterorhabditid nematodes: a handbook of biology and techniques*. Southern Cooperative Series Bulletin 331, 30 pp.

WOUTS, W.M. (1984). Nematode parasites of lepidopterans. In: Nickle, W.R. (Ed.). *Plant and insect nematodes*. New York, NY, USA, Marcel Dekker, pp. 655-696.

YAMANAKA, S., HAGIWARA, A., NISHIMURA, Y., HIROSHI, T. & ISHIBASHI, N. (1992). Biochemical and physiological characteristics of *Xenorhabdus* species, including *Steinernema kushidai* and their pathogenicity against *Spodoptera litura* (Lepidoptera: Noctuidae). *Archives of Microbiology* 158, 387-393.

YAP, W.H., ZHANG, Z. & WANG, Y. (1999). Distinct types of rRNA operons exist in the genome of the actinomycete *Thermomonospora chromogena* and evidence for horizontal transfer of an entire rRNA operon. *Journal of Bacteriology* 181, 5201-5209.

ZHOU, X., KAYA, H.K., HEUNGENS, K. & GOODRICH-BLAIR, H. (2002). Response of ants to a deterrent factor(s) produced by the symbiotic bacteria of entomopathogenic nematodes. *Applied and Environmental Microbiology* 68, 6202-6209.

ZIMBRO, M.J. & POWER, D.A. (EDS) (2003). *Difco and BBL manual. Manual of microbiological media*. Sparks, MD, USA Becton Dickinson and Company, 696 pp. http://www.bd.com/ds/technicalCenter/inserts/difcoBblManual.asp

Index